DARWIN'S UNIVERSE

EVOLUTION FROM A TO Z

DARWIN'S UNIVERSE

EVOLUTION FROM A TO Z

RICHARD MILNER

WITH A FOREWORD BY IAN TATTERSALL AND A PREFACE BY STEPHEN JAY GOULD

UNIVERSITY OF CALIFORNIA PRESS

BERKELEY   LOS ANGELES   LONDON

For my friends Susan and Bob Wilder, Gerry Ohrstrom,
Norman Shaifer, Bob Adelman, Henry and Gloria Jarecki,
and the late C. A. Tripp, with gratitude for all their support
and encouragement

And for Ed Knappman and Carl Sifakis, with appreciation

The present book, *Darwin's Universe*, has evolved from two ancestral forms titled *The Encyclopedia of Evolution*, published in 1990 and 1993. It has been updated, revised, and enhanced with many new essays and illustrations. The late Stephen Jay Gould's preface (pages 2 and 3) is adapted from his foreword to the first edition of that book. The entry "Cladistics" (pages 80–82) is by Gareth Nelson.

University of California Press, one of the most distinguished university presses in the United States, enriches lives around the world by advancing scholarship in the humanities, social sciences, and natural sciences. Its activities are supported by the UC Press Foundation and by philanthropic contributions from individuals and institutions. For more information, visit www.ucpress.edu.

University of California Press
Berkeley and Los Angeles, California

University of California Press, Ltd.
London, England

Library of Congress Cataloging-in-Publication Data

Milner, Richard, 1941–
  Darwin's universe : evolution from A to Z / Richard Milner ; with a foreword by Ian Tattersall and a preface by Stephen Jay Gould.
      p.   cm.
  "The present book, *Darwin's Universe*, has evolved from two ancestral forms titled *The Encyclopedia of Evolution*, published in 1990 and 1993. It has been updated, revised, and enhanced with many new essays and illustrations"—T.p. verso.
  Includes bibliographical references and index.
  ISBN 978-0-520-24376-7 (cloth : alk. paper)
  1. Evolution (Biology)—Encyclopedias.   2. Darwin, Charles, 1809–1882—Encyclopedias.   I. Milner, Richard, 1941– Encyclopedia of evolution.   II. Title.
  QH360.2.M55   2009
  576.803—dc22   2008035575

Manufactured in the United States of America

18 17 16 15 14 13 12 11 10 09

10 9 8 7 6 5 4 3 2 1

The paper used in this publication meets the minimum requirements of ANSI/NISO Z39.48-1992 (R 1997) (Permanence of Paper).

FRONTISPIECE: Statue of Darwin at the State Darwin Museum, Moscow.

CONTENTS

FOREWORD

Ian Tattersall

Curator of Anthropology, American Museum of Natural History

The title of this volume is *Darwin's Universe*, but *Milner's World* might well be more appropriate. This book is about Charles Darwin, and the way that his ideas transformed the world he lived in and are continuing to change ours. But, as befits a book whose author has created a unique niche in the history of Darwinian studies, it has Richard Milner's fingerprints all over it.

Alphabetical its arrangement may be; but this is no staid, humorless encyclopedia of natural history and evolution; rather, it is a joyous romp through a world inhabited by as eccentric and absorbing a cast of characters as was ever brought together between endpapers.

Aristogenesis and its egomaniacal inventor Henry Fairfield Osborn? You'll certainly find them in here. The anarchist Russian Prince Peter Kropotkin? He's here too, alongside the charlatan psychic Henry Slade, the Victorian "Gorilla Woman" Julia Pastrana, and the contraception pioneer Marie Stopes, who qualifies for inclusion—who knew?—as a paleobotanist.

But you will look in vain for a solemn exegesis of the Hardy-Weinberg equilibrium or of speciation theory. The rule is that if Milner is fascinated by it, you'll find it in here; if he finds it boring or overworked in other books, you'll search in vain. Fortunately, Milner's interests are wide, and there's not a whole lot of true general interest that he misses. And the result is a book that most readers will find it hard to put down without browsing through at least half a dozen entries more than he or she had intended to tackle.

While much in here is necessarily familiar—though always uniquely informed by Milner's distinctive way of looking at things—there is also a lot that is new, a result of his industrious delvings in obscure sources of Darwiniana. Milner has, for instance, exhumed a lot of fascinating material about Darwin's career as a magistrate and his war on the spiritualists, as well as a "lost" painting of a coral atoll he inspired. Milner even discovered that the forbidding Sir Richard Owen, Darwin's nemesis, was also a gifted cartoonist with a wry sense of humor.

But Milner is not just an idiosyncratic—though serious—historian of ideas and their inventors. He has a wonderful visual sense and a gift for sniffing out old and curious pictures that make the past and its denizens come alive. Like it or not, ideas of evolution have become woven into the fabric of modern society, and the wonderfully quirky selection of illustrations, some of them unseen in a century or more, brings home just how—and how much—our understanding of the world and our place in it have been revolutionized in the 150 years since the publication of the *Origin of Species*.

Personal though this view of the world is, Milner rarely takes sides on the many issues he discusses. Instead, he acts rather like a referee in the glorious game of science and life—and as an ambassador to evolutionary science as well, showing not only its fads and fallacies but its enormous ramifications in the development of our modern society, including law, politics, art, movies, and a host of other cultural nooks and crannies.

This book is crammed with the results of a lifetime of inquiry, and will become an indispensable reference for anyone interested in evolutionary thought and the changing milieus through which it has developed. But while, as a result, you will probably pick it up for information and insight, I guarantee that you'll read it for fun.

PREFACE

AN APPRECIATION OF MILNER'S *ENCYCLOPEDIA*

Stephen Jay Gould

"Fossilface" (Stephen Jay Gould) and "Dino" (Richard Milner), circa 1954.

We have, as a society, lost our bearings in so many ways. Perhaps other things are more important in a world of poverty and pollution, but I rank our growing preference for automated sameness over a personal touch as one of the greatest ills of our age. But primates are social creatures, and (for all its tragedies) perhaps the one great legacy of our cultural history resides in our stated respect for individuality. This principle applies with special force to scholarship.

With a misplaced definition of "objectivity," many people think that books of nonfiction, particularly reference works like encyclopedias, should be impersonal, and devoid of style or idiosyncrasy. No, and a thousand times no. The truly great books of reference have a personal stamp, as any work of passion worthy of our attention must. We still speak of Johnson's *Dictionary*, Roget's *Thesaurus*, and Robert's *Rules of Order*. Among the monuments in our intellectual history are Lyell's *Geology* and Marshall's *Economics*. Milner's *Encyclopedia of Evolution* has joined this noble tradition of reference books stamped with the vibrant idiosyncrasy of true scholarship. If camels are horses built by committees, then Milner's encyclopedia has provided Darwin with a full-blooded Thoroughbred.

Muscular and vibrant idiosyncrasy avoids crankiness and uses the personal touch both to impart a vision and to supply details that conventional accounts would never treat or even discover. This book finds just the right balance between the necessarily universal and the unique. It can be used as a work of reference in the conventional manner—to find the names, the dates, the leading concepts clearly and crisply defined, the major arguments and the chief actors.

But the strength of Milner's compendium does not reside in this worthy material, for others have worked in this mode. Milner's different approach follows the variety of his own life. He has loved and studied evolution and paleontology since we were boys together at a junior high school in Queens. (We were the only two dinosaur nuts at a time when such an interest marked a kid as a geek or a weirdo. He was called "Dino" on the playground; I was "Fossilface.") He completed five years of graduate work in anthropology, but then, through a series of interesting circumstances, spent fifteen years in the midst of pop culture before returning, in midlife, to his first (and longest) love for natural history.

From this vantage point both in and outside academia, Milner views evolution and its impact in a comprehensive way often ignored (through true ignorance, that is, not condescension) by more traditional scholars. Milner's uniqueness and idiosyncrasy (in the good sense) lie in his range of chosen topics—a full spectrum of influences on evolutionary thought, from the abstractions of high culture to the come-ons of sideshow hucksters. I see two special strengths in this ecumenical approach:

1. If we make an artificial division into high and vernacular culture, and consider just the former in a narrowly confined and misplaced concept of importance, then we will never understand the impact of science in society. Consider the role of garden clubs, of bird watchers, of telescope makers, of horseplayers who study probability. Academic knowl-

edge of science may be abysmal in America, but scientific styles of thought are widespread, if only we could harness them.

2. Science pursues an external truth, but only from a perspective inextricably embedded in social contexts. If scientists ignore the context, they not only act in an elitist way, they also preclude any real understanding of scientific change and utility. Social context is particularly important for evolutionary science, because this subject, more than any other, impacts the great myths and hopes of Western society—ideas of progress, God, human origins, the meaning of life, the basis of ethics, to name just a few.

Richard Milner has given us so much more than a conventional reference work. His book is a series of mini-essays on the highways and byways of this most socially contentious of all scientific ideas. It offers scores of essays on people, many unknown to almost all professionals but as important to the diffusion of the idea of evolution as the wisest professor: Henry Ward Beecher, the great Victorian liberal clergyman who defended Darwin; Carl Akeley and the canonical vision of taxidermy in the romantic mode; E. Ray Lankester, the head of the British Museum, who exposed the fraudulent "spirit-medium" Henry Slade; Chief Red Cloud, who befriended the great paleontologist O.C. Marsh while Custer advanced on the Black Hills; Willis O'Brien, who developed stop-motion photography and made the great dinosaurs of early movies (including the King Kong vs. pterodactyl scene); Charles R. Knight, whose paintings established the image of *the* dinosaur for professionals and amateurs alike, and who thereby (despite his invisibility in scholarly publications) has had as much influence in paleontology as any scientist on earth.

But Milner does not stop with historical personages. He treats the fictional (Tarzan) and the fossil, both real ("Lucy," *Zinjanthropus*) and fake (Piltdown Man). He also writes about individual nonhuman animals who have made a difference in evolutionary theory—all the leading personalities of apedom, from Digit the gorilla to Nim Chimpsky, and even Lonesome George, the last surviving (and mateless) saddlebacked Galápagos tortoise.

Milner's events are as fascinating, and diverse, as his people. I never knew that the trial of the medium Henry Slade featured a battle between Darwin and Wallace, as the rationalist (and wealthy) Darwin contributed funds to the prosecution, while the spiritualistic (and impecunious) Wallace testified directly for the defense.

I never heard about T.H. Huxley's arrest in connection with agitation by the Sunday Leagues against pervasive blue laws that had made almost everything but churchgoing illegal on the Christian sabbath. (He preached a "sermon" to 2,000 people on Darwinism one Sunday, surrounded with enough music and recitation to court definition as a "secular service." The baffled police knew that this wasn't kosher but couldn't devise an appropriate charge; they finally ran Huxley in for keeping a disorderly house.) I had never encountered "The Happy Family," a popular sideshow attraction of the 19th century, featuring animals calmly living together that would tear each other apart in the wild. Although started before Darwin, and meant (in part) as a commentary on Isaiah ("the wolf also shall dwell with the lamb"), these exhibits were later viewed as an argument against Darwin and the struggle for existence. (There was no particular secret in rearing such a group. Good trainers can recognize unusually docile animals, and most creatures raised together from birth and given adequate food on a regular schedule will grow to live in peace.)

Details in the hands of an anecdotalist or an antiquarian remain just that—interesting in themselves perhaps, but incoherent and without aim. Details in the hands of a scholar with a general purpose are the essence of understanding. Milner has given us details in the sublime and coordinated sense. He has also fused high with vernacular culture to show us the full impact of science's most seminal idea—evolution.

In so doing, he has produced a popular book worthy of Darwin's words to Huxley, as he importuned his old friend and combative champion to write a work on evolution for the general public: "I sometimes think that general and popular treatises are almost as important for the progress of science as original work."

INTRODUCTION

The artist Paul Gauguin, dying in Tahiti, scrawled onto his final masterpiece: *What are we? Where do we come from? Where will we go?* Simple, childlike questions, they have tantalized creative geniuses for centuries, inspiring Michelangelo's Sistine ceiling, Emily Dickinson's poetry, Stephen Hawking's cosmogony. When Victorian biologist Charles Darwin tackled those questions, he connected all three with one overarching concept: evolution. Although not the first to do so, he was by far the most persuasive, unleashing a tidal wave in Western thought.

"Nothing in biology makes sense," wrote geneticist Theodosius Dobzhansky, "except in the light of evolution." Every branch of the life sciences, from biochemistry and botany to zoology, has an evolutionary underpinning. Practical applications are so ubiquitous that we hardly notice their common yoke: DNA "fingerprinting" in the courtroom, bioengineering of livestock and plants in agriculture, medical strategies to combat resistant strains of bacteria, conservation programs for endangered species, growing awareness of planetary ecology. All are joined together by a basic, shared assumption: the idea of evolution, which only 150 years ago was considered unscientific and vaguely immoral.

Entire libraries are devoted to its scientific implications—histories of geology, journals of animal behavior and population genetics, ecological studies of jungle canopies and ocean floor vents, speculations about dinosaurs and "death stars." An encyclopedist of evolution could spend a lifetime exploring abundant riches without venturing beyond this scientific literature. But we are dealing with an impolite and unruly idea that simply refuses to stay put.

Vast and far-reaching in its effects, the theory of evolution by natural selection long ago leapt the boundaries of biology to influence an astounding variety of fields: science fiction novels, children's toys, Hollywood movies, even religious and political movements. Therefore, a comprehensive work devoted to this single, unifying idea has to include not only diverse scientific developments but also the major historical repercussions in literature, art, philosophy, and "pop culture" as well. And people, lots of interesting and unusual people.

During the colorful history of this great cultural enterprise, contributions to evolutionary knowledge have been applied (and misapplied) by a variegated spectrum of remarkable individuals: the world's richest man, with his strange evolutionary "religion" (CARNEGIE, ANDREW); a school dropout obsessed with collecting slabs of fossilized footprints made by animals no one had ever seen (BIRD, ROLAND THAXTER); a dinosaur expert who tried to become king of Albania (NOPCSA, BARON FRANZ); the occupational therapist from Louisville, Kentucky, whose love for African gorillas cost her her life (FOSSEY, DIAN); the artist who went blind painting prehistoric animals (KNIGHT, CHARLES R.); even the proverbial Doctor (MANTELL, GIDEON), Lawyer (LYELL, SIR CHARLES), and Indian Chief (RED CLOUD).

Even as evolutionists study change, their very perceptions of nature—and of change itself—are constantly evolving. Information about the web of life, animal species and their behavior, shifting theories and fads in science, all defy attempts to "freeze" them in a book. Upon close scrutiny, they seem to be perpetually transforming themselves.

Take chimpanzees, for instance: not long ago we thought of them as natural clowns, like Cheetah, the comic relief in Tarzan movies. In the 1970s, they became newly respected as hand-signing symbol users; next—during the ensuing "ape language controversy"—they were viewed as clever, manipulative brats that only *pretended* to know language in order to yank the human investigator's chain. During Jane Goodall's thirty years of African field studies, chimps went from peaceful, sweet-natured vegetarians to cannibalistic, infanticidal killer apes. The chimps remain chimps, but how drastically human perception of them has changed over a few decades. Our understanding of apes and language underwent another radical change in the 1970s, with the symbol-using achievements of a charismatic bonobo (or "pygmy chimp") called Kanzi (BONOBOS; KANZI). At the same time, an African grey parrot with a thimble-sized brain astounded us with his abilities to use and apparently understand English words (ALEX).

The status of fossil man is probably the most rapidly changeable area at present; the quest for "the missing link" became the search for the "oldest man," while our conception has shifted from "noble savages" to cannibals, from "killer apes" to food sharers. Species names change, genera are split or lumped. Experts attack each other's credibility, even while wriggling out of their own extravagant pronouncements of a few years earlier. More fossil hominids (and hominins) are known now than ever before, but paleoanthropologists cannot agree on just how many there are and which are ancestral to ourselves. (See AUSTRALOPITH-ECINES; CANNIBALISM CONTROVERSY; DMANISI HOMINIDS; FLORES MAN; FOSSIL HUMANS; "LUCY"; "MISSING LINK"; TAUNG CHILD; TOUMAI SKULL.) According to some anthropologists there may have been as many as seventeen different hominid species, and at least four of them alive at the same time!

I have attempted, in this book, to rescue many "unknown" incidents from oblivion. Did you know that Gregor Mendel never formulated Mendel's Laws? That arch "materialist evolutionist" Thomas Huxley liked to astonish audiences with his "supernatural power" to summon spirits? Or that Darwin said he could not really prove that any species had evolved—but didn't think that his theory should be judged on that basis? (See "CENTRAL DOGMA" OF GENETICS; HUXLEY, THOMAS HENRY; MENDEL'S LAWS; THEORY, SCIENTIFIC.)

Did you know that Captain Robert FitzRoy's personal mission aboard HMS *Beagle* was to return three young Indians to their homeland—part of an experiment dear to the captain's heart? The public neither knew nor cared that the young Charles Darwin was aboard as unofficial naturalist. But all English newspaper readers knew the outlandish names that sailors had given to the celebrated Fuegian passengers: York Minster, Fuegia Basket, and Jemmy Button. The "experiment" ended badly and tragically, as you will see (BUTTON, JEMMY).

Bitter conflicts over evolutionary ideas have frequently spilled out of scientific circles into the forums of popular culture. Everyone has heard of the 1925 Scopes "Monkey Trial" in Tennessee, but the Slade trial of Victorian England is almost forgotten. In 1876, British spiritualists and evolutionists fought an emotional courtroom battle over the authenticity of "Doctor" Henry Slade, an internationally celebrated spirit-medium (or channeler, in today's parlance). The issue was whether Slade had demonstrated that the human personality is inseparable from the biological brain. "Materialist" scientists insisted that Slade's communion with "departed spirits" had to be a criminal fraud to bilk the bereaved. Many may have privately agreed, but for a scientist to publicly insist that a human "soul" cannot exist apart from a body was deemed a shocking break from tradition—much worse than claiming apes for relatives. Evolutionists Charles Darwin and Alfred Russel Wallace (a prominent spiritualist) supported opposing sides in the Slade affair, each convinced that his own vision of evolutionary science was at stake. (See CREATIONISM; MATERIALISM; SCOPES TRIAL; SLADE TRIAL; SPIRITUALISM.)

Two years earlier, distinguished physicist John Tyndall had scandalized the British Association with the uncompromising "materialist evolutionism" of his Belfast Address, in which he openly declared that the church's "monopoly" on questions of human origins, nature, and destiny was over and done with. And in America, in 1925, Carl Akeley's statue *The Chrysalis*, commissioned by a New York church, became the center of another dustup. Akeley had depicted a man emerging from a cracked-open gorilla skin, offering a triumphant vision of ascent from the primates in place of a guilt-ridden Fall from Grace. Newspaper editorials denounced him as a pagan nature worshiper (see AKELEY, CARL; BELFAST ADDRESS; *CHRYSALIS*; MATERIALISM).

The mystery of mankind's origin, nature, and destiny obsessed and inspired not only the biologists Charles Darwin, Alfred Russel Wallace, and Thomas Henry Huxley, but a diverse "who's who" of great authors whose influence has permeated the far corners of our culture. George Bernard Shaw conceived *Back to Methuselah* (1920) as a sugarcoated manifesto for Lamarckian evolution, devoting the play's preface to an attack on Darwinism. H. G. Wells, whose novels include *War of the Worlds* (1898), was a student of Thomas Huxley's and wrote the first "Darwinian" science fiction. His extraterrestrials were products of an alien evolution, terrorizing Earth in a cosmic battle for "survival of the fittest." (See SHAW, GEORGE BERNARD; WELLS, H.G.) Sigmund Freud developed fanciful scenarios of Ice Age "primal hordes" to explain the "childhood" of the human race. Freud's theories were based on the belief, common among 19th-century biologists, that each individual's developmental stages recapitulated the evolution of the entire species. (See BIOGENETIC LAW; FREUD, SIGMUND.)

Twenty years before the publication of *Origin of Species* (1859), Charles Darwin's exciting account of the *Beagle* voyage (1839) instantly won wide fame, inspiring many younger naturalists to seek their own adventures. Alfred Russel Wallace remembered into old age how he and Henry Walter Bates dreamed of exploring the Amazon after reading Darwin's rhapsodic descriptions of its rain forests. (They also shared his passion for beetle collecting.) The same book, which was aimed at nonspecialists, entranced the great botanist Joseph Dalton Hooker during his own voyage to the Antarctic. He slept with pages of Darwin's published *Journal* under his pillow. (See BATES, HENRY WALTER; HOOKER, SIR JOSEPH DALTON; *ORIGIN OF SPECIES*; VOYAGE OF HMS *BEAGLE*; WALLACE, ALFRED RUSSEL.)

One of the most peculiar aspects of the worldwide spread and acceptance of evolution is the amazing range of perceptions and attitudes that have claimed it. Lamarckians equated evolution with progress, increasing complexity, and perfection. Darwin himself saw the possibility of degeneration as well as progress, of neutral "sideways" changes as well as adaptation. Some evolutionists, such as the German Ernst Haeckel and the Englishman E. Ray Lankester, became obsessed with degeneration, expressed in their rantings about nationalistic "racial hygiene." As playwright George Bernard Shaw put it, Darwin "had the luck to please everybody who had an axe to grind." (See DEGENERATION THEORY; HAECKEL, ERNST; LANKESTER, E. RAY; NEO-LAMARCKISM.)

For Haeckel, evolution meant the moral justification of ruthless competition, while for the Russian Prince Kropotkin it was altruism and social cooperation that ensured the survival of species. Some rejoiced in identifying their roots in the animal world and establishing a blood connection to all other living creatures. Others were shocked, as if they had discovered the revered founder of their family was, in the phrase of novelist D. H. Lawrence, "nothing but frog-spawn." (See KIN SELECTION; KROPOTKIN, PRINCE PETER; SOCIAL DARWINISM.)

Science views evolution as contingent history—a branching bush sending forth divergent twigs, as circumstance and variation allow, at each juncture. Mystics, theologians, and some philosophers have insisted that there must be an ultimate goal or unfolding of a preordained destiny: a ladder or escalator to the infinite. Evolutionary data have been used as evidence for the biological unity of mankind and also for its converse, dividing so-called "races" into "higher or lower" depending on their supposed time of evolution (a scientifically discredited idea that nevertheless lingers among racists). (See BRANCHING BUSH; CONTINGENT HISTORY; ORTHOGENESIS; RACE; TEILHARD DE CHARDIN, FATHER PIERRE.)

Certainly, Darwinism, natural section, "survival of the fittest," and other key concepts keep changing. Darwin tinkered with his theory from one edition of *Origin of Species* to the next, and no one since has left it alone. History has seen it evolve into Neo-Darwinism and later the Modern Synthesis, or Synthetic Theory of Evolution; the 1970s brought a significant and wide-ranging shift in emphasis regarding rates and pattern of change known as the theory of Punctuated Equilibrium (or "Punk Eek"). (See NATURAL SELECTION; NEO-DARWINISM; PUNCTUATED EQUILIBRIUM; "SURVIVAL OF THE FITTEST"; SYNTHETIC THEORY.)

Darwin knew he could not hope to foment a major revolution in thought by himself and constantly enlisted the help of top botanists (Joseph Hooker and Asa Gray), the best geologist (Charles Lyell), and his knight errant, zoologist Thomas Henry Huxley. Although they never coauthored papers (sharing credit for each morsel of knowledge is a 20th-century fashion), they were true collaborators in the evolutionary enterprise.

Darwin, of course, was compelled to accept Alfred Russel Wallace as his junior partner in the discovery of natural selection, which should properly be called the Darwin-Wallace Theory (See "DELICATE ARRANGEMENT"; WALLACE, ALFRED RUSSEL; "WALLACE'S PROBLEM.")

These Victorian naturalists were remarkable human beings whose personalities, friendships, and rivalries left their stamp on an era; often their private correspondence was even more interesting than their public contributions to science. In an 1874 essay on animal automatism, Thomas Henry Huxley remarked that when he was perplexed about a question, he liked "talking it over" with men "of real power and grasp" of a previous age. Despite our unquestioned conviction that present scientific knowledge far surpasses anything known a century ago, we can still learn a great deal by "talking over" the great questions of evolution and science with Darwin and his friends. There is also the added delight of wandering about in what Wallace aptly called "the wonderful century."

Contact with the personalities of Darwin, Wallace, Huxley, Hooker, and the rest imparts something else too: an intimate insight into lives extraordinarily well lived. All shared a fierce dedication to finding truth; a resigned humor about their own human limitations; a deep sense of loving wonder about the natural world; and, above all, a cheerful willingness to commit a lifetime's unstinting labor to a worthwhile quest. All this at a time when information couldn't be transmitted with a keystroke and words weren't "processed"—their overseas letters had to still be worth reading three months after posting, and their durable books were written with scratchy pens dipped in ink.

During most of the 19th century, there was little money to be made in science. It was carried forward on the shoulders of a volunteer army, in which great physical courage was taken for granted. Wallace habitually risked his life in the jungles, often dragging himself through bouts of malaria to capture new, undescribed butterflies. Darwin stayed aboard the *Beagle* for five years—seasick almost every day—though that ship was of a design and manufacture sailors had nicknamed "floating coffins," so commonly did they capsize. He also rode 400 miles on horseback through hostile Indian country in Argentina, where even the hardy gauchos feared to accompany him.

Perhaps the most delightful example of a naturalist's courage and resourcefulness concerns Darwin's friend and confidant Joseph Dalton Hooker, an expert on trees and flowers. Convinced that he was a spy for the British government, the Rajah of Sikkim sent a hundred armed tribesmen to ambush and kill the adventurous botanist at a remote mountain pass in the Himalayas, near the Indo-Tibetan border. Within a half hour, Hooker had turned them into a platoon of plant collectors. (See HOOKER, SIR JOSEPH DALTON.)

Darwin insisted that uncertainty is part of science. Does that mean evolutionary theory is worthless, as some have argued, because its truths are not eternal? While trading barbs about the uncertainty of scientific truth with Bishop Samuel Wilberforce at the Oxford meeting of 1860, Thomas Huxley asked his listeners to suppose they were lost in the countryside on a dark night, with no clue to the road. If someone came along offering a flickering lantern, "should I refuse it because it shed an imperfect light? I think not," Huxley answered himself, "I think not."

Looking back after a century's scientific "progress," you may be surprised to find in this book how well the old naturalists fought and debated many of the same questions offered as new in our own day. Often they argued more passionately than their intellectual descendants and with superior powers of expression. Darwin and his circle never disappoint. And no matter what hour you'd like to visit, they are always home.

ABANG
Orang Stone Toolmaker

Abang, a male orangutan in the Bristol zoo, mastered a skill once thought impossible for apes: he learned to make and use a stone knife.

Chimpanzees use rocks to smash hard nuts and fruit, but apes in the wild do not use stones to shape other stones. Humans craft stone tools by chipping, shaping, or flaking them with other stones; one of the commonest of ancient blades is a long, sharp-edged flake struck off a larger flint core. In 1971, British anthropologist R. V. S. Wright worked with a five-year-old orang called Abang in the Bristol zoo. Wright repeatedly showed him how to use a flint blade to cut a nylon rope tied around a food box. After about an hour of demonstrations, the orang learned to use the knife to get the food. Soon after, he learned to make his own knife by striking off a flint flake with a hammerstone.

Eventually, whenever Abang was given a tied-up box of food, a flint nodule, and a hammerstone, he quickly made a stone knife and cut the rope.

See also APES, TOOL USE OF; KANZI; ORANGUTAN

ACTUALISM
Continuity of Causality

Actualism is the assumption that the Earth's past can be explained in terms of natural processes observable in the present. Many historians credit James Hutton (*Theory of the Earth*, 1788) with first applying it to geology.

Half a century later, the works of Sir Charles Lyell, a meticulous field geologist, solidly established actualism. Lyell made it the keystone in his cluster of ideas later known as uniformitarianism—the foundation of modern geology.

Lyell's subtitle for his *Principles of Geology* (1830–1833) is *An Attempt to Explain the Former Changes of the Earth's Surface by Reference to Causes Now in Operation.* His systematic observations of erosion, sedimentation, and volcanic formations enabled him to clarify many long-standing mysteries about the Earth's features. Unlike his predecessors, Lyell did not fall back on miracles or divine intervention to explain ancient events. (He was stymied, however, by the origin of the Earth, since no comparable processes appeared to be observable now.)

Catastrophist geologists believed that most of the Earth's history was a series of cataclysms or upheavals unlike anything known today, involving drastically different processes. Though actualism eventually led to the downfall of catastrophist geologists, some of them initially embraced Lyell's approach. If more ordinary geologic features had been produced by known causes, they reasoned, then those that defied explanation could safely be assigned to forces outside the range of human knowledge.

By 1840, Lyell's uniformitarian principles had exerted a huge influence. Not only did they set the tone for the next century of geological research, but they also directly influenced Charles Darwin's view that species originated gradually through ordinary reproduction, rather than suddenly through supernatural agencies. Darwin had read Lyell during his voyage aboard HMS *Beagle*, and later claimed his theories about geology came "half out of Lyell's brain."

But while Lyell had no problem in imagining that great canyons or mountain ranges had been shaped slowly by natural forces over eons, he balked at accepting a similar process for the human species. It was not until 1863, in *The Antiquity of Man*, that Lyell publicly supported Darwin's ideas about the continuity of life in the natural world, though he still skirted the issue of humans. Perhaps, he grudgingly conceded, "community of descent is the hidden bond which naturalists have been unconsciously seeking while they often imagined that they were looking for some unknown plan of creation."

See also CATASTROPHISM; GRADUALISM; LYELL, SIR CHARLES; PROGRESSIONISM; STEADY-STATE EARTH; UNIFORMITARIANISM

ADAM AND EVE
Biblical Primal Couple

In the Old Testament, the Lord creates the first man out of clay, breathes life into him, and calls him Adam, which means "from the dust of the Earth." When Adam becomes lonely in his Garden paradise, the Creator fashions the woman Eve from the man's rib, as "an help meet [suitable] for him"—commonly misread as "a help-mate." (Despite common belief, men do not have one fewer pair of ribs than women.)

Until the first couple violates the Lord's instruction not to eat the fruit of the forbidden Tree of the Knowledge of Good and Evil, they live in total harmony with nature, over which they are given "dominion." Before the Expulsion, they spend their days enjoying the benevolent works of the Creator and inventing names for all the plants and animals in the Garden.

Carl Linnaeus, the great Swedish botanist who founded the system of biological classification, insisted Adam was "the first naturalist." To spend one's life studying, classifying, and giving names to plants and animals, Linnaeus wrote, was therefore sanctified in tradition as a manner of worshiping God (an excellent response, not incidentally, to bookish churchmen who scorned his practice of collecting dead plants and animals as vulgar).

Did Adam and Eve have navels? For painters and sculptors of religious subjects, this was a controversial question of great practical concern. During the Renaissance, the primal couple was sometimes represented without them, since neither was born of woman. Other artists avoided the question with strategically placed shrubbery. Some theologians argued the first man and woman were "finished" creations; they sported belly buttons though they never actually had an umbilicus—just as Eden's tree trunks were created complete with growth rings.

A popular 19th-century natural history writer, Philip Gosse, seized on that idea in an ill-conceived effort to finally reconcile scripture and science. If Adam and Eve's hair, fingernails, and navels were created complete in an instant, bypassing growth and development, he argued, then God must have created all the Earth's fossils and geologic strata as false remnants of a past that never actually happened.

Gosse's infamous 1857 book *Omphalos* (Greek for navel) was not credible to scientists and clerics alike. The Reverend Charles Kingsley, for instance, sadly confessed that if such mental contortions were really necessary to reconcile geology with the Bible, it shook his faith in scripture.

Mark Twain thought that Adam and Eve, the "founders" of the human race, deserve a memorial by their descendants. Twain campaigned for all the world's peoples to join in erecting a colossal statue of Adam and Eve—towering over all the divisive religious shrines—to be erected in the Holy Land. No one took his proposal seriously, but for once the celebrated humorist was not joking.

See also *OMPHALOS*; ORIGIN MYTHS

THE PRIMAL PAIR was often portrayed with navels by earlier church artists, but 19th-century illustrator John Tenniel avoided controversy by covering up the umbilical question.

ADAPTATION

Shaped for Survival

Fins and flukes evolved as swimming adaptations, wings for flight, and camouflage for defense. Structures and behaviors useful to an organism in a particular environment are adaptations, recognized as such long before Charles Darwin.

Woodpeckers (Darwin's favorite example) get their living by climbing tree trunks and extracting insects from bark. Adaptive features include a thick skull, "shock absorber" neck construction, chisel bill, long, barb-tipped tongue, claws like grappling hooks, and stiff tail feathers for stability. Admiring such adaptations—as narrators of television wildlife films often do—can lead to using them to "explain" evolution.

In fact, the concept of adaptation is one of the most troubling and puzzling in natural history. Since an animal is the product of a long history, its adaptation is relative at any given time. Feathers may now be adaptive for flight, yet they evolved before birds flew—possibly for retaining body heat. How could early wings have been adaptive? Or as paleontologist Stephen Jay Gould put it, "What good is forty percent of a wing?"

Many biologists believe the answer lies in a change or shift in function. A structure's eventual use may be quite different from its origin, a process known as "exaptation." Early flightless "wings" may have been used to stabilize swift-running birds or dinosaurs, as ostriches use them today, or they may have first functioned as heat regulators. Some living birds' wings are not adapted primarily for flying. Penguins use them to swim, and some wading birds curl them into glare shields while fishing in shallow water. For many structures, we cannot determine how they originated, nor what may be their future.

> **Adaptation is relative at any given time. Feathers may now be adapted for flight, but they evolved before birds flew.**

The explanation of the origin of adaptations is one of the most controversial areas in evolutionary biology. One persistent question is whether—as seems likely—new behaviors usually appear first and new structures subsequently evolve. Darwin thought behavior changed first, but rejected Lamarckian notions of organisms "willing" and "striving" in new directions.

Adaptation, which at first seems such an easy, commonsense concept, turns out to be slippery, sometimes even circular and paradoxical. A species is adapted if it survives in its environment, but how do we know it has not simply been dumb-lucky or neutral, while some calamity eliminated its competitors? Mass extinctions have wiped out 97 percent of all species that have ever lived on Earth. If a species becomes extinct tomorrow, how well adapted was it? And how can one explain such seemingly maladaptive structures as the peacock's glorious tail, which gets in the way of efficient flight or food-getting? (This last question led Darwin to propose his theory of "sexual selection.")

To complicate matters, many species that don't seem adapted to certain environments have moved into them anyway. When Darwin visited the Galápagos, he was fascinated by the plentiful marine iguanas. When not basking on rocky beaches, these large lizards spend their time underwater, grazing on seagrass. Although excellent divers and swimmers, they never (or have not yet) evolved webbed feet or streamlined forms. Their bodies give no obvious indications that they get their living on the ocean floor.

Biologists studying wildlife in a South American rain forest found fish that have recently become fruit eaters. With seasonal flooding of the river banks, several species of fish have learned to swim among submerged branches, feeding on the fruit. In Oregon, a local population of deer (usually vegetarians) now patrol a river's banks, eating beached fish that jump out of the water while spawning. And Galápagos gannets are often found perched on tree branches—bizarre behavior for webbed-footed birds.

One of Darwin's enduring demonstrations was that adaptations are usually not marvels of perfection at all, but historical compromises. On closer examination, they usually turn out to be jerry-built contraptions, products of a unique, opportunistic history.

Darwin explains his views on how adaptations evolve from previously existing structures in this passage from his 1863 book on the fertilization of orchids:

> Although an organ may not have been originally formed for some special purpose, if it now serves for this end, we are justified in saying that it is specially adapted for it. On the same prin-

ciple, if a man were to make a machine for some special purpose, but were to use old wheels, springs, and pulleys, only slightly altered, the whole machine, with all its parts, might be said to be specially contrived for its present purpose. Thus throughout nature almost every part of each living being has probably served, in a slightly modified condition, for diverse purposes, and has acted in the living machinery of many ancient and distinct specific forms.

With this approach, Darwin veered away from the natural theologian's concept of "perfect adaptation" by a "Designer." Adaptations are not perfect; they are often demonstrably makeshift. It was much more fruitful to focus on the "contrivances" and contraptions, evidence of made-over parts showing the pathways of an organism's specific, unique history.

Less sophisticated naturalists have, over the years, created fanciful "just-so" stories to explain the origins of particular adaptations. Appalled by these unbounded speculations, some biologists have suggested abandoning the concept of adaptation. They find it too vague to be useful and historically abused as a substitute for solid investigation.

See also CONVERGENT EVOLUTION; DARWIN'S FINCHES; DARWIN'S LIZARDS; EXAPTATION; "JUST-SO" STORIES; ORCHIDS, DARWIN'S STUDY OF; PALEY'S WATCHMAKER; PANDA'S THUMB

ADAPTIVE RADIATION

See ADAPTATION; DARWIN'S FINCHES; HAWAIIAN RADIATION

AFAR HOMINIDS
Ethiopian Fossils

Charles Darwin had written in *The Descent of Man* (1871) that he thought Africa would be the place to seek a sequence of apelike creatures ancestral to humans. "But not even Darwin," wrote paleoanthropologist Tim White, "could have imagined that a single geological deposition in the Horn of Africa would by 2005 have yielded a record of human evolution stretching across the last 6 million years."

Ethiopa's Afar Depression, a vast arid triangle between the Blue Nile and the Red Sea ("Ophir" in the story of Solomon and the Queen of Sheba), has turned out to be the hottest fossil field of the past quarter-century. Three of the Earth's gigantic plates pull and rub against one another in the Afar, creating a junction of rifts where earth movements and erosion of exposed sedimentary deposits constantly bring fossils to the surface. A particularly rich locality, Hadar, is near the Awash River, about 185 miles northeast of Addis Ababa.

Maurice Taieb, a French geologist, began surveying the region in 1971, with American Jon Kalb. The following year this international research team added an American graduate student, Donald Johanson, who discovered a fossil knee joint of an erect-walking hominid, which would later be called *Australopithecus afarensis*, so named for the Afar region. Hoping to find more, he returned the following year and recovered parts of a pelvis, ribs, arm and hand bones, and skull and teeth. With almost 40 percent of the original skeleton represented (after mirror-imaging the 20 percent that was actually recovered), it was the most complete fossil of an early hominid (three million years old) ever found. This skeleton was nicknamed Lucy, after the Beatles' "Lucy in the Sky with Diamonds," a camp favorite.

When the Johanson–Taieb team returned the following season, they surpassed the previous year's success by unearthing the largest collection of hominid skeletal bones ever found at that time depth (3.2 million years old). This extraordinary fossil trove, "The First Family," provided the first real glimpse of a population sample from that time. Thirteen individuals of various ages and both sexes are represented; the males appear to have been significantly larger

A CHILD'S SKULL was unearthed at Dikika, near the Awash River in Ethiopia, close to where "Lucy" was found in 1974. Dated at 3.2 million years, it had a chimp-sized brain. Reconstruction by Viktor Deak. © Viktor Deak, used by permission of Nevraumont Publishing Company.

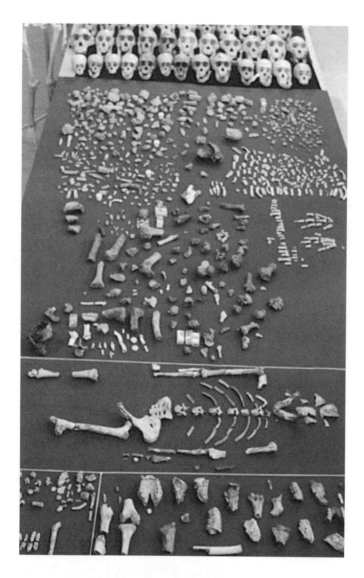

AFAR HOMINID FOSSILS from Don Johanson's early Ethiopian expeditions are spread out with chimpanzee skulls from the Cleveland Museum's collection. "First Family" fragments lie near the famous "Lucy" skeleton.

than the females. All were relatively short (about four feet tall) with bipedal hips, legs, and feet—fully upright walkers with chimp-sized brains.

Since the late 1990s, a succession of fossil discoveries by White and others has revealed a remarkable record of earlier and later human evolution in the Afar.

First, there was the Herto cranium, found with finely shaped implements of basalt and obsidian, a volcanic glass. It belonged to an ancient *Homo sapiens*, whose brain was slightly larger than our own, and is dated at 155,000 years ago. Near the Herto site, at Bodo, a 300,000-year-old cranium with a somewhat smaller brain capacity was found; however, it appears to be not fully human.

Nearby, in older sediments, White's team found a heavy-browed *Homo erectus*, along with stone tools and remains of extinct animals about a million years old. Further south, at Bouri, in 2.5-million-year-old deposits, White, then with Berhane Asfaw's Ethiopian team, found a creature they called *Australopithecus garhi*. It had an apelike muzzle, a brain the size of a gorilla's, and may be associated with cut-marked animal bones that were found nearby.

In 1992, while digging in 4.4-million-year-old sediments at a place called Aramis, White's team found the even more primitive-looking *Ardipithecus*. Four years later, Yohannes Haile-Selassie of the Cleveland Museum of Natural History reported teeth and bones from a 5.7-million-year-old ape that he named *Ardipithecus kadabba*.

Most recently, in 2006, at a site called Dikika, six miles from where Lucy was found, Zeresnay Alemseged of the Max Planck Institute discovered an excellent skull and partial skeleton of a 3.3-million-year-old juvenile female *Australopithecus afarensis*. More complete than Lucy, the three-year-old child was nicknamed Selam, the Ethiopian word for "peace." Over the past 25 years, according to White, the Afar has yielded the remains of 225 hominid individuals. Over several million years, they show a trend toward larger body size, larger brains, smaller teeth, and more sophisticated stone tools.

The importance of the Afar region has increased with each new discovery. Before the 1960s, no paleoanthropologists had ventured there. Taieb was the first to explore the Awash Valley area, with its sparse trees in the midst of Ethiopia's vast, arid desert region. He was interested in mapping its geological features but found himself in a remarkably rich fossil treasure trove. Steep ravines exposed sandstone layers, from which elephant bones, tusks, and extinct rhino remains stuck out as far as the eye could see. In 1964, he was joined by geologist Jon Kalb, who described a similar scene in Hadar, another locality in the Afar. Kalb wrote that he was awed by the sight of "an enormous, flat-lying encyclopedia of natural history with part of one page exposed on this hill, another in that ravine, another on the crest of a ridge. The formidable task ahead of us was to put together the pieces and see how much of any one page we could read."

See also AUSTRALOPITHECINES; JOHANSON, DONALD; "LUCY"

AGASSIZ, LOUIS (1807–1873)
Geologist, Zoologist

One of the most influential naturalists of the 19th century, Louis Agassiz (AGG-uh-see) was a comparative anatomist of the old school, who had studied with Georges Cuvier in Paris. His comprehensive, meticulous volumes established the specialty of fossil fish studies, and his work on European glaciations provided the foundation for all future research on ice ages.

Raised in the Swiss Alps, he suspected their glaciers were remnants of vast continental ice sheets, which had left such telltale evidence as isolated boulders, deeply etched grooves in surface rock, and characteristic rubble heaped by the moving ice. Geological studies in Scotland and Ireland produced supporting evidence for his glacial hypothesis. The world's leading geologist, Sir Charles Lyell, rejected the idea as fanciful, but within a few years Agassiz's discovery of periodic ice ages became universally accepted.

After university stints in Europe, in 1848 Agassiz was wooed by Harvard, where he became the most influential zoologist and paleontologist in America. He founded the Agassiz Museum of Comparative Zoology there. Completed in 1860, it is preserved today as a Victorian-style natural history museum, just as he designed it—a three-dimensional textbook of the Plan of Creation as reflected in classification.

As an educator, he hoped he had "taught men to observe." He advised students to "read nature, not books. . . . If you study nature in books, when you go out-of-doors you cannot find her." One of the standard ordeals he imposed on new students was to leave them alone in a room for hours with a reeking preserved specimen of a fish or bird, telling them they must not leave until they made a detailed written description of everything they could observe about it. Frequently, he would send them back for additional hours—or even days—of communing with it until he was satisfied with their observations.

Despite his studies of fossil animals and ancient climatic changes, Agassiz remained a staunch antievolutionist. A believer in divine plans and ideal forms, he thought that later species must have had separate, successive creations, unrelated to what had gone before. Each time the ice sheets retreated, the Earth was repopulated by divine creation of new species.

In his *Methods of Study in Natural History* (1863), Agassiz compared the idea of continuous evolution—then known as the development hypothesis—to medieval alchemy. "The philosopher's stone is no more to be found in the organic than the inorganic world," he insisted, "and we shall seek as vainly to transform the lower animal types into the higher ones by any of our theories, as did the alchemists of old to change the baser metals into gold."

Standing on his reputation as the greatest naturalist in America, he ridiculed the Darwinian theory when it appeared in 1859 and refused to reconsider his position to the end of his life. Agassiz was stunned, however, when his best students, including his own son Alex, a marine biologist, abandoned his system of thought and adopted Darwinian theory. Because his own teacher Cuvier had appeared to prevail over Lamarck's earlier evolutionism, Agassiz was convinced that he would "outlive this mania." Instead, his influence gradually eroded until he slipped off his pedestal as a leader of biological thought.

Years after his death, during the 1906 San Francisco earthquake, a huge marble statue of Agassiz actually did topple from the Zoology Building of Stanford University. Though unbroken, it was found upside down, head firmly planted in the cement. Stanford's first president, David Starr Jordan, recorded in his memoirs that one professor declared that he had always thought that Agassiz was "better in the abstract than in the concrete."

See also ICE AGE; MILLER, HUGH; PROGRESSIONISM

TOPPLED FROM HIS PEDESTAL as leading zoologist of the 19th century, Professor Louis Agassiz refused to recognize the Darwinian revolution in biology. After the San Francisco earthquake in 1906, a large statue of Agassiz was found upended, with its head stuck firmly in the ground.

AGNOSTICISM
Seeking Evidence for Belief

Evolutionist Thomas Henry Huxley (1825–1895) may not have been the first agnostic, but he was the first to call himself one. Comparative physiologist, innovative educator, and "Darwin's bulldog," Huxley's interests ranged widely over science, religion, and philosophy. He coined the term "agnostic" in 1869, when he joined London's Metaphysical Society, a group of theologians, scientists, and writers who met to explore questions of belief.

When asked whether he was an atheist, a Christian, a theist, a materialist, an idealist, a freethinker, or a pantheist, Huxley was at a loss. He hadn't "a rag of a label to cover [himself] with" and felt like the proverbial fox without a tail who was disowned by his fellows.

> The one thing in which most of these good people were agreed was the one thing in which I differed from them. They were quite sure they had attained a certain "gnosis"—that is, a revealed knowledge of the truth about existence.
>
> So I took thought, and invented what I conceived to be the appropriate title of "agnostic," (meaning without revealed knowledge). It came into my head as suggestively antithetic to the "gnostic" of Church history, who professed to know so much about the very things of which I was ignorant; and I took the earliest opportunity of parading it at our Society, to show that I, too, had a tail, like the other foxes. To my great satisfaction, the term took.

Agnosticism, Huxley took pains to point out, "is not a creed but a method," a skeptical, experimental approach to personal belief. "In matters of the intellect," he advised, "follow your reason as far as it will take you [and] do not pretend that conclusions are certain which are not demonstrated or demonstrable." Nevertheless, he had "a deep sense of responsibility" for his actions, and nurtured a profound religious feeling without relying on organized religion. When Huxley's young son died in 1860, the Reverend Charles Kingsley asked if he now regretted his lack of belief in the soul's immortality.

In an uncompromising and moving letter, Huxley replied:

> If a jeering devil asked me what profit it was to have stripped myself of the hopes and consolations of the mass of mankind . . . [I should answer] truth is better than much profit. . . . [I] refuse to put faith in that which does not rest on sufficient evidence, I cannot believe that the great mysteries of existence will be laid open to me on other terms.

See also HUXLEY, THOMAS HENRY; METAPHYSICAL SOCIETY; SECULAR HUMANISM

AKELEY, CARL (1864–1926)
Artist Who Saved African Apes

Gorillas, one of our close evolutionary kin, have yielded important clues to the roots of human behavior. But without the impassioned concern of artist-taxidermist Carl Akeley, there might have been no gorillas left in the mountains of central Africa for anyone to study.

Akeley became interested in animals while still a boy in the farming town of Clarendon, New York. A self-taught taxidermist by age 13, he practiced on neighbors' pets that had died, and then moved to Rochester to work at Ward's Natural Science Establishment. There he helped mount P. T. Barnum's famous Jumbo, when the elephant was killed in a circus train accident—a big step up from the neighbors' canaries.

Hired by Chicago's Field Museum to create a series of dioramas of North American mammals, he invented an entirely new method of taxidermy. After measuring muscles and bones when the animals were skinned, he sculpted clay models in realistic action poses. These were cast in plaster, and a light, hollow shell was made from the mold. Finally, the skin was carefully fitted over the sculpture. So startlingly lifelike were the results of his art that the Akeley Method was adopted by all world-class museums.

Akeley went to Africa for the Field Museum in 1896 and again in 1905. He fell in love with its wildlife—though the animals did not always return his affection. On his first trip, he was mauled by a leopard he wounded, which bit deeply into his left arm. Somehow, he managed to kill the enraged cat with his bare hands and escaped with his life.

After his African wildlife exhibits won fame and an invitation to dine at the White House with President Theodore Roosevelt, he was hired by the American Museum of Natural History in New York. In 1909, he joined Roosevelt's hunting safari in Uganda, where the president shot an elephant for the museum's mounted herd.

During the same field trip, Akeley had another close call when a bull elephant attacked. While recuperating, he conceived the museum's great African Hall: a wide-ranging depiction of the continent's ecology and wildlife. "My fondest dream," he called it, "the unifying purpose of my work."

Although it has endured as a world-class museum treasure, Akeley did not live to see his masterpiece completed. Arguments among museum officials and drawn-out quests for funding resulted in years of delay. Akeley's assistants and colleagues, whom he had trained, completed all the dioramas after his death.

Inspired by the accounts of explorer Paul du Chaillu, Akeley became increasingly attracted by the mountain gorillas of the Virunga volcanoes, then part of the Belgian Congo. When he returned to Africa in 1921, he sought the elusive apes in their remote forest home and was the first to take motion pictures of them in the wild. Although they were new to science, European "sportsmen" ruthlessly hunted them.

Despite intense feelings of affection and kinship for the great apes, Akeley shot five of various ages and sexes, took casts of their faces and hands, and brought their skins back to New York for his African Hall. His mounted family group, frozen in time as they browse in their lush mountain forest, is a masterpiece that still excites millions of visitors. Even Dian Fossey, the most fanatic of gorilla conservationists, said when she saw them for the first

If Carl Akeley had not convinced the King of Belgium to protect mountain gorillas, the apes would be extinct today.

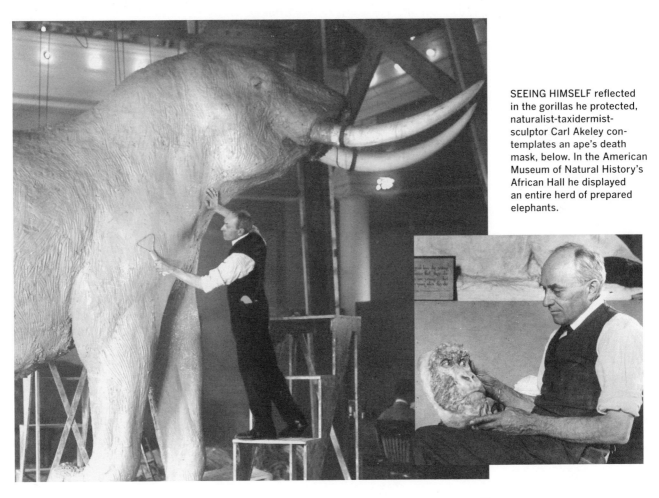

SEEING HIMSELF reflected in the gorillas he protected, naturalist-taxidermist-sculptor Carl Akeley contemplates an ape's death mask, below. In the American Museum of Natural History's African Hall he displayed an entire herd of prepared elephants.

time in 1983 that she didn't begrudge Akeley the taking of those gorillas—so "respectful" had he been in creating the classic diorama.

Starting in about 1922, Akeley became an insistent, lone voice calling out for the conservation of gorillas as a world treasure, and his campaign impressed the Belgian ambassador. Finally, he took his case directly to King Albert, who convinced the Belgian government to create the Parc National Albert in 1925.

In 1926, Akeley returned there to work with photographers, artists, and botanists on the background for his gorilla habitat group. Although afflicted with dysentery, he led his party through the soaking, misty forests to the heart of the gorillas' homeland, which he considered "the most beautiful spot in the world." There Carl Akeley died and was buried by Mary, his third wife, and a small group of friends.

Concrete was brought to the remote site, and Akeley's forest grave was turned into a permanent memorial, later visited by such noted naturalists as George Schaller and Dian Fossey when they followed him in observing wild goril-

MAN IN THE MIRROR
Carl Akeley in his field tent after tangling with a leopard, which he killed with his bare hands.

las. However, in 1979, at the peak of hostilities between Fossey and local gorilla poachers, the grave was broken into and Carl Akeley's bones were stolen. In 1990, some bones and a skull, presumed to be Akeley's, were anonymously returned, and Akeley's biographer, Penelope Bodry-Sanders, repaired the memorial.

But "his greatest memorial," writes Bodry-Sanders, "is that the mountain gorillas still walk the earth today. They are endangered . . . and under constant pressure from poachers and human encroachment, but they still exist." If it had not been for Carl Akeley, the gorillas' evidentiary link to human beings might have been lost forever, within a century of their discovery.

See also *CHRYSALIS; FOSSEY, DIAN; GORILLAS*

ALEX (1976–2007)
Language-Learning Parrot

That parrots can mimic human speech is no surprise, but an African grey parrot named Alex (an acronym for Avian Learning EXperiment) seemed to know what he was saying. Irene Pepperberg, a theoretical physicist turned cognition researcher based at Brandeis University, worked with Alex for more than 30 years. When shown two wooden triangles, one green and one blue, and asked, "What's the same?" Alex would reply, "Shape."

If asked, "What's different?" he said, "Color." He could identify 50 different objects, count quantities up to six, distinguish seven different colors and five shapes, understand "bigger," "smaller," "same," and "different," and understand the concept of absence ("none"). More talented than Pepperberg's other two parrots, Alex analyzed hundreds of shape and color combinations he had never seen before.

Until Pepperberg's first publication in 1981, scientists considered conceptual use of symbols unique to primates. Alex asked for things by name ("I want cork"

AFRICAN GREY PARROT
Alex astounded scientists for 30 years with his analytical and communicative skills, surpassing those of many primates. Photo by Arlene Levin-Rowe, courtesy of Irene Pepperberg.

or "Want key"), then demonstrated he really wanted the items by cleaning his beak with the cork or scratching his head with the key. If given a key when he asked for cork, he would toss it back. He knew more than 100 words and used them appropriately in simple sentences. He said "No!" ("nuh") or "I'm gonna go away" when he didn't want to do something, and sometimes demanded rewards by saying "Want a nut" loud and clear.

Alex's communicative abilities have yet to make anything like the popular impact of Jane Goodall's chimps "fishing" termite mounds with twigs—a possible example of our pro-primate bias. Yet the bird's accomplishments, with a brain the size of a thimble, may well revolutionize our estimation of conceptual capacities among nonhuman beings. In her book *Alex & Me* (2008), Pepperberg opined, "Clearly, animals know more than we think, and think . . . more than we know."

According to Pepperberg, Alex's final words as she returned him to his cage for the last time were his customary "You be good. I love you. . . . You'll be in tomorrow?"

See also APE LANGUAGE CONTROVERSY; CLEVER HANS PHENOMENON; KOKO; NIM CHIMPSKY; WASHOE

ALLELOCHEMICALS
Evolutionary Poisons

In the course of the evolutionary "struggle for existence," many animals and plants have evolved chemical defenses against predators. Such natural chemicals that affect the health, behavior, growth, or population biology of members of other species are called allelochemicals.

Many plants produce substances that are toxic to the insects, bacteria, rodents, and even viruses that try to feed on them. One of the most common allelochemicals, known as L-canavanine, has been identified in 1,200 legumes, including clover, wisteria, alfalfa, and some trees. When hookworm larvae were fed L-canavanine in experiments at the University of Kentucky, it interfered with their protein metabolism, thereby disrupting development and producing defective adults.

Some creatures utilize allelochemicals even though they cannot produce them. Certain caterpillars, for instance, have evolved an immunity to the poisonous leaves on which they feed. When attacked by ants, they repel them by regurgitating a toxic chemical.

Research into allelochemicals is providing a natural source of pesticides and other agricultural poisons—without the harmful side effects of DDT and other artificial chemicals.

See also NATURAL PRODUCTS CHEMISTRY

ALLOPATRIC SPECIATION
Evolution and Geographic Isolation

We enjoy a freedom of movement today that is new in the Earth's history. Until a century ago, most animals were often blocked in their travels by mountains, oceans, rivers, forests, or deserts—obstacles that created small, isolated breeding populations.

For instance, a river may divert and cut across a population's range, leaving breeding groups on either side. When populations are genetically isolated, they may diverge, first into geographic subspecies and, finally, species. These are known as "allopatric" ("different places") species. More rarely, new species may evolve from within a single, widely spread out population ("sympatric" or "same place" species).

Because individual subgroups encompass only part of the population's total variability, they take on different characteristics. As a result of sampling error (see GENETIC DRIFT), one cut-off group may have a higher proportion of the same genes or the presence or absence of particular alleles or mutations.

In small populations, such random differences can easily become established and passed on to a large number of descendants. Thus, the most common "origin of species" may occur on the edge of species ranges, with very small subpopulations that have become isolated from the main population.

After a time apart, if the barriers are removed and two populations should remingle, they may have evolved differences in behavior, color, or vocalization ("isolating mechanisms") that still keep them from mating.

AMERICAN PALEOZOIC MUSEUM
Victorian Dinosaur Disaster

A VICTORIAN VISION of monsters appears in this drawing (left) by Benjamin Waterhouse Hawkins of the ill-fated Paleozoic Museum he planned for New York's Central Park. Hawkins' earlier statues (right) survive in a park in Sydenham, England.

Photo on right courtesy of and © by Gryphon Productions Ltd.

After Benjamin Waterhouse Hawkins (1807–1894), the first great dinosaur artist, created a successful prehistoric park in England, Americans wanted a bigger, better one. Under anatomist Richard Owen's direction, Hawkins had sculpted life-size models of the giant saurians at London's Crystal Palace—inventing the first dinosaur theme park. His monsters were so popular with the public that he was invited in 1869 "to undertake the resuscitation of a group of animals of the former periods of the American continents" in New York City's Central Park.

Working with the paleontologist Edward Drinker Cope in Philadelphia, Hawkins absorbed the fossil hunter's dramatic tooth-and-claw view of prehistoric life. He took molds of American dinosaur skeletons, and then a year later he set to work at a studio in Central Park.

Hawkins began to model huge hadrosaurs being attacked by smaller dinosaurs while others fought over carcasses. Giant aquatic reptiles were to be half-submerged in real pools. Visitors would enter a huge, domed cavern, where mammoths and giant sloths awaited them—a "complete visual history of the American continent from the dawn of creation to the present time."

Unfortunately, the project ran afoul of the infamous Tweed Ring, a nest of corrupt, powerful politicians who controlled most municipal projects. In 1870, their leader, "Boss" William Tweed, managed to wrest control of all construction in Central Park from the commission that had hired Hawkins. With the false claim that it was costing the city too much money, the Tweed Ring ordered work on the project stopped. Hawkins was ridiculed in the press as an impractical dreamer.

By the spring of 1871, Hawkins had finished the plaster casts of seven major figures and was battling to keep the project alive. Late one night, Tweed's hired vandals smashed the molds with sledgehammers, destroyed sketches and small models, and taunted Hawkins about his "alleged pre-Adamite animals." To this day, Tweed's motives are unclear. Was it pure greed, because he couldn't make a huge profit from the project? An issue of control over turf? Or did religious zealotry fuel his implacable antagonism for the Paleozoic Museum?

Whatever Tweed's reasons, the result was a shattered studio and a Hawkins broken in spirit. He retreated to the Princeton Natural History Museum, where he spent a few years painting imaginative prehistoric landscapes, then returned to his native England in 1877. The smashed remnants of his work, his models, and his studio remain buried today near 63rd Street in the southwest corner of Central Park.

A few of his small models and sketches were rescued and give some inkling of what the magnificent exhibition might have looked like. His restorations were fanciful and inaccurate by present standards, but embody the unique style and viewpoint of their time. Hawkins's first prehistoric garden built for England's Crystal Palace survives at Sydenham

Park, in South London, where some of his splendid Victorian dinosaurs, still standing among streams and flower beds, remain to delight new generations.

See also HAWKINS, BENJAMIN WATERHOUSE

ANIMAL RIGHTS
Inter-Species Morality

Humanity's absolute right to use animals as it sees fit—hunt them, eat them, wear their skins, experiment on them—was for many years justified by the passage in Genesis where God gives man "dominion over all the beasts." Indeed, throughout the 19th century many scientists maintained that the purpose of animals was to be used by man. Horses were made for human transport, foxes for fur, and sheep and cows for meat. Anatomist Richard Owen of the British Museum even noted that there was a convenient gap in horses' teeth where the metal bit was "meant" to be inserted to anchor the reins.

When Darwin established an evolutionary kinship between species, he saw at once it might imply a moral obligation not to abuse animals. In one of his early private notebooks, he had jotted that "animals our fellow brethren in pain, disease & death & suffering & famine; our slaves in the most laborious work, our companion in our amusement, they may partake, from our origin in one common ancestor we may be all netted together."

His disciple in America, the Harvard botanist Asa Gray, pursued the point: "It seems to me that there is a sort of meanness in the wish to ignore that tie. I fancy that human beings may be more humane when they realize that, as their dependent associates live a life in which man has a share, so they have rights which man is bound to respect." The philosopher Bertrand Russell snorted that "such a philosophy could logically end with the demand of Votes for Oysters."

Yet many religions—including the Jewish and Christian faiths—have a tradition that grants animals the status of independent beings, created without reference to human wants, as in this rarely quoted passage:

> For the fate of the sons of men and the fate of the beasts is the same; as one dies, so the other. They all have the same breath, and man has no advantage over the beast for all his vanity. All go to one place; all are from the dust, and all turn to dust again. Who knows whether the spirit of man goes upward and the spirit of the beast goes down to earth? (Ecclesiastes 3:18)

Naturalist Henry Beston put it another way in his classic 1928 book, *The Outermost House:* "Animals," he wrote, "are not brethren. They are not underlings. They are other nations—caught with ourselves in the net of life and time, fellow prisoners of the splendor and travail of the earth."

The modern movement for animal rights, which was spearheaded by Princeton University philosopher Peter Singer, has made an impact. Through the efforts of activists, including Jane Goodall and the Great Ape Project, captive primates have been granted some protection from abuses. However, apes are better protected in Europe than in the U.S., where there are 3,000 chimpanzees in captivity—half of which are used in medical research. Such use of apes has been banned in the UK, Sweden, Spain, and the Netherlands.

Geneticist Steve Jones of University College London, however, has expressed concern that all animal experimentation could be banned. Taking a page from Russell, he told a BBC journalist, "Mice share around 90 percent of human DNA: should they get 90 percent of human rights?"

> Mallorca was first to guarantee apes legal rights, such as freedom from mistreatment or death at the hands of humans.

ANNING, MARY (1799–1847)
Discoverer of "Sea Dragons"

Mary Anning, a self-taught geologist, became world-famous for finding ancient "sea dragon" remains near her native Lyme Regis, a seaside village in southern England. At a time when virtually all geologists were wealthy gentlemen, Anning became expert at spotting fossils, extracting them from the rocks, identifying species, and analyzing their anatomy. Her 200-million-year-old plesiosaurs, ichthyosaurs, and pterodactyls (flying reptiles) became famous and helped to popularize the infant science of paleontology.

In the early nineteenth century, Lyme Regis was a poor little village, inhabited by rough fishermen and smugglers. Anning's father, a carpenter, supplemented his meager income by selling curious objects he had found at the water's edge to tourists. Young Mary assisted him, and seems to have been the inspiration for the famous tongue-twister: "She sells sea-shells by the seashore." (A popular rejoinder went: "Sam sells clam shells for a lot more.")

After her father's death in 1810, Mary and her brother Joseph made fossils their full-time business. Joseph discovered the head of a "crocodile" in 1811, and he and Mary found and restored its skeleton a year later. One of the first ichthyosaurs known to science, it still occupies an honored place in London's Natural History Museum.

A gentleman geologist, Thomas James Birch, bought many fossils from the Annings but became appalled at the family's poverty. When he visited them in 1819, Birch found the Annings "selling their furniture to pay their rent—in consequence of their not having found one good fossil for near a twelve month." Eventually, Birch sold off his entire fossil collection and donated the proceeds to the Annings, "who have in truth found almost all the fine things which have been submitted to scientific investigation."

Birch was not alone in his admiration of Mary across the class divide. Famous gentlemen called on her seeking fossils and her analyses of them—which they published under their own names. A female acquaintance wrote in her journal that Mary "says the world has used her ill and she does not care for it. According to her account these men of learning have sucked her brains, and made a great deal by publishing works, of which she furnishes the contents, while she derived none of the advantages."

Dr. Gideon Mantell, the premier discoverer of dinosaur fossils, visited Anning's "dirty shop, with hundreds of specimens piled around." He described her as "the presiding Deity, a prim, pedantic vinegar looking, thin female, shrewd and rather satirical in her conversation." Every day, in all kinds of weather, Mary could be seen combing the shores and cliffs of Lyme Regis with her collecting bag and geological hammer.

Another wealthy amateur geologist, Thomas Hawkins, related how Mary led him to discover a large plesiosaur skeleton (*Chiropolyostinus*) embedded at the base of a cliff, covered by seawater. He fell in love with the fossil, but Anning advised him against pursuing it, warning, "you will never get that animal, [because] the marl, full of pyrites, falls to pieces as soon as dry." Undaunted, Hawkins devised a chemical treatment that would preserve it. For days he "lay upon a thorny pillow listening the livelong night to the rumbling gale" and hoping against hope that his plesiosaur would not be destroyed. "The angry waters of the channel," he wrote, "are pent up by contrary winds and the relic of an incalculably remote generation sleeps on in his oozy bed secure beneath the main."

At Mary's urging, Hawkins positioned six workmen to be ready the moment the tide retreated. Within an hour, racing against the return of high tide, the men freed the fossil skeleton and got it into a horse cart. Hawkins gazed lovingly at his dragon as he secured it in a tight wooden case, encased in a shell of plaster of Paris. "I worshipped it for hours," he wrote, "in my mad intoxication of spirit." Such was the romance and excitement that Mary Anning had fostered in the quest for her "sea dragons," which are still the pride of London's and Oxford's Natural History Museums. At her death she was honored by London's Geological Society, a "gentlemen only club" from which she had been excluded in life by the very men who had built their reputations on her achievements.

APE LANGUAGE CONTROVERSY
Capacity for Symbolic Communication

Can animals learn human language? Dyak tribesmen told anthropologists that they thought wild orangutans could speak but pretended to be dumb when humans were around, because they were afraid men would put them to work. Charles Darwin's neighbor, Sir John Lubbock, tried in 1882 to teach his dogs the Sign Language for the Deaf, prompting Samuel Butler's remark: "If I was his dog and he taught me, the first thing I should

ONCE A FAMILIAR FIGURE combing the cliffs and beaches of Lyme Regis, England, Mary Anning, a self-taught paleontologist, remains a popular local hero.

"I worshipped [the sea dragon] for hours in my mad intoxication of spirit."

—Thomas Hawkins, *Book of the Great Sea-Dragons*, 1840

tell him would be that he is a damned fool." A century later—with Lubbock's dogs long forgotten—attempts to teach chimps human sign language became fashionable in research.

Robert Yerkes, the pioneering psychologist who began to study captive chimps around 1900, noticed they were more apt to mimic movements and facial expressions than the sounds of human speech. Early on, he suggested trying a visual language, but no one did until the 1960s, when a few psychologists began teaching chimpanzees American Sign Language (Amslan).

Sarah, one of the first famous language-learning chimps, was taught by psychologist David Premack to communicate with standardized plastic chips on a magnetic board. Another researcher taught a chimp called Lana to use a system of geometric symbols on a specially designed computer keyboard. By 1971, psychologists Roger and Deborah Fouts had taught Washoe, a young female chimp, 150 hand gestures in simplified Amslan. Next, Herbert Terrace's Nim Chimpsky mastered a vocabulary of 132 signs.

There were two basic questions. Could chimps learn to associate a given sign, symbol, or word with its referent? And were they capable of combining them into "sentences" according to some kind of rules (grammar)?

Herbert Terrace knew psychologists had been impressed before by "remarkable animals" in flawed experiments. In the nineteenth century, a horse named Clever Hans had amazed scientists when he tapped out answers to mathematical problems—until his trainer was completely removed from view.

In 1979, Terrace challenged his own methods and accomplishments. Reviewing the video-tapes, Terrace could not establish that his chimp had really used the symbols conceptually or mastered any of the grammatical rules that structure human language. Yes, Nim had learned scores of gestures, but he was often haphazard about their sequence. (He didn't seem to know the difference between "Nim eat banana" and "Banana eat Nim.") And he often produced a lot of irrelevant signs, which the experimenters had ignored.

Apes were playing a game all right, Terrace concluded, but not the language game; they were just running off various signs until they got what they wanted. No one had really demonstrated that chimps understood that signs carry definite meanings.

Those who wanted to preserve human uniqueness in the natural world were delighted, echoing Samuel Butler's sentiments: Researchers were deluded fools. Grant funding dried up and no new chimp language programs were begun. But the story was far from over; in fact, some surprises lay just ahead.

Duane Rumbaugh and Sue Savage-Rumbaugh, at the Language Research Center of Yerkes and Georgia State University, never doubted their apes had learned to use symbols. The challenge was to convince everyone else, so they redesigned their experiments.

The chimps were now given tasks that were impossible without real symbol use. Sherman and Austin would see a symbol flashed on a screen. The chimps then left the room, entered another room containing many objects, and were to bring back only the object named (symbolized) on the screen. To do this, they had to remember what the symbol stood for. Not only did the chimps retrieve the correct objects, but they came back empty-handed if the named object was not in the other room.

Going even further, Sherman and Austin began to communicate with each other using

AN OLD QUESTION about men and apes is expressed in this *B.C.* cartoon by Johnny Hart. Humans did not "come from" any existing ape. Chimps, gorillas, and humans share common ancestors. By permission of John L. Hart FLP and Creators Syndicate, Inc.

the signs. After being taught they must share food, the chimps used the symbols to ask each other for specific items. The Rumbaughs' work finally convinced many scientists that apes can use signs to convey meaning, though their ability to form grammatical combinations is extremely limited.

Sue Savage-Rumbaugh and Rose Sevik made a major breakthrough working with pygmy chimps, or bonobos, during the 1980s. While trying with little success to teach language behavior to a wild-born female, they inadvertently discovered that her captive-born son, Kanzi, "had learned everything we'd been trying to teach the mother." Extensive work with Kanzi resulted in his learning and demonstrating more sophisticated vocabulary and communication than in any previous experiments with apes.

During the 1980s, a growing body of research established that language-like abilities (conceptualizing and symbolizing) belong to some very diverse creatures: sea lions, dolphins, pigeons, parrots.

At the University of Hawaii, psychologist Lou Herman's dolphins not only associated hand signs with particular objects ("ball," "disk"), but also responded correctly to sentence-like instructions. Without having seen the sequence before, they were able to tell the difference between "Take the disk to the ball" and "Take the ball to the disk."

But perhaps the most astounding language learner of all was not a primate, but an African grey parrot named Alex. He could identify seven colors, five shapes, and quantities of up to six. And he needed no plastic chips, computers, or signing gestures—he did it in English! With a brain the size of a thimble, he appeared to have language capabilities that rivaled or even surpassed those of the vastly larger-brained monkeys and apes.

See also ALEX; CLEVER HANS PHENOMENON; KANZI; NIM CHIMPSKY

"APE WOMEN," LEAKEY'S
Primate Field Observers

Famous for his discoveries of early hominid fossils in East Africa, Louis Leakey wanted more than dry bones to help reconstruct early man—he wanted behavior. During the 1960s, he sought a few exceptional individuals to study our closest living relatives, the African apes.

His recruits would face dangerous, difficult years in the wilderness, trying to approach unpredictable animals capable of tearing a human apart—and they would not carry guns. From the first, Leakey believed that women were better suited for the job than men. Women, he thought, were more perceptive about social bonds and maternal behavior, more patient and capable of long-term dedication, and would perhaps appear less of a threat to the male apes.

> **Louis Leakey believed that women would be better suited than men to study wild apes without appearing to threaten them.**

Jane Goodall, a young Englishwoman, was the first of Leakey's three "ape women," whom he sometimes referred to as his "trimates." She came to visit him when he was curator of the National Museum in Nairobi, and was hired as assistant-secretary on the spot. "He must have sensed," she later wrote, "that my interest in animals was not just a passing phase." She accompanied Louis and Mary Leakey on their next paleontological expedition to Olduvai Gorge, where she absorbed their enthusiasm for understanding the roots of human behavior and evolution. Leakey asked her to study a group of wild chimpanzees he had seen near Lake Tanganyika.

Leakey was particularly interested in the lakeshore habitat because some East African early hominid sites showed evidence of lakeside living. When Goodall protested that she was inadequately trained to undertake such a study, "Louis [told me he felt] a university training was unnecessary, [and] even that in some ways it might have been disadvantageous," she later recalled. "He wanted someone with a mind uncluttered and unbiased by theory who would make the study for no other reason than a real desire for knowledge; and, in addition, someone with a sympathetic understanding of animals."

Next, Leakey wanted to find a "gorilla girl." Like Goodall, Dian Fossey had no university background in animal behavior, but had loved animals since childhood. She was working as an occupational therapist in Louisville, Kentucky, and had traveled to Africa on vacation to see wildlife. She met Leakey—who spoke of his search to find a woman to study wild gorillas.

In 1966, when Leakey gave a lecture in her home city, Fossey spoke with him about her continuing interest in studying gorillas. "After a brief interview," she wrote in *Gorillas in the Mist* (1983), "he suggested I become the 'gorilla girl' he had been seeking. . . . Our conversation ended with his assertion that it was mandatory I should have my appendix removed before venturing into the remote wilderness" of the Zaire volcanoes. She promised she would have the operation.

Six weeks later, on returning home from the hospital minus her appendix, Fossey found a letter from Leakey. It began, "Actually, there really isn't any dire need for you to have your appendix removed. That is only my way of testing an applicant's determination!" Was he kidding about having the operation, or about not having it? "This was my first introduction," she wrote, "to Dr. Leakey's unique sense of humor."

A few years later, inspired by the efforts of Goodall and Fossey, another young woman, Biruté Galdikas, undertook a prolonged field study of the orangutan in Indonesia.

Leakey had found and inspired three "unqualified" amateurs, who, though working outside of academia, conspicuously outperformed every established university department of anthropology in the world!

See also APES; FOSSEY, DIAN; GALDIKAS, BIRUTE; GOODALL, JANE; LEAKEY, LOUIS

APES

See APES, TOOL USE OF; BONOBOS; CHIMPANZEES; GORILLAS; KANZI; NIM CHIMPSKY; ORANGUTANS

APES, TOOL USE OF
A Puzzling Pattern

In an early series of experiments by the German psychologist Wolfgang Köhler published in 1917, a captive chimp named Sultan piled up several boxes, climbed atop this construction, and used a stick to knock down a banana dangling from the ceiling. When given a short stick, he used it to snag a much longer stick just outside his cage, then used the new one to reach up and knock down the prized fruit.

Over fifty years later, in 1971, British anthropologist R. V. S. Wright decided to settle the question of whether apes were capable of making stone tools with other stones. Working with Abang, a five-and-a-half-year-old orangutan in the Bristol zoo, Wright repeatedly demonstrated how to use a flint blade to cut a nylon rope tied around a food box. Abang learned to use the knife to get at the food. Next, he mastered the skill of making his own knife by striking off a sharp flint flake with a hammerstone, and used it to cut the rope.

In 1994, Carel van Schaik of Duke University documented the use of sticks and branches as tools among wild orangutans. In the swamps of the northwest corner of Sumatra, he watched the red apes prepare sticks to use for getting termites, breaking into ants' and bees' nests, and remove sharp spines from wild fruits.

Gorillas, though closely related to chimps and humans, had never been observed to use tools in the wild for getting or preparing food, though they do build night-nests out of sticks and leafy branches. In 2007, however, primatologists watched Congo gorilla families wade into swampy areas to forage on succulent plants. The apes used straight sticks or staffs to test the water's depth and the mud's capacity to support their weight, much as humans would do.

As is well known, chimpanzees strip the leaves from twigs to prepare them for poking into termite mounds to capture insects. After applying saliva to a thin twig, they push it into a termite burrow, where thirsty insects grab onto it. The apes then withdraw the probe and eat the termites. In 2004, Crickette Sanz of Washington University (St. Louis) and her colleagues videotaped Congo chimps using various kinds of specially prepared twigs—a veri-

JANE GOODALL'S CURIOSITY led her to seek out fossil-hunter Louis Leakey in Kenya, never dreaming he would steer her toward studies of chimpanzees that would bring her international acclaim.

Photo © by Kenneth Love.

table "tool kit"—to perform different tasks in gathering termites. They may prepare a heavy puncturing stick, a lighter perforating stick, or a light, flexible "fishing" stick.

Chimpanzees also chew leaves and use them as sponges for gathering water from tree hollows in the dry season, and have been observed using an anvil stone and a bashing stone to smash open nuts. Neither gorillas nor orangutans have ever been observed to use stones as tools in the wild.

In 2007, archeologists in West Africa discovered stone tools (simple bashing stones and anvils, like those chimps use today) that were made and used by apes four thousand years ago. Some stones—the first stone artifacts known from chimp prehistory—still contained nut protein residues.

Kanzi, a bonobo at the Georgia State University Language Research Center, has been spectacular at solving problems associated with making tools, including redefining the nature of the experiments. He acquired much of his advanced symbol-using abilities by observing other "educated" chimps, especially his adopted mother.

One primatologist tried to find a pattern. Was evolutionary closeness to humans correlated with proficiency of tool use? No such pattern could be discerned. To confuse matters further, in 2004 Cambridge University anthropologists Antonio Moura and Virginia Lee reported that they observed capuchin monkeys using stones to dig for insects, roots, and tubers and crack open seeds and twigs to probe for insects—although the monkey's brain is the size of a domestic cat's.

See also ABANG; CHIMPANZEE; KANZI

ARCHAEOPTERYX
Transitional Fossil

Charles Darwin, in a letter to Charles Lyell, predicted the discovery of an *Archaeopteryx*-like fossil more than two years before it was described. When the fossil was discovered in 1863, his friend Hugh Falconer excitedly informed him:

> Had the Solenhofen quarries [in Germany] been commissioned—by august command—to turn out a strange being à la Darwin—it could not have executed the behest more handsomely—than in the *Archaeopteryx*.

At Solenhofen, a fine-grained stone had been quarried for centuries because it is smooth, porous, and particularly suitable for lithography, such as in printing the famous Toulouse-Lautrec posters. Within these 150-million-year-old Jurassic stones, a beautifully detailed imprint of a single feather was found in 1861, and two years later a complete skeleton of the remarkable creature was discovered. It was named *Archaeopteryx lithographica*—ancient winged creature from the printing stone.

Coming just two years after the publication of Darwin's *Origin of Species*, it was hailed as a "missing link" between reptiles and birds, a proof of the theory of evolution written in stone. *Archaeopteryx* appeared to be a truly transitional creature, combining attributes of two classes of vertebrate animal. Its feet were suitable for perching, its pelvis seemed birdlike, yet it did not have the keeled breastbone to which flight muscles attach in modern birds nor their light, hollow bones. Feathers were completely birdlike, but the skeleton showed digits on the bifurcated wings, claws, teeth, and long vertebral tail. Darwin called it "the wondrous Bird . . . by far the greatest recent fossil."

Anatomist Richard Owen, a bitter enemy of the evolutionists, described it as an aberrant bird. But a sharp-eyed paleontologist, John Evans, noticed that Owen had overlooked the fine, perfectly formed teeth within the beak. Congratulating Evans on that crucial observation, Hugh Falconer joked that perhaps he would next find the creature's "fossil song."

Attempts have been made over the years, including one in the 1980s by astronomer Fred Hoyle, to prove that the fossil is a fake, cooked up by paleontologists to support their theories. But the microscopic structure of the feath-

THE THROWN-BACK HEAD in this classic *Archaeopteryx* fossil is consistent with death by trauma to the nervous system and can be seen in modern animals that die of brain injury or toxic bacterial infections.

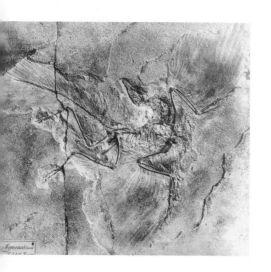

ers, preserved in fine-grained stone, is much too intricate to copy; subsequent discoveries of other specimens have passed every conceivable test for authenticity.

During the 1990s, hundreds of fossil discoveries from China established the existence of many kinds of early birds and "feathered dinosaurs." Many represent closer and much earlier affinities between dromaeosaurs—upright, meat-eating dinosaurs—and birds. *Archaeopteryx* is now thought to be a very early "bird," but only one of many transitional species that are now known from the late Jurassic and early Cretaceous.

See also BRANCHING BUSH; BRIDGEWATER TREATISES; CLADISTICS; DINOSAURS, FEATHERED; "MISSING LINK"; TRANSITIONAL FORMS

ARGUMENT FROM DESIGN
See also BRIDGEWATER TREATISES; INTELLIGENT DESIGN; PALEY'S WATCHMAKER; PANDA'S THUMB

ARISTOGENESIS
Evolution by the "Superior Few"

Paleontologist Henry Fairfield Osborn, the influential director of the American Museum of Natural History in its formative years, fancied himself an aristocrat. Upper crust in manners, education, and social standing, he believed evolutionary progress was driven by the "superior few"—his theory of aristogenesis.

Osborn extended his notion of superior and inferior to ethnic groups, never doubting that he and his circle were the highest products of evolution. To the lasting embarrassment of his admirers, he wrote enthusiastic introductions for his friend Madison Grant's notorious books *The Passing of the Great Race* (1916) and *The Conquest of a Continent* (1933). Grant warned of "Nordic debasement" by a flood of "alien" Italians, Jews, Asians, and Africans unless America maintained strict immigration quotas and laws against "racial intermarriage."

His theory of aristogenesis also led Osborn to disavow the "apish ancestry" of man—a strange position for America's leading evolutionist. Instead of Darwin's ape-man, he imagined a separate, superior line of "dawn-men," who never lived in forests (*Man Rises to Parnassus*, 1927). "We have all borne with the monkey and ape hypothesis long enough," he told an audience, "and are glad to welcome this new idea of the aristocracy of man back to . . . a remote period." Anatomist William K. Gregory claimed Osborn was "afflicted with pithecophobia—the dread of apes as relatives or ancestors."

See also EUGENICS

AUTOCRATIC PRESIDENT of the American Museum of Natural History and paleontologist, Henry Fairfield Osborn startled colleagues by disavowing an ape ancestry for humans. In this 1930 newspaper cartoon, a chimpanzee is greatly relieved at the news.

ARTIFICIAL SELECTION
Breeding Domestic Varieties

Natural selection is a familiar concept from Charles Darwin's *Origin of Species* (1859), but he also wrote several volumes on what he called artificial selection—the creation of domestic varieties by breeders.

In *Origin*, Darwin contrasts natural and artificial selection. He makes it clear that natural selection does not create variability—it merely acts on "the individual differences given by nature." Man selects the variations he can see, while Nature "can act on every internal organ . . . on the whole machinery of life."

Humans breed animals for what they find beneficial to themselves, while Nature selects

MICKEY MOO, a Holstein cow, was born on a farm in Whitefield, Maine. She was bought by Disneyland in 1988 and bore a calf there—which did not inherit the Mickey-shaped splotch that made its mother famous.

© Disney Enterprises, Inc.

for the benefit of the plant or animal species. Since artificial selection has produced the fastest horses, the most succulent fruits, and the most ornate pigeons, Darwin asked, "What may not natural selection effect?"

Characteristically, Darwin was not content to draw the comparison and rest; he methodically found out everything he could about plant and animal breeding. He built a dovecote behind his house, bred pigeons, and attended meetings of the local pigeon enthusiasts. He bombarded horse breeders, gardeners, and agricultural experts with his questions, and performed thousands of botanical experiments in his garden and greenhouse.

In 1868, he published a two-volume work on the subject: *The Variation of Animals and Plants under Domestication*. Here Darwin covers virtually everything that was then known about the breeding of dogs and cats, horses and asses, pigs, cattle, and other barnyard animals, pigeons, canaries, fowl, bees, fruit, flowers, and vegetables. Though little read today, its wealth of detail was invaluable to contemporary gardeners and breeders.

In fact, Darwin had gathered so much material on pigeons that the bird figured prominently in several of his books. When his publisher submitted the manuscript for *Origin of Species* to an independent evaluator for an opinion, the critic wrote back suggesting he cut the theoretical parts and turn it entirely into a book on pigeon breeding! "Everybody is interested in pigeons," he enthused. "The book would be reviewed in every journal in the kingdom, and would soon be on every library table."

See also NATURAL SELECTION

"ARYAN RACE," MYTH OF
Racial Supremacist Theory

One of the most infamous and disastrous attempts to trace the racial ancestry of Europe was born as a minor issue in comparative linguistics, developed into a pseudo-Darwinian theory of history, and became a cornerstone of Nazi ideology.

Originally, the term "Aryan" was applied to a language group also known as Indo-European. In the 1780s, Sir William Jones, an English Orientalist, compared languages and grammars and concluded that ancient Indian Sanskrit was related to Persian, Greek, Latin, Celtic, and Germanic languages. He thought that they all must have branched off from a lost mother tongue, which he called "Aryan" after the "Aryas," an ancient people who had supposedly invaded India and Persia. By the mid-19th century, German linguists and anthropologists developed "Aryan" studies into a major branch of inquiry.

It was but a short, illogical step from the notion of a single mother tongue to conjectures about a single original race that civilized Europe.

Count Arthur de Gobineau, a French journalist and historian who claimed aristocratic descent, believed that humankind was divided into three races, differing in degrees of superiority: black at the bottom, yellow in the middle, and white at the top. In his *Essays on the Inequality of the Races* (1850s), he asserted that within the white race, the Aryan branch was the highest of the high. Aryans originated in central Asia, he believed, and were tall, blond, alert, honorable, and powerful.

Gobineau wrote that he was "sure that everything great, noble, and fruitful that man has created on earth . . . issues from a single root, results from a single idea, and belongs to a single family—the Aryan race." In England, Max Müller, a professor of comparative philology at Oxford, championed the Aryan source of European civilization.

As Müller gathered more evidence, however, his belief that European culture was founded by a pure Aryan race evaporated. In *Biographies of Words and the Home of the Aryas* (1888), he about-faced, arguing that language has nothing to do with race and that a person of any race can learn to speak any language. "An ethnologist who speaks of Aryan race, Aryan blood, Aryan eyes and hair," he wrote, "is as great a sinner as a linguist who speaks of a dolichocephalic [long-headed] dictionary or a brachycephalic [broad-headed] grammar." But Müller's earlier teachings had been enormously influential and had already done their harm.

With the spread of European colonial empires and the inequities of economic domination came the rise of Social Darwinism as a convenient justification for conquest. If evolutionists had taught that it is "natural selection" for the "fittest" to survive, then it was only right that the "superior" white race should dominate and subjugate people with yellow or brown skin. And blond, blue-eyed people should rule over brown-eyed people, Germans over Jews, and so on. Darwin would have been appalled. Many times he had emphasized that he was not a Social Darwinist, that he detested slavery, and that his theories about the natural world were misapplied to commerce and politics.

In America, the most famous advocates of "Aryan" supremacy were Madison Grant, who in 1916 wrote *The Passing of the Great Race,* and Lothrop Stoddard, whose *Rising Tide of Color Against White World Supremacy* appeared in 1920. American racists propagandized against "mixing" with people of color, and also tried to bar immigration of "inferior" European types such as gypsies and Jews. Englishman Houston Stewart Chamberlain (*The Foundations of the Nineteenth Century*, 1899) and the German composer Richard Wagner (who published his anti-Semitic diatribe *Judaism in Music* in 1850) directed their venom at European Jews. In their popular writings, everything that was good, true, and pure was Aryan; everything that was low and degraded was Jewish.

As the Aryan hysteria continued to froth, any serious examination of language stocks or ethnic histories was now completely overwhelmed by polemics, hatred, and politics. Chamberlain wrote this prophetic statement in his *Foundations*: "Though it were proved that there never was an Aryan race in the past, yet we desire that in the future there may be one. That is the decisive standpoint for men of action." When asked to define an Aryan during the height of the Nazi madness, Josef Goebbels proclaimed, "I decide who is Jewish and who is Aryan!"

During the German Third Reich (1933–1945), the ideal of Aryan purity and supremacy became national policy. Adolf Hitler's program of herding "inferior" races into concentration camps and gas chambers was rationalized as making way for the new order of superior humanity. Meanwhile, S.S. officers were encouraged to impregnate selected women under government sponsorship to produce a new "master race"—an experiment that produced a generation of ordinary, confused orphans.

Hitler was furious when the black American Jesse Owens outraced "Aryan" athletes at the 1936 Berlin Olympics, contradicting his theories of racial supremacy. And when the "Brown Bomber" Joe Louis knocked out boxer Max Schmeling in 1938, German propaganda became even more vehement that white superiority would be vindicated. However, when Hitler needed the Japanese as allies in World War II, he promptly redefined the Asians as Aryans.

Historian Michael Biddiss has commented that "the history of the Aryan myth demonstrates the power of belief over the power of knowledge. . . . We may now hear more often of Caucasians than of Aryans, but the substance and errors of the belief in white supremacy linger."

See also LEBENSBORN MOVEMENT

ARYAN SUPREMACY POSTER, issued by the Nazi government in Germany (1932), shows a steely Nordic knight battling the many-headed reptile of "inferior races" within the Fatherland. It reads, "Fight the danger! Damage prevention is your duty! This concerns you!"

Heinrich Himmler, head of Hitler's Stormtroopers, believed the "Aryan race" originated separately from the rest of humankind, preserved as "living shoots" in ice crusts from outer space.

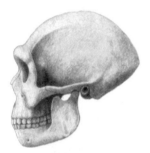

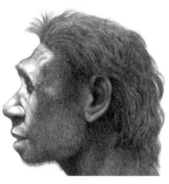

ATAPUERCA
"Pre-Neanderthal" Death Pit

The gently rolling hills known as the Sierra de Atapuerca in northern Spain hide a complex system of limestone caves where paleontologists and archeologists have uncovered abundant—if enigmatic—evidence of western Europe's earliest inhabitants.

Hundreds of bones and artifacts have been recovered from two principal Atapuercan sites over the past quarter-century. Although the caves are quite close to one another, their contents are believed to date from very different time periods—and apparently represent different species of hominids.

Railroad workers uncovered the first site, Gran Dolina, when they blasted a hillside to make way for train tracks in the 1890s, exposing a sequence of prehistoric deposits. Archeologists, alerted by the railroad company, immediately recognized it as an "early man" site, but excavations proceeded slowly for almost a century. During the 1990s, with an influx of government funding and the help of hundreds of student volunteers, the pace of excavation and discovery dramatically accelerated.

The site's eleven layers have yielded hominid fossils, 200 stone tools, and more than a thousand fossils of horses, deer, rhinoceros, bison, wolves, wild cats, elephants, and other animals. Most exciting for Spanish paleontologists José María Bermúdez de Castro, Juan Luis Arsuaga, and Eudald Carbonell, in 1994 their team unearthed a partial skull, jaw, and skeletal fragments of a new species of human, which they named *Homo antecessor*, "the man who came before." Because the layers were rich in marker pollens, small mammal fossils, and rocks amenable to dating by measuring their ancient magnetic fields (a technique first developed in the 1960s and 1970s for determining the age of tectonic plates), the team was able to date the skull at 800,000 years old. It appears to have belonged to an 11-year-old male, nicknamed the "Gran Dolina Boy." Bermúdez de Castro and his colleagues believe *H. antecessor* represents the last common ancestor of modern humans and pre-Neanderthals; many of the latter's remains were found together at the other, later Atapuercan site.

Also at Gran Dolina, the scientists discovered many bones of animals with telltale marks caused by cutting, scraping, and chopping with stone tools. Since the associated human bones show identical cuts and scrapes to those of the butchered animals, these hominids may have practiced cannibalism.

In 1976, half a mile away from Gran Dolina in a cave known as Cueva Mayor, spelunkers discovered the Sima de los Huesos (the Pit of Bones), a 42-foot-deep shaft filled with hundreds of pre-Neanderthal bones dated at from 400,000 to 600,000 years old and belonging to almost thirty individuals. Skeletons and skulls of cave bears were mingled with them, along with a well-formed quartzite hand axe. While there appears to be no direct connection between the two sites, they do demonstrate that these wooded hills, with their labyrinth of caves, have attracted hominid occupation during widely different time periods.

Ian Tattersall, curator of anthropology at the American Museum of Natural History, told *Natural History* magazine that the Sima de los Huesos is "the most astonishing concentration of human fossils that has been found anywhere in the world . . . but you have to descend down a vertical pit in a pitch dark cave to get to them." How these remains, belonging to various ages and both sexes, ended up in the Pit of Bones is a complete mystery—perhaps even a murder mystery. Were they sick or dying or already dead, or perhaps victims of a plague? Were they enemies or relatives of those who dumped them there so unceremoniously? Or is there some completely different explanation that we will never know? If humans did indeed deposit the bodies there, it would be the earliest known example of human funerary activity.

Atapuerca proves that Europe has been home to a variety of hominid species over the last million years. The earliest hominids in Europe whose bones are preserved lived 1.7 million years ago at what is now Dmanisi, in the Republic of Georgia. Atapuerca shows that early *Homo* species such as the proposed *H. antecessor* were also living in Europe 800,000 years ago. The hominids from the Pit of Bones were related to *Homo neanderthalensis*,

EXCAVATIONS AT ATAPUERCA since the 1990s have yielded scores of pre-Neanderthals (top) and the much earlier skull of a young boy (middle), called *Homo antecessor*, dated at 800,000 years ago.

which, beginning some 200,000 years ago, dominated Europe as well as parts of western Asia, and persisted for over 150,000 years—longer than our own species has existed. Then, around 40,000 years ago, a new species arrived in Europe, presumably from Africa: *Homo sapiens*. These modern humans were sophisticated artists, hunters, and toolmakers; by 30,000 years ago they had replaced Neanderthals throughout Europe.

See also *HOMO ERECTUS; NEANDERTHAL MAN*

AUSTRALOPITHECINES
Man-Apes of Africa

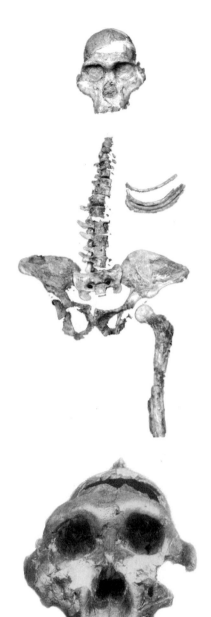

As African fossil hominids came to light over the past 80 years, one of the very surprising discoveries was that the early hominids were upright walkers, even though they had brains the size of a chimp's. Before these fossils were discovered, anthropologists assumed that upright walking had developed after the expansion of brain size. Australopithecines were bipedal apes. They had upright posture and walked bipedally on the ground, but they certainly lacked language and human cultural capacities.

Difficult to pronounce and misleading in meaning, *Australopithecus* refers to a genus of early hominids that were the closest evolutionary relatives to the genus *Homo* and our possible ancestors. Coined by South African anatomist Raymond Dart in 1924 to describe his famous "Taung child" skull, it means "southern ape." But subsequent discoveries showed they were neither southern nor close to modern apes. They ranged over the length of the African continent, from south and east to the northern areas of Ethiopia.

Raymond Dart and Robert Broom, two physicians with a passion for paleontology and anatomy, effectively began modern paleoanthropology with their discoveries of the australopithecine fossils in South Africa during the 1920s and 1930s. *Australopithecus africanus*, found by Dart, is the smaller, more lightly built hominid. A larger, more robust genus, known as *Paranthropus*, was discovered by Broom a few years later.

Several species of "australopith" are currently recognized. Some are more lightly built (gracile), such as *Australopithecus africanus*, from South and East Africa, and *A. afarensis* and *A. anamensis*, from East Africa and Ethiopia. They were about three to four feet tall, with arms longer than in humans but shorter than in modern apes, and were probably still good climbers. Their legs, and pelvises, however, indicate that they walked upright. Some of their faces were flatter than in modern apes, but jut out farther than in *Homo*. *Australopithecus anamensis* is the oldest of the australopiths, dated at around four million years, and *A. africanus* is somewhat younger at two to three million. Paleoanthropologists believe that *A. afarensis* may be ancestral to *Homo* as well as to other australopithecines.

Two larger, heavy-boned australopithecines are recognized: *Paranthropus robustus* and *P. boisei*, and a slightly earlier form called *Australopithecus aethiopicus*. Formerly known as *Zinjanthropus*, *Paranthropus boisei*, or "Nutcracker Man," had heavy grinding molar teeth and bony crests that supported strong jaw muscles. The heavy, large-toothed forms were at first thought to be a later side branch of the hominid family, but it now appears that these presumed "descendants" coexisted in the same times and places in Africa with some of the smaller, more lightly built "ancestral" species.

The long-standing tradition about the uniqueness of bipedalism in humans goes back to the ancient Greek philosophers. They knew that birds walked on two legs, but observed that man seemed to be the only other creature that did. Consequently, when one of Plato's students defined man as a "featherless biped," no one was able to offer a logical refutation. The next day, however, another student showed up for class waving a plucked chicken.

See also *AFAR HOMINIDS; DART, RAYMOND ARTHUR; TAUNG CHILD*

"MRS. PLES" was Dr. Robert Broom's nickname for the South African fossils he named *Plesianthropus* (top) but later lumped with *Australopithecus africanus*. "Zinj" or "Nutcracker Man" refers to the heavy-boned *Paranthropus boisei* (bottom) found by Mary Leakey in East Africa's Olduvai Gorge. It had huge grinding molars and powerful jaw muscles. Photos © by Margo Crabtree.

BABOONS
Evolved for Fighting or Finesse?

Baboons originally attracted the interest of anthropologists because these large monkeys had left the forest for the open plains of Africa, the presumed cradle of mankind. Some thought the baboons' social organization and behavior might offer clues about how the savanna shaped our ancestors' way of life as well.

Because males guard the group against leopards and other predators, an early study by Sherwood Washburn and Irven DeVore of the University of California, Berkeley, concluded that the biggest, fiercest males had their pick of food and consort females. The researchers also thought that all females were subordinate to the dominant males, and they did not bother to study the rank order of females. Dramatic, tooth-and-nail conflicts for patriarchal dominance, reminiscent of Freud's imagined "primal horde," became embedded in scientific literature as the central fact of baboon behavior—an image that bolstered the "killer ape" theory of human origins popularized by Robert Ardrey in his 1961 bestseller *African Genesis*.

Primatologist Shirley Strum studied olive baboons in the high-altitude savanna of the African Rift for more than 30 years, following them on foot and gradually acclimating the group to her presence. To her surprise, Strum could find no set dominance hierarchies among males, who might bluff-fight several times in a day, reversing position as winner and loser each time. Among the females, however, she discerned a very complex and long-lasting rank system, maintained by groups of close relatives. A low-ranking female, for instance, would not allow her infant to play with another whose mother was of much higher rank. In many cases, the rank is hereditary and maintained by tradition as much as by defense.

Olive baboons live in groups of 45–150 individuals. Of the 45 baboons in what Strum called the "Pumphouse Gang," only seven or eight were big males; the rest were females and juveniles. All spent a good part of their day foraging for plant food, insects, and anything else edible. Occasionally, a male would kill and eat a young antelope or small mammal.

In baboon society, females make friends and alliances with each other and with males. Males can befriend females or juveniles; adult males do not make friends with other adult males. A male new to the group does what Shirley Strum did—he sits on the edge, observes behavior, and tries to figure out relationships. He does not, as previous observers thought, attempt to move into the center of the group and take over from the dominant male.

Although the big males are certainly fierce fighters, what appears most interesting is the way other members of the troop get around them. A dominant male can compel a female to accompany him; if she doesn't like him, however, she will show it by moving away when he tries to mount her and gives him a hard time until he tires and loses interest.

When fights did occur between the big males, they frequently had unexpected outcomes. Several times, while large males fought over a female, a much less powerful male quietly walked up to the female, groomed her, and led her away to copulate while his more powerful rivals were still fighting.

Sometimes, males attacked by stronger ones will grab a female or juvenile to use as a living shield between himself and the aggressor. But this tactic only works if the baby or female is a friend who shrieks at the attacker. If the "shield" has no relationship with the defending male, it may shriek at him instead, causing double trouble—for then he will be mobbed by the whole troop.

Male baboons may hunt but always alone—never cooperatively. When a male makes a kill he does not offer to share it with the group, as chimps do, although he may give some meat to a favored female.

One of Strum's big males killed a small Thomson's gazelle and was soon approached by a female who liked meat. He didn't offer to share, so she walked behind him and began to groom his back until he was in an ecstatic "grooming stupor." Then she grabbed part of the carcass and ate it.

But this male would not be suckered again. Some days later, he obtained more meat, and she tried the same tactic. He allowed her to groom him, but every time she made for the meat, he slapped his hand down on it to block her. Finally, she walked away and attacked his favorite female.

This strategy put the male in a real quandary. He looked at the carcass, then at the female he should defend, then at the meat, then at the female. ("Fortunately for us," Strum says, "baboons think with their faces.") Finally, he went to the side of his favorite female; as he did so, the wily one bounded over and grabbed the meat.

According to Strum, the baboon social organization is one of reciprocity, understanding, friendship, and complex behavioral trade-offs. A London literary critic, reviewing her book *Almost Human* (1987), described Strum's preoccupation with the network of baboon relationships as "very California."

See also CHIMPANZEES; HUNTING HYPOTHESIS; KILLER APE THEORY

Only Clyde knew what a girl really liked.

© Pete Von Sholly.

BACTERIA
Most Ubiquitous, Resilient Life

We sometimes delude ourselves that we are living in an Age of Mammals and that warm-blooded creatures rule the planet. Yet all around us by far the most numerous and dominant life forms are bacteria; indeed, the Earth has been in the Age of Bacteria for the past three billion years.

The earliest forms of life that left fossil traces behind are prokaryotic (prenucleic) bacteria, whose remains are found in 3.6-billion-year-old rocks from Australia and South Africa. For a billion and a half years, they were the only living things on the planet. Then, about two billion years ago, eukaryotic cells—those containing nuclei and chromosomes—evolved and spread.

Bacteria are literally everywhere, from the rain forest floor to Antarctica, from the air around us to the insides of our intestines. In humans, the large intestine is home to some 500 to 1,000 kinds of bacteria, and there are ten times more of these microbes in our gut than the total number of all the cells in our bodies. There is even a remarkable species (*Helicobacter pylori*) that can live in the stomach lining, thriving in a bath of digestive acid. Unfortunately, this one causes ulcers and possibly gastric cancer. However, most bacteria in our bodies are friendly, ubiquitous, and beneficial.

One of them, known as *Bacteroides thetaiotaomicron*, helps its human host by breaking down carbs into sugars, and it even produces vitamins. In infants, it can turn on specific intestinal genes that promote the growth of blood vessels, and also stimulates production of chemicals that kill harmful bacteria. Experiments on rodents raised in a germ-free environment show that without a gut full of bacteria, they are much more prone to infections and must consume 30 percent more calories to maintain their body weight.

Outside our bodies, bacteria are even more abundant. The total mass of living matter above Earth's surface is dwarfed by what exists below. A handful of soil contains close to a billion organisms—including bacteria, fungi, and microscopic roundworms. Microbial organisms live by the uncountable billions two miles down into the Earth, in regions so hot they were once considered hostile to any form of life. Some bacteria thrive near the boiling point of water in the famous hot springs of Yellowstone National Park, where they produce colorful patterns on the surface, while others live in the subzero ice of Antarctica.

Archaea ("Ancient Ones") appear to be one of the three original branches on Earth's evolutionary tree of life—as old as the first bacteria. They are nearly indestructible, thriving in deep sea vents under crushing pressure and volcanic heat, and in rocks thousands of feet below the Earth's surface. Discovered only in the 1980s, they have demonstrated that life can persist in extreme conditions—and have raised the possibility that life might exist on other planets, or even in cosmic dust.

A single-celled microbe large enough to be seen with the naked eye (about three-quarters of a millimeter long) has been found by researchers sampling ocean dredgings in the South Atlantic. The bacterium is the largest ever identified. The microbe, discovered near Namibia, lives by absorbing sulfur and nitrates, and it swells as the chemicals are stored inside its cell walls. Researchers at the Max Planck Institute for Marine Microbiology, who made the discovery, said the microbes form chainlike colonies that glow from the absorbed nitrates and resemble a thin string of pearls. They named it *Thiomargarita namibiensis*, which means "Sulfur Pearl of Namibia," as it is found in great concentrations in Namibian coastal sediments that contain high levels of hydrogen sulfide—conditions that are toxic to most other forms of life.

Besides the 10 trillion cells in a human body, there are another 100 trillion bacterial cells living as symbiotes that help to sustain them.

Because unimaginable billions of bacteria waft easily between continents, scientists long assumed that the same kinds would be found all over the world. "Everything is everywhere; the environment selects" was a dictum coined by the Dutch microbiologist G. M. Baas-Becking in 1934, and widely believed by bacteriologists.

But field studies in 2003 revealed that not all microbes are world travelers. Local populations of bacteria are subject to the same processes of genetic isolation and evolutionary diversity as larger plants and animals; some species evolve in local environments and stay there.

David Ward of Montana State University, who studies the hot sulfur spring bacteria of Yellowstone, explains that microbiologists traditionally studied only bacteria grown and raised in laboratory dishes, rarely venturing into their natural habitats. In a 2003 interview for the *New York Times,* Ward opined, "We never had a natural history phase. And we're just now starting to see patterns of diversity. Sometimes it feels like we're bringing into microbiology thinking that's been in the rest of biology for 150 years."

See also PANSPERMIA; SULFUR-BASED LIFE

BANKS, SIR JOSEPH (1743–1820)
Natural History of "Oz"

Darwin's famous voyage of discovery was part of a British tradition that stretched back to the 17th century. The buccaneer William Dampier and the Prussian naturalist-explorer Alexander von Humboldt had published widely admired accounts of their travels, both of which inspired young Darwin. He was also following in the wake of England's first major scientific expedition, Captain James Cook's voyage on HMS *Endeavour* in 1768—only fifty years before the voyage of the *Beagle.*

Cook's crew of about a hundred included the best astronomers, naturalists, and artists he could find. His announced objective was to time the transit of Venus across the sun—from Tahiti, the best vantage point of that time. This bit of data would enhance the precision of nautical charts, preventing the loss of hundreds of sailing ships and their cargoes each year.

Under instructions from the Admiralty, however, Cook had a secret agenda: to search for the fabled fifth continent in the South Seas—whose existence he doubted. Just in case he were to find it, however, the Royal Society sent along twenty-five-year-old naturalist Joseph

Banks (1743–1820), whose responsibility was to collect and describe any plants and animals they might discover in the new land. Wealthy and well educated, Banks assembled a private staff of nine, a large natural history library, and extensive supplies for preserving specimens.

After the *Endeavour* had completed its astronomical observations in Tahiti, Cook sailed westward and first sighted "Terra Australis," the southern land, on April 19, 1770. Banks became the first European to discover (and dine on) a kangaroo. When the *Endeavour* returned to England, Banks presented the Royal Society with 500 previously unknown fish preserved in alcohol, 500 bird skins, innumerable insects, and 1,300 accurate and detailed drawings of plants and animals. He also brought back a thousand species of dried plants and seeds, including those of the giant eucalyptus, the acacia, and a large, diverse genus of odd flowering plants now known as *Banksia* in his honor.

Carl Linnaeus was so impressed with his plant collections and illustrations that he recommended the new southern land be named Banksia, but his suggestion was ignored. It was called Australia.

BARLOW, LADY NORA (1885–1989)
Darwin's Devoted Granddaughter

Lady Nora Barlow, granddaughter of Charles Darwin, cheerfully took on many tasks connected with collecting and editing her grandfather's notes, letters, and journals. She was the founder of the so-called Darwin Industry, which has occupied generations of scholars in sorting out and editing Darwin's literary and scientific legacy. Born Emma Nora Darwin, she was the daughter of Horace Darwin, an innovative manufacturer of scientific measuring devices at his Cambridge Instrument Company. She acquired her title by marriage to Lord John Barlow, a minister in the British government.

Lady Barlow was a good-humored link between her grandfather and 20th-century Darwin scholars; she spent years gathering, archiving, and editing family papers for several books, including republications of Darwin's *Diary of the Voyage of the Beagle* (1932), a collection titled *Charles Darwin and the Voyage of the Beagle* (1945), Darwin's notes on birds (1963), and the correspondence between Darwin and his Cambridge mentor, the botanist John Henslow (*Darwin and Henslow*, 1967). Her work was later taken up by her godson, the physiologist Richard Darwin Keynes, who continued to transcribe and publish Darwin's voluminous notes and manuscripts from the voyage.

In 1934, while working on the manuscript of the *Beagle Diary*, Lady Barlow visited Admiral Robert FitzRoy's daughter, Laura, who was then in her nineties. "Charles Darwin," said Miss Laura FitzRoy, "was a great man—a genius—raised up for a special purpose. But he overstepped the mark. Yes, he overstepped the mark for which he was intended." Lady Barlow dutifully recorded her words—a final summation of the fateful relationship between the commander of the *Beagle* and his ship's naturalist so long ago.

See also (HMS) *BEAGLE;* FITZROY, CAPTAIN ROBERT

BARNACLES
Darwin's Most Tedious Monograph

In July 1837, a year after returning from his world travels as ship's naturalist, Charles Darwin began a series of notebooks on "the species question." He knew his theories would not be taken seriously unless he first proved himself as a sober, serious scientist capable of detailed work in biology. So for eight years (1846–1854), he devoted himself to the study of barnacles.

Known to zoologists as Cirripedia ("curl-footed"), these marine invertebrates first attracted Darwin's attention when he discovered a new parasitic Chilean species that burrows into shellfish. He found no place for it in the established classification, then realized the zoology of the entire group was in complete disorder.

BARNACLE SEX and reproduction fascinated Darwin. He found one species in which the female was hundreds of times larger than the male, which was reduced to a parasitical sac clinging to her body.

Darwin's four volumes titled *A Monograph on the Sub-class Cirripedia* (1850, 1851, 1854, 1858) brought accuracy and order to the study of all fossil and living barnacle species. More than a century later, it is still a basic reference for barnacle specialists.

In the course of the long, tedious work, he became a master taxonomist (classifier) and sharpened his understanding of species and their variability—though he wondered whether it was really worth eight years of his life. About halfway through the work he wrote to a friend, "I hate a Barnacle as no man ever did before, not even a Sailor in a slow-sailing ship." After Darwin's death, his son Francis asked Thomas Henry Huxley if he thought the long barnacle project had been worthwhile and received this reply:

> Your sagacious father never did a wiser thing. . . . Like the rest of us, he had no proper training in biological science, and it has always struck me as [a] remarkable . . . insight, that he saw the necessity of giving himself such training, and [courageous], that he did not shirk the labor of it. . . . It was a piece of critical self-discipline, the effect of which manifested itself in everything your father wrote afterwards.

During his cirripedes work, Darwin wrote Captain FitzRoy that he was "for the last half-month daily hard at work in dissecting a little animal about the size of a pin's head . . . and I could spend another month, and daily see more beautiful structure."

After several years of this routine, the Darwin children accepted it as a normal part of life. One of his young sons was overheard asking a neighbor's child, "Where does your father work on his barnacles?"

BARNUM, PHINEAS T. (1810–1891)
Entrepreneur of Natural Hoaxery

America's great showman and shameless hoaxer, P. T. Barnum, knew exactly what the 19th-century public would pay to see: "fossilized" men, "primitive" savages, gorillas and orangutans, "unnatural" animals, and such extremes of human variation as dwarfs and giants. This same wellspring of curiosity also motivated serious students of evolution and natural science.

While he publicized the "educational" value of his exhibits, Barnum was often accused of "nature faking" by scientific experts. Thus he invented the classic ploy of urging the public to come see the disputed exhibit and "judge for yourself."

In 1869, a huge "fossilized man" was found buried on a farm in upstate New York. Known as the Cardiff Giant, it became a popular sensation. Thousands of people paid a dollar apiece to file past the pit where the "remarkable find" lay exposed. Actually, it was a figure carved in stone that had been artificially aged and buried a year before its "discovery" by a local hoaxer.

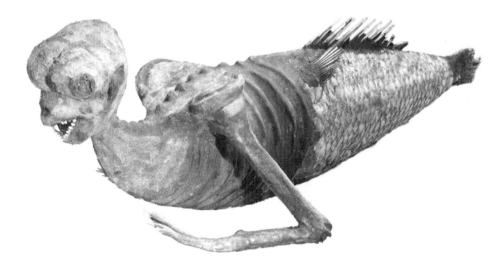

FAMOUS FEEJEE MERMAID was an outrageous fake made of various animal parts and fish bones, skillfully pasted together by Japanese craftsmen.

Recognizing a gold mine when he saw one, Barnum offered to lease the "petrified giant" for $60,000, but the owners were doing so well exhibiting it that they turned him down. Undaunted, Barnum commissioned a sculptor to carve a copy. After Barnum's brilliant promotion campaign, it proved to be even more lucrative than the original. He publicly admitted that both giants were frauds, but insisted his was a real copy of a fake original.

One of the classic early exhibits at Barnum's American Museum in New York City was the celebrated "Feejee Mermaid," which appeared to be a small human skull, torso, and arms, with the lower body of a fish. Actually, it was a monkey's partial skeleton skillfully joined to that of a large fish by Japanese craftsmen. Although it was denounced by zoologists, thousands paid to "judge for themselves."

Once, when a ship docked in New York harbor fresh from an African voyage and carrying a large primate aboard, newspapers described it as a rare gorilla, which stirred public interest. When Barnum went to see the animal, he recognized it at once as a baboon, but bought the animal anyway for his show.

Billed as a "The Most Remarkable Gorilla in Captivity," Barnum's new star attraction packed in the customers. A zoologist from Yale traveled to New York to see it and was outraged. "Mr. Barnum," he said, "it has a tail. Gorillas don't have tails." "Yes, but this one does," replied the master of humbug. "That's what makes him so remarkable."

See also CARDIFF GIANT; HAPPY FAMILY

"Come to the exhibition and judge for yourself."

—Phineas T. Barnum

BATES, HENRY WALTER (1825–1892)
Naturalist on the Amazon

Henry Walter Bates, one of the great pioneers in tropical biology, came from a Leicester family of hosiery makers, but refused to spend his life dyeing socks. While working 13 hours a day, he educated himself by night, reading classics and natural history. Young Bates developed a special passion for collecting beetles—the same enthusiasm that started Charles Darwin and Alfred Russel Wallace on careers as naturalists and evolutionists.

Beetle-mania brought Bates together with Wallace, who was then a young schoolmaster. On weekend bug hunts, the two friends talked of famed German naturalist and traveler Alexander von Humboldt and of the *Beagle's* Charles Darwin, whose writings had inspired them. They, too, wanted to see the South American rain forests and explore the Amazon. They decided to pursue that dream together, financing the journey by selling the specimens they would collect. And they had another, remarkable purpose—to gather data "toward solving the problem of the origin of species." (Darwin, in contrast, had no such plan when he set off for the tropics aboard HMS *Beagle*.)

Wallace and Bates reached Pará, at the mouth of the Amazon, in May 1848, where they stayed for a year and a half before venturing up a branch of the river; eventually, they parted company to continue independent explorations.

Wallace collected along the Rio Negro, about which he wrote the classic *Narrative of Travels on the Amazon and Rio Negro* (1853). He had to write it from unaided memory. On his way back to England, Wallace's ship caught fire, and all his notes and specimens, the product of four years' labor and hardship, were lost. Undaunted, the unsinkable Wallace set out for Malaysia, where he did much collecting and scientific work, and wrote the papers on natural selection that would link his name with Darwin's as cofounder of the new evolutionary biology.

Wallace sent Bates his 1855 paper on the "Sarawak Law" from Malaysia. "The theory I quite assent to," Bates replied from Amazonia, "and, you know, was conceived by me also, but I profess that I could not have propounded it with so much force and completeness."

Four years later, when Bates returned to England (after 11 years in the rain forests), he read Darwin's *Origin of Species* (1859) and became an instant convert. Bates had managed to return to England with almost 15,000 specimens,

MOBBED BY TOUCANS in the Amazon, naturalist Henry Walter Bates is surrounded after shooting one of the birds. He and Alfred Russel Wallace explored Brazilian rain forests seeking evidence for evolution.

mostly insects, including 8,000 species that were new to science. Darwin's theories gave him a unifying framework for interpreting his collections and observations.

In 1862, he published a paper ("Insect Fauna of the Amazon Valley") on his Amazonian butterflies in which he showed that good-tasting species mimicked the colors and forms of poisonous, foul-tasting ones. He explained this "mimicry" as "a most beautiful proof of natural selection," and Darwin praised the article as "one of the most remarkable and admirable papers I ever read in my life."

At Darwin's urging, Bates wrote his memoir *The Naturalist on the River Amazons* (1863). Although its composition was a long and painful process, Bates produced one of the great books about the South American rain forest and its peoples. One contemporary naturalist enthused, "Bates, I have read your book—I've seen the Amazons." Bates replied that he would rather spend another 11 years in the jungle than ever go through the ordeal of writing another book. He never did either.

See also BEETLES; MIMICRY; RAIN FOREST CRISIS; "SARAWAK LAW"; WALLACE, ALFRED RUSSEL

(HMS) *BEAGLE*
Vessel of Discovery

In his later years, Charles Darwin claimed he had acquired his "real education" aboard HMS *Beagle*—when he was not being seasick. Built in a shipyard at Woolwich-on-Thames in 1820, the *Beagle* was a 235-ton, 10-gun sloop brig of the Royal Navy. In whimsical contrast to her formidable artillery, she was named for a friendly little dog. It was not an isolated example of the shipbuilder's sense of humor: Other vessels in the series were named HMS *Porcupine,* HMS *Opossum,* and even HMS *Squirrel.*

The *Beagle* was of a design—the Cherokee class—that sailors had nicknamed "floating coffins" or "half-tide rocks" because they were difficult to navigate and went down so easily in foul weather. During her first surveying expedition along the coastline of Patagonia (1826–1830), she came close to capsizing in the tempestuous waters. By 1828, she was in poor repair, out of provisions, and her crew was wracked by scurvy and disease. Her captain, Pringle Stokes, retired to his cabin in despair for weeks, and then shot himself.

Lieutenant Robert FitzRoy, then a young officer stationed on HMS *Ganges* at Montevideo, was sent to replace the unfortunate Stokes. Shortly after assuming command, the 26-year-old FitzRoy almost lost the ship in a violent storm, but soon mastered the art of sailing her through difficult and jagged channels on the roughest seas. He managed to put the expedition back together and to accomplish its difficult mission of charting and exploring the coast of South America, establishing worldwide latitudes and even the exact longitude of Rio de Janeiro for the first time.

For the *Beagle*'s second survey of South America (1831–1836), FitzRoy was promoted to captain, and he selected 22-year-old Charles Darwin as his "gentleman's companion" and unofficial ship's naturalist.

Darwin worked in a tiny forward poop cabin preparing tons of natural history specimens for shipment to England, assisted by his servant Syms Covington. He attended mess with the officers and at night shared the captain's cabin. Frequently he went ashore on overland expeditions, sometimes for several weeks at a time.

Newly refitted, the *Beagle* was to chart the coast of Patagonia and Tierra del Fuego, then to circumnavigate the globe. She sailed from Devonport, England, on December 27, 1831, and returned to Falmouth almost five years later, on October 2, 1836.

In addition to her extensive work in cartography and hydrography, the *Beagle* was given the task of bringing Greenwich Mean Time all over the world. She had 22 precision chronometers, which were kept in special cases in order to protect them from the shocks of motion.

Barely 90 feet long, HMS *Beagle* carried 74 men, including eight marines. Also aboard were three lieutenants, a surgeon, and artists Augustus Earle and, later, Conrad Martens. Passengers included three Fuegian Indians known as York Minster, Jemmy Button, and Fuegia Basket, who were being returned to their homeland in Tierra del Fuego. Richard Mat-

"The Beagle was where I acquired my first real education."

—Charles Darwin

HMS *BEAGLE* APPROACHES James Island in the Galápagos on October 17, 1835, to pick up Charles Darwin and his party from their nine-day exploration of the island.
HMS Beagle in the Galapagos (detail), by John Chancellor. Courtesy of and © by Gordon Chancellor.

thews, a missionary, attempted to establish a colony with the Christianized Indians in Tierra del Fuego, but the disastrous experiment failed, and he was nearly killed by the Fuegians' relatives. He was rescued, however, and later disembarked in New Zealand.

After their voyage together, neither Darwin nor FitzRoy ever sailed on the *Beagle* again. In 1837–1843 she completed a third voyage, to New Zealand and Australia, under the command of Captain Wickham, who had served as an officer under FitzRoy.

Upon her return, the *Beagle* was retired from ocean duty, stripped of masts and gear, and used for Coast Guard patrols (1845–1870) on the River Roach in Essex, which was a smuggler's highway. After 25 years on the river, she was sold for scrap and eventually towed and abandoned in a Thames estuary.

A century later, in 1964, the Charles Darwin Research Station in the Galápagos Islands equipped a modern research vessel and christened her *Beagle II*. Over the years, there were also at least nine other ships called *Beagle*, including a steamer, a schooner, and two massive destroyers that saw action during both world wars in the twentieth century.

In 2003, a different sort of ship called the *Beagle 2* was launched by British astrophysicists led by Colin Pillinger—an unmanned spaceship shot into space to explore Mars. It reached the red planet, but crash-landed and stopped broadcasting information. That same year, the *Beagle*-obsessed Pillinger decided to use his sophisticated radar technology to try to find the remains of the original *Beagle*, which he deduced were buried in an Essex marsh.

The last mooring of Darwin's vessel is recorded in a Photographic Office Survey chart from 1847, which shows the ship anchored in the middle of the River Roach. However, it was dragged ashore in 1850, following complaints from oyster fishermen that it was blocking their passage. Old Coast Guard letters mentioned a particular spot on the shore where it was decided HMS *Beagle* should be dumped. Working with Pillinger, Robert Prescott of the University of St. Andrews completed an archeological survey of the site in 2003, and now believes they have identified her final resting place beneath tons of mud.

See also BUTTON, JEMMY; DARWIN, CHARLES; FITZROY, CAPTAIN ROBERT; VOYAGE OF (HMS) *BEAGLE*

BEECHER, HENRY WARD (1813–1887)
Evolutionist American Preacher

Reverend Henry Ward Beecher, the superstar of American preachers during the latter nineteenth century, reconciled science and religion with a metaphor his congregation of tradesmen could understand. If he found a watch in a field, he said (referring to William Paley's "argument from design"), he would not be as impressed with the watchmaker as he would with the factory designer who constructed the machines that turn out thousands of watches a day. It was inefficient for God to design each species separately, he explained, so He created laws and forces that generated all the world's creatures without divine intervention. Belief in evolution was not disrespectful to God because, in Beecher's words, "design by wholesale is grander than design by retail."

Once called "the most admired man in America after Abraham Lincoln," Beecher was on a pedestal of his own making. Crowds from three states ferried to Brooklyn Heights every Sunday to hear him hold forth at Plymouth Church. His series of sermons *Evolution and Religion,* originally delivered to capacity crowds, gained wide popularity. They were printed in newspapers across the country and republished as a book in 1885.

Beecher's biographers Lyman Abbott and S. B. Halliday claimed that his agenda of reconciling science and Christianity required courage, since "evolution had come to be identified in the public mind with infidelity. It does undoubtedly involve a recasting of the philosophic statements of creation, sin, revelation, and redemption [that appears] equivalent to an entire abandonment of these truths."

Indeed, for Beecher the Adam and Eve story was "a poem, not a treatise in cosmogony," his biographers asserted. An 1883 account of a Beecher sermon in the *New York Times* reports that he ridiculed the traditional view of divine punishment and Original Sin as unduly harsh. "If they had been your children [who ate the apple],' said Mr. Beecher, looking out over his audience, 'you would have spanked them and put them to bed.'" Instead, God "'told them he would make their descendants bad. Great Heavens!' said Mr. Beecher, 'they were bad enough without making them any worse.'"

In Beecher's interpretation, Christian belief was entirely compatible with evolution. God gave successive revelations to the evolving human race at each stage of its development, but only when it was prepared to receive them—an idea revived a century later in Stanley Kubrick's film *2001: A Space Odyssey.*

Heaven and hell were replaced by the choice between progressive evolution and degradation. Beecher described human moral degeneration in terms a naturalist might apply

to a parasitic worm. [See DEGENERATION THEORY.] Those who resist all divine elevating influences, he warned, "go down steadily lower and lower until they lose the susceptibility, the possibility of moral evolution, moral development; let them keep on, and in the great abyss of nothingness there is no groan, no sorrow, no pain, and no memory." Its opposite is "Regeneration," brought about by a willingness to live according to divine principles. Evolution, therefore, does not disprove design in nature but is part of a larger plan to bring man from the realm of animal physicality to a higher, spiritual realm.

In Beecher's upward-striving theology, the doctrine of Original Sin libeled a kindly, caring God. Sin was not Satanic, but a devolution into carnal animality. Beecher himself was guilty of just such a relapse when he had an adulterous affair with a woman in his congregation: Mrs. Tilton, who was also his best friend's wife. The resulting front-page scandal and uproar became known as "The Great Sensation" and preoccupied the nation for weeks. The elders of his church debated firing their minister but stopped short of doing so. His personality and charisma were still bringing millions of dollars to Plymouth Church.

During a civil suit brought by Tilton, Beecher's evasive testimony got him off. Although he escaped conviction and retained his pulpit, however, he lost his moral authority and never recovered it.

THE MOST INFLUENTIAL EVANGELIST in 19th-century America, Reverend Henry Ward Beecher fused Darwinism and religion into a new gospel of spiritual evolution. An abolitionist, he exhorted his parishioners to purchase the freedom of enslaved black children he brought up to the pulpit.

BEETLES
Evolutionist's Inspiration?

According to Alfred Russel Wallace, he and Charles Darwin shared a "childlike" passion for collecting beetles. That was one key similarity, Wallace suggested, that may have eventually led them independently to a theory of evolution by natural selection.

Wallace and Darwin recalled into old age their early love for their beetle collections; both could instantly recall the exact place and circumstance of each acquisition. So dedicated a collector was Darwin that he sought beetles between classes at Cambridge and even put one in his mouth so he could hold one in each hand. To his great disgust, the one between his lips gave off a noxious spray, causing him to lose all three prizes.

Both served as professional collectors of natural history specimens. Darwin spent a major part of his time on the *Beagle* collecting fish, insects, rocks, fossils, reptiles, and plants for shipment back to England. He later wrote a manual on collecting methods, including such practical matters as labeling, cataloging, and packing. When his voluminous, high-quality collections reached the British Museum, his reputation as a naturalist was assured.

For years, Wallace earned his entire living from collecting. He sold thousands of insects to the museum and to private collectors for two cents apiece—his sole means of financing his expeditions to the tropics.

Wallace thought beetle collectors were more likely to have developed an evolutionary perspective than anatomists or physiologists. Thomas Huxley, for instance, focused on living things as "mechanisms." If you dissected one crayfish, following Huxley's lab manual, you pretty much knew them all. Variations or anomalies were seen as departures from the normal type—curios, not sought-after prizes.

But collectors reveled in variations. They liked beetles because there are so many different species, subspecies, and seemingly endless varieties. Even a boy, as young Darwin discovered, could find, name, and publish a new beetle. (He remembered the thrill for the rest of his life.) When arranged in a collection, beetles formed series that ran from tiny to the size of a large potato, with hundreds of different colors, patterns, and wing shapes.

Eventually, a philosophical collector had to wonder how many good species he had and how many among them were varieties, or hybrids, and why there should be such a spectrum.

One might easily visualize a Creator forming several species of big cats, bears, or eagles, but separate creations for 150,000 kinds of beetles gave one pause. As Comte de Buffon wrote years earlier, it was hard to believe God "made a different kind of fold for each beetle's wing." Secondary causes appealed more to reason. Reverend Henry Ward Beecher saw no contradiction. In the 1870s, he told his congregation that belief in natural processes is no insult to God, "because design by wholesale is much grander than design by retail."

In Argentina, Darwin was bitten by a very dangerous beetle, the Benchuca, known as the "Great Black Bug of the Pampas." Some biographers attribute his chronic illness to a blood parasite these "assassin bugs" commonly carry. However, other historians believe the naturalist's mysterious malady was lifelong "psychosomatic" stress and neurosis.

During the 1930s, the great British geneticist and evolutionist J. B. S. Haldane was asked by a priest what his long study of nature had taught him about the attributes of God. "For one thing," Haldane replied, "He has an inordinate fondness for beetles."

See also BEECHER, HENRY WARD; BUFFON, COMTE DE; DARWIN, ILLNESS OF; "DELICATE ARRANGEMENT"; HUXLEY, THOMAS HENRY; SPECIES, CONCEPT OF; TRANSITIONAL FORMS; WALLACE, ALFRED RUSSEL

TIGER BEETLE.
Photo © by Marian Brickner.

ASTRIDE A GIANT BEETLE, young Charles Darwin was sketched by a fellow student at Cambridge pursuing one of his favorite hobbies: collecting insects. Redrawn by Pete Von Sholly.

Go it Charlie!

BEIJING MAN (PEKING MAN)
Homo erectus from China

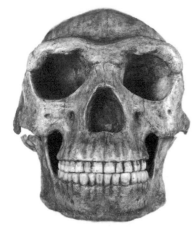

Homo erectus skulls were excavated from the "Hill of the Dragons" near Beijing. Casts were sent to New York in the crate shown below, but the originals were lost during World War II.

Beginning in the 1920s, an international team of scientists excavated the spectacular fossils of Beijing Man from a limestone quarry called "Hill of the Dragons" at Zhoukoudian, near Beijing. Although the discovery of this important hominid, now called *Homo erectus*, is often attributed to the Canadian Davidson Black, the German-American Franz Weidenreich, and the Chinese W. C. Pei, these investigators were only among the last and best funded, by Rockefeller grants, in a complicated history of discovery.

Germans, Swedes, Austrians, Frenchmen, and Americans were all poking around China in the early 20th century, when the idea of an Asian "cradle of mankind" was most popular.

The first major expedition came from Sweden's Uppsala University, at the instigation of Johan Gunnar Andersson, a mining expert advising the Chinese government on ore deposits. His field team—headed by Otto Zdansky, a young Austrian paleontologist—found the first two fossil teeth of Beijing Man in 1923 but Andersson cautiously kept the discovery under wraps until 1926, when he announced it at a scientific meeting.

A new round of Swedish activities began in 1927, headed by Birger Bohlin, a young expert on fossil giraffes, who was assigned an army of Chinese laborers and "ordered to find man." He did turn up a few more jaws and teeth, which began Bohlin's obsession with the image of Beijing Man (and Woman). He drew charming reconstructions of their faces and filled the walls of his home with endless variations of it.

After Bohlin's tenure, a succession of international paleontologists worked the site, including Americans working with Davidson Black and W.C. Pei, who found the first crania deep in a limestone cave. The heavy-boned hominid skull featured prominent brow ridges and a somewhat smaller braincase (about 1,000 cc) than is found in modern humans (1,500 cc).

After Black's death in 1934, the site was directed for a year by the French Jesuit Teilhard de Chardin, who was followed by Franz Weidenreich. By 1936, when all work had to stop because of the Japanese invasion of China, 14 skulls in varying conditions had been discovered, along with 11 jawbones, 147 teeth, and a couple of small armbone and femur fragments. There were also stone tools and ambiguous evidence of a layer of carbon ash, which led to the name "The cave of hearths."

But what did the excavated materials represent? Was it a living site? A burial ground? A place of ritual cannibalism? Theories ran rampant. Beijing Man was represented mainly by skulls—hardly any postcranial material. Not a pelvis or a rib. Just skulls. And the openings at their bases, the foramen magnums, had been widened and smashed, as if someone had wanted to scoop out the brains.

Some concluded that these individuals had lived in the caves, used fire, and were cannibals. Others thought that they may have been captured and their heads taken by *Homo sapiens*, who left no trace of themselves but the remains of their ghastly feast. Did they have language? Were these our ancestors, or a side branch of human evolution that our ancestors

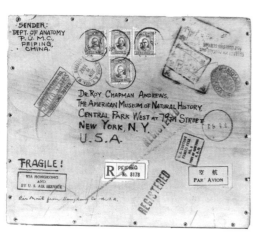

helped exterminate? Paleoanthropologist Russell Ciochan recently published his own conclusion, that the faces were bitten off by giant hyenas.

At the time of discovery, it was clear that Beijing Man (then placed in its own genus and species, *Sinanthropus pekinensis*) was virtually identical to the Java man (then called *Pithecanthropus erectus*), which had been found by Eugene Dubois in the 1890s. Over the next 50 years, similar remains turned up in the Middle East, Europe, North Africa, and East Africa, including a fairly complete skeleton of an adolescent [see TURKANA BOY]. During the 1950s, ornithologist Ernst Mayr reexamined the various classifications and "lumped" these widespread hominids into the single species *Homo erectus*.

The Beijing Man fossils also pose a much more recent mystery: No one knows what happened to them. When war broke out in 1936, Weidenreich hesitated to ship them out of the country, since they belonged to the Chinese government. Amidst the uncertainties of war-torn Beijing, however, it proved impossible to store them safely, so Weidenreich finally packed them for military shipment to the United States. They are believed to have been aboard the marine ship SS *President Harrison,* which was sunk in the Pacific in mid-November 1941. So Beijing Man's bones may now be resting on the ocean's bottom. Fortunately, Weidenreich had sent excellent plaster casts and notes back to New York. They are all that remain to science of the precious, hard-won fossils.

Tantalizing tips and rumors through the years have kept a tiny spark of hope alive that the priceless relics were not destroyed during the war after all and might yet resurface. The anthropologist Harry Shapiro spent much time and energy following up fruitless leads by cranks and pranksters.

For years the fossils were known as "Peking Man," but in 1979, China adopted the pin-yin system of transliteration, which more accurately reproduces Chinese sounds in English spelling. And so Peking Man became Beijing Man, but old linguistic habits linger on. Current textbooks still refer to the fossils as "Peking Man," just as American Chinese restaurants continue to serve Peking duck. And English-speakers adamantly refuse to refer to their Pekinese dogs as "Beijingese."

See also CHINA, DARWINISM IN; DUBOIS, EUGENE; GOBI DESERT EXPEDITIONS; *HOMO ERECTUS;* TEILHARD DE CHARDIN, FATHER PIERRE

BELFAST ADDRESS (1874)
Tyndall's Shocking Materialism

Among the leading lights of Victorian science, physicist John Tyndall (1820–1893) was an outspoken proponent of evolution who did not shrink from attacking religion's orthodoxies.

As president of the British Association, the nation's leading scientific organization, Tyndall delivered a speech on August 19, 1874, at the Belfast (Ireland) meetings that caused shock waves on both sides of the Atlantic. It summarized the advances of science, particularly evolutionary biology, and contained an explicit manifesto that scientific truth would replace religious revelation.

"We claim, and we shall wrest from theology," he trumpeted, "the entire domain of cosmological theory." His address began with a sweeping history of science from the Greek philosophers onward, and concluded with a detailed exposition of natural selection. Its major thrust was that the church's long monopoly on questions of human origins and nature was at an end and that all biology was now founded on evolutionary theory.

He reminded his audience that all of them, in the words of Darwin and Spencer, are shaped by interactions with their environment—a constant state of adjustment. The intellectual landscape around them, too, was changing, and it was best to adapt. Although his colleagues could "purchase peace at the price of intellectual death," he urged instead that they embrace "the leap of the torrent before the stillness of the swamp."

But what raised the most violent outcry was his assertion that one could discern in matter "the promise and potency of every form and quality of life." For focusing on the creative properties of matter rather than the Creator, critics branded him with the severe epithet of "Materialist Atheist."

Tyndall was impervious to the insult. Materialist science, he concluded, is the best hope for understanding nature. While "you are not urged to erect it into an idol," he told his audience, "science claims unrestricted right of search."

Most of Tyndall's ideas in "the notorious Belfast Address" were increasingly acceptable to many in the scientific community, but an English gentleman just didn't say such things in public. The Victorian attitude on public expression of unscriptural sentiments is epitomized by the apocryphal story of a genteel lady's famous remark about *Origin of Species:* "Descended from the apes? Well, let us hope that it is not true—but if it is, let us hope it won't become generally known."

See also AGNOSTICISM; MATERIALISM

"The church's long monopoly on questions of human nature and origins is at an end. Science claims unrestricted right of search."

—John Tyndall, 1874

BELT, THOMAS (1832–1878)
Naturalist, Geologist, Engineer

Charles Darwin thought "the best of all natural history journals which have ever been published" was *A Naturalist in Nicaragua* (1874), written by the self-taught ecologist Thomas Belt. Belt's work as a mine geologist and engineer took him to wilderness areas all over the world; during his off hours he recorded the natural wonders he found.

Among the adaptations Belt describes are those of insects that resemble leaves in shape, color, even veins. Some are faded and blotched as if dying; others appear brown and withered with "a transparent hole through both wings that looks like a piece taken out of the leaf." Even as he marvels at the realistic mimicry, he recalls anti-Darwinian objections that such detailed fidelity "could not have been produced by natural selection, because a much less degree of resemblance would have protected" the insect. What could account for such details as false holes?

Belt answers that natural selection not only protects mimetic (mimicking) prey animals, but also sharpens the perception of the predators, "a progressive improvement in means of defense and attack." On one side the disguise gets better, and on the other so does the ability to penetrate it, until the camouflage becomes very fine-tuned indeed.

This progressive refinement, he goes on to say, is no different from the way natural selection works on carnivores and their prey on open plains. From generation to generation, each species pushes up the other's running speed until both become more streamlined and fleet.

Daniel Janzen, a tropical biologist working to restore Costa Rican wilderness areas, read Belt for precise observations of what the forest was like more than a century ago. "Belt was working amidst an ocean of nature," Janzen lamented, "while all that remains to us are small and rapidly shrinking ponds. . . . Belt's book is [today] a litany of habitat destruction."

BENGA, OTA (1881–1916)
Man in the Monkey House

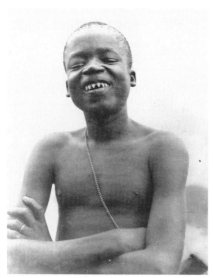

THE ONLY HUMAN ever exhibited in an American zoo, "Pygmy" Ota Benga shared a cage with an orangutan. Although the Bronx Zoo's director insisted he was free to leave whenever he wished, the incident created a public furor.

Visitors to the New York Zoological Park, better known as the Bronx Zoo, flocked to the monkey house in 1906. There they could see a living spectrum of their "lower" evolutionary cousins: monkeys, chimpanzees, America's first gorilla (Dinah), and an African pygmy tribesman. His name was Ota Benga, and he shared a cage with a parrot and an orangutan named Dohong.

According to the explorer Samuel Verner, he found Benga in the Belgian Congo held captive by another tribe and paid a ransom to free him. Then the little fellow helped Verner find other pygmies to accompany him to St. Louis, Missouri—to appear at the St. Louis World's Fair. But he couldn't have had the vaguest idea about what he was letting himself in for.

In 1904, an entire "Pygmy Village" was exhibited at St. Louis along with exotic African wildlife. Afterward, Verner handed Benga over to eccentric zoologist William T. Hornaday, the director of the Bronx Zoo during its early years.

Hornaday believed he could read the thoughts of zoo animals and considered apes almost human. He saw nothing wrong with keeping a little black man on view in the monkey house.

New York's black community was outraged, and church organizations insisted that Benga be freed from his captivity. White fundamentalist clergymen also expressed strong disapproval, but not because the pygmy was a human being caged like an animal. What these churchmen feared was that the exhibition in the monkey house would convince the masses of the truth of Darwin's theory of evolution.

Hornaday explained that he was merely offering an interesting exhibit and that Benga was happy and knew he was absolutely free to leave his cage at any time. This statement could not be confirmed, as Benga spoke no English. However, Hornaday finally bowed to the threat of legal action and insisted the pygmy leave his cage. Dapper in a white suit, Benga paraded around the zoo with crowds trailing him, but still slept in the monkey house.

After a while, Ota Benga began to hate being mobbed by curious tourists and mean children. He fashioned a little bow and a set of arrows and began shooting at zoo visitors, which caused his expulsion from the Zoological Park for good.

For a few years he was employed in a tobacco factory in Lynchburg, Virginia. Several individuals and institutions tried to care for him, but he grew hostile, depressed, and irrational, yearning for his native forests. Bitter, lonely, and tormented, Benga got hold of a gun and, in 1916, committed suicide.

BERINGER, JOHANNES (c. 1667–1738)
Victim of Fossil Fraud

There have been several famous attempts to test the credulity of learned men with forged fossils. One of the earliest of these geological hoaxes occurred in 18th-century Germany. The victim was Johannes Beringer, dean of the medical school at the University of Würzburg, court physician, and respected amateur collector of antiquities. It was not a harmless student prank, as some books still claim, but a deliberate attempt to destroy a man's reputation. The villains were two jealous academics: a professor of geography and a librarian.

The plotters eventually forged some 2,000 intriguing stones and hired three young student "collectors" to dupe Dr. Beringer into "finding" them. At first, they dug up what appeared to be petrified impressions of small animals and skeletons, but the "fossils" grew more and more improbable. All were carved with considerable skill and buried with great care in local fossil quarries. They included butterflies sipping nectar at flowers, spiders with their webs, and even letters of the Hebrew alphabet "fossilized" in stone!

Each successive find delighted the gullible doctor, and he wrote a learned treatise on the wonders of his stone collection and his "search for the truth" about how they might have been formed. His book, *Lithographiae Wirceburgensis,* was published in 1726. Shortly after it came off the press, he is said to have found the stone that finally gave the game away: the "fossil" had his own name on it.

Beringer furiously defended his honor by hauling the perpetrators into municipal court, where they were exposed and duly disgraced. Contrary to popular myth, Beringer himself did not die a broken man but went on to recover his reputation and position. The plotters confessed that they had sought revenge because of Beringer's insufferable arrogance and condescension. He had often proclaimed that he alone had been "chosen by divine providence" to discover the history of life.

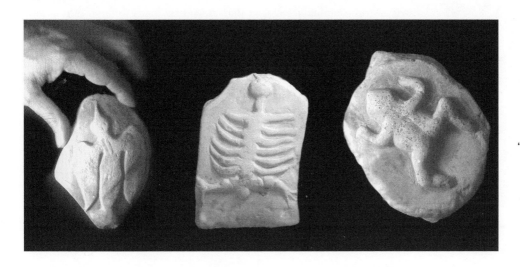

"LYING STONES" containing strange creatures and bizarre shapes were described in Professor Johannes Beringer's book. All were hand-carved hoaxes deliberately planted by malicious rivals. Photos courtesy of and © by William Schopf.

BINOMIAL NOMENCLATURE
Linnaean Scientific Names

In the 1750s, the Swedish naturalist Carl Linnaeus elaborated and popularized the system of scientific classification and nomenclature that is used throughout the world today. Latin or "scientific" names for animals and plants eliminate the confusion caused by the great variety of "common names" given to a single species. The binomial consists of two names, the first referring to the genus (a group of related species) and the second to the particular species within that genus.

In the Linnaean system, categories are nested within categories, like a series of ever-smaller boxes enclosed in larger boxes. For instance, in the class of mammals (warm-blooded, hair-covered, milk-giving animals), we can locate the order Carnivora (meat-eaters) and within that group the family of cats, Felidae. Within that family is the genus *Panthera,* comprising tigers (*Panthera tigris*), lions (*P. leo*), jaguars (*P. onca*), and leopards (*P. pardus*). A third name can be added to indicate a geographic subspecies, such as *P. leo persica,* for the Asiatic lion, or *P. leo krugerei,* the South African lion. (In the full listing, the name of the discoverer and the date are also attached, as in *Apatosaurus ajax* Marsh 1877.)

Scientific names must be precise in their classificatory function but otherwise can be utterly whimsical. For instance, two species of wasps were named *Polemistus chewbacca* and *P. yoda* by the entomologist who discovered them—apparently, a big *Star Wars* fan. One scientist named a small cave-dwelling arachnid that sucks the juices out of its prey *Draculoides bramstokeri.* Sometimes the new species name reflects the classifier's frustration in distinguishing it from look-alike species. Thus, we have the yellow daisy recently named *damnxanthodium* and the beetle called *Agra vation.*

To Linnaeus, similarities between groups expressed a divine plan of organization in which species were fixed and unchanging. After Charles Darwin published his *Origin of Species* in 1859, however, Linnaeus's static classification scheme took on a dynamic, evolutionary dimension. Similar groups are now thought to be genetic relatives descended from common ancestors.

See also BRANCHING BUSH; LINNAEUS, CARL; TREE OF LIFE

BIOGENETIC LAW
Ontogeny Recapitulates Phylogeny

Greek philosophers had noticed that an individual's development seems to resemble the larger history of its kind. Knowing that humans are born from a womb filled with liquid, the Ionian Anaximander (c. 600 B.C.) speculated that humans originated as fishy creatures that emerged from the ocean.

In the 19th century, German embryologist Karl von Baer made detailed studies of the evolutionary stages that individuals pass through during fetal development. His influential colleague, Ernst Haeckel, proclaimed the embryonic sequence to be nothing less than a retelling of the entire story of evolution. He called it his "Biogenetic Law," summarized by the slogan "Ontogeny recapitulates phylogeny."

That jaw-breaking phrase means that development of an embryo (ontogeny) is a speeded-up replay of millions of years of species evolution (phylogeny). Haeckel believed that a fetus relives all the evolutionary "stages": invertebrate, fish, amphibian, reptile, mammal, primate, ape, and finally human.

While there is some general resemblance between an embryo's development and its evolutionary history, Haeckel's theory was incapable of accounting for the many detailed differences. He had oversimplified, and equated embryonic stages with the *adult* forms of supposedly ancestral creatures. To make matters worse, Haeckel, who was a fine artist, drew attractive and convincing pictures of the forms his theory ideally predicted—but no one could actually observe them in their labs.

INFAMOUS EMBRYO DRAWINGS by Ernst Haeckel incorrectly depicted fish, amphibians, turtles, and chicks all starting out as identical, then replaying their evolution. Haeckel fudged the drawings to fit his theory.

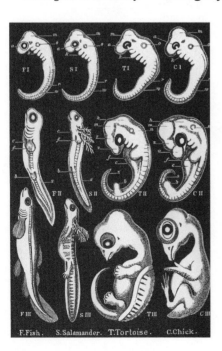

"Haeckel's biogenetic law was so extreme," Stephen Jay Gould wrote in *Ontogeny and Phylogeny* (1977), "and its collapse so spectacular, that the entire subject became taboo." Gould helped to revive interest, and in recent years a new generation of geneticists has carved out a research field known as "evolutionary developmental biology," nicknamed Evo-Devo.

Meanwhile, seemingly oblivious to the fact that Haeckel's law and his famous illustrations were discredited a century ago, creationists continue to cite them as evidence that evolutionary biology is based on "fraudulent science."

See also EVO-DEVO; FREUD, SIGMUND; HAECKEL, ERNST

BIOLOGICAL DETERMINISM
Possibilities and Constraints

Biological determinism is a hot potato in science and society. Broadly defined, it means that certain behaviors, potentials, even destinies are shaped by the genetic "cards" an individual is dealt at conception. Certainly all scientists can agree that our legs and feet are shaped to make us walkers, our brains process language, our sex determines whether we can give birth. But many apparent biological differences between individuals cannot be used to predict a human being's potential. Skin color, height, weight, or brain size—all are easily noticed features that have nothing whatever to do with an individual's character or capacity. There have been great short basketball players and brilliant scientists with small brains. The "genetically defective" bodies of Charles Proteus Steinmetz, the dwarf whose genius revolutionized the understanding of electricity, or the diminutive painter Toulouse-Lautrec did not keep them from lives of brilliant achievement.

Nineteenth-century criminologist Cesare Lombroso, among others, pushed theories of biological determinism far beyond what was justified by the facts. Lombroso, for instance, claimed he could tell a prostitute from a "normal moral woman" just by the shape of her skull, nose, arms, thighs, and facial wrinkles. Male criminals, he wrote, could be easily identified by their "feeble cranial capacity, heavy and developed jaws, projecting ears, and crooked or flat noses." For decades, such theories were taken very seriously by university professors, government policymakers, and police departments.

So much human suffering was caused by such ill-founded theories that later generations of biologists, reacting against facile determinism, refused to accept any conclusions about biological patterns and constraints of human behavior. Since the 1930s, attempts to measure intelligence of various ethnic groups have provoked bitter controversies; opponents argue that social, rather than genetic, factors are the key determinants of performance. Even the serious attempt to study biologically based human behavior (sociobiology) during the 1960s provoked hysterical attempts to defame entomologist E. O. Wilson and others.

Biological bases of behavior and genetic or sexual differences are legitimate areas for investigation. A major challenge to future researchers is whether such studies can be designed with full awareness and allowance for the researcher's ingrained cultural biases and expectations.

See also PHRENOLOGY

> "Male criminals are evolutionary throwbacks, identifiable by their small braincases, heavy jaws, and flat noses."
>
> —Cesare Lombroso

BIOLOGICAL EXUBERANCE
The Kingdom of Gay Animals

Televangelist Reverend Jerry Falwell told his flock, "God created Adam and Eve, not Adam and Steve." Homosexuality, in his fundamentalist view, was demonstrably "unnatural." Scientists who have looked into the matter, however, have reached the opposite conclusion: nature abounds in same-sex attractions, eroticism, and long-time companionships.

While still a graduate student in the 1970s, anthropologist Linda Wolfe (now at East Carolina University in Greenville, North Carolina) almost wrecked her career by reporting the unexpected behaviors she had observed in Japanese macaques. Many females paired off with each other, had sex to orgasm, and groomed each other exclusively. Wolfe's professors insisted that the female monkeys must have been mounting each other by mistake. They even questioned her own sexual preferences. "But it's part of what primates do," she insists, "part of their total sexual repertoire."

For years, many zoologists pointedly ignored any field research that documented homosexuality in animals. When the U.S. Marine Mammal Commission published its definitive report on killer whales, for instance, it deleted the passages its biologists had written about male orcas rubbing each other to orgasm, which was not government-approved.

In 1999, a landmark book by an independent American scholar, Bruce Bagemihl, challenged the prevailing view of animal sexuality. Bagemihl had spent a decade scouring the scientific literature, collecting documented cases of homosexuality and other mating behaviors among animals. His book *Biological Exuberance: Animal Homosexuality and Natural Diversity,* which profiles 470 species, mostly mammals and birds, assembled overwhelming evidence for homosexual or multisexual behavior up and down the animal kingdom.

Animal homosexuality, according to Bagemihl, is not a single, uniform phenomenon but an astonishing diversity: from "lesbian" gulls that share a nest and rear chicks together to male giraffes that rub and entwine each other's necks for hours, often ending in mounting and orgasm. Commenting on the book, primatologist Robin Dunbar says, "If you're looking for homosexual sex in vast quantities, forget humans, it's bonobos (the so-called 'pygmy chimps') you want. They'll have sex with anyone, never mind the sex or age."

In *Evolution's Rainbow: Diversity, Gender, and Sexuality in Nature and People* (2004), evolutionary ecologist Joan Roughgarden of Stanford University interprets these "diverse manifestations of gender and sexuality" as a challenge to Darwin's theory of sexual selection. Since homosexuals can't pass on their genes, she asks, Shouldn't they be an evolutionary dead-end? She offers instead her own theory of "social selection." "What if classic sexual ornaments such as the peacock's tail or a stag's antlers are there not to attract females," she asks, "but are badges of membership in power-holding cliques?"

Eric Idle, of Monty Python fame, seems to have anticipated her theory in this wonderful lyric from his *Rutland Isles* album:

In the *Origin of Species*, Charles Darwin liked to say
That dressing up and showing off was purely male display.
But sadly, scientifically he led us all astray
For what he failed to notice was—these creatures are all gay.

There's a tiny bird in Kent
Who's ridiculously bent
And a rutting stag who is a total dear
There are naughty big blue whales
And the sperm whale prefers males
And the dolphin is of course completely queer . . .

Gay, gay, gay, they all are madly gay
But in Victorian times that was not the thing to say. . . .

Lyric from "The Gay Animal Song," by Eric Idle and John Du Prez.
© 2003 Rutsongs Music. All rights reserved.

Of course, students of animal behavior have long noted the existence of same-sex "helpers" in social species like meerkats or acorn woodpeckers. Although they don't leave biological offspring, their successful participation in rearing the young of close relatives perpetuates their genes—a phenomenon known as kin selection.

Bagemihl, however, seems uncomfortable with such a cut-and-dried solution, for he would rather marvel at nature's enigmatic abundance—the "exuberance" of his title. He wants us to celebrate the fact that life appears in every conceivable shape, inhabits a wide range of environments, and manifests enormous sexual diversity—without human-imposed preconceptions or limitations. In a delightful poem about taking his grandchildren to the zoo, the great French author Victor Hugo once expressed his own puzzled delight at the exuberance of the natural world:

"If the master . . . should give us a hummingbird one day, and next day a mastodon— bad taste is one of his quirks."

—Victor Hugo

Personally, I don't expect God to keep himself under control, not always,
You have to put up with some vibrant excesses
From such a great poet, and not lose your temper
If the master who tinges peach-blossom so subtly
And arches the rainbow right over the ocean he pacifies
Should give us a hummingbird one day, and next day a mastodon.
Bad taste is one of his quirks,
He likes to add dragons to chasms and maggots to sewers,
To do everything on an astonishing scale . . .

See also SEXUAL SELECTION

BIOLOGY
Origins of the Discipline

Pioneer French evolutionist Jean-Baptiste Lamarck (1744–1829) coined the word *biologie* (from Greek meaning "the study or knowledge of life") in his *Philosophie Zoologique* (1804) to embrace the study of all living things. The term was rapidly adopted in Europe but did not become popular in England and the United States until the 1850s. Few biologists realize that their discipline was formalized so recently.

Until the mid-19th century, students of plants and animals were usually called "naturalists," and their subject "natural history"—which included not only botany and zoology, but geology and sometimes ethnology as well. Darwin came aboard HMS *Beagle* as "ship's naturalist," equally adept with a microscope, skinning knife, insect net, or geologist's hammer. (The fossils and rocks he collected on his travels were as influential as the living species in forming his ideas.)

By the late 1880s, biology was in vogue in England; natural history began to sound vague and old-fashioned. Geology was banished to a separate department, and a caste division arose between the professional biologist and the amateur naturalist, who was now relegated to Sunday butterfly-chaser and birdwatcher. However, some modern, unorthodox zoologists honor the old tradition by calling themselves naturalists, meaning they are curious about everything in nature.

The teaching of life sciences in schools and universities would not gain a foothold until the late 19th century. One of the major forces for its establishment was Thomas Henry Huxley, Darwin's friend and colleague. Previously, natural history was something one pursued outside of school. Classics, Greek, and Latin dominated academic studies, and it was considered vulgar to dissect frogs in a classroom. After all, how could tearing apart a plant or animal compare with the intellectual virtues to be gained by studying Homer? In his 1856 lecture "On Natural History as Knowledge, Discipline, and Power," Professor Huxley, as always, had an eloquent answer:

> Let those who doubt the efficacy of science as moral discipline make the experiment of trying to come to a comprehension of the meanest worm or weed—of its structure, its habits, its relation to the great scheme of nature. It will be a most exceptional case, if the mere endeavour to give a correct outline of its form, or to describe its appearance with accuracy, do not call into exercise far more patience, perseverance, and self-denial than they have easily at their command. . . . There is not one person in fifty whose habits of mind are sufficiently accurate to enable him to give a truthful description of the exterior of a rose!

Biology today encompasses botany, cell physiology, zoology, genomics, genetics, classification (taxonomy and cladistics), and attempts to construct a tree of life, showing how every living species is related to each other. Other fields in biology include ecology (the study of an organism's relationship to its environment), animal behavior (sociobiology), and the study of embryos and development. Since the Darwin-Wallace revolution, all biology is evolutionary biology.

See also HUXLEY, THOMAS HENRY; LAMARCK, JEAN-BAPTISTE

"There is not one person in fifty whose habits of mind are sufficiently accurate to enable him to give a truthful description of the exterior of a rose!"

—Thomas Henry Huxley, 1856

BIOSPHERE, EVOLUTION OF
The World of Life

> "The biosphere is the area of the Earth's crust occupied by transformations which convert cosmic radiation into effective terrestrial energy."
>
> —V. Vernadsky, 1911

Eduard Suess first used the word "biosphere" in 1875, in a geological treatise on the Alps, and the concept was developed around 1911 by the Russian scientist Vladimir Vernadsky: "The biosphere is the envelope of life . . . the area of the Earth's crust occupied by transformers which convert cosmic radiations into effective terrestrial energy." In other words, the biosphere is a complex web of plant and animal life that ultimately derives its energy from the sun. Recently, it has become clear that the biosphere itself evolves. According to the Gaia Hypothesis, life may play an important role in creating, maintaining, and expanding its own favorable environment.

Analysis of early microfossils and geochemistry show that the Earth's original biosphere contained no oxygen. But about two billion years ago, purple and green microbes began to transform water and sunlight into food. They used the hydrogen, and they expelled oxygen as waste into the atmosphere. To these anaerobic bacteria, oxygen was a deadly gas, but it turned out to be the staff of life to their successors.

According to biochemist Lynn Margulis of the University of Massachusetts at Amherst, "This toxic-waste crisis turned out to be a blessing, though, for it inspired further innovation among the microbes. Some bacteria . . . came up with means to detoxify and eventually exploit it; they invented oxygen breathing."

In her view, these new oxygen-breathing (aerobic) bacteria survived in local niches for hundreds of millions of years. As levels of oxygen in the atmosphere rose higher and higher, some aerobic and anaerobic bacteria joined together into a new form of life: the cell with a nucleus and cell membrane.

"This new piece of microbial technology," says Margulis, "was as different from the basic bacterium as the space shuttle is from a paper airplane." Known as the eukaryotic, or nucleated, cell, these composite creatures have dividing membranes, including one that cordons off a nucleus from the rest of the cell, or cytoplasm. Self-reproducing parts known as organelles perform specialized functions, such as respiration (breathing) and photosynthesis (food making). And their nuclei contain long twisted protein chains of DNA, libraries of information that enable them to produce copies of themselves.

Today the biosphere is relentlessly assaulted by pollutants introduced by automobiles and industry and by the unprecedented destruction of habitat. Oil spills, contaminating pristine Arctic waters, holes in the ozone layer caused by freon and other gases, reduction in oxygen caused by burning of vast rain forests, and the consequent "greenhouse" effect—all pose grave dangers to the biosphere. In ecologist Paul Ehrlich's metaphor, destruction of parts of the biosphere is like removing hundreds of the thousands of rivets from an airplane's wing. One never knows at which point the wing will fall off.

See also BACTERIA; EXTINCTION; GAIA HYPOTHESIS; GREAT DYINGS; RAIN FOREST CRISIS

BIRD, ROLAND THAXTER (1899–1978)
Dinosaur Tracker

> For years, R.T. Bird argued that dinosaur trackways were valuable records of behavior, but museum bosses only wanted bigger "monster" skeletons.

Junius and Roland Bird, brothers raised in New York's Catskill Mountains, were a most remarkable pair. Each spent a separate, individualistic career trekking through wilderness in search of clues to America's prehistoric past.

Roland (known as R.T.) was one of the last of the old-fashioned dinosaur hunters, best known for discovering and preserving the spectacular dinosaur trackways from the Paluxy River, near Glen Rose, Texas.

Junius, who became an archeologist at the American Museum of Natural History, sought traces of ancient man in the New World, from the Arctic to the tip of South America. (He found the famous Inca fabrics in Peru and, in Tierra del Fuego, proof that humans lived there 11,000 years ago, while giant ground sloths still roamed.)

R.T.'s true calling came after some aimless years of dabbling in cattle farming and land speculation and roaming the American wilderness in a motorcycle-camper he invented. (Recreational vehicles were still in the future.) His chance discovery of a fossil amphibian jaw sparked a lifelong passion. With his father's encouragement, he sent his fossil amphibian skull to Barnum Brown, then America's greatest dinosaur digger, who decided to take him on as an assistant.

Throughout the 1930s, Bird energetically excavated, prepared, shipped, and cataloged uncounted tons of dinosaur bones for the American Museum of Natural History, often so absorbed in his work that he let his paychecks pile up for months uncashed.

R.T. felt especially close to the ancient behemoths when following the trackways they had left in the mud, long since hardened into stone. R.T. cleared river banks and built dams to enable him to follow the tracks, stalking them even under rock formations by dynamiting away hills. (One local workman insisted R.T. was trying to make a fool of him, because "There's no way that big fella could have walked under there.")

In water, R.T. discovered, the creatures propelled themselves as modern hippos do, by pushing only their front feet against the bottom. Rarely, there was a single imprint of a hind foot, where the animal kicked into a sharp turn.

At a time when scientists believed the huge sauropods were slow-moving and had to stay half submerged to support their bulk, R.T. showed they could move quickly on land and traveled in herds. He uncovered one dramatic sequence where sauropods were being pursued by a hungry two-legged meateater. Forty years later, dinosaur specialist Robert Bakker noticed the tracks showed a social structure, with young and juveniles protected in the middle of the herd—which delighted R.T.

For years, he had argued that the trackways were really fossilized behavior, but museums (and sponsor Harry Sinclair, the oil magnate) were more interested in skeletons. Nevertheless, R.T. pestered Brown to let him work on the trackways until he got his way. Over several years, he meticulously chose, partitioned, and quarried stone blocks of petrified footprints and shipped them back to the Museum of Natural History in New York.

But while Brown's dinosaur skeletons were cataloged, prepared, and mounted, R.T.'s crates of fossil footprints languished in the museum yard. Stricken by illness, R.T. went into forced retirement and obscurity.

During the 1950s, Edwin Colbert, the museum's curator of vertebrate paleontology, was constructing a new hall of Jurassic dinosaurs and remembered the tracks. If they could be restored behind the giant dinosaur skeletons, he decided, it would make a marvelous exhibit.

By this time they had lain neglected outdoors through 14 New York winters, a "weathering mess." Crates were rotted, stones broken, labels washed away. More than 100 heavy blocks were heaped in jumbled disarray; a few more years of neglect and they would be lost to science. Colbert sent for R.T.—the only man in the world who could rescue the achievement of his youth.

Despite very frail health, Bird accepted the challenge, although he was so weak he was only able to work for two sessions a day of 20 minutes each, separated by a two-hour rest. Nevertheless, he doggedly persevered. In six months, with help from assistants, he had reconstructed the massive jigsaw puzzle; it was cemented in place behind Barnum Brown's dinosaurs.

R.T. died in 1978, but not before he had the satisfaction of seeing a permanent on-site museum built in Texas to preserve and display some of his beloved trackways. At his funeral, relatives lovingly placed in his pocket one of his prized, polished gizzard stones, which had once aided a dinosaur's digestion. His unique tombstone in the Grahamsville, New York, cemetery is carved with the likeness of a brontosaur and this epitaph: "R.T. Bird, discoverer of sauropod dinosaur footprints."

See also BROWN, BARNUM; "NOAH'S RAVENS"; SINCLAIR DINOSAUR

FOSSILIZED BEHAVIOR intrigued R.T. Bird, who collected these trackways at the Paluxy River in Texas and mounted them behind an apatosaur (brontosaur) skeleton years later. They clearly show a meat-eater (the three-toed prints) following and attacking a large plant-eating dinosaur.

BONOBOS
Pygmy Chimps

Bonobos are not called "pygmy chimpanzees" because they are small, but because they live in the same forests in central Africa as tribes of diminutive humans. Reclusive and rare, the bonobo is known as the "last ape" because the species was not recognized by science until 1929. Humans are as genetically similar to bonobos as we are to common chimpanzees, sharing 98 percent of our DNA. Some primatologists believe that intellectually and behaviorally, however, bonobos may be even more similar to humans than to other chimps.

Slimmer than the chimpanzee, the bonobo has long, graceful legs, a smaller head, and its back becomes straighter when it stands upright—presenting a decidedly more human-like posture. Anatomically, the bonobo resembles our ancestors, the australopithecines, more closely than any other living ape. Since the species has been evolving from ancestral hominids as long as we have, however, it cannot be considered a living "missing link."

Bonobos' behavior also sets them apart from other apes. While chimpanzees are known for their aggressive, male-dominated societies, bonobo communities seem to be egalitarian, female-centered, and relatively nonviolent. According to primatologist Frans de Waal of Emory University's Living Links Center, "The chimpanzee's sex life is rather plain and boring, while bonobos act as if they have read the Kama Sutra." Though they only reproduce every five years, bonobos engage in very frequent sexual activity, and in an astounding variety of combinations—males with males, males with females, females with females, and adults with juveniles. Their erotic behavior seems to be used mainly to ease (rather than promote) social tensions and conflicts. "The chimpanzee resolves sexual issues with power," according to De Waal, while "the bonobo resolves power issues with sex."

FAMILY PORTRAIT of bonobos in the Jacksonville, Florida, Zoo. Matriarch Linda, on the left, was a founding member of the colony. Photo © by Marian Brickner.

Bonobos and chimps differ greatly in other aspects of their behavior, such as communication and tool use. Wild chimps are famous for capturing termites with sticks and cracking nuts with rocks, but seem never to fabricate stone tools, even when captive animals are shown how to make them. Bonobos, on the other hand, readily use tools in captivity—and can even learn to chip stone flakes to use as cutting tools—but no one has ever seen them do so in the wild. Some researchers believe that bonobos also appear to be more empathetic, taking account of other individuals' feelings in ways that chimps do not, and appear to use more differentiated systems of vocal communication.

In captivity, bonobos have excelled in experimental language training. At the Language Research Center in Atlanta, Georgia, anthropologist Sue Savage-Rumbaugh has worked with both chimps and bonobos, teaching them printed symbols or "lexigrams" that each correspond to a single word. Her star bonobo pupil, Kanzi, astounded scientists with his abilities to make stone tools and to understand arrangements of symbols. He can apparently understand some English sentences, and is able to point out symbols that correspond to spoken words. Such comprehension, or "symbol-to-symbol transfer," had never previously been documented in nonhuman animals—and it was a feat Kanzi achieved on his own, without being specifically trained to do so.

Until the 1980s, local taboos protected bonobos from being hunted in their remote African rain forests, but today they are increasingly slaughtered for "bushmeat." Sadly, they are now endangered in the wild and are exceedingly rare in captivity, creating a dangerous situation for the "last ape" and a precarious future for acquiring better knowledge of our close relatives.

See also APE LANGUAGE CONTROVERSY; CHIMPANZEES; KANZI

BONOBO GESTURES and expressions seem uncannily human, whether showing a bruise to a chum, flossing one's teeth, or sharing a moment of **utter hilarity.** Photos © by Marian Brickner.

IMAGINING FACES AND FIGURES in every flint chip and flake, as in these illustrations from his book, Boucher de Perthes hurt his own attempts to convince scientists of the authenticity of prehistoric stone tools.

BOUCHER DE PERTHES, JACQUES (1788–1868)
Founder of Prehistoric Archeology

Scientists of the 1830s were not prepared to believe in the existence of 30,000-year-old man-made tools and artworks. That tons of such artifacts were sealed into rock strata, along with the bones of extinct European rhinoceroses and mammoths, seemed even more farfetched. So no one paid the slightest attention to their discovery by an eccentric French amateur in the rural village of Abbeville. Even his name sounded like an impossible joke: Monsieur Boucher de Crèvecœur de Perthes, roughly "Mr. Butcher Broken-Heart of Lost-Town."

Born to wealth, Boucher de Perthes also inherited his position as head of the Abbeville customhouse, which allowed him the leisure to attend social functions and to dabble in literature and politics. He wrote novels that didn't sell and plays that were never performed; he ran for parliament but was never elected.

In 1837, Boucher became intrigued by some unusual man-made objects turned up by workmen dredging the Somme Canal. They sold him what appeared to be an ancient polished stone axe hafted to a staghorn. During visits to several gravel pits in the Somme River terrace just outside Abbeville, he obtained a few bifacially chipped stone "hand-axes." Within a few weeks, he had become an enthusiastic collector.

After accumulating thousands of chipped flints and studying their contexts in excavations, he decided to publish an illustrated description of his discoveries. Since any pre-Roman artifact was then considered "Celtic," he entitled his book *Celtic and Diluvian Antiquities* (1847). Boucher asserted that the older tools must have been buried during the biblical Flood, which put off geologists, while churchmen rejected his insistence on the great antiquity of man. Many of the prizes in his collection turned out to be crude forgeries chipped by quarrymen, who thought it great fun to con a gullible gentleman out of a few francs.

Boucher was often the victim of such hoaxes, which was a major reason most scientists refused to take him seriously. Deception of the "toffs" (gentry) for profit by clever tricksters from the underclass was an international sport, for which science was just another arena.

In his classic *Antiquity of Man* (1863), Sir Charles Lyell, England's leading geologist, tells how each geologist, fearing all his colleagues had been duped, went into Boucher's excavation determined to "not quit the pits till he had seen one of the hatchets extracted" from a sealed deposit with his own eyes.

Lyell quotes the French geologist Albert Gaudry, who wrote in 1859 that "the great point" in making valid discoveries in prehistoric archeology "was not to leave the workmen for a single instant, and to satisfy oneself by actual inspection whether the hatchets were found *in situ*," sealed in a closed deposit with the bones and teeth of extinct horses and other mammals. Like detectives on stakeout, a scientist would have to watch closely for hours, or several would take shifts, never taking their eyes off the quarrymen. Exactly as at spiritualistic seances and carnival shell games, it was a contest of alertness and acuity between those who would fool the gentry and those who would catch the fraud.

One after another the British geologists came to Abbeville: the paleontologist Hugh Falconer, the clergyman-geologist Sir Joseph Prestwich, the evolutionist Sir John Lubbock, and Lyell. Acting separately, each man personally witnessed (or performed) the extraction of a flint tool from sealed deposits. (Prestwich even took a photo of one half uncovered.) Once they knew what to look for, they returned to England and promptly found similar artifacts in the gravels of the Thames Valley.

In later years, Charles Darwin was somewhat chagrined to recall how heartily he and his colleagues once scoffed at the reports of Boucher de Perthes. After publication of his *Origin of Species* in 1859, however, scientists were positively eager to find the "prehistoric man" themselves. Riding the wave of his success, Boucher decided to offer 200 francs to any workman who discovered human remains in the stone tool–bearing deposits. Sure enough, on March 26, 1863, he was shown a human jaw stuck in the gravels at a site called Moulin-Quignon.

After prolonged examination by an impressive committee of French and British scientists, the Moulin–Quignon jaw was accepted as authentic. Theories were spun about its resemblance to such "primitive peoples as the Lapps and Basques," and it received an honored niche among relics of the prehistoric past. Unfortunately, the jaw later proved to be a hoax and not at all ancient.

Boucher turned his huge ancestral mansion into a private museum, where he displayed his thousands of prehistoric artifacts and more than 1,600 Flemish, Dutch, and French oil paintings by old masters. After his death, Boucher's heirs had the picture collection appraised by art historians and received some bad news: Most of them were fakes.

See also FOUR THOUSAND AND FOUR B.C.; LUBBOCK, SIR JOHN; PRESTWICH, SIR JOSEPH

BRANCHING BUSH
Model of Evolution

Charles Darwin pictured evolution as a "Tree of Life." From a common trunk, branches diverge in several directions, each in turn sprouting numerous twigs. Paleontologist Stephen Jay Gould characterized Darwin's tree as "actually more like a branching bush." The trunk and heavy boughs of a tree are not very flexible; an intricate bush, with delicate twigs burgeoning in all directions, is closer to our current idea of evolutionary history.

Prior to Darwin, natural history was influenced for centuries by the medieval idea of the Great Chain of Being, which arranged living things in a hierarchy. The model was a ladder or staircase. Simple, lowly (base) creatures occupied the lower rungs; higher on the ladder were creatures successively closer to man, the pinnacle of all forms of life; only angels or other spiritual entities were higher. God was on the top rung.

The Chain of Being was a hard mental habit to break. Early evolutionists, such as Lamarck and Erasmus Darwin, transformed the ladder into an escalator. Animals were thought to aspire to the next higher rung, a constant striving upward. Even Charles Darwin realized he sometimes imagined development from "less perfect to more perfect," and reminded himself "never to say higher or lower."

Determining whether a clam is "higher" than a mussel or a hamster higher than a field mouse is impossible. Each species is the product of a unique history, influenced by origin, habitat, competition, predators, climate, opportunities, and luck of the draw. Galápagos island finches, for instance, diverged from mainland ancestors to become seed eaters, insect eaters, woodpecker-like species, etc. None can be said to be "higher" or "lower"; they simply adapted to various niches in the new environment.

Lamarck measured evolutionary "progress" in terms of closeness to man, an appealing idea from the human perspective but hardly fair to the rest of living things. (Old habits die hard; many texts still refer to the "higher primates" or the "manlike apes.")

This most common misunderstanding of evolution—a fallacy that distorts humankind's true place in nature—can be corrected by picturing the branching bush. Where it is very bushy, representing a cluster of many related species, we have a successful group that has radiated into many niches.

If most of the cluster became extinct, leaving only one surviving species and a few fossils, the usual procedure has been to bring in the old evolutionary "ladder." The fossils are arranged in a direct line, making "progress" or "leading up" to the single surviving twig. This is the illusion of "finalism"—that species evolve toward a final goal, be it the modern horse or man.

If many closely related species continue to survive, no one would dream of arranging them in a hierarchy leading to the "highest" one. For instance, there are many species of rodents on Earth

DARWIN'S GERMAN DISCIPLE Professor Ernst Haeckel had a special fondness for drawing evolutionary "trees of life," or phylogenies. Evolutionists today have modified the tree to a spreading bush, lacking the unilineal "trunk."

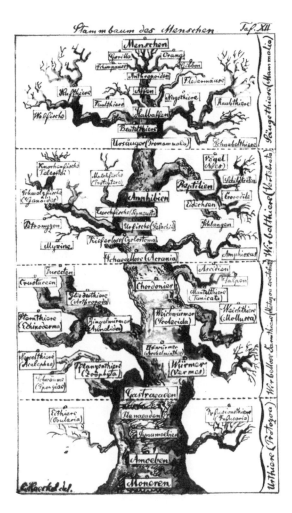

today and many different antelopes. No one wonders which is the "highest" antelope or rodent. But because only a single twig remains in the case of *Homo sapiens*, we commonly see ourselves as the goal or culmination of all hominid evolution.

See also DIVERGENCE, PRINCIPLE OF; GREAT CHAIN OF BEING; HORSE, EVOLUTION OF; ORTHO-GENESIS; TREE OF LIFE

BRIDGEWATER TREATISES (1830s)
Expositions of Natural Theology

When the Earl of Bridgewater died in 1828, he left £8,000 to sponsor a series of books illustrating the "Power, Wisdom, and Goodness of God as Manifested in the Works of Creation."

Seventeenth-century naturalist John Ray had begun the tradition of using natural history to discover God's designs, which later was elaborated and popularized by Reverend William Paley's classic *Natural Theology* (1802). Its cornerstone was the "argument from design"; whether analyzing the structure of the human eye or of an eagle's wing, natural theologians believed adaptations were conceivable only as products of divine workmanship.

Ignoring the clumsy structures and "made-over contraptions" that abound in the natural world, they focused on the tailoring of functional structure to prove the benevolence and wisdom of a caring God. Just as the intricate structure of a watch implies a watchmaker, these natural theologians argued that the complexity of living things proclaims the power of their Designer. (Almost two centuries later, the idea was resurrected as the "new" idea of Intelligent Design.) Several eminent scientists contributed volumes to the Bridgewater treatises, compiling endless lists of adaptations in nature.

The president of the Royal Society, the archbishop of Canterbury, and the bishop of London were appointed to find suitable authors. Among the first chosen was William Buckland, dean of Westminster Abbey and an accomplished geologist, whose Bridgewater treatise became a much-imitated classic of scriptural geology.

Buckland interpreted the history of the Earth on catastrophic principles. Evidence of earlier life forms was assigned to earlier creations, while "higher" species were introduced by the Creator when conditions on Earth had sufficiently improved. Geological phenomena were assigned to causes that were different in kind from those explainable by natural laws.

In 1838, the Cambridge mathematician Charles Babbage published his own, unauthorized *Ninth Bridgewater Treatise*, in which he argued that what seemed to be inexplicable miracles in the creation of the world and of living things might really be explained by some higher natural law, rather than by supernatural acts.

Babbage had invented the "calculating engine," a forerunner of the modern computer, which could be programmed to change its mode of operation according to a predetermined plan. Surely, he thought, God could program the universe to change the normal laws of nature from time to time, which would appear miraculous to the human observer.

His ideas were adopted by Robert Chambers in his influential and pseudonymous work *Vestiges of Creation*, published in 1844. Although it offered an evolutionary theory that in some ways anticipated Darwin, *Vestiges* put forward a very different notion of how things change. Rather than believing in a consistent, uniform operation of natural processes, Chambers, like Babbage, thought the rules of nature kept changing in a series of gear-shifting progressions.

This was no mere clockwork universe; admiration or explication of its mechanisms could never displace reverence for the guiding Cosmic Intelligence behind it. God had input a "law of progression," impelling the universe to undergo transformations that would ultimately unfurl the present laws of nature. Chambers and Babbage refused to speculate on the mechanism for creating the mechanism; that would be forever beyond human knowledge or investigation.

See also INTELLIGENT DESIGN; PALEY'S WATCHMAKER

Victorian mathematician Charles Babbage, inventor of the first computer, argued that God could reprogram the normal laws of nature at will.

BROOM, ROBERT (1866–1951)
Paleontologist Extraordinary

Robert Broom was a nonconforming individualist who made enemies and admirers throughout a long, productive, and adventurous life. He and his fellow scientist Raymond Dart changed the course of paleoanthropology by finding the first australopithecine fossils in South Africa—though it took them 40 years to convince the scientific world they had found the earliest hominids.

Born in Scotland in 1866, Broom trained as a surgeon but retained a strong interest in paleontology and comparative anatomy. He considered himself the "scientific son" of the British Museum director Richard Owen, who thought highly of him. Had Owen lived to see his protégé's discoveries, he would probably have been appalled that they supported the theories of Charles Darwin, his hated rival.

Broom established a successful medical practice in South Africa around 1900. After a few years, however, he became restless with city life and took off to explore the Karroo Desert region, where he prospected for fossils. Had he done nothing else in science, he would be remembered for his brilliant work on the mammal-like reptiles he discovered there, for which he won the Medal of the Royal Society in 1928.

When anatomist Dart found the first australopithecine fossil, the "Taung child," in 1924, Broom was overjoyed. Dart never forgot how Broom "burst into my laboratory unannounced. Ignoring me and my staff, he strode over to the bench where the skull reposed and dropped to his knees 'in adoration of our ancestor.'"

At the age of 69, Broom decided to find "an adult Taung ape" and searched in a limestone quarry near his home. This site, called Sterkfontein, near Krugersdorp in the Transvaal, yielded the first adult *Australopithecus* (originally named *Plesianthropus* by Broom).

A few years later, he heard that some teeth had been found by a schoolboy at a farm at Kromdraai, near his first site. He bought the teeth for a few shillings and five chocolate bars, and persuaded the boy to take him to the cave where he had found them.

The result was Broom's discovery of the large or robust form of australopithecine, to which he gave the name *Paranthropus* (meaning "to the side of humans") in 1938. His view, as expressed in the name, was that this genus was not directly ancestral to later hominids but was a distant side branch.

British scientists, notably Sir Arthur Keith, refused to go along with Dart and Broom, the renegades in South Africa. Their "near-men" did not at all fit the expected image of early humans: first an expanded brain followed by upright posture and then reduction of the teeth and jaw. Piltdown man, the fossil forgery, had a large brain and an ape jaw.

The australopithecines had chimp-sized brains and humanlike jaws and teeth. And they stood upright before they were anywhere near fully human. Broom triumphed, lived to see Piltdown exposed as a fraud, and continued searching for fossils. He was rewarded with a major trove of *Paranthropus* fossils at Swartkrans Cave, near Sterkfontein. (Later, he also found *Homo* fossils there—the first evidence for the contemporaneity of *Homo* and *Paranthropus*.)

At the age of 85, in failing health, Broom raced to complete his crowning monograph on *Paranthropus*. He corrected the final proofs on April 6, 1951, and wrote to his nephew: "Now that's finished . . . and so am I." He died that night.

See also AUSTRALOPITHECINES; DART, RAYMOND ARTHUR; KILLER APE THEORY; PILTDOWN MAN

BRILLIANT, ECCENTRIC physician Robert Broom devoted years to finding fossils of mammal-like reptiles (bottom) in South Africa, then later discovered spectacular australopithecine remains (top), helping to establish Africa as the cradle of humankind.

BROWN, BARNUM (1873–1963)
Dinosaur Fossil Collector

BONE HUNTER Barnum Brown, when almost ninety, supervised construction of Sinclair Oil's lifesize dinosaur models, shown below being barged down the Hudson River to the 1964 New York World's Fair.

Named after the great showman P. T. Barnum, indefatigable dinosaur digger Barnum Brown assembled his own version of "The Greatest Show on Earth": a parade of giant dinosaur fossils wrenched from the cliffs and arroyos of the American West. Brown's lasting contribution—hundreds of tons of dinosaur fossils—formed the nucleus of the American Museum of Natural History's world-famous collection.

During the 1960s, Brown, nearly 90, could still be seen leading visitors around the crammed dinosaur halls, announcing, "Here's another one of my children," as he pointed out the bones of a saurian giant. But when he began his career in 1897, the museum had not a single dinosaur.

As a child in Carbondale, Kansas, Brown collected fossils from freshly plowed fields. He attended the University of Kansas, then moved to New York City, where he studied paleontology at Columbia University and began working at the museum while he was still a graduate student.

For his first field assignment, the museum's director, Henry Fairfield Osborn, sent Brown to Como Bluff, Wyoming, to prospect its rich Jurassic deposits. Brown and his colleagues discovered new beds containing enormous quantities of fossils, including the *Apatosaurus* (then called *Brontosaurus*) that still dominates one of the museum's huge dinosaur halls. However, Yale paleontologist Othniel C. Marsh was furious about his former sites being worked and began a bitter feud with Osborn that lasted to the end of his life.

During the early years of the 20th century, Brown dug up fossils all over the West. One of his greatest discoveries, the first ever and a nearly complete skeleton of *Tyrannosaurus rex*, was blasted out of tons of sandstone near Hell Creek, Montana, in 1902. The fossils were then hauled by horse-drawn wagon to the nearest railroad 130 miles away.

As his exploits became known, Brown became nationally famous as "Mr. Bones." Crowds would meet his train and offer to help him find ancient monsters near their town. Now a celebrity, Brown dressed in expensive, fashionable outfits while exploring remote, dust-blown sites.

In 1909, Brown led an expedition along the Red Deer River in Alberta, Canada. The party navigated downriver on a large raft and found fossil deposits galore. "Box after box," he wrote, "was added to the collection till scarcely a cubit's space remained unoccupied on board our fossil ark."

Over the next decades, he searched for fossils and prospected for oil in India, South America, Ethiopia, and the Greek islands. Brown's second wife, Lilian, chronicled her adventures accompanying him on field trips in such books as *Bring 'Em Back Petrified* (1956) and *I Married a Dinosaur* (1950). When she first decided to join her bone-hunting husband in the field, the family maid expressed grave concern. "After all," she warned, "who knows what the beasts died of?"

One of Brown's most famous discoveries was the "great dinosaur graveyard" at the Howe Ranch, near the base of the Bighorn Mountains in Montana. After some preliminary work

SINCLAIR DINOSAURS ON WAY TO N. Y. WORLD'S FA

in 1933, he convinced the Sinclair Oil Company to put up the money for major excavations at the site. The team's efforts soon paid off when they uncovered a vast bone deposit—in Brown's words, "a veritable herd of dinosaurs." More than 4,000 bones (about 20 dinosaurs) packed in 144 crates weighing 69,000 pounds were shipped to New York.

Sinclair Oil, which used a "brontosaur" as its company logo, garnered a windfall of publicity from the public's interest in Brown's digs. During the 1930s and 1940s, the company gave free dinosaur stamps and booklets at its service stations, a promotion created and supervised by "Mr. Bones" himself.

In addition to being the world's greatest fossil hunter and a well-paid consultant to the oil industry, Brown had a clandestine career as a spy for the government—a story that was suppressed until 40 years after his death. He worked for the Office of Strategic Services, precursor of the CIA, which relied on his intelligence about the Aegean Islands as background for planning Allied invasion routes during World War II. During the 1940s, in between fossil-hunting expeditions, he assisted the Bureau of Economic Warfare.

In 1956, when he was 83, Brown explored a site at Lewiston, Montana, where he discovered and excavated a plesiosaur skeleton. Two years later he used a helicopter to prospect the Isle of Wight, where fossils abounded in the steep sea cliffs. After spotting skeletons from the air, he planned to strap himself into a bosun's chair and excavate while dangling above the English Channel.

While planning this expedition, he was approached by his old sponsor, the Sinclair Refining Company, to supervise the construction of life-size dinosaur models for the 1964 New York World's Fair. They were to be built in the town of Hudson, north of New York City, and transported to the fair via the Hudson River. Delighted at being offered a "new job" at the age of 89, Brown looked forward to startling Manhattanites with the bizarre sight of a bargeful of dinosaurs floating down the Hudson River.

Brown supervised the dinosaurs' construction but never did witness their journey to the fair. He died in February 1963, just a week short of his 90th birthday, and was buried beside his first wife, Marion. When Lilian died some years later, according to his daughter's memoirs, she "was buried on the other side of Barnum, who undoubtedly would have had a good chuckle over being sandwiched between his two wives."

See also BIRD, ROLAND T.; *FANTASIA*; OSBORN, HENRY FAIRFIELD; SINCLAIR DINOSAUR

Crowds would meet the train when "Mr. Bones" arrived in dusty Western towns to help them round up their ancient monsters.

BRYAN, WILLIAM JENNINGS (1860–1925)
Antievolutionist Crusader

During his lifetime, politician and great orator William Jennings Bryan won fame as a progressive reformer with a strong social conscience. Secretary of state under Woodrow Wilson, he had been the Democratic nominee for president three times. Bryan campaigned vigorously for women's suffrage, justice for the working poor, and curbs on corporate greed. He was also the architect of legislation prohibiting teaching evolution in the schools, thus leaving a legacy of continuing legal battles 60 years after his death.

Bryan has been vilified as an ignoramus and a demagogue who pandered to uneducated bigots in the backwaters of the United States. Movies and plays have portrayed him flailing and ranting as Clarence Darrow's adversary in the celebrated Scopes "Monkey Trial" of 1925, which was not his finest hour. Journalist H. L. Mencken depicted him as a religious fanatic, obstructing intellectual progress with a mulishly stubborn belief in the literal interpretation of the Bible.

In fact, Bryan had not always opposed evolutionary ideas, and had arrived at his reactionary position with the best of intentions for America's welfare. Convinced that the Darwinian theory, as many at the time understood it, was "a merciless law by which the strong crowd out and kill off the weak," Bryan preferred to believe "that love rather than hatred is the law of development." He also thought that "class pride and the power of wealth" were using Darwinism to justify exploiting the poor, just as European kings had once used the doctrine of Divine Right.

And his fears were justified. Industrial giants like John D. Rockefeller and Andrew Carnegie did indeed adopt Social Darwinist views about being "the fittest," their ruthlessness justified as part of a great law of nature. That this was a misreading of evolutionary theory occurred neither to Bryan nor the industrialists, since it was also taught by many biology professors of their day.

In addition, the Darwinian banner was being carried by militarists and, in Bryan's words, "was at the basis of that damnable doctrine that might makes right that had spread over Germany." He knew that during World War I, German intellectuals believed natural selection was irresistibly all-powerful (*Allmacht*), a law of nature impelling them to bloody struggle for domination. Their political and military textbooks promoted Darwin's theories as the "scientific" basis of a quest for world conquest, with the full backing of German scientists and professors of biology.

Bryan also perceived another evil resulting from the interpretation of Darwinism by the intellectuals of his day: an ill-conceived faith in eugenics as the wave of the future. It would paralyze the hope of social reform, Bryan realized, as "its only program for man is scientific breeding, a system under which a few supposedly superior intellects, self-appointed, would direct the mating and the movements of the mass of mankind—an impossible system!"

For these compelling reasons, as Stephen Jay Gould pointed out in *Bully for Brontosaurus* (1992), Bryan saw Darwinism as a many-faceted evil, quite apart from its conflict with biblical accounts of creation. Science had too easily lent respectability to political and social programs that went far beyond its proper sphere. Bryan "had the wrong solution," Gould wrote, "but he had correctly identified a problem!"

See also BUTLER ACT; *INHERIT THE WIND*; SCOPES TRIAL

> **"Taxpayers have a right to say what shall be taught. . . . The hand that writes the check rules the school."**
>
> —William Jennings Bryan

FLAG-DRAPED RELIGION was William Jennings Bryan's specialty. He was known as "The Great Populist" for his advocacy of overdue social reforms.

BUFFON, GEORGES-LOUIS LECLERC, COMTE DE (1707–1788)
French Naturalist

The orangutan, wrote the Comte de Buffon in the mid–18th century, "is a very singular brute, which man cannot look upon, without contemplating himself, and being convinced that his external form is not the most essential part of his nature."

Such tantalizing passages in Buffon's *Histoire Naturelle* (1749–1788) clearly hint at a common ancestry for man and apes. But although he appears to have glimpsed it, Buffon never could entirely break with the prevailing "chain of being" ideas that were sanctioned by his church.

Although Georges-Louis Leclerc, Comte de Buffon, was born to the wealth and power of the French aristocracy, his curiosity about nature drew him to a life of scientific pursuits. As a youth, he sought to please his family, by studying first law and then medicine. But neither could hold his interest as much as exotic species of plants or animals.

At the age of 33, he was made head of the King's Botanic Gardens (*Jardin du Roi*, later the *Jardin des Plantes*) in Paris; there he had ample opportunity to pursue his diverse interests. Soon after, he began his great work of natural history, the *Histoire Naturelle*, a compendium of everything then known about the natural world. It took him 39 years to complete. One of his innovative and controversial contributions was to include long chapters about humans along with the other animals, clearly anticipating Thomas Huxley's *Evidence as to Man's Place in Nature* (1863). Nearly a century before Charles Darwin, Buffon was groping for an evolutionary interpretation to tie together his observations.

He was struck time and again by such phenomena as offspring varying from their parents, the "fit" between organisms and environment, competition for resources, and development of races and varieties. His observations were astute, but a satisfying general theory eluded him.

Nevertheless, by 1766 Buffon had become convinced that related species could arise from a common ancestor. Although he proposed a principle of common descent, he thought that lineages could develop only from within a family, which was prevented from diverging widely by permanent constraints. For instance, Buffon thought that an ancestral cat could "degenerate" into tigers, lions, and leopards, but the original cat family was a "given." Buffon uses "degeneration" and "degradation" in a very general sense, similar to what later naturalists would refer to as "varieties"—departures from an earlier type or primal species. The species he describes are not diminished in complexity, as later degeneration theorists would say of snakes being "degraded" lizards, "losing" their limbs. Buffon simply believed that any change from a "pure" ancestral type was by definition a "degenerate" form.

Between species he often sought "intermediate gradations," because he believed that nature allowed no gaps. These linking species, he thought, came into being by gradual, regular processes, which he did not specify. But he asserted that God did not occupy Himself "with the particular fold in a beetle's wing," for he worked only through natural "Second Causes."

Buffon rejected Jean-Jacques Rousseau's view of man in the state of nature. [See NOBLE SAVAGE.] "In traveling over . . . the globe," he wrote, we do not find "human animals lacking words, deaf to voices as well as signs, dispersed males and females, abandoned children. . . . Children would die if they were not helped and cared for over several years. . . . It is not possible to maintain that man has ever existed without forming families."

Therefore, Rousseau's "natural man" was an impossibility. On the contrary, Buffon wrote, "the state of pure Nature is a known state; it is the Savage living in the wild, with his family, recognizing his children . . . using words, and making himself understood."

He also tackled geology and was the first European Christian to state openly that he thought the world much older than the 6,000-year limit imposed by church authorities. In 1788, he published *Les Epoques de la Nature*, in which he described the formation of the Earth's features by normal, currently observable processes. Some of his passages are strikingly similar to Sir Charles Lyell's statements of "uniformitarianism" that came 40 years later.

Buffon's musings on the relationship of apes and man seem very much ahead of his time. If we only study its body, he says,

> we might look on that animal as the one in which the ape species begins, or the human species ends. . . . The mind, thought, and speech, therefore do not depend on the form or organization of the body. Those are gifts bestowed on man alone. . . . Forced to judge by external appear-

"The mind, thought, and speech do not depend on the form of the body. . . . These are gifts bestowed on man alone."

—Comte de Buffon

ance alone, the ape might be taken for a variety of the human species . . . [but] the Creator has . . . infused this animal body [of ours] with a divine spirit.

It could have been Alfred Russel Wallace writing a century later.

Buffon also sounds surprisingly modern in his suggestion that distinctly human behavior is directly correlated with prolonged education of the young. Man is the only creature, he points out, where mother and infant continue an intense, intimate association well past the age of three years. It is a time of intense socialization and the learning of language, after which the child will still not be capable of surviving on its own for another ten years. Therefore, Buffon concludes that

> [A] state of pure nature, wherein we suppose man to be without thought and speech, is imaginary, and never had existence. This needful and long intercourse of parents with their children produces society in the midst of a desert . . . the parents communicate to [the child] not only what they possess from Nature but also what they have received from their ancestors, and from the society of which they form a part.

Buffon's *Histoire Naturelle* was extremely influential. By 1750, Buffon's books had outsold Diderot's *Encyclopedia,* and his popularity surpassed even that of Voltaire and Rousseau. The series was translated into many languages, and updated editions continued to be printed and read for 100 years after his death.

See also DEGENERATION THEORY; FOUR THOUSAND AND FOUR B.C.; GREAT CHAIN OF BEING

BURGESS SHALE
Earliest Known Animals

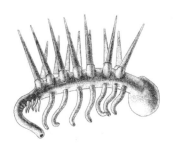

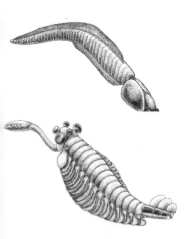

For years fossil hunters were stymied in their search for very early types of animals, since most remains were of creatures with hard parts. Fossils of early animals without shells or skeletons were extremely rare. However, in 1907 Charles D. Walcott of the Smithsonian Institution discovered an assemblage of previously unknown worms, jellyfish-like animals, and strange creatures near Burgess Pass in the Rocky Mountains of British Columbia.

Estimated to be more than 550 million years old, the Burgess Shale contains many soft animals preserved in rapidly deposited muds, the result of repeated underwater landslides. There are annelid worms (related to the living earthworms), Priapulidae (an archaic type of sea worm), and early chordates (animals with notocords, or primitive backbones.)

But the Burgess Shale is most interesting because it contains several animals belonging to phyla totally unknown on the Earth today. These lineages arose from unknown ancestors in Precambrian times, and all became extinct.

One segmented swimmer, *Opabinia,* has short stalks containing five compound eyes on its head and spiked pincers at the end of its flexible trunk. Another of these strange creatures has seven tentacles on its back and seven pairs of stiltlike legs. It was named *Hallucigenia* by its discoverer, who could scarcely believe his eyes. Fifteen of them were found clustered around a large worm, possibly preparing to dine when all were buried in the ancient mud.

Still another Burgess creature (*Pikaia*) is a slim, tentacled, cigar-shaped animal with no head, no eyes, no fins or limbs—but a rudimentary backbone. It may well be the ancestral vertebrate.

These early oceanic creatures make us wonder what turn of evolutionary history eliminated the extinct phyla, which seem so bizarre, and favored *Pikaia,* the early chordate from which mammals and birds may have eventually developed. Biologists cannot even begin to speculate on what kind of creatures might have colonized the land if *Hallucigenia* had represented the more successful phylum in Cambrian seas.

See also CONTINGENT HISTORY

THE STRANGEST CREATURES known to science are the fossil remains in Canada's Burgess Shale. Shown here are *Hallucigenia* (top), *Nectocaris* (middle), and *Opabinia,* with its clawed nozzle and five eyes—all unlike any known living phyla.

BURIAN, ZDENĚK (1905–1981)
Czech Paleoartist

One of the greatest—and fastest—of all painters of prehistoric life, Zdeněk Burian left a magnificent legacy of more than 15,000 artworks and an indefinite number of sketches and preparatory studies. In a few days, without sacrificing quality, Burian was able to turn out a finished painting that would take another artist weeks or months.

Born in Moravia, in what is now the Czech Republic, Burian graduated from the Academy of Fine Arts in Prague and started as an illustrator for book and magazine publishers. Fascinated with natural history since his boyhood and inspired by the work of Charles R. Knight, Burian turned his remarkable abilities to depicting the lost world of prehistory. Almost immediately, his ability to imbue extinct creatures with vitality, as well as the dramatic tension of his compositions, began to win recognition.

In 1932, paleontologist Josef Augusta saw some of his illustrations in an early book, *The Hunters of Mammoth and Reindeer* and sought out the painter. Together, they produced the most beautiful series of popular literature on prehistoric subjects ever created, including *Prehistoric Animals* (1960), *Prehistoric Man* (1960), *The Book of Mammoths* (1962), *Prehistoric Sea Monsters* (1964), and *The Age of Monsters* (1966). With another scientific collaborator, Burian produced three more profusely illustrated volumes on the life of prehistoric man. These books are now scarce and coveted by collectors; inexplicably, there have been no recent reprints of them.

In the Czech Republic, the government declared Burian's paintings national treasures, which cannot be sold or exported from the country. A small museum located in Dvůr Králové nad Labem, adjoining the city zoo, exhibits sixty of his works—a tiny fraction of his prodigious output.

A few years after Burian's death in 1981, his longtime student Vladimir Krb emigrated to Canada, where he created exciting dinosaur murals for the Royal Tyrrell Museum of Paleontology in Drumheller, Alberta, in the tradition of his teacher.

See also KNIGHT, CHARLES R.

BURNET, REVEREND THOMAS (c. 1635–1715)
Founder of Scriptural Geology

Many of the early geologists and naturalists were churchmen who wanted to reconcile their observations of nature with the word of God. During the 17th century, Reverend Thomas Burnet founded a style of scriptural geology that flourished in

England until the 19th century. His popular book *Sacred Theory of the Earth: Containing an Account of Its Original Creation, and of All the General Changes Which It Hath Undergone, or Is to Undergo, until the Consummation of All Things* (1691) was a pioneering attempt to explain geologic features by mechanical forces and principles and yet be true to scriptural teachings.

Following the lead of Isaac Newton in physics and René Descartes in cosmogony and geology, Burnet tried to explain the role of natural forces in shaping the Earth. His premise was that God had a plan for the Earth and for humankind, which He embedded into natural principles that could be known. If we can discover the natural laws, we would know how to fit into them and so further God's plan. That notion was later shared by Christian fundamentalists, Social Darwinists, eugenicists, environmentalists, and—minus the deity—Marxist socialists. All assume that those who are out of tune with natural laws are opposing the inevitable.

Burnet also believed that the Bible itself "providentially conserved . . . the Memory of Things and Times so remote, as could not be retrieved." From it he took his notions of the origin of man, the primal paradise, destruction of the ancient world by a "Universal Deluge or flood," and the "peopling of the second Earth." All would end, he believed, with "the fire next time."

In *Sacred Theory*, the world was created as a perfect sphere, containing a fluid mass. But as the crust dried, it cracked, allowing the inner waters to flow out and flood the land. Thus were created mountains, rivers, earthquakes, and all the untidy features of this "dirty little planet," which was pristine when first formed by God. Burnet wrote clearly of how ordinary processes of erosion, deposition, and volcanic activity had shaped the surface of the Earth. But he saw them mostly as destructive forces, a decline from original perfection.

His book was considered heretical, because Burnet described Noah's Flood more as a natural event than a punishment for human sin. Trying to have it both ways, he added that Earth crises were synchronized with human events—that it was "the great Art of Divine Providence to adjust the two Worlds, human and natural." Nevertheless, churchmen denounced him for trying to mesh sacred texts with his naturalistic theories.

In a later work, *Archaeologiae Philosophicae* (1692), Burnet tried to reconcile his account of earth history more closely with Genesis, but never succeeded in winning over his critics. If God was acting to punish man's sins, the critics preferred to read of direct divine intervention rather than such mundane secondary causes as erosion. In addition, they could not sanction Burnet's idea of a degraded Earth, ruined by the flood, at a time when most scholars sought evidence of "Divine Beneficence" in nature.

See also DIVINE BENEFICENCE, IDEA OF; "TWO BOOKS," DOCTRINE OF THE

BUTLER, SAMUEL (1835–1902)
Novelist, Polemicist

> "Meet we shall and part and meet again / Where dead men meet, on lips of living men."
>
> —Samuel Butler, 1898

Novelist Samuel Butler created two Victorian classics that continue to enjoy popularity almost a century after his death—the utopian satire *Erewhon* (1872) (an anagram for "Nowhere") and his prickly family saga *The Way of All Flesh* (1903). Practically forgotten now is the incredibly bitter and personal vendetta Butler launched against Charles Darwin in books such as *Life and Habit* (1877), *Evolution Old and New* (1879), *Unconscious Memory* (1880), and *Luck or Cunning?* (1887).

Once an admirer and correspondent of Darwin's (he had even visited Down House several times), Butler gradually turned against him. Darwin's theory of evolution wasn't even original, Butler argued. Buffon, Lamarck, and Erasmus Darwin had invented it years before, and natural selection was a gross error that added nothing. If habits and strivings of an individual's lifetime could not be passed on to offspring, there could be no progress, but only a "nightmare of waste and death," he wrote in his 1890 essay "The Deadlock in Darwinism."

But Darwin's star was riding high in the last quarter of the 19th century, and Butler's attacks antagonized a public newly enamored of science. Since Butler had also criticized and attacked the follies of religion, he stood in a no-man's land. George Bernard Shaw recalled "how completely even a man of genius could isolate himself by antagonizing Darwin on the one hand and the Church on the other."

It was Shaw who recognized that Butler's bile was not stirred by a mere argument about biology. Samuel Butler had declared "with penetrating accuracy that Darwin had 'banished mind from the universe.'" He extended the attack to Darwin's character, the playwright believed, because Butler was "unable to bear the fact that the author of so abhorrent a doctrine was an amiable and upright man." In *Evolution Old and New*, Butler described Erasmus Darwin's evolutionary theory and concluded that it was far superior to his grandson's.

At the time, Charles Darwin was publishing a translation of a German article, by a Dr. E. Krause, about Erasmus's life and work. Darwin sent a copy of Butler's book to Krause, who then added a few caveats to his admiring biography of Darwin's grandfather. Erasmus's theory was great in its time, Krause warned his readers, but "to wish to revive it at the present day . . . shows a weakness of thought and mental anachronism which no one can envy."

Butler was furious and insisted Darwin publicly admit that this passage (and others) had not appeared in the original German version but had been inserted as a devious attack on his own critical book. Darwin claimed innocence, though his son and biographer Francis later admitted that Butler "had some cause of complaint." Charges and countercharges flew by letter, though Darwin made no public response to Butler's published accusations. Thomas Huxley told Darwin to grin and bear it, for "every great whale has its louse."

The Darwin-Butler feud became convoluted and blown out of proportion; it was still the subject of argument and debate long after the principals were dead. Butler certainly appears to have been thinking of Darwin in the concluding lines of "The Life after Death," his prophetic poem about immortality:

> We shall not argue, saying "'Twas thus" or "Thus,"
> Our argument's whole drift we shall forget
> Who's right, who's wrong, 'twill be all one to us;
> We shall not even know that we have met,
>   Yet meet we shall and part and meet again
>   Where dead men meet, on lips of living men.

CONTENTIOUS NOVELIST
Samuel Butler accused Darwin of stealing his best ideas from his grandfather Erasmus Darwin.

BUTLER ACT
Tennessee Antievolution Law

On March 21, 1925, the Tennessee legislature passed the Butler Act, making it illegal to teach human evolution in the state's public schools. It provided

that it shall be unlawful for any teacher in any of the universities, normals and all other public schools of the State . . . to teach any theory that denies the story of the Divine Creation of man as taught in the Bible, and to teach instead that man has descended from a lower order of animal.

Two months after the law was adopted, John T. Scopes, a Dayton, Tennessee, high school teacher, was charged with teaching evolution and made national front-page news.

Scopes was convicted, but the verdict was overturned on a technicality. As a result, it was impossible to appeal and challenge the constitutionality of the law. Forty years later, antievolutionary laws were struck down in Tennessee, Arkansas, and Mississippi.

But antievolution laws were soon replaced in Arkansas and some other states by "balanced treatment" laws, which required "creation science" and evolutionary biology to be given equal time in classrooms. On January 5, 1982, the same day that Judge William Overton ruled that the Arkansas law was incompatible with the U.S. Constitution, the Mississippi Senate passed its own law requiring "balanced treatment" in the state's public schools.

See also BRYAN, WILLIAM JENNINGS; CREATIONISM; *INHERIT THE WIND*; SCOPES TRIAL

> "It is unlawful to teach any theory that denies the Divine creation of man as taught in the Bible."
>
> —Butler Act, Tennessee, 1925

BUTTON, JEMMY (1815–1864)
A Victorian "Experiment"

Had it not been for Captain Robert FitzRoy's promise to return Jemmy Button and two other Yahgan (Yamana) Indians to their homeland at the tip of South America, Charles Darwin's voyage of discovery might never have taken place.

"Jemmy Button" was the English name given to Orundellico, a 14-year-old Indian boy who was taken from Tierra del Fuego to England in 1830 aboard HMS *Beagle*. FitzRoy, according to his own accounts, had "paid" an adult Indian a large mother-of-pearl button for the lad; hence his nickname.

Imperious Captain FitzRoy "collected" three other Fuegians at the same time, not by force, but by describing the wonders of England to the naive tribesmen. On his own responsibility, the captain was attempting an experiment with human lives. He planned to take these "savages" from "brute creation" and expose them to the light of British civilization. They would be taught to speak and read English, learn Bible verses, eat with knife and fork, and then be returned to their homeland to spread the light of Britannia. Jemmy's companions were given the names Boat Memory, York Minster, and (a young girl) Fuegia Basket.

In FitzRoy's view, it was a philanthropic project to benefit both the British and the Fuegians. When Jemmy and his friends returned to their tribe, they would bring superior knowledge and prosperity to their benighted relatives. They could spread their new knowledge of clothing, English, Christianity, high morals, and the cultivation of food plants. When the next British sailors arrived at this remote, wild spot, FitzRoy believed, they would be greeted by friendly, English-speaking natives who would cheerfully supply food, wood, water, and other provisions—perhaps even a good English shepherd's pie.

Although Boat Memory died soon after arriving in England, Jemmy Button and his two other companions behaved remarkably well for people who were wrenched and dislocated from their culture and treated as specimens. They were bright, curious, and eager to learn the white man's ways. Most of the time they lived with missionaries in England, who took them on occasional excursions around London, where they even had a long audience with the king and queen. Jemmy enjoyed dressing as a dandy in highly polished boots, short oiled hair, and white kid gloves.

FitzRoy's Fuegian experiment was the original impetus for the *Beagle*'s second voyage to South America. When the Admiralty hesitated to finance another long surveying expedition, FitzRoy appealed to his wealthy relatives. Finally, fearing the surviving Fuegians would all grow discontented and die in England, he decided to finance their return himself.

After the captain had put up some of his own money, the Admiralty came through with funds for a new mission to chart the coast of South America. It was then that he decided to use the opportunity to advance another of his pet theories: he would gather evidence during the voyage that would prove the truth of Genesis.

JEMMY BUTTON, in English dress, 1833.

Both of FitzRoy's major projects became complete personal disasters. His experimental subject Jemmy Button ended up turning his back on British civilization and reverting to the "wild" ways of his naked tribe. And the naturalist he chose to help him prove the biblical creation theory turned out to be Charles Darwin. (FitzRoy committed suicide years later, in despair about what he thought was the failure of his grandest aims in life. Darwin had long feared that possibility, as FitzRoy's uncle had also been a suicide.)

Jemmy, 16 years old on the *Beagle*'s second surveying voyage of 1832, became a special favorite of the sailors. The Fuegian's eyesight was very keen, and he was able to spot distant objects before anyone else on board. When Jemmy couldn't get his way with the officer on watch, he would pout and say, "Me see ship, me no tell."

Many gifts had been given to the Fuegians with which to start their new civilized life in the wild country at the tip of South America. If the project's ludicrous Victorian smugness was unrecognized by the captain, it did not escape Darwin. He wrote in *Voyage of the Beagle*:

> The choice of articles [by the Missionary Society] showed the most culpable folly and negligency. Wine glasses, butter-bolts, tea trays, soup tureens, mahogany dressing case, fine white linen, beaver hats and an endless variety of similar things, show how little was thought about

the country where they [the Fuegians] were going to. The means absolutely wasted on such things would have purchased an immense stock of really useful articles.

A young missionary, Richard Matthews, had come along on the voyage to help Christianize the Fuegians and establish a settlement. Jemmy Button, Fuegia Basket. and York Minster were ferried in small boats up the Beagle Channel to Ponsonby Sound, along with supplies and a crew of sailors. They were met by a fleet of canoes manned by grease-covered Indians, naked in the cold except for skimpy capes of guanaco and otter skins. The sailors put up tents to house the cargo and constructed three wigwams: one for the missionary, one for Jemmy, and one for York and Fuegia Basket.

All hands began digging a vegetable garden, while 100 or so native tribesmen stood around, staring in wonder. Finally, Jemmy's mother, two sisters, and four brothers arrived. Darwin wrote that "it was laughable, but almost pitieable, to hear him speak to his wild brother in English, and then ask him in Spanish whether he did not understand him." Jemmy had forgotten his own language, although he was soon to regain it. Matthews stayed with the Indians, while FitzRoy and his men departed the camp.

When the *Beagle* crew returned 10 days later, everything was in a shambles. Matthews reported that as soon as the crew left, the natives had started to take everything in sight; when he tried to stop them, he was beaten and almost killed. The vegetable garden was trampled. York Minster had sided with the natives and was let alone, but Jemmy had been in several fights for defending the missionary and had been beaten as well. FitzRoy was shocked and disappointed with the results of his good intentions. He took Matthews back on board but left the Fuegians with their kinsmen and departed.

A year later, when the *Beagle* returned after further explorations, the camp was no more. York Minster and Fuegia Basket had taken all Jemmy's goods and departed with the other Fuegians. Jemmy had remained, but had replaced his European clothes with a loincloth, taken a native woman for his wife, and was hunting and fishing for a living. Although he visited his old friends aboard the *Beagle,* he told them he was finished with civilization forever. Darwin poignantly recorded his last view of Jemmy, standing on shore near his campfire, waving "a long farewell."

More than 30 years later, another ambitious crew set out for the region with the goal of converting the heathen, and once again Jemmy Button played a major role—but this time a much more sinister one. By dint of his bicultural expertise, Jemmy had emerged as informal leader and ambassador for his tribe of about 300 people. The new missionaries planned to build a church, establish a colony, and bring his kinsmen the light of English culture and religion. For a while he played along, persuading his tribesmen to cover their bodies, construct buildings, plant and tend gardens, and worship the Christian deity.

In *Evolution's Captain* (2003), Peter Nicols comments: "Here was the wildest Victorian pipe dream of colonization from the opium of faith and ignorance." In addition to their total misreading of the people and their culture, the colonizers callously imposed prisonlike restrictions on the islanders for their "improvement." Jemmy, in Nicols's conjectural view, became bitter at the years of capture, handouts, and humiliation, and was secretly brimming with anger at those who had dressed him in fine clothes and introduced him at their Royal Court, then cast him adrift to play the role of naked savage once more.

One day in 1860, during a routine prayer meeting in the church, the entire tribe attacked the eight missionaries, slaughtering them in cold blood. When authorities arrived at the scene some months later, Jemmy blamed "other" tribes for the attack and managed to talk his way out of receiving any blame or punishment. Some historians, however, think it probable that he was the leader and organizer of the bloody massacre.

In Jemmy's various personas as Victorian dandy in kid gloves, compliant savage, dutiful convert, and perhaps leader of a violent insurrection, we may glimpse a complex, cunning personality that belies Darwin's rose-colored image of a simple, endearing "child of nature."

See also FITZROY, CAPTAIN ROBERT; VOYAGE OF HMS *BEAGLE*

YAMANA INDIANS brought to England to be "civilized" were sketched by Captain Robert FitzRoy. Below: Fuegia Basket in 1833; Jemmy Button in 1834, after reverting to tribal life; and York Minster in 1833.

CAMBRIAN-SILURIAN CONTROVERSY
Rival Geological Periods

The Old Red Sandstone of England was the oldest system of fossil-bearing rock known until the 1830s. Packed with remains of an "age of fishes," it was called the Devonian, after Devon, where it was first found. But what came before the Devonian? Adam Sedgwick, a former theology student turned geologist, and Roderick Impey Murchison, an imperious aristocrat, traveled all over Europe attacking rock formations with their geologist's hammers to find the answer.

Once good friends, the two became bitter rivals in the quest to discover, describe, and name the oldest geological systems (and corresponding periods) in Earth's history.

After publishing many papers on the geology of Scotland, the Alps, and other parts of Europe—and gaining the presidency of the Geological Society in 1831—Murchison set out to study the Greywacke formations, which were thought to be very old.

His search led him to Wales, where in 1834 he found a formation of fossil-bearing rock that lay underneath the Devonian and was therefore more ancient. With romantic flair, he named it for the Silures, an ancient Welsh tribe that had resisted the invading Romans. Thus the Silurian period of Earth's history was established—and became Murchison's lifelong obsession.

While Murchison was working to establish the Silurian as the oldest rock system, his friend Sedgwick discovered an even more ancient one underneath it. Sedgwick gave it the archaic name for Wales, which is Cambria, and thus established the Cambrian period in geology.

Now began a great rivalry, with Murchison and Sedgwick trekking to rock formations all over Europe, each attempting to enlarge his own geological domain. Murchison extended the Silurian downward, while Sedgwick pushed his original Cambrian upward. The two men were completely unable to agree on where the natural boundaries occurred. Murchison, however, found a way to resolve the dispute. He got himself appointed director of the national Geological Survey and simply ordered that the name "Cambrian" be deleted from all government books and geological maps.

Science is not always fair—at least not while its contributors are still alive. But it also has a way of correcting itself in time, and after both men were dead, Sedgwick was vindicated. The Cambrian has grown in importance over the years, while the Silurian has diminished.

Since Murchison's time, his beloved rock kingdom has been shown to encompass several systems, one of which has been split off and renamed the Ordovician. Thousands of fossil finds of extinct early phyla, including those of the remarkable Burgess Shale, have literally put the Cambrian back on the maps for good. Sir Roderick Murchison, the self-proclaimed "King of Siluria," would not have been pleased.

See also BURGESS SHALE; MURCHISON, SIR RODERICK IMPEY; SEDGWICK, REVEREND ADAM

CANNIBALISM CONTROVERSY
Are Humans Man-Eaters?

In the history of anthropology, discoverers of almost every fossil manlike creature have rushed to announce they have found associated "evidence of cannibalism." Later, in most cases, their colleagues declare the evidence insufficient, and it is eventually dropped from discussion.

African australopithecines, Beijing Man (*Homo erectus*), Neanderthals, and Cro-Magnon were all at first believed to have had a taste for their fellows' flesh. Debate has gone on for years about whether our ancestors habitually dined on each other, or if the usual horrific interpretation reveals more about the minds of anthropologists than it does about prehistoric cannibalism.

In the case of *Homo erectus*, scratches on fossil hominid bones could have had other causes than cannibalism; they may have been gnawed and dragged by hyenas or other scavengers. *H. erectus* may have been the prey, not the predator. At the Beijing Man site, only skulls were found, which could mean that they were carried into the cave for a ritual. (Many contemporary tribal peoples use the skulls of dead relatives in ancestor worship.)

On the other hand, human bones of Atapuerca's "pre-Neanderthal" people, in Spain, exhibit cut marks that are identical with those on animal bones that were butchered there by ancient people. Other credible evidence of cannibalism has recently been found in the remains of 2,000-year-old Fiji Islanders, 100,000-year-old Neanderthals in France, ancient pueblo peoples of the American Southwest, and the Mayans of Central America.

Anthropologists still debate the nature and extent of cannibalism among tribal peoples in recent times. William Arens provoked an uproar among experts when he challenged the whole idea in his book *The Man-Eating Myth* (1979). Like most anthropologists, Arens had always taken for granted that 19th-century explorers had visited cannibalistic tribes in Africa, New Guinea, and South America. But when he sifted the massive literature on the subject, he could not find one satisfactory first-hand account of cannibalism as a socially approved custom in any part of the world. It was a myth among the anthropologists, he said, and challenged his colleagues to prove otherwise. Soon after his book appeared, several field-workers came forward to offer evidence that culturally sanctioned cannibalism was indeed part of the human repertoire.

Twenty years before Arens published his book, anthropoligist George Morren of Rutgers University lived in New Guinea, where older Miyanmin tribesmen gave him detailed recollections of the practice. He also studied records of a 1959 trial in which 30 Miyanmin were accused of killing and eating 16 people from a neighboring tribe. Documented instances continued into the 1980s.

When he pressed the Miyanmin about the possible religious or symbolic significance of the incident, they insisted there was none: "No, we just went after the meat." It appeared to be an authentic, documented account of recent culturally sanctioned "gourmet" (as opposed to ritual) cannibalism. There were many more to come.

A more informal account, this one from the nearby South Pacific islands of Vanuatu, was offered by journalist Mike Krieger in his book *Conversations with the Cannibals* (1994):

> I am sitting in a thatched-roof village on the island of Malekula listening to some old ex-cannibals as they have the time of their lives laughing and reminiscing about somebody they once ate. There is no guilt here. . . . To my astonishment, it becomes more and more apparent that with little coaxing, and if they could be sure of escaping detection, these old codgers would still love nothing better than to chew on another human once more.

In their youth, the Miyanmin tribesmen raided other villages. When an enemy warrior was killed, the victors would carry his corpse to their own village, where they would butcher it

CANNIBAL BUTCHER SHOP, supposedly in central Africa, was imagined in this old illustration from 1598, and reprinted in Thomas Huxley's 1863 book *Evidence as to Man's Place In Nature.*

and dine on the flesh. At the time, they had no pigs, cattle, or chickens. Eating human meat was a special occasion, which occurred three or four times a year among the men of the victor's village. The practice, which was already waning by the 1920s, ceased almost completely when American soldiers occupied the island during World War II.

Observers of primate behavior have also contributed to this debate. Among the Gombe Stream chimpanzees observed by Jane Goodall, a mother and daughter team suddenly began a series of cannibalistic infanticides. One would distract a new mother, another would snatch the baby, then they'd both kill and eat it. Goodall admitted that she had once thought chimps were "better" than humans, but now realized that a chimp's heart contains dark secrets as well.

See also ATAPUERCA; CHIMPANZEES; *HOMO ERECTUS*; KILLER APE THEORY

CARDIFF GIANT
Petrified Man Hoax

Fossil evidence for human evolution was still so scant in the 1870s that a bewildered public paid a fortune to see the most brazen scientific hoax in history. The Cardiff Giant, a crude stone statue, was successfully promoted as the petrified remains of a huge, extinct species of man that once inhabited upper New York State.

George Hull, a former cigar maker from Binghamton, New York, conceived the plot to create the giant and, in 1868, obtained a five-ton block of gypsum in Iowa and had it fashioned

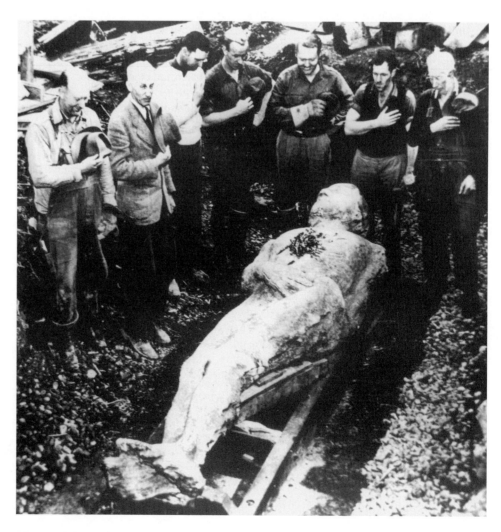

GIANT FRAUD being laid to rest at the Farmer's Museum in Cooperstown, New York, in 1948. Thousands had paid to see the fake "petrified man" that had supposedly been plowed up on a Cardiff farm; irreverent workmen pay their last respects with appropriately phony sentiments.
Courtesy of Fenimore Art Museum, Cooperstown, N.Y.

into the shape of an immense man by a Chicago stonecutter. He then shipped the statue to his cousin, William Newell, near Cardiff, New York, who supposedly discovered it while digging a well behind his barn a year later. It is not clear whether the hoax was originally planned as a swindle or if, as Hull later claimed, he built the giant to ridicule clergymen who insisted on the literal truth of every word in Genesis, including "there were giants in the earth in those days."

A Syracuse newspaper headlined the find as a wonderful discovery, and the pair pitched a tent on the farm and began exhibiting the giant in a ditch, charging a nickel a look. News of the find flashed around the world. Thousands swarmed to see it, and admission was raised to a dollar.

Meanwhile, experts argued about the fossil's authenticity. The director of the New York State Museum thought the giant was really a statue, but was indeed most ancient and "the most remarkable object yet brought to light in this country." Others, including Oliver Wendell Holmes and Ralph Waldo Emerson, concurred. Cornell's president pronounced the giant a gypsum forgery, and Yale paleontologist O. C. Marsh muttered it was "remarkable—a remarkable fake."

But the crowds, now arriving by special trains, continued to grow, and the great showman P. T. Barnum offered $60,000 to lease the giant from Newell for three months. The farmer refused. Undeterred, Barnum hired a sculptor, Professor Carl C. F. Otto, to make an exact copy of the giant.

When Hull and Newell came to exhibit their giant in New York City in 1871, they discovered that Barnum was already displaying his version in Brooklyn. While they hauled Barnum into court, newspapermen were investigating Hull's activities and uncovered his purchase of gypsum in Iowa. They located the stonecutter in Chicago, one Edward Salle, who admitted to carving the giant, aging it with sand, ink, and sulfuric acid, and punching pores into it with darning needles. Faced with the growing evidence of fraud, Hull confessed. Barnum now was able to avoid prosecution by claiming all he had done was show a fake of a fake—which could not be considered a forgery. His giant, after all, was guaranteed to be an authentic fake. In the end, Barnum's ripoff copy netted four times as much money as the fraudulent original.

Today, the Cardiff Giant—Hull's original phony, not Barnum's authentic copy—is displayed in an earthen pit at the Farmer's Museum in Cooperstown, New York.

See also BARNUM, PHINEAS T.; BERINGER, JOHANNES; PILTDOWN MAN (HOAX)

CARNEGIE, ANDREW (1835–1919)
Darwinian Industrialist

When steel was king, Andrew Carnegie, "the richest man in the world," was king of steel. A Scots immigrant from a poor working-class family, Carnegie rose in business to become a powerful, ruthless tycoon who exploited man and Earth, crushed competition, and justified his actions by a philosophy of Social Darwinism.

Entrepreneurial competition, he believed, does a service to society by eliminating the weaker elements. Those who survive in business are "fit," and therefore deserve their positions and rewards. Carnegie elevated the capitalist ethic to a law of nature.

Although he proclaimed himself a "Darwinist," Carnegie drew his inspiration from the English philosopher Herbert Spencer (1820–1903). Unlike Darwin, Spencer had sought to apply evolutionary thinking across a broad spectrum of political and social questions. "Before Spencer," Carnegie said repeatedly, "all for me had been darkness, after him, all had become light—and right." It was Spencer, after all, and not Darwin, who was the author of the phrase "survival of the fittest."

Carnegie believed that Spencer had revealed to humankind its own destiny:

ANDREW CARNEGIE, the richest man of his time, revered Herbert Spencer's evolutionary philosophy. Proud of having sponsored discovery of the huge *Diplodocus carnegii* dinosaur, he donated full-size casts to many museums.

social evolution toward a peaceful industrial world, whose mass-produced products and technologies would be available to all. Human striving had a purpose: progressive evolution to higher levels of efficiency and happiness. One biographer wrote, "Carnegie's happy little slogan, 'All is well, since all grows better' was for him the satisfying distillation of thirty volumes of Spencer's philosophy."

Yet Carnegie seemed blithely unaware that, in many crucial areas, his beliefs and actions were very un-Spencerian. For instance, he did not really favor laissez-faire capitalism, though he paid lip service to Spencer's enthusiasm for open competition. In fact, he fought for high protective tariffs and engaged in price fixing, patent monopolies, and other artificial reductions of free competition.

Unlike Spencer, moreover, Carnegie rejected the idea of eugenics and expected men of "genius" to rise from the poor classes, which Spencer had dubbed the "unfit." Carnegie, as an American immigrant, opposed the notion of hereditary social classes and believed that people who held vast wealth should give most of it away to foster the public good. "The man who dies thus rich, dies disgraced," he wrote. Carnegie's belief that it was the moral duty of the strong to protect the weak was another un-Spencerian notion.

Despite his loose reading of Spencer's teachings, Carnegie revered him as the greatest thinker alive. He believed Spencer had "proved" that evolution leads to progress and that ruthlessness can enable a man to acquire the means to promote the higher good.

Carnegie courted Spencer's friendship for several years and invited him to visit North America. In 1882, Spencer made the voyage at his publisher's invitation. Carnegie, who was visiting England at the time, hurriedly arranged his return home so that he could accompany his favorite philosopher on the journey across the Atlantic. After a brief tour of Canada, Spencer visited Pittsburgh, which Carnegie extolled as the model of the Spencerian future, the well-regulated industrial beehive. Historian Joseph Frazier Wall relates that Spencer

> did not recognize utopia when it was shown to him. [He complained that] in this smoky, polluted air a man would be fortunate if he could recognize his own hand held close to his face. [At the Bessemer steel plant] the heat and noise of the mills reduced Spencer to a state of near collapse. When the tour was over, he could only gasp out to Carnegie, "Six months' residence here would justify suicide."

Near the end of Spencer's visit, on November 9, 1882, Carnegie organized a great testimonial dinner for him in New York. Newspaper and railroad magnates, politicians, and such national celebrities as the Reverend Henry Ward Beecher gathered to honor Spencer at the peak of his popularity. The chairman welcomed them by stating that the glittering assemblage had been brought together "by natural selection."

Spencer's speech turned out to be not a clarion call to competitive battle, but a reclusive philosopher's reaction to the feverish activity of American industry. His disappointing message to the assembled capitalists was to slow down, learn to relax, and enjoy life—something he wished he could do.

In 1895, Carnegie endowed a museum in Pittsburgh; its most famous attraction was the fossil skeleton of a gigantic new species of sauropod dinosaur, measuring more than 87 feet in length, discovered in Colorado and Wyoming. Of all the scientific projects Carnegie sponsored, the long-necked behemoth—named *Diplodocus carnegii* in his honor—was his special favorite.

Full-size copies of the *Diplodocus* were cast, mounted, and sent all over the world at Carnegie's expense. His friend King Edward VII had admired it, so one was sent to the British Museum. Others went to the great museums in Germany, France, Austria, Mexico, and Argentina.

According to his biographer, Carnegie "delighted in this graphic proof of evolutionary development . . . an organism so successful in its physical development that it outgrew the brain and nervous system necessary to control it." Although biologists no longer characterize dinosaurs in this way, the metaphor has been applied to Carnegie's own Pittsburgh Steel Corporation, which became so vast it could no longer be controlled by its head.

See also SOCIAL DARWINISM; SPENCER, HERBERT

CARTESIAN DUALITY
Does a Frog Have a Soul?

Before Charles Darwin and Alfred Russel Wallace established an evolutionary perspective in Western thought, it was widely believed that humans and animals were two unconnected, utterly different orders of beings. A rationale for this duality was proposed by René Descartes, the French genius of the 17th century.

Descartes was an influential founder of modern philosophy and the father of analytical geometry. As a scientist, he adopted the mechanistic viewpoint: phenomena of nature were viewed as machines, whose parts and workings could be analyzed according to mechanical principles.

The human body also was a machine made of physical matter, as were animals' bodies. But Descartes had a problem. His traditional Christian religious beliefs maintained that human beings were more than matter, more than machines: they have immortal souls. How could Descartes continue to advocate scientific materialism, yet escape heresy?

His solution was to postulate a major difference in kind between humans and animals, which became known to philosophers as the Cartesian duality. Humans, he decided, must be the only conscious beings, because we alone are endowed with a soul by God. Animals are automata, robots, clockwork droids without consciousness or feeling.

Descartes thought that this duality between man and beast, spirit and machine, soul and soulless provided a reason for believing in life after death. If "the souls of animals are of the same nature as our own," he wrote, we would "have no more to fear or hope for after this life than flies or ants." That conclusion would lead to immoral conduct and social chaos.

Duality also resolved another disturbing problem. If animals can expect no justice in an afterlife and they are untainted by Adam's original sin, why would a just God allow them to suffer? Descartes's very unfortunate answer was that animals never suffer because, as mere machines, they are incapable of feeling pain.

Others were quick to embrace the Cartesian duality. Belief in a profound difference in kind between humans and animals relieved all guilt about killing or eating animals and enabled scientists to vivisect live creatures without paying any attention to their cries.

Descartes's ideas persisted for years. They were finally displaced by Darwin's demonstration that all life is related and connected. At last there was scientific support for what many nonscientists with common sense had known all along—that animals share many of our feelings, certainly including physical pain.

See also ANIMAL RIGHTS; MATERIALISM; MECHANISM

CATASTROPHISM
Upheavals and Cataclysms

Catastrophism (as a label) has been applied to many different theories about the history of life on Earth, all with the core idea of dramatic, fairly rapid, and discontinuous change.

Georges Cuvier, the influential 18th-century anatomist and zoologist, read a history of great "catastrophes" and "debacles" in the geologic record: widespread earthquakes, floods and inundations, and volcanic upheavals. In the classic version, there was a succession of extinctions, which wiped all life off the face of the Earth, followed each time by a new creation of a different set of plants and animals. Enlightenment Frenchman that he was, Cuvier called these violent convulsions "revolutions," sometimes (without conscious irony) translated by Englishmen as "catastrophes."

Until well into the 19th century, geologists thought the Earth too young to have acquired many of its features (including fossil-bearing rock layers) by a slow and gradual buildup. Clear and abrupt boundaries between strata, fossils of seashells on mountains, and successive extinctions suggested a picture of rapid and disastrous events, unlike anything observable in the contemporary world.

One common view was that the layers of fossil-bearing rock, with their different worlds of plants and animals, represented separate, "successive creations by the Author of nature." Some geologists thought this view blasphemous, as it implied the Creator was an inept craftsman, smashing his works again and again in an attempt to create a more perfect world. Others balked at accepting so many instances of life coming out of nonlife, since no such process could be seen currently operating on Earth.

Generally, catastrophism fit in well with such accepted biblical miracles as Noah's Flood, which lent authority to the concept. But Cuvier and others (notably the discoverer of ice ages, Swiss-American Louis Agassiz) were excellent scientists; to characterize them as religious zealots who ignored the evidence when it contradicted scripture is unfair. At the time, available evidence was still ambiguous, thus allowing room for many competing scientific interpretations.

Catastrophism is usually contrasted with Sir Charles Lyell's doctrine of uniformitarianism, which transformed geology and had a great influence on young Charles Darwin.

This battle between catastrophists (as champions of scriptural miracles) and uniformitarians (as knights of rational science) has been vulgarly oversimplified in hundreds of books. Lyell's methods of geology, particularly his insistence on understanding present processes as the key to interpreting the past, were admired by many geologists who have been labeled catastrophists; some invoked catastrophes only where no known (observable) processes could account for the geological facts.

As Thomas Huxley described the catastrophists of his youth, they went further than imagining great events like poisoned atmospheres or worldwide floods; they insisted these events operated according to different processes than those we consider natural today. In the past, they thought, nature operated according to different rules. As the mathematician Charles Babbage, inventor of the mechanical computer, insisted, if God could set up natural laws, he could also reprogram the universe at will.

See also ACTUALISM; AGASSIZ, LOUIS; LYELL, SIR CHARLES; NOAH'S FLOOD; STEADY-STATE EARTH; UNIFORMITARIANISM

CAVEMAN
Stone Age Icon

A cave in the Neander Valley, near Düsseldorf, Germany, inspired the term "caveman" in popular culture. There, in 1854, quarrymen discovered part of a skull and other bones of what became known as Neanderthal Man. The term seems to have been first used in *Pre-historic Times* (1865), written by John Lubbock, Darwin's protégé and only student, in which he also coined the term "Paleolithic," meaning Old Stone Age.

Despite the wide popularity of the caveman image, only very few early humans and hominids actually lived in caves. Thousands of stone tools have been found in river gravels and open-air sites all over the old world, and many works of Paleolithic art are found in the open. Portugal's Côa Valley, for instance, is famous for its many images engraved in rock overhangs and cliffs. However, several important early human sites have been found in caves, including the Beijing *Homo erectus* (a dozen skulls found in a Chinese cave in the 1930s) and the scores of pre-Neanderthal remains discovered since the 1990s in a cave at Atapuerca, Spain.

Spectacular discoveries of early human artifacts in caves helped to fuel public acceptance of Darwin's *Origin of Species* in 1859. For at least a half-century prior to its publication, discoveries of human fossils and stone tools with remains of extinct animals had been turning up. In 1858—a year before the *Origin* appeared—the archeologist William Pengelly began a decade of systematic excavations at Brixham Cave at Torquay, in Devon, England. His work there and in nearby Kent's Cavern over the next two decades yielded tens of thousands of fossil animal bones and early human artifacts, and established their association in time.

It turned out that some of the ancient "cavemen"—who did not live in the caves they decorated—were fine artists. In 1879, the first known painted cave was accidentally discovered

A WILD MAN OF THE WOODS, or wodewose, perches atop a Suffolk church, sporting wild hair, a beard, and a heavy club. Such enigmatic 15th-century icons may have influenced later images of "cavemen" in popular culture.
Photo courtesy of and © by John Telford-Taylor.

at Altamira, Spain; its images of extinct aurochs, bison, and horses stunned both the art and scientific worlds. Subsequent discoveries of European caves with sophisticated Paleolithic paintings of long-extinct animals fired the imaginations of scientists and the general public alike. Many of the spectacular paintings of animals and mystical symbols in France's Lascaux, Trois Frères, and Chauvet caves were done in almost inaccessible recesses that were probably used for mysterious rituals and initiation—certainly not for day-to-day living.

Many people continue to imagine our Ice Age ancestors mostly as wild-haired, club-toting clods, unmoved by the evidence that many cave paintings are extremely sophisticated, not at all childlike or "primitive." Perhaps that notion of the dim-witted hairy brute was carried over from a medieval tradition of "wild men of the woods," or wodewoses, in European art. These archaic 15th-century iconic figures, with their unkempt hair and ever-present clubs, prefigured and continue to influence our images of the caveman.

IN THE SILENT CLASSIC
The Three Ages (1923) comic genius Buster Keaton competes for a leopard skin–clad lady's attentions with a monstrous, club-wielding caveman.

Beginning in the 1870s and 1880s, a few European artists, notably Ernst Griset (commissioned by John Lubbock) and Emmanuel Fremiet turned their hands to painting prehistoric scenes. The undisputed king of the paleoartists, however, was Charles R. Knight, who painted many scenes of prehistoric humans for the American Museum of Natural History during the 1920s through the 1950s. Subsequent masters include the Czech artist Zdeněk Burian and the Americans John Gurche, Jay Matternes, and Viktor Deak.

Along with serious, science-based attempts to reconstruct our distant ancestors, a very popular genre of caveman spoofs and cartoons arose and flourished from the late 19th century to the present. In 1893, the British humor magazine *Punch* launched a delightful series of cartoons called *Prehistoric Peeps,* by Edward Tennyson Reed. One of his best-known drawings shows a tribe of cavemen playing primitive cricket at Stonehenge, using the giant stone arches as wickets. In 1905, a Hollywood live-action silent comedy based on Reed's work became the first dinosaur film, antedating the silent version of Sir Arthur

CRO-MAGNON CRICKETERS, drawn by Edward Tennyson Reed for the British humor magazine *Punch,* use the Stonehenge arches as wickets for their game.

Conan Doyle's adventure *The Lost World* (usually considered the first dinosaur movie) by two decades. In the movie version of *Prehistoric Peeps*, cavemen and dinosaurs were shown living together, a tradition that persisted in entertainment to the 1970s—perhaps fueling later Creationist fantasies of dinos living with prehistoric people. The 1930s comic strip *Alley Oop* featured a time-traveling caveman who rode his trusty saurian Dinny, and in the 1960s cartoon *The Flintstones*, the cave family's whining house pet was a dog-sized dinosaur named Dino.

Subsequent movie portrayals include the silent *The Three Ages* (1923), in which comic genius Buster Keaton appeared as a lovelorn caveman pursuing a fair lady through the Stone Age, ancient Rome, and the 1920s; his Paleolithic rival was a club-wielding, wild-haired, bearded giant. For sixty years, most Hollywood "caveman" films were low-budget stinkers showcasing curvy starlets clad in scanty skins (as in *One Million B.C.*, 1940, and its 1966 remake with Raquel Welch). Rubber dinosaurs terrorized the tribe in Ringo Starr's big-screen comedy *Caveman* (1981).

Part of the humor in caveman cartoons and films is that they satirize the eternal universality of the human condition. They are also perfect vehicles for absurd anachronisms. Contemporary cartoonist John Callahan, for instance, shows a caveman watching television, as the screen cautions: "The following program contains language."

See also ATAPUERCA; BEIJING MAN; FLORES MAN; NEANDERTHAL MAN

"CENTRAL DOGMA" OF GENETICS
From DNA Outward

A dogma is a doctrine or belief one accepts as true without question—an article of faith. How peculiar, then, that the scientific conclusion that genetic information only flows outward from DNA, and never into it, is called the "Central Dogma" of molecular genetics.

Francis Crick, codiscoverer of the structure of DNA, stated the "Dogma" thus:

> The transfer of information from nucleic acid to nucleic acid, or from nucleic acid to protein may be possible, but transfer from protein to protein, or from protein to nucleic acid is impossible.

The chemist Robert Shapiro, author of *Origins: A Skeptic's Guide to the Creation of Life on Earth* (1986), put the question directly to Crick about why the famous "Central Dogma" should have been so named. Astoundingly, the great geneticist cheerfully admitted that, at the time, he did not understand exactly what "dogma" meant. Later, a friend had explained to Crick that a dogma is a truth that must simply be taken on faith, without need for evidence.

"I didn't know it meant that," he replied, "I thought it meant a hypothesis, some arbitrary thing which was laid down for no particularly good reason. Otherwise it would have been called the 'Central Hypothesis,' and then nobody would have made all this fuss."

See also DNA, DISCOVERY OF; DNA IDENTIFICATION

CHAMBERS, ROBERT (1802–1883)
Evolution's Pioneer Popularizer

Fifteen years before Charles Darwin's *Origin of Species* (1859) appeared, a popular Scots author named Robert Chambers published his own highly controversial treatise on evolution. Chambers's *Vestiges of the Natural History of Creation* (1844) argued that all species had gradually developed according to natural laws, without direct intervention of a Creator. It became an instant bestseller.

Despite its continuing success (11 editions were published), Chambers never put his name on the book nor would even admit to writing it. With his brother William as lifelong partner, Robert, the self-taught son of a poor weaver, had worked his way up from street bookseller, to publisher, to popular author and man of letters. In 1848, he was a candidate for the post of Lord Provost of Edinburgh and seemed a shoo-in on the eve of the election. However, his opponent issued an ultimatum: Either face a public challenge that he was the anonymous "Mr. Vestiges," author of that "blasphemous, materialist book," or withdraw from the race. Thus ended his political career.

He believed his ideas would interest scientists, but that the mass of general readers would find them unpalatable. To his amazement, it was the scientists who ignored the book, while the wider public eagerly snapped it up.

Although scorned by most naturalists (including Charles Darwin and Thomas Huxley), *Vestiges* did have a profound influence on some younger men, helping inspire Alfred Russel Wallace and Henry Walter Bates to search for (and find) evidence of evolution in the Amazon rain forests. Wallace and Chambers became friends, sharing passions for evolution and Spiritualism.

Like Wallace, Chambers was impressed by the reports of friends who believed they had seen and heard spirit manifestations at séances, and were appalled that most scientists dismissed such accounts, just as they had at first rejected evolution. In 1859, the same year Darwin's *Origin of Species* appeared, Chambers published the pamphlet *Testimony: Its Posture in the Scientific World*, which examined scientists' criteria for accepting evidence of supernatural phenomena. Chambers challenged the physicist Michael Faraday, who had urged that "any extraordinary natural facts" reported by nonscientists should be routinely distrusted.

Too much reliance on "the skeptical method" leads to a "vicious circle" in which the search for knowledge cannot thrive, said Chambers. We cannot credit a fact until it accords with accepted laws of nature, but we cannot determine the laws of nature without accumulating facts. If we reject every novel fact that doesn't fit in with our conception of nature, "We can't know what is possible till we've learned everything."

In 1861, Darwin wrote Chambers, "You fulminate against the scepticism of scientific men. You would not fulminate quite so much if you had had so many wild-goose chases after facts stated by men not trained to scientific accuracy."

But Chambers shared Wallace's view that the thousands of eyewitness reports of supernatural phenomena, stretching back to biblical times, could not all have been false. Chambers invited Wallace to write the article on spiritualism for *Chambers's Encyclopedia*, and in 1867 wrote to Wallace: "I have for many years known that these phenomena are real, as distinguished from impostures." Even if most were rightly discarded as delusion, still there must be a few "golden grains" amid such a mass of earnest testimony. For Chambers and Wallace, it was this skepticism toward skepticism that allowed them to embrace both Spiritualism and evolution. Both phenomena were invisible to most scientists of the day. Yet both men shared an open-minded willingness to seek evidence of the unseen. As Chambers wrote Wallace, "We have only to enlarge our conception of the natural, and all will be right." Chambers wrote a book on Spiritualism that he never published. Fearing for his reputation even after a century and a half, his descendants still nervously keep the manuscript under lock and key.

Chambers entrusted two close friends and his wife and brother with the secret that he had written *Vestiges*. In a preface to the 1884 edition, a year after Chambers's death, his old friend Robert Cox at last revealed the truth. Chambers's wife and brother had died, both of them content to let future historians continue the guessing game. But Cox felt that would be an injustice to his friend's achievement. As the last man alive who knew the secret, he wrote, "I am unwilling that it should die with me."

See also SPIRITUALISM; *VESTIGES OF CREATION*

MYSTERIOUS AUTHOR of a controversial bestseller about organic evolution before Darwin's *Origin of Species*, Robert Chambers kept his identity secret. His *Vestiges of Creation* (1844) was a Victorian sensation.

ANCIENT ARTISTS of the Ardèche Valley, 32,000 years ago, observed and painted now-extinct European cave lions. Like smaller, modern lions, they stared at potential threats or prey. Left: © by Frans Lanting/Minden Pictures. Cave painting photos courtesy of Jean Clottes/French Ministry of Culture.

CHAUVET CAVE
Ice Age Gallery of the Lion King

On December 18, 1994, near the town of Vallon-Pont-d'Arc in the Ardèche Valley, three amateur speleologists led by Jean-Marie Chauvet, a guard who worked for the French Ministry of Culture, discovered a cave spectacularly decorated with ice age art. Although Southern France is world famous for its Paleolithic caves, this one proved unique.

Most painted caves are filled with images of aurochs (wild oxen), horses, mammoths, ibex, and bison. Among the 400 paintings of animals at Chauvet Cave are images of such predators as leopards, hyenas, bears, an owl, and many lions, all of which were rare or previously unknown in Paleolithic cave art. Moreover, when modern methods of radiocarbon dating were applied to the pigments and charcoals, researchers were amazed to find that the oldest drawings are 32,000 years old, and the youngest 27,000 years (compared with Lascaux's 17,000).

French rock art expert Jean Clottes, who had long served as director of prehistoric antiquities for the Pyrenees region, was among the first to visit the cave after its discovery. After Chauvet guided him through some narrow, tubelike passages, Clottes emerged into a spacious chamber and shone his miner's lamp on the amazing paintings.

"Most of the animals were recognizable as to species and sometimes even as to sex or behavior, " he wrote later. "Contrary to what can be seen in most painted caves, they seemed to be grouped into compositions." Two male rhinos seemed to be fighting; lions seemed to be courting or stalking bison. Overcome with emotion, Clottes "felt a deep and clear certainty that there was the work of one of the great masters, a Leonardo da Vinci of the Ice Age revealed to us for the first time. It was both humbling and exhilarating, underlining the immensity of our ignorance."

After months of intense competition over who was to study the remarkable site, in 1997 Clottes was put in charge of the research to be carried out in Chauvet Cave. His meticulous work there with teams of specialists has elucidated some of the cave's mysteries. However, it has been closed to the public and, for preservation reasons, will remain that way.

In October 2000, *Natural History* magazine asked Clottes to escort a lion expert, Craig Packer of the University of Minnesota, through the cave. Packer, who had managed the Serengeti Lion Project in Tanzania, wanted to examine the images from the point of view of a field zoologist. Cave lions, which became extinct about 12,000 years ago, were maneless and larger than their modern African counterparts, but Packer immediately recognized typical lion behavior in the Paleolithic sketches.

Modern male African lions, for instance, adopt a posture of hunkering down to the smaller female's height and walking closely alongside her during courtship—a behavior clearly recorded in a Chauvet drawing. Paintings also show lions lined up as they watch a herd of bison or react to threats. "Anyone who has spent time studying and photographing African

A MATED PAIR of modern African lions walk in tandem with tails curved, just as their huge Ice Age European counterparts are depicted doing in Chauvet Cave. Ancient artists depicted them as maneless. Right: © by Charles G. Summers, Jr./ Wild Images.

lions will be stunned by the cave artists' accomplishments in observing and recording the behavior of the big cats," Packer and Clottes wrote. "For the ancient artists to have made these [closeup] observations, the lions must have been very relaxed in their presence. . . . We can be sure [the artists] had courage and patience as well as a degree of curiosity that rivals that of the best naturalists of our own era."

See also ICE AGE; LASCAUX CAVE

CHIMPANZEES
Fascinating Forest Ape

Human perception of the African forest apes has changed drastically over the past few decades and keeps on changing. Chimpanzees were first described by early travelers as "wild" tribes of "Pygmie" men. Nineteenth-century explorers believed apes roamed the jungle at night bearing blazing torches and occasionally carrying off an African woman.

Victorian zoo-goers watched chimps drink from teacups, then make a shambles of the tea table. Later they became costumed circus clowns that rode motorbikes. To medical researchers, they were sullen and unpredictable stand-ins for humans. Psychologists treated them as retarded suburban children who couldn't speak.

Some tried to teach them language through hand signs. By the 1970s, we saw them either as symbol-learning prodigies or brats who deliberately gave false data for a few grapes. In movies they were Tarzan's antic jungle sidekick or Ronald Reagan's laboratory ape Bonzo. Will the real chimpanzee please stand up?

The greatest field observer of the natural behavior of chimpanzees is Jane Goodall, and even she finds her image of the chimp constantly changing. After her first 10 years in the forest, she thought them peaceful, gentle vegetarians who did little but socialize and pull up plants all day. A few years later, she redefined them as toolmakers and users of tools. After another decade of observation, she saw them as warring "communities," cooperative hunters of meat, and sometimes even cannibalistic baby killers.

Three communities of about 50 individuals each inhabit the Gombe Stream Reserve in Tanzania where Goodall watched chimpanzees for 35 years. Each group ranges over about 30 square miles, and its relations with neighboring communities are hostile. Males sometimes silently patrol the "border" and are now known to attack their neighbors, killing all males and infants from other groups. In such conflicts, females are taken into the marauding group, where they usually mate with the victors and produce new infants.

In 2003, Harvard's Richard Wrangham and colleagues videotaped a horrendous attack by a thuggish gang of male chimps on one male who had not properly established himself in their hierarchy, and on another who was from a neighboring group. Each attacker took his turn at bashing and kicking the downed, defenseless apes until they were beaten to death, despite their pitiful sobbing cries.

OUR CLOSE KIN, chim-
panzees are genetically
very similar to humans.
These African forest
apes have been variously
perceived by naturalists
as vegetarians and meat-
eaters, clowns and killers.
Photos © by Kenneth Love.

Although chimps feed mostly on wild fruits, vegetables, and palm fibers, they frequently hunt small monkeys, bushbuck, and piglets and sometimes kill and eat young baboons with whom they have regularly played and socialized. Only males hunt, and they never do so alone. Hunting is a social activity, preceded by several males working themselves up to a pitch of excitement. Prey are stalked, surrounded, and deliberately ambushed by several cooperating apes. They catch and kill with their bare hands, then simply tear up the carcasses.

Meat is distributed among the group in a leisurely food-sharing ritual, which is still not understood by anyone but the chimps. Some males are more generous than others, but all respond to the gesture of upturned palms or fingers pressed gently to the hunter's lips. Unlike their usual behavior, the more dominant individuals patiently wait their turn.

Chimps show regional cultures or traditions, especially in food-getting. They learn by observing, then imitate and practice. "Termiting" has become their most famous use of tools, where they strip leaves off twigs, chew them to the proper shape, and insert them in termites' burrows in the clay mounds. In 2004, Crickette Sanz of Washington University (St. Louis) and her colleagues videotaped Congo chimps preparing three tools: a heavy puncturing stick, a lighter perforating stick, and a light, flexible fishing stick. If the chimp is dealing with a below-ground termite colony, it brings the strong puncturing stick and a lighter digging stick. Using a foot and a hand, it forces the stick into the ground like a gardener using a spade. For an above-ground mound, the chimp pokes holes with a lighter stick, then uses a very slender twig to fish out the termites. The delicate tool is chewed at one end to make a brush, which is moistened with saliva.

Genetically, chimps are more like humans than any other creature; we have 98 percent of our genetic material in common. Their behavior includes embracing, back patting, open-mouth "kissing," and hand-holding. In addition to hunting, tool using, and food sharing, many researchers agree they demonstrate reasoned thought, memory, directed communication, and ability to plan for the immediate future. In captive experiments, chimps were able to learn at least 300 signs in computer language or sign language at an early age. Goodall believes they have more social complexity and intellect than gorillas.

Yet they are still hunted for food and cash by Africans, and many infants continue to be taken for biomedical research. Labs commonly lock them up in small, isolated cages until they literally go insane. With what we now know about them, says Goodall, it's time to treat chimpanzees with more kindness and respect, like the relatives they are.

See also APE LANGUAGE CONTROVERSY; "APE WOMEN," LEAKEY'S; APES, TOOL USE OF

CHINA, EVOLUTION IN
A Renaissance of Discoveries

During the 19th century, the West regarded China as a "sleeping giant," isolated and mired in ancient traditions. Few Europeans realized how avidly Chinese intellectuals seized on Darwinian evolutionary ideas and saw in them a hopeful impetus for progress and change.

According to the Chinese writer Hu Shih (*Living Philosophies*, 1931), when Thomas Huxley's *Evolution and Ethics* was published in 1898, it was immediately acclaimed and accepted by Chinese intellectuals. Rich men sponsored cheap Chinese editions so they could be widely distributed to the masses "because it was thought that the Darwinian hypothesis, especially in its social and political application, was a welcome stimulus to a nation suffering from age-long inertia and stagnation."

Within a few years, evolutionary phrases and slogans became accepted Chinese proverbs. Thousands named themselves and their children after them to "remind themselves of the perils of elimination in the struggle for existence, national as well as individual." A famous general called Chen Chiung-ming renamed himself Ching-tsun, or "Struggling for Existence." Author Shih himself adopted the name "Fitness" (Shih), from the phrase "survival of the fittest." He recalled that because of "the great vogue of evolutionism in China . . . two of my schoolmates bore the names 'Natural Selection Yang' and 'Struggle for Existence Sun.'"

Despite this early interest, China's development of evolutionary sciences was hindered for almost three-quarters of a century by political instability, invasion by Japan, civil war, and the Communist revolution and its aftermath. Since the latter part of the 20th century, however, Beijing's Institute of Vertebrate Paleontology and Paleoanthropology of the Chinese Academy of Sciences has built a world-class reputation. New generations of Chinese paleontologists, working with scientists from around the world, have produced a torrent of stunning discoveries.

Among the myriad new finds of invertebrate fossils, those of the Heilinpu Formation in Chengjiang, Yunnan Province, are among the most exciting. The Chengjiang fossils, from the Cambrian era, dated at 525 million years ago—10 million years older than the famous Burgess Shale fossils from British Columbia—have provided a remarkable window into the early evolution of multicellular life. Even more unexpected, during the 1990s, thousands of the earliest known embryos of sea creatures were found in the Doushantuo Formation of Southwest China; these tiny, planktonlike larvae that once floated by the billions in ancient seas predate the Cambrian "explosion" by at least 50 million years.

The Yixian Formation of Liaoning Province (northeast of Beijing) has recently yielded a treasure trove of previously unknown animals. Near the town of Sihetun, the Yixian Formation preserves a complete Cretaceous ecosystem: a lake and riverine environment from a warm, dry climate. It contains a wide range of animals and plants that lived from 130 to 125 million years ago. *Archaefructus*, a plant fossil found in the region, is thought be the earliest known flower.

Among the "firsts" from Liaoning are *Eomaia*, the earliest known placental mammal, and *Sinodelphis szalazi*, the earliest known marsupial. One small fossil mammal was found with a juvenile dinosaur inside its rib cage—the first known example of an early mammal feeding on a dinosaur. In Nanjing, a *Sinosauropteryx* was found that contained two unlaid eggs—never before seen in a dinosaur—and the remains of a shrew-sized mammal, its last meal.

But by far the most exciting discoveries from Liaoning are small carnivorous theropod dinosaurs, ranging from a couple of feet to about seven feet long. One species, *Dilong paradoxus*, is an early relative of *Tyrannosaurus rex*. Another theropod, *Psittacosaurus sinensis*, seems to have died while protecting a crèche of 34 chicks, all of whose fossil skeletons were found with it.

Many of the Liaoning theropod dinosaurs had body coverings of filaments, downy feathers, and even well-formed flight feathers. Many species of theropod dinosaurs evolved featherlike coverings, probably for thermal insulation, long before feathers were used in flight. Before the discovery of these Chinese fossils, paleontologists argued over whether birds evolved from dinosaurs. The massive evidence of the many varieties of Liaoning's feathered dinosaurs seems to have finally resolved that debate.

See also DINOSAURS, FEATHERED

CHRYSALIS, THE
Akeley's Evolutionary Sculpture

Two years before his final, fatal trip to his beloved East African wilderness, Carl Akeley (1864–1926) created an evolutionary sculpture for a church that caused a public sensation. The bronze depicted a handsome "modern" man emerging from a cracked-open gorilla skin; he titled it *The Chrysalis.*

Akeley is best remembered as the genius behind the magnificent African Hall at New York's American Museum of Natural History. He was also instrumental in saving the mountain gorillas when they were being slaughtered without limit, 70 years before Dian Fossey's famous battles with poachers. But although he was an accomplished taxidermist, anatomist, and naturalist, Akeley was first and foremost a sculptor, renowned for his wildlife bronzes.

The Chrysalis distilled in one sculpture Akeley's strong feelings of evolutionary kinship with the animal kingdom. The title refers to the cocoon in which caterpillars transform into butterflies; humans, he implied, emerged from apes. He hastened to add in a public lecture he knew full well that humans had not literally sprung from the gorilla. "They undoubtedly had a common ancestor. Science is on the trail of this ancestor and will locate it."

CONTROVERSIAL CHURCH SCULPTURE entitled *The Chrysalis* was considered very daring in 1925. Carl Akeley's man emerging from a cracked-open gorilla skin symbolized the evolutionary "ascent of man," rather than the traditional Fall from Grace.

The piece was commissioned for New York's West Side Unitarian Church, where it was on display for many years (the church no longer exists). Creationists were outraged and publicly criticized the Unitarians for placing it in their house of worship. "Could anything be more degrading," asked one newspaper editorialist, "than for a church to contain Akeley's statue and for the pastor of that church to say of it: 'I know of no concrete symbol which so well expresses the religious message which I am trying to preach every Sunday!'" The Unitarian pastor, Reverend Charles Francis Potter, was unperturbed by the fundamentalist tempest. "The point of the statue," he explained to the New York newspapers, "is not the gorilla, but the man, who has risen above his animal ancestry. This statue shows the rise of man as opposed to the fall of man."

When asked about his church affiliation, Carl Akeley replied, "Most of my worshipping has been done in the cathedral forests of the African jungles, with the voices of birds and animals as music."

See also AKELEY, CARL; GORILLAS

CLADISTICS
Search for the Sister Group

Guest essay by Gareth Nelson, Professorial Associate, School of Botany, University of Melbourne; Curator Emeritus, and former Chairman of Ichthyology at the American Museum of Natural History.

During the late 1960s to 1980s, the search for ancestors—the traditional quest in paleontology—was abandoned in favor of the search for the "sister group": the nearest related organisms, either extinct or alive. Known as *cladistics* (from the Greek *klados*, a branch), the new approach changed the focus and methods of evolutionary studies for almost two generations.

Cladistics aims to discover the interrelationships of species, genera, families, and larger groups rather than search for their specific ances-

INSECT EXPERT WILLI HENNIG clashed with old-guard systematists like George Gaylord Simpson in the 1960s over the issue of how to best classify living things from an evolutionary perspective. © 1989 Jorge Llorente.

tors or origin. The conflict between the old and new approaches was first highlighted by the Swedish entomologist Lars Brundin (1907–1993), who in 1966 applied the ideas of his German colleague Willi Hennig (1913–1976) to the geographical distribution of midges, tiny flies that live in streams of the Southern Hemisphere. Brundin collected and identified related groups and was able to correlate their distribution with the breakup of continents in Mesozoic times, leaving isolated "sister" populations in southern Africa, South America, New Zealand, and Australia.

Most evolutionary histories of sister groups are not as elegantly clear as that of the midges. Suppose we wanted to learn the nearest relative of the ostrich, now native to Africa. We could begin the search among the other flightless (ratite) birds, such as the South American rheas, the Australian cassowaries and emu, the New Zealand kiwis and recently extinct moas, and the also recently extinct elephant birds of Madagascar. Despite their far-flung geographical distribution, they have long been recognized as a group separate from other birds because of their flat, or raft-shaped ("ratite") breastbones, which do not have the "keel" typical of flying birds—a bony ridge that anchors powerful flying muscles.

We could either propose that the sister group of the ostrich includes only the two rheas of South America, which appear to resemble them most closely, or assign them to the larger group of all other living ratites—9–12 species in all. (Evidence from anatomy supports the former conclusion; DNA the latter.) What, then, would be the sister group of all the ratites? Ornithologists say that it is the tinamous, a group of small flying birds native to South America, of which there are nearly 50 species. Taken together, tinamous and the ratites form a group called *paleognaths* (old jaws).

And the sister group of these? You might suggest the group that includes all other birds, the *neognaths* (new jaws). How about the sister group of all the birds combined? Anatomy and DNA show that birds are most closely related to crocodiles, which could make their ancestors equally ancient—about 200 million years. And so on, through all of the interrelated groups until you come to the earliest life forms we know. There you find two groups of bacteria (Archaebacteria, Eubacteria) and a third group including all other living things (Eukarya). It is not clear which, if any, of these is the sister group of the other two, but thousands of scientists are working to clarify the relationships in the great tree of life.

The search for specific origins is another, much thornier—and perhaps insoluble—problem. Cladists question whether science can pinpoint an ancestor—be it for the ostrich, the ratites, all the birds, the birds and crocodiles combined, or any other particular group. While some paleontologists have searched for a suitable non-ostrich, non-ratite, or non-bird ancestor, how would you even recognize it?

Cladistics is a method of classifying living things while working out their evolutionary relationships with one another.

British paleontologist Colin Patterson (1933–1998), a leader in establishing cladistics, argued in his book *Evolution* (1978) that conjectural ancestors are literally non-entities. "Fossils may tell us many things," he wrote, "but one thing they can never disclose is whether they were [direct] ancestors of anything else."

Beginning in the late 1960s, Patterson expanded on one of Hennig's key ideas: *paraphyly*, meaning a mistaken, artifactual assemblage of organisms whose only shared attributes seem to be a *lack* of characteristics found in their presumed descendants. They are thus thought to be "primitive" in relation to other groups. But many paleontologists were deluding themselves, Patterson warned, by dealing in such "empty" or scientifically useless categories as nonvertebrate animals (invertebrates) or nonhuman anthropoids (apes).

One of my favorite examples of the paraphyly fallacy comes from the classical world. Ancient Greeks, secure in their cultural superiority, referred to non-Greek–speaking humans as "barbarians." That did not imply that they all had beards. Rather, their diverse languages, incomprehensible to Greeks, sounded to them like "bar-bar-bar"—the incomprehensible bleating of sheep.

More than two thousand years ago Plato (429–347 B.C.) used the empty category of "barbarians" as an example, warning his students not to fall into faulty thinking. In one of his dialogues, "The Statesman," he explained that it is

> the kind of a mistake a man would make who, seeking to divide the class of human beings into two, divided them into Greeks and barbarians. This is a division most people in this part of the world make. They separate the Greeks from all other nations, making them a class apart; thus they group all other nations together as a class, ignoring the fact that it is an indeterminate class made up of people who have no intercourse with each other and speak different languages. Lumping all this non-Greek residue together, they think it must constitute one real class because they have a common name, "barbarian," to attach to it.

In a series of papers and lectures in the 1970s and 1980s, Patterson argued that to state that vertebrates evolved from invertebrates (or humans from apes—or, we might add, Greeks from barbarians) means only that vertebrates evolved from nonvertebrates, or humans from nonhumans, or Greeks from non-Greeks. He urged paleontologists to seek sister groups with demonstrably shared characters, rather than those characterized by absences and presumed more primitive.

Even so, paleontology still poses questions such as "Are birds really dinosaurs?" and answers "Yes, because birds have evolved from dinosaurs." Interestingly, the evidence shows only that some dinosaurs relate more closely to modern birds than to other dinosaurs. One may even say that a modern bird relates more closely to one dinosaur (e.g., Dromaeosaurus) than to another (e.g., Allosaurus).

From a cladistic standpoint, this does not mean that birds are dinosaurs, or that birds have evolved from them, or that dinosaurs are ancestors of birds. It means that dinosaurs are not one group, but two or more: one, dromaeosaurs, is an extinct sister group of modern birds, and another, allosaurs, is an extinct sister group of dromaeosaurs + modern birds. Cladistically, some dinosaurs are extinct groups of bird relatives, but we must take care to reserve judgment as to whether we can pinpoint direct ancestral fossils.

Some old-school biologists reject Hennig's and Patterson's definitions of what constitutes a meaningful category in systematics or classification. They continue to value paraphyletic groups as ancestors and claim that such statements as "birds evolved from nonbirds and humans from nonhumans" are irrefutable. But what if the statements, irrefutable as they may be, are also without meaning? As Patterson put it, although such assertions have the appearance of knowledge, they contain no specific knowledge and are actually statements of antiknowledge. Nonbirds and nonhumans are not even mutually exclusive. After all, most nonbirds are also nonhumans and most nonhumans are also nonbirds.

If there is a lesson here, it is not simply the one from Plato that dinosaurs are to birds what barbarians are to Greeks. It is also about the impossibly high cost of "irrefutable" statements in science that, on examination, are devoid of any content.

See also *ARCHAEOPTERYX*; DINOSAURS, FEATHERED

CLEVER HANS PHENOMENON
Mystery of the "Talking" Horse

Evolution of the capacity for thought and speech has long fascinated anthropologists, but recent "ape language" experiments sparked heated controversy. Can Koko the gorilla really communicate in sign language? Why did Nim Chimpsky's longtime trainer decide he never really "spoke"? In these debates, scientists often cite the case of a famous "talking" horse who lived during the 1920s. His name was Clever Hans.

Billed as the smartest animal in history, Clever Hans could read, spell, do arithmetic, and work out musical harmonies. His trainer, Herr von Osten, posed mathematical and verbal questions, and the horse, with amazing accuracy, tapped out answers with his hooves.

Herr von Osten really believed in Hans. He swore he did not cheat by giving Hans the answers, and his sincerity was believable. To prove his point, he let strangers question the horse, and Hans still gave correct answers. Audiences were fascinated, and scientists baffled, until the mystery was unraveled by a psychologist named Oskar Pfungst.

In a series of systematic experiments, Pfungst rearranged elements of the question-and-answer proceedings. He soon discovered that if the human didn't know the answer to the question, the horse was also stumped. But when he searched for deliberate sound or hand signals by the trainer, he found none. Yet he also determined that the horse was baffled when the questioner was hidden from view. Eventually, Pfungst concluded that the animal responded to very minute cues the questioner wasn't even aware he was giving.

Hans performed best with men who began the session by leaning forward slightly in tense expectation, and then relaxed with barely perceptible movements when the horse had completed the correct number of taps—at which point Hans would stop. He was simply responding to human approval, not to the content of the questions.

Many of the "ape language" programs of the 1970s were greeted with initial enthusiasm but have since been shown to be tainted by the Clever Hans phenomenon. Involuntary human shaping of the animal's responses proved to be a major flaw and embarrassment. Experimenters now strive to eliminate human cues, however unintentional. When ape language researchers work with bonobos in the Language Research Center at Georgia State University, for instance, the scientists wear welders' masks to hide their eyes and facial movements. The "talking horse" of long ago is still telling us something.

See also APE LANGUAGE CONTROVERSY

A HORSE CALLED HANS astounded European audiences during the 1920s with his apparent ability to read, spell, and perform arithmetic calculations.

COEVOLUTION
Species Evolving Together

When Charles Darwin wrote of a "struggle for existence," he did not simply mean that an individual was pitted against all others in a contest for "fitness." From his studies of plants and insects carried out in his own greenhouse and garden, Darwin discovered that evolution also includes mutually beneficial partnerships between species. For example, he found that sexual parts of some orchids mimic the sexual colors and smells of the wasps that pollinate them. Structures of all organisms, he wrote in *Origin of Species* (1859), are "related in the most essential yet often hidden manner to that of all other organic beings with which it comes into competition for food or residence, or from which it has to escape, or on which it preys."

In the 1960s, population biologists Paul R. Ehrlich and Peter H. Raven coined the term "coevolution" to describe the reciprocal evolutionary changes that occur when unrelated species influence changes in each other. Their concept arose out of a combination of the older concerns of evolutionary biology with the newer ones of ecology and population genetics.

Coevolution may involve predation (animals that prey on others); competition for food, shelter, or other resources; or symbiosis, a close association between two species, which may be mutually beneficial or, as in parasitism, beneficial only to one.

Predation relationships act in nature like the human arms race. Each new weapon inspires a defense or a counterweapon. As carnivorous dinosaurs developed more powerful teeth and jaws, their herbivorous prey developed armored plates and spikes. Predatory cats developed greater ability to chase down their prey, even as antelopes evolved greater speed. High-speed chases between cheetahs and impalas at 70 miles per hour across the African plains approach the upper speed limits possible for four-legged animals. Similarly, hooves and legs of horses evolved in a feedback interaction with the speed of their carnivorous foes, just as the horses' teeth and digestive system coevolved with the grasses on which they fed.

In the 1920s, mathematical biologists Alfred Lotka and Vito Volterra showed how the populations of foxes and rabbits were affecting each other's numbers. Rabbits and foxes influence not only each other's population size but their physical evolution as well, and that of the vegetation in their habitats.

A classic example of coevolution involves the monarch butterfly and the milkweed plant. Milkweed long ago evolved a defense against most birds, insects, and mammals—a poison milky latex so deadly that South American Indians tip their arrows with it. But the monarch butterfly has evolved a defense to the poison. Females lay their eggs on the milkweed, where their larvae have adapted to feeding on the leaves while packing away the deadly, active ingredient in special sealed-off body cells. While the poison does the caterpillar no harm, it makes the insect distasteful to predators, even after it has matured into a butterfly.

Ant-acacia trees of Central America and their resident ant colonies are marvels of coevolution. Unprotected by poisonous leaves or sap, the trees would be vulnerable to the region's thousands of leaf-devouring insects. Each tree, however, has its own resident ant colony, acting as its immune system. In exchange, special structures at the bases of the leaves excrete proteins on which the ants feed. The ants also keep the tree clean, repelling intruders and destroying any climbing vines.

One acacia species has swollen thorns in which the ants hollow out nests. Ant protectors patrol their tree day and night, attacking any other insects they encounter and killing any nearby tree seedlings that might compete for sunlight. Eventually, the ant colony may number 30,000; when all the thorns on the tree are occupied, the colony may split in two, with one group migrating to another tree. Without their resident ant colonies, the coadapted acacia trees would be rapidly wiped out.

See also BELT, THOMAS; MIMICRY; ORCHIDS, DARWIN'S STUDY OF

Some acacia trees have coevolved with their own resident ant colonies. Antelopes and lions push up each other's running speed.

COMPARATIVE METHOD
Projecting from Present to Past

Scholars in various disciplines have referred to "the comparative method" as a means to reconstruct the past from the present.

During the mid-19th century, students of language (philologists) began to compare grammars and vocabularies of various existing languages, to group them in families and even to reconstruct words of a presumed "ancestral" language (Indo-European). In his *Antiquity of Man* (1862), Charles Lyell devotes a chapter to the similarity of reconstructing ancient languages with deducing the Earth's history by the comparative method.

In anthropology, a comparative method was established by Edward Tylor and Sir John Lubbock to reconstruct cultural evolution. The idea was that if one compared "primitive" tribes on Earth today, one could separate the commonalities from the special developments or rank the stages of development in ascending order. After more than a century of trying to make the scheme work, it became clear that modern groups neither are "primitive" nor represent our ancestors. They have developed specialized technologies that enable them to live in various environments, from tropical forests to the Arctic.

Darwin's ambitious book *The Expression of the Emotions in Man and Animals* (1872) compares infants, monkeys, and other mammals in an effort to understand the evolution of facial expressions, gestures, and vocalizations. Seventy-five years later, the ethologists (animal behaviorists) Niko Tinbergen and Konrad Lorenz picked up where Darwin left off, attempting to reconstruct the evolution of behavior. Lorenz, for instance, compared courtship behavior in many related species of ducks and postulated what the ancestral courtship rituals might have been like and how they became elaborated. However, such studies remain conjectural, despite continuing attempts by sociobiologists to refine the methods of comparison.

Molecular biologists now compare the "distances" of serum proteins, seeking to establish evolutionary relationships and times of divergence, using a biochemical version of the comparative method. Critics question its validity as a reliable evolutionary "clock," claiming it has not yet been demonstrated that the rates of change are uniform.

Darwin himself kept a sense of humor about the comparative method. In 1878, in a letter to George J. Romanes, who was then writing his book *Mental Evolution* (1883), Darwin slyly suggested, "You ought to keep an idiot, a deaf mute, a monkey and a baby in your house!" Romanes replied that the baby "would stand a poor chance of showing itself the fittest in the struggle for existence" in such a household.

Two years later, Romanes brought home a monkey from the London Zoo, "a very intelligent, affectionate little animal." He wanted to raise it in the same room as his own infant "for purposes of comparison," he wrote his mentor Darwin, "but the proposal met with [maternal] opposition. . . . I am afraid to suggest the idiot, lest I should be told to occupy the nursery myself."

See also CUVIER, BARON GEORGES; LORENZ, KONRAD; LUBBOCK, SIR JOHN; TINBERGEN, NIKOLAAS; UNIFORMITARIANISM

In comparing species, one seeks to separate the shared commonalities from specializations, the primitive condition from the derived.

COMPOSITE PHOTOGRAPHS
A Comparative Image

Charles Darwin's clever cousin, Sir Francis Galton, sought an objective way to picture a "typical" individual of any group. In his day, each learned professor worked out his own supposed criminal types or "racial" classifications, which prompted endless, insoluble arguments. Galton's interest in heredity also led him to ask what traits were shared in common among close relatives and which were variable or unique to the individual. In 1877, in response to a request from the director of prisons, he created the first composite photographs.

In *Inquiries into Human Faculty and Development* (1878), Galton explained his method.

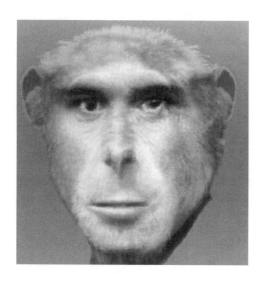

HUMAN-CHIMP COMPOSITE
Evolution II, by computer artist Nancy Burson, bears a striking resemblance to anatomical reconstructions from early hominid skulls. The technique of making such composite images was pioneered by Sir Francis Galton, cousin of Charles Darwin.

First, he collected frontal portraits of the faces to be compared, similar in attitudes and size. Next, he centered each picture on an easel, using pinholes through the eye pupils as fixed points. When other pins were punched through the photos and anchored in the board, each face could be exactly superimposed on all the others.

To find the "typical" features of a family group, Galton combined eight different faces: parents, sons and daughters, and close relatives. Using a slow photographic plate that took 80 seconds for full exposure, he shot each face for 10 seconds. (The lens was covered after each shot, while the next subject was put in place.) Only features common to all eight faces appear on the finished composite, because they receive the longest exposure. Individual variations are not exposed long enough to leave an image, and are eliminated.

American artist Nancy Burson (b. 1948) updated Galton's technique for the electronic age, using television cameras and digital computers. After an MIT engineer developed the basic system for her, special software devised by Richard Carling and David Kramlich permitted her to wed technology to art.

During the early 1980s, Burson made composites of the "typical" movie star, businessman, U.S. president, and "Big Brother" dictator. She sometimes got surprising results. For instance, in the composite *Three Assassins* (1982), "I had expected a portrait of evil and ended up with the boy next door."

For *Evolution II* (1984), she combined the faces of a man and a chimpanzee. The man's face was already a composite made up of a dozen popular film actors. Burson's intent, she explained, "was to see if by combining ape and man a credible image of early man would result." Burson's "early man" image appears strikingly similar to recent reconstructions of *Homo habilis* made by painters and sculptors using traditional anatomical methods.

CONGO
Chimpanzee Painter

As director of the London Zoo, Desmond Morris became intrigued with a chimp called Congo who loved to draw and paint with colors. In his book *The Biology of Art* (1962), Morris reproduced dozens of her productions, arguing that they show "a recognizable, personal style."

He reported that Congo was in no way trained; as soon as she was given a pencil, she began producing her characteristic designs, described as radiating fan shapes. She became famous in England for creating her artworks on *Zootime,* a popular 1950s television show.

An exhibit of Congo's paintings at a London art gallery in 1957 spread her fame worldwide. Many were sold for fancy prices. Salvador Dalí declared that Congo was a better abstract painter than Jackson Pollack; Miró and Picasso hung original "Congos" in their studios.

CONSPICUOUS COLORATION
Warning to Predators

Among Alfred Russel Wallace's many remarkable contributions to science was his demonstration that some of the most gaily colored animals in the world are also the most poisonous. For example, the so-called poison-dart frogs of South and Central America produce toxins so potent that forest Indians use them to tip their blowgun darts. Yet these lethal amphibians are jewel-like in gorgeous reds, yellow, and purples. A Victorian naturalist, Thomas Belt, wrote charmingly of a Nicaraguan species in his classic *A Naturalist in Nicaragua* (1874):

> The little frog hops about in the daytime, dressed in a bright livery of red and blue. He cannot be mistaken for any other, and his flaming vest and blue stockings show that he does not court

concealment. . . . I was convinced he was uneatable as soon as I made his acquaintance and saw the happy sense of security with which he hopped about.

When Charles Darwin first became interested in animals with bright colors, he thought the colors might be part of sexual display. But then he realized the evolution of gaudy caterpillars could not be connected with courtship because they are immature organisms. Why make themselves so conspicuous to predators? He appealed to Wallace for a solution.

As he did so often (to Darwin's delight and sometimes consternation), Wallace came up with an unexpected answer. Bright colors had evolved in many different types of poisonous or bad-tasting animals, he noted, so there must be some advantage to being easily noticed. Perhaps, he thought, the colors served as warning to predators. Maybe a hungry bird would learn to avoid certain insects or frogs more easily if it received a shock (a jolt of poison or disgusting taste) at the same time it saw a burst of bright colors. In the days before the development of behavioral psychology, this was not at all an obvious conclusion. "You are the man to apply to in a difficulty," Darwin wrote Wallace. "I never heard anything more ingenious than your suggestion." Wallace's idea has held up very well, and is still inspiring research today.

Bright color in prey can affect a predator in two ways: It can facilitate the association of a bad experience with a bold color, or it might invite more frequent attack at first, resulting in repeated negative lessons over a short period.

During the early 1980s, Paul Harvey at the University of Sussex set up domestic chicks (as predators) and bad-tasting crumbs (as prey) in a drab enclosure. Half the crumbs were dyed gray to match the floor and the other half were brightly colored. At first, chicks pecked more often at the colored crumbs, but learned to avoid them after a short time. By contrast, they never entirely learned to leave the equally bad-tasting, dull crumbs alone.

Building on this experiment, Harvey's colleague Tim Roper put in fewer bright crumbs and found that the chicks remembered to avoid them much more easily than the dull ones. He concluded that Wallace was right: that coloration could affect both rapidity of learning and ease of remembering.

Those unfamiliar with the older naturalists assume they simply were observers of nature, in contrast to the behavioral experimenters of our own day. Actually, Darwin and many of his contemporaries were tireless empiricists whose so-called observations were achieved through informal experimental designs. Darwin gave the following evidence on the function of conspicuous coloration in *The Descent of Man* (1871):

> [Mr. Wallace's] hypothesis appears at first sight very bold, but . . . Mr. J. Jenner Weir, who keeps a large number of birds in an aviary, informs me that he has made many trials, and finds no exception to the rule, that all caterpillars with smooth skins, . . . all which imitate twigs, are greedily devoured. . . . The hairy and spinose kinds are invariably rejected, as were four conspicuously-coloured species. When the birds rejected a caterpillar, they plainly shewed, by shaking their heads, and cleansing their beaks, that they were disgusted by the taste.

Yet puzzles still remain a century later. Researchers at Göttingen University in Germany gave starlings and domestic chicks a choice between mealworms painted dull green or painted boldly in black and yellow stripes. Inexperienced young birds consistently shunned the boldly striped worms. Why should these young birds avoid stripes without having learned that they taste bad? Can it be a hard-wired instinct?

Bold stripes, Roper theorizes, may perform a double function for prey animals. The stripes may act as both a camouflage and a warning: from a distance they break up the outline of a creature (as in zebras), and up close they may warn of poison. The black-and-yellow-striped cinnabar caterpillar is an example of a doubly protected species. Though highly visible up close, it is poisonous, while from a distance it is almost invisible against the bright flowers on which it feeds.

Of course, all of nature's defenses can backfire when used against a creature that is immune. Bee-eater birds of Africa, for instance, can counter a bee's toxicity by removing the stinger before eating the insect. In such a case, the warning coloration becomes an invitation to a meal.

POISON DART frogs of South and Central America secrete deadly toxins. Their bright, bizarre coloring serves as warning to predators. Top: © AMNH/R. Mickens; middle and bottom: © by Joe McDonald.

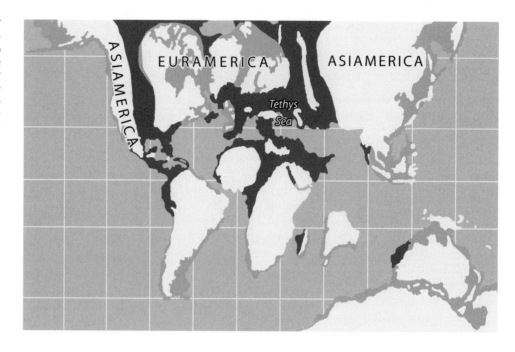

THE JIGSAW PUZZLE OF CONTINENTS led meteorologist Alfred Wegener to propose that they had once been joined, then somehow drifted apart. Here, the position of the continents is shown in the Late Cretaceous, 80 million years ago.

CONTINENTAL DRIFT
The New Geology

Alfred Wegener (1880–1930), a German meteorologist, studied masses of geological evidence from all over the world and came to a startling conclusion: All the Earth's continents were joined together at the beginning of the Dinosaur Age and had been slowly drifting apart for 225 million years.

Wegener noted that Cambrian rocks from Canada and Scotland are identical and that formations on Africa's west coast are the same as those in Brazil. Africa and South America, he pointed out, look like jigsaw puzzle pieces that fit into each other. (Brazil would nestle snugly into the curve of West Africa.) All had once broken away, he thought, from an ancient supercontinent he named Pangea, which means "all lands."

Although some geologists were intrigued by the evidence, most ridiculed the idea, as there was no known mechanism for how the continents could move around. They believed that continents were static and explained the similarity of widely separated fossil species by ancient "land bridges" between continents. The issue went unresolved until the 1960s, when geologist Harry Hess founded the modern study of plate tectonics, which is based on geophysical data that had accumulated since Wegener's day.

That the continents merged and drifted apart at least twice in the past, and are on their way to doing it again in another 200 million years, has now been established.

See also GONDWANALAND; JARAMILLO EVENT; LAURASIA; PANGEA; PLATE TECTONICS

CONTINGENT HISTORY
Quirky Pathway of Evolution

Science has traditionally sought to discover the fixed laws underlying nature, the reliable regularities that allow prediction of particulars. Evolutionary biologists, following the example of physicists and chemists, sought laws to explain the past and perhaps predict future trends. They thought if there were no such laws to be found, then the study of evolution might become just history—incapable of prediction, and therefore not a science at all.

For a century, many paleontologists adhered to the basic assumption that life followed an inevitable course of gradual progress from simpler to more complex forms, ultimately leading to humans as its highest production. Since the 1970s, however, there has been a quiet revolution in thought, stimulated by very technical and unglamorous studies of the strange little creatures, preserved in Canadian shale near the Burgess Pass, that inhabited the oceans some 550 million years ago.

Their discoverer, Charles Doolittle Walcott, head of the U.S. Geological Survey, described the Burgess Shale creatures as primitive arthropods (trilobites, insects), annelids (worms), and relatives of other known groups of sea creatures. He "made out" body segments and mouth parts, legs, and eyes in a manner consistent with the idea that they were early versions of modern phyla, prior to the "improvements" that came later.

With the pioneering work of invertebrate paleontologist Harry Whittington in the 1960s, carried forward by Simon Conway Morris and others a decade later, the Burgess fossils were reexamined and dissected for the first time with startling results. These creatures were unlike anything known in zoology. Some had five eyes, body plans radically different from arthropods, or mouth parts and spikes structurally unrelated to anything that came after them. They might as well have come from Mars.

Attuned to expect similarities with known phyla, Walcott had mistaken heads for legs and broken parts for whole organisms. What he thought were dry, objective descriptions turned out to be as wildly fanciful as anything in the emotional literature on fossil man.

Of the approximately 15 Burgess Shale phyla, primitive vertebrates and arthropods represented only a tiny, inconspicuous minority. There is no obvious reason why they should have survived while the majority of the flourishing Burgess fauna disappeared forever, leaving no descendants. Had conditions been slightly different and a tiny Burgess "worm," which possessed a central notocord (primitive spine), become extinct, animals with backbones (vertebrates) might never have developed—no birds, fish, frogs, or mammals. Instead, life on Earth might have sprung entirely from some of the other 14 body plans that thrived in those early seas.

A picture emerges of evolution as a unique, quirky history, whose details are chancy and contingent. To appreciate contingent history, paleontologist Stephen Jay Gould suggested a mental game of "what if." Gould's favorite example was Frank Capra's classic motion picture *It's a Wonderful Life* (1946), in which Jimmy Stewart's character is allowed to see what the world would have been like if he had never been born. Rescued from suicide by Clarence the Angel, the depressed hero believes his existence doesn't matter to anyone. But Clarence proves him wrong. If he hadn't saved his brother from drowning one summer, that brother wouldn't have later saved a shipful of sailors during the war. His wife might have become a lonely, aged spinster. Even the town's name would have been different. Gould applies the same principle to understanding evolution in his book *Wonderful Life: The Burgess Shale and the Nature of History* (1989).

"Contingency is both the watchword and lesson of the new interpretation of the Burgess Shale," writes Gould. "The fascination and transforming power of the Burgess message—a fantastic explosion of early disparity followed by decimation, perhaps largely by lottery—lies in its affirmation of history as the chief determinant of life's directions."

REWIND THE TAPE OF LIFE, change one fork in the road, and all else would change. In *It's a Wonderful Life* (left), the Jimmy Stewart character learns how much his own choices affected everyone in his town. Stephen Jay Gould applied that concept to understanding fossils from the Burgess Shale (right).

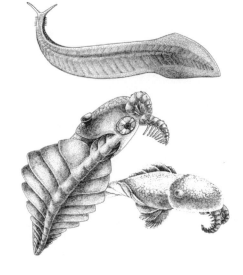

If the modern order was not guaranteed by basic laws such as natural selection or mechanical superiority in anatomical design but is largely a product of contingency, no particular outcome is inevitable. "It fills us with a new kind of amazement," wrote Gould, "at the fact that humans ever evolved at all. Replay the tape a million times from a Burgess beginning, and I doubt that anything like *Homo sapiens* (a tiny twig on an improbable branch of a contingent limb on a fortunate tree) would ever evolve again. It is, indeed, a wonderful life."

See also BURGESS SHALE; TELEOLOGY

CONVERGENT EVOLUTION
Caused by Similar Pressures

Convergent evolution is the process whereby species of differing ancestry independently evolve similar traits as a result of similar environments or selection pressures. For example, the streamlined torpedo shape for cutting through water has been evolved by sharks and bony fish, by mammals (whales and dolphins), and by the extinct seagoing reptiles (ichthyosaurs). Apparently, there is an optimum streamlined shape for gliding through water. Animals from widely divergent groups, with very different anatomies, have evolved amazing convergencies, even to the shape of tail flukes and position of dorsal fin. Similarly, the wings of bats, birds, and insects, though constructed very differently, show the effects of convergent or parallel evolution for enabling flight.

At least one paleontologist, Simon Conway Morris, believes that convergencies are inevitable, and that increasing complexity would inevitably have been reached from diverse origins. For instance, Conway Morris believes that if the human lineage had become extinct or never appeared, perhaps reptiles would have evolved bipedal creatures with serviceable hands, flat faces with eyes in the same plane (necessary for stereoscopic vision), and enlarged brains capable of advanced culture and language. In fact, one of the dinosaurs with the largest brains, Troodon, stood upright and had serviceable digits and eyes that faced somewhat forward.

Other evolutionists, however, believe that evolution proceeds by quirky, unpredictable turns along the kind of irreproducible path that Stephen Jay Gould called "contingent history." According to that view, if there had been a different turn at any fork in the road, there might well be no species on Earth that could lament the fact or ponder its own fate.

See also ADAPTATION; CONTINGENT HISTORY

If all hominids have gone extinct, would convergent evolution have produced a flat-faced, dextrous, upright, intelligent reptile?

COPE, EDWARD DRINKER (1840–1897)
Dinosaur Discoverer

For 30 years, Edward Drinker Cope was a driving force in American paleontology, possessed of an inexhaustible energy for research, discovery, and excavation of new dinosaur fossils. He was also jealous, combative, and downright underhanded. But without his raw egotism and ambition, the dinosaur halls of the great museums would not have become so quickly packed with Mesozoic saurians. His legacy also includes 1,400 books and articles describing and interpreting his fossil discoveries.

By the age of 21, the Philadelphia-born Cope had already published 31 papers on reptilian classification—this before he began his zoological education at the University of Pennsylvania under Joseph Leidy, the effective founder of American paleontology. He quickly made a name for himself, and was a museum curator and professor of zoology at Haverford College at age 24.

But a sedentary professor's life was not for the adventurous Cope. He made frequent trips to marl pits in New Jersey, paying quarrymen to notify him of interesting fossils. In the summer of 1866, a hadrosaur bone turned up, and Cope began to spend all his time at the pits. Soon after, he wrote his father that he had found "a totally new gigantic carnivorous Dinosaurian . . . the devourer and destroyer of Leidy's [his teacher's] *Hadrosaurus*, and of all else it could lay claws on." He named the beast "Laelaps," gave up his teaching post, and bought a house near the stone quarries.

Although he came from a respectable Quaker family, Cope was anything but quiet and peaceable. Almost a century after his death, his reputation still lingers as one of the most aggressive, belligerent scientists of all time. He was known to settle scientific arguments with his fists and did not shrink at outright thievery from his competitors.

Once he stole the entire skeleton of a whale. When the huge carcass washed up at Cape Cod, the Agassiz Museum at nearby Harvard sent out a team to collect its bones. They labored all day under a hot sun, hacking away tons of blubber and decomposing flesh, and finally loaded the bones onto a railroad flatcar. Cope, watching from a distance, waited until the workers left, then bribed the stationmaster to change the shipment's destination from Harvard's museum to his own in Philadelphia. For several years, the Harvard curator had no idea what had become of his whale skeleton.

In 1868, Cope began a series of digs in the American West that would lead to discovery after discovery. He held rebellious field crews together and charmed the hostile Cheyenne, who were warring on white intruders in the region.

His most acrimonious battles were with his chief rival, another wealthy and unscrupulously competitive paleontologist, Othniel C. Marsh, of Yale University. Marsh was equally adept at gathering dinosaur skeletons in hostile Indian country, bribing railway employees to reroute shipments, and claiming first discovery of species that Cope had already written up.

In later years, the feud died down, and the fortunes of both men declined. Cope had a bitter dispute with the government over a fossil collection he had made as a member of Ferdinand Hayden's expedition to the Yellowstone wilderness. The National Museum claimed total ownership, even though Cope had spent $75,000 of his own money to ensure the project's success. He had also made some bad investments, and finally had to sell most of his vast lifelong dinosaur collection to the American Museum of Natural History for only $32,000.

He ended his days in a Philadelphia brownstone that was packed floor to ceiling with fossils and papers but had virtually no furniture. Here the great dinosaur artist Charles R. Knight spent several weeks with Cope shortly before he died. Their intense discussions were fruitful, and Knight produced a remarkable series of paintings based on Cope's insights about the extinct monsters he loved.

See also COPE-MARSH FEUD; KNIGHT, CHARLES R.

STOGIE LOVER AND DINOSAUR DIGGER, paleontologist Edward Drinker Cope became sufficiently famous to front a brand of cigars.

Many dinosaurs were named twice by rival paleontologists Cope and Marsh— a mess that took thirty years to untangle.

COPE-MARSH FEUD
Dueling Dinosaur Diggers

One hundred years ago, the world's greatest collections of fossils from the American West were made by two rival paleontologists who loved dinosaurs and hated each other. Edward Drinker Cope of Philadelphia and Othniel C. Marsh of Yale's Peabody Museum were unscrupulous in competing for new discoveries.

Because both rushed to publish and never checked with each other, they often wrote up the same species at about the same time. Many dinosaurs therefore entered the scientific literature with two names: a Cope name and a Marsh name.

Once Cope even managed to insult his rivals by naming a mammal fossil *Anisonchus cophater,* the "jagged-toothed Cope-hater." He explained to a colleague, "It's no use looking up the Greek derivation . . . for I have named it in honor of the number of Cope-haters who surround me." The species name was officially accepted and still stands.

Often, when Cope had gotten what he'd wanted from a site, he would order his workers to dynamite the remaining exposed fossil fields to prevent others from working the area after him—a scientific crime. When he found that Marsh did not destroy fossil beds, Cope moved into his rival's abandoned excavations, which Marsh considered his own private domain.

One early incident, which caused years of enmity between the two former friends, occurred when an amateur fossil hunter offered some important bones for sale. Cope promised to buy them, wrote them up for publication, but didn't get around to sending a check. Marsh promptly wired the man a phony message canceling the deal and then bought the fossils himself.

DINOSAUR FINDER and evolutionist Othniel C. Marsh (right) of Yale discovered thousands of new species, often rushing to beat his chief rival Edward Cope (left) to the sites in their legendary "Great Bone War." Marsh's tough-looking gang of Western desperados (middle) was really a field team of Yale students.

Another clash came about when Cope had finished mounting his huge *Elasmosaurus* at Philadelphia. The creature had an enormous neck as well as a very long, tapering tail. When Cope proudly unveiled the skeleton, Marsh pointed out that he had put two tails on the creature, one at either end. The head was actually mounted on the end of one of the tails. Cope could not bear the thought that his rival had demonstrated that he didn't know one end of a dinosaur from the other, and tried to buy up all the copies of his monograph. He succeeded in finding and destroying all but two of the published volumes, which Marsh carefully tucked away in his own personal library.

Both men came from wealthy families, but Marsh had the advantage: His uncle was the fabulously rich George Peabody. When he sought a professorship at Yale, Marsh announced to the trustees that he had persuaded his uncle to contribute $150,000 to establish a major museum at the institution. He was promptly appointed professor, as well as director of the new Yale Peabody Museum.

In 1890, the Cope vs. Marsh feud went public, with the two acrimonious dinosaur diggers attacking each other in the newspapers. Cope said Marsh had plagiarized his work, and Marsh ridiculed some of the egregious errors in Cope's monographs. But their battle of the bones had been a boon to the new science of paleontology: The warring pair had discovered and named 1,718 new genera and species of fossil animals and packed several major museums with dinosaur skeletons.

In the bitter rivalry between Cope and Marsh, their mentor Joseph Leidy was all but forgotten. Leidy had been the first to discover and name a North American dinosaur and had founded American dinosaur studies, but he stepped aside in disgust when Cope and Marsh's acrimonious "bone wars" dominated the field. Henry Fairfield Osborn, director of the American Museum of Natural History, recalled that many of the Eocene and Oligocene animals had been given three names in the scientific literature: the original Leidy name and then the Cope and Marsh names. Speaking in 1924, Osborn lamented, "It has been the painful duty of Professor [William Berryman] Scott and myself to devote thirty of the best years of our lives trying to straighten out this nomenclatural chaos."

Cope once proposed that after their deaths, his brain should be weighed and compared with that of Marsh, to see which was heaviest—a finale to their lifelong competition that never took place.

See also COPE, EDWARD DRINKER; (CHIEF) RED CLOUD

COPROLITE INDUSTRY
How Dino Dung Saved England

Around 1840, an English farmer brought some strangely shaped stones to the Reverend John Henslow, who had been Charles Darwin's mentor at Cambridge. According to another of Henslow's students, Charles Kingsley, author of *The Water Babies:*

[Henslow] saw . . . that they were not, as fossils usually are, carbonate of lime, but phosphate of lime—bone earth. He said at once as by inspiration, "You have found a treasure—not a gold-mine, indeed, but a food-mine. This bone earth, which we are at wit's end to get [for our English crops] will increase immensely the food supply of England and perhaps make her independent of foreign phosphates in case of war."

What the farmer had found were coprolites from the Lower Cretaceous period: deposits of tons of dinosaur feces, mixed in with fossils of skeletons, claws, and teeth of many kinds of animals. Intrigued, Henslow sought and found a coprolite deposit not long after, and in 1843 published his first article on the subject. By the mid-1840s, large coprolite deposits were turning up in many parts of southeast England, with Cambridge near the center. William Buckland, the first professor of geology and mineralogy at the University of Oxford, went to see them, and in 1849 wrote that "in the remains of an extinct animal world England is to find the means of increasing her wealth in agricultural produce, as she has already found the great support of her manufacturing industry in fossil fuel."

The timing of these discoveries could not have been more fortunate. England's soil was badly depleted from heavy and continuous use for raising corn, pulses, and grains, and the country had begun importing expensive ground bone from abroad to use as fertilizers. According to one contemporary writer, "Great Britain was like a ghoul searching the continents for bones to feed its agriculture . . . [even mining] the battlefields of Leipzig, of Waterloo, and of the Crimea; already from the catacombs of Sicily she has carried away the skeletons of many successive generations."

Workers on the nearly played out farms were being paid abysmal wages for long hours of labor. When coprolite mining operations began thriving in town after town, however, England experienced an enormous economic and agricultural recovery.

Young men in the villages were now able to make many times their wages as farm laborers when they turned to digging coprolites. It was risky business, however, and many were killed in cave-ins during the precarious shovel and pickax operations. But now they had money in their pockets as well as leisure time, which many of them used for drinking. Pubs and "pot-houses" (bars) proliferated by the hundreds, and a new alcoholic village culture arose across the coprolite belt. One of their drinking songs went like this:

> Come listen you farmers to what I do say.
> We Coprolite diggers now can have fair play.
> You once did us grind down, but now it's our turn
> As we can get work and farm labour spurn!
> We are jolly young fellows, that do not work fear,
> We can work at the fossils, have a pot of good beer.
> With our spade and pickax we've no work to seek.
> We won't work for farmers for ten bob a week . . .
> Your sons and your daughters with all their fine clothes
> At the Coprolite diggers don't turn up your nose.

Coprolite mining had its heyday from the 1840s to the 1880s, during which it contributed greatly to increasing England's food supply and prosperity. Near the end of the century, deposits became played out and cheaper fertilizers from abroad became available to replace them. By 1885, the industry itself had become as extinct as the dinosaurs.

CORAL REEFS (1842)
Darwin's First Scientific Book

When Darwin set sail on his five-year voyage of discovery aboard HMS *Beagle* in 1831, he planned to concentrate his researches on geology and marine invertebrates, including corals. At the time, sailors and explorers were quite familiar with coral reefs and coral islands, but no one knew how they got there. Some geologists speculated that they were built by fishes carrying shell fragments in their mouths and cementing them in place. Others thought corals were like social insects, cooperatively building an enormous hive or nest.

Great Britain was like a ghoul searching the continents for bones to feed its agriculture, even mining Europe's catacombs and battlefields.

Young Darwin was a skilled geologist well before he had a comparable knowledge of botany or zoology. While exploring South America, he devoured the just-published *Principles of Geology* (1830–1833) by Sir Charles Lyell. Its uniformitarian approach inspired him to look for currently active "small causes" that may have created major geological features over time. Coral reefs perfectly filled the bill.

Darwin later recalled that he began thinking about coral reefs long before he had ever seen one, while still on the west coast of South America. For two years he had observed the shorelines of that continent, which seemed to show contradictory evidence of having been repeatedly built up and worn down. He began to form a theory, based on the effects of subsidence and uplift, of how coral reefs were formed.

At Tahiti he made his first field studies of reefs, and then gathered other evidence at Cocos, Keeling, and Mauritius that convinced him he was on the right track. Upon his return to England, he communicated his results to his mentor Lyell. The geologist was so overcome with delight that "he danced about and threw himself into the wildest contortions, as was his manner when excessively pleased." He wrote shortly afterward to Darwin: "I could think of nothing for days after your lesson on coral-reefs, but of the tops of submerged continents. It is all true, but do not flatter yourself that you will be believed till you are growing bald like me, with hard work and vexation at the incredulity of the world."

Lyell then helped his protégé get a hearing at the Geological Society of London in 1837. After 20 months' additional work on details, maps, and charts, Darwin published the study as his first scientific book: *The Structure and Distribution of Coral Reefs* (1842). Despite Lyell's fears, Darwin's explanation of coral reefs, backed up by complete maps and data, won the admiration of geologists and secured him instant recognition as an up-and-coming naturalist.

In the book, Darwin described three types of coral reefs: *atolls,* which are circular and enclose a lagoon; *barrier reefs,* which are long walls near a coastline, separated from the land by a channel; and *fringing reefs,* which stretch along a shoreline. He understood that all were built by millions of tiny colonial invertebrates, soft-bodied polyps (he called them "coral insects") that secrete protective limestone cells around themselves. But he wondered just how these huge structures were formed. How thick were they? How did their various shapes develop, and how long did it take for a reef to be produced? Why did they occur only in certain areas and not in others?

Darwin's theory of coral reefs starts with the fact that live corals grow only in shallow

CORAL LAGOON ISLAND, by Ernest Griset, was commissioned by John Lubbock in 1874, around the time that Darwin reissued his book on coral reefs.

© Bromley Museum Service.

water. He combines that with the observation that reefs seem to be associated with sea floors or submerged volcanoes that are subsiding. Accretions of rock build up as millions of coral animals secrete calcium carbonate atop the accumulated skeletons of older colonies. While the "basement" on which all are resting continues to settle, the "live areas" thrive only near the surface, where they can receive sufficient light to generate food. In other words, colonies keep reaching upward to receive sunlight, while the ocean floor beneath them keeps sinking. Given enough time, small forces can create major geological features; some reefs are thousands of feet thick and more than a mile long.

In 1874, more than 30 years after its initial publication, Darwin brought out a new, revised edition, but he saw no need to alter the theories he had published as a young man. (To commemorate the occasion, his protégé John Lubbock commissioned a painting of a coral lagoon island, apparently complete with HMS *Beagle* on the horizon—a long-lost watercolor that was discovered in a British museum storeroom in 2008.)

In his final book (1881), on earthworms, Darwin referred back to his first, on coral reefs, to emphasize how spectacular effects can be produced by "insignificant" forces acting over immense periods of time. In *Coral Reefs,* he had written: "We feel surprise when travelers tell us of the vast dimensions of the Pyramids and other great ruins, but how utterly insignificant are the greatest of these, when compared to these mountains of stone accumulated by the agency of various minute and tender animals! This is a wonder which does not at first strike the eye of the body, but, after reflection, the eye of reason."

Although Darwin's theory about reef formation long dominated the field, the question remained contentious. Alexander Agassiz, son of the Harvard zoologist Louis Agassiz, spent forty years and his considerable fortune visiting hundreds of the world's coral reefs. He planned to write a book that would challenge Darwin's explanation of reef formation but died before he could complete it. Meanwhile, several other geologists entered the fray with reef theories of their own. The problem remained unsolved for almost a century and a half.

Not until 1950—while attempting to destroy Eniwetok, a remote coral island in the Marshall group—did scientists finally find definitive answers. Preparing to test H-bombs there, the U.S. government sent geophysicists to drill test cores deeper than anyone had previously done. David Dobbs relates in *Reef Madness* (2005) that finally, at 4,200 feet, they hit "greenish basalt, the volcanic mountain on which the reef had originated. Dating of the tiny fossils in the bottommost layer of coral showed that the reef had gotten its start in the Eocene. For more than thirty million years this reef had been growing—an inch every millennium—on a sinking volcano, thickening as the lava beneath it subsided. Darwin was right, Agassiz wrong."

Over the next few years, deep drillings and echo soundings confirmed the subsidence theory all over the Pacific and Caribbean. As it turned out, Darwin's 1842 theory also fit perfectly with the discovery of plate tectonics, which was only developed in the 1960s. As Dobbs put it, "the movement of the earth's huge plates explains the subsidence of the Pacific and many other reef areas. . . . Darwin's theory was astoundingly correct."

See also CONTINENTAL DRIFT; EARTHWORMS; LYELL, SIR CHARLES

SHIP IN GRISET'S watercolor could be HMS *Beagle*. John Lubbock commissioned the painting for the 40th anniversary of Darwin's coral reef book. © Bromley Museum Service.

Darwin's theory of coral reefs remained controversial for a century, and was finally proven correct in the 1950s.

"CRADLE OF HUMANKIND"
Location of Human Origins

By the late 19th century, scientists no longer believed humans originated in a Garden of Eden somewhere in the Middle East. But what was the location of the actual "cradle of mankind"? And where were the fossils that could prove humans had evolved from apelike ancestors?

Charles Darwin, in his *Descent of Man* (1871), was an early advocate of an African origin. His reasoning was that man's closest living relatives, gorillas and chimpanzees, are found in Africa, perhaps still in the forests where they evolved. Therefore, although no early hominid fossils had yet been found anywhere in the world, he suggested that Africa was the place to look.

Writers celebrate Darwin's early conclusion but fail to mention that he had been scooped (again!) by his junior partner Alfred Russel Wallace. Seven years earlier, Wallace had attended a Geological Society meeting at which Sir Roderick Impey Murchison presented evidence that Africa was the oldest continent and was never submerged during Tertiary times.

Enthusiastically, Wallace related Murchison's views in a letter to Darwin. "Here then is evidently the place for finding early man," Wallace wrote in 1864. "I hope something good may [also] be found in Borneo and that the means may be found to explore the still more promising regions of tropical Africa, for we can expect nothing of man very early in Europe." Twenty-five years later, "Java Man" (Homo erectus) was discovered not far from Borneo. In one sentence, Wallace had predicted two major sites of early human discoveries. But the brilliant hunches of Darwin and Wallace seemed to die with them.

Some of the men who dominated evolutionary theory during the late 19th and early 20th centuries, imbued with visions of "Nordic" or "Aryan" racial superiority, wanted no part of an African origin for mankind. Africans were "retrogressive," said Henry Fairfield Osborn, president of the American Museum of Natural History, so he looked to the plains of Central Asia for humanity's roots. Despite his commitment to Darwin, Osborn later insisted there were no tropical forest apes in mankind's family tree.

After the discoveries of Homo erectus in Java, Osborn was convinced more than ever that Asia was the cradle of mankind. To prove it, he helped organize and fund one of the greatest paleontological expeditions ever mounted: the East Asiatic Expedition through the Gobi Desert, led by Roy Chapman Andrews. Its major objective was to find early hominid fossils. The expedition found many fossils—dinosaur skeletons, previously unknown early mammals, the first known clutch of dinosaur eggs—but no Asiatic "Dawn-Man."

After 1924, African anatomists Raymond Dart and Robert Broom started finding near-men (the australopithecines) in South Africa. No European anatomist or paleontologist was ready to accept them as anything but "aberrant chimpanzees," so sure were they that Africa could not have been the "cradle." Some British experts, with patriotic zeal, had even accepted the blatant hoax of an "earliest man" (Piltdown) perpetrated in the English countryside, within 30 miles of Darwin's home. And when Kenya-born Louis Leakey told his British professors in the 1930s he was going back to Africa to find early man, they told him not to waste his time looking there, since "everybody knew" that humans must have evolved in Asia.

During the next half-century, however, African evidence began piling up. Robert Broom, Louis Leakey, Mary Leakey, Richard Leakey, Donald Johanson, Tim White, and others turned up fossil after fossil. There were more new hominids than anyone could keep track of— "nutcracker man" (Paranthropus boisei), "Lucy" (Australopithecus afarensis), Homo habilis, Ardipithecus, and Homo erectus. No other continent yielded so many early hominids.

In recent years, the spotlight shifted from Leakey's East Africa to the Afar region of Ethiopia, where "Lucy" (3.3 million years old) was found, and, more recently, to central-western Africa, where Sahelanthropus lived six million years ago. With windows into hominid evolution now open in East Africa, Ethiopia, and Chad, it appears that hominids of many species were all over the African continent from the earliest times.

Darwin and Wallace, two Victorians who were not averse to having apes as ancestors, had been right about an African cradle of mankind. And they made a lot more sense out of a lot less evidence than many who followed them.

See also AFAR HOMINIDS; ARISTOGENESIS; "ARYAN RACE," MYTH OF; "DELICATE ARRANGEMENT"; PILTDOWN MAN

For decades, European scientists refused to accept australopithecines, denying that Africa could be the birthplace of humankind.

CREATIONISM
History of a Belief

A creationist is one who believes in a divine creator of Earth and living things. The belief does not necessarily imply anything about *how* the Earth and its inhabitants were brought into existence. Many Christians consider themselves creationists while believing that God has worked through the process of evolution. Others, particularly Protestant fundamentalists and Orthodox Jews, define creationism more narrowly, insisting on a literal belief in the Genesis account: fixity of species, six-day creation, young Earth, global Flood, and the impossibility of understanding any mechanisms of Creation.

Duane Gish of California's Institute for Creation Research defines the fundamentalist doctrine as a supernatural Creator "bringing into being the basic kinds of plants and animals by the process of sudden, or fiat, creation." In his textbook proposed for public schools, Gish argued that "we cannot discover by scientific investigation anything about the creative processes used by the Creator [because they] are not now operating anywhere in the natural universe." Henry Morris, Gish's colleague and founder of the Institute for Creation Research, disavows evolutionary biology because "we are completely limited to what God has seen fit to tell us, and this information is His written Word. The Bible is our textbook on the science of Creationism."

Historically, however, the manner in which creationists have interpreted the Bible has always been heavily influenced by secular scientific knowledge of nature and their own changing interpretations. Prior to the 18th century, for instance, church authorities did not insist on the fixity of type or permanence of species. "From Augustine through Thomas Aquinas to Francis Bacon and beyond," writes historian James R. Moore, "different `kinds' of plants and animals could be naturally generated; one kind could give rise spontaneously to another, and these kinds were by no means necessarily identical with those of the Book of Genesis."

It was not until the doctrine of fixity or permanence of biological species was asserted by the Cambridge botanist John Ray (1686) and reinforced by the great taxonomist Carl Linnaeus (1740s) that it became the prevailing view, synonymous with creationism. With the advent of Darwinism, many creationists, especially liberal Protestants, embraced theories of creative evolution--the idea that God had used evolution to perform His miracle of creation.

Creative evolution theories began to fall from scientific favor in the 1920s with the rise of Mendelian genetics. Protestant fundamentalists then broke from the majority of creationists to proclaim a return to the "fundamentals" of Christian doctrine. But in fact, the elements of their creed were distinctly modern innovations, unknown two centuries earlier.

The present creationist belief system owes much to a self-taught, armchair geologist named George McCready Price, who taught in Seventh-Day Adventist colleges. In 1923, he published a textbook, *The New Geology,* which explained all rock strata as dating from a worldwide catastrophic deluge; it also claimed the Earth was young, and that it had literally been created in six solar days. After the media drubbing it received at the Scopes trial in 1925, fundamentalist creationism remained for a time a quietly held, minority view.

The apparent quiescence was deceptive. Over the next half century, charismatic and dedicated creationists such as Henry Morris, a professor of hydraulics, Duane Gish, the Reverend Jerry Falwell, and others brought this once-minority movement to prominence in the United States, whence it has spread worldwide. According to James R. Moore, possibly as many as a quarter of the people in the United States

> live in a universe created miraculously only a few thousand years ago, and on an earth tenanted only by those fixed organic kinds that survived a global Flood. The Book of Genesis and its [fundamentalist] interpreters now command an audience unknown since before the time of Darwin.
> . . . The creationist cosmos of Protestant Fundamentalism has acquired an authority rivaling that of the established sciences.

In 1961 the religious movement received a tremendous impetus from a book called *The Genesis Flood,* coauthored by Morris and John C. Whitcomb Jr., an Indiana professor of Old Testament theology. Buoyed by a surge of support from their co-religionists, in 1963, Morris and a group of fellow creationists, many with doctorates in fields other than the life sciences, organized the Creation Research Society. By 1972 they had expanded it into the Institute for Creation Research, which they disingenuously described as a nonpolitical, nonreligious scientific organization.

However, this "scientific organization" requires a statement of faith from all members on the fixity of created species, the universality of the Flood, and the historical reality of the Genesis creation. Their chief activity seems to be mounting legal challenges to the teaching of evolution in public schools unless equal time is given to "creation science," though the courts have repeatedly held that their "science" is in fact a sectarian religion. They actively promote their view that "evolution science" (which they consider a religion) is a major cause of social problems. One of their books characterizes evolution as "not only anti-

Many Christians consider themselves both creationists and evolutionists, believing that God chose evolution as the method of creation.

Biblical and anti-Christian, but utterly unscientific and impossible as well. But it has served effectively as the pseudoscientific basis of atheism, agnosticism, socialism, fascism, and numerous other false and dangerous philosophies over the past century."

In 1981, Henry Morris obtained approval from the state of California for a graduate school run by his Institute for Creation Research, which offers degrees in science education, geology, astrophysics, geophysics, and biology—all from a creationist point of view. By 1986 he was able to move the school from the campus of Christian Heritage College in El Cajon, California, to its own campus. In its first catalogue, the institute's philosophy of scientific creationism is spelled out:

> Each of the major kinds of plants and animals was created functionally complete from the beginning and did not evolve from some other organism. . . . The first human beings did not evolve from an animal ancestry, but were specially created in fully human form from the start.

See also FUNDAMENTALISM

CREATIONISM, AMERICAN POLL ON
Consistent Split in National Beliefs

Only about a third of Americans believe that Darwinian evolution is well supported by the evidence.

According to a 2008 Gallup poll, Americans are divided between those who believe that God instantaneously created humans in their present form less than 10,000 years ago (44%), those who favor an evolutionary process guided by God (36%), and those who believe evolution has occurred without any divine intervention (14%). Public opinion is almost equally divided between those who believe that human evolution is well supported by evidence and those who accept the biblical account of creation as literal and infallible. Since 1982, when the Gallup organization began surveying Americans on human origins, these percentages have remained remarkably stable, varying little from year to year. Among Western nations, the United States consistently ranks among the lowest in public support for the evolutionary paradigm of mainstream science.

2008 Gallup Poll of Americans

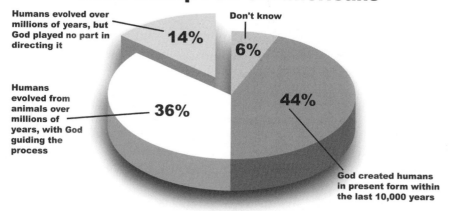

Humans evolved over millions of years, but God played no part in directing it — 14%

Don't know — 6%

Humans evolved from animals over millions of years, with God guiding the process — 36%

44% — God created humans in present form within the last 10,000 years

A CREATIONIST MUSEUM
Evangelical Darwin-Free Dinosaurs

Ken Ham, a Christian evangelist with a passion for dinosaurs, raised $27 million to build a lavish "biblically-based science museum" in Petersburg, Kentucky, 20 miles from Cincinnati. His Creation Museum, as it is now called, attracted about a half-million visitors within a year of its 2007 opening.

As director of the organization Answers in Genesis, Ham contends that every mainstream science museum, zoo, and national park in America is brainwashing children with "evolutionist propaganda." According to his website, secular science is based on a misplaced "faith in human reason," while "creation science" is "based on the only Eyewitness's revelation, as recorded in His own words."

Ham previously worked at the Institute for Creation Research, which founded a Museum of Creation and Earth History in Santee, California, during the 1980s. That museum is now dwarfed by the Kentucky museum, which includes a mile and a half of outdoor trails on its 47 acres and 70,000 square feet of indoor exhibitions. Ham has expanded on many of the smaller museum's concepts and themes, including the Six Days of Creation, Noah's Ark, the Tower of Babel, and Flood Geology.

The Creation Museum promotes unquestioning acceptance of the Bible's account of human origins (as interpreted by their ministry) as an antidote to "fallible human reason." Visitors are taught that God made the Earth and all its plants and animals in six days, and that major geological features were subsequently shaped by a Great Flood. Evolution is a fallacy and delusion; humans were created in their present form by divine fiat 6,000 years ago. A typical museum label reads: "*Velociraptor*. Means 'swift hunter.' Height: 4 feet. Length: 11 feet. Created on: Day 6."

A Hollywood theme park designer from Universal Studios was hired to build a section of Noah's Ark and life-size dioramas that dramatize biblical scenes. In the museum's Garden of Eden, dinosaurs are depicted living peaceably alongside Adam and Eve. Lions, tigers, and tyrannosaurs are shown as gentle vegetarians that never ate meat until humankind's sins brought violence into the world.

While Ham's ministry rejects almost every tenet of mainstream biology and geology, it embraces scientific evidence for the existence of dinosaurs. Dinosaurs have never previously appeared in Christian biblical imagery, probably because of their association with evolution. However, the Creation Museum's exhibits teach that Noah's ark had enough room for 16,000 "kinds" (not species) of animals, and that dinosaurs were indeed aboard. "Genesis says that the ark had two of every kind of creature that walked on the earth," according to a museum spokesman, "so we're taking the dinosaurs back from the evolutionists."

See also CREATIONISM; FUNDAMENTALISM; INTELLIGENT DESIGN; NOAH'S FLOOD; "NOAH'S RAVENS"

ANIMATRONIC DINOSAUR with cartoony spikes on its back greets visitors near the entrance to to the evangelical Creation Museum in Petersburg, Kentucky.

CREATIVE EVOLUTION
Vitalist Principle

French philosopher Henri Bergson had a rich literary style, clothing his arguments in emotionally affecting language. His influential book *Creative Evolution* (1907) was a treatise on evolution that purported to refute Darwinism on the basis of Bergson's intuitive feeling for a self-organizing principle he called the *élan vital*.

Scientists complained they had no way to work if Bergson denied them the possibility of finding causal explanations. Paleontologist George Gaylord Simpson argued in *The Evolution of Meaning* (1949):

> Such theories do not explain evolution, but claim it is inexplicable and then give a name to its inexplicability: *élan vital*, omega, aristogenesis, cellular consciousness, holism. . . . As Huxley has remarked, ascribing evolution to an *élan vital* no more explained the history of life than would ascribing its motion to an *élan locomotif* explain the operations of a steam engine.

MICKEY MOUSE EVOLVED to a "cuter" look (above) as his eyes and forehead got bigger and his snout shrunk. Betty Boop (below) has the large eyes and button nose of a human infant.

Top: © Disney Enterprises, Inc.; bottom: © King Features Syndicate.

CUTENESS, EVOLUTION OF
Biology of Infant Appeal

There seems to be a strong evolutionary basis for why we find the face of a baby cute and, by extension, the similar faces of baby animals, dolls, and cartoon characters. A certain kind of face seems to release protective, parental responses; we want to cuddle and care for helpless infants.

In 1950, the famous Austrian ethologist Konrad Lorenz published his theory that adult humans find baby animals "cute" because they share infantile features with newborn humans: "a short face in relation to a large forehead, protruding cheeks, maladjusted limb movements." He illustrated his point by comparing juvenile and adult faces of a duck, a dog, and a human. The adults have relatively flat, shortened foreheads; medium-sized eyes; long or protruding snouts (or beaks or noses); and an overall angularity. In contrast, the young have high, curved foreheads; larger eyes; and smaller noses and mouths.

These babylike features elicit affection; toys and pets that have the same infantile proportions are also perceived as appealing. Stephen Jay Gould wrote an essay on Mickey Mouse in 1979 in which he traced the evolution of Walt Disney's mischievous, angular rodent of the late 1920s through progressively more lovable versions. Mickey followed the Lorenzian formula exactly: his snout shrunk, his eyes got larger, and his head proportions changed through the years to conform increasingly to the popular image of "cuteness" accepted widely throughout the world. (For example, Mickey is now immensely popular in Japan, where all cartoon characters sport the large-eyed, large-foreheaded look.)

An infantilized version of feminine beauty is also widespread, as epitomized in the exaggerated facial proportions of geisha make-up and Betty Boop cartoons. In 1986, ethologists Robert Hinde and Les Barden wrote a paper on an unusual kind of "natural selection" in the only bear that reproduces asexually: *Ursus theodorus*, the Teddy bear. Named for President Theodore Roosevelt, the original toy bear looked like a long-snouted adult when it was created around 1900. Over the past 80 years, however, teddy bears have shown "a trend towards a larger forehead and a shorter snout," say the scientists.

As teddies with more babylike faces are bought each year, the ones that remain unsold are "selected out" of the "evolving" population by human taste. Teddy bear faces, like those of cartoon mice and kewpie dolls, are shaped by recognition patterns imbedded in the human brain, a form of parental behavior that has helped our species survive.

See also ETHOLOGY; LORENZ, KONRAD; NEOTENY; ROOSEVELT, THEODORE

CUVIER, BARON GEORGES (1769–1832)
French Comparative Anatomist, Paleontologist

A large block of limestone was delivered to "the father of comparative anatomy," Baron Georges Cuvier, at the Natural History Museum in Paris. Cuvier had announced to the crowd of students gathered in his laboratory that he could predict what kind of fossil animal was inside the rock. It was, he said, a very large kind of pig.

Technicians went to work with hammers and chisels, chipping away carefully at the block. Hours later, the spectators applauded when at last the fossil skeleton of an ancient pig emerged. Once again, Cuvier had lived up to his reputation as the man who could reconstruct an entire animal from a single bone.

Cuvier had studied thousands of skeletons and dissected carcasses of all kinds of creatures; his studies had shown him that there was an order to animal structures, a "correlation of parts." He wasn't "predicting," but using diagnostic features—some small foot bones that protruded from the rock.

Cuvier's lifelong interest in animals began in boyhood; by his teens he was already a well-read zoologist who had studied and hand-colored all the plates in a copy of Buffon's 36-volume *Histoire Naturelle*. In 1795, well schooled in sciences, languages, and administration, he accepted a position at the new Natural History Museum in the Jardin des Plantes, where he became fascinated with comparing the anatomies of many kinds of animals. His growing expertise enabled him to extend his methods to fossils, and he became a pioneer paleontologist as well. Soon he would produce his own magnum opus, *The Animal Kingdom*, published in four volumes in 1817.

Darwin's mentor, the geologist Charles Lyell, visited Cuvier in 1823, and wrote his sister a firsthand account of the great naturalist at work. Cuvier's "grand secret of [accomplishing his] prodigious feats," Lyell enthused, was his phenomenal ability at organization. Evidently he was one of the greatest multitaskers that ever lived.

His house adjoined the park's anatomy museum and was across from the Natural History Museum, where he was curator. In a huge, long room in his residence he had set up eleven desks. Lyell described the room as "like a public office for so many clerks. But all is for one man, who multiplies himself as author, and . . . moves as he finds necessary . . . from one occupation to another." Each desk was furnished with ink stand, pens, and everything needed for each project. It was like switching from file to file on a computer, only Cuvier would switch desks.

The first to include fossils in his classification of animals, Cuvier was fascinated with elephant bones found in Paris. He argued that fossils were the result of periodic global floods, after which new forms of life appeared. Opposed to the idea that species could change, he fought against Lamarck's notions of "transformation" (evolution) and proclaimed that human fossils would never be found: "Fossil man does not exist!"

Although he avoided considering naturalistic origins for humankind, Cuvier concluded that extinction was a fact for many species—a daring opinion for the time. The church taught that each animal was a link in God's chain of creation, which could not be broken. When mastodon bones were found, for instance, naturalists supposed that some of the animals must still be alive somewhere.

Cuvier wrote in his *Discourse on the Revolutionary Upheavals on the Surface of the Globe and on the Changes Which They Have Produced in the Animal Kingdom* (1825):

> It is vain for someone to seek in the forces which affect the surface of the earth today causes sufficient to produce the upheavals and catastrophes whose traces the earth's surface shows us. [For] a long time we thought we could explain earlier revolutionary upheavals by present causes, just as we readily explain past events in political history, when we know well the passions and the intrigues of our own time. . . . Nature's march has changed; any one of the agents which she uses today would not have been sufficient to produce these ancient works."

Even decades after Cuvier's death, his immense influence continued, and impeded the acceptance of Lyell's uniformitarian geology and Darwin's organic evolution.

See also CATASTROPHISM; EXTINCTION; UNIFORMITARIANISM

THE "MAGICIAN OF THE CHARNAL HOUSE," as Cuvier was known, was a masterful teacher, zoologist, and comparativie anatomist. His influential ideas delayed the acceptance of evolution.

"Life is the most beautiful spectacle and the most difficult problem presented to the curiosity of man."

—Baron Georges Cuvier

DART, RAYMOND ARTHUR (1893–1988)
Pioneering Paleoanthropologist

When Australian anatomist Raymond Dart patiently chipped the first known skull of an australopithecine child from a slab of limestone sent to him in 1924, he knew he had made the "early man find" of the century. But for decades he was alone in that knowledge. England's greatest anatomist, Sir Arthur Keith, proclaimed that Dart had merely found an "aberrant chimpanzee."

With patience and tenacity, Dart defended his view that the fossil braincase was unlike that of any known living or fossil ape. At a time when most anthropologists believed (on very little evidence) that humans had first appeared in Asia, Dart insisted that Africa was the cradle of mankind. Details of the shape of the brain, as preserved in a natural cast of the skull's interior, appeared hominid, not apelike. And the teeth, too, resembled the human pattern, with no large canines, which are characteristic of apes.

Pressing on alone and snubbed by European science, Dart continued to seek evidence of early hominids in Africa. He studied the compacted bone accumulations in rock cavities; to him, they indicated that early humans were aggressive hunters that frequently varied their diet of antelopes and other prey by dining on their own kind. Dart characterized nearby accumulations of antelope bones and horns as the raw material for hominid tools and weapons—which he dubbed the "osteodontokeratic [bone-teeth-horn] culture." After careful analysis of these deposits years later, however, others concluded that the bones were brought there by hyenas and other predators.

Robert Broom, a Scottish doctor and paleontologist who lived and worked in South Africa, later joined the search and discovered additional australopithecines (*Plesianthropus,* now lumped with *Australopithecus africanus*), as well as another, larger hominin (*Paranthropus robustus*) that is not part of the human lineage. European scientists continued to scoff, and for years Dart and Broom were, as Robert Ardrey described them in *African Genesis* (1961), "two wild men crying alone in the wilderness."

Raymond Dart lived to be 87. By that time, scientists had long acknowledged that his "Taung child" was indeed an epoch-making discovery. His old adversary, Sir Arthur Keith, proposed that the hominids be called "Dartians" (rhymes with Martians) in his honor, though the name never caught on.

See also AUSTRALOPITHECINES; BROOM, ROBERT; KILLER APE THEORY; TAUNG CHILD

THE FIRST TO RECOG-NIZE an australopithecine fossil, anatomist Raymond Dart chipped the skull of his famous "Taung child" out of a block of South African limestone. It took 30 years before scientists accepted his claim that it represented a previously unknown hominid.

DARWIN, CHARLES ROBERT (1809–1882)
Naturalist, Evolutionary Theorist

For 40 years in England's Kentish countryside, Charles Darwin, his wife Emma, and their seven surviving children lived quietly in a spacious old house, complete with gardens, fields, a patch of woods, greenhouse, later a clay tennis court, and about fifteen servants—the estate of a modest millionaire.

A semi-invalid who shunned social functions, Darwin spent his days writing, reading, strolling his grounds, dissecting barnacles or orchids, talking with local pigeon breeders, checking botanical experiments in his greenhouse, and observing the activities of bees in his garden. A casual visitor would never have guessed that from this sleepy, idyllic retreat he was shaking the world.

Conservative in lifestyle, Darwin dutifully contributed to the village church, helped organize local philanthropies and served on the local magistrate's bench as a justice of the peace. (He adjudicated such cases as rabbit poaching, "furious driving" of carriage horses, disputes over livestock road crossings, and vandalism to a farmer's fence.) So retired was his existence from the hubbub of London that when, around 1877, telephones became the latest rage among the well-to-do, he refused to have one installed.

Despite this penchant for privacy, Darwin was world famous as a prolific (and popular) author, naturalist, philosopher, botanist, geologist, explorer, and zoologist. Yet he had no professional training in biology, never passed a doctoral exam, accepted no formal students, and became nauseated at the thought of delivering a public lecture.

Darwin is still best known for the theory of natural selection, an achievement for which he had to share credit with Alfred Russel Wallace, a young naturalist who formulated it independently.

A much more impressive achievement was Darwin's application of the theory to thousands of specific problems in natural history, years of labor that created the new science of evolutionary biology. Although the Darwin-Wallace theory can be summarized in a few pages, how to use it to unravel nature's mysteries fills Darwin's 17 scientific books and more than 150 articles—an output that founded modern evolutionary biology. This incredibly productive life's work revolutionized every field he touched: botany, paleontology, physiology, taxonomy, comparative psychology, zoology, what we now call ecology, primatology, genetics, paleoanthropology, sociobiology, and all the life sciences.

These four decades at Down House may have seemed outwardly uneventful and tranquil, but Darwin's inner life was one continuous adventure. He had stored up more memories, impressions, and raw scientific data by age 27 than most people gather in a lifetime. On his epochal voyage of discovery aboard HMS *Beagle*, he explored the wilds of South America, Australia, Tahiti, and South Africa; observed and collected thousands of plant and animal specimens; chipped fossil skeletons of giant sloths out of cliffs; and discovered the secrets of coral islands and reefs.

It was an endless parade of wonders. For young Darwin it was a combination of adventure, hardship, scientific discovery, and unremitting hard work, all packed into five years. He had galloped on horseback alongside Argentine gauchos, ridden rough seas, survived killer storms and earthquakes, and wandered awestruck through creeper-laced rain forests teeming with gaudy birds and exquisite orchids. To the ordinary traveler, these events and experiences might register as striking but unrelated impressions. What was remarkable about the *Beagle*'s young naturalist was that he habitually sought underlying connections and regularities. Some years after returning home, he realized the key was the shared history of adapting to a changing Earth.

CHARLES DARWIN, shown in two different seasons of his life. As a young father, holding son William, his warmth and humor are apparent (bottom), but he is commonly remembered as a melancholic, wintry sage (top).

Gradually, he worked the notebooks and journals of his travels into a larger picture. His mind focused on a history measured in hundreds of millions of years, a puzzle that fascinated his brilliant grandfather, Erasmus Darwin: How do species come into being? What is man's place in nature? Is the "tangled bank" of plants and creatures an impenetrable jumble or is there an order accessible to the human mind?

Darwin's father, Dr. Robert Darwin, and grandfather Erasmus were both successful country physicians, and he was expected to follow in their footsteps. Medical school at Edinburgh, however, he found to be intolerable. General anesthetics had not yet been invented, and the practices of the day were often brutal. One day, unable to bear the

 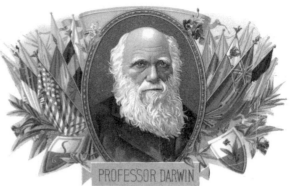

DARWIN REPLACED Charles Dickens on the 10-pound English bank note in 2000 (left). An early-twentieth-century cigar box label depicts him as an international icon (right).

screams of a strapped-down child during surgery, he ran out of the operating theater and gave up pursuing a medical career.

To his father's disgust, Charles's only real interests seemed to be collecting beetles, shooting birds, and studying rocks and plants—activities suitable only for an idle squire or a country parson. If he could not be a physician, Dr. Darwin urged his son to become a clergyman: Calling on parishioners was at least a respectable excuse to go tramping about the countryside. (Professional naturalists barely existed at all; many dedicated amateurs came from the ranks of rural churchmen.)

Charles agreed; he had read the Reverend William Paley's *Natural Theology* (1816) (parts of which he knew by heart) and was attracted to the idea of studying God's designs in nature. While he was waiting for the term to begin at divinity school, he went off for a few months of "geologising" with naturalist-clergyman Adam Sedgwick in Wales. But shortly after returning home, a fateful letter from Reverend John Henslow, his favorite Cambridge professor, was to change the course of his life forever.

Professor Henslow had been asked to suggest a ship's naturalist for a voyage around the world. Young Captain Robert FitzRoy (then 24 years old, only three years Darwin's senior) was taking the surveying ship *Beagle* to chart the coast of South America and then to explore the Pacific islands, Tasmania, Australia, and South Africa. Henslow offered to recommend Charles, expecting he would "snap at the opportunity." However, Dr. Darwin immediately put a damper on the young naturalist's enthusiasm, proclaiming it a dangerous, hare-brained scheme no "man of sense" would approve.

Charles appealed to his uncle Josiah Wedgwood, the famed ceramics manufacturer, a certified "man of sense" whose opinions carried considerable weight in family matters. "Uncle Jos" intervened, Darwin's father relented, and Charles went off to meet FitzRoy ("My beau ideal of a Captain"). Though some historians have painted Robert Darwin as a tyrannical ogre for opposing the voyage, he paid his son's expenses, bought equipment, and even provided him with a servant-assistant (Syms Covington) for the five-year trip.

On December 27, 1831 ("my real birthday"), HMS *Beagle* sailed from Portsmouth harbor, with 21-year-old Charles Darwin aboard as unofficial naturalist and "messmate" to the captain. His father was to take great pride in Charles's letters and accomplishments during the voyage. In retrospect, his opposition may have stemmed from a parent's legitimate fears for his son's life. Ships of the *Beagle*'s design had been nicknamed by sailors "floating coffins," and it was bound for the malarial tropics they called "the white man's graveyard."

Aboard ship, Darwin kept careful notebooks, journals, and diaries and read widely in natural history, including Sir Charles Lyell's *Principles of Geology* (1830–1833), which had just been published. Although he gathered, packed, and shipped home tons of natural history specimens, he began to believe "mere collecting" for its own sake was a vice of "the mob of naturalists without souls." His boyhood love of collecting insects had kindled his interest, as it had also for Alfred Russel Wallace, but "enlarged curiosity" eventually led both of them, independently, to join the ranks of philosophical naturalists—combining observation with a search for general explanations or laws of nature.

The Charles Darwin who returned five years later (1836) was a very different person:

seasoned, self-reliant, and with a mission to devote his life to natural science. His thousands of specimen rocks, fossils, birds, mammals, plants, and fish were to occupy naturalists at the British Museum for several years; hundreds of the species were new to science. His first two publications, *Journal of Researches Aboard the HMS Beagle* (1839) and *The Structure and Distribution of Coral Reefs* (1842), established him as a rising young scientist of talent and a popular author.

In 1839, Darwin married his first cousin Emma Wedgwood, whose traditional religious beliefs were opposed to his unorthodox inquiries into the origin of species. Soon after their marriage, she wrote him a letter begging him to reconsider challenging the Bible's account of Creation, lest they be separated for eternity in the hereafter. All his life he cherished her touching letter ("Many times have I kissed and cryed over this"), but remained committed to his scientific career.

After a couple of years spent in London organizing his collections at the British Museum and placing them in the hands of experts, Darwin and his wife decided to buy a country home. In 1842 they settled in the quiet little village of Downe, only 16 miles from London but—even today—a place that seems remote and timeless. ("I am now fixed on the spot where I will live my life to the end.") When they first moved in, the village still had a feudal flavor; country folk doffed their hats or curtsied when they encountered such "gentry" as the Darwins.

As he settled into the life of a country gentleman and father of 10 children (three would be lost to disease and genetic defects), he purposely turned his attention to a project as dull as the voyage had been exciting—describing and classifying the thousands of preserved barnacles he had gathered. For eight years he labored on these tiny, drab creatures, although he found the work numbing and tedious. ("I hate a Barnacle as no man ever did before, not even a Sailor in a slow-sailing ship.")

His purpose was to develop and establish his scientific judgment and credibility to deal with questions of species. Without paying his dues as a laborer ("hod carrier of science"), Darwin knew his theories on the "species question" would carry no weight with established zoologists.

And theories he had, notebooks full of questions, speculations, observations. Like his grandfather Erasmus, he had come to believe that all of living nature is related, that present species had evolved from ancient lineages. (He used the terms "transmutation" or "the development hypothesis." The term "evolution" was coined by philosopher Herbert Spencer and did not appear in Darwin's *Origin of Species* until the fifth edition in 1869.)

When he first met Thomas Huxley, later to become his great friend and champion, Darwin was examining some of his specimens at a laboratory table in the British Museum. Huxley, according to his recollections in *The Life and Letters of Charles Darwin,* brashly commented, "Isn't it striking what clear boundaries there are between natural groups, with no transitional forms?" Glancing up from the tray of preserved specimens, Darwin quietly replied, "Such is not altogether my view." Huxley later recalled that "the humorous smile which accompanied his gentle answer . . . long haunted and puzzled me."

To the sights and sounds of exotic wildernesses, Darwin added more commonplace observations, such as the unexpectedly wide range of variation in individuals, which he first learned from the thousands of barnacles he had dissected and classified. The domestic varieties of horses, hogs, and pigeons bred by "artificial selection" in the quiet Kentish countryside, too, were grist for his mill of evolutionary theory. ("If these varieties can be produced by man's hand, what might Nature not achieve?")

Darwin had only confided his ideas on "descent with modification" to a few people: the botanist Joseph Hooker, his mentor in geology Charles Lyell, and the American botanist Asa Gray. Although reticent about publication for years—as if testing

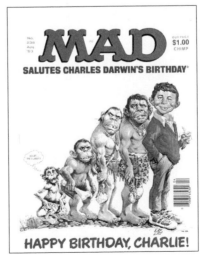

DARWIN POP CULTURE images include the cover of the April 1983 issue of *MAD Magazine.* Courtesy of and © by E.C. Publications Inc.

the waters before taking the plunge—his private correspondence reveals total dedication to revolutionizing the life sciences. Besides his tireless research, experimentation, and the writing of books and papers, he actively campaigned behind the scenes to convince influential scientists of the new paths to knowledge with an evolutionary perspective and even "grandeur in this view of life."

Despite having written a sketch of his theory in 1842 and expressing it in a letter to Hooker in 1844, the "great book" on evolution by natural selection was still unwritten in 1857. In the meantime, Wallace had published his paper "On the Law Which Has Regulated the Introduction of New Species" (1855), causing Lyell to warn Darwin that he must publish soon or risk losing his chance to be first with his theory. But despite Lyell's prodding, Darwin continued to work at his own careful (and procrastinating) pace.

Just as Lyell had feared, Wallace, on an extended field trip in the Moluccas (now Indonesia), had come up with exactly the same theory of evolution by natural selection. It was an independent discovery, which Wallace wrote out ("On the Tendency of Varieties to Depart Indefinitely from the Original Type," 1858) and sent to the one person he thought might appreciate its merit, Charles Darwin. Perhaps, wrote Wallace, if Darwin thought it worthy, he would pass it on to Sir Charles Lyell for publication. (Darwin had never told Wallace that he was working on a similar theory but had only vaguely mentioned his "long-term study" of the relationship of "varieties" to species.)

Wallace's paper threw Darwin into a panic, soon to be aggravated by a houseful of children quarantined with scarlet fever. "Your words have come true with a vengeance," he wrote Lyell in June 1858, "that I should be forestalled. I never saw a more striking coincidence. . . . Even his terms now stand as heads of my chapters."

Reeling from the possibility of being scooped by Wallace, he appealed for help to his friends Lyell and Hooker. He feared some might now accuse him of stealing from Wallace. ("I would as soon burn my whole book as any man believe I behaved in a paltry spirit.") He only hoped Wallace would not be aggrieved, and was relieved, on Wallace's return, to discover the younger man had a "generous and noble disposition."

Although Darwin was a careful and meticulous scientist, he was imaginative and empathetic as well. He related and reacted to plants and animals as fellow beings. He wrote of trees with their "noble heads raised high" and complained that climbing plants were "little rascals who never do what I want them to." He saw "life running riot" in rain forests, "antediluvian monsters" in harmless iguana lizards, and, on all creatures, "the indelible stamp" of their evolutionary origins.

Darwin is known as a solitary observer and field naturalist, but he excelled at two other modes of research that are less well known: collaboration and experimentation.

He experimented constantly—on the movements of plants, habits of earthworms, relationship of pollinating insects to flowers, digestion by carnivorous plants, cross-breeding of plant varieties, and seed germination. He did tens of thousands of experiments on which his later books were based. To test whether certain plants could have reached distant islands, he soaked seeds for months in barrels of brine, then planted them to see which could survive long immersion in salt water. He measured the activity of earthworms in his garden by calibrating the rate at which a heavy stone sank into the turf. He tested the reactions of the sundew plant to hundreds of substances, including dead flies, cobra venom, paper, atropine, nicotine, and human hair.

He found mystery and adventure while immersed in experiments with even the most commonplace and familiar forms of life, yet remained focused on such larger questions as the origin and evolution of humankind.

Darwin knew he could not hope to effect a major revolution in thought by himself and frequently consulted and collaborated at various times with Reverend John Henslow, Thomas Huxley, Charles Lyell, Joseph Hooker, Asa Gray, and his own son Francis. Although they almost never coauthored papers (sharing credit for each morsel of knowledge is a 20th-century fashion), they acknowledged intellectual debts by grandly dedicating their "great works" to one another.

Ironically, the one man with whom Darwin's name is forever linked as coauthor is Wallace, whose opinions and social standing differed greatly from his own. If he had had to choose a collaborator, surely he would have much preferred Hooker or Huxley, but Wallace was thrust upon him by his independent formulation of their theory of natural selection. (At first, Darwin was not very gracious about it, barely acknowledging Wallace in the first edition of *Origin of Species* [1859]. However, in later life, Darwin became one of the controversial Wallace's staunchest allies and helped to secure a government pension for him.)

Over the years, Darwin's books tackled the implications of his theory for human origins (*The Descent of Man*, 1871), behavioral evolution (*Expression of the Emotions in Man and Animals*, 1872), coevolution of insects and plants (*Orchids*, 1862), domestic breeding (*Variation of Animals and Plants under Domestication*, 1868), and botany and plant physiology (*The Movements and Habits of Climbing Plants*, 1865; *Insectivorous Plants,* 1875; *Different Forms of Flowers on Plants of the Same Species*, 1877). His first (*Coral Reefs*, 1842) and last books (*Earthworms*, 1881) were demonstrations of how the Earth's geological features may result from small, slow causes, acting regularly, over immense periods of time.

Darwin's scientific friends spoke of his intense honesty, the "central fire" of his character that never allowed him to cut corners in his work. He kept his vivid imagination disciplined by testing his ideas in thousands of careful experiments. Compulsively, he searched the scientific literature for facts that didn't fit into or contradicted his theories. Acknowledging all criticisms and suggestions with almost comically excessive gratitude, he carefully weighed every possible difficulty. In *Origin of Species,* he seems to be reluctantly driven to his conclusions, despite every possible objection, which he raises before his readers or critics can.

When he died of cardiac disease in 1882 ("I am not in the least afraid to die"), he expected to be buried in the old churchyard at Downe Village, but his powerful scientific friends petitioned for burial in Westminster Abbey, England's shrine of highest honor. Darwin's pallbearers included his old friends Huxley, Hooker, Wallace, Lubbock, and the presidents of the Royal and Linnean Societies. His final resting place is a few paces away from that of Sir Isaac Newton, another scientific immortal.

Eulogizing his old friend, the combative Professor Thomas Henry Huxley praised Darwin's lifelong restraint and diplomacy: "He delivered a thought-reversing doctrine to mankind with as little disturbance as possible to the deeply rooted sentiments of the age." Yet Huxley well appreciated that, notwithstanding Darwin's gentility and conventional lifestyle, he had managed to reshape and revolutionize Western thought:

> None have fought better, and none have been more fortunate than Charles Darwin. He found a great truth . . . [and] lived long enough to see it, chiefly by his own efforts, irrefragably established in science, inseparably incorporated with the common thoughts of men. . . . What shall a man desire more than this?

See also *CORAL REEFS;* DARWIN, CHARLES, ILLNESS OF; DARWIN, CHARLES, NICKNAMES OF; DARWIN, CHARLES, WORKS OF; DARWINISM; "DELICATE ARRANGEMENT"; *DESCENT OF MAN;* EARTHWORMS AND THE FORMATION OF VEGETABLE MOLD; ORCHIDS, DARWIN'S STUDY OF; *ORIGIN OF SPECIES;* TANGLED BANK; *VOYAGE OF HMS BEAGLE*

TRADING CARDS AND STAMPS bearing Darwin's likeness have appealed to collectors for more than a century.

DARWIN, CHARLES, DESCENDANTS OF

Emma and Charles Darwin had 10 children, of whom seven survived to adulthood. Two, a son and a daughter, were lost in infancy, and their eldest daughter, Annie, died at the age of 10. A bright and beautiful youngster, Annie had been the apple of Darwin's eye; her loss was one of the Darwins' greatest sorrows in life.

Darwin's eldest son, William, became a banker. His second eldest child, Henrietta, became the first family historian when she edited her mother Emma's letters, which were published in 1904. Her younger brother George became a noted astronomer and Plumian Professor of Astronomy at Cambridge who produced influential theories on the moon's effect on tides and Earth motions. Next came the youngest daughter, Elizabeth, followed by Francis (Frank), who became a botanist and edited *Life and Letters of Charles Darwin* (1887) and *More Letters of*

Charles Darwin (1903). Their younger brother, Major Leonard Darwin, had a military career and became a leading figure in the well-intentioned but ultimately disastrous eugenics movement. Finally, the youngest, Horace, was a designer of scientific instruments and founder of the Cambridge Instrument Company, which for years was known simply as "Horace's Shop." Horace also constructed the simple "wormstone" device, which still stands in the garden at Down House, to measure the rate at which earthworms bury stones by churning the soil.

George, Frank, and Horace were all elected members of the Royal Society, the fourth generation of Darwins in that prestigious scientific organization. The tradition extends through the fifth generation with Sir Charles Darwin and into the sixth with Richard Darwin Keynes and Horace Barlow.

Darwin had twenty-five grandchildren and has over a hundred living descendants.

Granddaughter Gwen Raverat wrote a sparkling recollection of her Cambridge childhood (*Period Piece*, 1952) illustrated by her own charming drawings. Another granddaughter, Lady Nora Barlow, edited Darwin's diaries of his *Beagle* voyage and his correspondence with the botanist Joseph Henslow. Lady Barlow was the founder of what is now called the "Darwin Industry"—scholars working to organize, archive, and analyze Darwin's voluminous letters, notebooks, and journals to reveal the development of his ideas. Grandson Bernard veered entirely away from science and became a noted golfer and sports journalist.

Richard Darwin Keynes, a great-grandson of Charles Darwin and a distinguished Cambridge physiologist, published a newly edited version of *Charles Darwin's Beagle Diary* (1988), as well as *Zoology Notes* (2000). He has assembled definitive accounts of the voyage from original documents, resulting in his books *The Beagle Record* (1979) and *Fossils, Finches, and Fuegians* (2006). During his travels in South America pursuing research on electric fishes, Keynes found a collection of the watercolor paintings Conrad Martens had made to document parts of the *Beagle* voyage before the invention of photography. Richard's brother Stephen Keynes formed the Charles Darwin Trust to promote science education and has worked tirelessly for the restoration and preservation of Darwin's home and its environs; his late brother, Quentin, amassed a remarkable collection of Darwinian books and letters, which he bequeathed to the Charles Darwin Trust.

Richard's son Randal Keynes has had dual careers—in the British Civil Service and as an author and historian with a deep interest in his family legacy. Intrigued and touched by a writing box that once belonged to Darwin's beloved daughter Annie, Randal Keynes wrote *Annie's Box* (2002), published in the United States as *Darwin, His Daughter, and Human Evolution*. In that book, he shows how Darwin incorporated his surroundings, including his garden, pets, and children, into his evolutionary studies. Randal has also been instrumental, with his uncle Stephen, in helping to restore Down House, which has been preserved by English Heritage as the Darwin Museum. They have also headed the effort to have the house and surrounding woods and fields designated a UNESCO World Heritage Site.

At the age of 14, Randal's son Skandar Keynes made an early mark as a popular movie actor, starring as Edmund in the Disney film *The Chronicles of Narnia: The Lion, the Witch, and the Wardrobe* (2005), which was based on a work by C. S. Lewis. Skandar has filmed sequels while completing his education; to a generation of youngsters, he is the most famous of Charles Darwin's descendants.

DARWIN'S LINEAGE includes his sons the astronomer George (left) and botanist Francis (middle). Great-grandson Richard Darwin Keynes (right) is a distinguished Cambridge physiologist and historian.

Ruth Padel, a Darwin great-great-granddaughter, is an acclaimed poet who began by studying Greek classics at Cambridge and became the author of *Summer Snow* (1990), *Angel* (1993), *Rembrandt Would Have Loved You* (1998), and *Voodoo Shop* (2002). Focusing now on poetry, her work has been described as "a fiercely idiosyncratic mix of erudition and the contemporary vernacular, bringing together references to Pushkin and Darth Vader, Odysseus and Iggy Pop."

With a University of London degree in creative writing, Emma Darwin—named for her great-great-grandmother—is the latest professional author in the family. Her well-received novels include *The Mathematics of Love* (2007) and *A Secret Alchemy* (2008).

Great-great-grandson Matthew Chapman was born in Cambridge but moved to the United States and became a screenwriter and director in Hollywood. In his darkly humorous book *Trials of the Monkey* (2000), Chapman recounts his bus journey through the American South, ostensibly to attend a reenactment of the Scopes Monkey Trial in Tennessee. In this "tragicomic accidental memoir," Chapman reports on his encounters with American evangelical creationists and his own inner turmoil as he seeks his identity as a Darwin.

Chapman attended the *Kitzmiller v. Dover Area School District* "intelligent design trial" in Dover, Pennsylvania, which he chronicled in *40 Days and 40 Nights: Darwin, Intelligent Design, God, OxyContin, and Other Oddities on Trial in Pennsylvania* (2007). The lawsuit was brought by parents who resented religious textbooks being foisted on students as a "scientific alternative" to Darwinian evolution by the local school board. His book's title was inspired by a lawyer who asked Judge John Jones whether he thought it a coincidence that the trial lasted 40 days and 40 nights, like Noah's flood. "Well," replied the judge, "it was certainly not by design."

Great-great-granddaughter Sarah Darwin Vogel, a gifted botanical illustrator, is an ecological botanist at the London Natural History Museum who donates some of her energies to the Galápagos Wildlife Conservation Trust. Her brother, Chris Darwin, a passionate environmentalist, lives in Australia, where he leads rock-climbing tours in the outback. On a 2007 BBC radio program, he recalled that his family was worried during his student days: "They didn't really mind me failing anything else, but not biology. I was doing badly and I got a tutor in it. . . . 'You're Darwin,' he said, 'I can't teach you anything about biology'—which I thought was really sweet. And then I went off and failed."

History is alive among the descendants of Charles Darwin. They still favor naming their offspring Erasmus, George, Emma, and Charles; one great-grandson even married a descendant of Thomas Henry Huxley. When Angela Huxley married George Pember Darwin in 1964, *Punch* magazine parodied the famous doggerel about the Lowells and the Cabots of Boston (". . . home of the bean and the cod / Where Lowells speak only to Cabots / And the Cabots speak only to God"):

A Huxley has married a Darwin
A fate that no Darwin escapes
For Huxleys speak only to Darwins
And the Darwins speak only to the apes.

CURRENT GENERATIONS include, from left to right, historian Randal Keynes; his son, film actor Skandar Keynes; Sarah Darwin Vogel with her son Leo Erasmus; and from top to bottom, poet and classicist Ruth Padel; novelist Emma Darwin; and author Matthew Chapman.

DARWIN, CHARLES, ILLNESS OF
A Mysterious Malady

A few years after returning to England from his five-year voyage of exploration, Charles Darwin became a semi-invalid who suffered daily for the rest of his life. Doctors were baffled; they could find neither cause nor cure.

As a young man Darwin had uncommon strength and stamina. During the *Beagle* expedition, he endured rough seas and primitive conditions on overland treks and rode spirited horses with the rough gauchos in Argentina. Whenever he encountered a mountain on his inland treks, he usually climbed it. Yet a few years later, he was afflicted with almost daily weakness, vomiting, and chronic fatigue.

One theory, proposed by Israeli parasitologist Saul Adler in 1959, identified the illness as Chagas' disease, which is acquired through the bite of the benchuca beetle, or assassin bug, also known as the "Great Black Bug of the Pampas." In fact, Darwin recorded in his journal that he was bitten by the insect in 1835 in Argentina, where the illness is common. It is caused by a blood parasite that became known to science only years after Darwin's death.

Other historians of biology suggest the illness was psychosomatic in origin. Dr. Ralph Colp Jr., a New York psychiatrist, has analyzed every scrap of evidence about Darwin's symptoms and the history of his condition and, in his 1977 book *To Be an Invalid: The Mysterious Illness of Charles Darwin*, made a meticulous case that the "mysterious malady" had a strong psychosomatic component. Colp traces the beginning of Darwin's illness to his work on evolutionary theory, over which he suffered extreme anxiety. From the first, his wife Emma worried that his scientific investigations were going to cost him his eternal soul. Darwin had nightmares of being beheaded or hanged; he thought that promulgating a belief so contrary to biblical authority was "like confessing a murder."

Emma nursed Charles, carefully scheduling his work and visitors so as not to tire him. Sometimes, he admitted, illness saved him from having to attend boring dinner parties or social engagements. When his doctors could offer no help, he pursued quacks and water cures, sometimes spending weeks at the spas that were then fashionable.

In her charming reminiscence *Period Piece: A Cambridge Childhood* (1952), Darwin's granddaughter Gwen Raverat concludes that "the attitude of the whole Darwin family to sickness was most unwholesome. . . . At Down [House], ill health was considered normal." She relates that her own father, George, as well as her aunt Etty and uncle Horace—all children of Charles and Emma Darwin—were "most affected by the cult of ill health. . . . [For Aunt Etty] ill health became her profession and absorbing interest. . . . She was always going away to rest, in case she might be tired later on in the day, or even next day." And she attributes the situation, to some extent, to Charles and Emma:

> The trouble was that in my grandparents' house it was a distinction and a mournful pleasure to be ill. This was partly because my grandfather was always ill, and his children adored him and were inclined to imitate him; and partly because it was so delightful to be pitied and nursed by my grandmother.

In 1989, however, Dr. Jared Goldstein of Randleman, North Carolina, published a new analysis of the Chagas' theory, in which he showed that a "sub-acute" form of the infection fit Darwin's symptoms very well. The incubation period and later stabilization, during which the patient improves, jibe with Darwin's medical history. Apparently, his immune system may have arrested the disease, which never progressed to organ impairment. Colp, who had been unaware that Chagas' could take that form, agreed with Goldstein's findings (and acknowledged them in the 2008 revision of his book, *Darwin's Illness*), while Goldstein endorsed Colp's psychoactive analysis as well.

It is remarkable that despite "never knowing a day of robust health" for 40 years, Charles Darwin managed to write his 17 scientific books and 155 articles—a lifetime output of more than 10,000 published pages—working no more than two or three hours a day. "I have always maintained," he said, "it is dogged as does it."

"It was . . . a mournful pleasure to be ill . . . because it was so delightful to be pitied and nursed by my grandmother."

—Gwen Raverat, 1952

DARWIN, CHARLES, MAGISTRATE
Part-time Justice of the Peace

On July 3, 1857, two years before publication of the *Origin of Species*, Darwin took on the part-time job of magistrate at Bromley, the largest nearby town to Downe Village. His neighbor, wealthy banker-scientist Sir John Lubbock (Senior), had talked him into accepting the honorary position, as part of a gentleman's duty to "help maintain order in the neighborhood."

In his oath of office, he agreed to "keep the Peace of one said Lady the queen in the said County, to hear and determine diverse felonies and also trespasses and other misdemeanours." In the same document, he was also enjoined from doing "anything to upset the religious values of the country." For almost two decades, his name remained on the list of active judges, and he often listed his occupation on official forms as Charles Darwin, J.P. (Justice of the Peace).

At times it was an exhausting responsibility. He wrote his friend Joseph Hooker in 1858, "I attended the bench on Monday and was detained in adjudicating some troublesome cases an hour and a half longer than usual, and came home utterly knocked up [exhausted] and cannot rally. I am not worth an old button."

The most common cases had to do with violations of the tough anti-poaching laws. Harsh penalties were given to trespassers for poaching game on private land. A man who killed a deer to feed his family might be sentenced to six months in the workhouse. In Darwin's court during March 1860, for example, a young man was fined for "catching a rabbit [in a wire snare] on his father's plantation." The father, it seems, was bent on teaching his son a stern lesson.

Poaching was considered an antisocial addiction, which, in the words of a Royal Commission, "must be cut at the root among young boys" because it "leads man, step by step, to almost every other crime." Several poachers of the period wrote books expressing their love of constantly trying to outwit animals, gamekeepers, and police alike. Some even became folk heroes, akin to Robin Hood—the classic English provocateur who defied the corrupt aristocracy by poaching the king's deer. Indeed, in Bromley there was an annual Robin Hood celebration at the White Hart Inn, where children dressed in forest green, played with bows and arrows, and eluded the vindictive Sheriff of Nottingham.

Other cases tried before Magistrate Darwin included domestic squabbles, public drunk-

> "I remember how troubled he was once when he had to punish some boys who had robbed his orchard. 'I do wish the police hadn't caught them,' he said to me."
>
> —George Sales, 1939

DOWNE VILLAGE children pose on High Street in the late 19th century, as a hay wagon rolls through town. Magistrate Darwin took an oath to "help maintain order in the neighborhood."

enness, vagrancy, vandalism of fences, boundary disputes, abuse of domestic animals, and fistfights in the local "pot houses," where pots of ale were served.

One man was charged with "furious driving" of a horse and cart in a public thoroughfare. The driver pleaded guilty to speeding and was fined, his second such offense. A newspaper's account of the case suggested that if the horse had been allowed to testify, the driver might have gotten off free: "The animal was well treated by his master, did not like staying out late, and was anxious to have a good feed in his stall. . . . If the little fat beast could have been examined in the matter, [perhaps] he would have . . . taken the blame on himself."

According to Darwin's son Francis, he was a sympathetic and often lenient judge, but on one transgression he was implacably harsh: cruelty to animals. Once he witnessed a man cruelly whipping his horse on the road. Pulling to a stop in his own carriage, Darwin angrily told the driver that he was a magistrate in the district and that if he caught the man abusing an animal again, he would personally haul him into court and throw the lawbook at him. Another time, a gentleman farmer had neglected his sheep, which starved to death. Darwin was ill, but roused himself to collect the evidence and got the man convicted.

One 1864 case that apparently amused Darwin and the other judges was that of a drunken man, a peddler by trade, who was accused by the police of being belligerent. He protested that he did not "create a disturbance" as charged. All he had done, he testified, was "to take off his coat and offer to fight the best man of the lot." Apparently, he thought that he was being prosecuted for removing his coat. The *Bromley Record* newspaper account noted that "this mysterious piece of logic was rather perplexing, and occasioned a good deal of laughter in the court, but failed to get the man off."

DARWIN, CHARLES, NAMESAKES OF

Darwin scholar Richard Freeman collected long lists of places, plants, and animals named for Charles Darwin to honor his achievements as explorer, geologist, botanist, and zoologist. There are probably close to 200. Among them are Darwin Bay, at Genovesa Island in the Galápagos, and the Darwin Channel, a westward continuation of the Aisén Fjord, in Chile. Two major mountains in South America are called Mount Darwin, one in Tierra del Fuego and one in Peru. The most northern of the Galápagos Islands (formerly Culpepper Island) is now officially Darwin Island. There is also an Australian city called Darwin and a Darwin district and mountain in South Africa. (There are also mountains named for Captain FitzRoy and his officer Stokes in Chile and Argentina and a Beagle Channel.)

About 100 living things bear Darwin's name as part of their official scientific label. Plants include a fungus, alga, orchid, potato, tulip, cactus, and tree. Animals include a ground beetle, legless lizard, jumping spider, iguana, fossil oyster, fossil giant sloth, tree frog, finches, and the ostrichlike South American bird known as Darwin's rhea.

DARWIN, CHARLES, NICKNAMES OF

Charles Darwin seems to have had more than his share of nicknames, reflecting the many facets of his complex career and personality.

When he was an old man, respectful journalists dubbed him "The Sage of Down," while ironic ones sometimes referred to him as "The Saint of Science." But his irreverent friend Thomas Henry Huxley satirized these overblown compliments by privately calling him "The Czar of Down" and the "Pope of Science."

As a small boy, Darwin and his brother Erasmus used to conduct amateur chemical experiments, and Charles once caused an explosion. For his part in this misadventure, his brother called him "Gas" for years.

At Cambridge, he took regular strolls in the countryside with his mentor, Professor Henslow, the botanist. He considered their rambling talks more valuable than any of his courses and became known to the other students as "the man who walks with Henslow."

On board the HMS *Beagle* as a young naturalist, he had two nicknames. One was "Philosopher," shortened to "Philos" by the captain and officers with whom he liked to discuss

"THE POPE OF SCIENCE" was one of Thomas Huxley's more acerbic nicknames for Charles Darwin. In a letter requesting a recommendation for a friend, Huxley lampooned Darwin as head of the "Church Scientific," with his favor-seeking friend as a kneeling supplicant.
Redrawn by Pete Von Sholly.

intellectual questions. The other was "Flycatcher," when his shipmates got tired of seeing the deck full of his natural history collections.

His own favorite nickname for himself was "Stultis the Fool." To Darwin nothing was obvious, and he would try experiments most people would prejudge to be fruitless. When he was investigating pollination, he put a female flower in a bell jar together with some pollen from a male plant to see if there was any way the two could get together without the help of insects. "I love a fool's experiment," he once told a friend, "and I am always making them." When describing his "fool's experiments" or new hypotheses in letters to scientific friends, he would wryly sign himself Stultis, a Latin name meaning one who is absurd, futile, ridiculous, insane.

DARWIN, CHARLES, WORKS OF

Charles Darwin's output was astounding, especially as he was a semi-invalid for most of his life after returning from his five-year voyage aboard HMS *Beagle*. Generally, he worked three hours a day, writing all his manuscripts by hand.

All told, he published 17 books in 21 volumes; these works contain more than 9,000 printed pages. Another 1,000 pages appeared in scientific journals. (Almost 10,000 additional pages of revisions were added to editions of various books over 43 years.) Also, several thousand of his letters have been published, many of them for the first time during the 1980s.

Many of Darwin's superficial critics focus on a few passages in *Origin of Species* (1859), unaware of the major body of his life's work. His wide-ranging contributions of new knowledge to geology, botany, animal behavior, reproductive biology, and dozens of other fields has never been equaled. And he established research programs in these fields that are still being profitably pursued. Following is a list of his books.

> "I love a fool's experiment, and I am always making them."
>
> —Charles Darwin

1839	*Narrative of the Surveying Voyages of His Majesty's Ships Adventure and Beagle (1826–1836).* 3 vols. London: Colburn. Volume 3 written by Darwin and reprinted as a separate work with the following title:
1839	*Journal of Researches into the Geology and Natural History of the Various Countries Visited by HMS Beagle.* London: Colburn.
1839–1843	*Zoology of the Voyage of HMS Beagle.* London: Smith, Elder. Edited by Charles Darwin, these were monographs by experts published in five parts; Darwin contributed notes and other material to each piece.
1842	*The Structure and Distribution of Coral Reefs.* London: Smith, Elder. First part of *Geology of the Voyage of the Beagle.*
1844	*Geological Observations on the Volcanic Islands Visited during the Voyage of HMS Beagle.* London: Smith, Elder. Second part of *Geology of the Voyage of the Beagle.*
1846	*Geological Observations on South America.* London: Smith, Elder. Third part of *Geology of the Voyage of the Beagle.*
1851–1855	Several volumes of monographs on fossil and modern barnacles (Cirripedes)
1859	*On the Origin of Species by Means of Natural Selection, or the Preservation of Favoured Races in the Struggle for Life.* London: Murray. 2nd edition, 1860; 3rd edition, 1861; 4th edition, 1866; 5th edition, 1869; 6th and last edition, 1872.
1862	*On the Various Contrivances by Which Orchids Are Fertilised by Insects.* London: Murray.
1865	*The Movements and Habits of Climbing Plants.* London: Linnean Society.

1868	*The Variation of Animals and Plants under Domestication.* 2 vols. London: Murray.
1871	*The Descent of Man, and Selection in Relation to Sex.* 2 vols. London: Murray.
1872	*The Expression of the Emotions in Man and Animals.* London: Murray.
1875	*Insectivorous Plants.* London: Murray.
1876	*The Effects of Cross and Self Fertilisation in the Vegetable Kingdom.* London: Murray.
1877	*The Different Forms of Flowers on Plants of the Same Species.* London: Murray.
1879	*Erasmus Darwin.* London: Murray. Text by Ernst Krause, with preliminary essay (longer than Krause's text) by Charles Darwin.
1880	*The Power of Movement in Plants.* London: Murray. Written with the assistance of his son Francis.
1881	*The Formation of Vegetable Mould through the Action of Worms.* London: Murray.

DARWIN, EMMA (1808–1896)
Charles's Wife, Partner, and Caregiver

EMMA DARWIN as a young woman just before her marriage to Charles, and a formal photograph in old age.

Soon after Charles Darwin's return to England, he married his first cousin Emma Wedgwood, who devoted her life to making his work possible even though she feared it might result in his damnation. Born into the famous family of prosperous potters, as Josiah Wedgwood's daughter she brought considerable means to the marriage. The Wedgwoods had been especially close to the Darwins for two generations, ever since Erasmus Darwin and the first Josiah Wedgwood founded their intertwined dynasties.

After their marriage, Emma wrote a deeply felt letter asking Charles to reconsider his defiance of biblical authority, so they might not be headed for two different places in the afterlife if he was wrong. Her new husband was touched when he received it. Years later, after his death, it was found with his jotted comment: "Many times have I kissed and cryed over this."

Emma soon grew used to having Charles's experiments about the house and was protective of his time and energies. She cautioned visitors to be brief and often cut them short by announcing it was time for his nap or his walk. She was nurse, companion, and mother of 10 children (seven of whom survived), and she supervised a large household staff.

She was tolerant of his friends but had little interest in their discussions. At one gathering, Charles saw her yawn and asked if she was bored by all the scientific talk. "Oh, no," she drily replied, "no more so than usual."

Emma regularly attended the village church on Sundays, and once heard a new pastor rail against Darwin's theories in a sermon. Although she held grave misgivings about her husband's philosophy, she promptly gathered her family and marched out of the church, to which she did not return until the tactless cleric no longer occupied the local pulpit.

DARWIN, ERASMUS (1731–1802)
Grandfather of Evolution

Erasmus Darwin, grandfather of Charles, was much more than a prosperous physician with a good bedside manner. He was also a famous poet, philosopher, and botanist and the first naturalist to publish a detailed theory of evolution—much of it in verse. His pioneering treatise on evolutionary theory, *Zoonomia; or, The Laws of Organic Life* (1794–1796), anticipated by more than 60 years the intellectual revolution his grandson would lead.

Born in the county of Nottingham in 1731, Erasmus showed a precocious intellect. Like his studious father, more interested in fossils than in sport, he later became the first in a line of six generations elected to membership in the Royal Society. After graduating from Cambridge, he studied medicine at Edinburgh, then moved back to England, where he established a medical practice in Lichfield. Some of the most brilliant and inventive men of his day lived in the area and enjoyed visiting together. Among these were inventor James Watt, metallurgist Matthew Boulton, chemist Joseph Priestley, and potter Josiah Wedgwood—all of them prime movers in creating the technology that would take England into the industrial era.

Dr. Darwin was the founder and ringleader of the Lunar Society, which brought these notables together regularly. The "Lunatics," as they called themselves, met only during full moons, so that they might find their way home by bright moonlight in their horse-drawn vehicles. Their gatherings lasted from early afternoon until night, and their talk was probably the most interesting in England at that time—of Watt's steam engine or Wedgwood's new ceramic methods, of Priestley's oxygen experiments or Darwin's evolution theories.

ERASMUS DARWIN, physician, botanist, naturalist, and poet, developed a theory of evolution, or descent with modification. Charles Darwin, despite his denials, clearly followed up on many of Erasmus's original ideas.

Although he was a great and expansive talker, with a wry sense of humor, Dr. Darwin stammered. When a rude young man asked whether his halting speech was frustrating, he replied, "N-n-no Sir . . . it gives me . . . time for reflection . . . and . . . saves me . . . from asking . . . impertinent questions."

In his books and letters, Erasmus Darwin discussed and anticipated many of our modern inventions and concerns, from eugenics and evolution to airplanes and submarines, from psychoanalysis to antiseptics.

In the late 18th century, he addressed two major evolutionary questions: whether all living things arose from a single common ancestor, and how one species could develop into another. He assembled observations from embryology, comparative anatomy, systematics, geographical distribution, and fossils. Overwhelming evidence, he thought, pointed to the development of all life from a single source, "one living filament."

Although he rejected the scriptural idea of separate creations for each species, that did not lessen his respect for "the Author of all things." It was just as wonderful, he argued, that "the Cause of causes" has set the whole evolving web of life in motion.

As for the means of evolution, Erasmus's writings touch on almost all the important topics except natural selection. He suggested that overpopulation sharpened competition, that competition and selection were possible agents of change, that humans were closely related to apes, that plants should not be left out of evolutionary studies, and that sexual selection could play a role in shaping species. (Of rutting stags, he wrote in *Zoonomia:* "The final course of this contest among males seems to be, that the strongest and most active animal should propagate the species which should thus be improved.")

In addition, he wrote on related problems in natural history, which remarkably parallel the subjects taken up by his grandson for whole books: twining and movement in plants, theory of descent, cross-fertilization in plants, adaptive and protective coloration, heredity, domestication of animals. Behind these investigations was Dr. Darwin's strong belief in progress toward "ever greater perfection in all the productions of nature."

See also DARWIN, CHARLES; *TEMPLE OF NATURE; ZOONOMIA*

"The final course of this contest among males seems to be, that the strongest . . . should propagate the species which should thus be improved."

—Erasmus Darwin, 1794

DARWIN COLLEGE
Cambridge Graduate School

Charles Darwin attended Christ's College at Cambridge during the 1820s, although he considered his stint there "an almost total waste of time." Nevertheless, one of his teachers, the botanist Reverend John Henslow, later recommended him as naturalist on HMS *Beagle*'s voyage, which he always referred to as his "real education."

Today Cambridge's ancient cluster of schools has a recent addition: Darwin College, founded in 1964 for postgraduate studies. Among its renowned alumni is Dian Fossey, who interrupted her study of African gorillas to take a zoology degree there in the 1970s.

DARWIN COLLEGE at Cambridge, England, founded in 1964 (left). This building, originally a granary, was later the home of Charles Darwin's son George, an astronomer and mathematician.

GROUCHO MARX takes over as dean of "Huxley College" (right) in the comedy *Horse Feathers* (1932). The school's football rival was "Darwin College," which at the time was equally fictitious.

Its first building was the "Old Granary," which had been converted into a private residence years before by Professor George Darwin, the distinguished astronomer and son of Charles. After his death, it was donated to the university by the Darwin family as the nucleus of the new college.

About 40 years before its actual founding, a fictional "Darwin College" was featured in the classic Marx Brothers comedy *Horse Feathers* (1932). The film's plot revolves around a football game between two rival schools: "Darwin College" and "Huxley College."

DARWIN CORRESPONDENCE PROJECT
Organizing Fifteen Thousand Letters

Darwin could not throw anything away. An inveterate collector of beetles and natural history objects since childhood, he also saved thousands of letters he received over the years. In addition, his family and friends kept nearly every scrap he wrote to them. As he developed his theories, he exchanged letters and requests for information with naturalists, travelers, and missionaries in every part of the world. About 15,000 letters to and from Darwin have become the core of one of the most ambitious scholarly projects ever undertaken.

In 1974, the American historian Frederick Burkhardt invited the Cambridge zoologist and literary scholar Sydney Smith to help with the immense task of gathering and organizing the thousands of letters for publication. A former president of Bennington College and president emeritus of the American Council of Learned Societies, Burkhardt gathered a team of scholars to preserve the correspondence and make it accessible to future generations.

The Darwin Correspondence Project has undertaken to retrieve, catalog, transcribe, annotate and publish both sides of the entire correspondence, of which about half were written by Darwin. It is located both at Cambridge University Library, England, its headquarters, and at the American Philosophical Society in Philadelphia. The late psychologist and Abraham Lincoln scholar C. A. Tripp contributed to the Project a program for immediately locating any word or phrase in Darwin's collected works and letters. In addition to their value as a history of Darwin's scientific work and as a snapshot of the time and culture in which he lived, many of the letters reveal the naturalist's humanity and sense of humor. Flashes of his wit are evident in his remark that a boring lecturer was "so very learned that his wisdom has left no room for his sense."

In the 1870s, the newly invented telephone became a fad among the well-to-do, and many rushed to have one installed. To our everlasting benefit, Charles Darwin refused to allow one in his home and continued writing and receiving letters to the end of his life.

> **About 15,000 letters to and from Darwin have become the core of one of the most ambitious scholarly projects ever undertaken.**

DARWIN FISH
Bumper Sticker Wars

Fishes are ancient symbols of Christ, whose disciple Peter—a fisherman by trade—was also a "fisher of men" for the new religion. The Greek word ΙΧΘΥΣ (*ichthys*), meaning fish, is also an abbreviation for "Jesus Christ, God's Son, Savior." During the 1970s, a simple fish logo began to appear on the bumpers of American automobiles, identifying the motorists as adherents to the literal truth of the Bible, especially its account of the Creation.

In 1987, during a discussion of the perennial creation-evolution controversy, Hollywood special effects artist Chris Gilman joked to friends that he was thinking of attaching a Darwinian fish to his bumper—one that had sprouted feet—to counter the blatant "religious advertising" he encountered daily on the freeways. Egged on by his buddies, Gilman created a prototype, affixed it to his car, and was soon hounded with requests for the stickers. The demand led to the creation of a company called Evolution Designs, which began to market them. Creationists responded by offering a "Truth Fish" swallowing the Darwin fish, and the battle was joined.

Soon the fish bumper stickers proliferated and evolved into the rotund Buddha fish, the (smoking) Rasta fish, the Jewish Gefilte Fish, and a dozen others. As the "fish wars" escalated during the 1990s, the company added such sticker slogans as "When teaching evolution is outlawed, only outlaws will evolve."

Today the "fish wars" have become a well-established phenomenon on American highways. Some drivers display the stickers in the spirit of light-hearted humor, while others regard them as part of a serious "culture war."

One California preacher has annoyed his fellow fundamentalists by proclaiming himself and his handful of followers the only "true Christians" in the country. He swears that his given name—and he doesn't like it—is Darwin Fish.

See also CREATIONISM, FUNDAMENTALISM

BUMPER STICKER WARS began in the 1980s, when secular motorists countered piscatorial religious symbols with satirical "Darwin fish."

DARWIN MUSEUM

See DOWN HOUSE; DARWIN MUSEUM OF NATURAL HISTORY (MOSCOW)

DARWIN MUSEUM OF NATURAL HISTORY (MOSCOW)

Moscow's Darwin Museum (now the State Darwin Museum of Natural History) was founded In 1907 by zoologist Alexander F. Kohts (1880–1964) and his wife, primatologist Nadezdha N. Ladygina-Kohts (1890–1963), as the world's first museum dedicated to evolution. Permanent exhibits cover Darwin's life, evolution, zoogeography, natural selection, biology, anthropology, and zoology.

INSPIRED BY DARWIN, Alexander Kohts and N.N. Ladygina-Kohts (bottom) taught biology to young women and founded the Moscow Darwin Museum. Their Hall of Early Man was packed with works of art but no actual fossils (top). Sitting among his sculptures is their creator, Basil Wattigan.

The core of the original museum was Kohts's private collection of preserved specimens of mammals, insects, and birds. A professor at the Moscow Higher Women's College, he used his own natural history specimens to train several generations of women biologists.

In 1935, the museum published Ladygina-Kohts's pioneering observations on the development of Joni, a juvenile captive chimpanzee. Following Darwin's lead from his book *The Expression of the Emotions in Man and Animals*, she compared the ape's instincts, emotions, play, and habits with those of her son Rudy as each developed. She worked with bold independence in the intellectual isolation of Stalinist Moscow, while American behaviorists were treating the human mind as a mechanical device rather than the product of primate evolution.

Owing to Kohts's interest in variation and selection, the museum's collecting philosophy has been different from those of American and British institutions. For instance, instead of showing only one or two "ideal" specimens of each species, as in American museums, the Moscow museum has specialized in collecting anomalies and variations within a species, to illustrate the principle of variation. For example, there are hundreds of individual stuffed foxes and wolves, showing the range within natural populations. The collection is unequaled for its examples of albinos, melanists, hybrids, and hermaphrodites—and the world's largest collection of aberrant mammals and birds. It also has a fine collection of recently extinct birds, including a dodo, great auk, passenger pigeon, and huia, which can be seen in only a few museums in the world.

A distinctive feature of the State Darwin Museum is its extensive use of wildlife art. Kohts believed that wildlife art facilitates the public interest in biodiversity and its conservation. He also commissioned sculptors and painters to recreate prehistoric animals, imaginative scenes of early humans, and statues of *Homo erectus* for a hall of early humans that was possibly the first of its kind.

Although the government had officially promised Kohts in 1926 that it would construct a new museum building, funding was repeatedly postponed on one pretext or another. For decades, a few devoted staff members stored the collections in their homes. Kohts spent most of his life hoping the government would come through and enduring years of broken promises. In 1960, near the end of his life, Kohts lamented that "because of sad reality, all these expectations had not been realized." However, ten years after his death, in 1974, construction of the building finally began. In 1994, the collection was moved to the new quarters on Vavilova Street, where the museum presently stands.

(CHARLES) DARWIN RESEARCH STATION
Conservation of the Galápagos

The Galápagos Islands, 600 miles west of Ecuador, have become famous not only for their profusion of natural wonders, but also for their pivotal role in inspiring Charles Darwin's theory of evolution. In 1836, when he explored these islands as an exuberant 26-year-old naturalist, Darwin encountered a giant tortoise, straddled its shell, and took a brief ride.

Galápagos means both "saddle" and tortoise in Spanish, because of the saddle-shaped carapace (upper shell) of the centenarian reptiles. Unhappily, in the century and a half since Darwin's famous visit, the islands' fragile ecosystems have been frequently damaged and threatened. Even in Darwin's day, thousands of giant tortoises were taken from the islands to stock ships' larders, resulting in the extermination of several species. Of the original 15 subspecies, five have recently become extinct. One variety is represented by only a single surviving individual at the Darwin Research Station. [See LONESOME GEORGE.]

HATCHING TORTOISE at the Darwin Research Station on Santa Cruz. The captive breeding program has returned thousands of tortoises to the islands on which they were almost eradicated.
© Heidi Snell/Visual Escapes.

Ecuador's government, which has designated most of the 3,000 square miles of the archipelago a national park, promotes conservation and attempts to stem the steady stream of immigrants from the mainland. Nevertheless, the ecosystem is presently burdened with 30,000 human inhabitants, most of them on San Cristóbal, Santa Cruz, Isabela, and Floreana Islands. Some 125,000 tourists visit each year, including many who travel on large cruise ships that impact the area.

The Charles Darwin Research Station, on Santa Cruz Island, was founded in 1964 to promote research, conservation, and the eradication of destructive plants and animals introduced by humans, such as goats and cats. Feral cats eat everything—sea birds and songbirds, eggs, lizards, and invertebrates. Donkeys, pigs, and goats graze on rare plants, causing loss of habitat. Since the 1960s, national park rangers and research station scientists have tried to eliminate these invaders that were inadvertently introduced by humans.

The Darwin Research Station also operates a famous tortoise nursery, where eggs are hatched and the young nurtured and raised in a sanctuary. Rare species have been rescued from the brink of extinction when hundreds of grown hatchlings were returned to their ancestral islands. In addition, the research station trains naturalist guides, who in turn teach tourists about Darwin's "eminently curious little world within itself." Visitors to the facility are permitted to get close to—but not touch or ride on—some of the venerable chelonians for which the islands are named.

See also DARWIN'S FINCHES; GALÁPAGOS ARCHIPELAGO

DARWINIAN MEDICINE
Ongoing Evolutionary Battle

One of the greatest medical discoveries of all time took place in 1929, when Alexander Fleming noticed that some *Penicillium* mold spores had blown into a laboratory dish of bacteria cultures and killed off all the surrounding bacteria. Realizing the importance of the accidental experiment, Fleming had discovered a powerful new group of medicines—antibiotics—that could cure many bacterial diseases.

In their enthusiasm for the new "wonder drugs," however, doctors prescribed them indiscriminately and in massive dosages. The eventual long-term outcome was not surprising. When a bacteria population is doused with massive doses of antibiotics, the medicine exerts a powerful selective force. Billions of bacteria are killed, curing the patient, but some resistant bacteria survive. At the point of "cure," the treatment is stopped, allowing the resistant strains to multiply and spread to another host.

Bacteria routinely evolved resistance to the antibiotic drugs, just as crop insects become immune to overused pesticides. After almost a century of being bombarded with antibiotics, virulent resistant strains of bacteria have become a common problem. Indeed, some diseases that were almost eliminated have become incurable again.

In a series of books beginning in the 1980s, R. M. Nesse and G. C. Williams have attempted to make the medical profession aware that absolute and final cures for bacterial diseases do not exist. They have promoted the idea that a physician must make intelligent, strategic choices of treatment—always mindful of the evolutionary phenomena that are crucial to long-term success. Some minor ailments should not be treated, as they stimulate natural immunities. One can only win temporary victories as the combatants (and their human hosts) continue to evolve immunities, resistances, and new toxins.

DARWINISM
Complex Cluster of Ideas

Darwin often spoke of "my theory" as if it were a single idea. Most historians of science agree that the core of Darwinism consists of two major ideas: the fact of evolution and the major mechanism of evolution he proposed, natural selection.

Other features implied in Darwin's thought are often neglected. As Harvard evolutionist Ernst Mayr pointed out, Darwinism also includes the concepts of common descent, gradualism, and multiplication of species.

Common descent is the idea that several related species trace back to a common ancestor from which they branched off. This idea cleared up puzzling similarities between various creatures, resolved the vague problem of archetypes or "common plans," and gave a new dimension to the classification of organisms into families, genera, and species. Formerly puzzling similarities of structure in families and genera "made sense" if they shared a common ancestor.

Gradualism is the idea that evolution proceeds slowly and continuously, rather than by leaps or jumps. Some of Darwin's supporters, particularly Huxley, did not wish to burden the theory by insisting on gradualism. After all, no one had actually seen a species formed, and the fossil record looked more jumpy than continuous.

Mayr believed Darwin's insistence on small changes working over a long period was necessary to counter prevailing ideas of special creation or supernatural intervention. In addition, Darwin's geological research on coral reefs and admiration for Lyell's uniformitarian geology came into play. If a great coral reef 1,000 feet thick could be produced over the centuries by tiny polyp animals, it was possible that species could be gradually transformed by small, steady forces acting slowly over immense periods of time.

Darwin was impressed with the "creative force" of species radiation in the Galápagos Islands, producing so many different but closely related animals and plants in a relatively short time. He was among the first to reject the view that "all places in the polity of nature are filled" and that new species only arose to replace those that became extinct. But although he helped establish that diversification of closely related species is a fact, the mechanism that caused it remained a puzzle to him. Fifty years after his death, population geneticists came up with plausible models of speciation, or how the multiplication of species occurs.

See also DIVERGENCE, PRINCIPLE OF; EVOLUTION; GRADUALISM; NATURAL SELECTION

POP DARWINISM inspired popular kitsch such as this pensive chimp seated on *Origin of Species*, once common to knick-knack shelves.

DARWINISM, NATIONAL DIFFERENCES IN

Charles Darwin's theory of evolution, as George Bernard Shaw observed in his preface to *Back to Methuselah,* "had the luck to please everybody who had an axe to grind." Interpretations of Darwinism were called on to support a bewildering variety of political beliefs, social theories, and conflicting ideologies, differing from nation to nation.

England was of two minds about Darwinism. On the one hand, natural selection justified

Britain's imperial colonial policies, but Huxley argued that a civilized people should rebel against the evolutionary past and seek a more humane and compassionate vision than "survival of the fittest." Natural selection itself was eclipsed for a while by the neo-Lamarckians but revived with renewed force after 1900 by neo-Darwinians in England, who were discovering it was compatible with the new population genetics.

German Darwinism was shaped by Ernst Haeckel, who combined it with anticlericalism, militaristic patriotism, and visions of German racial purity. He encouraged the destruction of the established church in Germany, with its sermons about "the meek shall inherit the earth" and compassion for unfortunates. Such a "superstitious" doctrine would lead to "racial suicide." During the 1930s, Adolf Hitler believed he was carrying Darwinism forward with his doctrine that undesirable individuals (and inferior races) must be eliminated in the creation of the New Order dominated by Germany's "Master Race." [See LEBENSBORN MOVEMENT.]

In the United States, Social Darwinism was advocated by laissez-faire economists like Yale's William G. Sumner, who found it compatible with free competition and rugged individualism. Philosopher Herbert Spencer, the prophet of progress, became more popular in America than Darwin. His conviction that unregulated competition is the key to progressive social evolution appealed to industrial capitalists.

Yet America also had a strong resistance to wholly "mechanistic" explanations of the origin of humans. Even Darwin's Harvard champion, botanist Asa Gray, could not imagine evolution without goals, without being directed by God. Many American biologists who retained their religious beliefs followed Alfred Russel Wallace in assigning to humankind a special spiritual origin not shared with animals.

French science resisted Darwinism for decades, perpetuating theories of Lamarckian evolution; national pride in the first evolutionary biologist being a Frenchman contributed to this stance. Influenced by the tradition of ideal types and classification, many scientists failed to see why Darwinism was so revolutionary outside France. Besides, how valid could it be if it had been become part of the ideology of resurgent German militarism?

Church influence remained strong in French education for much longer than in England. But France was also a treasure trove of human and animal fossils, prehistoric stone tools, and the stunning Paleolithic cave paintings. Before long, a few outstanding prehistorians and paleontologists arose from the ranks of churchmen, particularly from the Jesuit order. [See TEILHARD DE CHARDIN, FATHER PIERRE.]

Darwinism was welcomed in Communist countries, since Karl Marx and Friedrich Engels had considered *Origin of Species* (1859) a scientific justification for their revolutionary ideology. As far as socialist theorists were concerned, Darwinism had proved that change and progress result only from bitter struggle. They also emphasized its materialist basis of knowledge, which challenged the divine right of the czars.

An opposing branch of Russian Darwinism, led by Prince Peter Kropotkin, argued that "mutual aid" (and therefore socialism) was also a natural principle of evolution. The ideal society was based not on competition, but on cooperation. Individual needs were less important than those of the larger social entity.

Long after Lamarckian inheritance had been abandoned elsewhere, Russia stubbornly retained this 19th-century belief in the inheritance of acquired characteristics. Party theorists refused to accept that each generation must be educated anew, believing socialism would create permanent genetic transformations in the population.

Under Trofim Lysenko's dominance of Soviet science, "Mendelist" genetics was a forbidden doctrine, a bourgeois heresy. Lysenkoism was finally abandoned in the 1960s, but only after Lysenko's fraudulent research brought on agricultural disaster, which threatened the country with starvation.

Depending on a nation's culture, economic system and political history, Darwinian ideas shared the same fate as many religions and ideologies. They were interpreted so broadly that they could encompass anything. Identified with progressive change toward a final social good, Darwinism took on the qualities of each nation's history, self-image and aspirations.

See also CHINA, DARWINISM IN; ENGLAND, DARWINISM IN; HAECKEL, ERNST; KROPOTKIN, PRINCE PETER; LAMARCK, JEAN-BAPTISTE; LYSENKOISM; MARXIAN "ORIGIN OF MAN"; MONISM

Depending on a nation's culture and politics, Darwin's ideas were interpreted to support many contradictory causes.

"DARWIN'S BULLDOG"
Thomas Huxley Unleashed

Charles Darwin hated the thought of publicly defending his controversial theory of evolution. But his pugnacious friend, the zoologist Thomas Henry Huxley, loved nothing better than a war of wits. Temperamentally the opposite of Darwin, Huxley relished public confrontations with critics, a trait that earned him the nickname "Darwin's bulldog."

It was Huxley who first reviewed *Origin of Species* (1859) in the *London Times*, declaring it a "solid bridge of facts . . . [which] will carry us safely over many a chasm in our knowledge." His impromptu debate with Bishop Samuel Wilberforce at Oxford over the *Origin* has come down in history and folklore (somewhat inaccurately) as a milestone battle between "science and religion."

Certainly Huxley enjoyed deflating church authorities when they made pronouncements on scientific issues. He used to tell his anatomy students that they could remember the mitral valve, shaped like a bishop's miter, is on the left side of the heart "because a bishop is never known to be on the right."

Darwin was well pleased with Huxley's aggressive campaign to win over public opinion. In 1860, just after Huxley had bested Wilberforce, Darwin stressed the "enormous importance of showing the world that a few first-rate men are not afraid of expressing their opinion. . . . I see daily more and more plainly that my unaided book would have done absolutely nothing."

One student in Huxley's anatomy class at London College was the American Henry Fairfield Osborn, later to become the president of the American Museum of Natural History. One memorable day in 1879, Osborn recalled, the reclusive Darwin paid a rare visit to the classroom. The 22-year-old Osborn was thrilled when he was singled out to meet Darwin, who was then 70. After they exchanged a few words, Huxley hurried the great naturalist into the next room, saying, "I must not let you talk too much." Years afterward, Osborn recalled his sometimes formidable professor's touching solicitude toward his older friend. "You know, I have to take care of him," Huxley explained. "In fact, I have always been Darwin's bulldog."

See also HUXLEY, THOMAS HENRY; OXFORD DEBATE

DARWIN'S FINCHES
Classic Case of Speciation

When Charles Darwin explored the Galápagos Islands as a ship's naturalist in 1835, he was struck by the peculiar species he found there and made notes on their behavior and distribution. One group of birds, the Galápagos finches, is often cited as a striking illustration of the evidence for speciation he discovered; they have become famous as "Darwin's finches."

Adapted to exploit different niches and habitats in the islands, the Galápagos finches diverged from one mainland species into many. Ornithologist David Lack's study of Darwin's finches in the 1940s established the view that the birds were a crucial stimulus to Darwin, convincing him that divergent evolution occurs as a result of geographical isolation. Darwin, the well-known legend goes, realized these birds must have descended from a South American finch species that had somehow reached the islands. Adaptations to various niches in the new environment split the expanding population into varieties, then into species.

During the early 1980s, however, Frank Sulloway of the University of California at Berkeley reexamined the evidence and showed that even though the birds remain a classic example of island speciation, Darwin did not, at first, recognize them as such. When he collected them, Darwin thought the beak structures of the finches appeared so different that he did not recognize how closely related they were. After the *Beagle*'s return, it was ornithologist John Gould of the Zoological Society of London who identified all the species as a closely related group.

What is remarkable about the little gray or black finches is how minimal—yet how striking—are the differences between them. All are rather undistinguished sparrowlike birds

with thick pointed bills, but the various Galápagos populations have evolved very different feeding adaptations.

Some are adaptations in size or strength of beak to handle different kinds of foods, while others involve behavioral specializations as well. On Hood Island, a large ground finch with a stout beak and sturdy legs rolls over stones to find the food underneath. On Wenman Island, a sharp-beaked finch creeps up on nesting seabirds that are incubating their eggs, pricks their skin near the tail feathers, and, vampirelike, drinks their blood. Ferdinandina Island contains a local population of small ground finches that leap onto the backs of basking marine iguanas and clean them of ticks.

Perhaps the most remarkable of Darwin's finches is one of the only birds to use tools: the woodpecker finch. Since it lacks a woodpecker-like beak but feeds on bark insects, it holds a twig or thorn in its bill, which it uses to poke grubs out of their holes. Then it keeps its hunting tool, holding it in one foot, while it eats its catch.

Darwin did not really form his views on speciation until well after the *Beagle* voyage. Because he didn't recognize the significance of the finches, he neglected to label his specimens to document from which islands each bird had come. He later kicked himself for the omission and, similarly, for not bothering to note the islands from which he had gathered the giant tortoise shells in his collection. (It was only upon leaving the Galápagos that he was struck by the remark of the vice-governor, who claimed he could tell from which island any tortoise shell was taken by its pattern and shape.)

Back in England, after conferring with John Gould, Darwin belatedly realized the problem and tried to reconstruct the patterns of finch distribution from other collections. What made the situation even more difficult was the lack of a candidate for a hypothetically ancestral finch among the living birds of South America.

There was, however, a group of birds in the islands that did give Darwin a clue to evolution: the Galápagos mockingbirds. Although these birds were less dramatic in their adaptations than the finches, Darwin was able to identify several island species that clearly resembled their American cousins. Once he recognized that the islands harbored distinct varieties allied to a mainland form, he realized he had evidence in favor of speciation from a common ancestor—rather than separate divine creations for each small island.

Since the late 1970s, Peter and Rosemary Grant of Princeton University have made intensive, long-term studies of those Galápagos finches that live in very simple ecosystems—islands barren of almost all but finches, plants, seeds, and a few predators. They have caught and released more than 19,000 individual birds from 25 generations. The Grants measured their beaks and wings, and experimented with how much force it takes to crack the various seeds found on the island. Eventually, they were able to demonstrate that half a millimeter of beak could make the difference between survival and starvation.

Changes in beak size and form, it turned out, were highly heritable. The Grants were able to watch the beaks change in size and shape over the months and years, first in response to three seasons of drought, when only large, tough seeds remained on the islands, to years that saw the little islands buffeted by storms and soaked by heavy rainfall. After each climatic event, the differential survival and reproduction of birds on the islands changed dramatically, and their beaks changed in size and shape. The Grants also observed natural hybridization between species of the birds—a somewhat unexpected process that also fostered short-term evolution of new species.

In 2003, biologists Jeffrey Podos of the University of Massachusetts at Amherst and Stephen Nowicki of Duke University wondered: If beaks can change rapidly, wouldn't that affect the sounds of birdsong? And wouldn't changes in birdsong in turn affect breeding patterns? Their preliminary studies showed that Galápagos finches, like other songbirds, do indeed produce different sounds and songs according to the size and shapes of their beaks. Since the songs are used to attract mates, the researchers' preliminary work has led them to question whether "the functional link between beaks and song may have contributed to the process of speciation and adaptive radiation in these birds."

See also ADAPTIVE RADIATION; GALÁPAGOS ARCHIPELAGO

IN A CLASSIC EXAMPLE of divergence of closely related species, some finches of the Galápagos Islands are adapted for feeding on large seeds, others on small seeds, and some for digging insects out of bark. Darwin at first missed their evolutionary significance. © Tui de Roy/ Roving Tortoise Photos and Edward S. Barnard.

"DARWIN'S LIZARDS"
Adaptive Radiation in Quadruplicate

Galápagos finches often serve as the classic example of adaptive radiation—many kinds of related creatures evolving rapidly from a small founding population on small islands. Darwin said he could scarcely believe that islands so close to each other as the Galápagos could be "so differently tenanted." In the 1990s, an even more striking case of adaptive radiation was found among the little anole lizards, which are sometimes called American chameleons.

Lizards of the genus *Anolis* have evolved many species on the Greater Antilles islands, which include Cuba, Jamaica, Puerto Rico, and Hispaniola. One anole that inhabits the tree-tops developed large toe pads with which to grip leaves, and changes color to blend into the foliage. Another, shorter variety has evolved a slender tail and legs and balances easily on narrow twigs. Yet another, with a stocky body and long limbs, is a swift ground runner. Just as the finches of the Galápagos evolved beaks suited to various foods, so these lizards have adapted to feeding at different heights in the forests.

What makes them spectacular examples of evolutionary adaptation, however, is that each of the four islands has its own parallel species of habitat specialists. Each island has its own unique species of lizard with big toe pads living in the canopy, its own twig-dwelling slender lizard, and its own long-limbed sprinters on the forest floor.

When Todd R. Jackman of Villanova University, Allan Larson of Washington University, and Kevin de Queiroz of the Smithsonian Institution analyzed the DNA of the Antillean lizards, they made an intriguing discovery. Lizards in a particular niche are not closely related to the lizards filling that same niche on the other islands; the habitat specialists evolved independently on each of the four islands. The anoles demonstrate convergent evolution of an entire ecological community—in quadruplicate.

Genetic tests showed that these differences in body form or phenotype are not hard-wired into the species' genome. Lab studies confirmed that the various kinds of lizards' bones and muscles develop differently during their lifetimes, depending on the strains and stresses they experience during growth and development. This phenotypic plasticity seems to anticipate an adaptive radiation like those of Hawaiian honeycreepers or Galápagos finches.

A major question, of course, is how these phenotypic adaptations become incorporated into the genome and passed on to future generations—an unsolved riddle in our present understanding of the origins of adaptations.

See also ADAPTATION; CONVERGENT EVOLUTION; DARWIN'S FINCHES; HAWAIIAN RADIATION

DAVID GRAYBEARD (CHIMPANZEE)
First to Contact Humans

A chimpanzee with gray chin-hair, David Graybeard was the first wild ape to fully accept the presence of a human scientist with tolerance and friendly curiosity. In 1960, Jane Goodall had already spent months tracking chimps at the Gombe Stream Reserve in Tanzania, but they only let her get close enough to observe them through binoculars.

David Graybeard changed that by deciding to trust her. He was the first to voluntarily visit her camp and to let her touch him. Soon after, others followed suit until she was accepted by the group.

Carrying on with his normal activities despite Goodall's presence, he allowed her the first observations of meat eating by apes. From David Graybeard she also got her first look at the now famous "termiting" behavior. He stripped leaves from a twig, wet it in his mouth, then poked it into the mound of hard clay to capture the thirsty insects on his "termite popsicle."

Inspired by Goodall's experiences with chimps, Dian Fossey sought out mountain gorillas on their forested peaks and, in time, became similarly accepted, beginning with a gorilla she called Digit. But it was David Graybeard, no less than Jane Goodall, who pioneered changing the entire relationship between humans and free-living apes from fear to trust.

See also "APE WOMEN," LEAKEY'S; CHIMPANZEES; DIGIT

DAWKINS, RICHARD (b. 1941)
Darwin's Rottweiler

Richard Dawkins, one of England's preeminent lecturers and writers on Darwinian evolution, claims that as a child he had no interest in the natural world nor in science. Whereas many child naturalists grew up collecting beetles or fossils, Dawkins's imagination was captured instead by a classic children's book: the whimsical, fantastic tales by Hugh Lofting about the remarkable Dr. Dolittle.

The good doctor was a Victorian gentleman who regularly conversed with mice and parrots and whales. An adventurous sort, he traveled the world to learn the secrets of faraway places. When the adult Dawkins encountered the life and works of Charles Darwin, he welcomed him as an old friend and hero of his youth. Dr. Dolittle and Darwin, he opines, "would have been soul brothers."

Raised in Kenya by English parents, Dawkins studied at Oxford, graduating in 1962, then continued his studies there with the Dutch ethologist Niko Tinbergen. A pioneer in developing the new science of animal behavior, Tinbergen was concerned with how animals communicate and how their instinctive behaviors had evolved—and how these behaviors constitute adaptations for survival. Dawkins later characterized animals as "survival machines" for the genes they contain, replicate, and pass on.

In his international bestseller *The Selfish Gene* (1976), Dawkins focused on the primary role of genes in evolution. He questioned how genes communicate, why they cooperate, and how they compete. His view of biology was as a flow of information, which he later applied to human culture. Like genes, ideas (he called them "memes") could also compete, mutate, and replicate and are subject to a form of natural selection.

For several years Dawkins engaged in debates with the Harvard paleontologist Stephen Jay Gould in book reviews and articles. Aside from their prima donna competition as audience-pleasing writers and speakers, they raised some substantive issues. Dawkins came from a tradition of British evolutionists whose most important topic was adaptation. Gould criticized those biologists who uncritically attributed everything to adaptation, accusing them of concocting "just-so" stories. Also, he emphasized contingent history and the role of chance, such as mass extinctions, in determining which species survived in the long run. To Gould, the individual organism was the target of selection; to Dawkins it had always been the genes, first and foremost.

They also clashed in their views of science. Gould considered science incomplete, embedded in social context, and respected religion as a "non-overlapping" realm of knowledge and ethics. Dawkins, an uncompromising atheist, considered Gould a hypocritical "appeaser" of religious sensibilities and promoted his own Enlightenment belief in science as the only method capable of tackling the enigmas of nature. In recent years, Dawkins has extended his naturalistic skepticism to exposing fraudulent psychics, telepaths, and quack healers.

See also GOULD, STEPHEN JAY; "JUST-SO" STORIES; "SELFISH GENE"

RICHARD DAWKINS

DEAN, BASHFORD (1867–1928)
Collector of Fishes in Shining Armor

Bashford Dean became the world's greatest expert on both armored fishes and military armor. As curator of fishes at the American Museum of Natural History and curator of arms and armor at the Metropolitan Museum of Art, Dean built world-class collections in both institutions. His energy was prodigious. While acquiring and cataloging the collections and preparing exhibits, he also taught ichthyology to a generation of students at Columbia University, where he was professor of zoology.

Dean's unique career resulted from his pursuit of two childhood interests. While growing up in a well-to-do New York City family, he had been entranced with books about King Arthur's Round Table, and at age eight was drawing pictures of spirited knights in combat. His father, a skilled amateur birdwatcher and collector of fossil invertebrates, imparted to him a fascination with the natural world as well.

HEAVY METAL fan Bashford Dean was an expert in the evolution of armored fishes.

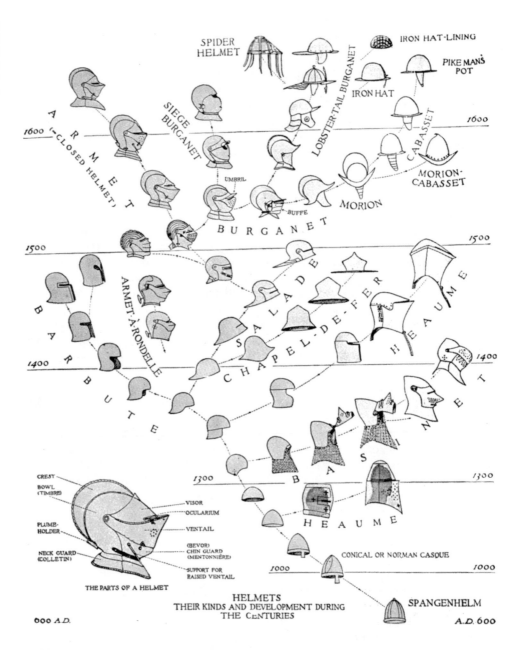

THE EVOLUTION OF ARMOR was worked out by Bashford Dean, who was also an expert on armored fish. He saw strong parallels between biological and cultural evolution.

After a productive career in ichthyology, during which he published over 100 journal articles, Dean gradually allowed his interest in armor to take over his life and devoted his last years to building the armor collection at the Metropolitan Museum. He could spot the characteristics of each armorer and was assiduous in establishing the history and authenticity of every object. In a field plagued with forgeries, he strove to keep the museum's collection free of counterfeits. He scoured Europe for new acquisitions, and persuaded several wealthy patrons to donate their armor collections to the museum. An expert artisan himself, Dean taught himself to repair old arms and armor with the finesse of a master.

Dean wrote hundreds of scientific and historical journal articles on both fish and armor, and compiled massive bibliographies on both. In search of the early developmental stages of the hagfishes and sharks, he traveled to Alaska, California, Japan, Norway, Russia, Italy, and Portugal. His work in ichthyology included monographs on sharks' embryology and development. On weapons, he wrote on the historical development of helmets, breastplates, gauntlets, shields, pole arms, swords, crossbows, guns, and spurs.

After working out "family trees" in the evolution of armored fish, Dean sought to find similarities in the history of armor. The anatomist William King Gregory noted that "his drawings of the evolution of various lines of helmets and pole arms in Europe strongly recall the phylogenetic tree of certain groups of fossil cephalopods." In both cases, according to Gregory, an ancient, primitive pattern gives rise to gradually diverging series that become "extreme" in old age, then go extinct. (Evolution of species used to be likened to an individual's cycle of birth, maturity, senescence, and death—a model that has long since been abandoned in science.)

In Gregory's view, Dean's jump from organic to technological evolution was convincing. "Dean showed," wrote Gregory, "that the evolution of arms and armor is only a sort of biological evolution once removed, the conservatism of armorers and purchasers taking the place of heredity in tending toward continuity, and the stress of environment evoking mutational changes in type."

During the First World War, in 1917, Dean was asked to lend his unique talents to the war effort. As a major in the Ordnance Department, he invented effective body armor that was adopted by aviators. His knowledge of armor enabled him to design a classic piece of battle gear: the U.S. infantry helmet.

DECEPTION, EVOLUTION OF
Origin of Lying

Because humans have the unique ability for spoken language, it is commonly thought that our species is the only one capable of lying. But deception is not confined to speaking falsehoods. Other strategies of deception have evolved in many diverse organisms, including plants, insects, birds, dogs, and apes. New observations of baboons, chimps, and gorillas especially seem to show deliberate intentions to deceive.

Misleading colors and structures are common in nature, though they have evolved by selection, not as conscious deception. Viceroy butterflies have come to resemble poisonous monarchs as a defense against predators, just as harmless king snakes mimic venomous coral snakes. Some orchids fool male wasps into "mating" with them by imitating female wasp odor, which ensures the pollination of the next flower the males visit. Other plants attract flies by giving off a false scent of rotting meat. [See MIMICRY.]

Many creatures raise appendages or feathers, or puff themselves up with air, to appear larger and more fearsome when attacked. When shrimps are molting—and defenseless without their shells—they rush straight at predators as if to attack, though they are incapable of biting. Bluff charges are also common in rhinos and gorillas, and these animals are perfectly capable of following through. But they would rather not expend more energy or risk possible injury, when a bluff is usually enough to scare away an intruder.

The stilt, a long-legged wading bird, acts wounded to lure predators away from her nest and chicks. Using a strategy adopted by several kinds of birds, she flaps around near the ground as if her wing is broken, always just out of range of the foe. When she has led the enemy a safe distance away from the nest, she suddenly takes off and flies away. Just how conscious the bird is of this deception is unknown. Some naturalists have called it an automatic "fixed-action pattern," while others give more credit to the animal for a deliberate and skillful performance.

Robert Mitchell, a researcher at Clark University, is particularly intrigued by animals' misleading behavior that appears to be deliberate. For instance, one chimpanzee in a free-ranging group sometimes gave a "food call," indicating that there were bananas nearby. When the others ran off in the indicated direction, the "lying" chimp went the opposite way, where it knew the food was really hidden. "To my amazement," wrote Mitchell, "I learned that in some animal species deception may be more common than truth-telling."

Mitchell also observed that animals in contact with people sometimes fake injuries to get attention, just as a child might. A female zoo gorilla pretended to have her arm hopelessly stuck between the bars of her cage—a ploy to attract the keeper over for a hug. He also observed a dog that had broken its leg and became used to extra sympathy from its master, which ended when the leg healed. Thereafter, it faked a pitiful, convincing limp when it wanted attention. While Mitchell describes that behavior as "deception," others might see

Animals in contact with people sometimes fake injuries to get attention, just as a child might.

in it only a "conditioned response." Limping behavior was learned and reinforced by petting, like the association Pavlov's dogs made between food and the dinner bell.

Richard Byrne and Andrew Whiten, psychologists at the University of St. Andrews in Scotland, found what they believed to be deliberate deceptive behavior among baboons. They observed a youngster who waited for an older animal to dig up a juicy root, then screamed as if it were being attacked. The juvenile's mother came running over and drove the industrious animal away, and then the juvenile ate the root. According to the investigators,

> [We both] saw this young baboon go through the same routine with different "victims" on different days. . . . The behavior was not a coincidence but a tactic. [He] was not genuinely scared [but his] deception of his mother gave food that he could not have gotten in any other way.

Another of their observations of "tactical deception" among baboons concerned a hefty adolescent they named Melton, who had bothered a juvenile until it screamed. Several adults, including the young one's mother, came tearing over, heading straight for Melton.

> Instead of fleeing or showing submission, he immediately stood on his hind legs and scanned the distant hillside—in exactly the way that these baboons behave when they have seen a predator or a group of foreign baboons. The pursuers skidded to a halt and looked intently in the direction Melton was staring; they never resumed the chase. . . . [There was] no genuine cause for alarm; the "outside threat" was a fiction.

Jane Goodall observed a chimpanzee move out of sight of hidden food, so that a second chimp couldn't even see him look at where the food might be hidden. When the second chimp went away, the first immediately went over and retrieved his banana.

Goodall's Gombe chimps produced an even more striking incident, observed by Frans Plooij. A male was zeroing in on some bananas he had hidden, when a second showed up. The first walked away, sat down, and looked all around—everywhere but where the food was hidden. Then the second pretended to leave, but hid behind a tree to watch the deceiver. When he uncovered the stashed food, the hiding male showed himself and grabbed it away.

Byrne and Whiten suggest that if "fixed action patterns" or "conditioning" cannot explain such an incident, we may have to conclude apes are capable of thinking such devious thoughts as "X thinks I think he doesn't know where the banana is, but I think he really does know, so I'll just wait and see where he goes."

Ability to deceive or mislead is therefore neither new nor unique to our species, but has evolved in a variety of creatures. Because it appears to have a "natural" origin doesn't make lying justifiable; it is not always effective, either. As Abraham Lincoln once observed, "No man has a good enough memory to be a successful liar all the time."

See also *EXPRESSION OF THE EMOTIONS*

DEGENERATION THEORY
Evolution To "Lower" State

A peculiar byway of evolutionary theory in the 1870s held that life forms could evolve in three directions. Species could remain relatively unchanged over time; they could become more elaborate and complex; or they could become simpler and "degenerate."

German zoologist Anton Dohrn was a strong champion of the last idea; his English friend E. Ray Lankester, director of the British Museum of Natural History, dedicated a little book on the subject to him in 1880. Their prime examples of degeneration were microorganisms and parasites, which they thought had "lost" the structures to move about or capture their own food. Others applied the concept to more complex animals perceived to have become simpler. Whales and snakes, for example, were described by one authority as "degenerate quadrupeds."

Charles Darwin's protégé Sir John Lubbock, who made pioneering studies of social insects, was intrigued by ants that kept aphids as "slaves." The ants had "lost" some of their mouth parts and were now totally dependent on the aphids for the sugary food they produced. A similar fate, he warned, was in store for degenerate human slaveowners, who would soon be unable to feed or care for themselves.

Lankester did not hesitate to apply the degeneration principle across the board, to everything from pond organisms to human history. "At one time," he wrote, "it was a favourite doctrine that [all] the savage races of mankind were degenerate descendants of the higher and civilised races." Although that idea "has been justly discarded . . . it yet appears that degeneration has a very large share in the explanation of the condition of the most barbarous races, such as the Fuegians, the Bushmen, and even the Australians." There was a lesson here, Lankester thought, for "the white races of Europe."

> It is well to remember that we are subject to the general laws of evolution, and are as likely to degenerate as to progress. . . . It is possible for us—just as the Ascidian throws away its tail and its eye and sinks into a quiescent state of inferiority—to reject the good gift of reason with which every child is born, and to degenerate into a contented life of material enjoyment accompanied by ignorance and superstition.

Lankester's antidote to human degeneration was knowledge of man's place in nature, so that "we shall be able by the light of the past to guide ourselves in the future. . . . The full and earnest cultivation of Science—the Knowledge of Causes—[is the best] protection of our race—even of this English branch of it—from relapse and degeneration."

Well-meaning though it was, Lankester's hysterical, quasi-religious faith in science as salvation, together with his fear of human degeneration, was to have many sinister echoes. Restrictive immigration laws and sterilization of the "unfit" in America were founded on a similar belief; it was also Nazi Germany's rationale for the extermination of "inferior races."

See also BUFFON, COMTE DE; LEBENSBORN MOVEMENT

"DELICATE ARRANGEMENT"
The Darwin–Wallace "Joint" Publication

Although Charles Darwin is usually credited with originating the theory of organic evolution by natural selection, he shares the discovery with a man few educated people could name.

Alfred Russel Wallace, a naturalist 14 years Darwin's junior, was ready to publish a crisp summary of the theory in 1857, before *Origin of Species* (1859) was written. How they came to publish together has become known as the "delicate arrangement."

Darwin had been working in secret for years, confiding his theory to only a few close friends. "Outsiders" (including his sometime correspondent Wallace) knew only that he was interested in the relationship between varieties and species.

As a field naturalist in the Malay Archipelago, Wallace studied the distribution of plants and animals; he was also struck by competition for resources among the native tribal populations. Having read the same book (Thomas Malthus's *On Population*, 1798) that inspired Darwin, he conceived the idea of evolution by natural selection during a malarial fever. Later, he wrote a paper on the subject (the so-called "Sarawak Law" paper, 1855) and mailed it to Darwin. If Darwin thought it worthy of publication, Wallace requested he pass it on to the influential geologist Sir Charles Lyell.

Lyell knew that Wallace was close to an evolutionary theory and had been warning Darwin to publish soon or lose the chance to be first. Receipt of Wallace's paper threw Darwin into a panic. He wrote his mentor Lyell,

> Your words have come true with a vengeance—that I should be forestalled. . . . So all my originality . . . [is] smashed. . . . I never saw a more striking coincidence. If Wallace had my MS sketch written out in 1842, he could not have made a better short abstract! Even his terms now stand as heads of my chapters.

It could not have come at a worse time. One of his children had recently died and another was sick with scarlet fever. Darwin, who was usually generous and fair, admitted he was filled with "trumpery feelings." He could not bear the idea of being scooped by anyone publishing "his" theory first. Yet he would rather burn his book, he said, than for anyone to believe he stole his ideas from Wallace or behaved in a "paltry spirit."

Without consulting Wallace, Lyell and Hooker arranged for a joint announcement of the

"Your words have come true with a vengeance. . . . So all my originality [is] smashed."

—Charles Darwin to Charles Lyell, 1858

theory at the Linnean Society of London in 1857. Since Hooker was England's leading botanist and Lyell its most eminent geologist, the pair had great influence in scientific circles. In 1858, they persuaded the Linnean Society to publish Wallace's essay along with extracts from two letters by Darwin (1842, 1844) in which he had sketched his views to friends. Darwin and Wallace were to share credit as codiscoverers of the theory of evolution by natural selection.

When Lyell wrote Wallace about their proposed course of action, Wallace wrote back with gratitude for his "favourable treatment."

The phrase "delicate arrangement" comes from a passage by Thomas Huxley's son Leonard in his *Life and Letters of Sir Joseph Hooker* (1918): "Wallace's paper had come like a bolt from the blue. . . . Yet . . . when this delicate situation had been arranged [Darwin wrote Hooker], 'You must let me once again tell you how deeply I feel your generous kindness and Lyell's on this occasion, but in truth it shames me.'"

Darwin then went furiously to work on the *Origin of Species* and completed it in 13 months, after almost two decades of putting off the writing. Years later, in his *Autobiography* (1876), Darwin claimed he had "cared very little whether men attributed most originality to me or Wallace"—a statement that still strains the credulity of even Darwin's staunchest admirers.

Darwin was apprehensive about Wallace's reaction on his return from Malaysia and was greatly relieved to discover his "noble and generous" disposition. Wallace modestly stated that his work on the problem was short compared with Darwin's decades of painstaking groundwork and never questioned his priority. ("I shall always maintain it is actually yours and yours only.") However, after Darwin's death, Wallace admitted he had "no idea" his paper had thrown the senior naturalist into such a panic.

To the end of his long life, Wallace insisted that being first to publish is meaningless if an idea makes no impact, saying that "my paper would never have convinced anybody." In *The Wonderful Century* (1898), he lauded Darwin's achievement in having established widespread acceptance of the evolutionary origin of species:

> The whole literary and scientific worlds were violently opposed to all such theories . . . [but] the greatness and completeness of Darwin's work [caused a] vast change in educated public opinion. . . . Probably so complete a [reversal] on a question of such vast difficulty and complexity, was never before effected in so short a time. It . . . places the name of Darwin on a level with that of Newton.

Journalist Arnold C. Brackman's notorious 1980 book *A Delicate Arrangement* accused Darwin of lifting his theory from Wallace and alleged that the joint papers were presented at the Linnean Society of London without Wallace's knowledge or approval, since he was then deep in the jungles of Malaysia. However, Darwin's great-grandson, the late Quentin Keynes, said that Brackman later told him he would never have written that book had he had access to a crucial letter from Wallace that had long been tucked away in Keynes's private collection.

That letter, sent from Ternate, Moluccas, to Sir Joseph Hooker on October 6, 1858, proves that the younger naturalist heartily approved of the joint publication and was grateful to be named as Darwin's co-author of the theory of natural selection. He wrote:

> I cannot but consider myself a favoured party in the matter, because it has hitherto been too much of a practice in cases of this sort to impute all the merit to the first discoverer . . . and little or none to any other . . . [who may have] arrived at the same results a few years or a few hours later. . . . I must again thank you for the course you have adopted, which while strictly just to both parties is so favourable to myself.

Though published in *The Correspondence of Charles Darwin* (vol. 7, 1991), the letter went largely unnoticed by historians until Michael Shermer called attention to it in his 2002 book *In Darwin's Shadow: The Life and Science of Alfred Russel Wallace*. Shortly before his death in 2003, Quentin Keynes bequeathed the letter to the Charles Darwin Trust, where it is now archived.

Historians now agree that Darwin's written sketches of the theory in 1842 and 1844 amply document his priority, and that there is no question of him ever having appropriated any of its content from Wallace.

See also "CRADLE OF MANKIND"; LINNEAN SOCIETY; "SARAWAK LAW"; WALLACE, ALFRED RUSSEL

DE MAILLET, BENOÎT (1656–1738)
Cosmological Theorist

Benoît de Maillet thought planet Earth had developed by natural processes, still observable today, and considered fossils "libraries" of ancient plants and creatures that "exist no more." He even had a rough idea about the succession of strata, the kinship of man and apes, and the origin of life from tiny organic atoms like those seen under the new microscopes—more than a century before Darwin.

However, his fumbling attempts were also mixed in with stories of mermaids and other fantastic myths. And though he proposed the "development" of species, his notions of evolution were crude. He saw flying fish, for example, as being well on their way to becoming birds.

Yet he insisted that if God were a watchmaker (in the favorite metaphor of the period), He did not have to continually intervene to keep his machines running. Instead, de Maillet wrote, the Creator "had skill enough to make a clock so curiously, that by the Disorder which Time should produce in her Parts and Movements, there should be new Wheels and Springs formed out of Pieces, which had been worn and broken."

His book *Telliamed: Or, Discourses between an Indian Philosopher and a French Missionary on the Diminution of the Sea, the Formation of the Earth, the Origin of Men and Animals, etc.* (1750), published posthumously, was translated from the French and widely read in English. Of course, there was no Telliamed. Philosophers who challenged church teachings often attributed them to fictional Oriental sages. "Telliamed" is "de Maillet" spelled backwards.

THE DESCENT OF MAN (1871)
Darwin on Human Origins

Only a tiny hint about human evolution appeared in Charles Darwin's *Origin of Species* (1859): "Light will be thrown on the origin of man and his history." Although Darwin had often thought about man, for 30 years his published writing stuck to animals and plants. With *The Descent of Man*, he finally let the other shoe fall.

By this time, evolution had won over the majority of botanists and zoologists from the idea that each species appeared instantly and separately. Others, more bold than he, had published pioneering works on human evolution in the wake of the *Origin*. Thomas Huxley's *Evidence as to Man's Place in Nature* (1863) and the German Ernst Haeckel's *Natural History of Creation* (1868) were among the most influential.

But the public was eager for Darwin's own version of human origins from apelike creatures, though he fully expected the book would "be denounced by some as highly irreligious." Still, he wondered, why is human evolution from "some lower form" any more shocking or incredible than the development of individuals from sperm and eggs? The fact that man has "risen to the very summit of the organic scale" instead of having been placed there from the beginning, Darwin wrote, "may give him hope for a still higher destiny in the distant future." But hopes and fears aside, Darwin concluded that the time had come to acknowledge that "with all his noble qualities . . . god-like intellect [and] exalted powers—Man still bears in his bodily frame the indelible stamp of his lowly origin."

Three great "groups of facts" could no longer be denied: similarities in structure (and, we now know, biochemistry) among members of the same groups, patterns of geographical distribution, and the worldwide succession of one group by another in the fossil record: "It is incredible that all these facts should speak falsely. He who is not content to look, like a savage, at the phenomena of nature as disconnected, cannot any longer believe that man is the work of a separate act of creation."

Darwin noted that man is not descended from any existing monkey or ape but from ancestors that would be recognizably apish. Darwin thought human ancestors had dagger-like canines—similar to those found today in male gorillas and baboons—which became reduced in size as stone tools and weapons were mastered. That idea, found in *The Descent of Man,* inspired bitter controversies a century later. A distorted view of baboon social behav-

> "With all his noble qualities . . . and exalted powers, man still bears in his bodily frame the indelible stamp of his lowly origin."
>
> —Charles Darwin, *The Descent of Man*, 1871

ior, based on the primacy of males with large canines, was promoted and widely accepted as the model for early man. [See BABOONS; HUNTING HYPOTHESIS.]

The Descent of Man and Selection in Relation to Sex (its full title) is actually two books in one. The first argues that humans evolved from apelike ancestors. Aside from a Neanderthal skullcap, fossils of early hominids were not yet known to science. Darwin's "fossil-free" conclusions about human origins were drawn entirely from embryology, comparative anatomy, animal behavior, and biogeography.

Darwin also published his famous guess—correct, it now appears—that Africa was the cradle of humankind. He reasoned that it was a center of radiation, because our closest surviving relatives, chimps and gorillas, still live there. Over the following century, fossil hunters scoured Europe, Asia, and North America in vain for evidence of the earliest hominids. Eventually the rich fossil record unearthed in Africa confirmed Darwin's hunch.

The Descent's second part deals entirely with Darwin's theory of "sexual selection": competition for mates as an important factor in evolution. In the final chapter, he draws together both themes in discussing the role of sexual selection in shaping the local variations ("races") of humankind.

Why did Darwin combine these two seemingly disparate topics in one work? In his attempt to identify uniquely human adaptations, he became baffled. Certain traits that humans do not share with apes, such as relative hairlessness (more pronounced in women) and musical ability, appear to have no adaptive advantage whatsoever. Then there was the puzzle of human racial differences. Could adaptation explain the evolution of so many different hair textures, nose shapes, skin colors, and bodily proportions among the world's peoples? He had to admit such traits might be unnecessary or neutral in making humans better fitted to survive.

Therefore, he took a long look at animal species that evolved traits that were "unnecessary" or even detrimental in relationship to their environments. Peacock's tails, for instance, only get in the way of feeding or escaping enemies. But they certainly may help the cock compete in attracting peahens. Similarly, Darwin thought, smoother, hairless skin in human females might have evolved to enhance attractiveness to males. Many other disparate characteristics, including musical ability, hair texture, or skin pigmentation, might also have evolved through sexual selection. Women, in choosing their mates, may be inclined to select strong, smart hunters and good providers, but perhaps they are also attracted by such "nonadaptive" traits as artistic or musical ability.

Unfortunately, like some modern practitioners of "evolutionary psychology," Darwin based parochial, unfounded assumptions about universal human attributes on his own Victorian society. He did not take into account, for instance, that in some tribes men, and even whole families (not just women), were in charge of mate choice, or that women were not always selected by men for their beauty. (He thought that the best hunters would get the prettiest women, just as in 19th-century European culture the richest men, even if physically unattractive, could attract young, beautiful wives.) A related point he missed is that, among some peoples, physically strong, hard-working women are much more desirable as wives than great beauties.

However, he did send questionnaires to missionaries and travelers all over the world to gather information about ideals of beauty among various tribes. He found out that local standards of attractiveness vary greatly and often are exaggerations of the tribe's special characteristics. These differences notwithstanding, mate selection frequently appeared to be based on physical attractiveness. To complete his case, Darwin presented the results of these inquiries alongside a compilation of evidence for sexual selection in antelopes, monkeys, birds, and other animals.

Historians of science James Moore and Adrian Desmond argue that it was Darwin's hatred of slavery that impelled him to try to understand "racial" differences such as skin color and hair texture that evolved within a single, unified human species. His horror when he witnessed "heart-sickening atrocities" of slavery in Brazil and genocide against Argentinean Indians by the Spaniards confirmed his commitment to the "monogenic" doctrine that all humans are members of the same species. That view of the unity of mankind, which

A peacock's tail slows down his escape from enemies, but it attracts females— an advantage in leaving more offspring.

Abolitionists believed was supported by the story of Adam and Eve, was central to their moral revulsion against slavery.

Pro-slavery partisans, on the contrary, argued for a "polygenic" origin of mankind. God had created different "kinds" of people—including superior varieties that were intended to be masters of the "inferior" kinds. One classic excuse for dominating and enslaving tribal peoples was that their colonial masters would "improve" them and their primitive cultures. "As far as I can see," Darwin wrote sarcastically in his journal, "we are improving them right off the face of the earth."

Subsequent authors not only explored and dissented from the book's ideas, but also played endlessly with the wording of its title. Theologian Henry Drummond wrote a popular work, *The Ascent of Man* (1894), in which "upward" biological progress is only the prelude to higher "spiritual evolution." In Darwin's day, one cartoonist depicted a man "descending" an evolutionary staircase toward the worms! In 1972, Elaine Morgan's *The Descent of Woman* offered a feminist corrective to the "ascent of man."

See also SEXUAL SELECTION

DE VRIES, HUGO (1848–1935)
Instantaneous Speciation

Hugo de Vries is remembered in the history of biology chiefly as the first of three rediscoverers of Gregor Mendel's long-neglected paper on plant hybrids in the spring of 1900. With hindsight, we see de Vries as a midwife to the birth of genetic science, helping to usher in the modern synthesis of Darwinian and Mendelian thought, the dual foundation of evolutionary biology.

De Vries himself would have been appalled and outraged, for he considered his recognition of Mendel's work only a minor incident in an illustrious career. It was his mutationism—an antigradualist theory of evolution by large, discontinuous jumps—that he thought would ensure his enduring fame.

De Vries's story begins in 1886 when he found, in a field near Amsterdam, large numbers of the evening primrose growing wild. Among the plants were two new varieties that differed distinctly from the normal form of the species: "Both [these new species] come perfectly true from seeds. They differ from [the parental stock] in numerous characters, and are therefore to be considered as new elementary species."

There, in a field in the Netherlands, was something that no one steeped in the Darwinian tradition of slow, gradual change expected: brand-new species! And unlike what the Darwinian gradualists had envisioned, they did not grade away imperceptibly from their progenitors, nor was any selection necessary to produce them. In a single step, in one generation, the new species had come into existence and, when self-fertilized, remained absolutely constant.

During the 1890s de Vries conducted a prodigious number of plant crosses, not only between species but between varieties of the same species. Sometimes an apparently new species was created, as he had hoped, but it was a rare occurrence. He kept careful records of his crosses and noticed that certain ratios of traits (such as the classic three-to-one) in hybrid generations came up again and again, but the results made no sense to him. Then, in early 1900, Professor Martinus Willem Beijerinck (1851–1931) of Delft sent de Vries a copy of Mendel's paper with the note: "I know that you are studying hybrids, so perhaps the enclosed reprint by a certain Mendel which I happen to possess is still of some interest to you."

In a flash, all of the notebooks full of numerical ratios that de Vries had recorded over the past decade of crossbreeding experiments fell into place, and he immediately set about writing not one but three Mendelian papers.

De Vries's enthusiasm soon evaporated, however, when he saw that the simple, classic Mendelian ratios only appeared in special cases. Since the complexities of Mendelian theory had not yet been worked out, de Vries concluded that Mendel's laws had nothing to do with the origin of species. "It becomes more and more clear to me," he wrote to Mendel's champion in England, William Bateson, in 1901, "that Mendelism is an exception to the general rules of crossing. It is in no way the rule!"

> "The new species originates suddenly from the existing one, without preparation or transition."
>
> —Hugo de Vries,
> *The Mutation Theory*, 1903

Between 1901 and 1903, de Vries published the two volumes of his magnum opus *Mutationstheorie (Mutation Theory),* in which he spelled out his conclusions. "The new species originates suddenly," he wrote; "it is produced by the existing one without any visible preparation and without transition."

For approximately two decades de Vries's theory eclipsed Darwin's, and during this period most biologists were confident that de Vries had laid Darwin to rest. Erik Nordenskiöld's *History of Biology,* written in the early 1920s, claimed that "Darwin's theory has long been rejected in its most vital points." And as late as 1932, Clarence Ayres, in his biography of Thomas H. Huxley, said: "All of Darwin's 'particular views' have gone down wind: variation, survival of the fittest, natural selection, sexual selection, and all the rest. Darwin is very nearly, if not quite, as outmoded as Lamarck."

De Vries's theory of speciation by sudden jumps or mutations was finally eclipsed, however, as Mendelian principles proved to be the soundest basis for understanding evolutionary change after all. By the 1930s, most of the early difficulties in applying them to more complex results had been ironed out. By then, it was clear that de Vries's "new species" were not new at all but only variations created by abnormal chromosomal pairings. (Darwin himself had once warned a correspondent that nature often presented ambiguities to the scientist and that "she will lie to you if she can.")

Today, scarcely anyone remembers either that Darwin's theory fell on hard times around the end of the 19th century or that de Vries's mutation theory dominated evolutionary thinking during the early 20th century.

DEWEY, JOHN (1859–1952)
American Philosopher, Educator

When the Darwinian revolution shook up Western thought during the latter half of the 19th century, many perceived the uproar as a conflict between science and religion. (Some still do.) Shortly after 1900, American philosopher John Dewey realized that the most profound rumblings were coming from within science itself. Long before the physicist Werner Heisenberg had made it respectable in science, Charles Darwin had introduced a principle of uncertainty.

In a famous lecture at Columbia University in 1909, "The Influence of Darwinism on Philosophy," Dewey recalled that the history of science had been a search for certainties and fixed laws that represent "the truth" about nature. Species were thought of as both fixed forms and the divinely ordained patterns that caused them to exist. That conception of species, said Dewey, "was the central principle of knowledge as well as of nature. Upon it rested the logic of science." He explained that

> for two thousand years, the conceptions that had become the furniture of the mind rested . . . upon treating change and origin as signs of defect and unreality. In laying hands upon the sacred ark of absolute permanency, in treating the forms [species] that had been regarded as types of fixity and perfection as originating and passing away [becoming extinct], the *Origin of Species* introduced a mode of thinking that in the end was bound to transform the logic of knowledge, and hence the treatment of morals, politics, and religion.

Dewey also pointed out that Darwin's way of looking at nature was pluralistic, seeking different explanations with various kinds of questions, including a historical approach, which considered nature over long periods of time. (History previously had not been part of the scientific enterprise.)

Darwin was interested in function, how specific things worked in the living world: whether seeds could sprout after soaking for weeks in salt water, or how orchids attracted pollinating insects. He asked nature thousands of limited questions with his experiments. Darwin was not concerned with who made the world or why, said Dewey, but with what kind of a world it is: How does it work? Darwin was unconcerned with proving any "absolute truth"; he thought the best theory was the one that could link up the most facts. Darwin expressed delight with an obscure reviewer in an April 1861 letter. The reviewer, he wrote, was "one of the very few who see that the change of species cannot be directly proved, and that the doctrine must

"Origin of Species . . . was bound to transform the logic of knowledge, and hence the treatment of morals, politics, and religion."

—John Dewey, 1906

sink or swim according as it groups and explains phenomena. It is really curious how few judge it in this way, which is clearly the right way."

Even today, the deepest argument between religious fundamentalists and scientists is not about conflicting accounts of how the world (or species) began, but over the nature of certainty. Can a creationist be sure of his answers? Certainly, since he's also convinced there is only one "right" interpretation of scripture (his own). When he asks a scientist directly if Darwin's theory could be disproved tomorrow by a new theory, the scientist must answer yes. To the fundamentalist, this admission that Darwin's truth is relative and less than eternal "proves" its worthlessness. Dewey realized that "old ideas give way slowly; for they are more than abstract logical forms . . . [they are] deeply engrained attitudes." Questions in science often cannot be answered within their own frame of reference:

> In fact intellectual progress usually occurs through sheer abandonment of questions. . . . We do not solve them: we get over them. . . . Old questions are solved by disappearing, evaporating, while new questions . . . take their place. Doubtless the greatest dissolvent in contemporary thought of old questions, the greatest precipitant of new methods, new intentions, new problems, is the one effected by the scientific revolution that found its climax in the *Origin of Species*.

Many supposed that Darwinian evolution established a new law that must be followed: We must try to conform to nature's "goals" of evolution for "improvement of the species." John Dewey thought Darwinism implied the opposite. Because there are no predetermined goals or necessary progress in Darwinian evolution, Dewey saw new possibilities for human freedom. "Conformity with laws," no matter how noble and perfecting, was still a belief in determinism—a new way of saying man had no real influence over his destiny. Part of the Darwinian revolution, Dewey believed, was the acknowledgment that we do have a newfound freedom, and consequent responsibility, to understand and meet challenges that have direct bearing on the future of our species.

See also ESSENTIALISM; SPECIES, CONCEPT OF; THEORY, SCIENTIFIC

DIGIT
Gorilla Martyr

Dian Fossey first encountered the young male gorilla she called Digit (because of a twisted middle finger) in 1967, when he was about five years old. He was always the first member of his group to come forward to investigate human visitors brought by Fossey. She enjoyed his company regularly for a decade, during which she watched him mature into the leader of his own family group.

African poachers murdered Fossey's gentle giant on New Year's Eve day in 1977 as he tried to protect his family from their

DIGIT AND DIAN FOSSEY, circa 1970. © The Digit Foundation.

onslaught. His head and hands had been hacked off, sending Fossey into a state of shock, anger, and "withdrawal into an insulated part of myself" from which she never recovered. "Digit gave his life," Fossey wrote, "so his family group might survive for the perpetuation of his kind."

The tale of Digit's heroic self-sacrifice has since become the stuff of pop culture legend. In Disney's animated feature *Tarzan* (1999), for instance, the old silverback male, Kerchek, is killed as he defends the lives of his infants and females from hunters.

A preposterous thriller, *Instinct* (also 1999), starred Anthony Hopkins as an anthropologist who joins a group of forest gorillas and becomes their dominant male when their own silverback leader is shot. When armed humans attack his adopted family, he kills some poachers, but is captured and subsequently tried for murder.

The movie is a far cry from the unique, thoughtful novel that inspired it. *Ishmael* (1992), by Daniel Quinn, consists of a long Socratic dialogue between a man and a telepathic gorilla, who instructs him on how we humans can renounce our role as conquerors and save the planet from ourselves.

See also FOSSEY, DIAN; GORILLAS

Poachers killed Digit the gorilla as he tried to protect his family— sending Dian Fossey into a deep, debilitating depression.

A MAMMOTH-SHAPED TRAVELING CASE was built to transport "Dima," the mummified baby mammoth. In 1979, paleontologist Andrei Kapitsa accompanied her to a special exhibition in London, where she was kept under police guard.

DIMA
Celebrated Infant Mammoth

A complete frozen carcass of a small baby mammoth was found by Soviet miners on June 23, 1977, as they thawed patches of frozen ground (permafrost) in Siberia searching for gold. It was a six-month-old female, unusually complete and exceptionally well preserved.

Found near a tributary of the River Kolyma, the ancient infant was immediately flown to Leningrad for study by Professor Nikolai Vereshchagin. Little "Dima" proved to be very similar to a modern African or Indian elephant, except that she had long reddish-brown hair and very small ears—just as Paleolithic cave artists had recorded.

A special traveling case was designed and built for Dima's remains, which were sent round the world during the late 1970s. From Moscow to London, she proved a great attraction; museum-goers formed long lines and waited patiently to see her.

Dima was one of the first frozen mammoths to be investigated with modern biochemical techniques. Electron microscopes revealed that her white and red blood cells were intact and that her albumen was still genetically active after thousands of years in the natural deep-freeze. When tested, her blood serum showed close kinship to that of modern elephants.

While these results came as no surprise, it was a dramatic confirmation that biochemical techniques told the same story of age and relationship to modern elephants that had been determined from bones, anatomy, and Ice Age artists' drawings. Lab scientists were ecstatic; it isn't every day they get to test the blood of an animal that disappeared from the face of the Earth 10,000 years ago.

A decade after Dima's discovery, in Western Siberia, another nearly intact frozen infant mammoth was discovered as it emerged from the ice during a spring thaw.

DINOSAURS
Ruling Reptiles of the Mesozoic

Dinosaurs were a successful and varied group, ranging in size from 80 tons to creatures no larger than a chicken. Thousands of different species filled all the available niches or habitats: browsers, swamp and river dwellers, herbivores, two-legged runners, and the largest, most fearsome meat-eaters ever to walk the Earth.

The world's biggest and heaviest known dinosaur is *Argentinasaurus,* a 120-foot-long, 80-to-100-ton herbivore from South America, whose largest vertebrae are six feet tall. (An eight-foot fossil vertebra from a creature called *Amphicoelias fragillimus* was pulled out of the Morrison Formation of North America in 1877 by Edward Drinker Cope, but has been lost.

All told, dinosaurs ruled the planet for about 150 million years before their comparatively sudden worldwide extinction. Their heyday, comprising the Triassic, Jurassic, and Cretaceous periods, 213–65 million years ago, is known as the Mesozoic ("middle animals") era.

Although descended from common ancestors—the small, swift, bipedal pseudosuchins—dinosaurs formed two great groups (orders), which were anatomically distinct. Based on different structures of the pelvis, the first are the Saurischia (lizard-hipped) dinosaurs, while the second are Ornithischia (bird-hipped). (Oddly, birds are descended from the lizard-hipped dinosaurs, not the bird-hipped group.) Two famous saurischian species are the long-necked plant-eater *Apatosaurus* (formerly called *Brontosaurus*) and the bipedal meat-eater *Tyrannosaurus.* Among the ornithischians were horned and armored dinosaurs, including *Stegosaurus, Triceratops,* and *Iguanodon.*

Certain prehistoric reptiles that are popularly included with dinosaurs are not dinosaurs at all but are classified with other groups. For instance, sail-backed reptiles (pelycosaurs), dolphinlike reptiles (ichthyosaurs), flying reptiles (pterosaurs), and paddle-flippered plesiosaurs are not members of the dinosaur group.

Everything we know about dinosaurs has been discovered only since about 1830, when the first known specimens led to a bone hunt that accelerated throughout the 19th and 20th centuries. In the 1960s, the old view of dinosaurs as slow, dim-witted creatures gave way to new interpretations of them as quick-moving, warm-blooded, intelligent, and social animals, closely linked to birds. Since the 1970s, so many new fossils and new species have come to light that knowledge of dinosaurs has more than quadrupled in four decades.

Three-quarters of all known dinosaur genera have come from the United States, China, Mongolia, Canada, England, and Argentina. As of 1990, when a comprehensive reference book, *The Dinosauria,* was published, scientists had described about 285 genera (related groups) of dinosaurs. By 2004, paleontologists had found another 222 genera—about 120 per decade. According to an analysis and projection by Peter Dodson of the University of Pennsylvania in Philadelphia, before long scientists will bring the total to almost 2,000 genera of dinosaurs, containing thousands of new species.

> Every decade, more than 100 new species of dinosaurs are discovered.

DINOSAURS, EXTINCTION OF
Why the "Great Dying"?

Because humans are the species writing evolutionary history, we inflate the importance of the last two or three million years, which produced mankind. But from a larger perspective, the conquest of the land by hundreds of kinds of dinosaurs, lasting 150 million years, is the major triumph of land vertebrates.

Yet at the close of the Cretaceous period, about 65 million years ago, something devastating happened. Every large land animal then in existence was wiped out, as well as the flying pterosaurs and sea-going plesiosaurs and mosasaurs. The cause must still be considered unknown, although there is no lack of clever theories. Here are a few of those that have enjoyed some prominence and popularity over the past century and a half.

1. *Pre-Evolutionary Theories.* Dinosaurs were antediluvian or "pre-Adamite" monsters that were wiped out in a great flood. Some early naturalists speculated they were part of a previous creation. Others thought that there may have been no room for them on Noah's Ark, and so they were doomed to drown.

2. *"Racial Senility" Theory.* Some paleontologists, like W. E. Swinton in the 1930s, thought that species, like individuals, may become old and senile. Extinction was therefore inevitable, like the death of an individual. Evidence of species "senescence" was found in "overossified" dinosaur skulls with bony frills and grotesque horns—supposedly produced

by growth hormones run wild. Now these "monstrous" appendages are seen as gradually developed features, adaptive for defense and reproduction. Dinosaurs were a progressive group, constantly developing new niches and specializations. "Senility of species" is a false analogy with the life cycle of individuals.

3. *Too Stupid and Slow vs. Mammals.* Recent evidence and new interpretations show that many dinosaurs were fast-moving, perhaps warm-blooded, and in some cases quite intelligent. Late Cretaceous wide-eyed "ostrich" dinosaurs (coelurosaurs) and dromaeosaurids like *Deinonychus* and *Sauronithoides* are examples of "bright" dinosaurs, which were developing just before the whole group became extinct. Some of these had such human-like characteristics as stereoscopic vision, correlated with opposable thumbs and large brains. Adrian Desmond writes that they were "separated from other dinosaurs by a gulf comparable to that dividing men from cows." Although these agile, alert creatures did not survive their comparatively dim-witted giant relatives, they were quite spectacular in contrast to the small and not very bright or prepossessing mammals of the time.

4. *Mammals Ate the Dinosaurs' Eggs.* No doubt they did, and so did other dinosaurs. The first dinosaur eggs found, which captured worldwide attention, were collected by Roy Chapman Andrews on the Gobi Desert (Central Asiatic) Expedition during the 1930s. Fossils of small early mammals were found nearby, leading to the popular view that the mammals hastened the dinosaurs' demise. It is more likely that the mammals survived unobtrusively amid the giants until the dinosaurs had died off from other causes.

5. *Dinosaurs Were Poisoned.* Various scientists have suggested that the development of poisonous defenses by land plants did the dinosaurs in. Alkaloids in the newly developed flowering plants have been suggested, and one scientist even fed poisons to tortoises to prove they could not taste them. In fact, flowering plants were around for quite a few million years before the dinosaurs' demise and may even have been beneficial to them. Such explanations still would not account for the disappearance of seagoing plesiosaurs and for the mass extinction of many invertebrates and other nondinosaurian species during the same time period.

6. *Failure to Adapt.* Perhaps the most common belief about dinosaur extinction is that when an animal becomes overspecialized and fails to adapt to changing circumstances it becomes extinct. First, this is a form of circular reasoning—like saying that those that do not survive were not fit. Second, it doesn't specify what circumstances changed. (No creature is adaptable enough to survive all changes, except within a certain range.) Finally, it has the ring of morality rather than science. It teaches the lesson of flexibility, like the Zen story of the bamboo that can bend in the wind, while the sturdy oak is felled by a storm. But some of the most archaic and least flexible creatures of the Cretaceous—tortoises and crocodiles—managed to survive and are still with us.

7. *Cosmic Cataclysms.* Did extraterrestrial events cause sweeping changes on Earth? While earlier theories postulated an exploding supernova, which altered Earth's climate, giant asteroid impacts are now in vogue. Such impacts could have raised immense clouds of dust, which would block solar rays and cool the planet. Even a small change in worldwide temperatures could account not only for the extinction of the dinosaurs, but for the disappearance of many other species as well, including the seagoing reptiles, shellfish, ammonites, and plankton that vanished at the same time. Geophysicists Walter and Luis Alvarez were the first to discover layers of iridium at the Cretaceous-Tertiary boundary—a substance not normally found in the Earth's crust except at impact sites. Periodic episodes of such impacts would also explain other mass extinctions in the fossil record. These astronomical theories, the iridium evidence, and the interest in "punctuated equilibrium" have fostered a new respectability for admitting some catastrophism back into uniformitarian geology.

See also GREAT DYINGS; PUNCTUATED EQUILIBRIUM

DINOSAURS, FEATHERED
Are Birds Evolved Dinos?

Why do a chicken's feet resemble those of a bipedal dinosaur: trident-like toes, hind claw, and scales? The answer, supported by many recent discoveries in northeastern China, is that birds and dinosaurs are close cousins. Indeed, paleontologists have concluded that birds evolved from a diverse group of carnivorous running dinosaurs.

Some similarities have been obvious for many years. The first dinosaur remains found in America were neither skulls nor skeletons, but four-toed footprints preserved in stone, known as the "tracks of Noah's ravens." These petrified imprints were thought to represent similarities between bird's feet and those of some lizards. Famed 19th-century evolutionist Thomas Henry Huxley carefully compared the skeletons of birds and dinosaurs and concluded that the two groups were indeed closely related. Few followed up on Huxley's insight, but more than a century and a half later his views have been vindicated, to say the least. Birds are now thought to belong to a clade called Maniraptora, a branch of the theropod dinosaurs.

Until recently, the oldest known bird was a creature called *Archaeopteryx* ("ancient wing"), the 150-million-year-old fossil found in a Bavarian limestone quarry in 1861. While its wings sported fully developed feathers, *Archaeopteryx* also had a lizardlike jaw filled with teeth rather than a beak.

For years, birds were defined by their feathers, as well as by breastbones and wishbones—and, often, winged flight. Some paleontologists thought that feathers must have appeared along with wings, but the question became: How could wings have evolved in the first place? After all, what good is half a wing? (See EXAPTATION.) Some experts thought that *Archaeopteryx* appeared too late in the fossil record to have been a founding avian.

In 1996 and '98, Chinese paleontologist Ji Qiang of the Chinese Academy of Geological Sciences published two previously unknown species of "feathered dinosaurs," *Sinosauropteryx* and *Caudipteryx*. The fossils Ji unearthed in Liaoning Province were surrounded by fine volcanic ash that had settled in an ancient lake, allowing detailed preservation of their downy plumage. They are about 130 million years old, younger than *Archaeopteryx*.

These dromaeosaurs, as they are called, represent a lineage of small, meat-eating, fast-running theropod dinosaurs, related to velociraptors, that had begun to develop feathers long before their descendants evolved the power of flight. Feathers may have been useful in regulating the animals' body heat. Stumpy arms, the precursors of wings, may have helped the creatures balance when running. Some dinosaur fossils have tested positive for beta keratin, the main protein in bird feathers.

During the 1990s and early 2000s, the fossil-rich Yixian Formation has yielded fifteen genera (different groups) of dinosaurs with preserved fossil feathers. Other birdlike dinosaurs and dinosaurlike birds have been found in Madagascar, Mongolia, Patagonia, and Spain. Many types of theropods may have had feathers, not just those that are especially similar to birds.

Thousands of specimens have been found in China recently, ranging from the size of pigeons to that of ponies, and with plumage ranging from fluff to feathers. In addition to all the fossils, in 2008 the bird-dinosaur link was given an unexpected boost. Molecular biologist Chris Organ of Harvard and colleagues compared collagen proteins from a 68-million-year-old *Tyrannosaurus rex* leg bone with those of living animals. The result: dinosaur proteins turned out to be most similar to those of ostriches and chickens, not lizards or alligators. The *Washington Post* headlined an account of the story: T. REX CLOSER TO GIZZARDS THAN LIZARDS.

See also *ARCHAEOPTERYX*; CHINA, EVOLUTION IN; "NOAH'S RAVENS"

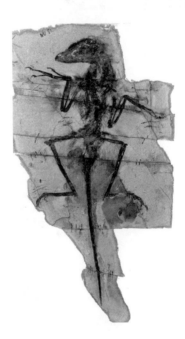

THE FOSSIL OF A YOUNG FEATHERED DINOSAUR, *Microraptor zhaoianus,* was discovered in 1998 in China's Liaoning Province. About two feet long, it lived about 130 million years ago. Reconstruction courtesy of and © by Mick Ellison.

DIVERGENCE, PRINCIPLE OF
"Keystone" of Darwin's Theory

Evolution is often pictured as a family tree or branching bush, bristling with divergent forks. Each lineage repeatedly splits and differentiates, and lines splay out, in Alfred Russel Wallace's image from his Sarawak paper (1855), "like the twigs of a gnarled oak or the vascular system of the human body." Some of the early evolutionists, such as Ernst

Haeckel, spent years working out "trees of life," showing divergence of families, genera, and species.

But though they seem inseparable today, evolution and divergence have not always been associated. Erasmus Darwin drew no family trees, nor did Jean-Baptiste Lamarck. Although Charles Darwin sketched such a tree in an early notebook, the principle of divergence occurred to him much later, about 15 years after he had developed his basic theory of natural selection. Divergence was a crucial missing piece during the writing of *Origin of Species* (1859), yet he called its last-minute inclusion "the keystone" of his book.

In the later views of both Darwin and Wallace, divergence serves a dual function in evolution. First, it enables a given species under selection pressure to survive in modified form by exploiting new ecological niches. Second, the gain in diversity boosts the habitat's carrying capacity, enabling it to support a greater total amount of life. Typically, small isolated habitats (like Darwin's beloved Galápagos) exhibit a startling diversity of closely related species, adapted for exploiting different foods or parts of the habitat. (Long before he formulated his theories, Darwin was impressed by the "abundance of great creative force" in those tiny, isolated islands.)

Darwin's notebooks and papers show he was stimulated by Wallace's preliminary publication on evolution but worked out divergence independently, as he later claimed. Wallace's 1855 paper (the "Sarawak Law") began with the question: If one examines the numbers of closely related species within genera, geographic distribution of natural groups, and differences between species in a given area, what overall pattern (he called it a "law") would emerge? The answer was that the largest number of species seemed to be produced when genera were confined to a small area (such as islands), and their differences were related to feeding adaptations (sharp beaks, blunt beaks, long beaks, etc.).

When Darwin read this paper, he scrawled on it, "Why should this law hold?" The answer, he later realized, was that under selective pressures, organisms evolve to fill "vacant places in the natural economy." He compared it to a division of labor, with efficient specialists exploiting the various food sources in a limited area. In one of his garden experiments, Darwin counted 20 species of wild plants (belonging to eight orders and 18 genera) that had sprouted on a cleared piece of turf measuring three by four feet.

What is now taken for granted was then a startling insight: contrary to "common sense" expectations, the fierce struggle for existence does not reduce the overall number of related species in an area with limited resources. Instead, it has the paradoxical effect of allowing many more species and individuals to thrive there, constantly evolving "endless forms most beautiful and most wonderful."

See also BRANCHING BUSH; DARWIN'S FINCHES; "DELICATE ARRANGEMENT"; HAWAIIAN RADIATION

TELEVISION'S SHERLOCK HOLMES, as played by Jeremy Brett, ponders the reasons for roses in *The Adventure of the Naval Treaty*.
Photo © ITV/Granada Television.

DIVINE BENEFICENCE, IDEA OF
Evidences of God's Goodness?

During a brief lull in a criminal investigation (*The Adventure of the Naval Treaty*, 1893), Sherlock Holmes took a moment to smell a red rose. The flower, he remarked to his friend Dr. Watson, clearly was evidence of divine beneficence. "Our highest assurance of the goodness of Providence seems to me to rest in the flowers," proclaimed the world's greatest detective. "All other things, our powers, our desires, our food, are really necessary for our existence in the first instance. But this rose is an extra. Its smell and colour are an embellishment of life, not a condition of it. It is only goodness which gives extras, and so I say again we have much to hope from the flowers."

Holmes was not the first to follow that line of reasoning. In the late 17th century, the English clergyman-naturalist John Ray popularized the venerable school of thought known as natural theology.

By studying plants and animals, Ray believed, one could find proof of divine beneficence and providential design.

A century later, another parson-naturalist, William Paley, expanded on Ray's theme. Paley's *Natural Theology; or, Evidences of the Existence and Attributes of the Deity, Collected from the Appearances of Nature*, first published in 1802, was studied by many aspiring natural-ists, including the young Charles Darwin. Paley argued that just as a watch implies the existence of a watchmaker, so the intricate mechanisms of living things must be the handi-work of a divine artificer. Through anatomical studies of eyes, wings, and hands, the natural theologians attempted to prove that biology confirmed their theology. Or, as Holmes once declaimed to Watson, "Religion . . . can be built up as an exact science by the reasoner."

Divine beneficence was a blanket explanation for natural phenomena in Victorian natural history. A dramatic example is Dr. David Livingstone's description of his near-fatal encounter with a lion, in *Missionary Travels and Researches in South Africa* (1858):

> Starting, and looking half round, I saw the lion just in the act of springing upon me. I was upon a little height; he caught my shoulder as he sprang, and we both came to the ground below together. Growling horribly close to my ear, he shook me as a terrier dog does a rat. The shock produced a stupor similar to that which seems to be felt by a mouse after the first shake of the cat. It caused a sort of dreaminess, in which there was no sense of pain nor feeling of terror . . . like chloroform . . . the shake annihilated fear, and allowed no sense of horror in looking round at the beast. This peculiar state is probably produced in all animals killed by the carnivora; and if so, is a merciful provision by our benevolent Creator for lessening the pain of death.

Of course, his explanation charmingly sidesteps the question of why a benevolent deity would have allowed Dr. Livingstone (or other prey) to be mauled by the lion in the first place.

Why, indeed, do flowers exist? In the 17th century, the German botanist Rudolf Jakob Camerarius—and later the Swedish taxonomist Carolus Linnaeus, among others—had come up with a shocking answer: sex.

In 1737 Linnaeus published his sexual system of botanical classification. The number of a plant's stamens, or male parts, determined the class to which Linnaeus assigned it, while the number of its female parts, or pistils, determined the subgroup, or order. He character-ized some species as diandria—those that have two stamens (or "husbands") on a flower with one pistil ("wife")—others as polygamia (one "husband" with multiple "wives"), and so on. He even named a genus of pea plant *Clitoris*. Although Linnaeus's sexual system was widely accepted, some embarrassed professors dubbed it botanical pornography. William Smellie, who worked on the first *Encyclopaedia Britannica*, denounced Linnaeus for going "far beyond all decent limits" and exceeding the most "obscene romance-writer."

The English physician and poet Erasmus Darwin (Charles's grandfather) not only ex-panded on the Linnean sexual system but also popularized it in his long poem *The Loves of the Plants* (1791). Erasmus's lusty lyric to nature became a best-seller, but many of his contemporaries found it unthinkable that beautiful blossoms, to perpetuate themselves, required vulgar insects to crawl among their sexual parts.

In 1793 Christian Konrad Sprengel, a German schoolteacher and botanist, published a triumphantly titled book, *The Secret of Nature Revealed in the Structure and Fertilization of Flowers*. The "secret" he revealed was that blossoms and fragrances do indeed function to attract insects that carry and disperse pollen. Sadly for Sprengel, his book was ignored for decades.

Twenty-five years after Sprengel's death, however, Robert Brown, first keeper of botany in the British Museum, sent a copy to his friend Charles Darwin. "It may be doubted," wrote Darwin's son Francis years later, "whether Robert Brown ever planted a more fruitful seed than in putting such a book in such hands." Darwin was delighted and amazed, and he began a research program, based on Sprengel's ideas, to discern the fertilization mechanisms of orchids. He also independently concluded that in gathering nectar for their honey, bees confer upon each species of plant—including those with hermaphroditic flowers—the great advantage of cross-fertilization. His own botanical treatises would have to wait, however, until he completed his magnum opus on evolution. "Who has ever dreamed," exclaimed

Deducing the purpose of flowers was anything but elementary.

Darwin's friend Thomas Huxley in admiration, "of finding a utilitarian purpose in the forms and colors of flowers?"

Darwin's orchid book was intended as a "a flank movement on the enemy"—that is, on teleology or purposive design. In 1862, when he published *On the Various Contrivances by Which British and Foreign Orchids are Fertilised by Insects, and on the Good Effects of Intercrossing* (his first book after the *Origin of Species*), Darwin demonstrated that flowers resulted from neither happenstance nor design: they coevolved with the insects that pollinate them. Insects had, in effect, selected the scents and colors of flowers. In species after species, the orchid's tubular bags of nectar and the mouthparts of the pollinating insects had evolved a precise lock-and-key fit.

So convinced was Darwin that the structures of flowers had coevolved with insects that he boldly predicted the discovery of a bizarre moth that no one believed could exist. The white Christmas Star orchid of Madagascar sports foot-long containers that have nectar only at the bottom. "What can be the use," Darwin wondered, "of a nectary of such disproportionate length?" Then, with Holmesian logic, he concluded: "In Madagascar there must be moths with proboscides capable of extension to a length of between ten and eleven inches! This belief of mine has been ridiculed by some entomologists." Forty years after he wrote those lines, a night-flying moth with a twelve-inch coiled tongue was indeed found on Madagascar—a subspecies of the African sphinx moth—which was dubbed *Xanthopan morganii preaedicta*: the moth that was predicted.

But it remained for his protégé and neighbor, Sir John Lubbock, to follow his mentor's lead in establishing how insects had shaped flowers. A banker, anthropologist, and entomologist, Lubbock had devoted himself to solving the puzzles posed by Darwin's theories. In *Ants, Bees, and Wasps* (1882) and *On the Senses, Instincts, and Intelligence of Animals* (1888), Lubbock demonstrated by careful experiment that bees have good color vision and olfaction. Lubbock wrote in the *Journal of the Linnean Society* in 1898 that it is to insects "we owe the beauty of our gardens, the sweetness of our fields. To them, flowers are indebted for their scent and color; nay for their very existence [and for] the form, size, and position of the petals."

By establishing the existence of color vision in bees, Lubbock had paved the way for Karl von Frisch to pursue his elegant investigations into the dance of the honeybee. During the middle years of the 20th century, von Frisch was able to show how scout bees communicate the location of flowers to other members of their hive. Sprengel was finally and fully vindicated, as Darwin had hoped. Scientists could no longer doubt that nature's glorious flower show is indeed directed at insects, not humans.

We read in the later reports of Dr. Watson that Sherlock Holmes eventually retired from solving London's crimes and devoted himself to beekeeping on the Sussex Downs. As all Sherlockians know, the insatiably curious Holmes even produced a technical monograph, *The Practical Handbook of Bee Culture, with Some Observations upon the Segregation of the Queen*. In *His Last Bow* (1917), set during the detective's last years, Holmes described this treatise on apiculture as "the fruit of my leisured ease, the magnum-opus of my latter years! . . . the fruit of pensive nights and laborious days when I watched the little working gangs as once I watched the criminal world of London."

For such an authoritative work, the meticulous Holmes must surely have consulted Darwin's books on orchids and insect pollination, as well as Lubbock's classic works on color vision in bees. However, the good doctor left no clue as to whether the great detective ever changed his views on the reasons for roses.

See also COEVOLUTION; LUBBOCK, SIR JOHN; ORCHIDS, DARWIN'S STUDY OF; SEXUAL SELECTION

DMANISI HOMINIDS
Earliest European *Erectus*

While excavating in the medieval town of Dmanisi in the Republic of Georgia (in the Caucasus, between the Caspian and Black Seas) in 1992, Antje Justus received the shock of a lifetime. Amazed by a well-preserved hominid jaw beneath the skeleton of a saber-tooth cat, the German paleontologist had inadvertently opened a new window on the earliest appearance of humans in Europe.

Dmanisi was once an important center of trade along a branch of the Silk Road. A church was built there in the 6th century, and the city prospered for many years under various rulers. Between the 10th and 14th centuries, it was attacked, sacked, and rebuilt several times.

But Dmanisi's rich medieval history pales beside the unprecedented hominid fossils that are being found beneath the old city. During the past fifteen years, they have added an unexpected chapter to the story of humanity's emigration from its African birthplace.

The first jaw, now dated at 1.8 million years, appeared similar to that of the Turkana boy or *Homo ergaster* from Nariokotome, East Africa. Its discovery caused a sensation among anthropologists, although some were cautious about inferring too much from an isolated jaw. By 2007, however, five crania had also been excavated at the site, establishing Dmanisi as a major, world-class treasure trove. The sample of this ancient population showed variability from very light, small individuals to one whose jawbone was massive and robust—probably a big male.

Nearby, researchers found over 3,000 crude stone tools, shaped in the simple early East African (Oldowan) manner—not the more sophisticated teardrop-shaped hand axes that first appeared about 1.6 million years ago. Along with the hominid finds are fossils of many long-extinct mammals, whose known dates have helped to pinpoint the age of the hominid fossils. Intensive work at the site continues by a crew of about 30 people, under the leadership of Georgian paleoanthropologist David Lordkipanidze.

The discoverers of the Dmanisi hominids classified them as a distinct species, *Homo georgicus*, although they now prefer to consider them a subspecies, *Homo erectus georgicus*. Among the many carnivore remains associated with them are those of giant hyenas, which may have caused the hominins some sleepless nights. When alive, they were between four and five feet tall, walked upright, and resembled an intermediate between Africa's *H. habilis* and *H. ergaster*, which is thought to be a closely related predecessor of *H. erectus*. Their cranial capacity, which indicates brain size, ranges between a gorilla-sized 600 and 775 cc—just within the range of early *Homo*.

If *Homo* evolved in Africa about two million years ago, as many believe, the Dmanisi fossils indicate that early humans did not linger exclusively on that continent for very long before spreading out into Eurasia. Indeed, they appear to have arrived on the fringes of Europe long before the ancestors of Neanderthals and modern humans got there.

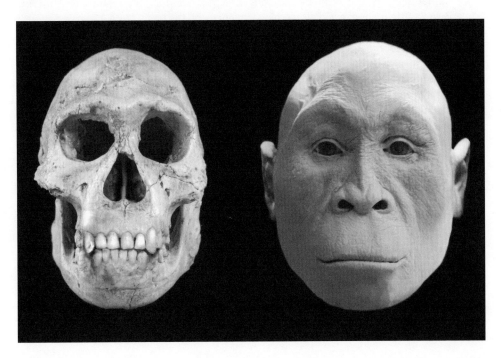

A DMANISI SKULL of a young male, reconstructed by paleoartist Viktor Deak, represents one of the earliest humans found in Eurasia.
Reconstruction © by Viktor Deak.

DNA (DEOXYRIBONUCLEIC ACID), DISCOVERY OF
Watson and Crick's Precursors

DNA, the biochemical code of instructions for living things, appears in the nucleus of every cell in plants, animals, and bacteria. The two complementary strands that entwine to form the molecule's double helix are composed of phosphates and sugars bonded together at intervals across the molecule's center.

Three molecular biologists, the American James Watson and the British Francis Crick and Maurice Wilkins, shared a Nobel Prize in 1962 for their work in establishing this structure as the physical basis of heredity. Watson and Crick's celebrated paper on DNA, published in 1953, marked a turning point in 20th-century biology. It ushered in an unprecedented expansion of molecular research, as well as a new biotechnology.

Watson and Crick's contribution was immense, but they did not actually discover the substance DNA, which had first been extracted from human cells in 1869 by the German scientist Friedrich Miescher. He called the substance "nuclein," thought its function was phosphorus storage, and never suspected its role in heredity.

By 1944 biologists had defined the molecule's role as a transmitter of heredity. Then, in 1953, Rosalind Franklin, working with X-ray crystallography, discovered its characteristic patterns on photo plates.

According to historian of genetics Edward Yoxen of the University of Manchester, the Watson-Crick breakthrough did not arise out of any startling new laboratory discoveries of their own, nor out of original experiments. Rather, Watson and Crick took it upon themselves to review all the known experimental work with the aim of extracting new significance from it. Their hypothetical model of the three-dimensional structure of DNA was achieved by reasoning and imagination alone.

It immediately appealed to those working in the field, but left much data unexplained and could have been wrong, since it was a hypothesis rather than a proof. Decades of experimental research have since provided abundant confirmation.

See also "CENTRAL DOGMA" OF GENETICS; FRANKLIN, ROSALIND; GENETIC ENGINEERING

DNA IDENTIFICATION
Biochemical Tool for Criminal Law

Animal DNA, as well as that of humans, has been used to catch murderers. In 1996, two Seattle men received long prison sentences on the basis of DNA from a dog—the first criminal trial in which animal DNA was used to win a conviction. Kenneth John Leuluaialii and George Tuilefano were charged with killing a young couple after breaking into their home and demanding drugs and cash. When the victims said they had neither, the thieves shot them dead. Leuluaialii also shot their German shepherd when it lunged at him to defend its owners.

DNA tests of blood stains on the suspect's jacket did not find a match with either of the human victims, but the blood stains did match their dog. When asked at the trial how blood got on his jacket sleeve, Leuluaiailii said he had been in a bar fight and someone had struck him. "What did your assailant hit you with," asked the prosecutor, "a dog?"

In the past, prosecutors have nailed criminals by analyzing bits of hair or body fluids left at the crime site. But the outward form, color, or thickness of hair or the type of blood or semen may leave room for doubt. Analysis of the DNA from protein molecules within the blood, hair, or semen produces "absolute identification." (Identical twins, who developed from the same egg and share the same DNA, would be a possible exception.) Thousands of more specific variable features than exist in fingerprints can be compared, making the older methods seem crude by comparison.

Among the first to be convicted by DNA was Timothy Spencer, for a double murder in Virginia in 1988. His attorney tried to discredit the new DNA test but could not find one biochemist who would challenge its validity. Shortly thereafter, the Seattle police department began a project to catalog the DNA of all known violent criminals in its files.

DOBZHANSKY, THEODOSIUS (1900–1975)
Evolutionary Geneticist

Harvard's Ernst Mayr, one of the architects of the modern Synthetic Theory of evolution and a historian of biology, pointed to one particular book that heralded the beginning of the new understanding of evolution: Theodosius Dobzhansky's *Genetics and the Origin of Species*, published in 1937.

It was Dobzhansky's first book, and he claimed he would never have found the time to write it if he hadn't been forced to spend weeks in bed after a horseback riding accident. A Russian immigrant who left the Soviet Union when Mendelian genetics was forbidden by Stalinist dogma, Dobzhansky came to work at Thomas Hunt Morgan's "fly room" at Columbia University in 1927. He brought to America the innovative techniques developed by Russian geneticists before their science was crushed by the tragic madness of Lysenkoism. Versed in the research problems of both field naturalists and lab men, he was able to make connections between the two approaches

During the first 20 years of the 20th century, Darwin's theory of natural selection had fallen out of favor among scientists. Many thought it insufficient to explain the origin of adaptations, while new discoveries of gene mutations seemed to be incompatible with Darwinian models of change. Dobzhansky's book was the first systematic overview encompassing organic diversity, variation in natural populations, selection, isolating mechanisms (a term he coined), and species as natural units. Later, working with geneticist Sewall Wright, he went on to demonstrate how evolution can produce stability and equilibrium in populations rather than constant directional change.

He discovered that successful species tend to have a wide variety of genes that may not be useful to the organism in its present environment but provide populations with genetic diversity—which enables them to adapt effectively to changes.

Dobzhansky's studies of isolating mechanisms identified nongenetic barriers to reproduction, such as behavior or vocalizations, which may keep populations distinct even after geographical barriers to interbreeding are removed. Dobzhansky also helped demonstrate that a population arbitrarily divided into two subpopulations can diverge into two species (genetic drift) even in the absence of any selection pressure. Among other projects, he studied the mechanisms of genetic variability within natural populations and then showed how detrimental genes can spread in certain combinations with beneficial ones. The classic example is the sickle cell gene, which increases resistance to malaria while also fostering anemia.

Dobzhansky was particularly fascinated with unraveling the multiple effects of a single genetic change (pleiotropy) and with the complex role of gene arrangement and chromosomal structure in producing evolutionary change. Never content with laboratory studies alone, he repeatedly stalked wild populations of fruit flies in the mountains of Arizona, New Mexico, and California and the rain forests of Brazil. Dobzhansky's intimate familiarity with the processes of variation and evolution in these fast-breeding insects also enabled him to apply his methods to understanding variation and change in human populations.

See also ALLOPATRIC SPECIATION; FLY ROOM; GENETIC DRIFT; ISOLATING MECHANISMS; LYSENKOISM; MAYR, ERNST; SYNTHETIC THEORY; WRIGHT, SEWALL

> "Nothing in biology makes sense except in the light of evolution."
>
> —Theodosius Dobzhansky

DODO
Icon of Extinction

Dodos were large-beaked, bulky ground birds from the island of Mauritius; their name has become synonymous with creatures that are "dead, gone, failed, vanished." The dodo has become an icon of extinction.

Historical accounts depict the dodo as cumbersome and defenseless, as it had no natural predators on the island. When ships from Holland began arriving at Mauritius in 1598, sailors slaughtered the birds indiscriminately, even though they reportedly tasted bad. By 1861, the last dodo was killed—the victim not only of humans but also, it has long been supposed,

AN UNHAPPY DODO is ferried by Charon the boatman across the River Styx "to join the Archaeopteryx" in a charming poem and drawing by Oliver Hereford (1901).

of ship-borne rats that sought out its rudimentary ground nests and devoured thousands of eggs and chicks.

Dodos were easy to hunt, but hunting alone probably did not wipe them out. Recent research indicates that the early Dutch settlers rarely ate dodo meat. Nor did the deforestation of the island doom the dodo, since it began after the dodo became extinct.

No one was quite sure what sort of fowl it was, nor what its closest relations were. It looked a bit like a turkey designed by a committee. Eventually, however, scientists decided that it was a kind of giant pigeon. DNA studies show that its closest relative was the solitaire, another extinct flightless bird that lived only on the nearby island of Rodrigues.

In 2005, a team of scientists from the Geological Survey of the Netherlands working on Mauritius discovered a trove of dodo fossils at a site called Mare aux Songes that may be as old as 3,000 years. As they reconstruct the island's ecology, they are seeking clues to the dodo's true fate.

Children's books featuring dodo images have proliferated over the years, Lewis Carroll's dodo in *Through the Looking Glass* being perhaps the most famous of them all. One memorable poetic eulogy appears in a 1901 book, *More Animals*, written and illustrated by Oliver Hereford:

This Pleasing Bird, I grieve to own
Is now Extinct. His Soul has Flown
To Parts Unknown, beyond the Styx
To Join the Archaeopteryx.

What Strange, Inexplicable Whim
Of Fate, was it to banish him?
When Every Day the numbers swell
Of Creatures we could spare so well:
Insects that Bite, and Snakes that Sting,
And many another Noxious Thing.
All these, my Child, had I my Say,
Should be Extinct this very Day.
Then would I send a Special Train
To bring the Do-Do back again.

A film called *A Flock of Dodos*, made by biologist-turned-filmmaker Randy Olsen in 2005, documents the contemporary debate between evolutionary scientists and creationism or intelligent design proponents. Which side in this public controversy are the dodos? In the filmmaker's view, both groups are proving inflexible and unable to adapt to the changed cultural environments in which they find themselves.

DOLLY
First Cloned Sheep

In February 1997, Scottish scientists captured the world's attention when they announced that they had successfully cloned a sheep. The young lamb, named Dolly, set off a polemical firestorm over the scientific potential and ethical concerns of the new technology.

What made Dolly so special wasn't that she was the first animal to be cloned—that distinction went to a set of genetically identical tadpoles created in 1952. The tadpoles, and later some mammals including sheep and cows, had been cloned from cells taken from early embryos. But Dolly sprang from the udder cells of a mature, six-year-old ewe, even though many scientists believed it would be impossible to clone a mammal from the body cells of an adult.

To achieve this feat, Ian Wilmut and his colleagues at Scotland's Roslin Institute removed the nuclei (which contain an animal's genetic material) from unfertilized sheep eggs. In their place, they inserted the nuclei from adult udder cells that had spent the last three years in the lab freezer. The altered eggs were then implanted into a surrogate sheep. Of the 277 nuclear transfers the team performed, only Dolly was carried to term. The scientists tested Dolly's DNA to confirm that she was a genetic duplicate of the donor ewe. But even before the results were in, they were confident of their success: Dolly, like the donor of the frozen udder cells, was a Finn Dorset sheep, bearing no resemblance to either her surrogate mother or her egg donor.

With this one successful experiment, the possibility of cloning humans became real, and self-appointed moralists rushed to ban further research. President Bill Clinton banned the use of federal funds for research into human cloning and asked private institutions to adopt a "voluntary moratorium." Many scientists were disappointed, convinced that the door was being closed on a great potential for human benefits. When the procedure was perfected and made more cost-efficient, livestock could be genetically "improved," animals could be made to secrete beneficial drugs in their milk, and organs could be altered for transplant into humans without fear of rejection by the body's immune system.

DOLLY'S SURROGATE MOTHER was this Scottish blackface sheep. Dolly's genes were taken from a Finn Dorset female's udder.

Since Dolly, a number of other mammals have been cloned, including mice, goats, pigs, and a kitten named "cc," for "carbon copy." Attempts to clone dogs have been unsuccessful, although some hopeful owners have paid thousands to have their beloved pooch's DNA stored.

Unfortunately, Dolly's health deteriorated from a young age. When she died in Edinburgh on February 14, 2003, at the age of six, Dolly received an obituary in the *New York Times*. She was survived, it said, "by six lambs, produced in the customary way, with a ram."

DOWN HOUSE
Home of the Darwins

Soon after Charles Darwin returned to England from his voyage around the world, he married his cousin Emma Wedgwood. After a few years in London, the Darwins decided to move permanently to the English countryside. In 1842, they bought a Georgian home in the tiny village of Downe, 16 miles from London.

Despite its closeness to the city, it is an isolated area, nestled in the gently sloping chalk downs from which it takes its name. The Darwin home was (and is) called Down House. In 1842, during paranoia about Irish rebels, the village became "Downe," to distinguish it from County Down, but Darwin refused to go along with the ridiculous spelling change.

Down House sits on 18 acres, including fields, gardens, and woods. A sand-covered path, the Sandwalk, was laid out winding through the shady woods and then returning toward the house along a sunny straightaway bordered by hedges. Darwin called it "my thinking path." Work on his 17 books and scientific articles was punctuated daily by several brisk turns around the Sandwalk, often with his terrier Polly trailing along.

THE DARWIN HOMESTEAD still stands in the village of Downe, near Bromley, about 16 miles south of London, and is open to the public as the Darwin Museum.

Sometimes, when deep in thought, he would stack up a few flints around the turn in the walk. He might have a "three flint problem," just as Sherlock Holmes had "three pipe problems." Darwin would walk around the loop, knocking away a flint with his walking stick each time he passed. When all the stones were gone, it was time to head back home.

Here Darwin lived and worked for 40 years. At various times he added to the house; he put up a dovecote out back for breeding pigeons, and added a greenhouse (which still stands), a laboratory (now in ruins), and, later, a clay tennis court for his children. Charles and Emma had 10 children, of which three did not survive childhood.

Down House was sold after Emma Darwin's death and was used as a private school for girls until the 1930s, when Sir Buxton Browne arranged to purchase it for the Royal College of Surgeons. Relatives and friends of the Darwin family contributed original furnishings, which had been scattered, and the house was restored to its appearance when Charles and Emma lived there. Today it is a museum, managed by English Heritage, where one can walk along Darwin's Sandwalk and visit his study.

In recent years, the garden, orchard, and greenhouse have been restored and replanted with the trees, shrubs, and even insectivorous plants that were there in Darwin's day. In the garden is the "wormstone" device that Horace Darwin constructed at his father's request to measure the rate at which earthworms could bury a large stone.

See also SANDWALK

DOYLE, SIR ARTHUR CONAN (1859–1930)
Materialist and Spiritualist

ARTHUR CONAN DOYLE as Professor Challenger, above. In *The Lost World,* below, Challenger/Doyle closely resembles the ape-man king.

Sir Arthur Ignatius Conan Doyle, the Scottish-born author, is best remembered for his 56 stories and four novels about the detective Sherlock Holmes, one of the most durable and compelling characters in fiction.

Holmes was not Doyle's favorite alterego, however. He identified most strongly with his less famous character, Professor George Edward Challenger, who starred in five tales, including Doyle's classic evolutionary fantasy, *The Lost World* (1913).

Although both heroes represent the scientific quest for truth through logic and physical evidence, the two could not be more different. Challenger is burly and hirsute where Holmes is wiry and clean-shaven; loud and obnoxious, in contrast to Holmes's genteel introversion; a public provocateur, while Holmes shuns the limelight. Professor Challenger has little use for his scientific colleagues, whom he regards as timid, mediocre intellects, constitutionally incapable of making important discoveries. He is physically imposing, charismatic, and bombastic—described by one of his detractors as "a homicidal megalomaniac with a turn for science." His very name is emblematic of both a gadfly who takes on the scientific establishment and a fearless explorer. (Britain's first geophysical research vessel, launched in 1872, was HMS *Challenger*, a name reappropriated for a U.S. Space Shuttle more than a century later.)

The Lost World chronicles Challenger's expedition to a Venezuelan valley, a backwater of evolution, where dinosaurs and ape-men still roam. In the final Challenger story, *Land of the Mists* (1926), the professor discovers and explores the spirit world, and is guided by the spiritualist-scientist Alfred Russel Wallace. Like Wallace, real-life codiscoverer of the theory of evolution by natural selection, Challenger believes that scientific evidence has led him from materialism to spiritualism.

Doyle donned full makeup to appear as Professor Challenger in his book illustrations, and played practical jokes while in costume—something he would never dream of doing as Sherlock Holmes. When he meets the king of the ape-men in *Lost World,* the primitive tribal chief turns out to be a mirror-image of Doyle in his Challenger get-up.

During the last decades of his life, Doyle campaigned relentlessly on several continents to convince audiences of the reality of the Spirit World. Two major conflicting philosophical views of his time—materialism and Spiritualism—drove his life and his art.

See also DIVINE BENEFICENCE, IDEA OF; *LOST WORLD;* PILTDOWN MAN (HOAX); SPIRITUALISM

DRAGONS
Myth That Came True

Dragon legends have persisted for centuries in Norse epics, medieval English ballads, Wagnerian operas, Japanese art, and Chinese folktales. To modern scientists, gigantic reptiles were pure myth until about 1825. At about that time, marveling European geologists began turning up huge fossil teeth and bones. Huge reptilian monsters, it turned out, had once lived and breathed, though long before there were pure-hearted knights to battle them.

Why had this ancient tradition of dragons persisted in so many of the world's cultures, long before science knew of dinosaurs? Perhaps it was more than a coincidence of fact and fancy; fossils were found and collected among many peoples long before the rise of paleontology. Fossils (not only of dinosaurs, but of any large animal) were regarded in many parts of the world as "dragon bones" or "dragon's teeth." In China, farmers collected them for sale to apothecaries, who displayed bins full of them, ready to sell and be ground up into traditional medicines.

In Asia, dragons have been prominent in art and legend for thousands of years; they were considered wise, beneficent, and bringers of good luck. Over the years, they took on attributes of other animals (elongated as snakes, antlered like reindeer) and symbolized the whole Chinese nation. Even today, huge dragon puppets manned by dozens of people are trotted out in American Chinese communities as the highlight of New Year celebrations.

LINDWURM FOUNTAIN at Klagenfurt, Austria, depicts a traditional European dragon. Its peculiar head, sculpted in 1590, may have been inspired by fossil rhino skulls found in the nearby countryside. It is perhaps the first known attempt to reconstruct a prehistoric beast from fossil evidence.

In European folklore, dragons were evil monsters that guarded great treasures or imprisoned damsels. Ordinary folk were not sufficiently pure of heart to slay them, but great warriors like England's St. George sought and killed them. A famous dragon sculpture—a large stone fountain—dominates the town square of Klagenfurt, Austria. Carved in 1590 by the sculptor Ulrich Vogelsang, in body it is a traditional representation of a European dragon. But several paleontologists have been intrigued with its head proportions. The facial shape, with its pointed, angular jaws, seems extraordinarily similar to the extinct European rhinoceros. According to local accounts, the artist was indeed inspired by fossil rhino skulls found in the nearby countryside.

In 1899, K. A. Heberer, a German naturalist, brought back a large collection of the Chinese "dragon bones" purchased from druggists and had them examined by museum experts. Ninety fossil species were identified, including extinct bears, giraffes, rhinos, and lions—but no reptiles. During the 1920s, anatomists G. H. R. von Koenigswald and Franz Weidenreich, realizing these sellers of folk medicine might be enlisted in the cause of science, began a systematic search of the shops for unusual fossils.

It was in such a druggist's bin that von Koenigswald recognized a few enormous hominid teeth. Thus, he discovered the giant extinct ape he named *Gigantopithecus*. Later, von Koenigswald's conclusion was confirmed after more teeth and fragments were found, though paleontologists are still seeking a skull or skeleton.

A particularly rich fossil site near Beijing was long known as Zhoukoudian, "Hill of the Dragons." There, in quarries and limestone caves, the famous Beijing Man skulls were found by Davidson Black, Franz Weidenreich, and W. C. Pei during the 1930s. Originally named *Sinanthropous pekinensis*, they are now lumped with *Homo erectus*, an ancient species of human whose remains have been found in East Africa, Java, and Europe.

In Beijing, there is a famous "Wall of the Dragons," an artistic treasure of old China carved with magnificent Oriental dragons. In 1987, modern Chinese paleontologists from Beijing's Institute of Vertebrate Paleontology and Paleoanthropology quarried and exhibited a rock slab showing six fossil synapsids, or mammal-like reptiles. They nicknamed their spectacular prize "The Scientific Wall of the Dragons."

See also BEIJING MAN; DINOSAUR; KNIGHT, CHARLES R.; MANTELL, GIDEON ALGERNON

DRYOPITHECINES

Miocene Apes

During the Early Miocene period (about 23 to 16 million years ago) a group of ancestral medium-sized apes was spread widely over Africa, Asia, and Europe. In 1856, the first fossils were discovered in southern France by Eduard Lartet, who named them

dryopithecines, meaning "oak forest apes." Poorly known for a century, the fossil representation has been substantially filled in. One site in Kenya alone yielded 13 skeletons, and collections of remains have been found recently in Spain and Hungary.

The best known of the dryopiths is *Proconsul*, discovered by Mary Leakey at Rusinga Island, Kenya, and recognized as an early representative of the hominoid lineage. It was tailless, had large canines (as do modern apes), and appears to have had a monkeylike type of locomotion. Its back teeth showed the hominid (Y-5) dental pattern. A later, related genus, *Sivapithecus*, found in India and Pakistan, may be ancestral to the orangutan.

The relationship of the dryopithecines to the living apes is not yet known, but since 1990 new discoveries of Miocene apelike fossils have begun to fill in the picture.

See also AUSTRALOPITHECINES

DUBOIS, EUGÈNE (1858–1940)
Discoverer of Java Man

With a stubborn insistence that appeared entirely unreasonable to his associates, a Dutchman named Eugène Dubois set out to find a needle in an immense haystack when hardly anyone thought there was a needle there at all. His discovery of Java Man remains one of the most incredible stories of self-confidence and determination in the face of seemingly impossible odds in the history of science.

Since his boyhood in Roermond, Holland, Dubois had always been an amateur naturalist whose pockets were bulging with stones and bits of animal bone. In 1877, at the age of 19, he entered medical school. There he was exposed to the exciting ideas of Darwinian biologists, who were revolutionizing every field from embryology to paleontology. In particular, he was enthralled by a visiting German lecturer, the famous Ernst Haeckel, who was trying to fill in the "missing links" in the chain of evolution.

Compelling as Darwin's theory was, there was still only a scrap or two of fossil evidence for human ancestors, most notably the Neanderthal skullcap found in 1856 in Germany. Anatomists agreed it was certainly human and, therefore, no "missing link."

Professor Haeckel insisted a creature intermediate between men and apes must have existed and gave it the Latin name *Pithecanthropus alalus*, meaning "ape-man without speech." Young Dubois made up his mind that he was going to find the creature Haeckel had named. It would be the final "proof" that Darwin was correct.

Dubois completed his medical studies and became a teacher of anatomy at the University of Amsterdam but was restless with academic life. He was increasingly consumed by the idea of discovering the "missing link," but in what part of the world would he find it?

Haeckel had thought that "Lemuria," a supposed continent that had sunk beneath the Indian Ocean, might have been the cradle of humankind, from which early hominids could have spread westward to Africa, northward to Asia, and eastward via Java. Dubois eagerly read Alfred Russel Wallace's 1869 book *The Malay Archipelago: The Land of the Orangutan and the Bird of Paradise,* which sketched Sumatra as a fascinating, prehistoric-looking landscape—and it was conveniently under Dutch colonial rule. Wallace had written:

> It is very remarkable that an animal, so large, so peculiar, and of such a high type of form as the orang-utan, should be confined to so limited a district—to two islands. . . . With that interest must every naturalist look forward to the time when the caves of the tropics be thoroughly examined, and the past history and earliest appearance of the great man-like apes be at last made known.

Dubois left his medical practice, quit his teaching job, and took off for Sumatra to find the "missing link."

Dubois wanted to explore those caves for himself and announced to his colleagues that he was leaving Amsterdam for Sumatra to find the missing link. They thought he was throwing away a brilliant medical career for an impossible, crazy scheme, but his new, young wife believed in him. In 1887, he quit his teaching post and spent many frustrating months trying unsuccessfully to find backing for an expedition.

When he realized that neither private nor governmental sponsors were to be found, Dubois enlisted as a doctor in the Dutch Indian Army. He, his wife, and their small children made the seven-week voyage aboard a mail boat, then found their way to a small hospital in the interior of Sumatra. His duties were light, as he had hoped, and he spent his time and money investigating many limestone caves and deposits. Once, he crawled headfirst into a tiger's den while its occupant was out hunting, and got stuck. Fortunately, his helpers returned before Dubois himself became a skeleton in a Sumatran cave.

After an attack of malaria in 1890, he was transferred to the neighboring island of Java, placed on inactive duty, and finally given assistance in his search by the colonial government, which supplied him with a crew of convict laborers. His workers secretly sold the fossils as fast as they dug them up to Chinese merchants, who ground them up for use as "dragon bones" in folk medicines. Despite such difficulties, crates of rocks started to pile up at Dubois's house in Tulungagung, and by 1884 he had shipped 400 cases of interesting new fossils back to Holland—but no early hominids or humans.

Dubois had a hunch that the most promising site was an exposed, stratified 45-foot embankment along the Solo River, near Trinil village. His workmen combed it inch by inch and in August 1891 came up with a molar tooth which he thought came from an extinct ape. Months later, three feet away, his diggers found a brown object resembling a turtle shell. It proved to be a fossil skullcap with thick brow ridges, and Dubois wrote "that both specimens come from a great manlike ape."

Rains came, the river rose, and Dubois pondered his skull for a year before he could dig again. When he did, he unearthed a left femur, or thighbone, at the same site. Although thicker and heavier than that of modern humans, it certainly came from a bipedal primate. Then another tooth turned up, and in 1893 Dubois formally announced discovery of the creature Haeckel had predicted only seven years earlier. Because of the shape of the leg bone, Dubois dubbed his find *Pithecanthropus erectus*—ape-man who walked erect. Haeckel cabled him, "CONGRATULATIONS TO THE DISCOVERER OF PITHECANTHROPUS, FROM ITS INVENTOR." After that incident, zoologists announced a new rule of classification. Henceforth, one could not name a new species until it had actually been found.

Pithecanthropus became Dubois's constant companion, to which he grew greatly attached. Returning from Java in 1895 during a storm at sea, he went rushing to the ship's hold to embrace his precious cargo. "If something happens," he told his wife, "you see to the children. I've got to look after this."

Dubois carried the bones to scientific meetings all over Europe, presenting his case. Some pronounced it a primitive man and no "missing link," while others, such as the anatomist Rudolf Virchow, said it was an extinct giant gibbon.

For several years, Dubois made his fossils accessible and produced numerous photos, casts, and reconstructions, but the scientific world would not accept his interpretation of their antiquity or their pivotal role in human evolution. Several scientists whom he allowed to examine and measure the fossils published lengthy and detailed papers on them, effectively scooping their discoverer. Deeply hurt, he withdrew them from further study and became a recluse.

By 1930, the search for early humans had entered a different phase. However, more scraps of Java Man had been found, and several spectacular skulls had been discovered near Beijing, China. Despite Dubois's insistence that his Java fossils were quite unlike the Chinese finds, *Pithecanthropus* was becoming well defined as a widespread population of an early type of human. Eventually, more remains turned up, scattered on several continents.

Today, Java Man has been assigned to the species *Homo erectus*, along with similar fossils from East Africa, Asia, the Middle East, and Europe. *Homo erectus* appears to have reached Southeast Asia by 1.6 million years ago and to have persisted as recently as 40,000 years ago; it is now considered a widespread early human species, not a "missing link" between men and apes.

See also HAECKEL, ERNST; *HOMO ERECTUS*; WALLACE, ALFRED RUSSEL

"Congratulations to the discoverer of *Pithecanthropus,* from its inventor."

—Ernst Haeckel to Eugène Dubois, 1873

EARTHWORMS AND THE FORMATION OF VEGETABLE MOLD
The Subject of Darwin's Last Book

One of the secrets of Charles Darwin's greatness was his pleasure in finding wonder in the small, commonplace features of the Earth that most of us ignore. At the age of 72, the year before his death, he published his very last book, *The Formation of Vegetable Mould through the Action of Worms* (1881). This simple farewell treatise, researched in his own backyard, tells us as much about the man and his methods as about worms.

While his wealthy neighbors in the English countryside amused themselves with social gatherings, bird shooting, and cricket matches, Darwin loved strolling in his woods and fields, lifting rocks and poking logs, exercising his "enlarged curiosity" as he had done since childhood. To the end of his life, he never tired of watching the bees in his kitchen garden, the climbing plants in his greenhouse, and the earthworms in his fields.

Today we take for granted the industriousness of worms and their important place in the ecology. In Darwin's day, these facts had still to be proved. Indeed, it was startling to some when he claimed for worms an important role in world history. "In many parts of England," he wrote, "a weight of more than ten tons of dry earth annually passes through their bodies and is brought to the surface on each acre of land; so that the whole superficial bed of vegetable mould [topsoil] passes through their bodies in the course of every few years." (He had first expressed this view in a short paper in 1837, more than 20 years before the *Origin of Species* appeared.)

Worms had originally been brought to Darwin's attention by his uncle Josiah Wedgwood, the same benefactor who had convinced his skeptical father to let him go on the *Beagle* voyage. Darwin had noticed that stones scattered in fields tended to sink and become buried with time, and Wedgwood suggested that they were being covered by worm castings.

Darwin then began adding to his thousands of other observations the condition of several fields near his home. One in particular had been plowed in 1841 and was called by his sons "the stony field," as it was thickly covered with small and large flints. In the earthworms book, Darwin noted:

> When [my sons] ran down the slope the stones clattered together. I remember doubting whether I should live to see these larger flints covered with vegetable mould and turf . . . [but] after thirty years (1871) a horse could gallop over the compact turf . . . and not strike a single stone. . . . The transformation was wonderful [and was] certainly the work of the worms.

Once Darwin got hold of a subject, he pursued it for all it was worth. He had his son Horace fit a round, heavy stone, known as the "Wormstone," with a metal gauge in its center so he could measure precisely how far it sank in soil continuously churned by earthworms. One of Horace's wormstones can still be seen in the garden at Down House.

Darwin, even took the long rail journey to Stonehenge to observe how far worms might have buried the ancient "Druidical stones." His sons helped dig test holes near the monoliths' bases, and Darwin duly recorded how far the great stones had sunk—though he was somewhat disappointed in the worms' lack of industriousness in that region.

In his search for nature's truths, Darwin was a man who literally left no stone unturned. He even brought pots full of worms into the house to measure their activity at various temperatures and had his son play notes on the bassoon and the piano to them in the drawing room to see whether they reacted to higher- or lower-pitched notes. (How his wife Emma reacted to worms on her piano is not recorded.)

Punch magazine greeted the earthworm book with a caricature of Darwin pondering a giant worm coiled like a question mark, next to this verse:

I've despised you, old Worm, for I think you'll admit
    That you never were beautiful, even in youth;
I've impaled you on hooks, and not felt it a bit;
    But all's changed now that DARWIN has told us the truth
Of your diligent life, and endowed you with fame—
    You begin to inspire me with kindly regard:
I have friends of my own, clever Worm, I could name,
    Who have ne'er in their lives been at work half so hard.

A CARICATURE OF DARWIN and his earthworms appeared in *Punch* magazine in 1881, the year before his death. Darwin's study of worms stressed a major theme of his life's work—that small causes produce major effects in shaping the Earth.

In the last paragraph of the book, Darwin refers to another animal of "lowly organization," the coral. One of Darwin's very first treatises was *The Structure and Distribution of Coral Reefs* (1842), in which he suggested that the world's great ocean reefs were the work of millions of tiny creatures. The progress of science has often been retarded, he wrote, by "the inability to sum up the effects of a continually recurrent cause . . . as formerly in the case of geology, and more recently in that of the principle of evolution."

"Clever old man," wrote Stephen Jay Gould in a centenary appreciation of the worm book, "he knew full well. In his last words, he looked back to his beginning, compared those worms with his first corals, and completed his life's work in both the large and small."

See also CORAL REEFS; DARWIN, CHARLES; ECOLOGY; UNIFORMITARIANISM

ECOLOGY
Organism-Environment Relationship

Darwin had stressed the importance of adaptation, the evolutionary adjustment of an organism to its environment. It was his German disciple Ernst Haeckel who suggested the creation of a whole new field: the study of the "homes," or niches, that animals and plants occupy. In 1866, he coined the term "oekology," from the Greek word *oikos*, "a home."

Ecology has since become a scientific specialty, analyzing the relationship between organisms and their internal and external environments. The ecology of a particular bird, for instance, would include the kinds of trees it inhabits; the food it eats; the air it breathes; the parasites in its blood; the bacteria in its digestive system; its predators; its nesting materials; its relationships with other species; the extent of its range and territory; its population density; social dynamics; and so on.

Often, the word "ecology" is misused in a narrow political sense to mean environmental or conservation issues.

See also ADAPTATION; HAECKEL, ERNST

EDIACARAN FAUNA
600-Million-Year-Old Ecosystem

Almost five-sixths the way up the timetable from the very earliest traces of life to the present—some 600 million years ago—multicellular animals first appear in the rocks. Fossils of these sea creatures were first discovered in 1946 at a place called

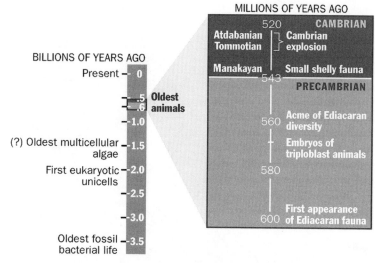

BILLIONS OF YEARS AGO

Present — 0

—.5 **Oldest animals**
—.6

—1.0

(?) Oldest multicellular algae —1.5

First eukaryotic unicells —2.0

—2.5

—3.0

Oldest fossil bacterial life —3.5

MILLIONS OF YEARS AGO

520   CAMBRIAN

Atdabanian Tommotian   Cambrian explosion

Manakayan   **Small shelly fauna**
543
PRECAMBRIAN

560   **Acme of Ediacaran diversity**

Embryos of triploblast animals

580

First appearance of Ediacaran fauna
600

THE OLDEST ANIMALS known now go back beyond the Cambrian "explosion" of life forms 530 million years ago to a period between 800 and 543 million years before the present. These fossil Ediacaran life forms were unlike any creatures with which we are familiar.

Ediacara (ee-dee-AK-ara) in southern Australia, and have since been found on all continents except Antarctica. ("Ediacara" is an Aboriginal word meaning "foul waters.") Scientists began to be aware of their enormous significance only during the 20th century's closing decade.

Ediacaran organisms were flat, composed of only two sheets of cells, like a balloon inside a balloon, and some were built of numerous tube-like sections that were joined together. Some scientists think they resemble early jellyfish, anemones, or corals, while others consider them dead ends or "evolutionary experiments" that left no descendants. Most living things with which we are familiar are called triploblasts, because they develop embryologically from three layers of cells. They, too, first appear in the Ediacaran deposits.

Cambrian rocks (543–580 million years old) famously show an "explosion" of many different body plans, or phyla, including the first appearance of our familiar animal groups, but the "flat" Ediacarans dominated the seas right up to the dawn of the Cambrian period. Some were jellyfish-like, had no shells or backbones, and ranged from tiny to a few feet in length. Fossils show neither complete internal organs nor a mouth or anus. Others, known as rangeomorphs, superficially resemble fern fronds anchored to the seafloor by stalks.

A Canadian site discovered in 2004 has revealed a preserved ecosystem that was buried under a layer of volcanic ash. Just as in Pompeii, every organism died where it lived. "When you walk on a rock surface," says Guy Narbonne of Queen's University in Ontario, "it's like walking on a 565-million-year-old seafloor." Narbonne hopes to study the positions and spacing of the organisms, which may show how they lived and fed, by comparing the fossils to the layout of known seafloor ecologies.

Ediacaran organisms dominated the seas for at least 50 million years and then became extinct during the Cambrian, when they were replaced by a host of new phyla, all triploblasts.

In 2004, the International Union of Geological Sciences named a new period, the Ediacaran—the first time since 1891 that the geological timescale was amended. The Ediacaran period lasted from 620 million years ago to the onset of the Cambrian period 543 million years ago. The identification of this period has begun intense work on understanding further divisions of the so-called Archean and Proterozoic eons, an enormous time span that covers 90 percent of geologic history.

ENDOSYMBIOSIS
A Beast with Five Genomes

Symbiotic organisms—animals or plants that have a mutually beneficial relationship—are a well-known marvel of evolution. Clownfish hide safely in sea anemones, scavenging remains of its meals in the midst of its poisonous tentacles. The fish cleans the invertebrate; the anemone's stingers protect the fish from predators. Ant colonies that live in acacia trees are another striking example: the insects repel all invaders that might harm the tree, in exchange for food and shelter.

Over the last few decades, however, researchers have found evidence that many types of cells (once thought to be the smallest unit of life) are actually combinations of originally different creatures living together as a cooperative entity. It is even possible that all complex cells evolved in this manner.

A well-studied case in point is the one-celled animal (protozoan) *Paramecium*. The tiny "hairlike" filaments (cilia) by which it moves with coordinated whiplike motions are actually separate creatures moving in unison. Similarly, tiny termites have incorporated thousands of even smaller microorganisms, which have evolved in their guts to break down wood; they are, in effect, the termite's digestive system. Component organisms of this kind are called endosymbionts, helpful creatures that live inside (endo).

Equipped with their own DNA (sometimes called nonchromosomal genes), these tiny life forms reproduce asexually along with their host (or fellows). One extraordinary animal, *Mixotricha paradoxa*, which lives in the sealed micro-universe of a termite's gut, is a compound organism that has been called "the beast with five genomes." Although at first it appears under the microscope as a species of one-celled animal, studies have shown that it is made up of five distinct kinds of smaller creatures, each with its own DNA. Reef-building corals are also now known to have five different genomes of once independent organisms.

See also EUKARYOTES; MITOCHONDRIAL DNA; PHYLA

ENGLAND, DARWINISM IN

Within two decades after publication of the *Origin of Species* (1859), Darwinian evolution was completely accepted by British science, though his theories of natural and sexual selection, among others, were not pursued by experimental biologists until six decades later.

Yet England was deeply conflicted about integrating Darwinism into its national belief system. On the one hand, a crude interpretation of "survival of the fittest" was used to justify British imperial rule over "backward" nations and peoples. But a tradition of fair play and Christian values urged both kindness and social justice. Thomas Huxley argued fervently that if survival of the fittest is truly the law of nature, then civilized people must "turn their backs on the law of the jungle" and seek a higher ethic based on compassion and generosity.

Evolutionary themes began to permeate English literature. Poet laureate Alfred, Lord Tennyson, had infused his popular poem *In Memoriam* (1850) (a favorite of Queen Victoria's) with evolutionary ideas even before Darwin's *Origin of Species* had appeared. One of Huxley's students, H. G. Wells, was the first to write popular fantasy novels on time travel, worlds of alternative evolutionary results, and other scientific themes. Even Arthur Conan Doyle, the creator of Sherlock Holmes, got into the act with his fantasy-adventure *The Lost World* (1912), the tale of a land where ape-men and dinosaurs still roam.

By the end of the 19th century there was no longer any uproar over teaching Darwin in British schools. (Ironically, when Thomas Huxley became minister of education he insisted that the Bible still be taught—but as "great literature" rather than science or religion.)

MIXED FEELINGS ABOUT DARWIN caused English officialdom to withhold honors while he was alive, but he received a national hero's funeral. London's Natural History Museum installed his statue on the main stairway but later moved it to a café.

London's Natural History Museum, once the stronghold of its antievolutionist founder, Richard Owen, became England's bastion of evolutionary biology. In 1885 a large marble statue of Darwin was installed on the stairway of the Great Hall, greeting every visitor who entered, and replacing one of Owen. (Some years ago, Darwin was moved to a small upstairs cafe, which he shares with a companion figure of Thomas Henry Huxley. More recently, the museum opened a permanent Darwin wing to exhibit thousands of preserved specimens from the museum's research collection.)

During the 1890s Darwinism began to fade from the attention of British science. However, at Oxford, Darwin's disciple E. Ray Lankester promoted the work of the little-read German geneticist August Weismann and encouraged the study of population genetics, which revitalized Darwinian theory. In 1898, Lankester became director of London's Natural History Museum.

Biologist Julian Huxley, grandson of Thomas, helped develop ethology, the evolutionary study of animal behavior. He was also a founder of the modern Synthetic Theory (which he named), combining Darwinism with population genetics, paleontology, and other fields. A succession of Oxbridge evolutionists, including R. A. Fisher, J. B. S. Haldane, and John Maynard Smith, devoted their careers to the evolution of adaptations.

Julian Huxley also churned out popular essays on his grandfather's favorite themes: social and ethical issues arising from evolution, science, and religion. His brother Aldous's novels, such as *Brave New World*, explored such kindred issues as eugenics and mass production of test-tube babies.

> The British government never honored Darwin in life, but in 2000 his face replaced Charles Dickens's likeness on the 10-pound note.

In the 1990s, Down House, Darwin's home near Bromley, in Kent, was restored and turned into a charming museum by English Heritage, barely in time to save it from the ravages of time and the wrecker's ball. The British government never officially honored Charles Darwin during his lifetime (he was, however, buried in Westminster Abbey), but in 2000 his image appeared on the Bank of England 10-pound note, replacing that of Charles Dickens. Two years later, in a BBC public poll to determine the "Ten Greatest Britons," Darwin ranked fourth, somewhere between Winston Churchill and Horatio Nelson, the naval hero who won the battle of Trafalgar. (Further down the list were William Shakespeare, Princess Diana, and John Lennon.) In September 2008, a vicar of the Anglican Church offered a heartfelt public apology to Darwin on behalf of England's state religion.

The Darwin Correspondence Project at Cambridge has taken on the task of publishing some 15,000 letters to and from Darwin, in a series that will eventually consist of some thirty volumes. It is the largest scholarly project ever attempted about one person.

See also DARWIN CORRESPONDENCE PROJECT; DARWINISM, NATIONAL DIFFERENCES IN; DOWN HOUSE; HUXLEY, SIR JULIAN SORELL; LANKESTER, E. RAY

ESSENTIALISM
Belief in Ideal Types

One of the great shifts in conceptualizing the natural world, according to Harvard evolutionist Ernst Mayr, has been the change from "essentialism" to population thinking. Charles Darwin and Alfred Russel Wallace helped create this modern outlook, which departed from 2,000 years of intellectual tradition—although Darwin himself sometimes appeared confused as to whether it was individuals or species that should be the units of study.

Essentialism is a way of looking at nature that goes back to the ancient Greek philosophers. Plato and his student Aristotle believed an "ideal" reality or "essence" lies behind everything we perceive in the world. Plato's famous allegory of the cave was meant to point up the illusory nature of what we think we see. Viewers in his cave, like the audience in a movie theater, react to flat, projected shadows on the wall, and not to reality.

In human art, essentialism does work. A craftsman makes a chair, for instance, which is only an imperfect representation of the ideal "Chair," a type he has in his mind. But it does not therefore follow, as Plato believed, that plants and animals in nature are also "imperfect" models of ideal types. That conclusion implies they began as distinct ideas in the mind of God, which is something we cannot know by deduction, inference, or observation. (We can believe it on faith, but then we leave the realm of natural science.)

Linnaeus and other early naturalists were typologists: they ignored variations of individuals in their constant search for the perfect "type specimen" of each species. Despite their efforts to accept the evidence of observation, their view of nature was really a continuation of the old essentialist outlook. (Even today, in such trivial matters as so-called beauty pageants, judges still seek Miss World or Miss America—the individual who comes closest in their opinion to the "ideal.")

In his essay "Darwin and the Evolutionary Theory in Biology" (1959), Ernst Mayr argued that essentialists believe

> There are a limited number of fixed, unchangeable "ideas" underlying the observed variability [in nature], with the *eidos* (idea) being the only thing that is fixed and real, while the observed variability has no more reality than the shadows of an object on a cave wall. . . . For the typologist the type (*eidos*) is real and the variation an illusion, while for the populationist, the type (average) is an abstraction and only the variation is real. No two ways of looking at nature could be more different.

So long as naturalists accepted the premise of essentialism, the idea of the fixity of species did not seem at odds with nature. Today's creationists still hold on to a notion of "kinds," which is the same as "ideal types." Divine "ideas," they believe, shape each species, which reproduce "after their kind." When variations are shown to them that fall between "types," they insist on assigning them to one "kind" or another and thus never confront the phenomenon of blurred boundaries, which more accurately describes nature.

Darwin and Wallace, after systematically acquainting themselves with the enormous range of variability in organisms, pointed out that nature was not so tidy. During the first few decades of the 20th century, mathematicians developed new tools for analyzing population phenomena and describing variability. Population geneticists and biometricians made variability—not ideal types—the focus of their studies. In the 1930s, these newer methods were combined with the older disciplines of paleontology, comparative anatomy, and systematics (taxonomy) to shape the modern approach to evolutionary biology. Because it fused together several diverse sciences, it is known as the Synthetic Theory.

See also *NATURPHILOSOPHIE;* POPULATION THINKING; SYNTHETIC THEORY

> "For the populationist, the type is an abstraction and only the variation is real."
>
> —Ernst Mayr

ETHOLOGY
Biologically Based Behavior

Charles Darwin initiated a new approach to understanding animal behavior in his book *The Expression of the Emotions in Man and Animals* (1872). At a time when zoologists focused on the classification and anatomy of preserved specimens, he studied the behavior of live animals—including his own dogs and cats—and tried to work out how their behavior could have evolved.

Curiously, Darwin's exciting lead in watching animals' natural behavior was not followed for 60 years, but instead had the opposite effect of sending scientists into their laboratories to work on dead or captive specimens. In the 1930s, the Austrian Konrad Lorenz and later the Dutchman Niko Tinbergen began to study animals behaving naturally in the field, with the aim of understanding behavior from an evolutionary perspective.

They were especially successful at observing the interaction between innate, species-specific behavior and learning. They also gained new understandings of animal signals and communication, a line of research Darwin had begun in *Expression of the Emotions.* Their work with the territorial behavior of stickleback fish, courtship displays of ducks and geese and "imprinting" of ducklings and goslings on their mothers, and social and maternal behavior of herring gulls gave new impetus to the study of behavioral evolution.

After several decades of innovative work, however, during which Lorenz and Tinbergen shared a Nobel Prize (with bee expert Karl von Frisch), the ethologists got bogged down in reductionist explanations of "drive," "motivation," and "releasing mechanisms" that stretched their explanatory models to the breaking point. A spate of facile books—including Lorenz's *On Aggression* (1966), Robert Ardrey's *The Territorial Imperative* (1966), and Desmond Morris's *The Naked Ape* (1967)—also hurt their credibility by prematurely applying observations of fish and birds directly to humans. Several used ethological studies to "prove" that humans are genetically hardwired to be aggressive, incorrigible, territorial killers.

Those scientifically questionable—but wildly popular—books helped to launch a subfield known as evolutionary psychology, which purports to elucidate the biological underpinnings of human learned or cultural behavior. Although the search for a science of human nature is worthwhile, it still remains a quest more than a body of knowledge. One notorious attempt at evolutionary "explanation," *A Natural History of Rape: Biological Bases of Sexual Coercion* (2000) by Randy Thornhill and Craig T. Palmer, concluded that men who rape women are following an ancient "adaptive" strategy for spreading their genes. The book is a prime example of how this offshoot of ethology has degenerated into junk science and fanciful "just-so" stories.

During the 1970s, many ethological concepts were abandoned as investigators of animal behavior redefined their subject. First, there was "sociobiology," proposed by ant expert E. O. Wilson as a comprehensive evolutionary approach to all behavior, including that of humans. "Behavioral ecology," a more circumscribed study, focuses on how animals interact and evolve with their environments. Scientific fads and fashions come and go, but the evolutionary roots of human nature, language, and social behavior remain tantalizingly elusive.

See also EXPRESSION OF THE EMOTIONS; "JUST-SO" STORIES; LORENZ, KONRAD; TINBERGEN, NIKOLAAS

EUGENICS
Breeding Better Humans

I n 1883, Charles Darwin's cousin Sir Francis Galton coined the word "eugenics" from a Greek root meaning "noble in heredity" or "good in birth." He had first advanced his theories in 1865 in a series of magazine articles, which were later expanded into his book *Hereditary Genius* (1869).

Pursuing the social implications of Darwinism, Galton wanted to create a science that could improve humanity by giving "the more suitable races or strains of blood a better chance of prevailing speedily over the less suitable." He believed natural selection should be given an artificial assist, much as the domestic stockbreeder improves horses or cattle. According to historian Daniel J. Kevles, Galton

> suggested that the state sponsor competitive examinations in hereditary merit, celebrate the blushing winners . . . foster wedded unions among them at Westminster Abbey, and encourage by postnatal grants the spawning of numerous eugenically golden offspring. . . . The unworthy, Galton hoped, would be comfortably segregated in monasteries and convents, where they would be unable to propagate their kind.

As historian James Moore has pointed out, Darwin himself was worried about the effects on families of producing offspring with close cousins—such as his own. (His wife Emma was his first cousin.) In 1871, he strongly suggested to Parliament that they include a survey of close cousin marriage in the next British census, but they flatly turned him down. In *The Descent of Man,* published that same year, Darwin railed against "ignorant members of our legislature [for] rejecting with scorn a plan for ascertaining by an easy method whether or not consanguineous marriages are injurious to man."

Once almost obligatory in all biology textbooks, the promotion of eugenic programs became anathema after the disastrous, barbarous attempts to create a "master race" in Nazi Germany. However, the notion lingered on for several decades in America, bolstered

CHARLES DARWIN'S SON LEONARD (below, right) co-chaired the Third International Congress on Eugenics in 1932 at the American Museum of Natural History with museum president Henry Fairfield Osborn, at left. At bottom, the tree of eugenics is shown rooted in many sciences.

LIKE A TREE
EUGENICS DRAWS ITS MATERIALS FROM MANY SOURCES AND ORGANIZES THEM INTO AN HARMONIOUS ENTITY.

by miscegenation laws in southern states, immigration quotas, and legislation of sterilizing criminals and the "feeble-minded." Eventually, it became clear that the presumed scientific rationale for these social policies did not, in fact, exist.

But although Galton's term for the genetic betterment of humans is tarnished, a new field of quasi-medical practices has arisen that involves genetic testing, parental selection, manipulation of embryos, and human genetic engineering. Over all these, Galtonian eugenics casts a long shadow.

See also ARISTOGENESIS; GALTON, SIR FRANCIS; LEBENSBORN MOVEMENT

EUKARYOTES
Cells with Nuclei

Eukaryotes are living cells that have a central nucleus suspended in cytoplasm, the whole wrapped in an outer cell membrane, like an egg yolk surrounded by protein, enclosed in a shell. Nucleated cells make up trees, starfish, fish, insects, birds, mammals—what we usually think of as our fellow organisms. But for millions of years before cells with nuclei appeared, living prokaryotes (cells without nuclei) dominated the Earth. Their close relatives are still with us in the form of cyanobacteria, archaea, and other microbes.

According to biologist Lynn Margulis, "the sudden appearance of eukaryotes in the fossil record, and the absence of any intermediate forms, suggest that they were not the gradual result of genetic mutation but a technical innovation forged by communities of symbiotic bacteria." In her view (built on the work of Lemuel R. Cleveland and others), nucleated cells originated when non-nucleated bacteria engulfed one another, forming compound organisms.

Cleveland studied the tiny universe in a termite's gut: millions of microscopic creatures swallowing and merging with others in remarkable ways. Some of those that are eaten are not digested and destroyed, but live cooperatively within other microbes, producing parasitic offspring to live within their host's descendants.

In many cases the relationship is not really parasitic, but mutually beneficial. Cells within cells become interdependent (endosymbiotic) and form stable organisms—new wholes greater than the sum of their parts.

Biology has long taught that we are complex, coordinated colonies of cells, but in the new view each of our cells may in turn be a colony made up of cooperating organisms.

See also ENDOSYMBIOSIS

EVO-DEVO
Evolutionary Developmental Biology

Evolution of development (Evo-Devo for short) is a relatively new biological field that seeks to analyze the roles of genes and embryos in evolution. Charles Darwin hinted at that pivotal connection when he wrote the botanist Asa Gray in 1860: "Embryology is to me by far the strongest single class of facts in favor of the change of form." Evolution of humans from marine invertebrates, he believed, was no more far-fetched than the commonplace miracles of transformation that occur within every maternal womb. For the better part of a century, however, the relationship between embryos and ancestors seemed intractable.

Embryology, despite a promising start during the nineteenth century, seemed to have reached a dead end by the twentieth, with the collapse of Ernst Haeckel's theory that developing embryos replay the "stages" of evolution. While it seems plausible in a general way, the Biogenetic Law ("ontogeny recapitulates phylogeny") turned out to be unworkable and misleading when applied to specifics. When Haeckel's theory went down in flames, it put a damper on attempts to link evolution and development for almost a hundred years.

In the 1960s, genes—the nucleotide sequences in DNA that encode amino acid chains and proteins—became a major focus and touchstone of biological research. Geneticists first sought to understand how the genetic code was passed on to successive generations, rather than how genes express themselves to build anatomy in individuals. Since then, the emphasis has turned to gene expression.

Huxley liked to remind people that every individual begins as a tiny aquatic creature in the maternal womb.

In their current quest to understand the elusive "translation" from DNA "blueprints" into living organisms, developmental biologists believe they have achieved a preliminary understanding of how bodies are formed. According to Sean B. Carroll of Howard Hughes Medical Institute, the key is the genetic regulatory networks that orchestrate pattern formation and cell fate decisions, whose first wiring principles are now known.

During the 1980s, new studies began to reveal some startling surprises. First, evo-devo specialists found that a small number of master genes are responsible for all basic body plans and parts throughout the animal kingdom. These few ancient genes, recycled and repurposed, control embryonic development in organisms of striking diversity.

Within the DNA (of which only 1.9 percent codes for proteins that build bodies) there is a hierarchy, in which a very few master, or "toolkit," genes control the expression of those at lower levels. Some of these master genes act upon genetic "switches" in DNA, telling other genes to turn on or off, thereby producing thousands of different species from relatively small and very similar sets of genes. Other recently discovered genes, known as *Hox* genes, regulate the number and kinds of repeating parts, such as segments in insects or vertebrae in mammals.

Very different-looking animals, it turns out, often share genes that are very similar. "There aren't new genes arising every time a new species arises," Brian K. Hall, from Dalhousie University in Nova Scotia, told the *New York Times* (June 26, 2007). "Basically you take existing genes and processes and modify them, and that's why humans and chimps can be 99 percent similar at the genome level." Complex new species, rather than requiring the "small, insensible" changes Darwin postulated, or the many new mutations some biologists had expected, instead often evolve by using already existing genes in different ways.

In addition, it has long been known that changes in the rate of development can produce the hallmarks of different species. (*Heterochrony*—literally, "different timing"—is the general term for such processes.) One common alteration of timing, known as neoteny, preserves the characteristics of juveniles into adulthood—as in the axolotl, a Mexican salamander that retains external gills throughout its life. Many anthropologists over the years have remarked on the unmistakable resemblance of adult human facial proportions to those of juvenile or infant chimps or gorillas. With our lack of heavy brow ridges and snouts and our comparatively flat faces, we may well be neotenous apes.

Evo-devo studies have also dislodged the commonly held belief that large, complex creatures must have many more genes than smaller, supposedly simpler animals. In fact, recent studies have shown that mice and humans have a surprisingly similar number—some 20,000 genes—and even much simpler invertebrates have tens of thousands. When comparing mouse and human genomes, biologists have found that at least 99 percent of all our genes have counterparts in mice.

In his 2005 book *Endless Forms Most Beautiful*, Carroll reports an even more remarkable comparison than mice and humans: "Most of the genes first identified as governing major aspects of fruit fly body organization were found to have exact counterparts that did the same thing in most animals, including ourselves." One of the most exciting of these discoveries, according to Carroll, is that such basic structures as eyes are regulated by the same ancient eye-genes (known as *Pax6*) in widely different lineages—and were not reinvented many times by convergent evolution, as had been thought.

The eye-gene controllers present in humans are identical to those in such extremely distant relatives as horseshoe crabs, octopuses, centipedes, and flies, with their multifaceted compound eyes. Thus, toolkit genes go back beyond primates, mammals, and even vertebrates in our common evolutionary history—leaving genetic footprints in our DNA from the long, winding pathway of our immense journey.

> "To think heredity will build organic forms without mechanical means is a piece of unscientific mysticism."
>
> —Wilhelm His, "On the Principles of Animal Morphology," 1888

EVOLUTION
Species Change through Time

The word "evolution" in the 17th century meant an "unrolling" or "unfolding" of a plan that was already in place, as in the theory of preformation. Similarly, it also meant embryological development of an individual.

When Darwin began turning his attention to "that mystery of mysteries, the origin of species," the idea that species change through time was called "transmutation" or the "development hypothesis." In the Darwin and Wallace papers read before the Linnean Society in 1858 announcing the principle of natural selection, neither author used the word "evolution," nor did it appear in the 1859 first edition of *Origin of Species*. (It first appears in the fifth edition of 1869.)

As with "survival of the fittest," it was British philosopher Herbert Spencer who originated the term "evolution" in this context. Wallace and Darwin later adopted it, but in the public mind, both phrases have become completely identified with the theories of Charles Darwin.

By "evolution" biologists mean that the change in gene frequencies of populations over the generations produces biological changes that can result in new species. Darwin called it "descent with modification": a slow process, usually operating over hundreds of thousands, even millions, of years.

There are four commonly confused meanings of evolution, which should be kept separate and distinct: (1) the general process of populational and species change, which is an established scientific fact; (2) inevitable "progress" from lower to higher life forms, a discredited notion; (3) the particular history of the "branching bush" of life and the origin of various clades, or phylogenies, which are interpreted from the fossil record and biochemical studies; and (4) the main "engine" of evolution, which Darwin and Wallace proposed as "natural selection," along with other possible mechanisms.

Here are some of the major arguments and objections that opponents of evolution never tire of raising—and some answers from the perspective of evolutionary biology:

1. *Fact or Theory?* Evolution became established as fact not because it won debates among armchair philosophers or logicians, but because it unified thousands of disparate observations by comparative anatomists, field naturalists, geologists, paleontologists, botanists, and (later) geneticists and biochemists. Without the overarching concept of a changing world in process over eons of time, what we consider modern science would not exist. That species are related through common ancestry is supported not by one argument or chain of reasoning, but by scores of interlocking research fields, each of which feeds into and supports the rest. Evolution is as well established a fact as gravitation. To paraphrase a leading paleontologist, apples are not going to stop falling in midair while scientists debate whether Newton's law of gravitation has been superseded by Einstein's theories. And species keep on changing over time, while we continue to search for the why and how of evolution. If one insists evolution is merely one interpretation of nature, what is the alternative? That the thousands of dinosaurs and the species that came before and after them appeared full-blown and had no common connections? Such a model, whether called religion or "creation science," cannot lead to fruitful inquiry. It is an "answer" that stops all further questions.

2. *"General" Evolution vs. Speciation.* While some critics concede that new species (of fruit flies, for instance) have been produced in laboratories, they claim that general evolution has never been experimentally demonstrated. By this, they mean breeding a succession of progressively higher or more complex species. But there is no such theory of general evolution; Victorian notions of inevitable progress in biology are outmoded.

3. *Transitional Forms.* The oft-repeated claim that there are no transitional forms is demonstrably false. The Karroo region of South Africa, for instance, is a vast graveyard of the remains of mammal-like reptiles, a whole array of species whose anatomy was intermediate between reptiles and mammals. There is the famous *Archaeopteryx* and the dromeosaurs, or "feathered dinosaurs," from China, demonstrating the transition between dinosaurs and birds. The past few decades have brought to light early limbed fishes as well as "walking whales." And the African hominid fossils represent creatures with humanlike dental patterns, small apelike brains, arms longer than those of humans but shorter than those of modern apes, with pelvis, feet, and legs for upright walking.

> There are four commonly confused meanings of evolution, which should be kept separate and distinct.

Transitional fossils are notably rare because, according to current theory, most species remain stable for long periods. When change occurs it is fairly rapid (in geologic terms) and often among small, isolated populations. The fossil record has been compared to freezing a multilevel parking garage in time. Most cars would be found on the various floors, with very few on the ramps. The amount of time each car spends on the ramp is short compared to the length of time it remains parked, yet each must have traveled the ramp.

Other evidence of transition is found in the geographical distribution of living species. On Pacific island chains, for instance, biologists have tracked widespread populations of intergrading species over thousands of miles, discovering intermediate forms from one end of the island chains to the other.

Darwin himself was so impressed with a series of such geographic variations in Amazonian butterflies over a vast area of rain forest, he was moved to remark, "We feel to be as near witnesses, as we can ever hope to be, of the creation of a new species on this earth." Among living creatures, there is a series of gradual, intermediate species between lizards and snakes, thrushes and wrens, sharks and skates, and many grasses and trees.

4. *Evidence and "Proof."* There is a common misconception that Darwin thought he had "proved" by logic that species had evolved. He was, in fact, a much more subtle thinker and philosopher of science. "The change of species cannot be directly proved," he wrote a friend, "and . . . the doctrine must sink or swim according as it groups and explains [disparate] phenomena. It is really curious how few people judge it in this way, which is clearly the right way." (A few years later he wrote that he was "weary of trying to explain" this point, since most could not grasp it.)

5. *"Holes" and Questions.* That there are "holes" and unanswered questions in evolutionary theory (just as there are in particle physics) is undeniable, which is normal for a healthy science. At the Oxford debate in 1860, Thomas Henry Huxley asked the audience to imagine being lost in the middle of the countryside on a dark night. If someone offered a dim, flickering lantern, would they reject it on the grounds that it gave an imperfect light? "I think not," said Huxley, "I think not."

6. *Tautology of "Survival of Fittest."* This old chestnut that evolutionary theory is built on the circular reasoning that "the survivors survive" was laid to rest long ago. Critics argue that "only the fittest survive" is an untestable proposition without a uniform definition of fitness and, therefore, meaningless as an explanation. Fair enough, and it is true that some fuzzy-thinking scientists have concocted fanciful "just-so" stories about the origins of particular adaptations.

But whatever the fate of those archaic catch phrases "natural selection" and "survival of the fittest," the heart of the Darwin-Wallace theory remains sound: overproduction of offspring in nature, genetic variability, and a sorting process, which result in both long-term stability and episodic divergence of populations. Increasingly, new research is focusing on gaining a deeper understanding of these mechanisms of genetic variation and differential sorting as they occur on various levels within the same organism.

7. *"Just History," Not Science.* Some assume that research and inquiry into living things must lead to the formulation of fixed "laws" like those of chemistry or physics. Dissecting the anatomical structures of extinct creatures, working out their distribution in space and time, and reconstructing past climates and ecologies may seem to be "just history" to a physicist or chemist, but to most biologists it is certainly science.

The kind of scientific illiteracy that rejects evolution, regarding it as a humanist religious belief, can result in serious errors in understanding and even loss of human life. For example, in 1984, Dr. Leonard L. Bailey of the Loma Linda University School of Medicine tried to save the life of "Baby Fae," an infant born with a severely malformed heart. He surgically implanted a baboon's heart, but the organ was quickly rejected and the child died.

Soon after, he was asked why he hadn't used a chimpanzee's heart instead, which would

"The change of species cannot be directly proved, and [the theory] must sink or swim as it groups and explains [disparate] phenomena."

—Charles Darwin

have offered a far better chance of success because of the chimp's much closer evolutionary propinquity and genetic fit. Dr. Bailey replied that he "didn't believe in evolution," and, in any case, couldn't see what it had to do with the practice of medicine.

See also ADAPTATION; ALLOPATRIC SPECIATION; BRANCHING BUSH; COEVOLUTION; CONVERGENT EVOLUTION; DARWINISM; DARWIN'S FINCHES; EXAPTATION; FITNESS; GENETIC DRIFT; HOMOLOGY; NATURAL SELECTION; NEO-DARWINISM; ORTHOGENESIS; OXFORD DEBATE; PIGGYBACK SORTING; SYNTHETIC THEORY; TANGLED BANK

EVOLUTION, MYSTERIES OF
Major Unsolved Problems

Charles Darwin candidly admitted there were "great difficulties" and unsolved puzzles in the study of evolution. Science teachers often make the grave mistake of glossing over these areas of ignorance, attempting to provide an answer for everything. Creationists rightly puncture such pretensions to knowledge, but then go on to scoff at evolutionary theory for what it hasn't solved, as if that disproves all of modern biological knowledge.

The following unresolved questions pose the most profound challenges to evolutionary science. Despite years of research and discussion (and, in some cases, promising directions), there are still no settled answers.

1. *Origin of Life.* How did living matter originate out of nonliving matter? Was it a process that happened only once or many times? Can it still happen today under natural or artificial conditions? Did it evolve out of the kind of growth and replication processes we see in organic chemical compounds or on an entirely different basis?

2. *Origin of Sex.* Why is sexuality so widespread in nature? How did maleness and femaleness arise? If it is necessary to maintain genetic variability, why can many microorganisms and even some vertebrates (e.g., whiptail lizards) do without it? How can one account for such phenomena as parthenogenesis: frog eggs, for instance, can produce tadpoles if they are pricked by pins or stimulated by electric current, without having been fertilized by male sperm.

3. *Origin of Language.* How did human speech originate? We see no examples of truly primitive languages anywhere on Earth today; all are evolved and complex. Can the answer be sought in the structure of the brain, experiments in teaching apes to use symbols, analysis of natural animal communication systems? Or is there no way to find out?

4. *Origin of Phyla.* What is the evolutionary relationship between existing phyla and those of the past? Why did so many different body plans arise during the Cambrian "explosion"? There is still no agreement on how many there are today, how many we know from the fossil past, and which may have come out of which. Transitional forms between phyla are almost unknown.

5. *Cause of Mass Extinctions.* Asteroids are currently in vogue, but far from proven, as a cause of worldwide extinctions. And though punctuated equilibrium theory helps account for the so-called sudden appearance of new groups and long persistence of others, it has raised many new questions about the stability and extinction of species.

6. *Relationship between DNA and Phenotype.* Can small, steady changes (micromutations) account for evolution, or must there be periodic larger jumps (macromutations)? Is DNA a complete blueprint for the individual? What are the influences and constraints in its expression? How are the presumed "instructions" in DNA translated into living forms? Are there any circumstances under which environment or behavior can work "backwards," influencing changes in the DNA? Is most DNA less a "blueprint" for organisms, or a series of "switches" that can allow or block certain expressions? Are there physical or mechanical properties of matter that determine the forms of cells and embryos?

Evolution still presents many unsolved puzzles in the ongoing quest to understand nature.

7. *How Much Can Natural Selection Explain?* Darwin never claimed natural selection is the only mechanism of evolution. Although he considered it a major explanation, he continued to search for others, and the search continues. Natural selection is an eliminative process that does not explain the generation, proliferation, and direction of variations.

To dismiss these open questions with pseudo-answers just to fill in unanswerable gaps is intellectually dishonest and no service to science. How much better to admit and identify areas of profound ignorance and challenge the next generation to explore them.

See also DINOSAURS, EXTINCTION OF; PHYLA

EVOLUTIONARY PROGRAMMING
Computer-Based Natural Selection

Evolution has been hijacked from the world of the living and is now a proven method of designing complex machines. The Boeing 777 airliner, for instance, has an engine whose turbine geometry evolved inside a computer. Using special software, the computer can combine a daddy jet engine and a mommy jet engine and select the "offspring" that combines the best features of both—which can then be "mated" with other hybrids. Deere & Co. uses computer natural selection programs to evolve optimal daily schedules for factory assembly lines. Natural Selection, Inc., in La Jolla, California, creates evolutionary programming that produces the "fittest" designs for medical testing.

John Holland of the University of Michigan has been a central figure in evolutionary computation since the early 1960s, when he invented genetic algorithms. These mathematical procedures for solving problems, he found, can use computers to manipulate potential solutions as if they were living organisms. You can mate them, breed them, crossbreed them, introduce random variations, and set up "rules" for what constitutes the "fittest." For 20 years, existing computer hardware could not accommodate generations of cyber-evolution, but recent leaps in their memory and capacity have at last given cyberorganisms a universe in which to evolve.

In 1993, engineering professor Andrew Keane, from the University of Southampton in England, tackled the problem of designing a better space-station girder for American astronauts. He devised strings of numbers expressing the girder's thickness, material, angle of attachment, and other requirements. The numbers were analogous to genes, with each string of numbers representing a chromosome.

Keane copied this "genome" until he had produced a diverse founding population, then ran his evolution program on eleven networked computers. According to *U.S. News and World Report* (July 27, 1998), "For several days the truss designs had cybersex—they swapped digital genes with random abandon. . . . Those [designs] that suppressed vibration best yet remained lightweight and strong were rewarded with greater fertility. Generation by generation, the fittest got fitter. The program threw occasional random mutations among the competing genomes to provide a little extra variety."

Fifteen generations and 4,500 different designs later, an optimal truss emerged, looking vastly different from the ones conceived of by NASA's human engineers. According to Keane, the lumpy, knob-ended result looked somewhat like a leg bone. Tests on the evolved models proved them superior to those designed by humans.

In London, in 2004, university scientists developed a technique for "evolving" a superior "breed" of Formula One racing cars. Researchers were able to knock crucial tenths of a second off lap time. Using genetic algorithms, Peter Bentley and his colleagues "evolved" a faster car by a process akin to natural selection. They introduced variability of parts into a "natural selection" program for optimal overall speed and performance. The computer generated small "adaptations" in wing height, suspension stiffness, gear ratio, tire rubber type, and about 70 other parameters. This simulation of the evolutionary process resulted in slight—but winning—increases in speed.

Evolutionary selection reliably produces highly efficient designs.

In a computer, one can mate a daddy jet engine to a mommy jet engine and select the best of the offspring.

EVOLUTIONARY PSYCHOLOGY

See ETHOLOGY

EXAPTATION
New Uses for Old Structures

One of the great puzzles of adaptation is how certain structures that work so well developed by stages. If birds' wings were not designed to fly from the first, for instance, why would their ancestors have started growing them? As Stephen Jay Gould phrased the problem, "What good is half a wing?"

Studies of anatomy, fossils, and the behavior of living animals suggest an answer to this brain-teaser. Anatomical structures used in one kind of behavior often can be shifted or co-opted for another function.

A favorite example of paleontologist Elisabeth Vrba's is the African black heron, a wading bird with well-developed wings that is perfectly capable of flying. But the bird has found another major function for its wings. While fishing, it spreads them around its head to eliminate glare from the surface of the water. Thanks to this built-in sunshade, it can get a much better view of the fish and tadpoles it seeks. (Also, the prey may seek the shade.) If most other birds should become extinct, someone might look at the African heron and conclude that flight was only a stage in the evolution of glare shields.

When birds first evolved from reptiles, it seems likely that their half-wings served an entirely different purpose than flight. They may have been useful for heat regulation, for instance, or for steadying the animal as it ran along the ground.

Animals often find themselves in different environments than those to which they have become adapted. Sometimes they "make do" with the old structures under the new conditions. For instance, marine iguanas of the Galápagos swim underwater, grazing on seaweed, but they have never developed flippers or webbed feet. Observations of certain populations of deer in Oregon report them habitually feeding on dead fish that are abundant on the riverbanks, although their teeth and digestive systems are supposedly adapted to herbivorous diets.

Over time, new behaviors may create new selection pressures. Old structures may be "made over" as certain variations are preserved and modified by natural selection. Parts of the old reptilian jaw became the ear bones in mammals. In elephants, an organ of smell became a powerful, flexible "hand." And of course, Gould's favorite example, the giant panda, retains a meat eater's teeth, while a wristbone has evolved into a crude "thumb" for holding the bamboo stalks on which it now feeds.

Science writers formerly used the term "preadaptation" for structures that were present earlier, then switched function. The term led to confusion, however, since it suggests foresight, design, or "preplanning" of the ultimate use of that organ. Exaptation is consistent with the idea that each species is the result of a special, unique history—something that cannot be predicted.

See also ORTHOGENESIS; PANDA'S THUMB; TELEOLOGY

FANNING OUT WINGS, these African black herons create glare shields to see fish and tadpoles. These birds illustrate how existing structures can take on new functions during evolution. Photo © by Alan C. Kemp.

EXPRESSION OF THE EMOTIONS (1872)
The Evolution of Behavior

At a time when most evolutionists were dissecting and comparing muscles and bones, Charles Darwin was still ahead of the pack. In 1872, he published the first modern book on the evolution of behavior, *The Expression of the Emotions in Man and Animals*, on which comparative psychology, ethology, and sociobiology are based.

Darwin thought that studying the evolution of bodies without also studying behavior was meaningless. He also realized, as Konrad Lorenz wrote in his introduction to a 1965 reissue of the book, "that behavior patterns are just as conservatively and reliably characters of species as are the forms of bones, teeth, or any other bodily structures."

Darwin described movements specific to species, which were later called "fixed action patterns": a species' particular way of courting, fighting, resting, or feeding. He mentions, for instance, how dogs may turn circles and scratch the ground before going to sleep on a carpet, "as if they intended to trample down the grass and scoop out a hollow, as no doubt their wild parents did." (He credits his grandfather Erasmus as the first naturalist to describe such repeated, stereotyped behaviors in animals.)

In *Expression of the Emotions*, he also discusses general principles that seem "to account for most of the expressions and gestures involuntarily used by man and the lower animals, under the influence of various emotions and sensations." Some, he thought, originated in habits that became "fixed" somehow in the species, like the dog preparing the ground for sleeping even when there was no need to do so.

Another was his principle of "antithesis": Opposite emotions produce opposite expressions. When a dog is aggressive, his ears and tail stiffen and his body strains forward. When he is submissive, he crouches, with ears and tail folded down. Darwin's astute observation is still valid, though today's investigators focus on "reduction of ambiguity" in the messages. Opposite signals for opposite intentions are easy to read.

With characteristic thoroughness and curiosity, Darwin explored the effects of electric stimulation on human facial muscles, expressions of the insane, emotional communication by actors and in works of art, and the development of facial gestures in infants.

If human emotions and their expression formed an unbroken continuity with the natural world, it was only a step to the next conclusion: Animals have minds and thoughts as well as feelings. He did not tackle that question directly, but served as mentor to young George Romanes, who published *Mental Evolution in Animals* (1883) shortly after Darwin's death. (That volume contains a chapter on instinct written by Darwin.)

For *Expression of the Emotions*, Darwin not only collected voluminous observations of animals, but he also wrote to many travelers and missionaries to find out if different peoples in remote places of the earth had the same facial expressions for grief, shock, or anger. He discovered, somewhat to his surprise, that "the same state of mind is expressed throughout the world with remarkable uniformity."

But the main thrust of this book was the attempt to carry the evolutionary argument a step further. Using the comparative method, he sought to show that a human smile was derived from a monkey's grimace of submission, that animals and humans both tremble with fear. Knocking down the wall between human and animal behavior, it was a demonstration of the grand view he had written in a notebook years earlier: "We may all be netted together." And a challenge to future generations to ultimately understand the evolution of human behavior.

See also COMPARATIVE METHOD; ETHOLOGY; LORENZ, KONRAD; ROMANES, GEORGE

SIGNALS OF FEAR, AGGRESSION, submission, and pleasure in animals, and the evolution of these behaviors, were the subject of Charles Darwin's book *The Expression of the Emotions* (1872). He founded the research tradition that led to ethology and sociobiology.

EXTINCTION
Destruction of Species

The history of the past few hundred years includes the extinctions of hundreds of species of plants and animals. Among them, quite a few—including the great auk, dodo, and passenger pigeon—were exterminated by humans who just didn't care. American bison were pulled back from the brink of extinction when only a few hundred were left alive, out of a population that had numbered 40 million. Between 1870 and 1875, buffalo hunters were slaughtering 2.5 million of them annually.

Extinction has always been a fact of life. According to ecologists Paul and Anne Ehrlich in their book *Extinction* (1981), 98 percent of all species that have ever lived have become extinct. There are probably about 10 million species alive on the Earth today, one million species in the Amazon basin alone, of which only 1.5 million have been discovered and given scientific names. Many are now disappearing before they are even discovered, particularly in the tropical rain forests.

Until the late 18th century, naturalists did not imagine that extinction was possible. Each species was believed to be a distinct idea in the mind of God, a link in an unbroken cosmic chain that allowed no gaps. When fossils of strange animals, like mastodons, were discovered, it was assumed that there must be some still living in the vast wilderness areas that had not yet been explored.

As more and more fossils were found (the first dinosaur teeth were discovered only in 1825) and more wilderness settled, the evidence of extinct creatures began literally to pile up. Late in his career, the eminent French anatomist Georges Cuvier had to admit that fossil bones were the remains of extinct species.

Paleontologists have documented several mass extinctions, which wiped out the majority of life on Earth, allowing new forms to radiate and develop. One such "mass dying" occurred after the Cambrian period, eliminating the once-numerous trilobites. Another took place at the end of the Permian period, eliminating most living things on the Earth. Still another was the famous and much-pondered Cretaceous extinction, which ended the 150-million-year reign of dinosaurs as the dominant form of life and ushered in the Age of Mammals.

Human activity, with its destruction of habitats as well as hunting, has been devastating, precipitating what has been called the sixth mass extinction. Although we have yet to see a new species evolve in nature—a very slow process—we often see them end, which can happen very quickly. As British naturalist Sir Peter Scott put it, speaking at a 1972 conference on breeding endangered species, "Living species today, let us remember, are the end products of twenty million centuries of evolution; absolutely nothing can be done when the species has finally gone, when the last pair has died out."

See also DINOSAURS, EXTINCTION OF; FULLER, ERROL; GREAT DYINGS; LONESOME GEORGE; RAIN FOREST CRISIS

IVORY-BILLED WOODPECKER,
extinct since 1950

THYLACINE (TASMANIAN WOLF),
extinct since 1936

YANGTSE RIVER DOLPHIN,
extinct since 2007

QUAGGA,
extinct since 1885

FANTASIA (1940)
Disney's Epic of Dinosaurs

With his innovative feature film *Fantasia*, Walt Disney hoped to elevate animated cartoons to high art. Against a background of classical music, he filled the screen with such grand themes as the unleashing of hellish forces, man's eternal attempt to gain power over nature, and the evolution and extinction of the dinosaurs.

An orchestral score conducted by Leopold Stokowski provided the framework for a series of imaginative visual interpretations. A remarkable vision of evolution and extinction, oddly enough, was set to Igor Stravinsky's stirring and dramatic *Rite of Spring*. (To Disney's dismay, the irascible maestro later called the musical interpretation "execrable" and "the visual complement . . . an unresisting imbecility.") In *Fantasia*'s most famous sequence, Mickey Mouse's meddling nearly destroys the natural order of the universe, to the accompaniment of Paul Dukas's scherzo *The Sorcerer's Apprentice*.

Fantasia's dinosaurs are mighty but slow-witted behemoths caught in a change of climate. The Earth is heating up, and thirsty dinosaurs undertake a vast migration across bleak, parched landscapes in search of water. Actually, geologic evidence points to world climates having gotten cooler, not hotter, at the time of the dinosaurs' demise. Disney's apocalyptic vision of dinosaurian dehydration had been inspired by the discovery of a spectacular "saurian graveyard" in Montana, where hundreds of sauropods apparently had become trapped in mud as their marshes and wetlands dried out. It was a local catastrophe, but in the Disney version it appeared as a global cataclysm.

Renowned conceptual artist-essayist Robert Smithson traced the look of the Disney dinosaurs back to museum painter Charles R. Knight, who, in his 1946 book *Life through the Ages*, described *Tyrannosaurus rex* as "an enormous eating machine."

Fantasia's giant, bipedal "tyrant lizard" is shown with three clawed digits on its small "hands." Years later, Disneyland technicians fashioned a life-size animatronics tyrannosaur based on the same mistaken restoration. When it was almost completed, Disney invited a distinguished paleontologist to the park to check its accuracy. "It's very good, Mr. Disney," said the scientist, "except for one thing. You'll have to change his forelegs; any schoolboy knows tyrannosaurs had two clawed digits, not three."

Disney stared at his dinosaur, thought for a long moment, then shook his head. "Nope," he replied, "I really think it looks better with three."

About thirty years later, in 1990, it turned out that Disney wasn't entirely wrong. Paleontologists discovered a new species of bipedal dinosaur, *Nanotyrannus*, which were smaller relatives of *Tyrannosaurus*. They greatly resembled their larger cousins except for one thing: their forelimbs had three-clawed digits.

See also BROWN, BARNUM; KNIGHT, CHARLES R.; SINCLAIR DINOSAUR

"FIGHT TO THE FINISH: Survival of the fittest is demonstrated by two dinosaurs in the 'Rite of Spring' sequence of Walt Disney's animated classic *Fantasia*." This sentence appears on the Disney publicity still, illustrating a popular misconception of Darwinian evolution.

© Disney Enterprises, Inc.

FITNESS
The Secret of Survival?

Erasmus Darwin, grandfather of Charles, had speculated that in early tribes the biggest, most powerful males would attract more females, father the most children, and thus perpetuate their attributes. Philosopher Thomas Hobbes had thought that in a "state of nature" each man would be "at war with every other man," the stronger directly eliminating the weaker. Outside scientific circles, that is still the popular idea of "survival of the fittest."

Charles Darwin used "fitness" in several ways: (1) in his grandfather's sense of robust individuals; (2) in terms of populations better adapted to particular environments, such as specialists at obtaining particular foods; and (3) those with more complex nervous systems or "improvements in organization." Today geneticists prefer phrases like "high representation of gene frequencies in descendant gene pools." While gaining a precision useful in certain kinds of research, however, this mathematical definition offers no sense of the living creatures.

Alfred Russel Wallace, in his classic *Darwinism* (1889), defined "fitness" from the viewpoint of a naturalist who had spent years observing animals in tropical forests. To Wallace the fittest are "those which are superior in the special qualities on which safety depends":

> At one period of life, or to escape one kind of danger, concealment may be necessary; at another time, to escape another danger, swiftness; at another, intelligence or cunning; at another, the power to endure rain or cold or hunger; and those which possess all these faculties in the fullest perfection will generally survive.

In *Evolution: The Modern Synthesis* (1942), Julian Huxley differentiated between "survival fitness" and "reproductive fitness." The first concerns the individual's success at growing to maturity, finding food, and living long enough to reproduce. "Reproductive fitness" concerns efficiency in leaving offspring—such things as clutch or litter size, ratio of sexes in a population, and competitive attractions for the opposite sex.

Fitness can also mean the range of variability in a population's gene pool, which allows flexibility in crisis situations, such as changes in climate, famine, or disease. Long-term fitness is relative. Species may persist for a very long time, like the dinosaurs, which were certainly "fit" enough when they ruled the earth for 140 million years. But when mass extinctions claim more than 90 percent of the planet's life forms, fitness may be reduced to a matter of sheer luck. By far the great majority of all species that have ever lived have become extinct.

See also "SURVIVAL OF THE FITTEST"

FITZROY, CAPTAIN ROBERT (1805–1865)
Explorer, Meteorologist, Administrator

It was Captain Robert FitzRoy's idea to take a young naturalist aboard HMS *Beagle* during its second voyage, a five-year surveying expedition for the British Admiralty. After interviewing several applicants, he chose 21-year-old Charles Darwin, a well-recommended but inexperienced amateur who had never been to sea. This enthusiastic Cambridge-educated gentleman, FitzRoy thought, would make an intelligent companion and "messmate." Darwin was equally impressed with FitzRoy, describing him as "my beau ideal of a captain."

A staunch believer in the literal interpretation of scripture, FitzRoy could not have imagined that his vessel would become famous as the birthplace of evolutionary biology. Years later, he vehemently repudiated his old friend and cabinmate Darwin, cursing him as "a viper in our midst." Although he fully acknowledged Darwin's great achievements, he could not forgive him for leaving a Creator out of his theories.

In 1828, FitzRoy was given command of HMS *Beagle*, a 10-gun brig, after the previous captain had shot himself. The ship's mission was to map the frigid and desolate southern coasts of South America, including Patagonia and Tierra del Fuego. Weather was rough,

ARISTOCRATIC NAVAL OFFICER Robert FitzRoy (which means "son of the king") was a brilliant navigator and commander. Darwin admired him, but their friendship did not survive the evolution controversy.

supplies short, and the crew was depressed and sick with scurvy. During this first command, before Darwin joined the expedition, FitzRoy became obsessed with the Indians of Tierra del Fuego. Despite the harsh climate, these sturdy people went nearly naked, lived in rough huts, eked out a living by hunting and fishing, and appeared to the Englishmen almost as beings from another planet.

When a group of the natives invaded the English ship and made off with one of its small whaleboats, which they much preferred to their own canoes, FitzRoy retaliated by taking some Indians hostage. The tribesmen, however, refused to exchange the boat for their kinsmen. Finally, FitzRoy released them all except for a nine-year-old girl who wanted to stay aboard the ship. He named her Fuegia Basket and decided to teach her English.

Soon after, the captain "acquired" three other Indians, young men whom he considered "scarcely superior to the brute creation [animals]." As an experiment, he brought all four to England, "civilized" them at his own expense, then returned them to their homeland several years later—a major objective of his second *Beagle* voyage. To FitzRoy's dismay, this philanthropic attempt to spread the light of Britannia ended in tragic failure. [See BUTTON, JEMMY.]

His other personal agenda was to find geological evidence confirming the biblical account of Creation; he became increasingly upset when Darwin's researches seemed to be taking a quite different direction.

FitzRoy was imperious and arrogant; he could not bear to be contradicted about anything. Once he told Darwin that slavery was justifiable, as black people were childlike and better off under the plantation system. Darwin, an abolitionist, insisted that no man or woman would choose to be a slave.

During their visit to a Brazilian plantation, the owner gathered many of his slaves and asked whether they would rather be free. All agreed they were content with their lot, and FitzRoy crowed to Darwin that his case had been proved. "I then asked him," Darwin wrote in his *Autobiography* (1876), "perhaps with a sneer, whether he thought that the answer of slaves in the presence of their master was worth anything. This made him excessively angry, and he said that if I doubted his word, we could not live any longer together."

Despite these occasional flare-ups, relations between the two were good, and Darwin was chosen to accompany FitzRoy on many of his overland expeditions. Under FitzRoy's driving perfectionism, the ship made its way around Cape Horn and circumnavigated the globe, pausing in the Galápagos, Tahiti, New Zealand, and Australia. Upon returning to South America, the *Beagle* expedition made new explorations of its coasts, then returned to Falmouth, England, with an immense body of new information.

FitzRoy's maps and charts formed an accurate basis for those still in use today. He wrote a volume of the *Narratives* (1839) of the *Beagle* voyage and edited the others, but was somewhat jealous when Darwin's journal of the voyage became far more popular and well-known than any other volume in the series.

His interest in meteorology led him to devise the popular, easy-to-use FitzRoy Barometer, which was widely distributed to British coastal fisherman, and he invented the system of raising storm warning flags, which was adopted all over the world. He also pioneered a system of gathering widely scattered reports into "synoptic" maps (the term still used for them today). As head of the Admiralty's Meteorological Office, he got the *Times* to begin publishing newspaper weather forecasts and maps, which had never before been done. For this he was widely ridiculed, as few believed weather could be predicted with any accuracy. His persistence carried the day; royal messengers occasionally asked him for private weather forecasts for Queen Victoria's outings.

In later years, his imperious and moody personality—and conviction that he was right about matters he didn't understand—led him into many scrapes and difficulties. Nevertheless, he was promoted to admiral, and later in life served as governor of New Zealand. Unfortunately, Fitzroy's temperament was ill-suited for cross-cultural politics and diplomacy, and he left that country in worse turmoil than he found it.

Back in England, he nearly fought a duel of honor with a rival for political office but trounced the fellow with an umbrella instead. This bizarre scene took place at the Admiralty, where his opponent—brandishing a whip—said, "Captain FitzRoy, I shall not strike you, but consider yourself horsewhipped." FitzRoy replied by knocking the man down with his umbrella, to the astonishment of onlookers.

Feeling guilty that his expedition had played a part in "undermining" scripture, several times over the years he appealed to Darwin to recant his evolution theory. At the famous Oxford debate between Thomas Huxley and Bishop Wilberforce, FitzRoy was said by some to have held a Bible over his head and shouted, "The Book, the Book."

With his glory days long behind him, FitzRoy became increasingly despondent and withdrawn. Like his uncle, Viscount Castlereagh, before him, FitzRoy became deeply depressed and took his own life at the age of 60.

Perhaps FitzRoy's unfortunate end was hastened by his self-invented position as a weather forecaster. As Sir Roderick Murchison, president of the Royal Geographical Society, noted in his eulogy: "If FitzRoy had not had thrown upon him the heavy and irritating responsibility of never being at fault in any of his numerous forecasts of storm in our very changeful climate, his valuable life might have been preserved." In 2002, however, on the 137th anniversary of his death, the British Government's Meteorological Office (which FitzRoy founded) named a reporting station after him. Three times a day, sea captains still tune in to their radios to hear the weather reports from FitzRoy.

See also (HMS) *BEAGLE*; DARWIN, CHARLES; OXFORD DEBATE; VOYAGE OF THE HMS *BEAGLE*

FLAT-EARTHERS
Oldest "Bible Science" Sect

Hebrew scribes wove into the Old Testament their concept of a flat, stationary Earth—an ancient view they shared with older Babylonian and Egyptian cultures. Centuries later, when early scientists concluded that the Earth was a spinning globe orbiting the sun, church authorities attacked astronomy as a satanic attempt to destroy religion and morality. Long before scriptural literalists fought against the theory of evolution, they battled scientists over the shape of the Earth.

Right down to our own time, a handful of die-hard "flat-earthers" believe the Apollo moonwalk and astronaut photos of a round planet Earth are government lies fostered by a conspiracy of atheistic scientists.

Round-earth theories were often debated in the ancient world, even while the older notions prevailed. Aristotle had proposed a reasonable proof that the Earth is a globe, based on observations of lunar eclipses, differences in star patterns seen by travelers, and the manner in which ships "descend" at the horizon.

Ptolemy's *Almagest* (A.D. 140) established the global idea in Western culture, although he considered Earth the center of the universe. It remained for Copernicus (1473–1543) and Galileo (1564–1642) to destroy that geocentric notion—and for the Catholic church to destroy Galileo.

In 1849, a modern flat-earth movement was founded in England by Samuel Birley Rowbotham. Under the pseudonym "Parallax," Rowbotham published a pamphlet in that year titled *Zetetic Astronomy: A Description of Several Experiments Which Prove That the Surface of the Sea Is a Perfect Plane and That the Earth Is Not a Globe!* ("Zetetic" is from the Greek *zetetikos*, meaning "to seek or inquire.") In his view, the world is a vast flat circle, something like a pancake. The North Pole is at its center, and the perimeter is walled in by high sheets of ice that are mistakenly called the South Pole. Sun, moon, and planets circle above at an altitude of about 600 miles. Phenomena like the rising and setting sun and the apparent disappearance of ships over the horizon are explained by optical illusions and the "zetetic law of perspective."

"How many times did you flee from the Indians? How often were you carried away by floods? And how many times were you kilt?"

—Capt. Robert FitzRoy to Charles Darwin, 1833

About 1860, just after Darwin's *Origin of Species* (1859) appeared, the flat-earth movement and "Zetetic Astronomy" became household words in England. In America, flat-earthism became a central doctrine of the Christian Catholic Apostolic Church in Zion, which claimed 10,000 members during the 1920s and 1930s. An entire town of about 6,000—Zion, Illinois—was founded by the sect and ruled by the iron hand of Wilbur Glenn Voliva. Voliva even had his own radio station, on which he thundered against "the Devil's triplets: Evolution, Higher Criticism [historical biblical scholarship], and Modern Astronomy."

Voliva taught that the stars are much smaller than the Earth and rotate around it, the moon is self-luminous, and the sun is not a giant star but a much smaller light a few thousand miles away. In a 1930 issue of his sect's periodical, *Leaves of Healing,* he wrote: "Where is the man who believes [if the earth really rotates] that he can jump into the air, remaining off the earth one second, and come down to the earth 193.7 miles from where he jumped up?" When novelist Irving Wallace interviewed him for a newspaper in 1932, Voliva proclaimed that the sun was only 3,000 miles away and 32 miles in diameter. He attacked the conclusions of astronomers with homespun ridicule: "God made the sun to light the earth, and therefore must have placed it close to the task it was designed to do. What would you think of a man who built a house in Zion and put the lamp to light it in Kenosha?"

After Voliva's death in 1942, the popularity of his flat-earth brand of ultra-fundamentalism started to wane, and his town of Zion has long since caught up with the present century. Still, some of its elderly members still believe in their hearts that, in their founder's words, "the [majority of] so-called fundamentalists . . . strain at the gnat of evolution and swallow the camel of modern astronomy."

See also FUNDAMENTALISM; HAMPDEN, JOHN; "SCIENTIFIC CREATIONISM"

FLINT JACK (c. 1815–c. 1880)
Greatest Faker of Prehistoric Tools

ON THE MOVE in Scotland, England, and Ireland, finding suckers for his ersatz antiquities, Edward Simpson made excellent fake prehistoric tools, which he sold to museums and wealthy collectors.

By the time Darwin published *The Descent of Man* in 1871, many people had already accepted the fact that the human past far predated biblical accounts. Once "antiquarians" and their workmen began combing Europe's caves and gravels for ancient stone tools, they found them by the thousands—sometimes mingled with mammoth or cave bear bones. Victorian collectors, who already had a mania for rare and beautiful minerals, orchids, shells, and bird's eggs, now went mad for prehistoric artifacts.

Dealers in stone tools prospered, and with them so did the inevitable producers of fine fakes. The so-called "Prince of the Counterfeiters" was England's Edward Simpson, alias "Flint Jack," who had begun producing clever replicas by 1850 and for 17 years purveyed his freshly fabricated antiquities to museums and private collectors.

Flint Jack was also known to dealers and scientific men as "Fossil Willy," "Old Antiquarian," and "Cockney Bill." For six years he assisted a fossil-collecting physician at Whitby and became expert in finding fossils and stone tools. Eventually he mastered the art of copying chipped flints, "the fabrication of which," according to one account, "rendered him subsequently so dextrous as to succeed in gulling, not merely the public, but learned ones who had spent the whole of their lives in archaeological pursuits."

Always a short jump ahead of the law, he plied his trade all over England, then moved on to Scotland, and finally set up shop in Ireland, where, according to his own account, he "left behind him many a fine celt, arrow-head, hammer, and spear" (not to mention his freshly made "ancient British urns, a Roman breastplate fashioned from an old tea tray, an old silver coin made from the handle of a German spoon, and many fake fossils). In 1862, Flint Jack was persuaded by some of his antiquarian friends to give demonstrations of flint knapping at meetings of both the Archaeological Society and Geological Society in London. Before the assembled gentry, "in a few minutes he had produced a small arrow-head . . . with a facility and rapidity which proved long practice." The gentlemen were delighted and lined up to purchase every stone tool he could fabricate on the spot. Unfortunately, that stint at respectable celebrity was the high point of his career. A few years later, as more and more genuine stone tools entered the marketplace and all the dealers and collectors had become familiar with his notorious handiwork, Flint Jack found himself without repeat customers. A penniless alcoholic, he resorted to stealing and wound up serving time in Bedford jail, where he died.

Historian of science Anne Secord, speaking about Flint Jack at the Society for the History of Natural History meetings (2004) at Darwin College, Cambridge, explained how the prince of prehistoric forgeries managed to fool his learned clients for so long: "In mid-Victorian Britain, it was the presumed ignorance of a workingman like Flint Jack that had fueled his trade. Collectors could dispel any doubts about the authenticity of his specimens by reassuring themselves that a ragged itinerant was incapable of creating such things himself."

FLORES MAN
Extinct Dwarf Cave-Dwellers

On the small Indonesian island of Flores, 370 miles east of Bali, a dwarf species of hominin—a three-and-a-half-foot-tall early *Homo* with a small skull and heavy brow ridges—appears to have evolved and lived for tens of thousands of years. Isolated from the mainstream of Asian and European *Homo sapiens* populations, these diminutive archaic humans, *Homo floresiensis,* were hunters who lived in caves and made stone tools for butchering carcasses.

Several independent methods of dating their bone fragments and tools found in Liang Bua cave place them over a long range of occupation, between 95,000 and 13,000 years ago. The most complete Floresian remains are a skeleton dated at 18,000 years, which is surprisingly recent, given that the last Neanderthals disappeared 24,000 years ago. (Modern *Homo sapiens* reached Asia about 50,000 years ago.)

Small groups of the tiny Floresians may even have survived into historic times; locals tell handed-down tales of little people who still inhabited forest caves when Portuguese traders arrived in the 16th century. According to these traditional stories, the Floresians were troublesome to the local villagers and frequently tried to steal their food. Although the Floresian skulls are small, they do not resemble those of modern human pygmies; instead, the facial structure and heavy brow ridges are reminiscent of the much larger *Homo erectus* that roamed Africa, Asia, and Europe as early as two million years ago.

Flores lies just beyond Wallace's Line, the zoological boundary that separates Asian from Australian animals, in an area where species that originated elsewhere were free to migrate and mingle. [See WALLACE'S LINE.] Because of limited food supplies and absence of large predators on small islands, it is a general rule that tiny immigrants tend to evolve into larger ones, while big animals often become smaller. The Floresians inhabited just such a "lost" island world, which they shared with pony-sized dwarf elephants (*Stegodon*), rats as big as goats, and Komodo dragon–sized predatory lizards.

Ancient charcoal deposits indicate that the Floresians used fire for cooking. Some of their stone tools have been found between stegodon ribs that show cut marks, suggesting that they hunted and butchered the small elephants. Cooperative hunting also may imply the existence of language and social cooperation.

Scientists were unaware of the Floresians until 2003, when the skeletal remains of a 30-year-old adult female (nicknamed "Flo") were discovered and excavated from a limestone cave by a team of Australian and Indonesian archeologists. Paleoanthropologists Peter Brown and Michael J. Morwood, of the University of New England in Armidale, New South Wales, immediately recognized that they had found a new, previously unknown species in the human family. Not a hardened fossil, the soft, waterlogged, fragile mass required great skill and patience to retrieve intact. Bones from six more individuals were found in the cave in 2004. Over the objections of his partner, Morwood insisted on nicknaming them "hobbits," after J. R. R. Tolkien's mythical little folk, which became their popular moniker in the press.

Previous excavations by Morwood and a team of archeologists established that full-sized

On a remote Indonesian island, a tribe of little hunters lived in a lost world of giant lizards and dwarf elephants.

early humans had arrived on Flores by 840,000 years ago and had left many characteristic stone tools. However, the tools associated with the Floresians, who arrived much later, were more sophisticated, including small blades that may have been mounted on wooden shafts.

What began as Brown and Morwood's triumphal discovery of this spectacular, unexpected hominin took a sharp turn toward disappointment, however, when their precious prehistoric bones were sequestered for study by a senior Indonesian anthropologist. Teuku Jacob, of Gadjah Mada University in Yogyakarta, was accustomed to receiving first look at the country's paleontological finds; an Indonesian anthropologist sent him the fossils without consulting his distraught Australian colleagues. After months of bitter wrangling over academic turf, the materials were finally returned to their discoverers—but not before some of the rare fossils had reportedly been damaged. One of the jawbones was accidentally broken when it was pried from a mold, then clumsily and incorrectly glued together; the only pelvis was compromised beyond repair. Morwood, in a bitter moment, offended Indonesian decorum by publicly accusing the respected senior professor of greed and irresponsibility.

Discovery of the Floresians poses many more questions than it answers. For instance, were these little people, like the miniature elephants they hunted, examples of evolutionary "dwarfing" in response to their island habitat? Or did they arrive on Flores as already diminutive immigrants from elsewhere? One thing is certain: the longstanding correlation of large brains and refined stone tool-making doesn't hold here; with perfectly formed human brains no larger than those of chimps or australopithecines (about 380 cc) the little Floresians flourished quite efficiently and successfully for the better part of 100,000 years.

FLY ROOM
Thomas Hunt Morgan's Lab

In a 16-by-23-foot room at New York City's Columbia University, a makeshift laboratory known as "the fly room," Thomas Hunt Morgan and his associates used tiny flies to understand the workings of heredity.

Morgan had little faith in Mendelism when he came to Columbia in 1904. At first, he tried to study genetics in mice and rats, but without success. He made progress only after he chose the humble fruit fly, *Drosophila melanogaster*, as his subject. The fly could be bred by the thousands in milk bottles. It cost nothing but a few bananas to feed all the experimental animals. Their entire life cycle lasts 10 days and they have only four chromosomes.

Between 1907 and 1917, Morgan and his team of seven worked in the fly room, attempting to encourage mutations with heat, X-rays, and chemicals. Using the most laughably simple equipment, but lots of scientific talent and perseverance, they gradually provided evolutionary biology with a new foundation—and, to Morgan's surprise, showed Gregor Mendel to be correct. In 1933, Morgan won the Nobel Prize in Physiology or Medicine "for his discoveries concerning the function of the chromosome in the transmission of heredity."

FOSSEY, DIAN (1932–1985)
Gorillas' Greatest Defender

From an ordinary existence as an occupational therapist in Louisville, Kentucky, Dian Fossey's life was transformed into one of the great dramas of modern Africa. For 19 years, her beloved "family" became a band of mountain gorillas, with human poachers her sworn enemies. Ultimately her passion to preserve the great apes cost her her life.

Fossey's lifelong interest in animals prompted her to visit Africa in 1963, where she met Louis Leakey at his fossil digs in Olduvai Gorge. The crafty paleoanthropologist suggested that she study wild gorillas in the Congo.

Rare and elusive, the mountain gorillas live only on six extinct volcanoes, located in the border area of Rwanda, the Democratic Republic of the Congo, and Uganda. Fossey began her field work in 1967, when about 600 animals were left of the many thousands that had been discovered in the 1860s. When she founded the Karisoke Research Center, the nearest village was several hours' walk down a mountain trail.

Her study area was within the Parc National des Volcans in Rwanda, which is a wildlife preserve, but the local people illegally farmed cattle, poached wildlife, and cleared trees wherever they pleased, and the few park guards were easily bribed.

Fossey found the dense foliage booby-trapped with wire nooses for the capture of antelope and forest hogs; often they snared gorillas, causing gangrene of hands and feet. Poachers killed entire troops to capture infants for sale to zoos, and cut off the adults' massive hands and feet to sell as souvenirs. Alone with her love of gorillas in a war zone between apes and local hunters, she feared the apes would all be exterminated before she had a chance to study them.

At first she hiked many miles tracking groups, unsure of what to do when she got close. Her courage was tested when huge silverback males ran at her, roaring and chest-beating. Eventually, through patience and perseverance, she was accepted by the gorillas, and spent months sitting within them as they fed, played, and conducted their social lives. At Leakey's urging, Fossey brought her data to Cambridge University, where she earned a doctorate in zoology from Darwin College.

TRIUMPH AND TRAGEDY characterize the life of Dian Fossey, who entered into the world of mountain gorillas, then defended them from extermination. On a rare visit to New York City, Fossey grudgingly admired Carl Akeley's mounted gorilla group at the American Museum of Natural History. Neil Selkirk/Time & Life Pictures/Getty Images.

Fossey trained her staff to go on patrols, dismantling traps and nooses and attempting to scare off poachers. Some poachers quit, but others continued to hunt as their forefathers had done for millennia.

Fossey was traumatized when her favorite gorilla, Digit, was slaughtered in 1977. She had spent hundreds of hours with Digit, had watched him grow from infancy to maturity, and regarded him as a personal friend. According to Fossey's account in *Gorillas in the Mist* (1983), her assistant had "found Digit's mutilated corpse lying in . . . blood-soaked . . . vegetation. Digit's head and hands had been hacked off; his body bore multiple spear wounds." She agonizingly reconstructs his last stand:

> Digit, long vital to his group as a sentry, was killed in this service by poachers on December 31, 1977. That day Digit took five mortal spear wounds into his body, held off six poachers and their dogs in order to allow his family members, including his mate Simba and their unborn infant, to escape. I have tried not to allow myself to think of Digit's anguish, pain, and the total comprehension he must have suffered in knowing what humans were doing to him.

After that incident, Fossey suffered a nervous breakdown and "came to live within an insulated part of myself." Digit's death became a rallying point for conserving the gorillas.

Like a soldier whose buddy is killed by the enemy, Fossey became bitter. She joined with police in military-type raids of local villages, rounding up poachers. She helped capture three men responsible for Digit's death, who were tried and given prison sentences, but several others eluded capture.

Publicity brought more assistants, researchers, and interested tourists, but she shunned them. Withdrawn and obsessed, she desired only to be left alone with her gorillas and for her gorillas to be safe. Despite the growing success of local efforts to protect the apes, Fossey refused to take part in conservation programs in Rwandan schools, claiming that it was too little, too late. On December 27, 1985, Dian Fossey was murdered by a machete-wielding intruder in her forest cabin.

Dian Fossey's violent death remains unsolved. Initially, suspicion fell on Wayne McGuire, the graduate student who reported finding her body. He continued his research for months but returned to America upon learning that he was about to be arrested. Despite his protestations of innocence, McGuire was later indicted and convicted in absentia by the Rwandan government. A decade later, authorities focused on Protais Zigiranyirazo, a government official whom Fossey had threatened to expose for trafficking in endangered wildlife, according to some press reports. In 1994, he was indicted as a genocidal murderer of thousands of Tutsis, but he fled the country before he could be charged with Fossey's murder.

Dian Fossey is buried in the Karisoke gorilla cemetery, next to her beloved Digit.

See also AKELEY, CARL; "APE WOMEN," LEAKEY'S; DIGIT; GORILLAS

FOSSIL HUMANS
Relics of Relatives

When Darwin published *Origin of Species* in 1859, scarcely a fossil of early humans or near-humans had yet come to light. His hypothesis that humankind evolved from African primates (represented today by lemurs, monkeys, and apes) was criticized as airy speculation. Opponents taunted, "Where is the missing link?" Today, there are enough fossils of early hominids to fill an exhibition hall.

Darwin's friend, the zoologist Thomas Henry Huxley, included the first known fossil of a Neanderthal, a skullcap found in Germany in 1856, in his influential *Evidence as to Man's Place in Nature* (1863). Although convinced this was the most ancient humanlike fossil known at the time, Huxley thought the Neanderthals so similar to modern people that they could not be the sought-after "missing links" or "ape-men." Also, between the 1840s and 1860s, archeologists found thousands of prehistoric stone tools in English caves and French river gravels, sometimes associated with extinct rhinoceroses, bears, and mammoths.

In 1891 the Dutch physician Eugène Dubois found the Java Man in Indonesia (the first known *Homo erectus*, originally called *Pithecanthropus*), which some hailed as the "link." Although other hominid fragments turned up in Europe and Africa, many anthropologists thought they would eventually hit the jackpot in Asia, which they believed was the true cradle of humankind. But despite a seven-year search (1921–1928) by the Gobi Desert Surveys, known as the "Missing Link Expeditions," no "dawn man" fossils turned up.

By the mid-1920s, there were two other finds: one revolutionary, the other an outrageous forgery. The fraud was Piltdown Man, "discovered" in 1912 in Sussex, England, and exposed 40 years later. The authentic discovery, *Australopithecus africanus,* or "Southern African man-ape," found by anatomist Raymond Dart in 1924, was the first solid clue to an African genesis of humankind and the beginning of modern paleoanthropology.

THE HOMINID SPECIES called *Homo rudolfensis* lived in East Africa between 1.8 and 1.9 million years ago—one of many previously unknown extinct relatives of humans that have been discovered since the 1990s. Reconstruction by Viktor Deak. © Viktor Deak, used by permission of Nevraumont Publishing Company.

While European scientists continued looking to Asia, Dart, later joined by Robert Broom, unearthed more African man-ape fossils, including some larger, heavier-boned species, such as *Paranthropus robustus*, that were not ancestors of humans. During the 1930s, paleoanthropologists became excited about excavations in China that yielded several skulls of Beijing Man, another *Homo erectus* (then called *Sinanthropus*). Later specimens of *Homo erectus* eventually also turned up in North Africa, Europe, East Africa, and Israel.

During the mid to late 20th century, the search intensified in East Africa, led by Louis and Mary Leakey and later by their son Richard. At Olduvai Gorge, Koobi Fora, and other sites in Kenya and Tanzania, the record was enriched with some still-bewildering discoveries: the enigmatic, larger-brained *Homo habilis*, some two million years old, as well as both smallish (gracile) and heavier (hyper-robust) australopithecines. Richard Leakey found an almost complete skeleton of a *Homo ergaster* boy at Lake Turkana, estimated at 1.6 million years old.

Three or four different species of later hominids (and perhaps several more) now appear to have existed at about the same time, particularly in Africa and Eurasia. Although originally arranged in some ancestor-descendant series, it became clear that there was an adaptive radiation or "branching bush" of early hominids, of which only one "twig," *Homo sapiens,* managed to survive.

During 1974, Donald Johanson is credited with making remarkable discoveries in the "Afar triangle" of Ethiopia. The most famous is "Lucy," the oldest partial skeleton of a hominid yet found, which Johanson assigned to a new species (*Australopithecus afarensis*). Estimated at over three million years old, these upright-walking, small-brained hominids had apishly long arms not previously known from bipedal hominid skeletons.

Johanson collected 13 individuals, nicknamed "The First Family," at Hadar the following season; whether or not they represent more than one species is still controversial.

No fossil australopithecines, *Homo erectus, Homo habilis,* or any other very early hominid, not even the comparatively recent Neanderthals, are found in the Americas, although remains of early AmerIndians may go as far back as 15,000 years.

Anthropologists debate endlessly about which fossil hominid species were ancestral to us, and which were "side branches" of "the main line" leading to ourselves. The details of this complicated family tree are not likely to become clearer anytime soon. But the broad picture is encouraging, for as far as "missing links" go, there is not one but many. We have fossil remains of at least four early upright-walking manlike creatures with brains the size of modern apes dating back to the period between six and four million years ago, and perhaps more than 20 ancient kinds of hominids in all.

Some (*Homo habilis*) appear to have been closer cousins of ours than others. Some (*Homo erectus*) had a brain two-thirds the size of our own, and others (Neanderthals) had crania as large as ours. We also have the remains of humans like ourselves dating from as far back as 190,000–160,000 years ago. The earliest left rather archaic, crude stone tools, but by 30,000 years ago their presumed descendants were making remarkable cave paintings.

Discovered in the 1990s, the Dmanisi hominids from Georgia in the Caucasus, are currently the most ancient undisputed hominid fossils outside of Africa. At 1.8 million years old, they almost doubled the accepted date for the first African émigrés.

In 2003, skeletons of a group of small hominids that lived as recently as 18,000 years ago turned up on the Indonesian island of Flores. Nicknamed "hobbits," the Floresian people stood three feet tall and apparently used stone tools to butcher game, including miniature elephants (*Stegodon*) and giant lizards.

See also AFAR HOMINIDS; AUSTRALOPITHECINES; BEIJING MAN; *DESCENT OF MAN;* DMANISI HOMINIDS; FLORES MAN; GOBI DESERT EXPEDITIONS; *HOMO ERECTUS; HOMO HABILIS;* "LUCY"; NEANDERTHAL MAN; PILTDOWN MAN (HOAX); TURKANA BOY

FOSSILS
Reading the Rocks

Fossils (from the Latin *fossa,* "hole") originally meant any curious natural object that was dug up from the ground. That included not only petrified plants, shells, or animal skeletons, but also crystals, gems, and odd mineral ores. Early collectors debated whether the stones shaped like organisms were really once alive and how they might have been formed.

Conrad Gesner (1516–1565), an industrious early naturalist, wrote a book *On Fossil Objects, Chiefly Stones and Gems, Their Shapes and Appearances* (1565) that described and illustrated for the first time a large and varied collection of these mysterious objects. Some of the most familiar shapes seemed to be mineralized seashells, though the puzzle was that they were found on mountaintops. Others appeared to be fish or animal skeletons, but did not resemble any creatures that now exist. Many had strange and bizarre geometric shapes, unlike any living things.

For several hundred years after Gesner, attempts to understand fossils included theories that (1) they are the remains of creatures that once lived; (2) they are a manifestation of "pure forms" that look like, but are not, once-living creatures; (3) they are some kind of mysterious, devilish trick to disturb the natural order; or (4) they are a divine test of faith from God. Intelligent investigators, at various times, have believed all these things.

Easily identifiable fossils, such as oyster shells, appeared even to the ancient Greeks as having once been living creatures. Five hundred years ago Leonardo da Vinci suggested a friend study fossil shell deposits to understand past changes in the Earth. But early philosophers were puzzled by skeletons of unfamiliar organisms. Some imagined that a mysterious force, or *vis plastica*, shaped rocks into forms that resembled strange animals or plants that had never been alive. Like geometric crystals or unusual gemstones, fossils were considered natural marvels and were prized by collectors.

It was only when an ever-increasing number of large animals and previously unknown fish were discovered that the older questions faded away, as paleontologists realized they had to explain a vast prehistoric zoo. By the early 19th century, the questions had become: How did so many creatures' fossil remains get into the cliffs and outcrops? How old were they? How did they fossilize? And what was the world like when these creatures were alive?

In one of the most bizarre Victorian attempts to explain them, respected naturalist Philip Gosse argued that fossils were placed in the ground by God as part of the spontaneous creation of an ongoing world, just as he had included growth rings on trees that had never been saplings and navels on Adam and Eve, who were never born. Fossils were put in the Earth to create it whole, complete with a past that never occurred. [See OMPHALOS.]

Gosse's views notwithstanding, most naturalists came to realize that fossils are the remains of once-living creatures, including many species that have become extinct. By the 1870s, fossil hunters of the "great bone rush" filled museums with dinosaurian giants and scoured the world for "missing links" between humans and apes.

Today, less spectacular fossils are yielding important information about the past. Microfossils and drab remains of tiny sea creatures have provided keys to understanding such phenomena as cycles of global extinction, shifts in ancient climates, movements of continents, changes in the composition of sea water, and the advance and retreat of glaciers.

See also BERINGER, JOHANNES; BURGESS SHALE; COPE-MARSH FEUD; FOSSIL HUMANS; ICE AGE

FOUR THOUSAND AND FOUR B.C. (4004 B.C.)
Earth's Biblical Birthdate

Prior to the publication of Charles Darwin's *Origin of Species* in 1859, the first words in most English Bibles were "4004 B.C."—the date fixed for the creation of the world in the official view of the church. Until recently, many Bibles were printed with a special column of dates running alongside the text, which was meant to give the reader a precise chronology of events. These dates were not part of traditional scripture but had been added in the 17th century. Most readers assumed, however, that these recently added dates were part of the ancient text.

This "official" ecclesiastical timetable dated the planet Earth as about 6,000 years old. In our present understanding, that is about the time of the First Egyptian Dynasty, a complex civilization created by humans like ourselves. As long ago as 30,000 years, human artists created remarkable drawings of mammoths and lions on the walls of European caves. Stone tools and artifacts are well known from as far back as 2.5 million years ago. We can now date fossil hominids like *Homo erectus* well beyond a half-million years, while some African australopithecines left their remains in the earth at least four million years ago.

As geneticist Karl Pearson wrote in *Charles Darwin, 1809–1882: An Appreciation* (1923), "If we have ceased to believe that the world and all its forms of life were created in 4004 B.C. it is because Darwin freed us from that cramping doctrine." For instance, Englishman John Woodward (1665–1728), who founded a geological museum at Cambridge to house the fossils he spent a lifetime collecting, might have groped his way toward great geological truths, but he was stymied by the scientific establishment of his day, which refused to publish his papers or lend credence to any human prehistory going back further than the church's date of 4004 B.C.

In the 18th century, some thought fossils were no older than the Creation itself, while others believed the Devil inserted them in the Earth to tempt man to disbelieve the church's teachings. "With a complete history of the world supposedly known from its creation in 4004 B.C., no real solution was possible" to the problems posed by fossils and flint tools. This "bondage of 4004 B.C.," as Pearson called it, continued in force even through the 1860s.

One of the reasons Darwin was so hesitant and cautious in presenting his views on the great age of the Earth was that he—like most Victorians—assumed that 4004 B.C. was part of the original scripture itself. In 1861, Darwin was astonished to learn that the "official" biblical chronology had been calculated and added to the Bible by Ireland's Archbishop Ussher in 1650—just 200 years earlier.

See also CREATIONISM; USSHER-LIGHTFOOT CHRONOLOGY

FOX-FARM EXPERIMENTS
Domesticating Wild Canines

Charles Darwin's theory of evolution was heavily affected by his observations of domestic animals, including dogs. In the *Origin of Species*, he wrote:

> If . . . it could be shown that the greyhound, bloodhound, terrier, spaniel, and bull-dog, which we all know propagate their kind truly, were the offspring of any single species, then such facts would have great weight making us doubt about the immutability of the many closely allied natural species—for instance, of the many foxes—inhabiting the different quarters of the world.

We now know from genetic experiments that all the hundreds of dog breeds have, in fact, evolved from the wolf. And thanks to a decades-long experiment in Siberia, we have a good understanding of the mutability of foxes as well. In 1959, the Russian geneticist Dmitry Belyaev undertook an impressive experiment in artificial selection, with the goal of breeding tame foxes. Darwin had noticed that different domestic species share many physical traits: "not a single domestic animal can be named which has not in some country drooping ears," he wrote. They also often have wavy hair, curly tails, and a piebald coat with white patches of fur. Belyaev theorized that these particular traits may not have been selected for necessarily, but were actually a side effect of a desirable behavior—specifically, tameness.

Belyaev and his colleagues began the experiment with 130 silver foxes that they bought from a commercial fur farm. Since they'd been caged for generations, the animals were already tamer than wild foxes, but only those that earned the very highest tameness rating—foxes that wagged their tails, whined, and let themselves be petted—were allowed to breed. By the sixth generation of selective breeding, the foxes had changed enough that an even higher tameness rating had to be added, dubbed the "domesticated elite."

Elite foxes begged for human contact, whining for attention and licking the researchers' faces before they were even one month old. After 30 to 35 generations, the foxes didn't just act like dogs, they also began to look more like dogs and other domestic mammals, just as Belyaev had predicted. Tame foxes began to display shorter, curly tails; floppy ears; and piebald fur. Also like dogs, their skulls got smaller, their snouts got broader, and the differences between males and females became less pronounced.

How is it that simply selecting for a particular behavior can cause all of these physical changes? Hormones and other neurochemicals control aggression and tameness, but they also play a pivotal role in an organism's development. Selecting for tameness is, in effect, selecting for particular levels of different neurochemicals and hormones. Many of these changes involve the retention of juvenile traits in an adult, a process known as neoteny.

Darwin and his colleagues would have been delighted, though not surprised. Thomas Huxley,

EVOLVED FROM THE WILD in just forty years, foxes bred for tameness crave human company, do not bite, and often develop piebald fur, floppy ears, and curly tails.

for instance, in his essay "Evolution and Ethics" (1893), wondered why man could not do a better job of domesticating himself. "The intelligence which has converted the brother of the wolf into the faithful guardian of the flock," he wrote, "ought to be able to do something towards curbing the instincts of savagery in civilized men."

See also ARTIFICIAL SELECTION; CUTENESS, EVOLUTION OF; NEOTENY

FRANKENFOOD
Giant Salmon and Suicide Seeds

n the introduction to his 1868 work *The Variation of Animals and Plants under Domestication*, Charles Darwin wrote:

From a remote period, in all parts of the world, man has subjected many animals and plants to domestication or culture. . . . It is an error to speak of man "tampering with nature" and causing variability. If a man drops a piece of iron into sulphuric acid, it cannot be said strictly that he makes the sulphate of iron, he only allows their elective affinities to come into play. If organic beings had not possessed an inherent tendency to vary, man could have done nothing.

In the 9,000 years since humans first domesticated plants and animals, we have undertaken a most successful experiment in genetic engineering. By selecting for favorable traits, crossbreeding, and hybridizing, we have literally designed most of the foods we eat today. Darwin believed that while man had the power to select for variations present in nature, he couldn't actually introduce novel variations himself. But times have changed. Today, scientists can splice genes from virtually any living organism into another no matter how distantly related.

Is bioengineering a natural next step or a reckless lurch, veering like Dr. Frankenstein's monster into the realm of the unnatural? While proponents tout genetically modified foods

THE FARM, by Alexis Rockman. Courtesy of and © by Alexis Rockman, 2000.

as potential salvation for the world's underfed, critics fear that "Frankenfoods" have the potential to become a runaway menace.

The first genetically engineered food hit U.S. stores in 1994 with the Flavr Savr tomato, which was designed to stay on the vine longer, develop more flavor, and ship longer distances without spoiling. Because of production and distribution problems, the tomato is no longer on the market, but plenty of other high-tech foods are. In fact, by 2002 two-thirds of the processed foods lining U.S. store shelves contained genetically modified ingredients. Such gene splicing is usually performed to render the crops resistant to herbicides or diseases, toxic to insect pests, or more nutritious.

One of the most common biotech foods is Bt corn, which contains a bacterial gene, *Bacillus thuringiensis,* that makes it poisonous to the European corn borer and other insects. A variety of salmon has been engineered to grow twice as quickly as natural varieties, and researchers are attempting to genetically modify pigs to be less aggressive, allowing them to be raised at higher densities.

Critics of genetic engineering worry about food safety or are uncomfortable about tinkering with nature. As Manhattan chef Eric Ripert told the *New York Times* in 1998, "I don't want a cow gene in my cabbage." Others don't like to see modern agriculture fall into the hands of a few powerful companies. One of the largest producers of biotech crops, Monsanto ("Monsatan" to critics), produces an effective herbicide called Roundup to eliminate pesky weeds—which further induces farmers to buy Roundup Ready crop seeds that can survive it.

The company also came under fire as one of several prominent seed distributors that patented so-called terminator seeds in 1997. These seeds were genetically programmed to be sterile after being sprayed with a chemical signal, ensuring that farmers would have to buy new seeds from the company year after year. In developing countries, where farmers grow about 80 percent of their crops by saving seeds from the previous year or exchanging with others, use of such seeds would lead to economic ruin.

Another major concern is that genes inserted into agricultural crops for herbicide tolerance will be transferred to wild plants through pollen, creating "superweeds" that run rampant. And if super fast–growing salmon were to enter wild waterways, they could threaten natural ecosystems by outcompeting the species that grow at normal rates.

Possible benefits of genetically engineered foods are impressive, however. Planting crops that produce their own pesticides would mean less dumping of toxic chemicals into the environment. As the world population continues to balloon, crops that are designed to survive poor soil and drought conditions could prevent famine, and vitamin-rich foods could increase nutrition for both the developing and the industrial world.

In the works are plants that produce biodegradable plastics in their leaves and seeds, goats whose milk contains a malaria vaccine, potatoes that offer immunity against toxic strains of *E. coli* bacteria, and many more. But designing these plants is only half the battle. They still have to get planted—and that can involve major delays. What was supposed to be the hero of biotech foods—golden rice—is still waiting to prove itself.

In many regions of the world where rice is a staple, children don't get enough vitamin A, causing blindness and even death for thousands. Enter golden rice, invented by the German geneticist Ingo Potrykus and colleague Peter Beyer in 1999, which contains bacterium and daffodil genes, making it rich in vitamin A. The rice's inventors hoped to send it around the world free to farmers in need. But patent disputes and political issues prevented its release for a decade. Despite pleas by the angry and hearbroken inventors, tons of seeds were locked inside a grenade-proof greenhouse near Zurich, pending resolution of all the lawsuits.

Since then, scientists produced a new version (GR2) that incorporates a maize gene and produces 23 times more beta carotene. That strain, too, for scientific and economic reasons, has become stalled in the pipeline.

See also GENETIC ENGINEERING

FRANKLIN, BENJAMIN (1709–1790)
A Darwinian Precursor

Every American schoolchild knows Benjamin Franklin as a printer, statesman, scientist, diplomat, first postmaster-general, and the author of *Poor Richard's Almanac*. One of America's great founding fathers, he was also a keen observer and experimenter, inventor of the lightning rod, the Franklin stove, and a pioneer in harnessing electricity. But it is not so well known that, like his friend Thomas Jefferson, his observations in natural history place him among the early intellectual forerunners of Charles Darwin.

In 1750, Franklin wrote a letter about some pigeons he had observed nesting on the side of his house. It struck him that organisms increase to fill all the available space:

> I had for several years nailed against the wall of my house a pigeon-box that would hold six pair; and, though they bred as fast as my neighbours' pigeons, I never had more than six pair, the old and strong driving out the young and weak, and obliging them to seek new habitations. At length I put up an additional box with apartments for entertaining twelve pair more; and it was soon filled with inhabitants, by the overflowing of my first box and of others in the neighborhood.

In a short treatise written the following year, Franklin concluded that there is:

> no bound to the prolific nature of plants or animals but what is made by their crowding and interfering with each other's means of subsistence. Was the face of the earth vacant of other plants, it might be gradually sowed and overspread with one kind only. . . . And were it empty of other inhabitants, it might in a few ages be replenished from one nation only; as for instance, with Englishmen . . . one million English souls in North America [doubling every 25 years] will in another century be more than the people of England.

Benjamin Franklin was one of the founding fathers of the theory of natural selection.

Franklin's pamphlet was read by Thomas Malthus, an English clergyman and economist who took the ideas much further. People increase at a geometric rate, he wrote, while their food supply increases only arithmetically. The result is that populations expand to the limits of their resources, followed by a crash, which could mean famine, war, or disease—man's tragic "struggle for existence," he called it.

Malthus's *Essay on the Principle of Population* (1798) in turn directly inspired Darwin and Alfred Russel Wallace, both of whom independently applied its principles to the natural world. Thus, Benjamin Franklin was not only one of the founding fathers of the United States, but of the theory of natural selection as well.

See also JEFFERSON, THOMAS; MALTHUS, THOMAS; NATURAL SELECTION

FRANKLIN, ROSALIND (1920–1958)
Codiscoverer of DNA

By 1952, scientists believed that the DNA molecule was the genetic blueprint for living things, but no one had yet identified its structure. A few talented British physicists and chemists (and Linus Pauling in America) were racing to be first to solve its molecular structure, a key to the physical basis of life.

A year later, the puzzle was solved. American James Watson and British scientists Francis Crick and Maurice Wilkins, both of King's College, Cambridge, but working in rival labs, published their model: The long-sought structure was a double helix—resembling a twisted spiral staircase. Rosalind Franklin, a low-ranking young chemist at Wilkins's lab, supplied the crucial evidence for the discovery, but her name did not appear as a coauthor.

In 1962, the three men shared a Nobel Prize in Physiology or Medicine for the discovery, but Franklin was not accorded the honor of a codiscoverer. (She died in 1958 and was technically ineligible—but it is doubtful they would have included her had she been alive.) Since then, however, two major biographies have attempted to set the record straight, and Franklin has become something of a feminist icon: a woman who made major contributions to science but was denied recognition because of her sex.

Franklin, the child of a wealthy, highly cultivated British family, was brought up in an atmosphere of intellectual freedom. After she pursued graduate studies at a crystallography laboratory in Paris, she accepted a position at King's College London, with the understanding that she would be allowed to pursue her own research into DNA. She was to head up a new X-ray diffraction unit, and assumed she would be free to discuss ideas freely with her colleagues, as she had in Paris. To her dismay, she found that Wilkins treated her as his assistant, and refused to honor the terms under which she was hired. Only males were permitted to enter the university dining rooms, and after hours her colleagues talked over their daily work in men-only pubs.

Her response was to retreat into shyness and a shell of introversion as she focused intently on her work. Early in 1953, she discovered that there were two forms of DNA, and photographed them with advanced X-ray crystallography techniques that she had perfected. One of the photos, which she duly handed in to Wilkins, showed clearly the double helix structure of the molecule.

Meanwhile, Watson and Crick, working in their own lab at Cambridge, had come up with a three-stranded structure. Wilkins and Franklin opined that the three-stranded model was incorrect, and explained to them why it would not work.

Watson and Crick redoubled their efforts and eventually theorized that a double helix structure might be the correct solution. Two days after Franklin had completed a paper on the subject, Watson and Crick received a paper from Linus Pauling that showed he was also on the right track. Shortly thereafter, in an attempt to help Watson and Crick prove their hypothesis, Wilkins showed them Franklin's extraordinary photograph—without her permission and without bothering to tell her about it. Watson and Crick were stunned; Franklin's X-ray image proved that DNA was, in fact, a double helix. "My jaw fell open and my pulse began to race," Watson later recalled. They rushed to publish without including Franklin, who did not complain.

Shortly thereafter, Franklin left King's College and went on to explore the structure of viruses, including the polio virus. Even as she was dying of ovarian cancer over the next five years, she published 17 papers on her continuing research. When Watson, Crick, and Wilkins received the 1962 Nobel Prize for the double helix, they pointedly ignored the importance of her contributions to the discovery, except for a brief mention in Wilkins's acceptance speech.

In March 2000, King's College London gave Franklin posthumous coequal recognition with Wilkins by erecting the Franklin-Wilkins Building for pharmaceutical research. Her old nemesis Maurice Wilkins attended the dedication ceremonies, along with James Watson.

See also DNA, DISCOVERY OF

> In the 1950s, biochemist Rosalind Franklin was not permitted to eat or drink with male scientists at Cambridge.

FREEMAN, RICHARD BROKE (1915–1986)
Darwin Chronicler

Today there is a full-fledged "Darwin industry," made up of academics who pore over and write about Charles Darwin's life and works for a living. But English zoologist Richard Freeman, one of the brilliant founders of the field, immersed himself in Darwiniana for the sheer joy of it.

An expert on insects with a doctorate from Oxford, he joined the teaching staff of University College London in 1947 and taught zoology and taxonomy there for more than three decades. Over the years he developed a passion for natural history books in general and books by and about Charles Darwin in particular. His book collecting and worldwide correspondence with scholars and book dealers turned into a career.

Freeman's *The Works of Charles Darwin: An Annotated Bibliographical Handlist* (1965; 2d ed., 1977) lists every edition of every one of Charles Darwin's works. He set a new standard for scientific bibliography in the completeness and meticulousness of his survey of almost 2,000 items, including all editions, translations, and reprints. With the increasing interest of book collectors in Darwin, Freeman's work is the first place to turn to determine the rarity and historical value of a Darwin work.

But Freeman's most remarkable compendium is *Charles Darwin: A Companion* (1978), an alphabetical listing of every person, place name, and important topic mentioned in Darwin's books and letters. His *Darwin Pedigrees* (1984) brought information about the family up to date.

Although debilitated in his later years by a partial stroke and confined to a wheelchair, Freeman continued his vigorous Darwin scholarship to the end. Those who venture into the immense Darwin literature find their path easier and more enjoyable as a result of Freeman's labor of love.

See also DARWIN, CHARLES; DARWIN, CHARLES, DESCENDANTS OF; DARWIN, CHARLES, WORKS OF

FRERE, JOHN (1740–1807)
First Archeologist of Prehistory

The first published scientific evidence for prehistoric man was John Frere's *Account of Flint Weapons Discovered at Hoxne in Suffolk*, read before the London Society of Antiquaries in 1797, a dozen years before Charles Darwin was born. Finely worked flint hand axes had been discovered in association with several large bones and a massive jawbone, probably of a mastodon, all sealed together in the same strata.

Frere was attracted to the site by workmen, who had been excavating clay for bricks. He took careful notes on the position of the bones and artifacts in the deposit and made detailed drawings of the worked stones, which were published in *Archaeologia* in 1800.

These shaped flints were man-made tools, Frere concluded, "fabricated and used by a people who had not the use of metals. . . . The situation [depth] at which these weapons were found may tempt us to refer them to a very remote period indeed, even beyond that of the present world."

Unfortunately, this pioneer of archeology was a lone voice, and the significance of his work went unnoticed until long after his death. John Frere's great-great-great granddaughter was Mary Leakey, whose mother's maiden name was Frere. Along with her husband, Louis, and son, Richard Erskine Frere Leakey, Mary Leakey was a major force in the discovery of early human ancestors and their stone tools in East Africa.

See also LEAKEY, MARY; LEAKEY, RICHARD

FREUD, SIGMUND (1856–1939)
Evolution of Neurosis

On a table in his consulting room, Dr. Sigmund Freud kept a glass case full of the "antiquities" he enjoyed collecting: Egyptian scarabs, paleolithic stone hand axes, ancient figurines. What did this eclectic curio cabinet have to do with psychoanalysis? Freud was exploring how events of his patients' childhood had shaped their adult personalities. These prehistoric artifacts, he often told visitors, were clues to the "childhood of the human race."

Freud was strongly influenced by the theory that each individual replays the evolution of the species.

Freud was trained as a biologist during the heyday of classical Darwinism, when German medical schools taught recapitulation in the womb as an accepted part of evolutionary theory. Promoted in Germany by Ernst Haeckel, the idea was that the development of each individual (ontogeny) is a speeded-up replay of the whole history of the species (phylogeny).

Based on general comparisons of developing embryos, recapitulation theory became extremely influential outside of science. Pushed to explain many social phenomena, the misapplied analogy caused a great deal of mischief. Starting with the inaccurate notion that a human embryo at various times resembles an adult fish, reptile, and monkey, it asserted that all individuals undergo the same evolutionary stages of development.

According to the extension of this idea, the minds of European children passed through a stage similar to that of adults of existing "lower" races (savages) or our primitive prehistoric ancestors. Tribal peoples who had been enslaved or colonized by Europeans were regarded as "childlike primitives," requiring the firm supervision of a paternal missionary or colonial administrator.

To the triad of children, savages, and early man, Freud added a fourth: the neurotic adult. (Other theorists of his time, such as Cesare Lombroso, had suggested that "criminals" and "morons" are stuck at an earlier stage of human development, or are "throwbacks" to a primitive type.)

Most writers have treated Freud as if his theories had arisen fully formed out of his own system of thought, with no scientific precedents. In fact, as University of California historian of science and psychologist Frank Sulloway demonstrated in *Freud, Biologist of the Mind* (1979), this "absolute originality" is a myth; 19th-century evolutionary ideas had an enormous influence in shaping Freud's thought.

Freud began his *Introductory Lectures on Psychoanalysis* (1916) with the statement of Haeckel's premise, which he thought self-evident: "Each individual somehow recapitulates in an abbreviated form the entire development of the human race." And in 1938 he explained that "with neurotics it is as though we are in a prehistoric landscape—for instance in the Jurassic. The great saurians are still running around; the horsetails grow as high as palms."

Stephen Jay Gould noted that "recapitulation was both central and pervasive in Freud's intellectual development." His "oral" and "anal" stages not only represent the infant's early experiences, but also hark back to a four-legged animal ancestry of all primates.

Freud's famous *Totem and Taboo* (1913) is subtitled *Some Points of Agreement between the Mental Life of Savages and Neurotics*. In this imaginative classic, he concocted his own psychoanalytic myth of "the primal horde," the Freudian version of Original Sin. Freud imagined that the first prehistoric society was a patriarchal clan, ruled by a dominant father who monopolized food and sex. In order to get at the women, the sons murdered their father. But then they were too guilty to enjoy the women, which Freud thought was the origin of the incest taboo.

CHILDHOOD STAGES of development could underlie adult neuroses, taught Sigmund Freud, just as the "childhood" of the human species might explain certain religious practices and taboos.

Later, the sons assuaged their guilt by merging the memory of their father with a symbolic totemic animal that it was taboo to kill. However, once a year, the sacred totem was symbolically slain and eaten. Freud's new "origin myth" was daring, imaginative, and influential, but there is really no evidence that these events ever took place.

Even more far-fetched speculations surfaced in Freud's manuscript *A Phylogenetic Fantasy*, written in 1915 but forgotten and stored in an old trunk for 70 years and finally published in 1987. This strange work traces hysteria, obsessions, anxiety neurosis, and other modern disorders to the harsh life of our ancestors during the Ice Age. Anxiety, for instance, arose because "mankind, under the influence of the privations that the encroaching Ice Age imposed upon it, has become generally anxious. The hitherto predominantly friendly outside world, which bestowed every satisfaction, transformed itself into a mass of threatening perils."

In Freud's view, we've been plagued by anxiety ever since. In addition, under harsh conditions, people had to limit their numbers, which caused redirection of libidinal urges to other objects and resurfaces today as "conversion hysteria" or fetishism (sexual desires directed at objects, like shoes or leather, rather than to the opposite sex). His idea was that behaviors that make no sense in today's world must have had a utility in the past, and have been handed down as a sort of inherited memory.

Freud's conclusions about the origins of dysfunctional behaviors are therefore based on two antiquated theories in biology: recapitulation and Lamarckian inheritance. Most present-day Freudians, unfamiliar with the history of evolutionary theory, cannot appreciate how deeply Freud's theories rest on these two major 19th-century scientific fads, which have long since been abandoned by biologists.

See also BIOGENETIC LAW; COMPARATIVE METHOD; HAECKEL, ERNST

FULLER, ERROL (b. 1947)
Hunter of Extinct Birds

Errol Fuller, a British painter, pugilist, and self-taught ornithologist, lives in two conjoined houses in Southborough, Kent, that are packed floor-to-ceiling with Victorian taxidermy. His unique natural history collection includes bones, skins, and mounted specimens of birds that have gone extinct.

Since childhood, Fuller has been haunted by the mystery and irrevocability of extinction. "The death of any individual has always seemed a desperately profound thing to me, and the death of the last individual of a species greatly magnifies the sense of loss," he says.

Fuller brings a scientist's knowledge, an artist's passion, and a boxer's tenacity to his quest to commune with extinct birds. For his idiosyncratic book *The Great Auk* (1999) he visited scores of museums, private collections, and archives, seeking sketches and paintings of auks, as well as every known mounted specimen and egg (78 still exist). That book is itself a museum, containing photos of bones, beaks, letters, newspaper clippings, and even a rare, battered tin that once contained Great Auk cigarettes—an extinct English brand. He has also written and illustrated *Extinct Birds* (1987), *The Lost Birds of Paradise* (1995), and *The Dodo* (2003). In its World of Birds exhibition, the Bronx Zoo in New York has built a shrine to extinct birds and covered the walls with many of Fuller's illustrations, ironically etched in stone.

"I read at an early age the story of the last two great auks that were killed," Fuller recalled in a 1999 interview with *Natural History* magazine, "and the longing and intrigue grew over many years into a full-blown obsession." On June 3, 1844, on the remote Icelandic island of Eldey, three infamous sailors killed the last pair on earth.

Known as the "penguins of the North Atlantic," these flightless relatives of murres and puffins once ranged widely over subarctic seas. Gathering in great numbers on the remote, windswept rocks they used as rookeries, great auks were easy to catch and kill. Mariners would set up camp at Funk Island, off Newfoundland, herd thousands of the trusting birds together, throw them into boiling cauldrons while still alive, skin the carcasses, and render them for their fat and oil. But after that June morning in 1844—one of the best-documented finales of an entire species—there were no more auks.

By the end of the 19th century, the few preserved remains of the birds began to fetch fancy prices. Fuller has spent about $200,000 on his auk mania, including $45,000 for his own mounted specimen. When he discovered a stuffed auk at a French château, where it had been kept since 1858, he purchased it on the spot. "I just had to make contact with this poor little thing, to be able to touch it," he says. "It's all an illusion, of course, but this sad bundle of feathers did actually fire me up to create the whole project."

Some of Fuller's paintings deal with the auk's extinction. His *Evening of June 2nd, 1844*, depicts the last two birds—a lively, somewhat comical little couple—dwarfed by an enormous, cosmic sunset on the evening before their last day on earth. "The jauntiness of the birds as they face eternity represents the spirit of life," says Fuller, "the idea that we can never give up the battle until it is truly over."

See also EXTINCTION

AMONG ERROL FULLER'S SOUVENIRS of vanished birds is this battered tin of now-extinct Great Auk cigarettes, c. 1905. Below: *The Last Stand*, by Errol Fuller. © Errol Fuller, 1996.

FUNDAMENTALISM
Conservative Religious Movement

Christian fundamentalism reached its zenith of popular support in the United States during the 1920s, when 37 antievolution resolutions were brought before the legislatures of 20 states. Seven voted them into laws forbidding the teaching of evolution in schools. The World's Christian Fundamentals Association and the Anti-Evolution League of America packed considerable power in their war on Darwinism.

The term "fundamentalist" was coined in July 1920 by Curtis Lee Laws, editor of the *Baptist Watchman-Examiner,* to denote "believers who cling to the great fundamentals and who mean to do battle royal" for them. Before the 1920s, a "fundamentalist" movement existed in America, but it was not actively attacking science.

In 1910, the General Assembly of the Presbyterian Church adopted a creed of "five fundamentals": (1) inspiration and infallibility of scripture, (2) deity of Christ, including virgin birth, (3) substitutionary atonement of Christ's death, (4) literal resurrection of Christ from the dead, and (5) the literal return of Christ in the Second Coming.

To these, the Reverend Jerry Falwell later added that modern fundamentalists also believe in a literal heaven and hell, the depravity of mankind, the importance of soul-winning (evangelism), the Holy Spirit, and the existence of Satan.

A series of 12 booklets called *The Fundamentals* was published by Presbyterians in Chicago between 1910 and 1915 to "set forth the fundamentals of the Christian faith." They were sent free to ministers and others engaged in "aggressive Christian work." Although they contained some antievolutionary essays, the booklets also included writings of early fundamentalists who attempted to reconcile Christianity with Darwin, which may seem surprising today. Augustus Hopkins Strong (1836–1921), a leading Baptist theologian, had no quarrel with evolution and said so in *The Fundamentals.* Another eminent theologian, James Orr (1845–1913), wrote in *The Fundamentals* that the Bible should not be read as a science textbook, that the world is much older than 6,000 years, and that "evolution is coming to be recognized as a new name for 'creation.'"

By the 1920s, however, the movement had become a militant crusade, based on the fear that American children were being taught to deny God, subvert family authority, destroy the credibility of the Bible, and base their morality on a crass Social Darwinism. The Devil's work was being done by Darwin, science, and secular humanism.

It was not until the confrontation with lawyer Clarence Darrow at the Tennessee "Monkey Trial" of 1925 that fundamentalists began to lose ground in the public arena—even though they won that battle. High school biology teacher John Scopes was convicted of unlawfully teaching evolution, but Darrow's defense of science was a landmark case. As historian James Moore puts it, Darrow "made [William Jennings] Bryan talk nonsense, confess ignorance, and, most important of all, admit that he did in fact 'interpret' the Bible." Under Darrow's questioning, Bryan gave the lie to his own position that the "word of God" was unambiguous and not subject to human interpretation. Fundamentalists still claim to believe there is only one possible interpretation of God's word.

Since then, a succession of state statutes banning the teaching of evolution have been struck down. Nevertheless, fundamentalists have made periodic shows of strength: In the 1970s, a conservative Christian constituency led by Falwell succeeded in influencing the presidential campaign. In the 21st century, vocal, well-funded advocates of "intelligent design," some of whom acquired doctorates in science, took over the fundamentalist cause to replace evolution, though they deny that they have a religious agenda.

In 1987, the Supreme Court struck down the last of the "equal time" laws for teaching "creation science," ruling that it was an intrusion of religion into the public educational system. In 2005, a federal judge in Dover, Pennsylvania, made a similar ruling about the introduction of creationist textbooks. Undaunted, fundamentalist groups and their "intelligent design" allies have vowed to continue using the courts to oppose the teaching of evolutionary biology in public schools—the legacy of William Jennings Bryan.

See also BUTLER ACT; CREATIONISM; INTELLIGENT DESIGN; SCOPES TRIAL; SCOPES II

GATHERING THEIR FORCES, the Fundamentalist movement became militant during the 1920s, when they organized to introduce legislation prohibiting the teaching of evolution in many states.

GAIA HYPOTHESIS
Mother Earth's Stabilizers

Gaia, an ancient Greek name for the Earth goddess, has been applied to a hypothesis about how our planet supports life. The controversial idea was developed in the 1970s by two scientists: James Lovelock, an organic chemist, and Lynn Margulis, a molecular biologist. They believe that our planet is not simply an insensible rock hurtling through space, but a sensitive system that operates according to its own laws to support, develop, and preserve life.

The first scientific experiments on ecological life-support systems were done by the great English chemist Joseph Priestly in 1772. Exploring the question of whether plants breathe, Priestly placed a plant in an airtight bell jar; after a few days, it suffocated and died. Next, he enclosed a plant and mouse together in the same airtight chamber and found that both could breathe and live! Soon after, he realized that the animal inhales oxygen (O_2) and exhales carbon dioxide (CO_2), while the plant obligingly does just the opposite. It was the first scientific demonstration of a mutual life-support system.

In our own time, every schoolchild learns that the Earth's atmosphere consists of a certain mixture of oxygen (21%), nitrogen (78%), carbon dioxide, and other gases. This ocean of air extends to a height of about 60 miles and envelops the earth with a mantle of the same gases in the same proportions. But in 1972, James Lovelock wondered why these ratios remain constant over the whole planet and persist for vast periods of time with very little fluctuation.

He had found some exceptions, which help establish the more usual rule of stability. Evidence in the rocks reveals that eons ago, during the Cambrian era, there was an imbalance in the atmosphere. Oxygen levels dropped tremendously, causing mass extinctions among sea animals. However, by some unknown means, the balance was corrected, and the biosphere was rebuilt and stabilized once more for millions of years.

What caused the gases to make the necessary adjustment? We don't know. But Lovelock postulates the existence of a vast and delicate mechanism whose workings are still mysterious, but which seems in some way to react as if it were an organism. Earth's air, living matter, oceans, and land surfaces are all parts of the mechanism, the complexities of which are only now beginning to be unraveled.

According to Lovelock, in his book *Gaia* (1979), "the most essential part is probably that which dwells on the floors of the continental shelves and in the soil. The tough, reliable workers composing the microbial life of the soil and sea-beds are the ones who keep things going."

Lovelock is impressed, for example, by the millions of microbes that live in the guts of termites. Without this miniature universe of diverse creatures, termites would be unable to digest the wood (cellulose) on which they feed. But because an incredible living world has evolved inside—and along with—termites, the forests of the world are continually recycled, instead of being stacked miles high with billions of dead shrubs and trees.

Lovelock sees the entire range of living matter on Earth, from viruses to whales, from oaks to algae, as contributing to the Gaia system, which maintains and manipulates the Earth's atmosphere to suit its overall needs. It is also, Lovelock points out, a concept that is in tune with many of the world's ancient religious traditions, such as the Hindu, Buddhist and Native American.

Some scientists embrace Gaia as the world's only hope: If a tough, self-regulating mechanism exists in Mother Earth, they reason, we can all rest easier, as it may compensate for such human tampering as destruction of rain forests or spread of radioactive waste.

It seems more likely, however, that we cannot expect such a far-reaching, delicate system to heal any and all wounds inflicted by humans, as Mary Midgley, a philosopher of science, has warned in her book *Evolution as a Religion* (1985). "If Gaia finds her system getting out of kilter because one element in it is insatiably greedy," Midgley writes, "she simply ditches that element as she has done so many others before. But she herself can no doubt be killed with all her children. No universal fail-safe mechanism protects either her or us."

See also COEVOLUTION; ECOLOGY; ENDOSYMBIOSIS; EXTINCTION; RAIN FOREST CRISIS

GALÁPAGOS ARCHIPELAGO
Darwin's "Enchanted Islands"

The Galápagos Archipelago, now officially called the Archipiélago de Colón, is a cluster of volcanic islands in the Pacific Ocean astride the equator, 600 miles west of Ecuador. Early Spanish sailors called them Las Islas Encantadas, the Enchanted Isles, miniature self-contained worlds of desert, scrub, misty forest, and black volcanic beach. "Here, both in space and time," Darwin wrote in his *Journal* (1839), "we seem to be brought somewhat near to that great fact—that mystery of mysteries—the first appearance of new beings on this earth."

Darwin and HMS *Beagle* spent only five weeks exploring the islands in September and October of 1835, but they made a lasting impression on him. The Galápagos Archipelago is a compact model of how the whole planet may have been populated by radiating species, adapted to different ways of life and descended from common ancestors. In the first edition of his *Journal* he only hinted at such a conclusion, but in subsequent editions—as he sorted and pondered his collections—he began to realize that the Galápagos were key to the explanation he was seeking for the origin of species.

> The archipelago is a little world within itself, or rather a satellite attached to America, whence it has derived a few stray colonists. . . . Considering the small size of these islands, we feel the more astonished at the number of their aboriginal beings, and at their confined range. . . . The different islands to a considerable extent are inhabited by a different set of beings. . . . One is astonished at the amount of creative force, if such an expression may be used, displayed on these small, barren and rocky islands.

The nature and diversity of animal life was a real delight to the ship's young naturalist: there were sea lions, dolphins, penguins, prehistoric-looking iguana lizards, bright red crabs, incredibly gigantic tortoises (Darwin rode one like a pony), and a most interesting series of mockingbirds and finches. Large reptiles, rather than mammals, dominated the landscape.

Despite their fearsome appearance, the marine iguanas are vegetarians that dive deep in the ocean to graze on seaweed. (Darwin cut one's stomach open to discover this.) As for the tortoises, they were so docile that sailors were in the habit of carting off hundreds of the hapless beasts, to be stored live in ships' holds awaiting their turn to become soup. One tortoise could yield 200 pounds of meat (with a 400-pound shell); it took six or eight men to lift one and carry it away.

Because of their long isolation, the Galápagos animals had no fear of man. "A gun is here almost superfluous," Darwin wrote, "for with the muzzle I pushed a hawk off the branch of a tree." He was astonished at the contrast between this trusting disposition and the birds back in England and Europe, which had apparently learned to keep their distance from people thousands of years ago.

But it was the diversity of closely related species on the several islands that Darwin said "strikes me with wonder." Each island had its own species of tortoise,

"A gun is here superfluous, for with the muzzle I pushed a hawk off the branch of a tree."

—Charles Darwin, on the tameness of Galápagos animals

© Tui de Roy/Roving Tortoise Photos.

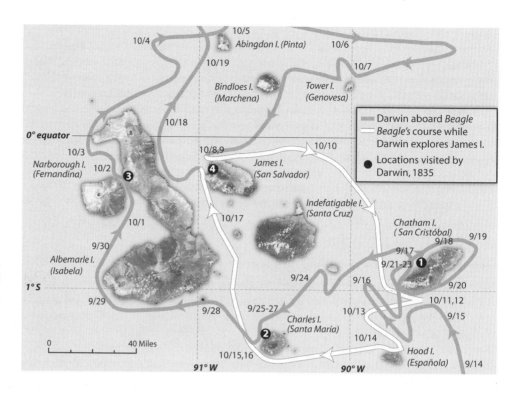

10/5
10/4
Abingdon I. (Pinta)
10/6
10/19
10/7
Bindloes I.
(Marchena)
Tower I.
(Genovesa)

Darwin aboard *Beagle*
Beagle's course while
Darwin explores James I.
Locations visited by
Darwin, 1835

10/18
0° equator
10/8,9
10/10
10/3
Narborough I.
(Fernandina)   10/2
James I.
(San Salvador)
Indefatigable I.
(Santa Cruz)
Chatham I.
(San Cristóbal)
9/19
3
4
10/17
9/18
9/17
10/1
9/21-23   1
9/30
Albemarle I.
(Isabela)
9/20
1°S
10/11,12
9/24
9/16
9/15
9/29
9/28   9/25-27
Charles I.
(Santa María)
10/13
10/14
0   40 Miles
2
Hood I.
(Española)
9/14
10/15,16
91°W
90°W

THE ENCHANTED ISLES
seemed to take young
Charles Darwin back in
time as he explored the
Galápagos Archipelago in
1835. Darwin's routes are
superimposed on a NASA
satellite photo.
© John Woram.

and various mockingbirds and finches abounded, although he did not bother to record from which islands he took his specimens. Later, back in England, working with preserved collections at the British Museum under the guidance of ornithologist John Gould, he realized that the islands contained 13 types of finches, each with a different type of beak, from delicate to heavy, small to large, and that they occupied different niches in the ecology. "Seeing this gradation and diversity of structure in one small, intimately related group of birds," he wrote, "one might really fancy that from an original paucity of birds in this archipelago, one species had been taken and modified for different ends."

He had thought the tortoises on all the islands were the same species until just before he left, when Mr. Lawson, vice-governor of one of the islands, told Darwin "that the tortoises differed from the different islands, and that he could with certainty tell from which island any one was brought. I did not for some time pay sufficient attention to this statement, and I had already partially mingled together the collections from two of the islands."

Geologically, the Galápagos islands present a mystery that Darwin never confronted. Until the late 20th century, geologists had supposed that the islands were about two to three million years old. Two biochemists at the University of California, Berkeley, however, thought that was far too young for evolution to have produced so many different and unique life forms there. During the 1970s, biochemists Vincent Sarich and Jeffrey Wiles compared the serum albumin of Galápagos land iguanas with those of the Galápagos marine iguanas, and concluded that they must have diverged from a common ancestor close to 10 million years ago. In 1983, Sarich and Wiles published a paper titled "Are the Galápagos Iguanas Older Than the Galápagos?" in which they predicted that submerged islands would be discovered that were above water when the ancestral iguanas originally landed.

The mystery was solved in 1992, when geophysicist David Christie of Oregon State University dredged the seafloor for a month. His team found convincing evidence that the Galápagos chain once included several much older islands—about nine million years old—that long ago sank beneath the waves.

See also DARWIN RESEARCH STATION; ISOLATING MECHANISMS; LONESOME GEORGE; VOYAGE OF HMS *BEAGLE*

TOURISTS EXPERIENCE
the curious seals, iguanas,
and giant tortoises of the
Galápagos, which is now an
Ecuadorian national park.

GALTON, SIR FRANCIS (1822–1911)
Eugenicist, Statistician

Charles Darwin's cousin Sir Francis Galton was the epitome of the compulsive Victorian genius. By the age of 22, he had degrees in both medicine and mathematics, then went on to become a famous travel writer (on Africa) and statistician. Galton is chiefly remembered today as the founder of the eugenics movement—a well-intentioned attempt to bring about social improvement through selective breeding. If horses and cattle could be improved through artificial selection, he asked, why not people?

Eugenics was, of course, an unmitigated disaster, an ideology that caused untold suffering for more than 60 years. Eugenic principles were called upon as the "scientific" justification for restrictive immigration laws, compulsory sterilization in the United States, racist social policies in several countries, and the nightmarish mass exterminations in Nazi Germany.

Galton had so many diverse interests, such an active mind, and was so convinced of his own greatness that in later life he hired secretaries to follow him everywhere, lest some of his thoughts escape unrecorded. Few did.

Among his almost 200 scientific publications are some intriguing oddball titles: "On a New Principle for the Protection of Riflemen" (based on the trajectory of spherical bullets) (1861), "Statistical Inquiries into the Efficacy of Prayer" (1872), "The Visions of Sane Persons" (1881), "Measurement of Character" (1884), "Note on Australian Marriage Systems" (1889), "Head Growth in Students at the University of Cambridge" (1889), "Arithmetic by Smell" (1894), *Fingerprint Directories* (1895), "Three Generations of Lunatic Cats" (1896), "Cutting a Round Cake on Scientific Principles" (1906).

Profoundly inspired by Darwin's evolutionary theories, Galton spent much of his life attempting to follow out their practical, mathematical, and social implications.

HEREDITARY GENIUS was more than the title of one of Sir Francis Galton's books; Galton—a cousin of Charles Darwin—saw in the concept his own family's destiny. But Galton's plan to breed superior humans proved to be not an intelligent design.

Searching for patterns of hereditary talents among close relatives, he began to gather background information on a sample of accomplished Cambridge college students. Eventually, he pulled together data on 1,000 men from 300 families for his study *Hereditary Genius* (1869). Some families, he found, produced an inordinate number of scientists, musicians, painters, judges, military commanders, or even oarsmen or wrestlers.

These special talents, he believed, were innate. Galton thought the advantages of tradition, wealth, opportunity, and education didn't count for much, compared with hereditary abilities present in the individual at birth. Thus, a person born with superior character and intelligence should inevitably rise to eminence, even if reared in the lowliest slum. Since few of the poor became eminent professionally, he concluded that the lower classes remained poor because of genetic "inferiority."

Charles Darwin, on reading part of *Hereditary Genius*, wrote Galton: "You have made a convert of an opponent in one sense, for I have always maintained that, excepting fools, men did not differ much in intellect, only in zeal and hard work; and I still think this is an eminently important difference." Galton respectfully disagreed, insisting that "character, including the aptitude for work, is heritable like every other faculty."

Eugenics, the "science of breeding the best," became Galton's obsession. As he wrote in *Memories of My Life*, he believed it was "within [man's] province to replace Natural Selection by other processes that are more merciful and not less effective. This is precisely the aim of Eugenics. Its first object is to check the birth-rate of the Unfit. . . . The second object is the improvement of the race by furthering the productivity of the Fit by early marriages . . . of the best stock." He feared the lower classes would take over society by having the most children, which would result in the "degeneration" of the "British race." His ideas were taken seriously, attracting such thinkers as playwright George Bernard Shaw, novelist H. G. Wells, and socialist Sidney Webb to join the Eugenics Education Society, which Galton founded in 1907.

The United States proved to be even more fertile ground than England for the adoption of eugenic policies. Sterilization laws for "imbeciles" were passed, and the usually compassionate justice Oliver Wendell Holmes upheld its constitutionality in a 1927 decision: "We have seen more than once that the public welfare may call upon the best citizens for their lives. It would be strange if it could not call upon those who already sap the strength of the State for lesser sacrifices. Society can prevent [the] unfit from continuing their kind."

Galton also established the tradition in England of biometrics, the application of statistical and populational techniques to the study of heredity and variation. His journals and laboratories laid the groundwork for population genetics, which years later helped to provide Darwinian theory with a mathematical basis. He also took up the unfinished work of Captain Robert FitzRoy, reestablishing his meteorological office and its practice of furnishing weather maps and forecasts (a term FitzRoy had coined) to the newspapers.

Galton was once asked, when in his seventies, why he had no deep wrinkles or furrows on his forehead. "Oh, that is easy," he replied. "It's because I am never puzzled."

See also COMPOSITE PHOTOS; DEGENERATION THEORY; EUGENICS; FITZROY, CAPTAIN ROBERT; GALTON'S POLYHEDRON

GALTON'S POLYHEDRON
Equilibrium and Constraints?

In his influential book *Hereditary Genius* (1869), Darwin's eccentric cousin Sir Francis Galton presented an unusual metaphor for the evolution of species. Suppose, he wrote, that you had a stone with many smooth facets that was tumbling across a slanted table. It would rest in a stable equilibrium on one facet until considerable energy was exerted to push it over onto the next facet, where it would again be quite stable.

He imagined a kind of "buildup" in organisms, which could erupt in seemingly sudden changes of form. As Stephen Jay Gould put it in his essay "A Dog's Life in Galton's Polyhedron," "the polyhedron's response to selection is restricted by its own internal structure; it can only move to a limited number of definite places . . . [an] interaction of external push and internal constraint."

In *On the Genesis of the Species* (1871), naturalist and theologian St. George Mivart used Galton's polyhedron as the basis for his hypothesis of Specific Genesis, wherein an organism—though adapted to particular conditions—contains within itself certain well-defined possibilities for adjusting quickly to sudden environmental changes. Decades later such large rapid changes in individuals would be called "macromutations," but Mivart and Galton were writing years before the rise of genetics.

A pattern of rapid changes in species followed by long periods of stability is now seen by some paleontologists as a more accurate description of the fossil record than Darwin's insistence on slow, steady rates of change. In their contribution to *Models in Paleobiology* (ed. T. J. M. Schopf, 1972), Gould and Niles Eldredge called attention to this model of evolution, which they named "punctuated equilibrium."

GARNER, RICHARD LYNCH (1848–1920)
Pioneer Primate Observer

In America during the 1890s, the only primates zoologist R. L. Garner had ever seen were locked in cages. Fascinated with human evolution, he decided to make pioneering studies of monkey and ape behavior in the wild. Fearing attacks by wild apes, Garner hit on a unique fieldwork method: He built a large cage in the deep forest and locked himself inside!

Garner's cage was no flimsy hunter's blind; it was strong enough to hold off any animal except an elephant. Armed with a rifle, revolver, and bush-knife, he sat patiently for days and weeks on end, emerging each evening to return to base camp. Almost from the start, monkeys and other wildlife gathered in nearby trees to observe the caged human. Garner did see passing gorillas and chimpanzees, but never tracked them or observed normal group behavior.

Nevertheless, he learned to imitate 10 chimpanzee sounds and identified 20 more. He also observed the whooping, frenzied vegetation-tearing displays accompanied by drumming, which he described as a "chimpanzee carnival." Relying on local accounts, Garner

correctly described the gorilla family group, with its several females clustered about a large male leader, in *Gorillas and Chimpanzees* (1896).

Later, he became the Bronx Zoo's representative in Africa, buying animals and shipping them to New York. In 1906, Garner sent the first live gorilla to reach an American zoo (England's first had arrived in 1860). Another was sent in 1911, but neither survived for long. Zoo directors despaired of ever keeping gorillas alive.

Garner also studied captive monkeys, particularly some capuchins from South America. An overzealous Darwinian, he imagined he could discern distinct calls for "milk," "apple," "banana," "bread," and "carrot"—a whole "monkey language," which no one else could later verify. He transcribed the calls as "melodies" in standard musical notation (*Apes and Monkeys*, 1900), concluding that monkeys had "all the characteristics of speech."

Despite his limitations, Garner made the bold first attempt to observe wild apes, which inspired others (including Carl Akeley) to follow him. If the heavily armed, pith-helmeted Victorian scientist sitting in his jungle cage seems quaint, his successors did little better. For the next 75 years they carried guns, tracked baboons from Land Rovers, and watched gorillas through binoculars.

It was only in the 1960s that Jane Goodall began mingling with chimps and George Schaller tracked gorillas unarmed and on foot. By the 1970s, mountain gorillas had accepted Dian Fossey into their groups, Schaller had walked among tigers, Shirley Strum had abandoned her van to mingle with baboons, and Jacques Cousteau's divers had swum alongside whales, completely dependent for their safety on the animals' curiosity, gentleness, and friendliness.

See also APE LANGUAGE CONTROVERSY; "APE WOMEN," LEAKEY'S; BABOONS; CHIMPANZEES; FOSSEY, DIAN

GENETIC DRIFT
A Chance Factor in Evolution

Geneticist Sewall Wright identified the phenomenon of genetic drift, which he called the "founder effect." It operates when a very small group of individuals becomes the basis of a new, larger population.

In humans, genetic drift may occur when a small tribal band splits up and a few families migrate to colonize a new territory. Founder populations can also be the remnants of harsh winters, wars, or famines. Examples in living human populations include some isolated Alpine villages, which have unusually high percentages of albinism, and the South Atlantic island Tristan da Cunha, whose population—descended from a single Scots family and some shipwrecked sailors—has a high proportion of a rare hereditary eye disease. As our species became dispersed over a wide range, irregular and nondirectional changes in gene frequencies within small populations probably played an important role in human evolution.

GENETIC ENGINEERING
"Inventing" New Life Forms

In 1987, the U.S. Patent Office decreed that living organisms produced by recombinant DNA are patentable inventions. That same year, a biologist patented a strain of mice that is guaranteed to develop cancerous tumors, for use in cancer research.

Biotech companies have induced bacteria to make human insulin and dissolve dangerous blood clots, and will soon turn microbes into living factories for the production of serum and vaccines. Some are trying to evolve bacteria that will eat up oil spills at sea or provoke human antigens to destroy organisms that decay teeth.

In one experiment, a modified microbe was injected into corn to see if it could give the plant immunity from its archenemy, a moth larva known as the corn borer. The microbe had been genetically altered to attack alkaline chemicals, and it attacked the corn borers' stomachs, giving them ulcers—and protecting the plants.

In the 1980s, tobacco plants—the experimental "white rats" of botany—received genetic material from fireflies. The result: a tobacco plant that glows in the dark. It is now much in demand for various studies of cell physiology, since the glowing "marker" cells can be easily traced.

At the University of Idaho, a state where fish farming is a major industry, genetic engineers are attempting to create the ideal food fish of the future. So far, they have found fish much more easy to manipulate than mammals. Experimenters have been able to reverse the sexes of fish, multiply their chromosomes, and even produce an "invented" variety made up of cells from two different species of trout.

Cattle embryos have been successfully cloned at Granada Genetics, Inc., a leading livestock genetic engineering firm in Marquez, Texas. Researchers there have already raised several small herds of genetically identical cattle and are preparing for the day when many cloned embryos of a single desirable animal can be routinely implanted in ordinary cows.

Genetic engineering has its vehement critics, however. Environmental advocate Jeremy Rifkin, for example, has challenged scientists' right to unleash "new" organisms into the environment for profit, and sued the U.S. Patent Office to challenge its policy that living things can be legally patented as "inventions."

See also DNA; FRANKENFOOD

> **Today, genetic variability among humans is rapidly increasing, while the variability of many animals and plants is shrinking.**

GENETIC VARIABILITY
The Raw Material of Evolution

The incredible range of variation within species has always fascinated naturalists and frustrated classifiers. Anyone can see that dogs and cats are different species, but the population biologist, or practical breeder, becomes entangled in the great variations within each species, sometimes even within a single litter.

Although Darwin saw thousands of different species in the wild during his voyage on HMS *Beagle,* it was only after detailed scrutiny of a single group of organisms that he considered himself a finished naturalist whose theories could command respect. Upon his return home, he embarked on an eight-year study of barnacles, focusing on the variability among thousands of individuals.

Later, Darwin sought out local farmers who bred different types of pigs, gardeners who experimented with varieties of flowers, and hobbyist breeders of fancy pigeons. He questioned them endlessly about strains, crosses, and hybrids and also conducted his own garden experiments. He compiled the results in his two-volume work *Variation of Plants and Animals under Domestication* (1868). In the 1870s, he corresponded with botanists who were breeding potatoes and encouraged them to seek varieties immune to the blight that was ravaging Ireland, causing mass famine and emigration.

It was one of Darwin's great insights that if certain genetic characteristics could be combined or eliminated in domestic plants and animals by deliberate human selection, then selection could similarly work on genetic variation in the wild. In *Origin of Species* (1859),

Darwin's argument for natural selection was based on this amazing variability within species and on the analogy with artificial selection of domesticated organisms.

Today there is probably greater human genetic variability (polymorphism) than ever before. Since culture, diet, and medical technology have lengthened our lives, more people are protected from being "weeded out" before they have a chance to reproduce. Therefore, today's human gene pool contains an immense range of biochemistries, skin colors, body structures, sizes, proportions, and, unfortunately, genetically carried diseases. At the same time, however, the genetic variability of the animals and plants with which we share the world, especially the species we use for food, is rapidly shrinking.

Several hundred years ago, each area of the world produced many kinds of rice, wheat, beans, corn, cattle, sheep, or chickens, which local farmers nurtured and bred from the wild stock over thousands of years. Andean Indians, for instance, had scores of potato varieties adapted to different altitudes and levels of moisture.

During the past century, however, much of this variation has been deliberately bred out of domesticated plants and animals to produce "superior" strains. We have all benefited from the worldwide production of large, tasty, high-yield varieties of staples. But when farmers universally adopted the new, "improved" types to increase their yields—often in response to government programs—they stopped cultivating their ancient local varieties. Indian corn and Texas long-horned cattle became rare; they are now propagated only as curiosities.

By the mid-1970s virtually all corn, wheat, rice, and many other grains were each grown from a single strain. While the new "super" varieties offer great advantages, it is dangerous to put all our eggs into one genetic basket. For instance, the better-tasting, faster-growing food crops are defenseless against wheat rust, corn blight, and other fungal invasions. If they are genetically uniform, they can all be wiped out in a single epidemic, as happened in the 1970s to millions of acres of corn, from Iowa to Texas. (Near the end of the 19th century, a similar catastrophe destroyed the vast British coffee plantations on Ceylon—now Sri Lanka—which had to be replanted with tea, turning the English from a nation of coffee drinkers to tea drinkers almost overnight!)

In nature, variability within a species allows the population to continually adapt. During changing conditions, many individuals may be wiped out, but some will survive. As a hedge against climatic change and epidemics, plant geneticists belatedly realized that some of the original, nearly lost varieties of food plants had to be preserved. In the late 1970s, various governments and the United Nations Food and Agriculture Organization began creating seed banks to preserve worldwide genetic variability of agricultural plants. By the time they started gathering rare seeds, however, many important varieties were close to extinction.

Today there are more than 1,400 seed banks around the world, containing about 6.5 million plant samples. At the federal gene bank in Fort Collins, Colorado, America stores plant tissues in liquid nitrogen, while millions of seeds, representing thousands of plant varieties, are stored in refrigerated vaults. These varieties must periodically be grown out in fields as well, because long storage and laboratory germination kills a certain percentage of the plants. Seeds with valuable properties can be lost simply because of archival conditions, which may create an unintended selection for survival in the artificial environment.

Of the thousands of crop varieties that farmers raised during the 19th century, 95 percent have been lost. Only 150 kinds of agricultural plants now feed the world's peoples, and of those only 12 provide 80 percent of the total food energy. The top four are rice, corn, potatoes, and wheat, which supply at least 50 percent of the total. In 2004, many nations ratified an international "seed treaty" as a kind of group insurance policy against agricultural disasters caused by war or climate change.

Although the United States did not ratify the 2004 treaty, private American philanthropies have donated millions toward the maintenance of the Svalbard Global Seed Vault, which opened in 2008. The $7 million dollar facility, built by the Norwegian government and funded by the Rome-based Global Crop Diversity Trust, is nestled in Norway's frozen tundra and set into a mountain of rock. Nicknamed the "doomsday vault," it is an extraordinary repository to protect humanity's agricultural legacy—the seeds of more than 200,000 crop varieties from all over the world.

At a major seed bank at Fort Collins, Colorado, millions of seeds, representing thousands of plant varieties, are stored in refrigerated vaults.

GENOGRAPHICS
Tracing the Odyssey of Humankind

Paleoanthropologists believe that the six billion people on earth today are descended from a founding African population of fewer than 10,000, some of whom migrated to Europe and Asia about 100,000 years ago but left few archeological traces along the way. They did leave evidence, however, in the form of genetic markers in our DNA.

To gather evidence, geneticist Spencer Wells traveled the world in the 1990s to collect and analyze thousands of samples of modern human blood. His mentor, the Italian geneticist Luigi Luca Cavalli-Sforza, now at Stanford University, had begun the work thirty years earlier by collecting and comparing samples from far-flung tribes and ethnic groups. Their efforts show that only a few hundred individuals originally left Africa, perhaps as late as 50,000 years ago, and that we are separated from them by only 2,000 generations.

While the self-replication of DNA usually takes place with high accuracy, rare mutations ("copying mistakes") do occur. Wells and his colleagues sought mutations within Y chromosomes, which are carried only in males, and used them as markers to identify movements of ancient populations. After sorting results from thousands of individuals throughout Africa, Asia, India, Europe, and the Americas, he concluded that there had been migrations over thousands of years, but only by a few major routes. During many millennia-long odysseys, our ancestors braved scorching deserts and arctic ice with phenomenal strength and resilience.

Wells confirmed Cavalli-Sforza's suggestion that the !Kung-San of the Kalahari Desert in southern Africa are the direct descendants of the world's earliest people and have remained in the Kalahari for over 50,000 years. Die Kelders Cave, in Cape Province, South Africa, may contain remains of their ancestors from 80,000 years ago, but the !Kung-San seem to have left that area 50,000 years ago as glacial ice moved northward, destroying pasturelands and making game animals scarce. The oldest known human remains in Africa (195,000 years old) have been found at Omo-Kibish, in Ethiopia, and the next oldest are from Herto (Ethiopia) and Jebel Irhoud (Morocco), both dated at 160,000 years. These populations are apparently from "stem" *Homo sapiens* and predate recognizable "racial" diversification.

> **Africa may have been the cradle of mankind, but genetic markers point to Central Asia as its longtime nursery.**

Outside Africa, the oldest fully human (*H. sapiens*) remains have been found in Australia, at a desert site called Lake Mungo, an ancient lakebed in New South Wales dated at >45,000 years. In Southeast Asia, the oldest archeological evidence of humans also goes back to around 40,000 years. Because a particular genetic marker is common in both Southeast Asian and Australian males and descends from an older African lineage, it appears that the traffic from Africa to Australia was one-way only, by way of Southeast Asia.

Some present-day Aborigines scoff at Wells's scientific "story" and insist that their ancestors emerged from sacred rock overhangs and caves in their land at the beginning of time. One man told Wells that if a blood connection to Africa really exists, then some of his ancestors must have migrated to Africa, not the reverse, as he was quite certain that Australian Aborigines were the first and original people.

When Wells began his study, he realized there was a gap of almost 50,000 years separating African and Australian populations and that scientists had no physical evidence along the coastline of South Asia. Forty millennia ago, the Indian and southern Asian coastlines were very different from today, and sea levels were much lower. Thousands of years later, as ocean levels rose dramatically, whatever trail of human migrations existed (bones or archeological remains) was likely washed away.

Wells sought genetic evidence connecting Africa and Australia. After taking blood samples from many indigenous groups along the possible route from Africa, his team finally found an individual in South India with the African genetic marker: the long-sought link between African and Australian genomes. Apparently, from genetic evidence, the South Indian group had branched off from the African about 50,000 years ago—a migration that appears to have involved the ancestors of only about 10 percent of the world's present non-African population. The other 90 percent took a different route—through the Middle East—and populated most of the interior of Eurasia and the Americas.

One subpopulation of the main migrating group split off, winding up in Central Asia, while another made its way to India. Later, a second wave of migrants from Africa also reached China, then split again into two distinct populations that settled in the north and south. Humans ventured into Europe only 35,000 years ago, during a period of cold and glaciation.

In Central Asia, after collecting blood from over 2,000 people, Wells and his colleagues turned up an ancient DNA marker that Asians share with Europeans. Populations carrying this marker went from the Middle East through Central Asia during worldwide droughts, which were followed by rainy climates that nurtured rich grasslands. Genetic markers show they arrived on the plains of Asia about 45,000 years ago and spent several thousand years there, where game was plentiful, before moving on to Eurasia and North America. The trek from the Middle East to Europe via Central Asia took 10,000 years. If Africa was the cradle of humankind, according to Wells, Central Asia was its longtime nursery.

Wells found one Central Asian man, named Nysazov, whose Y-chromosomal DNA carried the most ancient form of the genetic "copying mistake" that spread to Europe, India, Russia, and the Americas. His DNA indicates that his family has been living in Central Asia for nearly 2,000 generations; Wells calculates that he descended from a man who lived in the area 40,000 years ago.

Wells next traced the same marker to the Chukchi, reindeer herders who inhabit northeastern Russia. Living for the most part without heat or electricity, they are culturally and physiologically adapted to the brutal cold. A tiny group of Chukchi, Wells found, carry a genetic marker that links them with Native North Americans—evidence that their close relatives crossed over the Beringia land bridge from Russia to North America 13,000 years ago.

Within 800 years, incredible as it seems, an ancestral founding population of as few as 10 people who migrated across that frozen land bridge could have expanded to populate both North and South America, from the Mayans, Incas, and Aztecs to the Navajo, Sioux, and Iroquois. Like the Australian Aborigines, however, some traditional Native Americans are unconvinced that their people originated outside their present home territories. Navajos, for instance, asserted that they are the world's original people, and that their ancestors emerged from the Earth near their sacred rocks—almost exactly the same tale told by Australian Aborigines and by Africans on the other side of the world.

In 2005, Wells's pioneering work resulted in the formation of the Genographic Project, an internationally coordinated effort to read the human past in our blood chemistry. Ultimately, the goal is to reconstruct the story of human diversity, ancient migrations, and present variability. Funded by private and corporate sponsors, the project maintains field researchers in a dozen countries and enlists public participation in collecting hundreds of thousands of human blood samples. One of its studies recently concluded that at various times, tiny human groups in Africa and elsewhere had become rare, and that our species nearly went extinct during harsh climatic changes. "But they came back from the brink to reunite and populate the world," reports Wells on the project's website, "until the fragile populations eventually reunited to populate the entire planet. Truly an epic drama, written in our DNA."

GENOME
Makeup of a Species

Each individual's genetic information, embedded in the chromosomes, is its genotype. The expression of that code is the physical body, or phenotype. However, many unexpressed or recessive characters in the genotype do not show up in the phenotype.

The genome—the sum of all genotypes in a species—is the current focus of gene-mapping projects. A major surprise in comparing the human genome with the genomes of other organisms is that ours does not contain markedly more genes than do those of "simpler" creatures. The fruit fly has about 13,600 genes, mice have 22,000, and sea urchins 23,300. Humans have between 20,000 and 25,000 genes, making us more similar to urchins and mice than to flies. The lack of obvious correlations between genome size and organismal complexity has become a contentious topic among geneticists.

GERTIE THE DINOSAUR
First Animal Cartoon

I f animated cartoon creatures evolve, it seems appropriate that years before Mickey Mouse, the first animal "toon" star was a dinosaur. *Gertie the Trained Dinosaur* (1914) was the brainchild of innovative artist Winsor McCay, who had first introduced the friendly brontosaur in his Sunday newspaper comic strip *Little Nemo in Slumberland*. She proved so popular that he decided to feature her in an animated film.

Gertie is considered the first great animated film. Twenty years before the Disney studio existed, McCay worked alone, drawing 24 pictures for each second of screen action. He was first to give his cartoon character personality and emotion.

But Gertie was more than a screen cartoon; she was also McCay's partner in a clever vaudeville act that toured for years. McCay himself would appear on stage, dressed in a formal suit, to put Gertie through her paces. Seemingly at his command, her projected image would eat, dance, or laugh at his jokes. When she misbehaved, he scolded her and she cried. For the public, it was the first time a cartoon character had been "brought to life" with a believable personality, marking the creation of a Hollywood industry.

Near the end of the stage act, McCay would ask Gertie to lower her long neck so he could climb on. Then he would walk toward Gertie and, with split-second timing, slip backstage through a slit in the screen. At the same instant, a life-size image of McCay in his tux would appear to be hoisted up by Gertie, smoothly crossing over into the cartoon universe. With her human friend perched on her back, Gertie would then turn and walk away into the movieland Mesozoic.

See also *FANTASIA*

A SINGULAR SENSATION, cartoonist Winsor McCay's vaudeville act featured "Gertie," an animated dinosaur who appeared to respond to his every command. She was the first big hit as an animal cartoon star, a niche later occupied by toon ducks, mice, and rabbits.

"GHOST" SPECIES
Plants with Missing Partners

I n the 1977 movie *Oh, God!* the John Denver character marvels that everything in nature is perfect. "Oh, I've made some mistakes," replies George Burns as God. "For instance, the avocado: I made the pit too big."

There may be a good evolutionary reason for the apparently oversized, tough-shelled avocado seed. It seems to be an example of an "evolutionary anachronism," a fruit adapted for relationships with large mammals that have gone extinct. Paleoecologists Paul Martin and Daniel Janzen believe that avocados were once dispersed by giant ground sloths, gomphotheres (a kind of extinct elephant), or hippolike *Toxodon*, who gulped down the fleshy fruits whole and then excreted their large seeds miles away. No living North American mammals have sufficiently large guts to do the job.

Some plants seem to evoke the lifestyles of vanished animals from the Pleistocene with which they coevolved. Martin expressed his fascination with such seemingly puzzling adaptations in a 1992 essay, "The Last Entire Earth":

> In the shadows along the trail, I keep on eye out for ghosts, the beasts of the Ice Age. What is the purpose of the thorns on mesquites in my backyard in Tucson? Why do they and [the trees known as] honey locusts have sugary pods so attractive to livestock? Whose foot is devil's claw intended to intercept? Such musings add magic to a walk and may help to liberate us from . . . the hubris of the present, the misleading notion that nature is self-evident.

Science writer Connie Barlow, who has expounded on Martin's and Janzen's ideas in her book *The Ghosts of Evolution: Nonsensical Fruits, Missing Partners, and Other Ecological Anachronisms* (2000), sees the present American landscape as incomplete. In his foreword to her book, Martin has written, "We lost our American Serengeti . . . only 13,000 years ago, long after the extinction of the dinosaurs. It was more than an injury; it was an ecological and evolutionary amputation."

Mango, papaya, tamarind, osage orange, Kentucky coffee tree, gourd, peach, and watermelon—all appear to have catered to fruit-loving animals that once dispersed their seeds. Some of these plants continue to prosper today only because humans and their domestic animals now spread their seeds. Horses and cows, for instance, now snag the devil's claw's large burrs with their hooves (the feet of native pronghorns and deer are too delicate), while gourd vines (once the food of ground sloths) hitch rides in the blades of road graders and in tire treads.

For more than 20 million years, the honey locust tree, native to the central United States, lured mammoths to disperse its tough, tooth-resistant seeds by embedding them in sweet, protein-rich pods. But this tree also evolved formidable thorns on its trunk and lower branches. Elephants could browse for seed pods with their trunks among the upper branches but were prevented from stripping bark, which would kill the trees. Sturdy and drought-resistant, honey locusts are now commonly planted in city parks and streets. While the male trees, podless and thornless, are most common, older plantings include female trees, with their seed pods and armor—just in case any mastodons should return to eat them.

Despite George Burns's theological apology, the avocado was not the product of a divine mistake. By the very structure of the fruits, seed coats, and armaments of certain plants, they seem to be proclaiming that they are adapted for life in a vanished world.

See also COEVOLUTION

ICE AGE MASTODONS once dispersed seeds of the honey locust tree, whose sharp spikes protected its bark and lower branches. Today the trees are propagated by humans, who have allowed the now-superfluous spikes to remain on male trees. Top: Edward S. Barnard; bottom: © Michael Rothman, 2002.

GOBI DESERT EXPEDITIONS (1921–1930)
Search for Origins in Asia

Charles Darwin thought Africa was probably the "cradle of mankind," a clever hunch that would have been forgotten long ago except that science now agrees with it. But Darwin was in the minority; from his day until more than 70 years after his death, the search for human origins focused on Asia. Between 1921 and 1930, the most audacious and expensive land expedition in history combed the far reaches of Outer Mongolia, then as little known to science as the surface of the moon. Its mission: to find the birthplace of the human species.

Led by Roy Chapman Andrews of the American Museum of Natural History, the remarkable Central Asiatic Expedition ranged over China and Mongolia, resulting in the first systematic study of the Gobi Desert (a redundancy, since *gobi* means desert). There were five seasons of exploration and collecting, which cost a total of $700,000, an outlandish sum during those years of the Great Depression. At the project's zenith, in 1925, it had 14 scientists and technicians, 26 native assistants, five touring cars, two trucks, and a support caravan of 125 camels.

Among the expedition's accomplishments were archeological excavations of Paleolithic and Neolithic sites, the first accurate maps of the Gobi, collection of thousands of new plants and animal specimens, detailed geological surveys, and the gathering of vast collections of fossils—including important early mammals and the first known dinosaur eggs. Amazingly, the mineralized eggs still lay in concentric rings, just as their mother had arranged them in her nest 150 million years before.

ADVENTUROUS ARCHEOLOGIST Roy Chapman Andrews, leader of the Gobi Expeditions, was a real-life "Indiana Jones," battling bandits and sandstorms in search of "Dawn Man" fossils. Although he found none, the Central Asiatic Expedition obtained many scientific treasures, including the first known fossilized dinosaur eggs.

Although these exotic eggs, vast graveyards of "shovel-tusked elephants," and other wonders brought fame and honor to the Gobi team, the expedition did not accomplish what it set out to do above all else: to find evidence of the "Dawn Man," which museum president Henry Fairfield Osborn believed was the "missing link." Just a few years before, some truly early hominid remains had been found by anatomist Raymond Dart in South Africa, but European and American scientists, convinced of Asiatic origins, had ignored them.

Andrews and his team faced dangers and adventures worthy of the fictional archeologist Indiana Jones. On more than one occasion, they had to shoot their way out of ambushes or race their motor cars against brigands on horseback. Bands of armed bandits demanded cash for safe passage, corrupt politicians required payoffs, and the scientists frequently wandered into local civil wars.

Reasoning that the foreigners were not crazy enough to spend a fortune collecting mere rocks, Chinese officials concluded they were looting fabulous national treasures and seized about 100 huge crates. To their disappointment, they found nothing but fossils, which they eventually released for shipment to the museum in New York.

After his return to New York, Andrews decided to auction off a dinosaur egg at a fundraising event to help pay for the expedition. One wealthy bidder paid $5,000, then donated the fossil to Colgate University's museum.

When the news reached the Chinese, they concluded that each and every dinosaur egg must be worth $5,000. Their worst suspicions were confirmed: Andrews was no seeker of knowledge, but a greedy treasure hunter plundering the country's riches. Eventually, the flap over the high-priced dinosaur egg and other misunderstandings led to the American Museum of Natural History fossil hunters being barred from the Gobi for 61 years.

See also "CRADLE OF HUMANKIND"

GONDWANALAND
Paleozoic Supercontinent

In the late 19th century, Austrian geologist Eduard Suess noticed close similarities in the geology of central India, Madagascar, and southern Africa and concluded that they must have once been joined. He claimed a supercontinent had once existed and named it for the Gonds, an ancient Dravidian people located in central India near the fossil beds. (Gonds still survive in India today as isolated forest tribes.)

In 1912, a meteorologist named Alfred Wegener formulated a theory of continental drift, which remained controversial for many years. As more evidence accumulated, geologists began to form a picture of gigantic, movable continental plates that shaped the current geography of the planet. South America, Australia, Africa, India, and Antarctica were all

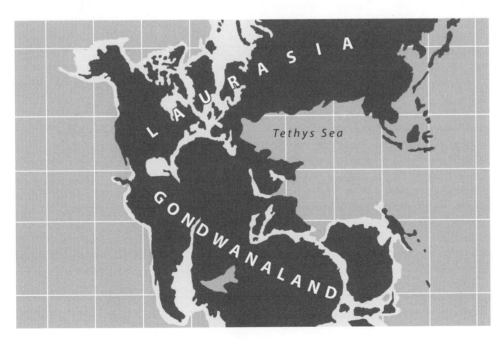

ANCIENTLY CONNECTED, present-day South America, Africa, India, and Australia share many of the same geological features and fossils. This map represents the later Triassic times, about 200 million years ago, during the early dinosaurs' heyday.

once part of the vast supercontinent of Gondwanaland, which existed during Paleozoic times and broke up during the Mesozoic.

In the 1960s, Edwin H. Colbert of the American Museum of Natural History found fossil evidence that confirmed the existence of the vast supercontinent. He found fossils of the tusked reptile *Lystrosaurus* first in South Africa, then in central India (in the land of the present-day Gonds), and finally in Antarctica. *Lystrosaurus* had wandered over thousands of miles without barriers.

Gondwanaland's breakup into the recognizable continents of today presented a puzzle when and in what sequence it occurred. During the last half of the 20th century, geologists believed that Africa split off first, around 130 million years ago, and subsequently two land bridges—between Antarctica and South America, and between Antarctica and Australia—broke off between around 90 and 85 million years ago. But in 2004, paleontologist Paul C. Sereno's team, working in Niger, uncovered a 95-million-year-old dinosaur skeleton that bears a strong resemblance to recently found fossils from South America. These allied species suggest that a transatlantic land bridge between Africa and the rest of Gondwana, via South America, persisted as late as 95 million years ago. At that time, during the mid-Cretaceous period, continuous populations of some of the same kinds of dinosaurs spread across South America and Africa while they were still connected.

See also BROOM, ROBERT; CONTINENTAL DRIFT

GORILLAS
Largest Living Primates

Elusive creatures of the African forest, gorillas were practically unknown to Europeans a century ago. Often they were confused with chimpanzees in travelers' reports; both apes sometimes lived in the same forests.

During the 1870s, French explorer and hunter Paul du Chaillu drew huge lecture audiences, hungry for his tales of encountering and killing wild gorillas in the Belgian Congo. The stuffed specimens with which he shared his lecture platform created a sensation at just about the time Darwin and Huxley put out their books on human evolution. Three of du Chaillu's mounted family, a male, female, and juvenile—the first gorillas ever seen in Britain—are still on view at the Ipswich Museum in England, where they have been displayed for more than a century.

Du Chaillu described the great apes as creatures "from a hellish nightmare." The way to kill one was to walk up to it and stand your ground while it made the most hideous roars and mock charges. Then you waited until it grabbed hold of the muzzle of your rifle and lifted it to its mouth—that was the time to fire. If you waited a moment too long, du Chaillu warned, the gorilla would bite the barrel closed or bend it into a curve, then kill the hunter. Thus was born the durable image of the gorilla as Gargantua and King Kong, which lasted for 70 years.

As a few real gorillas reached zoos and comparative psychology labs, another picture of the gorilla emerged. Wrenched from their forest homes and their social groups, gorilla individuals appeared sullen and intractable; they were often described as "introspective"—meaning they would rather stare at the wall than participate in "intelligence tests."

By the late 1950s, popular notions about gorillas were reshaped again by the field naturalist George Schaller, who tracked them from a distance. He carried no gun and was never attacked. He found the apes to be shy, skittish, unaggressive, and rather boring. They were peaceful vegetarians who had few fights, made no tools, and hunted no prey. Despite their great strength and the male's massive canines, they mostly sat around eating wild celery.

While their days are spent feeding and resting on the forest floor, gorillas often build night nests in tree crotches where they can safely sleep. Intensely social, they typically live in groups consisting of several females and their young, a few subadult males, and one large, experienced "silverback" male—the group's leader, protector, and defender. Large silverbacks can weigh 400 pounds, and they intimidate intruders with terrifying charges and vegetation-tearing displays, though they rarely follow through with an attack.

After Schaller's pioneering efforts, study was continued by the remarkable and tragic

Du Chaillu said the only way to kill a gorilla was to wait until he lifted your rifle barrel to his mouth, then fire it before he could bite the end shut.

GENTLE GIANTS, wild gorillas are shy vegetarians who rarely attack other creatures. The great apes' ferocious appearance and bluff charges terrified 19th-century hunters, who described them as monstrous killers, as in the image at top from an explorer's 1870 book. Decades later, the image in a 1938 Ringling Bros. circus poster (below) was even more terrifying. Poster courtesy of Circus World Museum, with permission from Ringling Bros. and Barnum & Bailey®.

Dian Fossey, who spent 17 years among mountain gorillas, until her (still unsolved) murder at Karisoke, Rwanda, in 1985. Fossey not only studied these largest of living primates, but also succeeded in drawing world attention to their precarious situation. Thanks to her local "war" on poachers in the 1970s and improved park security, the mountain gorilla populations were saved from extinction.

During the 1990s, groups of lowland gorillas in the immense, pristine forest of Gabon were studied by Michael Fay and others. These apes frequently emerged from the dense woodlands to wade through the swamps of Mbeli Bai, where they foraged among succulent water plants. From a high tower built in the swamp, observers could comfortably spend the day watching the behavior of entire groups that were out in the open, unobstructed by forest vegetation. Some gorillas were observed to carry stout sticks or staffs with which to test the depth of the mud and water, exactly as a human might do.

Today, a score of gorilla groups have become so accustomed to human visitors that they have become important economic resources in Rwanda, Central African Republic, and Democratic Republic of Congo, attracting tourist dollars. Their celebrity status insures government measures for their continued protection, as they are much more valuable to the local economy alive than as hunting trophies. Yet they survive under problematic conditions: Constant exposure to humans may irrevocably disturb their normal behavior. Science's understanding of "natural" gorilla behavior, ecology and evolutionary adaptations may still forever be lost, even if the apes themselves are given a reprieve from extinction.

See also APES; "APE-WOMEN," LEAKEY'S; GARNER, RICHARD LYNCH; DIGIT; FOSSEY, DIAN

GOULD, STEPHEN JAY (1942–2002)
Paleontologist, Essayist, Science Historian

When five-year-old Stephen Jay Gould first laid eyes on the towering *Tyrannosaurus* skeleton in the American Museum of Natural History, he decided to spend his life studying fossils. The tyrant lizard, he later recalled, followed him home and into his nightmares. Decades before dinosaurs became a staple of American childhood, and almost alone among his peers in Queens, New York, young Gould never considered any other career but paleontology.

For most of his professional life, Gould was a professor at Harvard University and a curator of its Museum of Comparative Zoology. He had attended Antioch College, and studied paleontology at Columbia University. His thesis focused on variation and evolution in an obscure Bermudian land snail. Like Darwin with his barnacles, Gould pursued his later theorizing only after intense scrutiny of a single group of organisms.

He had hoped to find correlations between variation and different ecologies within the mollusk's range, but the snails' sizes, colors, and shell shapes varied quite independently of local environment. Impressed with the importance of nonselectionist factors in evolution, Gould became interested in structural constraints and limitations as organisms change.

Gould also became interested in distinguishing incidental features from adaptive ones. He and geneticist Richard Lewontin published an influential paper about "spandrels"—angular wall spaces on structural supports for medieval cathedral domes. Often these surfaces are decorated with paintings that have interested art historians. But when analyzing these paintings, they ignored the spandrel's humble origin as an unavoidable consequence of stress distribution—a structural byproduct of the dome's construction.

TWO PRODUCTS OF EVOLUTION contemplate one another. Stephen Jay Gould and a giraffe in Kenya, 1990.
Photo © Delta Willis.

In their paper, Gould and Lewontin explain how slight changes in one feature can alter others without reference to adaptation—what Darwin had called "correlation of parts." Using spandrels as a metaphor, they pointed out that the human chin—often cited as "advanced" in comparisons with the chinless primates—holds no special correlation with higher intelligence. Chins, like spandrels, are the result of stress and growth factors in the human jawbone.

Gould's fellow graduate student at Columbia, Niles Eldredge, had studied thousands of trilobites that revealed a pattern that had impressed Thomas Henry Huxley a century earlier: The fossil record shows long periods of stability, punctuated by "bursts" of speciation. Darwin's explanation for this seeming absence of gradual transitions was that the fossil record was then too fragmentary and incompletely known to provide evidence of steady rates of change. It was like a book with pages and even whole chapters missing.

Looking at a much more complete fossil record more than a century later, Gould and Eldredge thought it was time to acknowledge that such episodic patterns in the rocks, separated by long periods of stability, probably reflect the reality of life's history. By the 1980s, "punctuationalism" had become widely adopted and was fruitful in generating new insights and research.

Darwin was one of Gould's lifelong heroes, whose achievements he celebrated in such books as *Ever Since Darwin* (1977) and *The Panda's Thumb* (1980). Nevertheless, he was irreverent toward the orthodox Synthetic Theory of evolution that has prevailed in biology since the 1940s. Dissatisfied with the limits of its explanatory power, he often championed other possible mechanisms and approaches to supplement traditional natural selection—to the dismay of more conservative colleagues.

One of his approaches was to emphasize the hierarchy of levels on which evolution operates: biochemical, genetic, embryological, physiological, individual, societal, species, lineages. Sorting or selection on any of these levels, he believed, produces significant effects on the level above or below it—a largely unexplored area for future research.

He also believed that heterochrony—evolution that speeds up or retards stages in the individual's life cycle—was an important force in generating new species. A new species could result, for instance, if the adults remained stuck at an early stage of their development, which could be programmed by regulatory genes. The classic example of such "neoteny" is the axolotl—a salamander that retains its infant gills into adulthood and never leaves the water. Another possible example is that adult humans seem to preserve the characteristics of juvenile apes, such as a flattened face and greatly reduced eyebrow ridges, a condition known as pedomorphism.

Gould did not shrink from public controversy. He appeared before congressional committees on environmental issues, was a courtroom witness in the Arkansas Scopes II trial about teaching evolution in the public schools, and spoke out against pseudoscientific racism and biological determinism.

His fatal bout with cancer at the age of 60 cut off a brilliant intellect in its prime of productivity. During his last years, Gould raced to produce his magnum opus, *The Structure of Evolutionary Theory,* in which he defined his views over the whole range of evolutionary thought. He likened its intellectual edifice to a Spanish cathedral that had changed and evolved over the centuries, adding sections that were in tune with the fashions and temper of the times. The core structure of the cathedral remained in place, however much its extensions and facades might vary or become obsolete over the years. Gould viewed Darwinian evolutionary theory as sound, even as it changes and itself evolves.

See also BARNACLES; BIOLOGICAL DETERMINISM; CONTINGENT HISTORY; NEOTONY; PANDA'S THUMB; PUNCTUATED EQUILIBRIUM; SCOPES II

GRADUALISM
Slow and Steady Change

One key feature of Darwin's original theory was that evolutionary change must have proceeded by "slow, insensible degrees"—a progression of tiny changes adding up to produce new species over immense periods of time. It was Darwin's attempt to apply Sir Charles Lyell's uniformitarian geology to the world of life.

Lyell had been a voice of reason at a time when geologists invoked imagined violent and sudden catastrophes, convulsions, floods, and supernatural forces to explain the features of the Earth. Presently observable processes of wind, water, volcanoes, erosion, and deposition, Lyell thought, could account for them all.

Darwin went so far as to adopt "Nature makes no leaps" as an axiom, or basic assumption. But from the first his friend and supporter, Thomas Henry Huxley, thought it "unnecessary to burden the theory" with an unproven gradualism, which he later described as an "embarrassment" when he noticed that some patterns of fossils over time showed little change, and then relatively rapid replacement.

When critics asked why the fossil record, though it showed change over time, did not demonstrate this smooth succession of small, gradual transitions, Darwin replied that it was very "imperfectly" known and that subsequent discoveries would fill in the picture. In fact, many transitional forms have since come to light, though they are still comparatively rare.

By the 1970s, the concept of Darwinian gradualism came under increasing attack by biologists. Apparent discontinuities, or "jumpiness," in the fossil record led to theories of "punctuated equilibrium" and intense scrutiny of Cambrian and pre-Cambrian fauna, since the basic body plans or phyla first appeared and proliferated during that time.

See also "HOPEFUL MONSTERS"; PUNCTUATED EQUILIBRIUM; UNIFORMITARIANISM

HANGING OUT WITH DARWIN, Stephen Jay Gould strikes a casual pose at Down House, now the Darwin Museum.
Photo © Delta Willis.

Top: A computer portrait of Gould by Pat Linse.
© Pat Linse/Skeptic.com.

GRAY, ASA (1810–1888)
American Botanist, Evolutionist

Asa Gray was, in Charles Darwin's words, "a complex cross of lawyer, poet, naturalist, and theologian." He was also one of the greatest botanists of the 19th century and for years supplied Darwin with detailed information on the distribution of plant families and species throughout the world.

Although he was Darwin's greatest champion in America, Gray had a theological bent and tried to defend evolution from charges of atheism. He wanted to make things easier for liberal clergymen, arguing that natural selection did not wholly do away with the "argument from design"; God accomplished his plan through evolution. When some insisted that new species could only be created by the direct intervention of God, Gray countered that they were putting limits on scientific investigation without enlarging the understanding of religion.

Despite his brilliance, Gray was very much in the shadow of Louis Agassiz, the most influential anti-Darwinian naturalist in America. (Both were professors at Harvard.) But Agassiz really did not understand Darwin's theories. Gray wrote that "he growls over it, like a well-cudgelled dog—is very much annoyed by it," but was just too old to change the entire view of nature he had learned from his teacher Georges Cuvier.

In 1857, Darwin had given Gray a sketch of his theory in a letter, making him one of the intimate circle sworn to secrecy about the coming bombshell. After *Origin of Species* (1859) was published, Gray tried to please all sides and continued to equivocate for another dozen years. Finally, he published a clear statement of his position in a series of essays called *Darwiniana* (1876). Darwin liked it so well that the following year he dedicated his book on the forms of flowers to Gray.

Darwin was grateful for Gray's championing of evolution in America, but they parted company on the question of design (teleology). Gray thought variations in nature were provided by a good-natured God to be used by species for their benefit, like a friend who hands you pieces of a jigsaw puzzle in the proper order for you to fit them in.

In *Variation of Animals and Plants under Domestication* (1868), Darwin said natural selection worked more like a builder putting together a structure with odd bits of stone and rubble from previous houses. The fragments were not shaped for particular ends, but were selected to fit in where they could be useful. In a letter to Gray on May 22, 1860, Darwin

BOTANIZING IN STYLE: An 1876 "expedition" in the Colorado Rockies was led by Asa Gray of Harvard (seated on ground) and Joseph Hooker of Kew Gardens (to Gray's right). Posing with tools of their trade are the camp's shooter, angler, and cook.

said that although he had "no intention to write atheistically," he simply could not see "evidence of design and beneficence on all sides of us [because] there is too much misery in the world." He could not believe, he continued, that a kind and loving God could purposely design the ichneumid wasp, whose young feed on the insides of living caterpillars, "or that a cat would play with mice."

In a similar vein, Darwin wrote Lyell in August 1861 that he could not imagine that variation of a particular species was "ordained and guided" any more than an astronomer would suggest that a particular meteoric stone or an individual sparrow should fall according to any "preconceived and definite plan."

See also AGASSIZ, LOUIS; DIVINE BENEFICENCE, IDEA OF; NATURAL SELECTION; TELEOLOGY

GREAT CHAIN OF BEING
Scala Natura

For hundreds of years before evolution became a part of the common thought of our culture, the concept of the "Great Chain of Being" dominated the Western view of nature. Everything in the universe was arranged in a hierarchy, from low to high, from base to noble. There were "base" metals like lead, and "noble" ones like gold and silver. Man, as humankind used to be known, was the noblest of creatures; all other living things occupied different links below him in the chain. Above were various spiritual beings, with God at the top of the cosmic order.

Arthur Lovejoy traced the history of the idea in his classic work *The Great Chain of Being* (1936), in which he called it "one of the half-dozen most potent and persistent presuppositions in Western thought," from medieval times right up until about a century ago.

The chain carried through to the social order, too: peasants at the bottom, then servants to the gentry, then various grades of nobility, and the monarch at top. Racism is embedded in the Chain of Being, which ranked the various races into "higher" and "lower." Of course, the white Europeans who devised it were at the top.

Darwin realized the chain concept simply didn't describe nature very well, and resolved not to use "higher" and "lower" in his descriptions of animals. Nevertheless, the mental habit was hard to break. The original title of his book on comparative psychology was to be *Expression of the Emotions in Man and the Lower Animals*, but he changed the title to *Expression of the Emotions in Man and Animals* (1872).

See also ORTHOGENESIS; CONTINGENT HISTORY

THE ASCENT OF MAN from worm through monkeys to Charles Darwin himself was satirized in this 1882 *Punch* cartoon by Edward Linley Sambourne.

GREAT DYINGS
Mass Extinctions

Extinction is a normal event in the history of life. Every so often a species will be faced with the loss of its habitat or the introduction of a new predator, or it may be outcompeted for food and resources. Mass extinctions, or "great dyings," in contrast, are rare and catastrophic events that wipe out tens of thousands of species at once when whole ecosystems collapse.

Mass extinction events can be used as hallmarks to define geological history. When geologist John Phillips first drew the horizontal lines that separate the Paleozoic, Mesozoic, and Cenozoic eras in 1840, he was indicating that in the rocks above each line were fossils of different life forms than in those below it. The major "jumps" or "punctuations" record the wholesale extinction of a majority of living things, followed by a new radiation of different families and species.

Of the half-dozen or so mass extinctions, two of the most extensive and devastating were at the end of the Permian, just before the Age of Reptiles, and the Cretaceous "great dying" that ended the dinosaurs' days of dominion. Paleontologist David M. Raup has estimated that during the Permian extinction of 250 million years ago, 90 percent of all then-existing species (about five million kinds of living things) vanished. About 250 million years later, during the Cretaceous extinction, thousands of kinds of dinosaurs were wiped out along with flying reptiles, plesiosaurs, ichthyosaurs, and many once-abundant sea creatures such as ammonites.

Current opinion among geologists is that both mass extinctions were triggered when massive meteorites or asteroids hit the Earth with enormous impacts. Immense dust clouds raised by the collisions may have filtered the sun's rays, cooling the Earth's climates.

Recent studies suggest that an asteroid that hit the Earth at the end of the Permian was about the size and mass of Mount Everest, leaving a buried 125-mile-wide crater called Bedout, off northwestern Australia. A massive crater off Yucatán, known as Chicxulub, is associated with the Cretaceous extinctions. Both the end-Permian and Cretaceous impacts may also have been responsible for setting off massive and widespread volcanic activity, which further devastated the habitats of millions of plants and animals.

At present, the world is experiencing a "sixth mass extinction" of species caused by another kind of devastating collision, as human industrial activity and population growth eliminate forests, warm the atmosphere, and pollute and deplete the oceans. As our species population shoots up to 10 billion during the next hundred years, thousands of other species will join the ranks of the dodo and the passenger pigeon. Since mankind appeared on Earth, we have increased the rate of extinction of other species by 1,000 percent.

In Paul Ehrlich's metaphor, it's like taking rivets out of the wings of an airplane. We may still be perfectly safe if a dozen or even two are removed. But if we keep on taking out rivets, how can we know at what point the wings will suddenly fall off?

See also BIOSPHERE; DINOSAURS, EXTINCTION OF; EXTINCTION

GROUP SELECTION
Survival of the Social Unit

Cooperating groups of creatures may survive and reproduce better than individual members could on their own. Social insects are the classic example. Hives of wasps or bees function almost as if they were themselves organisms, with individuals specialized to fulfill the functions of an immune or defense system, reproductive and food-gathering system, and maintenance and repair system. Some bees even station themselves at air passages and beat their wings, thus air-conditioning the hive.

Even those individuals that do not directly reproduce have a genetic stake in the next generation because they share many genes with the breeding individuals—a process known as kin selection.

The Japanese primatologist Kinji Imanishi documented that monkey groups differ in their habits of defense, food preferences, and even food preparation traditions (such as the washing of vegetables to remove grit). Imanishi thought that such social traditions could have a selective advantage for the group as a whole.

Most biologists agree, however, that there is no evidence that individual animals deliberately act "for the good of the group" or "the good of the species." Although certain behaviors may have evolved to function for the good of the group, it appears that individuals respond, even as asocial creatures do, purely in their own perceived interests.

Recent experiments with chimps have shown that they do help others reach something or find something without promise of reward—establishing the existence of generosity and altruism in the natural world. There are more than enough examples in nature, as in scripture, to establish the "naturalness" of any possible kind of human behavior.

See also FITNESS; KIN SELECTION; NATURAL SELECTION; "SELFISH GENE"

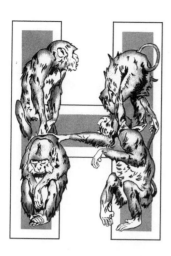

HAECKEL, ERNST (1834–1919)
German Evolutionist

Born in Potsdam in 1834, the son of a government lawyer, Ernst Haeckel was trained as a physician, graduating in 1859, shortly before the appearance of Darwin's *Origin of Species*, the book that dramatically changed his life. Here, he thought, was the answer to everything he had been seeking in science, philosophy, ethics, religion, politics—a unified, or monist, view of the world based on the creative properties of matter. His own fanaticized version of evolution—the "non-miraculous history of creation"—became an obsession and guiding passion, with Darwin his greatest hero.

Abandoning his medical practice, he set out for the University of Jena, Germany, to embark on a career as an anatomist. In 1859–1860, he undertook a zoological expedition in the Mediterranean, during which he discovered 144 new species of radiolaria, a marine invertebrate. His monograph *Die Radiolaren* (1862), and others on sponges, worms, and medusae, established his scientific credibility.

In 1862, Haeckel became professor of zoology and comparative anatomy at Jena, and turned from studying marine invertebrates to human evolution. His *Generelle Morphologie* (1866) contains the first of his famous drawings of "evolutionary trees of life," as well as his exposition of the monist philosophy. That same year he made the pilgrimage to England, visiting Darwin at Down House. He was not disappointed: "I fancied a lofty world-sage out of Hellenic antiquity . . . stood alive before me," he reminisced in a lecture shortly after Darwin's death. Darwin, too, was impressed. In his *Descent of Man* (1871), Darwin claims he would not have bothered to write the book if he had known that Haeckel had already begun one on the same topic.

Haeckel invented many scientific terms that are still in common use. Among them are "ecology" (the study of the interrelationship between organisms and their environment), "ontogeny" (the study of embryological development), and "phylogeny" (the study of evolutionary descent or lineage). He also became convinced that he had discovered a basic law of biology, "Ontogeny recapitulates phylogeny": that is, the development of an embryo (ontogeny) is a speeded-up replay of the evolution of the species (phylogeny). It was an enormously influential idea, utilized by both Darwin and Huxley, who were impressed with Haeckel's detailed illustrations comparing development in various animals and humans.

According to Haeckel's drawings, embryonic pigeons, dogs, and humans look identical in their early stages, and differentiate only later on in their development. However, Haeckel deliberately fudged his pictures to support his theory. As critics pointed out during his lifetime, obvious differences between birds, mammals, and reptiles are present from very early in their development.

EVOLUTIONARY ZOOLOGIST
Ernst Haeckel, known as the "Ape Professor of Jena," became a celebrity and revered national hero in 19th-century Germany.

HAECKEL'S GOTHIC decor, for a lecture in Jena in 1907, used stuffed apes, skeletons, and posters of human embryos to create an evolutionary theatrical spectacle.

Recapitulation theory enjoyed a tremendous vogue for a few decades, but eventually proved too vague to be useful in research. Before it was abandoned, however, it shaped scientific thought of the period, including the psychoanalytic theories of Sigmund Freud.

In one of his bolder public pronouncements, Haeckel stated that the "missing link" between men and apes must have lived in Java or Borneo. Although there was no known fossil evidence for such a creature, he gave it a Linnaean name anyway—*Pithecanthropus alalus* (meaning "ape-man without speech")—and encouraged his students to go out and find it. It is one of the amazing stories of science that a young Dutch doctor who attended Haeckel's lectures, Eugene Dubois, became so infused with his confident enthusiasm that he went to Java and found what he called *Pithecanthropus erectus* (now *Homo erectus*).

Elated, Haeckel promptly sent a congratulatory telegram "TO THE DISCOVERER OF PITHECANTHROPUS FROM ITS INVENTOR." Zoologists have since made a rule that scientific names cannot be given to species that have not yet been discovered.

In his *Natural History of Creation* (1868), Haeckel argued that the church, with its morality of love and charity, was an effete fraud, a perversion of the natural order:

> If we closely examine . . . the mutual relations between plants and animals (man included), we shall find everywhere and at all times, the very opposite of that kindly and peaceful social life which the goodness of the Creator ought to have prepared for his creatures—we shall rather find everywhere a pitiless, most embittered struggle of all against all. . . . Passion and selfishness . . . is everywhere the motive force of life. . . . Man . . . forms no exception to the rest of the animal world. . . .
>
> The whole history of nations . . . must therefore be [a physiochemical process] explicable by means of natural selection.

Haeckel's lectures drew huge, enthusiastic crowds at international conventions of freethinkers and monists. After one such gathering, held in Rome, the pope ordered a "divine fumigation," as one journalist put it, of the Holy City to clear the air.

Haeckel remained enormously popular in Germany, even when his scientific reputation was all but gone. Because he steadfastly refused political office, attacked entrenched church authorities, and promoted German nationalism, he was adored by the populace and honored by government. As a national hero he convinced masses of his countrymen to accept their evolutionary destiny to "outcompete" inferior peoples. It was right and natural that only the "fittest" nations should survive. He believed that "politics is applied biology"—an idea that later became a Nazi slogan. But despite his disastrous political agenda, Haeckel is remembered as a charismatic biologist and a world-class popularizer of evolution.

See also "ARYAN RACE," MYTH OF; BIOGENETIC LAW; DUBOIS, EUGENE; FREUD, SIGMUND; MONISM

CANONIZING HIS HERO, Haeckel is shown in this 1882 cartoon offering a halo to Darwin as he ascends into heaven, embracing his beloved apes.

HALDANE, JOHN BURDON SANDERSON (1892–1964)
Geneticist

One of the great rascals of science, J. B. S. Haldane was independent, nasty, brilliant, funny, and totally one of a kind. Son of an Oxford professor of physiology, he began as his father's assistant. Eventually he taught genetics and biometry at University College, London, and helped to shape the Synthetic Theory of evolution.

He learned Mendelian genetics while still a boy by breeding guinea pigs, and often served as one himself when he helped his father. Once, the elder Haldane made him recite a long Shakespearean speech in the depths of a mine shaft to demonstrate the effects of rising gases. When the gasping boy finally fell to the floor, he found he could breathe the air there, a lesson that served him well in the trenches of World War I.

A physically courageous 200-pounder, Haldane continued the family tradition of using his own body for dangerous tests. In one experiment, he drank quantities of hydrochloric acid to observe its effects on muscle action; another time he exercised to exhaustion while measuring carbon dioxide pressures in his lungs.

Haldane had mastered Latin, Greek, French, and German while still a student. Later, he wrote extensively on history and politics, and made important contributions to chemistry, biology, mathematics, and genetics. Along with E. B. Ford and R. A. Fisher, he pioneered population genetics, which reshaped modern evolutionary biology.

During World War I, Haldane volunteered for the Scottish Black Watch and was sent to the front. There he found, to his shock and dismay, that he liked killing the enemy. Twice wounded, he personally delivered bombs and engaged in sabotage behind enemy lines; his commander called him "the bravest and dirtiest officer in my Army."

In 1924, Haldane published a remarkable work of fiction, *Daedalus,* the first book about the scientific feasibility of "test-tube babies." He predicted their feasibility, but became an outspoken critic of eugenicists, calling them "ferocious enemies of human liberty."

Daedalus was a popular and influential book, inspiring Aldous Huxley's novel *Brave New World* (1932), in which a well-intentioned society based on test-tube babies turns out terribly wrong. Huxley also put Haldane in another of his novels, *Antic Hay* (1923), as Shearwater, "the biologist too absorbed in his experiments to notice his friends bedding his wife."

Shortly before his death in 1964, the irrepressible Haldane wrote an outrageous comic poem for his friends, mocking his own incurable disease:

I wish I had the voice of Homer
To sing of rectal carcinoma,
Which kills a lot more chaps, in fact,
Than were bumped off when Troy was sacked.

See also EUGENICS; HUXLEY, ALDOUS; KIN SELECTION; SYNTHETIC THEORY

HAMPDEN, JOHN (1819–1891)
Flat-Earther, Antievolutionist

John Hampden, the self-proclaimed champion of biblical "flat-earthism," drew the great evolutionist Alfred Russel Wallace into one of the most bizarre episodes in the history of science. Son of a rector and nephew of a bishop, Hampden had sworn to destroy the "infidel pagan superstitions" of modern science. A zealous disciple of Samuel Birley ("Parallax") Rowbotham, author of *Earth Not a Globe* (1869), Hampden warned that if schools continued to teach the "satanic globular theory," it would mean the destruction of all morality and religion. Evolution, he believed, was only the latest wrinkle in the more basic blasphemies of Copernicus and Sir Isaac Newton.

In 1870, Hampden offered to bet any scientist £500 that no one could prove the earth is a globe. To collect, the challenger had to demonstrate the existence of a convex railway track or curving surface on a large body of water. Hampden's preferred targets—wealthy, influential gentlemen-scientists like Charles Darwin or Sir Charles Lyell—ignored Hampden's bait. But Alfred Russel Wallace, with his openness, naiveté, and perpetual near-poverty, thought he might win a few points for science and an easy five hundred quid. After all, he had been trained as a surveyor.

As he recalled later in his autobiography, *My Life* (1905), Wallace hesitantly asked the great geologist Lyell whether he thought it wise to accept such a challenge. "Certainly," Sir Charles replied, "it may stop these foolish people to have it plainly shown them." Wallace had chosen the worst possible advisor to protect his interests. Lyell had pushed Charles Darwin to publish his natural selection theory quickly, warning that Wallace was also developing a theory of evolution and might beat him into print. Lyell, Darwin wrote in his *Autobiography* (1876), "strongly advised me never to get entangled in a controversy, as it rarely did any good and caused a miserable loss of time and temper." Yet Lyell encouraged Wallace to lock horns with one of the most malicious, abusive crackpots in all of England.

Confident of his scientific prowess, Wallace entered the contest unconcernedly, like a lamb to the slaughter. Hampden asked him to pick an umpire, and Wallace chose a man named J. H. Walsh, who was a stranger to him. As the well-known editor of *The Field, the Farm, and the Garden: The Country Gentleman's Newspaper,* Walsh had a reputation for fairness and objectivity. Graciously, Wallace offered to let Hampden appoint another impartial judge of his own choosing. He named a printer, who was a close friend and dedicated flat-earther, which Hampden did not think it necessary to mention. At the last minute Walsh was unable to attend Wallace's demonstration, but he remained as umpire while Wallace selected another man as his on-site referee.

On Saturday morning, March 5, 1870, the parties met near the north end of Old Bedford Canal, about 80 miles from London. The waterway ran straight and unobstructed for six miles between two bridges. Wallace affixed large cloth rectangles to the facing sides of each bridge, both bearing a bold black stripe running parallel to the ground and both placed exactly the same height above the water.

Halfway between the two bridges, Wallace stuck a tall pole in the bank, bearing two large disks as height markers. He measured the height of the lower disk, placing it exactly the same distance above the water level as the black stripes mounted on the bridges. After carefully lining up a surveyor's level-mounted telescope at the same height as the markers, Wallace asked the referees to sight through it from either bridge. From whichever vantage point, the pole's disks at the midpoint appeared much higher than the bars mounted on the bridges, demonstrating the convexity of the water's surface.

Both referees agreed on what they had seen but were completely conflicted in their interpretations of the results. Wallace's referee announced that the round-earth case was clearly proven, but Hampden's referee "actually jumped for joy" as he proclaimed the experiment proved precisely the opposite. Wallace was stunned. He had not reckoned with the strange logic of "Zetetic" astronomy, as set forth in Rowbotham's book, which explains away such a result as a mere optical illusion, to be expected if the Earth is flat. Hampden immediately claimed victory and demanded Wallace's money.

> Lyell counseled Darwin to avoid public controversies, while encouraging Wallace to lock horns with a malicious crackpot.

Walsh, the umpire who could not be present at the experiment, was sent the drawings and interpretations of what was seen through the telescope. After studying them for a week, he announced in his newspaper that Wallace was clearly the winner. Furious, Hampden demanded the return of his money, but instead Walsh sent it to Wallace with congratulations. For this "perfidy," Hampden directed a barrage of letters and pamphlets at Walsh, Wallace, and all their friends and colleagues, calling them liars, thieves, cheats, and swindlers. He harangued the officers of all the scientific societies to which Wallace belonged and sent Mrs. Wallace this note:

MADAM—If your infernal thief of a husband is brought home some day on a hurdle, with every bone in his head smashed to a pulp, you will know the reason. Do you tell him from me he is a lying infernal thief, and as sure as his name is Wallace he never dies in his bed.

You must be a miserable wretch to be obliged to live with a convicted felon. Do not think or let him think I have done with him.

JOHN HAMPDEN

In response to that written threat, Wallace brought Hampden up before a police magistrate, who ordered him to stop his flow of venom and keep the peace for three months. "As soon as the three months were up," Wallace wrote in his autobiography, "he began again with more abuse than ever." Hoping to end the harassment, Wallace took Hampden back to court, where he was fined and spent a week in jail. Nonetheless, over the next four years the vituperative crank kept up his nasty attacks and was convicted three times. He got several months in jail the next time and six more when he harassed Wallace again; he also was directed to pay damages and court costs, which he never did. Instead, he brazenly hid all his assets under a relative's name and declared bankruptcy.

Next, Hampden attempted to turn the tables in court. In 1876, his barrister advised him to take a new tack: English law did not recognize gambling debts as legal or enforceable. So this time he sued Walsh, the stakeholder, for the return of his £500 that had been paid to Wallace. Since losers of wagers had no legal obligation to pay, Hampden argued that he had asked for his money back before Walsh turned it over to Wallace, and that the stakeholder was therefore liable. This time Hampden won, and Wallace was stuck with returning the money, plus Walsh's court costs. Although he had been declared winner of the challenge and wager, and had prevailed several times in the courtroom, Wallace ended up the loser, deep in debt, and thoroughly disgusted with British justice.

Wallace expressed amazement at Hampden's unabated virulence, as if he were some strange specimen of natural history. "Seldom has so much boldness of assertion and force of invective been combined with such gross ignorance," he wrote. "And this man was educated at Oxford University!"

See also FLAT-EARTHERS; WALLACE, ALFRED RUSSEL

HAPPY FAMILY, THE
Survival of the Friendliest

The Happy Family is "one of the most wonderful and unique sights to be seen in the world," trumpets the 1896 *Guidebook to the Consolidated P. T. Barnum Greatest Show on Earth*. Pictured is a large cage containing many different species of animals, some of them "natural enemies," living together in peace and harmony.

Evoking a zoological Garden of Eden rather than a Darwinian "struggle for existence," the famous exhibit demonstrated

the possibility of that millennial time when the lion shall lie down with the lamb, and the wolf beside the sheep. In this cage will be found cats playing harmlessly with mice; dogs, foxes, monkeys and rabbits peacefully eating food out of the same dish without harming each other; owls, eagles, vultures watching the frolics of squirrels, bats and small birds, their usual prey,

"THE RAREST CURIOSITY EVER SEEN," according to its proprietor's handbill, was a London exhibit where "revengeful passions are not known, the weak are without fear, and the strong without the desire to injure. The rats and mice go to sleep under the cats."

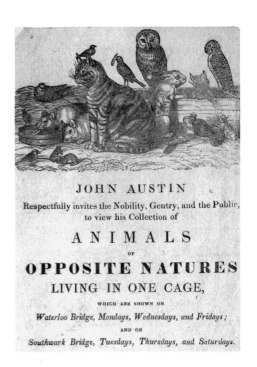

JOHN AUSTIN

Respectfully invites the Nobility, Gentry, and the Public, to view his Collection of

ANIMALS
OF
OPPOSITE NATURES
LIVING IN ONE CAGE,

WHICH ARE SHOWN ON

Waterloo Bridge, Mondays, Wednesdays, and Fridays;

AND ON

Southwark Bridge, Tuesdays, Thursdays, and Saturdays.

without making the least attempt to capture or kill them. In fact such a collection of animals in one cage was never seen since the days of Noah and the animals in the Ark.

Thousands of Barnum's paying customers found the exhibit appealing as well as amazing. It seemed a reversal of natural law, a contradiction of the popular conception of "survival of the fittest." How did predators lose their cruelty and prey their fear? Barnum proclaimed that "a secret," but assured viewers that the animals were not unhealthy or drugged.

"We were the first to originate the idea of a Happy Family and our success caused numbers of our competitors to imitate us . . . but they lamentably failed," Barnum crowed. In fact, the great master of humbug shamelessly lifted the idea, which seems to have originated in England. In 1852, Francis T. Buckland wrote of two exhibitions of "Happy Families" in London. "One stands at Charing Cross and about the streets, the other remains permanently at Waterloo Bridge. They both claim to be the original 'happy family,' but I think the man at the bridge has the greatest claims to originality. He is the successor to the man who first started the idea, Austin by name," who opened his exhibit to the public around 1816.

There was no special trick to the effect. Combining animals in "unnatural" groupings takes great patience and requires choosing docile individuals, acclimating them gradually to each other, and keeping them all well fed, or in some instances obtaining the various species when young and rearing them together. Animals that cannot get along are excluded by the trainer until a workable balance is struck.

Happy Family exhibits proved so popular that Mark Twain could not resist penning a parody ("The Lowest Animal," c. 1900)—his own attempt, he claimed, to prove that man was of the "Lower Animals." First, he wrote, he experimented by placing some "Higher Animals," such as dogs, cats, and rabbits, in a cage, and taught them to be friends.

In the course of two days I was able to add a fox, a goose, a squirrel and some doves. . . . They lived together in peace; even affectionately.
Next, in another cage, I confined an Irish Catholic from Tipperary, and as soon as he seemed tame I added a Scotch Presbyterian from Aberdeen. Next a Turk from Constantinople; a Greek

"LAW OF THE JUNGLE" seemed suspended in this famous live exhibit of peaceful predators and prey, as illustrated in *P. T. Barnum's Greatest Show on Earth Souvenir Book* for 1882. Barnum's claim that he invented the "Happy Family" was humbug.

HAPPY FAMILY

Christian from Crete; an Armenian; a Methodist from the wilds of Arkansas; a Buddhist from China; a Brahman from Benares. Finally, a Salvation Army Colonel from Wapping.

[When I checked on the experiment after a couple of days] the cage of Higher Animals was all right, but in the other there was but a chaos of gory odds and ends of turbans and fezzes and plaids and bones and flesh—not a specimen left alive. These Reasoning Animals had disagreed on a theological detail and carried the matter to a Higher Court.

See also BARNUM, PHINEAS T.; TWAIN, MARK

HAWAIIAN RADIATION
Diversity from Isolation

I f Charles Darwin had explored the Hawaiian Islands (rather than the Galápagos, which so impressed him) he would have seen much more striking examples of diversity among closely related species. Evolutionists after him have found in these volcanic islands, long isolated from the major continents, an extraordinary natural laboratory of adaptive radiation.

Most famous are the 23 remaining species of finchlike birds known as Hawaiian honeycreepers and the more than 500 different fruit flies that have evolved, diverging from island to island and adapted to different habitats or foods on the same islands.

During the several million years since their ancestors reached the islands, some honeycreepers evolved into seedeaters with heavy beaks; others developed straight, thin beaks for spearing insects; while still others diverged into parrot-beaked species and delicate nectar-feeders with long curving bills and tubular tongues for probing flowers. More than half of the original 47 honeycreeper species have become extinct during the past 1,500 years, since the advent of humans and imported predators. (Some were wiped out by the original Polynesian settlers and others only during the past few hundred years by Europeans.)

Under the former conditions of isolation, fruit flies (with their very short reproductive cycle) radiated far beyond the honeycreepers. Among the hundreds of spe-cies, some have become specialized for feeding on nectar or sugar; others eat decaying leaves; some are parasites on spider eggs; and some live only in a single valley on one island. They show a spectacular diversity in their body shapes, but even among those that appear pretty much the same (even to other fruit flies), they can be told apart by their sounds and behavior.

Hawaiian fruit fly populations have evolved scores of different courtship behaviors by which to recognize members of their own species, including elaborate airborne "dances." During the late 1980s, researchers also found the "songs" of Hawaiian fruit flies are as amazingly varied as their bodies. Some species make pulsing cricketlike sounds, while others sound more like cicadas than flies. Like body shapes or genes, these "songs" are providing more clues about how the various species diverged and spread throughout the islands.

See also ADAPTATION; DARWIN'S FINCHES; DIVERGENCE, PRINCIPLE OF; ISOLATING MECHANISMS

RAPID EVOLUTION of honeycreepers from fairly recent common ancestors has produced scores of closely related but divergent species in the Hawaiian Islands. Honeycreepers with sharp, heavy beaks can penetrate bark, while delicate, elongated, curved beaks evolved in nectar-feeders. Species with parrotlike beaks crush seeds and pits.

OBSESSED BY EXTINCT MONSTERS, British artist Benjamin Waterhouse Hawkins is surrounded in his studio by full-sized models of ancient animals, including reconstructions of the few dinosaur species then known.

HAWKINS, BENJAMIN WATERHOUSE (1807–1894)
First Evolutionary Artist

Best remembered for his giant models of dinosaurs sculpted for the Crystal Palace Exhibition (1853), Benjamin Waterhouse Hawkins pioneered the art of reconstructing extinct animals. Although some of his creatures are now considered fanciful and anatomically inaccurate, Hawkins was the first to stimulate tremendous public interest in life of the past.

Before he turned his hand to prehistoric monsters, Hawkins was an accomplished animal artist who contributed many illustrations to zoology monographs published by the British Museum. He drew the plates (and transferred them to litho stone) for the volume *Fish and Reptiles* of *The Zoology of the Voyage of HMS Beagle* (1839–1845); his models were the specimens collected by the young Charles Darwin.

Hawkins made his dinosaur reconstructions under the supervision of Richard Owen, the irascible anatomist and director of the museum. An enemy of Darwin's, Owen coined the word "dinosaur" to describe the newly discovered fossil creatures but did not accept the idea of evolution. Hawkins's monsters were presented as "antediluvian." Later, Hawkins worked with leading evolutionists, including Thomas Huxley and Edward Drinker Cope.

The opening of his Crystal Palace exhibition was celebrated with a festive dinner held inside the huge *Iguanodon* model, attended by England's leading scientists. Hawkins's dinosaur park was such a huge success that he was summoned to America to build a grandiose Paleozoic Museum in New York's Central Park, a project that ended disastrously.

During his years in America, he reconstructed huge fossil skeletons for the great paleontologist Edward Drinker Cope. He also painted haunting, nightmarish landscapes with grinning saurians reminiscent of gargoyles and hellish demons. Some of these works are still exhibited at the Princeton Natural History Museum, and his Crystal Palace statues have been restored to their former glory at Sydenham Park, in south London.

In their recent biography of Hawkins, Valerie Bramwell and Robert M. Peck reveal Hawkins's secret life: the artist wed two women and had separate families living a few miles apart in London. He didn't see either family while working in the United States for ten years.

Despite Hawkins's close working relationships with Darwin, Huxley, Cope, and Joseph Leidy during the early years of his career—and despite his pioneering visualizations of extinct animals—Hawkins did not believe in evolution. After returning to London in 1878, he spent his last years lecturing against Darwin's "mistaken hypothesis."

See also AMERICAN PALEOZOIC MUSEUM; IGUANODON DINNER

HENSLOW, REVEREND JOHN STEVENS (1796–1861)
Botanist, Geologist, Teacher

When the Reverend Professor John Henslow joined the teaching staff of St. John's College, Cambridge, in 1814, he had (at his father's insistence) given up his childhood dream of exploring African jungles. He was orthodox and devout in his religious beliefs, not very original as a thinker, and unambitious for advancement—an unlikely figure to help spark a scientific revolution.

Yet Charles Darwin said this unassuming cleric-naturalist was the greatest single influence on his career.

Henslow was a self-taught all-around naturalist, especially knowledgeable in geology, botany, and entomology. When he came to Cambridge, no degrees were given in natural history subjects, a situation he determined to change. He revived the university's moldering herbarium and natural history museum, replanted its botanic garden, and helped make scientific learning respectable in an institution long dominated by classics and theology.

Professor Henslow favored "hands-on" experience with living things. He took his students on high-spirited field trips into the countryside, where he showed them how to ask nature, rather than books, for answers. "Nothing could be more simple, cordial, and unpretending than the encouragement which he afforded to all young naturalists," his former student Charles Darwin later recalled. Never flaunting his vast knowledge, he put the young men perfectly at ease, treating their questions and blundering "discoveries" with the same respect he gave accomplished senior colleagues. Earnest mistakes were corrected "so clearly and kindly that one left him in no way disheartened, but only determined to be more accurate the next time."

By the time young Darwin returned home from college in 1831, Henslow's encouragement had transformed him from a boyish beetle collector into a budding naturalist. Although his father was urging him toward a life as a minister in the Church of England, he was busily making plans for a geologizing trip to Wales and the Canary Islands with Professor Henslow—which did not please Mrs. Henslow.

Everything changed, however, when Darwin received a fateful letter. Henslow had been asked to recommend a naturalist for a long round-the-world voyage of exploration. He considered signing on himself, but this time his wife put her foot down. The professor's dream of exploring exotic lands and collecting strange plants and animals was thwarted again. Still, he hated to see the opportunity go to waste and urged young Darwin to seize it, even though he wasn't "a finished naturalist." Henslow would coach him from half a world away.

Just before the *Beagle*'s departure, Henslow advised Darwin to take along the just-published first volume of Charles Lyell's *Principles of Geology* (1830–1833). By all means read it for its facts, he advised, "but on no account to accept the views therein advocated." Lyell's book became a crucial influence on the development of Darwin's thought as he geologized throughout his explorations.

During the entire five years of the voyage, Henslow corresponded regularly with Darwin, offering detailed advice and guidance to the inexperienced young naturalist. It truly became their voyage. From Brazil, Darwin wrote in 1832: "The delight of sitting on a decaying trunk amidst the quiet gloom of the forest is unspeakable and never to be forgotten. How often have I then wished for you." His teacher replied, "Your account of the Tropical forest is delightful, I can't help envying you."

Henslow took on the burden of receiving the endless crates and boxes of rocks, plants, and preserved mammals, birds, insects, and fish that Darwin shipped home. He carefully opened each, removed damaged specimens, and arranged and stored the precious collections.

"I firmly believe that, during these five years," wrote Darwin in

Henslow took on the burden of receiving the endless crates and boxes of rocks, plants, and preserved animals that Darwin shipped home.

MENTOR TO DARWIN, the Reverend John Henslow became young Charles Darwin's landbound partner during his voyage of discovery but was later unhappy about his student's advocacy of evolution.

1862, "it never once crossed his mind that he was acting towards me with unusual and generous kindness. . . . I owe more than I can express to this excellent man." Similarly, it seems never to have occurred to Darwin that he was giving his mentor great joy by allowing him to vicariously live out his boyhood dream of exploring the tropics.

Although the two men continued to correspond until Henslow's death, Darwin rarely discussed his evolutionary views, in an effort to spare Henslow's sensibilities. For his part, as the botanist Joseph Hooker wrote, Henslow was "a man who, with strong enough religious convictions of his own, had the biggest charity for every heresy so long as it was conscientiously entertained." He could not accept Darwin's conclusions, but never doubted his integrity.

When Richard Owen launched a vituperative, mean-spirited attack on Darwin's *Origin of Species* (1859), Henslow wrote his old pupil: "I don't think it is at all becoming in one Naturalist to be bitter against another—any more than for one sect to burn the members of another." To the gentlemanly Reverend Henslow, torching heretics was simply not good manners.

See also DARWIN, CHARLES; LYELL, SIR CHARLES; VOYAGE OF HMS *BEAGLE*

HOAXES, SCIENTIFIC

See BARNUM, PHINEAS T.; BERINGER, JOHANNES; BOUCHER DE PERTHES, JACQUES; PILTDOWN MAN (HOAX)

HOLMES, SHERLOCK

See DIVINE BENEFICENCE, IDEA OF; DOYLE, SIR ARTHUR CONAN

HOMININ
Classifying the Human Animal

The term "hominin," a recent addition to anthropological lingo, was coined to describe evolutionary classification of humans and our primate relatives with greater precision. Instead, it has further obscured them.

According to the Linnean system of classifying animals and plants, each species belongs to a genus, which in turn belongs to a family, order, class, phylum, and kingdom. Darwin gave that system an evolutionary dimension: species that share the most recent ancestor should be in the same genus, the most closely related genera should be grouped in the same family, and so on.

Humans are related to chimps, gorillas, and to various extinct upright apes. But how closely? Scientists formerly placed humans and our fossil antecedents, such as Neanderthals and *Homo erectus*, in the family Hominidae (the hominids), while the apes were called pongids, members of the family Pongidae.

Over the past few decades, however, genetic testing has shown that the various apes branched off at different split-off points (with the orangs most distant) and should not be placed in the same family after all. The term "hominid" is now applied only to the most closely related living apish trio—humans, gorillas and chimps—and their immediate ancestors.

About 10 million years ago the gorillas split off to form the subfamily Gorillinae. After another three million years, the chimps diverged into several species that are assigned to a separate subfamily called Panini. Modern humans and their direct ancestors are now placed by themselves in the subfamily Homininae—hence the term "hominin."

Both hominins and hominids are still within the more encompassing superfamily Hominoidea (homin*oids*), along with chimps, gorillas, gibbons, and orangutans. How do the pros keep them all straight? Harvard anthropologist Daniel Lieberman was quoted in *New Scientist* as saying, "I prefer to use 'hominid' as a colloquial, but use 'hominin' when I am being cladistically formal." So wear your tuxedo when you use "hominin." That same publication predicts that "it's only a matter of time before we reach hominaaargh!"

> There are now hominoids, hominids, and hominins. "It's only a matter of time before we reach hominaaargh!"

HOMO ERECTUS
Men of the Mid-Pleistocene

Homo erectus, a species of human somewhat different from ourselves, roamed eastern Asia up until the late Pleistocene, about 400,000 years ago, when it disappeared.

Dr. Eugene Dubois found the first *H. erectus* fossils in Java (1891) but named the species *Pithecanthropus erectus*, or "ape-man who walks upright" (Java Man). Four decades later, in 1929, excavations in an ancient Chinese cave yielded *Sinanthropus pekinensis*, or "Chinese man from Peking" (Beijing Man). A score of fragments from other Asian sites are now all considered members of the single species, *Homo erectus*.

Adults are thick boned, with massive jaws and heavy eyebrow ridges. Although they were first called "ape-men" or "missing links," there is really nothing apelike about them. They were a species of humans, with brains a bit smaller than our own: about 1,000 cc versus 1,400 for *Homo sapiens*. Pelvis and legbones show they were fully upright and bipedal, as we are.

Estimated at 1.6 million years old, the Turkana Boy's skeleton from East Africa is among the oldest-known hominids of this kind, though this fossil is now assigned to its own species, *Homo ergaster*, which means "work man," an acknowledgment that this hominid was a stone tool-maker. Most experts believe this species originated in East Africa and from there dispersed northward and eastward. Its descendants, among them *Homo erectus*, ranged throughout the Old World during the next million years.

Homo erectus remains have been found at widely scattered sites. Each discoverer gave his find a new name, convinced that it was very different from anything previously known to science. (A tendency to proclaim hominid fossils "the oldest," a "missing link," or a "previously unknown species" still pervades paleoanthropology.)

In 1950, evolutionary biologist Ernst Mayr, an expert systematist (classifier of organisms), decided to tackle the mess of multiple names for these fossil bones. Unlike most anthropologists, for whom slight variations justified new species, he had a zoologist's wide-ranging familiarity with the animal kingdom. Mayr had unraveled complex relationships between families and species of birds at the American Museum of Natural History and was able to review the hominid fossils with some detachment.

After comparative study, Mayr concluded there had been no scientific basis for splitting the so-called pithecanthropines into a dozen invented genera, and instead lumped them all into the established genus *Homo*. The species name *erectus* was taken from the original moniker by Eugene Dubois.

Anthropologists gratefully accepted Mayr's reclassification, which resolved years of confusion. But later fossil finds demonstrated even more complexity in the record than Mayr's schema allowed; far from being the "middle" species in the evolving *Homo* lineage, *Homo erectus* is increasingly viewed as a local East Asian variant of *Homo*, which eventually became extinct.

Ian Tattersall and Niles Eldredge of the American Museum of Natural History have suggested that the slenderer, lighter-skulled *H. ergaster* may have been ancestral to both the heavier-boned, beetle-browed *Homo erectus* and a second line that led ultimately to ourselves, *Homo sapiens*.

See also BEIJING MAN; DUBOIS, EUGÈNE; MAYR, ERNST; TURKANA BOY

First found in Java and then in China, *Homo erectus* seems to have originated in East Africa, migrating to Eurasia by 1.8 million years ago.

HOMO HABILIS
Controversial Ancestor

In the late 1960s through the 1970s, when most anthropologists thought a convincing picture was emerging that traced the human lineage back through *Homo erectus* to an ancestral australopithecine, the Leakey family dissented.

Back in 1960, Louis Leakey's son Jonathan had found a lower jaw and some cranial fragments at Olduvai Gorge, dated at about 1.75 million years by the potassium-argon method, which seemed to belong to an "advanced" hominid. Louis Leakey named it *Homo habilis*,

A GROUP OF HABILINES butcher a zebra carcass on the African grasslands. Some drive off wild dogs that made the kill, while others make stone tools for cutting up the meat.
© 1995 J.H. Matternes.

or "handy man," because he believed it was associated with chipped stones found at the site. "This was the first evidence," wrote Richard Leakey and Roger Lewin, "that early members of the human lineage were contemporaries of the australopithecines, not descendants as was generally believed."

Debate swirled over whether this "habiline" was really a distinct species of *Homo* or a variety of australopithecine. In 1972, Bernard Ngeneo, of Richard Leakey's team of fossil finders, discovered a much more complete skull at East Turkana, which received the specimen number "1470" at the Kenya National Museum.

The 1470 skull was pieced together by Richard Leakey's wife, anthropologist Meave Leakey, and several anatomists from dozens of fragments—a jigsaw puzzle that took six weeks to assemble. Dated at 1.89 million years, with a cranial capacity of 750 cc, it was first thought to be be the oldest fossil of a true human ancestor. It became the de facto best example of *Homo habilis,* prompting some paleoanthropologists to relegate the australopithecines to side branches of the hominid family tree. However, others suggested that the new fossil should be classified as a separate human species, *Homo rudolfensis*; still others favored lumping it with the australopithecines. Although *Homo habilis* is still acknowledged to be the earliest named species of *Homo*, its precise characteristics remain unclear. Other problematic fragments have been asigned to it—a situation that Richard Leakey has called "a mess."

In 2007, Meave Leakey and Fred Spoor announced their discovery of two new hominid fossils from Turkana, northern Kenya: a *Homo habilis* upper jaw dated at 1.44 million years and a *Homo erectus* jawbone dated at 1.55 million years.

These two fossils, found in the same deposits, showed that the two species lived side by side in eastern Africa for more than 500,000 years. Whatever their correct allocations to species ultimately turn out to be, they add to the growing body of evidence that several species of early humans typically lived at the same times and places. They also provide further evidence for demolishing the unilineal "ladder" of human evolution, according to which anthropologists had posited a progressive sequence from australopithecines to *Homo erectus* to *Homo sapiens.*

See also *HOMO ERECTUS;* LEAKEY, LOUIS; LEAKEY, RICHARD; "LUCY"

HOMO SAPIENS, CLASSIFICATION OF
Defining the Human Species

W hen, in 1735, the great Swedish naturalist Carl Linnaeus attempted to classify every living thing, he was not sure just how many species of humans existed. There were garbled reports from Africa and Asia about apes, "Jackos," "Pygmies," "wild men," and other near-men, but reliable information was scarce. In fact, few European scientists had even seen the carcass of a gorilla or chimpanzee, and the apparent variety of so-called savages was bewildering.

Of one thing Linnaeus was pretty certain: The earth contained creatures so anatomically similar to humankind that there was no reason to put them in a separate genus. (After Linnaeus, zoologists did separate them, placing humans and each of the apes in distinct genera. Now, after 200 years, biochemical and genetic tests show they are much closer than anyone had thought. Humans and apes are now placed together in the family Hominidae.)

Despite the striking similarity of structure, Linnaeus was more concerned with how apes and men behaved and communicated. In fact, Linnaeus's definition of man was based not on his usual description of form (morphology) but on mind and behavior, particularly the use of language. Where he usually listed anatomical traits, he wrote an enigmatic line next to *Homo sapiens* ("Man the wise"): *Homo nosce Te ipsum* (Man, know Thyself.) In other words, look in a mirror.

Under Linnaeus's system, one individual of each species is described in detail, then preserved as the "type specimen," or holotype. Museum drawers are full of the skins, bones, and pressed plants that represent each of the known living species. (The Linnean Society of London still has thousands from Linnaeus's own collection.)

In the case of ourselves—a polytypic species where populations vary greatly—what individual shall represent us? Shall he or she be tall as a Watusi or short as an Mbuti pygmy? Dark-skinned or fair? With woolly hair or straight? In Africa, Asia, and New Guinea, local populations vary more among themselves than averages of the various groups differ from each other. More variation exists among black Africans, for example, than between Africans and Europeans.

Historian Wilfred Blunt argues that the type specimen of mankind must be the individual described in greatest detail by the namer of the species. Having written five versions of his autobiography, the man Linnaeus described most completely was himself. Blunt therefore concludes that Carl Linnaeus is the legitimate type specimen of *Homo sapiens*. Entombed in the floor of Uppsala Cathedral, his skull and bones represent us all.

See also BINOMIAL NOMENCLATURE; LINNAEUS, CARL

Which individual represents our entire species— the type specimen of *Homo sapiens*?

HOMOLOGY
Similarities of Common Descent

One meaning of *Homo*, from Greek, is "the same." The seven bones in the human neck correspond with the same seven, much larger, neckbones in the giraffe: They are homologues. Seven neckbones, or cervical vertebrae, are a trait shared by thousands of creatures descended from a common ancestor. Sparrows also have the same seven neckbones. Distantly related species share corresponding structures, though they may become modified in various ways.

All mammals—from mice to whales—share a common ancestor. All have at least some body hair, are warm-blooded, have nucleated blood cells, suckle their young with milk from the female, and have four-chambered hearts. A closely related subgroup, or family, shares more specialized traits, which evolved more recently. Humans' close kinship with the great apes was first demonstrated by Thomas Henry Huxley in *Evidence as to Man's Place in Nature* (1863), in which he showed that, muscle for muscle and bone for bone, humans are more similar to apes (gorillas, chimps, orangs) than apes are to monkeys. As Darwin wrote in *Descent of Man* (1871), the evidence of homologies fairly shouts that "man still bears in his bodily frame the indelible stamp of his lowly origin."

Some similarities between distant species, known as analogous structures, may result from adaptation to similar environments rather than close common descent. Development of streamlined fins in fish (teleosts) and flippers in dolphins (mammals) are analogous: they function alike, but are very different in underlying structure. Wings of birds, insects, and bats are also analogous: although similar in form and function, they have independent evolutionary histories—known as convergent or parallel evolution.

Linnaeus's original classification of animals does not distinguish between analogous and homologous structures. Creatures were often put in the same groups if they seemed similar enough to express an imagined "divine plan" or "design." Since Darwin's impact on biology, species are classified to reflect the relative closeness or distance of their common ancestry.

See also CLADISTICS; CONVERGENT EVOLUTION

HOOKER, SIR JOSEPH DALTON (1817–1911)
Botanist, Evolutionist

One of the most remarkable men of the 19th century, Joseph Dalton Hooker had wanted to be just like Charles Darwin (eight years his senior) when he grew up. As a young man, he kept advance proofs of Darwin's travel journal of 1839 (obtained by his botanist father) under his pillow. Then, like his hero, he half-heartedly pursued a medical career, but left it in 1839 to sign on as unofficial naturalist on the surveying ship *Erebus*. During the expedition to the Antarctic, he made a large collection of previously unknown plants.

Soon after the young botanist's return to England in 1842, he was elated when Darwin sought him out. Within a few months, he became the great naturalist's closest friend, confidant, lieutenant, and research partner—a relationship that was to last for almost 40 years.

Their coming together was no accident; throughout the *Erebus* voyage, long before they knew each other, Darwin had followed Hooker's adventures and achievements through the letters he sent home. Darwin got hold of them by the same route Hooker had received his *Journal* proofs: through the benign meddling of the older generation of naturalists.

After reaching England from the south polar seas, Hooker's letters followed a complicated path. Their intended recipient was his father, William Jackson Hooker (1785–1865), the famous botanist who was founder-director of the Royal Botanic Gardens and herbarium at Kew. Delighted with his son's accounts of strange places and wonderful new species, the elder Hooker proudly showed them to his wealthy amateur botanist friend, Charles Lyell of Kinnordy, who passed them on to his son, famed geologist Sir Charles Lyell—friend and mentor of Charles Darwin.

Hooker had taken Darwin's *Journal* along on his travels, consulting it constantly, just as Darwin had taken Lyell's *Principles of Geology* (1828–1832) on his own voyage. Upon Hooker's return, the usually reclusive Darwin sought him out and offered him the task of classifying the *Beagle*'s plant specimens, still languishing undescribed in the British Museum. Hooker jumped at the opportunity.

Darwin liked Hooker immediately and admired his vast botanical knowledge. (Hooker had, after all, been helping his father with thousands of plant specimens since the age of 12.) He knew much more than Darwin about identification and traditional botany, but Darwin's way of looking at plants ("philosophical botany") was a revelation to him. Unlike professional botanists of the day, Darwin focused on the living plant: its physiology, ecology, distribution, growth, and reproduction.

It was not long before Darwin decided to take this excellent young man into his confidence about the theories he had secretly been developing for years. After all, Hooker could not help him marshal supporting evidence if he didn't know what to look for.

On January 11, 1844, Darwin revealed to his new friend that he had long been engaged in

> a very presumptuous work . . . [most would] say a very foolish one. I was so struck with the distribution of the Galápagos organisms . . . I determined to [tackle the problem of] what are species . . . and I am almost convinced (quite contrary to the opinion I started with) that species are not (it is like confessing a murder) immutable. I think I have found out (here's presumption!) the simple way by which species become exquisitely adapted to various ends. You will now groan, and think to yourself, "On what a man have I been wasting my time in writing to." I should, five years ago, have thought so.

With this hesitant, self-satirizing letter, Darwin confided to Hooker the agenda of his life's work, which he had kept even from Lyell. Soon after, some 15 years before publication of the *Origin of Species* (1859), he entrusted him with a preliminary essay (the "1844 Sketch") of his theory of natural selection.

Darwin had picked the right man; Hooker was to become his indispensable collaborator and ally in establishing the theory of evolution, backing it up with pioneering studies of plant distribution throughout the world. In 1848, before Hooker left on grand new expeditions to India and Tibet, Darwin loaded him with hypotheses to test, samples to be taken, questions to be answered. Hooker was becoming the world's greatest expert on the distri-

> "I am almost convinced . . . that species are not (it is like confessing a murder) immutable."
>
> —Charles Darwin to Joseph Hooker, 1844

bution of plants, and all his knowledge, research, maps, and specimens were at Darwin's disposal. Sometimes Darwin feared he was selfishly taking all his friend's ideas, but Hooker always maintained that he received more knowledge from Darwin than he gave.

Hooker made three expeditions to India during 1850 and 1851, all of them conducted in the grandest Victorian manner. On one exploration, he traveled by elephant 5,000 feet up a sacred Himalayan mountain, attempting to reach Tibet. But the rajah of Sikkim, paid by the Chinese to keep Englishmen out of the area, sent 100 men to capture him—which didn't faze Hooker a bit. He had taken the precaution of hiring his own private security force. In an 1848 letter to his father, he described his bodyguards: 56 Gurkhas, who were "immense fellows, stout and brawny, in scarlet jackets, carrying a kookry [dagger] stuck in the cummerbund and heavy iron sword at their side." Within two hours, he had turned both his and the rajah's men into a combined army of plant collectors.

On Hooker's expedition to the Sikkim Himalaya in 1848, he brought back hardy species of rhododendrons, which became an integral part of British gardens thereafter, an expression of the romantic Victorian notion of exotic beauty. He wrote feelingly of these danger-filled quests for exquisite, rare flowers amid rivers "roaring in sheets of foam, sombre woods . . . crested with groves of black firs, terminating in snow-sprinkled rocky peaks."

Hooker's massive party included armed guards, bird and animal shooters, cooks, porters, plant collectors, and an herbarium crew. Weather was so wet in the Khasia Hills that it took special efforts to dry specimens and press them in papers. He rented "a large and good bungalow, in which three immense coal fires were kept up for drying plants and papers, and fifteen men were always employed . . . from morning till night." His discovery of the rare blue vanda orchid in the Khasias, unfortunately, resulted in an invasion by ruthless commercial collectors who stripped the forests bare.

Hooker succeeded his father as director of the Royal Botanic Gardens at Kew, serving for 20 years (1865–1885), at which time his son-in-law, William Thistleton-Dyer, took over. All three directors were knighted for making this institution a world center of plant knowledge. (His father's own collection of specimens had formed the herbarium's core, occupying 13 rooms of his private house before it was moved to Kew.) When Darwin's long hesitation to publish threatened his place in history, it was Hooker (with Lyell) who rescued his claim to fame. In the introduction to his *Flowers of Tasmania* (1857), Hooker became the first respected naturalist to publicly take a stand with Darwin on natural selection.

And during the legendary Oxford Debate over evolution, it was Hooker—not Thomas Huxley—who won the day with a sober recitation of evidence after Huxley and the bishop had traded barbs.

In the end, Joseph Dalton Hooker had gone his youthful daydreams one better—he had become one of Charles Darwin's heroes.

See also *INDEX KEWENSIS*; LYELL, SIR CHARLES; OXFORD DEBATE

HOOTON, EARNEST ALBERT (1887–1954)
American Physical Anthropologist

An early student of mankind's ways, the Roman philosopher Horace coined a maxim justifying his wide-ranging curiosity: *Homo sum: humani nil a me alienum puto*, or "I am human; nothing in human nature is alien to me." With characteristic humor, Professor Earnest Hooton of Harvard revised it to read: "I am a primate; nothing in primate nature is alien to me."

For 40 years, Hooton was the premier gatherer of material on monkeys and apes, popularizing knowledge about them and encouraging students to learn about our own species by studying the context of our evolutionary lineage. His very readable books, *Man's Poor Relations* (1942) and *Up from the Ape* (1946), were widely influential, helping inspire the wave of primate field studies that transformed the following generation's physical anthropology from an anatomical to a behavioral science.

Hooton had begun his career as a classicist at the University of Wisconsin, where he won a Rhodes scholarship to Oxford in 1910. Although he had expected to continue his studies

FOUR STRANGE FIGURES (left), inspired by Professor Earnest Hooton's dubious claim that if fossil men walked the street today, no one would notice them. At right, showgirl Sherry Britton bones up on Hooton's *Apes, Men, and Morons.* Photo by Weegee (Arthur Fellig)/© ICP.

of ancient Greek and Roman literature, he became fascinated in London with the work of the British archeologist R. R. Marett (1866–1943) and the anatomist Sir Arthur Keith (1866–1955), and returned to the United States a physical anthropologist. He taught the subject at Harvard from 1914 until his death in 1954, training the first generations of professional physical anthropologists in the United States.

His approach was quirky and entertaining, including such novel studies as comparing the measurements of a gorilla with those of a professional wrestler. (Hooton insisted he meant no offense to the wrestler, who was a "real gentleman.") Among zoological descriptions of the primates, he inserted such doggerel as the following, in *Up from the Ape* (1946):

> The lemur is a lowly brute;
>   His primate status some dispute.
> He has a damp and longish snout,
>   with lower front teeth leaning out.
> He parts his fur with this comb-jaw,
>   and scratches with a single claw
> that still adorns a hinder digit,
>   whenever itching makes him fidget.
> He is arboreal and omnivorous;
>   From more about him, Lord deliver us!

Often he involved his students in his studies of anatomical variation. His "Harvard Fanny Study," for instance, involved measuring buttock spread and buttock-knee lengths of hundreds of students in order to design more comfortable chairs for the Pennsylvania railroad. Such studies led to the creation of applied physical anthropology.

Hooton was beloved by his students for his sometimes unintentional eccentricities as well. For instance, he so admired the great English physicist Sir Isaac Newton that he named a son after him, Newton Hooton. The boy took a lot of ribbing, so Hooton tried a more innocuous combination when he named his daughter Ima. Eventually, she married a Mr. Goody and became Ima Hooton Goody.

"HOPEFUL MONSTERS"
Macromutations

Richard Goldschmidt (1878–1958), a brilliant but unorthodox geneticist, did not believe that Charles Darwin's idea of slow, gradual changes could account for the origin of species. Forced out of his native Germany by the Nazis, he continued to develop his research at the University of California at Berkeley, where he wrote his magnum opus *The Material Basis of Evolution*, published in 1940.

Although he recognized the constant accumulation of small changes in populations (microevolution), he believed they did not lead to speciation. Between true species he saw "bridgeless gaps" that could only be accounted for by large, sudden jumps, resulting in "hopeful monsters."

Goldschmidt tried to explore possible genetic mechanisms by which rapid change might occur in lineages of organisms. He suggested that a relatively small change might have a large effect on the phenotype, especially through "controlling" genes that mediate the expression of the organism's blueprint. (Such genes have since been discovered.) Later, he thought macromutations, or mutants (which used to be called "monsters"), might arise in a single generation, and that this biological novelty might enjoy a selective advantage under changing environmental conditions.

That was where the "hopeful" came in. One hope was that the mutation would prove so useful in the newly changed environment that it would become selected as a new norm. Another hope was that the variant would appear often enough in the population to allow several similar "monsters" to find one another and produce offspring. There is a grotesque humor about the unfortunate phrase "hopeful monsters" that lent itself to caricatures and ridicule of Goldschmidt's ideas and obscured the important theoretical issues.

See also DE VRIES, HUGO; GRADUALISM; SALTATION

HORSE, EVOLUTION OF
Saddled with Errors

Professor Othniel C. Marsh of Yale, one of America's greatest paleontologists, set out to confirm Charles Darwin's evolutionary hypothesis by working out the evolution of the horse.

He collected a magnificent set of American fossil horses and published a paper in 1874 tracing equine development from a small three-toed animal "the size of a fox" through larger animals with progressively larger hooves, developed from the middle toes. Darwin thought Marsh's sequence from little *Eohippus* ("dawn horse") to the modern *Equus* was the best evolutionary demonstration anyone had produced in the 15 years since *Origin of Species* (1859) was published.

When Darwin's friend and "general agent" Thomas Henry Huxley toured America, he visited Marsh at Yale and was mightily impressed with his progressive series of fossil horses. "Whatever I ask for, you can produce," he enthused.

Marsh's classic unilineal (straight-line) development of the horse became enshrined in every biology textbook and in a famous exhibit at the American Museum of Natural History. It showed a sequence of mounted skeletons, each one larger and with a better developed hoof than the last. Almost a century later, paleontologist George Gaylord Simpson reexamined horse evolution and concluded that generations of students had been misled. In his book *Horses* (1951), he showed that there was no simple, gradual, unilineal development at all.

In fact, there were three complex radiations in the course of horse evolution, as they shifted from browsers on forest leaves to grazers on the grassy plains. Sixteen different genera developed, of which 15 became extinct. Several lineages developed in each adaptive zone, producing complex branches rather than a single line. Some grazers had well-developed hooves; other retained their toes. Rates of development were not gradual, but "jerky." Teeth, toes, and body size varied independently in different lineages.

Marsh's mistake was an easy one to make, since only one genus of horse is left today, *Equus*. He arranged his fossils to "lead up" to the one surviving species, blithely ignoring many inconsistencies and any contradictory evidence. Ironically, his famous reconstruction of horse evolution was copied by anthropologists. They, too, thought they saw a straight-line lineage "leading up" to *Homo sapiens* as the sole surviving species of a once-varied group. Anthropologists have since replaced that model with Simpson's model of the "branching bush."

See also BRANCHING BUSH; COPE-MARSH FEUD; ORTHOGENESIS

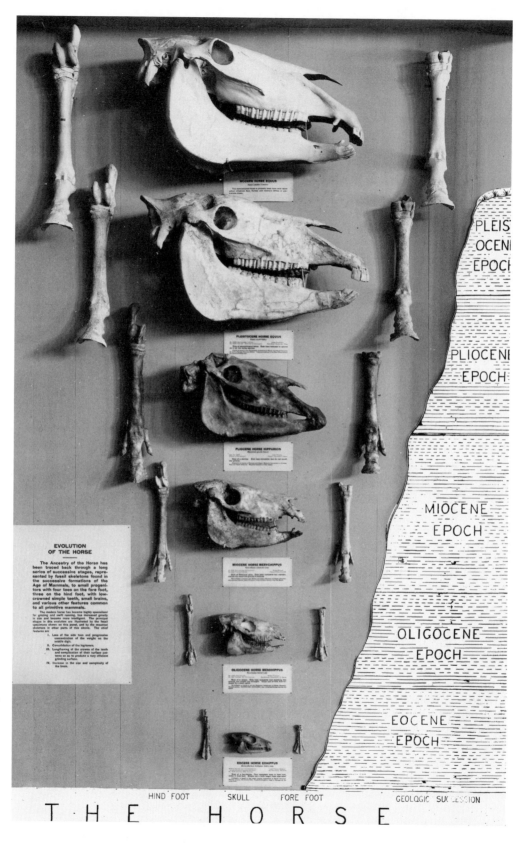

THIS OUTMODED EXHIBIT depicted straight-line evolution of the horse from small, five-toed animals, "leading up" to the progressively larger-hoofed *Equus* of today. Paleontologists later demonstrated that the real story was much more complex: a branching of many divergent populations over time and space.

HUNDREDTH MONKEY PHENOMENON
"Critical Mass" for Social Change

The story went like this: In the 1950s, Japanese primatologists had watched a new "pre-cultural" behavior emerge in a captive group of macaques on Koshima Islet. Scientists there left sweet potatoes on the beach for the monkeys. After one youngster discovered that dunking her potatoes in the sea got rid of the sand, she taught it to her mother. Soon, other macaques in her group learned the trick and it began to spread.

After the hundredth individual adopted potato-washing, according to this tale, the rest of the colony immediately began doing it. A few days later, macaques on the mainland and other islands—having absolutely no contact with their Koshima brethren—spontaneously began washing their food. The "scientific" conclusion: When an innovative idea reaches a "critical mass" in a population, the entire species instantly adopts it.

The "hundredth monkey phenomenon" became well established in New Age literature as a strategy to bring about social change. If thousands of individuals concentrate, say, on a just world free of hatred and weapons, the idea could reach a "critical mass" and suddenly humankind would end all warfare. Human evolution is therefore within our conscious control. No doubt the yearnings for peace are universal, but coercive attempts to foster a species-wide consensus have always led to greater divisiveness and conflict.

In point of fact, the phenomenon of the hundredth monkey does not exist. When asked by a journalist in 1984 whether he could verify the facts, Masao Kawai, director of the Primate Research Institute, said the story was hokum. His island monkeys had learned to wash sweet potatoes through the normal process of observing and learning, but there always remained a minority (mainly older individuals) who were never "converted" to the new ways.

There was no "critical mass" that spiked the learning curve overnight, and no spontaneous and rapid sweep of the behavior to groups of macaques on other islands or on the mainland. The idea of some kind of species telepathy, Kawai added, was introduced by Western writers.

British writer Lyall Watson seems to have originated the story. A best-selling author, designer of animal exhibits, and producer of television nature programs, Watson studied with the anthropologist Raymond Dart (author of the "killer ape" theory) and the English zoologist Desmond Morris, who taught chimpanzees to create abstract paintings for galleries. Both mentors, undeniably brilliant, were blessed with overactive imaginations, and encouraged Watson to pursue phenomena that, as he put it, "are on the borderline of soft and hard science."

Watson said that he originally published the "hundredth monkey" idea as a "speculation" in his book *Lifetide* (1979), and was astounded at its widespread, instant acceptance as fact.

> "The idea of species telepathy in Japanese monkeys is the creation of Western writers."
>
> —Masao Kawai, director of Primate Research Institute, 1984

HUNTING HYPOTHESIS
How Apes Killed a Theory

While thousands of Berkeley students protested the Vietnam War outside his window, Professor Sherwood Washburn of the University of California taught that aggressiveness had been the impetus for humans to evolve tools, technology, intelligence, and social complexity.

Studies of chimpanzees and gorillas in the African forest up to the 1960s had described the apes as peaceful vegetarians who neither hunted nor made war on their fellows. To Washburn, they posed a major puzzle for anthropology: How could "aggressive" humans have been derived from gentle, fruit-eating ancestors? His solution: "adaptation to a hunting way of life."

With the disappearance of forests, Washburn believed, early hominids had to venture out onto the savanna grasslands. In that new ecology, there would be selection pressure for bipedalism—standing upright—to scan the flat ground for enemies or prey. There

How could
aggressive
humans have
been derived
from gentle,
fruit-eating
ancestors?

humans became aggressive hunters. Cooperative hunting would shape social behavior, and tool use would lead to the development of weapons, as well as stone knives and cleavers for butchering prey.

During the 1970s and 1980s, field studies of apes took a dramatic and unexpected turn. Continuing observations proved that chimpanzees, for all their reliance on plant food, do hunt on occasion; their favorite prey are monkeys, baby antelopes, and the young baboons with which they frequently socialize. Moreover, they hunt cooperatively; several chimps go out together to stalk and surround their prey, driving it into ambush.

Killing and rending carcasses with their bare hands, chimps use no weapons or tools for butchering. Even more startling, the kill is not kept by the hunters; it is divided and shared among other members of the group as well. Chimps of all ages and sexes surround the hunters and beg for morsels with outstretched palms. Meat-sharing is a leisurely and elaborate social ritual lasting several hours; dominant males wait patiently with the others for their doled-out portion.

Then came the biggest shocker of all: "war" within species. Jane Goodall, in 1976, reported a "chimpanzee holocaust": the deliberate slaughter of all the males and infants in a neighboring group. Between three and five male chimps ganged up on each individual and brutally bashed him to death with fists and feet. Although they used no weapons, they did slam their victims against tree trunks. (They spared the females, who then joined the group and later mated with the victors.)

Goodall was heartbroken. She had thought chimps behaved "better than humans," but now realized her apes were no angels. "They are even more like us than I imagined," she sadly admitted. Later, she reported on chimp infanticide and cannibalism: two females kidnapped, killed, and ate several infants within their own group.

Hunting, murder, and mayhem can no longer be considered human specialties among the hominoids. Murderous aggression does not require upright posture, weapons-making, or a life on the open plains; stooped-over, knuckle-walking chimps, mostly vegetarian, living in the tangled forest without weapons or man-sized brains, can be as vicious and bloody as any carnivore.

Science remains ignorant of the basic adaptation that shaped the human species. Despite our bloody history, it is still possible that humankind is malleable, educable, and no more aggressive by nature than most other primates.

See also CHIMPANZEES; KILLER APE THEORY

HUXLEY, ALDOUS (1894–1963)
Novelist, Philosopher

A BRAVE NEW WORLD of high tech and low virtue is the best-remembered vision of Aldous Huxley. Novelist, essayist, futurist, mystic, and sometime Hollywood screenwriter, he entertained, provoked, and infuriated readers with consummate literary skill.

Evolutionist Thomas Henry Huxley, friend and champion of Charles Darwin, founded an intellectual tradition in Western thought and in his own family. One of his grandsons, Sir Julian Huxley, became a distinguished evolutionary biologist and essayist in his own right and a founder of the modern Synthetic Theory of evolution. Julian's brother, Aldous, followed a literary career, a poet's quest for new values and perceptions in the "brave new world" science had wrought.

Yet echoes of his evolutionist grandfather's major concerns reverberate throughout the fictional creations of Aldous Huxley: What kind of creatures are we? How can we know what is real? On what values should we base our actions? What can we honestly claim to know or believe? Into what kind of being will we allow ourselves to evolve?

Huxley's novels, casually saturated with the classical and historical traditions in which he was steeped, were addressed to an educated elite. But like his grandfather, who enjoyed lecturing on evolution to working men, Aldous Huxley had no snobbery about writing for *Vogue*, *Life*, or *Playboy* to popularize the results of his search for a view of life.

Huxley was a working writer, proud that he could support himself by the pen. During his early days the going was rough; his parents were "impecunious but dignified," as he recalled, "putting on dress clothes to eat . . . out of porcelain and burnished silver—a dinner of dishwater and codfish, mock duck and cabbage." From early youth he suffered from poor eyesight, which made a literary career especially difficult. He was blind by middle age.

He wrote everything from drama reviews to ad copy, plays, poetry, articles for the Sunday papers, even screenplays for Hollywood; but his fame rested on the essays and novels. During his last years he assumed the mantle of philosopher, mystic, and guardian of the public conscience. His pacifism during World War II and "inner explorations" with psychedelic drugs (*The Doors of Perception*, 1954) brought controversy and a measure of notoriety.

Among his best-known novels are *Chrome Yellow* (1922), *Antic Hay* (1923), *Those Barren Leaves* (1925), *Point Counter Point* (1928), and *Brave New World* (1932). The last is perhaps his most popular, despite its widely acknowledged flaws of one-dimensional characters and an unsatisfying "solution." It remains a vivid projection into the future of the eugenic ideas current in the early 20th century, when evolutionary biologists spoke seriously of the urgent need for mankind to take hold of its own evolution. Huxley saw and savagely satirized the kind of society that could result from biotechnology run wild; and the specter of his "brave new world" continues to loom as an ominous warning, especially since the technology he predicted has actually arrived.

See also EUGENICS; HALDANE, JOHN BURDON SANDERSON; HUXLEY, SIR JULIAN SORELL; HUXLEY, LEONARD; HUXLEY, THOMAS HENRY

HUXLEY, SIR JULIAN SORELL (1887–1975)
Evolutionary Biologist, Administrator

Thomas Henry Huxley said of his three-year-old grandson, Julian Sorell Huxley: "I like that little chap; he'll look you straight in the eye and deliberately disobey you." The great Victorian evolutionist did not live to see it, but biologist Julian (and his brother, the novelist Aldous) were to extend the family's fame and achievements well into the 20th century.

While still a boy, Julian decided to follow his grandfather into the study of evolutionary biology. Later in life—also like Thomas Henry—he found a conventional scientific career too limiting and entered the public arena. Early on, Julian thought a commitment to achievement "implies constructiveness not only about one's own life, but about society, and the future possibilities of humanity." As Thomas's evolutionism had led to an interest in exploring new views of religion and ethics, Julian's own study of biology expanded into a rethinking of humanist values and the application of science to social problems.

As a young student, Julian observed the courtship performances of the crested grebe. His description of their "plesiosaur-like dance," in which they entwine their long necks, jump out of the water, and utter a crescendo of excited cries, became a classic in the field study of animal behavior. His analysis of how various "normal" behaviors seem to become "ritualised" in the courtship sequence was later taken up as one of the influential concepts in the new study of animal behavior (ethology), developed by Konrad Lorenz and Niko Tinbergen. [See RITUALIZATION.]

Huxley attracted attention with his experiments on the axolotl salamander, a permanently juvenile species that retains gills throughout life and never leaves the water. [See NEOTENY.] Huxley and a colleague fed two five-inch axolotls extracts of cattle thyroid. After two weeks of ingesting hormones, the animals changed color, absorbed their fins and gills, and began walking on dry land. This "artificial metamorphosis" created wild speculation in the press about its applications for humans.

Huxley was invited to pursue research at Woods Hole Marine Biological Laboratory in Massachusetts. Studying fiddler crabs, which develop an extraordinarily large claw for

AT HIS GRANDFATHER THOMAS HENRY HUXLEY'S KNEE, Julian Huxley imbibed a tradition of scientific integrity, evolutionism, public service, and the search for "religion without revelation."

signaling and courting, Huxley tried to understand the rates of growth and proportions of the creature. "I was here able," he wrote, "to find a definite mathematical expression relating the weights of the claws and the rest of the body. The two behaved like two sums of money put out at two different rates of compound interest; in other words, the ratio of their growth rates was a constant."

Later, Huxley weighed and compared deer antlers as he had fiddler crabs claws, and worked out their constant differential growth rates. His first major treatise, *Problems of Relative Growth* (1932), was followed by *The Elements of Experimental Embryology* (1934), coauthored with British biologist Sir Gavin de Beer.

Although his academic career was going well—he then occupied the chair of zoology at King's College, London—around age 40 Julian decided to abandon it. He wrote several influential books, including *Religion without Revelation* (1927), *The Individual in the Animal Kingdom* (1912), and *Essays of a Biologist* (1923). With novelist H. G. Wells, who had been a student of Thomas Huxley's, he collaborated on a monumental textbook, *The Science of Life* (1929–1930), to make biology comprehensible to the general reader, as Wells had tried to do with historical studies in his famous *Outline of History* (1919–1920). (Wells died during the collaboration, and Huxley finished the book with Wells's son.)

During the 1930s, Huxley was influential in shaping the modern, or synthetic, theory of evolution: an update of Darwinism incorporating 20th-century developments in paleontology, Mendelian genetics, population genetics, systematics and other branches of biology. Along with Theodosius Dobzhanksy, Ernst Mayr, George Gaylord Simpson, J. B. S. Haldane, and others, Huxley was one of the architects of the new understanding, as explicated in his book *Evolution: The Modern Synthesis* (1942).

For some years, Huxley served as director of the London Zoo, then took on other administrative posts and was eventually asked to serve as the first director-general of the United Nations Educational, Scientific, and Cultural Organization (UNESCO).

Huxley thought the idealistic internationalism of the early United Nations could lead to an "evolutionary step up" (culturally, not biologically) from parochial warfare and the conflicts of antagonistic national governments. In a controversial pamphlet outlining his vision for UNESCO (*UNESCO: Its Purpose and Its Philosophy* [1946]), he went on record as a biologist turned social activist:

> UNESCO must constantly be testing its policies against the touchstone of evolutionary progress. . . . The key to man's advance, the distinctive method which has made evolutionary progress in the human sector so much more rapid . . . is the fact of cumulative tradition, the existence of a common pool of ideas which is self-perpetuating and itself capable of evolving.

This attempt by an "evolutionist-humanist" to bring together the human species created immediate polarization. Huxley's notion that man was in charge of his own destiny shocked many religious leaders as impious. A reference to Marxism as "the first radical attempt at an evolutionary philosophy" made Americans bristle, while another about the value of contraception in world social planning angered Catholics and others. And Communist ideologues demanded to know "whether acceptance of UNESCO required the abandonment of their philosophy in favour of Dr. Huxley's."

See also HUXLEY, ALDOUS; HUXLEY, LEONARD; HUXLEY, THOMAS HENRY; SYNTHETIC THEORY; WELLS, H. G.

HUXLEY, LEONARD (1860–1933)
Biographer, Editor, Poet

L eonard Huxley, a quiet and unassuming literary man, not only chronicled the accomplishments of his father, Thomas, but also inspired and shaped the continuation of a remarkable family tradition.

His own gifts were considerable, but he devoted them mainly to preserving the achievements of the previous generation in his classics *Life and Letters of Thomas Henry Huxley* (1900) and *Life and Letters of Sir Joseph Hooker* (1918).

Perhaps inhibited by living under the shadow of the great Victorian evolutionist, Leonard never exhibited the flamboyant Huxley inventiveness or intense commitment to the pursuit of new truths. In *Sheaves from the Cornhill* (1926), Leonard wrote:

> Unreason in every form is the enemy of scientific method, and the victory of science . . . means the gradual banishment of . . . many fancies . . . which survived to form a beautiful if misty background to everyday thought. . . . Beauty does not rest in untruth, nor is the loveliness of a landscape less appreciated by [understanding] perspective. The knowledge which destroys false beauties enthrones new ones.

Leonard's sons, to whom he passed on the family quest, were the biologist Julian and the novelist Aldous. Their grandfather had helped to locate man's place in nature, and now they were asking what humankind was going to do about it.

See also HUXLEY, ALDOUS; HUXLEY, SIR JULIAN SORELL; HUXLEY, THOMAS HENRY

HUXLEY, THOMAS HENRY (1825–1895)
Evolutionary Biologist

Thomas Henry Huxley is remembered mainly as Darwin's combative champion and defending knight: a forceful speaker, witty writer, and debater who gladly battled bishops and scientific foes to establish evolutionary biology. He is also known as the founder of an intellectual lineage, which includes his grandsons novelist Aldous and biologist Julian. Not so well known is his lifetime of solid accomplishment in many fields.

Huxley was a first-rate comparative anatomist and paleontologist, an educator who established lab courses in colleges, a philosopher of science, invertebrate zoologist, and popular essayist. He was also a medical doctor (though he never practiced), an anthropologist, and a tireless campaigner for freedom of thought. Huxley often said he would rather not have students hold his ideas if they believed them for the "wrong reasons," such as uncritical conformity. "Every great truth," he wrote, "begins as heresy, and ends as superstition."

Born in Ealing, a suburb of London, on May 4, 1825, he was the youngest of seven children. His father was a schoolmaster, but the family moved when Thomas was young, and he received little guidance in his early education. Although he wanted to be a mechanical engineer, his two brothers-in-law were doctors, and the family persuaded him to follow suit. One got him interested in human anatomy, and he soon became a highly skilled dissector and teacher. However, he believed he was infected while watching his first dissection (though he wasn't cut) and suffered afterward from what was then called "a form of chronic dyspepsia."

After a stint at London University, he won a scholarship to Charing Cross Hospital medical school, where he gained a knowledge of comparative anatomy far beyond what was required of a physician. Later, he remarked that his boyhood interest in mechanical engineering had been transferred to organisms. He wanted to take them apart and analyze their structures as living machines. About that time (1845), he discovered a layer of cells in the root-sheath of hair, which is still known as Huxley's layer.

He applied for an appointment in the Royal Navy, and was made assistant surgeon on the HMS *Rattlesnake*, which was about to survey the waters between Australia and the Great Barrier Reef. During the voyage (1846–1850), Huxley studied delicate ocean creatures

"DARWIN'S BULLDOG," the zoologist Thomas Henry Huxley debated bishops and lectured working men on evolution. His writings on science, ethics, and religion were enormously infuential.

(hydrozoa, tunicates), which float near the surface but do not last long for study. On the basis of his morphological work, he reclassified hydrozoans with sea anemones and corals, establishing them in a separate class.

Huxley's paper on the family of medusae (1849) also contained the important and original observation that two layers of cells in these primitive, hollow sea creatures were similar to an early stage of vertebrate embryos. This discovery took on added importance some years later, when evolutionists sought evidence for the "recapitulation" of evolution (phylogeny) in the development of individuals (ontogeny).

On his return to England, Huxley gained recognition and was made a Fellow of the Royal Society. Although he continued to make important discoveries in comparative anatomy, he could not earn a living in science, and at times he despaired. Until the late 19th century, science was mainly the province of the wealthy or those subsidized by them, and few salaried positions existed.

Finally, Huxley was hired as a lecturer in natural history at the Royal School of Mines, and also became a paid naturalist on a geological survey. In 1855, he felt settled enough to marry, and sent for Henrietta Heathorn, whom he had met in Sydney, Australia, while on the *Rattlesnake* voyage. She had been waiting for him for seven years, during which they conducted a long-distance courtship by letter. Her health was so poor that they both wondered if she'd still be alive by the time he was able to support a family. Once they were married, however, Henrietta proved a strong and able wife and a full partner.

Huxley first crossed swords with Richard Owen, the reigning anatomist and paleontologist at the British Museum, over the structure of the skull. Owen insisted the skull was developed from an enlarged vertebra, according to the archetypal theory he had learned from Lorenz Oken and Johann Wolfgang von Goethe. In a brilliant demonstration (1858), Huxley showed that the skull was entirely different in structural development from the spinal column and that Owen had distorted the facts to make his case. By decisively proving the senior anatomist wrong, he earned Owen's lasting enmity.

Huxley's life was transformed by a book: Darwin's *Origin of Species*, published in 1859. Prior to that time, he had not thought there was enough evidence to support a "development hypothesis" (evolution) or a plausible theory of how species could change. Darwin's work hit him like a revelation. "How extremely stupid not to have thought of that myself," he remarked. Fortunately for the theory, it was Huxley who was asked to review Darwin's book for the influential *London Times*. His enthusiastic reception got *Origin* off to a promising start, despite harsh critics—including the spiteful Owen.

That same year, Owen read a paper at Cambridge declaring that the human brain was structurally different from that of apes. Man's brain, he said, has a backward projection of the cerebral hemispheres that covers the cerebellum and wraps around a "hippocampus minor," utterly different from other primates. Huxley rose at the meeting and stated that Owen was clearly wrong and that he would soon prove it.

In 1861, Huxley published two essays that showed the human brain has no structures setting it apart from those of gorillas and chimps. Huxley expanded the work into *Evidence as to Man's Place in Nature* (1863), his most important book, which brought him into the public eye, where he remained for the rest of his life.

He became even more famous that year for his part in the famous "Oxford Debate" with Bishop Samuel Wilberforce, who had attempted to ridicule the new evolutionary ideas by asking Huxley if he was descended from an ape on the side of his grandfather or grandmother. Huxley's reply that he'd prefer an ape ancestor to a man "who used great gifts to obscure the truth" caused a sensation. (The inaccurate news went out that Huxley said he would rather be an ape than a bishop.) He once wrote of the chemist Joseph Priestley, one of his heroes: "There are men to whom the satisfaction of throwing down a triumphant fallacy is at least as great as that which attends the discovery of a new truth . . . and who are even more for freedom of thought than for mere advancement of knowledge." It was equally a description of himself.

An admirer of the philosopher René Descartes, Huxley believed doubt should have a high place in the life of the mind, not be condemned as "grievous sin," and thought it appropriate that his given name was that of the doubting apostle. He wrote that it was "a man's most sacred act" to say what he believed to be true and that he would not lie where there was no clear evidence for belief. Accordingly, he coined the term "agnostic" to describe his lack of proof for or against the existence of God, and the term became part of the language. Some have called agnosticism "a coward's way of being an atheist," but few have taken the trouble to appreciate Huxley's precision of thought. He well understood the limitations of the mechanistic philosophy and was not an advocate of crude "godless materialism," as some have characterized him.

He maintained that materialism was not an absolute truth but simply the most workable assumption that science could adopt for its purposes.

Even in the depths of grief when his infant son died, he refused to take comfort in "the hopes and consolations of the mass of mankind." In a famous letter to the Reverend Charles Kingsley, who had asked if he wasn't now sorry he didn't have religion, he replied, "My business is to make my aspirations conform to the facts, not to make the facts conform to my aspirations. . . . I refuse to believe that the secrets of the universe will be laid open to me on any other terms."

Yet while he denounced literal belief in demons or miracles, Huxley clearly recognized that some eternal truths lay behind the prevailing religious mythology. "It is the secret of the superiority of the best theological teachers to the majority of their opponents," he wrote in his 1892 essay "An Apologetic Eirenicon," "that they substantially recognize [certain] realities of things, however strange the forms in which they clothe their conceptions." He included among them

> the doctrines of predestination, of original sin, of the innate depravity of man and the evil fate of the greater part of the race, of the primacy of Satan in this world. . . . Faulty as they are, [they] appear to me to be vastly nearer the truth than the "liberal" popular illusions that babies are all born good and . . . that it is given to everybody to reach the ethical ideal if he will only try . . . and . . . that everything will come right (according to our notions) at last.

PORTRAITS and calling cards of Thomas Henry Huxley from his youth to old age.

Huxley's anatomical studies led him to unite reptiles with birds, a recognition of similarities between the two groups that was confirmed more than a century later. He also worked on dinosaurs and on classification of mammals, identified "crossopterygian" fishes, and was active in ethnology and physical anthropology.

He also gave much attention to writing and thinking about the implications of evolutionary theory, not only for science but for philosophy and ethics. At a time when "Social Darwinists" insisted "natural law" sanctified elimination of the weak by the strong, Huxley came out with a opposing view. Because we are human, he argued, we have the intelligence and duty to fight against accepting "survival of the fittest" as our ethic. We have the option of being a nurturing, protective, and compassionate species, deliberately turning our backs on "the law of the jungle" to create a humane society.

Debating the bishop of Oxford shaped Huxley's early reputation as a feisty, articulate scientist, but that was not his favorite type of contest. What he really liked was to master specialties far removed from his own fields of zoology and physiology and beat the experts at their own game. When theologian-naturalist St. George Mivart became one of Darwin's most adept and troublesome critics, Huxley took him on. But instead of arguing from a scientific base, he attacked with a virtuosic command of Roman Catholic theology, trouncing Mivart on his own turf.

Similarly, with William Gladstone and others, he delighted not in biological demonstrations, but in refuting their fuzzy biblical, historical, and philosophical scholarship.

True to form, when Huxley responded to claims of communication with the dead, he outfoxed "table-rappers" at their own tricks. He had learned that the Spiritualist movement started with two teenage sisters who produced phony "spirit raps" with their feet. Through long practice, Huxley became adept at loudly snapping his big toes inside his boots. Whenever his sense of humor moved him, he announced that he, too, had the power of summoning the spirits. Then, as if from nowhere, would follow a staccato of mysterious knocking sounds. Psychics were confounded, and true believers astounded.

In later years, Huxley took on administrative positions in universities and government and even served as director of the Royal Fisheries. Increasing ill health forced him into retirement, and he died at Eastbourne on June 29, 1895. His witty letters were collected by his son Leonard in *The Life and Letters of Thomas Henry Huxley* (1900), and his classic essays in the nine-volume *Collected Essays of Thomas H. Huxley* (1893–1894).

Although Huxley is remembered as "Darwin's Bulldog," spreading the secular gospel of Darwinism, he actually had a few difficulties with Darwin's ideas—not the basic idea of evolution, which he regarded as established fact, but Darwin's assertions about the universality of natural selection. Also, he told Darwin he thought his idea of evolutionary change occurring everywhere at a slow, gradual rate was an "unnecessary burden" for the theory to bear. His reading of the fossil record impressed him with the fairly sudden appearances and rapid divergence of various classes and families followed by long periods of stability. Recent reexaminations of the evidence, in the light of "punctuational" theory, have confirmed the acuity of Huxley's observations.

See also AGNOSTICISM; MATERIALISM; OXFORD DEBATE; PUNCTUATED EQUILIBRIUM; SPIRITUALISM; WELLS, H. G.

HYBRIDIZATION
Crossbreeding of Species

Crossbreeding is well known to be common among plants, but many naturalists still believe that hybridization is unusual and even "unnatural" in animals. However, according to geneticist James Mallet of the Galton Laboratory, University College, London, and other biologists, not only do 25 percent of plant species interbreed with others, but at least 10 percent of animals species (more in some families) produce viable hybrid individuals in the wild—and sometimes even hybrid species.

Around 1956, the "biological species" concept, promoted by influential evolutionist Ernst Mayr, became the prevailing definition of a species—a "real" entity that was reproductively isolated because its members mate exclusively with others of their own kind.

Mayr argued that in nature, when small populations are prevented from interbreeding by geographic barriers such as mountains or rivers, the result is commonly the origin of new species—an idea he elaborated from such earlier writers as Moritz Wagner, E. B. Poulton, and Alfred Russel Wallace. If geographic barriers to gene flow later disappear, the long-isolated populations may evolve "isolating mechanisms"; they no longer find colors, odors, or behavior from the long-separated population sexually attractive.

Charles Darwin made thousands of tests on hybrids of different varieties of plants and studied the results of crossing in pigeons and other domesticated animals. Although he never figured out the mechanisms of heredity, he published three books on his own experiments with variation and hybridization. Without an understanding of how genetics works, however, he believed in the "blending inheritance" of traits from father and mother, an idea later disproved. Genetic traits, as Mendel's classic pea plant experiments showed, cannot blend together; they are "particulate" and combine as dominant or recessive characters.

It is not known just how "distant" two related species can be and still produce viable hybrid offspring. When viable hybrids are produced, the male offspring are sterile, and females either are sterile, are less fertile, or produce offspring that do not live long. Yet that is not always the case, depending on the amount of divergence between the species. Sometimes hybrids of both sexes are fertile and viable, and can cross back to the parental species with no ill effects.

One of the most familiar of hybrid domesticated mammals is the mule, a cross between a mare and a male donkey. Mules, as sure-footed as donkeys and as strong as horses, possess a uniquely bright and stubborn personality all their own. However, they have to be crossbred from horses and asses each time and (with very rare exceptions) cannot reproduce themselves. One often thinks of them as the "classic" animal hybrid; indeed, in the 19th century, a synonym for "hybrid" (even in plants) was "mule."

When human breeders cross domestic animals, successful hybrids have been produced between cattle and bison ("beefalo"), turkeys and chickens ("turkens"), and horses and zebras ("zhorses"). A compilation of *Mammalian Hybrids* (1972), by Annie P.

A HORSE-ZEBRA HYBRID, or "zhorse," at a zoo near northern Gütersloh, Germany, shows an unusual combination of the coats of both parents. Photo: Martin Meissner/Associated Press.

Gray, lists 573 crosses between different species and varieties of mammals that have produced offspring. Many that do not occur in the wild sometimes take place in zoos, such as the "liger" or "tigon." In the 1930s, a male polar bear (*Ursus maritimus*) and a female kodiak bear (*Ursus arctos middendorffi*) at Washington's National Zoo produced cubs of both sexes that in turn birthed their own healthy male cub. More recently, DNA testing authenticated a wild-caught bear cub in Alaska as the offspring of a grizzly and a polar bear. Its white fur was interspersed with brown patches, and it sported long claws, a humped back, and other grizzly characteristics.

Botanists since the time of Linnaeus have known that many plant species commonly hybridize; willows, periwinkles, and many vascular plants, for instance, refuse to stick to clear species boundaries. According to geneticist Mallet 25 percent of vascular plants produce natural hybrids, as do passion flower butterflies, birds of paradise, and American warblers and tits. Three-quarters of British duck species, as well as partridges, grouse, and quail, hybridize with at least one other species. Over the past few decades, field data have accumulated on natural hybridization among populations of trout, rattlesnakes, and Galápagos finches.

Influenced by the "biological species" concept, conservationists adopted an almost eugenic view about preserving the "integrity" of species. During the 1970s, the Endangered Species Act in America stigmatized hybrids as unworthy of conservation.

A contentious case was the "red" wolf. Some scientists convinced the U.S. Fish and Wildlife Service that this smallish, ruddy canid was the indigenous wolf of the southeastern United States. Although a true species, they argued, it resembles a hybrid because it mates with the much more abundant coyotes. To prevent the endangered population from being "swamped" by coyotes, in 1980 the government put the presumed last "pure" red wolves into a captive breeding program and released some into the wild.

However, with increasing knowledge about genetics and the "naturalness" of hybridization, in 1990 that policy was changed. Intensive DNA research produced no distinctive genetic markers that would set the red wolf population off from gray wolves or coyotes. From a genetic standpoint, there is no such thing as a "pure" red wolf. Mallet, however, feels that the red wolf deserves protection precisely *because* it *is* a hybrid. "There is no such thing as a 'real' species anyway," he argues, and "hybrid zones are some of the most interesting populations in the world."

Studies have shown that at least 1–3 percent of all plant species have been formed by hybridization. When closely related species come together in the wild, the result is sometimes an array of hybrids, upon which natural selection can operate. Hybridization, it now appears, plays a significant role in continuing to generate biodiversity and the evolution of new species.

ICE AGE
Great Climatic Shifts

Harvard zoologist Louis Agassiz, the leading light of American biology, scorned Charles Darwin's theories of evolution as "impossible" when they were published in 1859. Twenty years earlier, however, Agassiz himself had launched a major unorthodox theory, to which Darwin was an early convert—the astounding idea that great continental glaciers had periodically advanced and retreated during prolonged ice ages.

Although at first skeptical, young Darwin was soon won over by Agassiz's radical idea. He later wrote in his *Autobiography* (1876) that, as a student on holiday, he had gone "geologising" with Adam Sedgwick in Wales and "had a striking instance of how easy it is to overlook phenomena, however conspicuous, before they have been observed by anyone." Following Sedgwick's lead, he had spent the day examining rocks for fossils.

> Neither of us saw a trace of the wonderful glacial phenomena all around us; we did not notice the plainly scored rocks, the perched boulders, the . . . terminal moraines [long heaps of pebbles and gravels]. Yet . . . a house burnt down by fire did not tell its story more plainly than did this valley. If it had still been filled by a glacier, the phenomena would have been less distinct than they now are.

As the ice age theory began to take hold in geology, it also posed a great mystery. What drives glacial cycles? In 1842, only a few years after Agassiz had announced his conclusions, Joseph A. Adhemar, a French mathematician, suggested that cold periods might be caused by long-term changes in Earth's position relative to the sun. But it was not until the 1920s that Milutin Milankovitch, a Yugoslav astronomer, worked out the idea in detail and came up with a plausible pattern.

Backed by a long and intricate series of calculations, Milankovitch proposed a curve of fluctuations in global climate governed by three factors: the slant of the spinning Earth relative to the sun (it regularly tilts a couple of degrees back and forth every 41,000 years), the shape of its orbit (which elongates and contracts every 100,000 years), and the precession, or wobble, of the planet (a circular movement completed every 23,000 years). Milankovitch showed that these three factors could vary the amount of sunshine reaching Earth by about 20 percent—enough to account for the formation and melting of great ice sheets. It was an intriguing theory, but for years no one was able to test it against a complete record of past glaciations. Over much of the Earth, the most recent advances and retreats of the ice had obliterated the evidence of previous episodes.

Beginning in the 1950s, however, micropaleontologist and geologist Cesare Emiliani found a new way of obtaining such a record—not on land, but in seafloor sediments, in the shells of tiny marine organisms called foraminifera. Emiliani showed that the proportion of oxygen isotopes in their remains reveals the composition of sea water at the time they were alive. (Emiliani is also considered the "father" of paleoceanography.)

Fluctuations in isotope levels within fossil forams, it turns out, correlate with the percentage of the world's water that is locked up in ice sheets. Therefore, cores drilled from the seafloor showed an isotopic pattern that correlated with Milankovitch's curves, and as others demonstrated in the 1970s, the regular fluctuations in the Earth's orbital movements fit right into this emerging picture.

Many kinds of evidence corroborate existence of the cyclic pattern: Global ice volume has peaked every 100,000 years during the past million or so. After each peak, the ice rapidly retreats for a few thousand years, causing a warming or interglacial period, such as the one we are in right now that began about 14,000 years ago.

However, the mechanisms that cause ice ages are not as simple to work out as the interaction of regular planetary movements or differential sunlight or rainfall. Scientists are only beginning to discover the dauntingly complex relationships between oceans and atmosphere, living things and circulation of global currents, and the distribution of carbon and oxygen.

We now know that Ice Age air (some of which is still trapped in ice bubbles formed thousands of years ago) contains one-third less carbon dioxide than interglacial air. Also, scientists have recently discovered that most (60 percent) of the world's carbon dioxide is in the ocean, captured by tiny ocean plants at the surface and "pumped" into the depths by the food chain. Since ocean currents profoundly affect the gas exchange process, it seems that, as geochemist Wallace Broecker and geologist George Denton put it, "only a major shift in the ocean's operation could account for such a dramatic change in atmospheric composition."

One geochemical study at MIT (again of foraminifera fossils) provided independent confirmation that ocean currents must have circulated differently during the last glaciation. Currents act like conveyer belts for heat, transferring vast quantities of solar energy gathered far from land to the continents, where oceanic winds strongly affect terrestrial climate and vegetation. These currents, in turn, are set in motion by complex relationships between the distribution of sea salt, differential formation of water vapor, and the transport of fresh water from melting glaciers.

In Broecker and Denton's view, transitions between glacial and interglacial conditions represent jumps between two stable but very different modes of ocean-atmosphere operation. Evidence from different parts of the world all seems to tie in. Warmer North Atlantic surface water, the melting of northern ice sheets as well as of Andean mountain glaciers, the reappearance of European forests, and changes in plankton around Antarctica and the South China Sea all point to around 14,000 years ago as the end of the last ice age.

Many scientists have expressed concern that the present interglacial phase may be artificially extended by the warming effects of human industrial activity, burning of fossil fuel, and destruction of rain forests.

See also LASCAUX CAVES; MAMMOTH; NEANDERTHAL MAN

EXCELLENT ARTISTS who lived 35,000 to 10,000 years ago left paintings and carvings of extinct bison, mammoths, woolly rhinos, and other Ice Age animals, as well as sculptures of obese or pregnant women.

ICEMAN
Alpine Neolithic Mummy

Encouraged by a bout of warm weather, two German hikers in the Ötztal Alps, Helmut and Erika Simon, ascended the Tyrolean trails on September 19, 1991. Meandering through a glacial area almost directly on the border of Italy and Austria, they saw a dead body with raised hand rising from a pool in the melting ice.

The Simons called the police, who arrived with a forensic anthropologist to examine the body. Although the corpse seemed peculiarly dehydrated, its antiquity was unsuspected until a flint knife, copper axe, quiver of arrows, and pair of grass-lined shoes came to light. The Iceman was also wearing leather leggings and a sturdy cape made of woven grasses. Radio-carbon dating at the University of Innsbruck established his age at about 5,300 years.

DNA tests of intestinal contents revealed that shortly before his death he had dined on ibex meat, dicots, red deer, and einkorn wheat. (Prehistoric stone engravings of red deer are found all over the central and eastern Alps.)

Nicknamed Ötzi for his native mountains, the Neolithic chap stood five feet three inches tall, and was between 25 and 45 years of age. Many small skin punctures surrounded his arthritic joints, suggesting that he may have used needles to relieve pain. Pathologists found that he also suffered from hardened arteries and intestinal infections—and serious wounds as well.

Eduard Egarter Vigl, the Iceman's new caretaker at the South Tyrol Museum of Archeology, defrosted a hand and determined that it had a deep gash, suggesting that Ötzi had met a violent end. Further examination showed that he had been stabbed in several places (including on the hand he had raised to defend himself) and had taken an arrow in his back; its stone tip was still embedded there. His assailants had left him on the mountain to bleed to death.

One of Ötzi's stone weapons contained traces of blood DNA from two different individuals, suggesting that the Iceman had killed an enemy, retrieved his stone arrowhead or knife, then brought down another man as well, before he himself was struck down. His cloak also contained blood from two other humans.

Many who view Ötzi are amazed at the preservation of this 5,300-year-old corpse and wonder what special circumstances allowed it to mummify when most human bodies rot, even in the cold. The answer may lie in the lack of bacteria, which normally decompose a body. Bacteria are always found in the blood of the living, and after death they begin to multiply in the corpse. Since all the Iceman's blood drained out of his multiple wounds, so did these bacteria. In other words, the very same cause that killed the Iceman—massive loss of blood—also helped create the conditions that would preserve his body for millennia!

Soon after the discovery, local authorities fought over whether Ötzi belonged to Austria or Italy, since the site was near their common border. Eventually, the courts determined that it was on the Italian side of the line. The legal ruling fortunately matched the archeological verdict: chemicals in Ötzi's body (remnants of vegetable foods) were identical to the proportions found in the soil of a nearby Italian valley. Ötzi now reposes in a humidified and refrigerated chamber in the Museum of Archeology at Bolzano, the capital of the mountainous Tyrol region. Visitors to the Iceman spend millions of euros there each year.

Cognizant of Ötzi's significant contributions to the State coffers, Helmut Simon, the German tourist who discovered him, sued the local government for a substantial finder's fee. While the courts agreed that payment was in order, litigation over the amount dragged on for years. Eventually, an agreement was reached for 25 percent of Ötzi's cash value to the town, but the case was not settled until 2008—four years after Helmut Simon's death.

While hiking alone in the same mountains on a snowy day in 2004, Simon, 67, died of an apparent heart attack and fell off a 300-foot cliff. After a rescue party searched for a week, they found his frozen body—just 100 miles from where he had discovered the Iceman 13 years before.

> **Ötzi died after being repeatedly stabbed and speared by several attackers. Police arrived at the scene of the murder 5,300 years later.**

IGUANODON DINNER
Victorian Dinosaur Celebration

To celebrate New Year's Eve 1853, 21 famous Victorian scientists were served a seven-course meal inside the belly of a giant iguanodon statue. The occasion was the opening of the first dinosaur exhibit at the Crystal Palace exhibition at Sydenham in South London, where the statues can still be seen. The dramatic dinosaur replicas stirred public interest in extinct fossil animals, though at the time many viewed the monsters as casualties of Noah's Flood.

The life-sized iguanodon, constructed of metal and concrete, was the work of Benjamin Waterhouse Hawkins, the first great reconstructive dinosaur artist. Dinosaurs had been known only for a few decades; they had been discovered in Kent and given their name by the cantankerous chief anatomist of the British Museum, Richard Owen. At the dinner, with his usual charm and tact, Owen made a speech attacking Hawkins for "getting the iguanodon wrong," since recently discovered fossil tracks suggested the animal really walked upright on its two hind legs. Yet Owen himself had supervised Hawkins while he built the dinosaur models, and so earned his place literally at the head of the table inside the iguanodon's head.

Illustrated invitations were drawn by the artist, and a special song was composed for the occasion, with the chorus: "The jolly old beast / Is not deceased / There's life in him again." The dinner was a great success, although it was not the first of its kind. In 1801, the American artist and museum owner Charles Willson Peale had assembled the first skeleton of a mastodon at his museum in Philadelphia. In December of that year, Peale invited 12 distinguished guests to join him for dinner inside the immense skeleton, where they made patriotic toasts and sang "Yankee Doodle."

See also HAWKINS, BENJAMIN WATERHOUSE; JEFFERSON, THOMAS; OWEN, SIR RICHARD; PEALE'S MUSEUM

AT THE IGUANODON DINNER of New Year's Eve 1853 (top), anatomist Richard Owen (in head), hosted a meal to celebrate the completion of his dinosaur reconstructions. Invitations (bottom) were drawn by the dinosaur sculptor Benjamin Waterhouse Hawkins. Courtesy of M.E. Korn.

INDEX KEWENSIS (1892–1895)
Greatest Plant Catalogue

Charles Darwin's last written contribution to the study of natural history was not a book but a bank check—to underwrite the "botanist's bible," the *Index Kewensis*.

For years the Royal Botanic Gardens at Kew, near London, had kept up their lavishly illustrated catalogs of all known species of flowering plants. By 1880, however, they were hopelessly out of date. Victorian naturalists and botanists were adding newly discovered species so rapidly that a completely revised world inventory was needed.

Darwin's good friend Sir Joseph Hooker, director of the institution, could not raise government backing and appealed to the public for financial assistance. In January 1882, Darwin sent him £250 to begin the project and told his children to continue supporting the work with annual payments if he didn't live to see its completion. He died soon after, in April of that year, at the age of 73. His estate continued to underwrite costs of the work, and within a few years the manuscript of the *Index* weighed two tons.

The *Index Kewensis* is part of Darwin's lasting scientific legacy. Now computerized, it is still in use today and is continually kept current for more than 200,000 species. The great evolutionist—who convinced the world that species are unfixed, changeable entities—funded an immense, definitive species list as his final gift to science.

See also HOOKER, SIR JOSEPH DALTON

INFANTICIDE
Nature's Murder Mystery

Baby-killing is commonplace in nature. Fathers, mothers, siblings, close relatives, parents' new mates—all may destroy infants or eggs under certain circumstances. Infanticide has been observed among rotifers, ants, guppies, swallows, lions, rats, monkeys, acorn woodpeckers, apes, prairie dogs, and humans, to name just a few. Since it removes genes from the next generation's population, the widespread phenomenon must affect evolution—but how?

Only a few years ago, scientists thought infanticide was an abnormal behavior produced by crowding or stress in captive animals. During the 1970s, however, field studies showed it to be common among wild populations. Can this behavior be beneficial to species? What is the effect on sex ratios, population structures, reproductive strategies?

One pattern has already emerged in species where social groups consist of "harems" of one or two males and many females, such as lions and chimpanzees. When males from an outsider group attack, they attempt to drive away or kill resident adult males. Invading males, after killing unweaned infants, end up siring many more, thus perpetuating their genes.

Lactating females usually become fertile again and quickly conceive. Also, they accept the situation in a manner foreign to human empathy, never rising up in outrage against the invading males who killed their offspring nor resisting their sexual advances. Often offspring of the invaders seem more vigorous than those produced by a closely inbred social group.

Infanticide in nature takes many forms. When a female poison dart frog comes upon a male brooding a nest for another female, she crushes the eggs, then mates with him. In black eagles, the first-hatched (larger) chick harasses its sibling, monopolizes the food, and the second starves to death. Among hyenas, a dominant female may harass a subordinate nursing mother until her cubs die, freeing her to help nurse the dominant female's litter.

In many species, the evolutionary mechanisms favoring infanticide are extremely puzzling. A seven-year study of black-tailed prairie dogs by John Hoogland found that more than half of all litters born were destroyed by lactating females killing their own sister's pups. Walter Koenig observed that female acorn woodpeckers—another animal that lives in large colonies—smash eggs laid by close relatives. Whether there may be evolutionary or genetic benefits to the colony of such "kin-directed ovicide" is still unknown.

> When a female poison dart frog finds a male brooding another's nest, she crushes the eggs, then mates with him.

INHERIT THE WIND (1955)
Dramatization of Scopes Trial

A tense courtroom drama based on the famous "Tennessee Monkey Trial" of 1925, *Inherit the Wind,* fascinated New York City theater audiences and went on to international fame. Based on a science teacher's clash with local religious beliefs, its theme was freedom of thought.

At the historic trial, John T. Scopes was prosecuted for presenting Darwin's theory in a high school biology class in defiance of a state antievolution law. He was defended by Clarence Darrow, a flamboyant attorney with a passion for freedom of thought and the separation of church and state. An equally charismatic figure, the popular politician and creationist William Jennings Bryan represented militant fundamentalists for the prosecution. National press attention was focused on the little town of Dayton, Tennessee; the *Baltimore Evening Sun* sent journalist H. L. Mencken.

Inherit the Wind was written by Jerome Lawrence and Robert E. Lee. After a brief, successful tryout in Dallas, it opened in New York City on April 21, 1955, to rave reviews and settled in for an extended run. The Bryan character (called Brady) was played by Ed Begley, the journalist by Tony Randall, and Darrow (Drummond) by Paul Muni.

Stanley Kramer made a classic motion picture version (1960) starring Spencer Tracy as Drummond, Fredric March as Brady, and Gene Kelly as the cynical newspaperman. In one of Drummond's most memorable speeches, he argues that a loss of cherished beliefs may be the price of new knowledge:

> Progress has never been a bargain. You've got to pay for it. Sometimes I think there's a man behind a counter who says, "All right, you can have a telephone, but you'll have to give up privacy, the charm of distance. . . . Mister, you may conquer the air, but the birds will lose their wonder, and the clouds will smell of gasoline!"

The play's title is from Proverbs 11:29: "He that troubleth his own house shall inherit the wind." Although the characters and conflict were clearly based on the Scopes case, only a few bits of dialogue were taken directly from the trial transcript. Playwrights Lawrence and Lee insisted in a preface that their play "does not pretend to be journalism." The stage directions set the time of the trial as "Not too long ago. It might have been yesterday. It could be tomorrow." Indeed, it turned out to be the first of many similar judicial challenges, which appear to have no end in sight.

See also BUTLER ACT; CREATIONISM; FUNDAMENTALISM; SCOPES TRIAL

HOLLYWOOD VERSION (1960) of the Scopes Monkey Trial of 1925 starred Spencer Tracy and Fredric March as the battling barristers. The drama focuses more on freedom of inquiry than on evolution.

INSECT SOCIETIES, EVOLUTION OF
Most Ancient Social Communities

Ants, bees, wasps, and termites have evolved societies that rival those of humans in complexity. The most highly social (eusocial) species support many nonreproductive helpers in the nest while one or more queens and a few males monopolize reproduction. Not all wasps and bees are social, but all termites are, as well as some beetles, thrips, aphids, spiders, caterpillars, and a tiny Caribbean shrimp.

To build their nest, 5,000 ants, weighing a total of seven-tenths of an ounce, moved 44 pounds of sand in five days.

The stinging, biting social insects—ants, bees, and wasps—all descended from wasplike insects without stingers that lived 100 million years ago. Most injected their eggs beneath the bark of trees, using a long ovipositor. This ancestral egg-laying appendage developed into the stinger that bees, wasps, and some ants use against large predators. Termites arose independently more than 200 million years ago from social cockroaches, which, like termites, carried thousands of microorganisms in their guts that can collectively digest wood.

Social insects live together in colonies that can range in size from a few dozen to millions of individuals. In 2002, European scientists reported a supercolony of Argentine ants stretching thousands of miles from the Italian Riviera along the coastline to northwest Spain. Accidentally introduced to Europe around 1920, this colony—the largest known to science—is composed of billions of insects living in millions of communities. They are all close enough genetically to recognize one another as kin, despite being from different groups with different queens, so they do not attack each other's nests.

Unlike some human societies, in ants there is no chain of command or control from the top of the hierarchy. The queen, after a short phase of colony founding, becomes a helpless egg-laying machine. Males generally die after impregnating the queen; when the first generation matures, the offspring tend the queen and assume all non-egg-laying duties in the colony.

The community's economy is based on the exchange of food. Ants, bees, and wasps solicit partially predigested liquid nutrients from "caretaker" individuals. They provide food to each member of the colony, and to some "hangers-on" of other species that mooch a living by staying within the colony. Often these resident strangers mimic solicitation signals of the host species.

Each member of the colony belongs to a "caste" that specializes in childcare, construction, foraging outside the nest, or soldiering. Workers in some species can, like insect Cinderellas, transform into reproductive queens if the original queen dies. In some cases individuals may start out taking care of the young, later go out to forage for food, and then transform into fighters as changing requirements of the colony dictate. In some species, "soldier" caste members have become specialized, sporting saberlike mandibles and organs that squirt toxic chemicals.

Differential nutrition and environmental variations such as temperature can cause genetically similar individuals produced by the same parents to develop into members of widely different castes. Some social insects have the ability to regulate sex ratios. Fertilized eggs develop into females, while the unfertilized ones become males.

Charles Darwin, in *Origin of Species* (1859), wrote that at first he was stumped by the evolution of sterile castes. He then introduced the concept of the entire family as being the unit of selection. Such families, which have the capacity to produce sterile altruistic individuals, become the units of evolution—an idea that was the foundation of modern kin selection theory.

Some ants and wasps are predators on other insects, while bees subsist on flower pollen and nectar, and seed-collector ants gather and maintain stores of grain. Some ants herd domesticated "ant cows" (aphids) from which they extract chemical foods. Other species have evolved complex symbiotic relationships with plants, the insects providing defense for plants in exchange for food and shelter. Several species of ants and termites are farmers that collect and nurture various types of fungi, maintaining them in just the right conditions of nutrients, moisture, and temperature. Termites, with the aid of microorganisms living in their guts, are specialized for eating dead wood and leaves, which they turn into soil.

Colonies often last as long as the queen lives—over twenty

THIS COLONY IN SPACE is a plaster cast of a Florida harvester ant nest, meticulously excavated and pieced together by entomologist Walter Tschinkel of Florida State University. It is 30 feet long, 10 feet deep, and has 135 chambers, storing a quarter million seeds. Photo by Charles F. Badland.

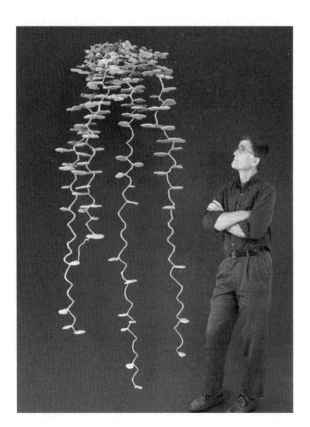

years in some species. Most social insects live below the ground, though some inhabit wood (especially termites and carpenter ants), which is closest to the lifestyles of their distant ancestors. Some nests, like the termite mounds in Africa and Australia, are earthen structures that rise 20 feet above the ground. Wasp queens may lay eggs in hexagonal cells constructed out of wax or of paper made from chewed wood. Certain ants live in trees, where they weave nests out of leaves with silk exuded by larvae, which adult workers hold and use like glue guns to stitch leaves into shelters.

Army ants of the Central and South American rain forests and African legionary ants send out fierce predatory columns, which roam the countryside and devour animal and plant life in their path. Some nomadic species carry their queen in a living fortress of ants borne aloft by the plundering stream of tiny predators. Other legionary species have a home base out of which swarms stream during the day to pillage, returning to the nest at night. Ant species will attack other colonies in order to defend and expand their territories.

Early observers described ants as if they were little people who carried out such human activities as warfare, slavery, animal domestication, and gardening. Scientists now try to stay away from such language, which is loaded with human values. It is clear, for instance, that ant "slavery" and "domestication" are very different phenomena than superficially similar human activities.

Social insect societies observed today evolved to their present range of complexity by about 50 million years ago—without major changes since then as far as we can tell—while we humans have been evolving complex societies for less than 20,000 years.

INSECTIVOROUS PLANTS (1875)
Darwin's "Disguised Animal"

> During the summer of 1860, I was surprised by finding how large a number of insects were caught by the leaves of the common sun-dew (*Drosera rotundifolia*) on a heath in Sussex. I had heard that insects were thus caught, but knew nothing further on the subject. I gathered by chance a dozen plants, bearing fifty-six fully expanded leaves, and on thirty-one of these dead insects or remnants of them adhered.

So begins Charles Darwin's 14th book, *Insectivorous Plants* (1875), the first scientific investigation into the behavior of carnivorous plants. Although his work in this area is sometimes neglected among his many achievements in natural history, Darwin was the first to prove that Venus flytraps, sundews, and pitcher plants are actually meat-eaters. Like so much of his research, this "fine new field for investigation" began with "idling" observations while "resting" from other projects.

Darwin soon developed a special affinity—even defensive affection—for these remarkable organisms. When American botanist Asa Gray suggested he might better occupy his time, Darwin wrote (1863): "Depend on it you are unjust on the merits of my beloved *Drosera*; it is a wonderful plant, or rather a most sagacious animal. I will stick up for *Drosera* to the day of my death." Venus flytraps, with their hinged, spiked leaves that snap closed on insect victims, Darwin thought "the most wonderful plant in the world." His wife, Emma, wrote a friend, "He is treating *Drosera* [the sundew plant] just like a living creature, and I suppose he hopes to end in proving it to be an animal." Emma was not exaggerating. "By Jove," he wrote to Joseph Hooker, director of the Royal Botanic Gardens, "I sometimes think *Drosera* is a disguised animal."

Darwin's was the first extensive physiological-behavioral study of insectivorous plants, covering 30 species belonging to eight genera. He wanted to get at three basic mechanisms: the plants' power of movement when capturing prey, the secretions of glands for digesting insects, and the absorption of the digested matter. His thousands of experiments resulted in the first definitive evidence—where before there had been only speculation—that these plants really were getting nutrition by capturing animal life.

No novice to plant movement (he had published his book *The Movements and Habits of Climbing Plants* [1865] some years before), Darwin found both the general sensitivity of

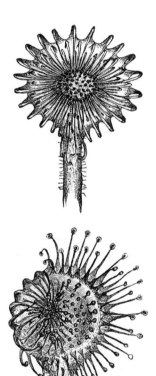

A ROUND-LEAVED SUNDEW plant (*Drosera rotundifolia*) closes its tentacles over a meat morsel in Charles Darwin's book *Insectivorous Plants*. "At the present moment," he wrote a friend, "I care more about *Drosera* than the origin of all the species in the world."

these plants and their finely coordinated movements to be fascinating. After all, they had neither muscles nor nervous system. "It appears to me that hardly any more remarkable fact than this has been observed in the vegetable kingdom," he marveled.

While admitting that the coordinated movement of plants was inferior to that of animals, Darwin envisioned them as a possible early stage in the evolution of animal nervous systems. "But," he says, "the greatest inferiority of all is the absence of a central organ, able to receive impressions from all points, to transmit their effect in any definite direction, to store them up and reproduce them." In other words, they lacked a brain!

Experiments on the adaptive significance of carnivory were assigned to Darwin's son Frank, who grew two groups of plants. One he fed insects; the other had to take its sole sustenance from the soil. The insect-fed plants were more vigorous and produced more flowers and seeds. "The results," Frank concluded, "show clearly enough that insectivorous plants derive great advantage from animal food."

Carnivory in plants is now recognized in eight families, 15 genera, and roughly 500 species among an approximate 422,000 total flowering plant species on earth. Thus, it is a very rare occurrence; why this is so is not understood, and the best that has been done is to correlate it with habitat. In all cases, insect-eating plants are found in moist, nutrient-poor, usually acidic soils, often in sunny locations.

Perhaps the most exotic habitats of carnivorous plants occur on the "Lost World" tepuis (tep-POO-eez) of Venezuela: high, flat-topped stone mountains or mesas, whose long-isolated plants and animals helped inspire Sir Arthur Conan Doyle's famous adventure tale. But they are also commonly found in north temperate peat and fen lands, the little-known hillside seep bogs of western Louisiana and east Texas, and the *Cephalotus* bogs of southwestern Australia—all of them characterized by a poor supply of nitrogen and other plant nutrients.

Sir John Lubbock's first wife, Lady Ellen Lubbock, upon reading *Insectivorous Plants*, composed a delightful poem and sent it to Darwin. It was purportedly written by insects who had "buzzed all over" his book and suddenly realized the horrific circumstances of how several of their friends had disappeared. Some sample verses:

(I never trusted Drosera,
Since I went there with a friend,
And saw its horrid tentacles
Beginning all to bend.

I flew away, but he was caught,
I saw him squeezed quite flat—
I don't go any more to plants
With habits such as that.)

We are very much obliged to you,
For now, of course, we shan't
Be taken in and done for
By any clever plant.

But this has to be considered:
It isn't much we need,
But if we daren't go to any plant,
On what are we to feed?

See also DARWIN, CHARLES; GRAY, ASA; HOOKER, SIR JOSEPH; *LOST WORLD*

INTELLIGENT DESIGN
Creationism's "Trojan Horse"

The "argument from design" was the widely accepted explanation of the natural world until around 1859, when Darwin's *Origin of Species* convinced scientists that life's complexity and diversity was best explained by evolution. In the early 21st century, however, this old fossil of intellectual history was resurrected and given a new lease on life by a small group of dedicated Christian fundamentalists, later joined by some orthodox Jews and Muslims. Reverend Barry W. Lynn, a Christian theologian who is director of Americans United for the Separation of Church and State, expressed an opposing view. "Intelligent design science," he wrote, "has as much to do with science as reality television has to do with reality."

In the late 18th century, the "argument from design" for the existence of God was famously promoted by the Reverend William Paley, an English theologian and popular writer on natural history. If we find a stone lying in a field, Paley wrote, we might assume that natural forces, acting blindly, produced it; but if we find a pocket watch, we immediately infer the existence of an intelligent watchmaker who designed it. Likewise, he reasoned, the complex forms of living things are "evidences" of a supernatural Creator.

Beginning in the 1990s, a few "intelligent design" proponents, some newly armed with doctorates in science, vowed to "topple Darwinism" from within the academic establishment. Microbiologist Michael Behe of Lehigh University and mathematician-philosopher William A. Dembsk of Southwestern Baptist Theological Seminary both insist that their version of neocreationism is not the same old Paleyism, but is based on new "discoveries" in mathematics, logic, and microbiology.

One of Behe's arguments is the "mousetrap" metaphor, which he proposed in 1996. At the subcellular level of molecular machines and chemical reaction pathways, he wrote, systems are "irreducibly complex," like mousetraps. Mousetraps have four working parts—a platform, a hammer, a spring, and a catch—each of which is essential for the trap to function. If any of these parts are missing, it cannot work, and so the system "must have been designed" with all parts in place. The argument ignores the fact that many so-called irreducible systems in single cells can be demonstrated to have evolved in other contexts, and are often appropriated, through evolution, for new functions.

Behe was outraged when his departmental colleagues at Lehigh publicly disavowed his views, stating: "It is our collective position that intelligent design has no basis in science, has not been tested experimentally, and should not be regarded as scientific."

Dembski believes he can always tell "empirically" whether or not a structure is the product of intelligent design. Lumping birds' plumage patterns or wing structures with the presidential portraits sculpted on Mount Rushmore, for instance, Dembski argues that they are inexplicable without invoking an intelligent designer. Boiled down, his "specified complexity" test comes right back to Paley's watchmaker. Yet he asserts that his criteria for detecting design should be accepted as a proven scientific procedure. In fact, there is no experimental ("empirical") basis for his assertion.

Intelligent design proponents claim not to be interested in who the "intelligent designer" might be, or in the mechanisms by which the "designs" are produced. They are only gathering evidence that some kind of conscious design has produced all living things. When pressed, they insist, somewhat disingenuously, that it does not matter to them whether the designer is an alien life form or a team of cosmic designers. In their writings outside of scientific journals, however, they are not reticent in asserting that the "designer" is the Judeo-Christian Creator god, as interpreted by evangelical sects.

These antievolutionists differ from their immediate predecessors, the fundamentalist, "young Earth" creationists, by their acceptance that some species do change—for example, that bacteria acquire resistance to antibiotics. They assert, however, that small changes cannot lead to new species. While they admit, for instance, that the beaks of Galápagos finches have been shown to evolve, they like to point out that "no finch was seen turning into a vulture or a duck!" Also, some proponents of intelligent design no longer believe that the Earth is only 6,000 years old, as the "young Earth" creationists do.

Historian and philosopher of science Barbara Forrest of Southeast Louisiana University and biologist Paul Gross of the University of Virginia have tracked the movement's history over the past 20 years, and have concluded that "its proponents invest most of their efforts in swaying politicians and the public, not in convincing the scientific community." Also, they

"This isn't really . . . a debate about science. It's about religion and philosophy."

—Phillip E. Johnson

"We seek nothing less than the overthrow of materialism and its damning cultural legacies."

—Mission statement, Discovery Institute

documented that pushing "intelligent design" into school science classes was part of the infamous "Wedge Strategy" propounded by Phillip E. Johnson, a Berkeley law professor who campaigned "to replace naturalistic science with 'theistic science.'"

Johnson launched the Wedge campaign with his book *Darwin on Trial* (1991), in which he treats evolutionary biology as an adversarial courtroom "theory" rather than a scientific edifice built on decades of experiment and observation. Johnson sought "nothing less than the overthrow of materialism and its damning cultural legacies."

Blaming secular or "atheistic naturalism" for all political and social ills of our time, Wedge strategists seek to introduce their belief in a supernatural Higher Being into the public schools—in Dembski's words, "defeating naturalism, and its consequences." In 1996, Johnson declared: "This isn't really, and never has been, a debate about science. It's about religion and philosophy."

Dembski, in an unguarded moment, has written that intelligent design "is just the Logos of John's Gospel restated in the idiom of information theory." His colleague Jonathan Wells, of the privately funded Discovery Institute, who is a member of Sun Myung Moon's Unification Church, admits that "Father Moon" encouraged him to earn Ph.D.'s in both religious studies and biology, specifically "to devote my life to destroying Darwinism." Behe, too, says he wants to "make room for religion" in science. At heart, proponents of intelligent design seek to change the definition of science itself. According to Dembski, any "systematic theory," including astrology, would qualify as science.

It is certainly reasonable—and Darwin expected it to happen one day—that his theory of evolution by natural selection would be superseded. But the old cosmic Watchmaker is not a testable scientific alternative. In calling their doctrines "science," neo-Paleyists are trying to play tennis without the inconvenience of a net. It can be done—but then you really can't call it tennis.

In an 1871 letter to Herbert Spencer, science publisher Edward L. Youmans reported that theologian and president of Princeton University James McCosh told an audience of concerned clergymen "not to worry, as whatever might be discovered [by science] he would find design in it and put God behind it."

See also CREATIONISM; FUNDAMENTALISM; MATERIALISM; PALEY'S WATCHMAKER; SCOPES TRIAL; SCOPES II

> "Whatever science may discover, I will find design, and put God behind it."
>
> —James McCosh, as reported by E. L. Youmans to Herbert Spencer, 1871

ISLAND LIFE

See WALLACE, ALFRED RUSSEL

ISOLATING MECHANISMS
Species Mate Identification

By their choice of mates (and rejection of others in closely related species), animals reveal that their breeding population is a real species, not something arbitrarily defined by the zoologist. Some of the most unusual, spectacular structures and behaviors in the animal world seem to function mainly for species recognition and sexual attraction. The riotous color patterns on the faces of African guenons (masked monkeys), the hundreds of different sounds and aerial "dances" of Hawaiian fruit flies, and the elaborate courtship rituals among birds of paradise help keep related breeding populations genetically distinct. Hence, they are known as "isolating mechanisms," a term coined by geneticist Theodosius Dobzhansky.

Such signals are not fail-safe, however. Hybrid crosses between varieties or species do occur. Some creatures even seem totally heedless of these genetic "traffic signals." Male dogs, for instance, are notoriously indiscriminate as to where they direct their procreative energies. And twice during the 1980s, male moose in New England attracted national media attention for their unrelenting attentions to dairy cows. Among plants, such as willows and grasses, there are many more instances of loose boundaries between species.

See also DOBZHANSKY, THEODOSIUS; GENETIC DRIFT; HYBRIDIZATION: SPECIES, CONCEPT OF

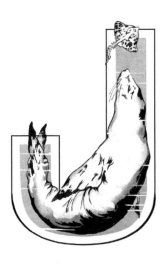

JARAMILLO EVENT
Evidence for Plate Tectonics

In 1966, earth scientists first identified the Jaramillo Event, an ancient reversal of the Earth's magnetic field that took place 900,000 years ago.

Recorded in rocks gathered at Jaramillo Creek in New Mexico's Jemez Mountains, it was only the latest discovery in a long sequence of known magnetic polarity changes. But Jaramillo, as it turned out, was the crucial bit of evidence that triggered a revolution in geology as profound as that of Einstein in physics or Darwin in biology.

At this point, American, Canadian, English, and French researchers realized there was a pattern to the sequence that exactly matched another they had seen: a recently compiled magnetic profile of the northern Pacific seafloor. It was like fitting in the one jigsaw puzzle piece that makes the whole picture understandable.

Geophysicist Fred Vine, quoted in William Glen's *The Road to Jaramillo* (1982), recalled that he "realized immediately that with the new time scale, the Juan de Fuca Ridge [in the Pacific, southwest of Vancouver Island] could be interpreted in terms of a constant [rate of seafloor] spreading"—the extension of the crust when molten magma rises through cracks between moving plates. "And that was fantastic," Vine continued, "because we realized that the record was more clearly written than we had anticipated. Now we had evidence of constant spreading."

Discovery of the Jaramillo Event, in conjunction with data gathered in the East Pacific two years earlier by the research vessel *Eltanin*, was the spark that set off a complete transformation in geology. If seafloors regularly expand and move, then "continental drift" was confirmed. As evolutionary theory had overthrown the fixity of species, so a century later plate tectonics overturned the ancient belief that continents and oceans were permanently locked into their present positions.

The Jaramillo/*Eltanin* pattern convinced many geologists that continent-sized slabs or plates of the Earth's crust are actually in constant motion. Since 1966, a rapidly growing body of research has documented that the Earth's crust is made up of about 25 huge plates, which open and close ocean basins, thrust up mountains where they collide (as in the Himalayas), and even submerge under each other to be melted back into magma.

Plate tectonics is the grand unifying theory that made sense of a great body of previously unconnected geological data. It explains, for instance, why the shapes of continents—such as the outlines of eastern South America and western Africa—seem to fit together. (They were once a single landmass, but later split and drifted apart.) Similarly, it explains why some continents, now widely separated in different climatic zones, contain identical kinds of rocks and fossil remains, such as the tropical *Lystrosaurus* found in India, South Africa, China, and Antarctica.

See also CONTINENTAL DRIFT; *LYSTROSAURUS* FAUNA; PEER REVIEW; PLATE TECTONICS

> Earth is made up of huge plates that create ocean basins, thrust up mountains, and sometimes melt back into magma.

JAVA MAN

See DUBOIS, EUGÈNE; *HOMO ERECTUS*

JEFFERSON, THOMAS (1743–1826)
Presidential Paleontologist

THE FIRST PALEONTOLOGIST PRESIDENT, Thomas Jefferson published descriptions of mastodon teeth and giant sloth's claws that were found on his Virginia estate.

Forty-nine Nobel Prize winners gathered in the White House on a spring evening in 1962 to be honored for their achievements in the arts and sciences. "I think this is the most extraordinary collection of human talent, of human knowledge, that has ever been gathered at the White House," said President John F. Kennedy, "with the possible exception of when Thomas Jefferson dined alone."

Jefferson was indeed a man of wide-ranging talents. He could design a graceful building, calculate an eclipse, tie an artery, argue a legal case, break a horse, survey an estate, dance a minuet, or give one of the first scientific descriptions of a fossil mastodon skeleton found in America.

In the late 1700s, fossil bones of elephantine proportions were discovered in various sites in Europe and America, creating a sensation among intellectuals. What could they mean? How old were they? Benjamin Franklin and George Washington both expressed interest in the massive tusks and grinding teeth, but were puzzled. Could these bones, as Quaker naturalist Peter Collinson thought, be mixed remains of elephants and hippopotamuses, perhaps drowned in the Flood?

News of the finds greatly interested Jefferson, who had just published a description of ancient elephant teeth dug up near his Virginia property in his *Notes on the State of Virginia* (1781–1782). Then living in Paris as America's ambassador, he boldly contradicted the great French naturalists Georges Buffon and Louis Daubenton, who thought the bones were mixed up from different species. Jefferson argued that they represented one animal, in some ways similar to an elephant, but also in some ways different. However, he stopped short of concluding it was extinct. The irrevocable disappearance of one of God's creatures was still too irreverent a conclusion, even to a free-thinking deist. "Such is the economy of nature," Jefferson wrote in his book *Notes,* "that no instance can be produced, of her having permitted any one race of her animals to become extinct; of her having formed any link in her great work so weak as to be broken."

Even while engaged in his many activities as a planter, politician, and statesman, Jefferson continued to seek additional fossils. In 1796, a year before taking office as vice president under John Adams, he acquired some remains turned up by workmen excavating saltpeter in a western Virginia cave. These bones proved that still another previously unknown giant animal had existed in the area—a huge, clawed creature quite unlike the elephantine monster. Jefferson thought it a kind of lion and named it *Megalonyx*, or great-claw, because of its huge hooked nails.

A few years later, Dr. Caspar Wistar and Jefferson jointly described the find in the *Philosophical Transactions of the American Philosophical Society*. Wistar correctly concluded that the animal was a giant sloth, similar to one that had been found in South America, and named the creature *Megalonyx jeffersonii*. Although Jefferson again chose to believe that the creature still had to be alive somewhere, extinction would soon become an unavoidable conclusion for him and all other fossil hunters.

Jefferson's fossils, among others, excited top anatomists. Similar finds inspired Georges Cuvier to tackle his comparative study of elephants, living and extinct (1796). The great French anatomist reviewed the evidence of fossil crocodiles, bears, rhinos, and deer and concluded that "they have belonged to creatures of a world anterior to ours, to creatures destroyed by some revolutions of our globe; beings whose place those which exist today have filled, perhaps to be themselves destroyed and replaced by others some day."

The craze was on for new digs, and farmlands across the country were eagerly dug up by many hands in search of "the great American incognitum," often with the interest and encouragement of then-President Jefferson. Finally, in 1801, Charles Willson Peale, curator of Peale's Museum in Philadelphia, successfully unearthed the first complete skeleton of a mastodon in New York State.

It was a patriotic triumph more than a scientific one. For years, American naturalists had smarted from the slurs of their European colleagues about the "degenerate" New World. Frenchmen claimed that in America, dogs lost their bark and men their virility. Buffon wrote that "all animals are smaller in North America than in Europe."

In reply, Jefferson had compiled a list of weights and measures of American bears, beavers, and otters—all larger than their European counterparts. He had even sent Buffon a huge stuffed American moose, but the Frenchman remained unimpressed. In 1803, his friend Peale crated up the complete, immense mastodon skeleton and took it to Europe for public exhibition. The awe-inspiring "mammoth" elephant came to symbolize American pride. Now let the European naturalists dare to call North America a place of small, second-rate inhabitants! President Jefferson was well pleased.

See also EXTINCTION; MAMMOTHS; PEALE'S MUSEUM; SECULAR HUMANISM

JOHANSON, DONALD (b. 1943)
Discoverer of "Lucy"

A Chicagoan born to Swedish immigrants, Donald Johanson was determined to become a professional scientist, probably a chemist. Then one day, while still in high school, he read an article by Louis Leakey in *National Geographic* that was to change his life forever.

Leakey's account of his search for early-man fossils at East Africa's Olduvai Gorge stirred Johanson's sense of adventure. "The name Olduvai, with its hollow, exotic sound," he later recalled, "rang in my head like a struck gong." And the bold quest of the ambitious, competitive, audacious Leakey also struck a chord. It was the beginning of Johanson's intense love-hate relationship with the Leakey family, which was to have a dramatic effect on paleoanthropology for the next several decades.

REMARKABLE FOSSILS catapulted young Donald Johanson (left) to fame before he had completed graduate school. He and Tim White (right) established their 3.2-million-year-old "Lucy" skeleton as a milestone in hominid evolution.

Johanson decided to major in anthropology at the University of Illinois and went on to graduate work at the University of Chicago. While still a student, he wangled his way onto an international fossil-hunting expedition to Ethiopia and within two years had become world famous for discovering the spectacular "Lucy" fossils—hailed as the "oldest and most complete skeleton of an upright-walking hominid." By the time he completed his Ph.D. and became a curator at the Cleveland Museum of Natural History, he had discovered another treasure trove of hominid fossils ("The First Family"), and had taken some of the limelight from the Leakey family's long dominance of African paleoanthropology.

Ironically, the Ethiopian expedition, originally led by French geologist Maurice Taieb and American Jon Kalb, had gained funding on Louis Leakey's recommendation, well before Johanson joined their team. And in his early seasons in Ethiopia, Johanson had sought and won the cooperation of Mary and Richard Leakey. But Johanson's published views and statements instigated an ever-widening rift with the first family of paleoanthropology. This split was partly caused by his insistence that his Lucy (*Australopithecus afarensis*) was a more ancient human ancestor than anything the Leakeys had found, his knack for grabbing press attention at conferences, and his compulsion to play scholarly one-upmanship with Richard Leakey, a past master of that game.

When Johanson announced the proposed new australopithecine species *afarensis*, he did not use "Lucy" as the type specimen, but instead identified it with a jaw Mary Leakey had found in East Africa—a fossil she considered not an australopithecine at all, but more likely a *Homo habilis*. Offended at the "appropriation" of her discovery by an upstart with whom she did not agree, Mary made her displeasure with Johanson known.

With his talent for fossil finding, fund-raising, and political infighting, Johanson became codirector of the Ethiopian project, along with Yves Coppen and Maurice Taieb, and emerged as a rival of Richard Leakey, at least in the popular press. In 1979, he announced his own family tree of the hominids, on which Lucy was ancestral both to other australopithecines and to the human line.

The Leakeys stuck to the beliefs of Louis (who had recently died) that *Homo habilis* was an "oldest" human ancestor and that australopithecines were merely side branches. When more fossils were discovered, Richard Leakey argued, they would show a distinct lineage leading through his *Homo habilis* to *Homo sapiens*. But beneath the scientific disputes was another question: Who would be king of the hill?

Press accounts, including major articles in *Life* and the *New York Times*, eagerly seized on the growing rivalry. Leakey tried to avoid responding to Johanson's challenges, instead promoting his own reading of the fossils, by this time embedded in his own book *Origins* (1977) and his "Making of Mankind" television series.

In 1981, matters came to a head before millions of people in a televised confrontation that had been suggested by Richard Leakey. Just before they were to begin filming, Leakey told both Johanson and newscaster Walter Cronkite that he did not want to debate their differences after all. Because of increasing creationist attacks on evolution, he said, it was more important to give the public some scientific education.

It was not to be. No sooner were the cameras rolling, than Cronkite asked each man for his view on what the fossils meant. Johanson at once pulled out a prepared chart of a "family tree" with Lucy as its matriarch and humankind evolving from the little australopithecine, with Leakey's *Homo habilis* emerging much later as an ancestral *Homo*.

On national televison, Richard Leakey drew a big "X" over Johanson's chart of human evolution, replaced it with a question mark, and then stormed out of the studio.

When asked for his own view, Leakey protested that he had brought no artwork, so he was handed a marker and asked how he would modify Johanson's chart. With a nervous chuckle, Leakey drew a big bold "X" through Johanson's entire scheme. "And what would you replace it with?" he was asked. Leakey then drew a large question mark. A few minutes later, he stormed angrily out of the building.

"When next I saw Richard," says Johanson in *Lucy's Child* (1989), "he looked right past me. That was four years ago. We haven't spoken since. In public, he has continued to insist that our 'rivalry' is largely the media's creation, and that beneath the hype there lie nothing more than minor professional disagreements. I wish it were that painless."

In 1987, Johanson and his colleague Tim White sought fossils at Olduvai Gorge in Tanzania and found a fragmentary skeleton they identified as a female *Homo habilis* (OH65), estimated at 1.8 million years old. Seeing resemblances between this new hominid and his famed *Australopithecus afarensis* from Ethiopia, Johanson nicknamed the find "Lucy's child" and made her the subject of his 1989 book. He argued that this Olduvai female was descended from the older *Australopithecus afarensis* population, and that both were direct ancestors of the genus *Homo*.

In 1981, Johanson founded the Institute of Human Origins, now at Arizona State University, of which he remains director. The institute has served as a clearinghouse for information on fossil hominids and a launching base for many fossil-hunting expeditions, particularly in Ethiopia.

See also AFAR HOMINIDS; "LUCY"

JURASSIC PARK (1993)
Spielberg's Cinematic Saurians

For millions of film buffs, the definitive view of how dinosaurs looked, moved, and behaved is based on Steven Spielberg's 1993 movie *Jurassic Park*. In the film, John Hammond (Richard Attenborough), a well-meaning but arrogant entrepreneur, funds the cloning of live dinosaurs from ancient DNA for his island amusement park. Just prior to the park's opening, however, the dinosaurs escape their electrified paddocks. Visiting scientists and Hammond's young grandchildren come terrifyingly close to becoming saurian snacks. Spielberg and author Michael Crichton moralize that scientists risk disaster if they tamper with the elemental forces of nature, especially if the motivation is greed—a theme common to *Faust*, *Frankenstein*, and the 1933 cinematic masterpiece *King Kong*. The giant gorilla, like *Jurassic Park*'s dinosaurs, was exhibited by an arrogant impresario to the ticket-buying public.

Crichton wrote *Jurassic Park* as his seventh novel. In seeking a plausible scenario for reviving the dinosaurs, he seized upon recent advances in biotechnology. In the 1990s, scientists learned how to amplify DNA from a tiny sample, and also found some in very ancient bones [see MULLIS, KARY B.; POLYMERASE CHAIN REACTION]. Where would one find dinosaur DNA? Crichton came up with the idea of dinosaur blood preserved in the gut of an ancient mosquito trapped in amber. Perhaps the dinosaurs could be reconstituted from their DNA.

If such a technological feat were possible, Crichton wondered, who would pay for it? The cost would be too much for research labs, but he thought that the funds would most likely come from the entertainment industry. In *The Making of "Jurassic Park"* (1993), he is quoted as saying, "Commercialization of genetic engineering [is] a very serious problem, and one that we are still not facing. . . . All this amazing technology is being used for essentially commercial and frivolous purposes."

Ironically, the film itself is a triumph of technology that was created to entertain. David Koepp, who wrote the final version of the screenplay, mused in the same book,

> Here I was writing about these greedy people who are creating a fabulous theme park just so they can exploit all these dinosaurs and make silly little films and sell stupid plastic plates and things. And I'm writing it for a company that's eventually going to put this in their theme parks and make these silly little films and sell stupid plastic plates. I was really chasing my tail there for a while trying to figure out who was virtuous in this whole scenario—and eventually gave up.

Originally, the dinosaurs were to be created by a combination of puppets operated by cables and hydraulics, stop-motion miniature models, and a few enormous, full-size robots. Computer generated imaging (CGI) was in its infancy, and a few of the artists in the project were experimenting with it on their own. When they showed their test footage of a walking dinosaur to Spielberg's team, they agreed it seemed more real than anything the best stop-motion animators, like Ray Harryhausen or Phil Tippett, had ever been able to produce.

Spielberg soon realized that his team was on to something extraordinary, and decided to switch from his models and giant puppets to computer graphics for the major dinosaur scenes. That decision would revolutionize movies, sparking a new generation of special effects and "virtual" characters.

Jurassic Park also changed the public's ideas about dinosaur behavior, as it drew heavily on paleontological research of the 1970s and 1980s. "The new view," wrote paleontologist John R. (Jack) Horner, who served as a consultant on the movie, was that dinosaurs were not sluggish cold-blooded reptiles, but "were avian in the ways that mattered most, anatomically and metabolically, more akin to chickens and ostriches than geckos and monitor lizards. If the films had been made as little as fifteen years earlier, many scenes, especially those dealing with eggs, embryos, and nests; parental care; and herd behavior, would not have been possible."

Tyrannosaurus rex had stood erect for a century in museum displays on its hind legs, but Horner pointed out that those reconstructions were now thought to be incorrect. *T. rex* was balanced horizontally across its legs like a roadrunner, in Horner's view. It was a long-

VELOCIRAPTORS ARE CUNNING creatures (top) who hunt cooperatively in *Jurassic Park*. Paleontologist consultants provided current scientific interpretations of the behavior of *T. rex* (bottom) and other dinosaurs.

distance walker, not a runner, and its tiny arms would not have been much use in grabbing prey. Although its eyes were small, implying poor vision, the olfactory lobe of its brain, where odors were processed, was huge—similar to that of vultures, which can detect the scent of rotting flesh from miles away.

The film inspired two sequels, *Jurassic Park: The Lost World* (1997) and *Jurassic Park III* (2001). Of course, the former title was in homage to the granddaddy of all dinosaur adventure yarns, *The Lost World,* published in 1912 by Sir Arthur Conan Doyle and made into a silent film in 1925. That film also introduced a new technology—stop-motion animation—which went on to dominate cinematic special effects for 70 years.

See also KING KONG; LOST WORLD; O'BRIEN, WILLIS

"JUST–SO" STORIES
Fanciful Evolutionary Explanations

Rudyard Kipling, author of *The Jungle Book* and *The Man Who Would Be King*, and chronicler of the British Raj in India, is also famous for a series of charming children's tales about the origins of animals. The *"Just–So" Stories* (1902) are fanciful explanations of how the elephant got his trunk (a crocodile bit it and stretched it) and how the rhinoceros got his loose skin (he took it off on a hot day and it never fit correctly afterward). Modeled on folktales, they express humor, morality, and whimsy in "explaining" how various animals gained their special characteristics.

For more than a century after Darwin, scientific journal articles were full of "just-so" stories, speculations, and imaginative "explanations" about structures that had evolved "for the good of the species." Armchair biologists would construct logical, plausible explanations of why a structure benefited a species or how it had been of value in earlier stages. Unfortunately, many of these clever conjectures were unsupported by evidence or proof.

Some of the most humorous, egregious examples of "just-so" stories can be found in Harold O. Whitnall's children's book *A Parade of Ancient Animals* (1936). Here, for instance, is his account of how birds evolved from dinosaurs:

> We can imagine this agile little reptile, about the size of a crow, running over the ground on two feet faster than any animals had ever run before. For many long years, he was the champion racer. Now and then he leaped to a low-lying branch of a tree and caught it with his clawed forefeet. . . . All this running around and leaping did queer things to the long scales that covered their skins. The pressure of the air as they rushed about began to fray the scales into thin shreds. The same thing would happen to our clothing if we stood for a long time in front of an electric fan . . . our clothes would be frayed and fuzzy all over.

Another gem from Whitnall's book is his "explanation" for why the largest known extinct mammal, *Baluchitherium* (now known as *Indricotherium*), ultimately become extinct. The immense animal, he wrote, "did not live on the earth long, for his unwieldy body kept him from moving fast. That meant that when new pastures had to be found, he was the last to arrive and the 'cup-board was bare,' because the swifter ones had come before him. He starved to death because he was always late to his meals."

To be charitable, one wants to assume that Whitnall, a professional geologist, was just joking while attempting to stimulate children's imaginations, or perhaps was satirizing the "just-so" stories his colleagues were proposing in earnest. If parody or whimsy were not his aims, however, he produced the worst book on evolution ever written.

Stephen Jay Gould pointed out three basic fallacies in the attempt to reason out "logical" explanations for structures in organisms based on a vague "survival value." First, it is "hyperselectionist"—seeking adaptation in every minute structure, when some traits could well be neutral. Second, plausible logic is no substitute for gathering sufficient supporting evidence. And finally, reconstructions cannot be made according to supposed "stages" of progression. Each organism is the product of a special, unique history, and we usually have no way of knowing how early adaptations meshed with past environments.

See also ADAPTATION; NEUTRAL TRAITS

> **The largest land mammal "starved to death because he was always late to his meals."**
>
> —Harold O. Whitnall,
> *A Parade of Ancient Animals*, 1936

KAMMERER, PAUL (1880–1920)
Controversial Evolutionary Biologist

n 1926, shortly after his research on Lamarckian inheritance was denounced as a fraud, Austrian biologist Paul Kammerer committed suicide. To this day, scientists speculate on his tragic life and argue about the integrity of his evolutionary experiments.

Kammerer is most often dismissed as a neo-Lamarckian who believed he had proved that acquired characteristics could be genetically passed to offspring. In fact, for most of his career he was a Mendelian who could find no satisfying explanations for his unusual experimental results. If anything, he thought they demonstrated not novelty but atavism (reversion or throwback), an expression of existing genetic characters that had only been "suppressed" by the animal's mode of life.

Before World War I, he conducted experiments with salamanders, newts, and an amphibian known as the midwife toad. Salamanders raised on different colored gravels changed their colors and breeding habits and passed the changes on to their offspring. Blind newts, bred in infrared light, "redeveloped" eyes. And male midwife toads regained the "nuptial pads" on their thumbs with which they grasp females when mating in water. Male midwife toads do not normally have the special thumb pads that other pond-spawning toads use for hooking onto swimming females. But when several generations are forced to live and breed in water, Kammerer claimed, the males (and their offspring) develop the pads.

After the war, Kammerer traveled to England and America to continue his work, but was met with cold hostility by Mendelian geneticists. They could not reproduce his results, and Oxford's Gregory Bateson accused him of incompetence or fraud.

Critics examined some of his pickled toads and announced that their nuptial pads were more prominent because they had been injected with India ink. Kammerer protested that he had not tried to "fake" anything, but Bateson cried "hoax," the experiments were discredited, and the depressed biologist killed himself a few months later.

Kammerer's story was the focus of *The Case of the Midwife Toad* (1971) by Arthur Koestler, who suggests it was not Kammerer, but an overzealous assistant who attempted to restore faded markings by inking the 15-year-old specimens. Others have blamed an insanely jealous rival (there was one) or Nazi saboteurs displeased with Kammerer's politics.

In the 1920s, his popular lectures about evolving ourselves into a vastly superior species caused a sensation. According to Koestler, when many German, British, and American scientists were using Mendelism to justify racist eugenics—preventing "inferior" groups from breeding—Kammerer championed a more benign program. Now that he had proven the reality of Lamarckian inheritance, he believed, all peoples could make evolutionary progress by improving their social environments. Achievement of "supermen" was possible without culling the "unfit." His optimistic, egalitarian program for planned future evolution made front-page news.

Such political statements did not sit well with the Nazis, but Soviet scientists welcomed him with open arms. His views were compatible with socialist ideals of progress for the masses. Soon they would cite his work to buttress the disastrous pseudoscience of Lysenkoism. Kammerer accepted a research position in Moscow, but died shortly before he was to begin.

According to Koestler's version, Kammerer was simply hounded to death: a victim of the narrow-minded scientific establishment. His work could not be replicated because few had his remarkable flair for breeding amphibians under artificial conditions. Other supporters insisted his earlier body of work could not all have been faked; it had been examined and praised by leading biologists. Yet geneticists ceased quoting any of his experiments, and most were never reattempted.

Historians of science generally consider Koestler's account of the troubled biologist unreliable. Some even claimed that Kammerer's suicide was triggered by an unhappy love affair, rather than by acrimonious scientific controversies.

To this day, the debates and uncertainties continue. Historians and scientists agree on one fact about Paul Kammerer: He was one of the most ambiguous, enigmatic, and tragic figures in the history of biology.

See also LYSENKOISM; NEO-LAMARCKISM

KANZI
The Brilliant Bonobo

Shortly after his birth on October 28, 1980, at the Georgia State University Language Research Center field station at Lawrenceville, Georgia, a bonobo ape named Kanzi was taken from his apathetic mother by another adult female. A few weeks later, Kanzi and his adoptive mother, Matata, were brought to the university's Language Research Center in Atlanta, to begin experiments that would change science's evaluation of ape language capabilities. Sue Savage-Rumbaugh and her colleagues were trying to teach Matata single-word symbols on a computer keyboard. Although Matata appeared to be a bright ape, she made little progress in learning the lexigrams.

After two years of training and little to show for it, Matata was transferred to a nearby field station to breed again, and the following day, Savage-Rumbaugh set up the keyboard to begin training Kanzi. To her amazement, he correctly used many of the symbols immediately. On that first morning in 1982, he pressed the symbols for "apple" and "chase," then picked up an apple and ran away, grinning. Could he really have learned the lexigrams by attentive observation of his mother's daily sessions—the same way human children learn language? A series of controlled experiments proved that he had indeed.

Kanzi had more shocks in store for his human companions. As they continued to work with the pygmy chimp, they began to suspect that he could understand spoken English as well as printed symbols. In the past, some scientists had claimed that their apes could understand spoken words, but they were subsequently proven wrong by controlled experiments. When the exchange of subtle glances between apes and humans was excluded, the apes could not give correct responses. [See CLEVER HANS PHENOMENON.]

Kanzi, however, seemed to listen to the researchers' conversations and understand words he had not been taught. When he overheard humans utter the word "light," for instance, he ran to the light switch and flipped it on and off.

At first, Savage-Rumbaugh didn't believe Kanzi's level of comprehension was possible, but a series of tests soon had her convinced. By speaking its name or showing Kanzi its lexigram, Savage-Rumbaugh asked him to select photographs of various objects. He had not been trained for the test and did not receive rewards for correct answers. Nevertheless, Kanzi correctly identified 95 percent of the objects in the lexigram trials; in the spoken trials, he scored 93 percent. Researchers concluded that Kanzi understood 150 spoken words.

Critics were quick to attack their results, and most colleagues dismissed Kan-

KANZI THE BONOBO excels at learning language and symbols. When researchers tried to teach him to chip stones by hand, he preferred producing sharp flakes by throwing rocks against a concrete floor.

zi's purported abilities as impossible. Savage-Rumbaugh took the skepticism in stride. "Science is not about doing things people will believe," she wrote in *The Ape at the Brink of the Human Mind* (1994). "It must explore the phenomena that are out there, believable or not." Today Kanzi uses over 200 words to communicate, and comprehends around 500; and he is not an anomaly. Savage-Rumbaugh has since shown that other bonobos and even common chimps can learn to use lexigrams spontaneously, after observing others who can.

Paleoanthropologist Nick Toth of Indiana University at Bloomington began working with Kanzi to investigate the bonobo's ability to make and use stone tools. Could Kanzi learn to produce simple stone knives, as australopithecines had done 2.5 million years ago?

Toth showed Kanzi how to strike two rocks together to chip off sharp flakes, and then use the sharp edges to cut the cord on a box containing tasty treats. Kanzi quickly learned how to use flakes to cut the twine, but never adopted the human method of flint-knapping. Kanzi did learn to bang one stone against another to make small flakes, but decided it was much easier to shatter a whole rock and then choose a large, sharp fragment. Suddenly standing upright, he would forcefully fling a rock onto the concrete floor of his enclosure, squealing with delight as he produced a shower of chips.

Toth tried covering the hard floor with carpeting so that Kanzi's hurled rocks would not fracture. Would Kanzi now be compelled to take a stone in each hand and strike them together? Not if he could help it. After examining the floor carefully, Kanzi found the seam in the carpet, pulled it up to expose the concrete, and then threw stones against the bare floor as before.

Toth was frustrated, but Sue Savage-Rumbaugh was delighted. Not only had Kanzi managed to fracture his stones, but he had thought out an ingenious way around the carpeting problem. When warmer weather arrived, Kanzi was taken outdoors to continue his tool-making sessions. He soon discovered that rocks would not shatter on the soft earth any better than on the carpet. So Kanzi placed a large rock on the ground as an anvil, stood back, and threw another directly at it with unerring aim, producing a shower of flakes as he had done before. In contrast to Abang, the orangutan at the Bristol Zoo who had been trained to make tools by a rote, step-by-step process, Kanzi's solutions were instantaneous, conceptual, and original.

> **Kanzi's solutions to problems were instantaneous, conceptual, and original.**

See also ABANG; APE LANGUAGE CONTROVERSY; BONOBOS; CHIMPANZEES; KOKO

KELVIN, LORD (WILLIAM THOMSON) (1824–1907)
Champion of a Young Earth

William Thomson, Lord Kelvin, codiscovered the second law of thermodynamics, and his Kelvin temperature scale is still the standard in physics labs. But evolutionists remember the great 19th-century physicist chiefly for an arrogant blunder, which Charles Darwin called "one of my sorest troubles."

Kelvin believed that not enough time had elapsed since Earth's formation for evolution by natural selection to work. The immense time spans Charles Lyell and Charles Darwin imagined as the canvas of evolutionary creativity by slow, gradual degrees did not exist, said Kelvin. Earth was simply too young, and he could prove it with physics and mathematics.

In 1866, Lord Kelvin published a very short paper titled "The 'Doctrine of Uniformity' in Geology Briefly Refuted." The Earth began as a molten body and has been steadily losing heat, he argued. Interior temperature measurements (from mines) could establish the rate of heat loss from Earth's crust. Then one could reason back to the time when the Earth was hot enough to be molten, which would give the age of the planet. In a series of elegant calculations, Kelvin then assigned an age of about 100 million years to the Earth, which he later whittled down to 20 million. (Today we believe the earth is 4.5 billion years old.)

Kelvin's assumptions were that heat dissipates at a constant rate, that the Earth began as a molten entity, that its composition is fairly uniform, and that no renewable sources of heat exist in the Earth's crust. This last assumption ultimately proved to be his undoing, when decades later it was found that radioactivity in the crust constantly generates new heat. Rate of heat loss, therefore, became meaningless as a measure of age.

For 40 years, Kelvin campaigned for a young Earth and brought the prestige of the "hard" sciences to bear on geologists' thinking. To some, his "proof" presented an insuperable difficulty to evolutionary ideas. But most, including Thomas Huxley and Alfred Russel Wallace, accepted Kelvin's verdict on the age of the Earth and tried to cram all of evolutionary history into it, arguing that evolution must proceed at a quicker pace than Darwin thought.

In 1903, Pierre Curie discovered that radium salts constantly emit heat, and the following year Ernest Rutherford presented a paper on radium that shattered Kelvin's "proof" forever. Rutherford later recalled that he hesitated when he spotted Kelvin in the audience.

> To my relief, Kelvin fell fast asleep, but as I came to the important point, I saw the old bird sit up, open an eye and cock a baleful glance at me! Then a sudden inspiration came, and I said Lord Kelvin had limited the age of the earth, provided no new source of heat was discovered. That prophetic utterance refers to what we are now considering tonight, radium!

The discovery of radioactivity not only destroyed Kelvin's argument. "It also," wrote Stephen Jay Gould, "provided the clock that could then measure the earth's age and proclaim it ancient after all!" The constant rate of radioactive decay proved to be an excellent measure of elapsed time, giving dates going back billions of years. Darwin's hunch about the great age of the Earth, based on his broad knowledge of geology, fossils, and biology, was finally vindicated by the physics and mathematics that had so worried him.

See also FOUR THOUSAND AND FOUR B.C.; USSHER-LIGHTFOOT CHRONOLOGY

KENNEWICK MAN
Early American Ancestor—But Whose?

The bizarre tug-of-war over the bones of Kennewick Man began on July 28, 1996, when county police were called to investigate a possible homicide on the banks of the Columbia River, near Kennewick, Washington. Two student boaters had found a human skull and skeleton, which the coroner's office sent to anthropologist James Chatters, a local consultant on forensics, for identification. After noting that the facial structure resembled a "Caucasoid" individual (he thought perhaps it was an early white settler), Chatters had the bones radiographed.

He was in for a surprise. The radiograph revealed there was a broken spear point buried in the man's hip, which was not visible from the surface. Chatters then became excited about the find and had the skeleton tested for age. Radiocarbon analysis revealed that the dead man had last drawn breath about 9,000 years before Columbus's arrival in America. What's more, the skeleton turned out to be one of the oldest and most complete Late Pleistocene human remains ever found in North America.

Working with sculptor Tom McCleland, Chatters spent three months reconstructing the face of Kennewick Man, who was between 45 and 50 years old when he died. When the clay sculpture of the head was unveiled to the world, it closely resembled the British actor Patrick Stewart, who played a futuristic starship commander on the popular television series *Star Trek: The Next Generation*. Chatters also reported that the skeleton showed evidence of injuries that may have caused chronic pain: The man's chest had once been crushed, and he had lived for years with a stone spear point lodged in his pelvis. He may have died from repeated infections of the festering wound.

A bitter battle ensued between archeologists, the U.S. Government, and the Umatilla Tribe of northeastern Oregon and other ethnic tribes. The Army Corps of Engineers, which has jurisdiction over the Kennewick site, wanted to turn the skeleton over to the Native American tribe for reburial, under the 1990 Native American Graves Protection and Repatriation Act. Anthropologists had raided Indian graves for over a century, and the tribes wanted to rebury the "Ancient One" among their own sacred dead. As the Pawnee historian Walter Echo-Hawk put it in 1990, "Desecrate a white grave, and you get jail. Desecrate an Indian grave, and you get a Ph.D."

Eight prominent anthropologists sued the Army Corps of Engineers to hand over the bones for scientific study. They argued that this individual could not be considered an ancestral Umatilla because Amerindians did not inhabit the area 9,000 years ago; archeologists place them there no earlier than 5,000 years ago. Umatilla spokesmen replied that they did not believe the archeologists' version of prehistory, nor that their ancestors migrated from Asia to North America over a land bridge. According to their traditional beliefs, Indian peoples have occupied North America "since the beginning of time." Besides, according to the repatriation law, any human remains over 500 years old were automatically assumed to be Native Americans. Archeologists contended that reburying the bones would destroy crucial evidence about the peopling of America.

Within months, four other tribes joined to claim return of the bones under the Graves Protection and Repatriation Act. The U.S. Army Corps of Engineers locked up the bones pending the outcome of their lawsuit. In the midst of the dispute, an organization of self-proclaimed Vikings claimed Kennewick as their ancestor; the bones were proof, they argued, that Scandinavians peopled America first. They held religious ceremonies for Kennewick Man, raising cups of mead to Odin and Thor. Finally, a Samoan chief also vied for custodial kinship to the Ancient One, whom he also claimed as an ancestor.

Chatters said he had no idea his characterization of "Caucasoid" features would provoke such highly charged reactions. "Racial" descriptions of bones are impressionistic at best, particularly when applied to an individual who lived almost ten thousand years ago. Why should an individual from that period be recognizable as belonging to any ethnic group alive today? And with modern understanding of genetics, the 19th-century typological approach to racial features has been abandoned by science for at least 50 years.

In 2004, a Federal Court finally ruled that any skeletal remains found in the United States that are over 3,000 years old cannot be claimed as a direct ancestor by any living ethnic group—a victory for the scientists. "The Indian activists see it as another defeat at the hands of the white man," says Chatters. "They simply cannot stand to lose the case—and refuse to see it as a gain in knowledge for everyone. We may have won this battle, but they will persist, and ultimately, I think, they will win the 'war.' Many bones and artifacts crucial to reconstruct the history of this continent will be destroyed."

For over a decade, Kennewick Man's remains have been claimed by contending factions. About the only person who hasn't claimed the Ancient One as his ancestor is Patrick Stewart.

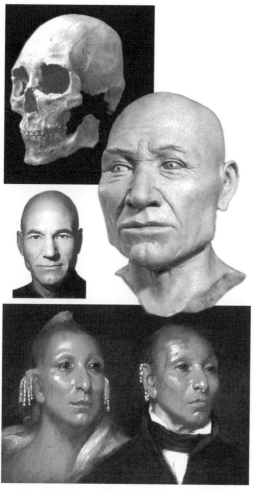

FACIAL RECONSTRUCTION (middle) of the Kennewick Man's skull (top) seemed to resemble actor Patrick Stewart (middle left) of *Star Trek: The Next Generation* fame. A Native American critic pointed out a striking resemblance to the 19th-century Sauk chief Black Hawk and his son, Whirling Thunder (bottom).

KEYSTONE SPECIES
Shapers of Environment

Some species are not merely adapted to environments—they create optimum living conditions for themselves and others. One of the productive results of James Lovelock's Gaia hypothesis—that life helps create the conditions for its own survival and evolution—has been to stimulate new research into how animals and plants shape the planet rather than simply occupy it.

The keystone species concept has been a mainstay of ecology and conservation biology since its introduction by University of Washington zoology professor Robert T. Paine in 1969. A keystone species is one whose impact on its ecosystem is much greater than would be expected from its relative abundance. Working with the coastal ecology of Washington State, Paine removed thousands of a single predator species, the *Pisaster ochraceus* starfish, over three years, and studied the resultant changes.

As Paine had suspected, the *Pisaster* starfish is the keystone that stabilizes the intertidal ecosystem. Other species that play similar roles include the freshwater bass and the sea otter. The return of the otter to southern California, where it had been extirpated, has resulted in restoring the giant kelp beds. Other species may have impacts that are large and important but not disproportionate to their total abundance, and therefore they are not considered keystone species.

During the 1980s, marine ecologist Mark D. Bertness of Brown University demonstrated the environmental impact of a small saltwater snail on the Atlantic coast of the northeastern United States. Bertness noticed that the advance of the periwinkle snail, which was introduced from Europe a century ago, seemed to be correlated with the conversion of coastal marshes to rocky beaches, drastically changing the living community. He wondered whether the little snail was a keystone species, and how it could function as one.

Coastal marshes support distinctive plants and the creatures that use them for food and shelter. But the marsh grasses need to be rooted in mud and silt, which adheres to algae-covered rocks. Periwinkle snails, which feed on algae, keep beach rocks clean and smooth. Mud and algae layers cannot adhere or build up, and the marsh habitat disappears. When the beach consists only of cleaned rocks and pebbles, a different set of organisms thrives: hermit crabs, for instance, often move into the empty shells of periwinkle snails.

To test his theory, Bertness performed a simple but ingenious experiment. On one rocky shore at Bristol, Rhode Island, he constructed eight bottomless cages, each enclosing several square feet. All periwinkles were picked out of the caged areas. After several months, within the test cages, algae formed on the rocks, silt clung to it, and marsh plants took root. A small marsh habitat developed in each of the cages, distinct from the surrounding area. Eventually, mud worms, fiddler crabs, and mussels thrived there.

Perhaps the most spectacular kinds of keystone species are the modifiers that can actively change their environments. The North American beaver, for instance, dams streams to create ponds and lakes. These new habitats enrich biodiversity by providing niches for entire communities of plants and animals.

In East Africa, decimation of elephant herds by ivory poachers has had profound and unexpected effects on many organisms. Important modifiers of environments, elephants transform their habitats. Herds clear thousands of acres of trees and underbrush, creating new grasslands. During droughts elephants dig waterholes, supporting many other species.

Knowledge of keystone species is crucial for understanding how to preserve habitat and perhaps even to restore some that have been destroyed. However popular and reassuring it may be, the idea that all nature is in balance is incorrect. Many ecologists now view nature as unstable and in flux. Extinction of some species may cause little harm to the survivors, while others are absolutely crucial to the web of life. Ecologists are trying to discover which particular microorganisms, for instance, may be essential for renewing the air and water upon which all life depends.

See also CORAL REEFS; ECOLOGY; GAIA HYPOTHESIS

> **Extinction of some species may cause little harm to survivors, while other species are absolutely crucial to the web of life.**

KIDD, BENJAMIN (1858–1916)
Survival Value of Religion

For many social thinkers during the late 19th century, Darwinism was linked with optimism about social progress. Society, they thought, was evolving toward a more rational, progressive way of life. So when Benjamin Kidd, in his book *Social Evolution* (1894), concluded that natural selection actually favors the preservation of "nonrational" institutions—such as religion—he provoked a critical storm.

Kidd's controversial idea was that the nonrational factor of religious belief had been immensely important in the evolutionary survival of societies. Reason and logic, he thought, were socially disintegrating forces, since each individual would act and plan for his own self-interest. Such mystical or "nonrational" beliefs as reward in an afterlife for selfless acts, the efficacy of prayer and ritual, and eternal punishment for immorality were the underpinnings of social solidarity and efficiency, and thus an adaptive trait.

Darwinians had insisted morality did not need to be backed by religious belief and that the evolution of life was toward ever-increasing rationality. Kidd's book provoked wounded cries from secularist scientists, who accused him of reviving religion in a new guise, whereby belief in a deity, even if false, promotes group survival and overrides individual selfishness. Kidd's point was that religion had been naturally selected as essential to the survival of societies.

See also HUXLEY, ALDOUS; SECULAR HUMANISM; SOCIAL DARWINISM

KILLER APE THEORY
Our Violent Heritage?

Not in innocence, and not in Asia, was mankind born," begins the best-seller *African Genesis* (1961) by dramatist Robert Ardrey. We evolved in Africa "on a sky-swept savannah glowing with menace," he wrote, where "man was born with a weapon in his hand," descendant of a "killer ape" who hunted for a living and often turned murderous weapons upon his own kind.

On Ardrey's "personal odyssey into the human past," he visited Africa, where he met anatomist Raymond Dart and paleontologist Robert Broom—scientific mavericks who had been finding the world's oldest hominid remains. They were snubbed by their European colleagues, who looked to Asia as humankind's probable birthplace.

Ardrey was fascinated by Dart's theory that man was descended from a carnivorous African ape. In 1953 Dart had published a paper called "The Predatory Transition from Ape to Man," in which he argued that a line of African apes branched off from fruit-eating cousins to become bipedal carnivores. Hands were freed to fashion weapons, and hunting led to success. The new human, Dart thought, soon began turning his weapons on his own kind. Those with the better weapons, sharpest minds, and strongest stomachs prospered by killing their rivals—and thus man was born with the mark of Cain upon him.

© 2004 www.mattcioff.com.

No doubt, the lingering popular fantasies of Hollywood's *King Kong* (1933) and the killer ape of Edgar Allan Poe's "Murders in the Rue Morgue" (1841) lurked somewhere in the background. From famed 19th-century gorilla hunter Paul du Chaillu to P. T. Barnum and his Gargantua, the great apes were always depicted as fearsome killers, often with the implication that they also carried off women for interspecies sex.

An antelope horn, a piece of long bone, a chipped rock—all were imagined by Dart as weapons, used for murder and warfare as well as hunting. Years later, South African paleontologist C. K. Brain reexamined the hominid and animal bone, but concluded that the holes and punctures in the australopithecine remains were the work of leopards and hyenas.

The killer ape captured the academic as well as the public imagination, perhaps offering an escape from guilt. If man was born aggressive, it is part of our biological nature, and we must accept killing and warfare as a fact of life. Soon, the pendulum of anthropological fashion swung the other way, and a rash of books were produced on such phenomena as the harmless, peaceful Tasaday forest tribe (which later turned out to be a hoax) and cooperative social behavior in animals. A few years later, the women's movement would influence new interpretations of primates and human evolution; women primatologists began to study maternal behavior in langurs and rank among females in baboon troops. In prehistoric scenarios, man the hunter was joined by woman the gatherer.

In the years since Ardrey's book was published, many early hominid fossils have been discovered, and the vast majority lack evidence of violent deaths. Meanwhile, George Schaller's decades of fieldwork among such true predators as lions, leopards, and tigers show that humans are, by comparison, a very peaceful species. In the United States, for example, there are about 10 killings for every 100,000 people. A field observer would have to watch 10,000 people for a full year in order to witness a single murder. If human beings were as violent as lions are among themselves, there would be a dozen or two slayings every year on every city block!

Richard Leakey, who has viewed and discovered more than the lion's share of early hominid fossils, has written:

If human beings were as violent among themselves as lions, there would be two dozen murders every year on every city block.

The evidence for a predatory early hominid is perfectly valid, but a predatory ape is not a killer in the sense that the "killer ape" was introduced and is popularly conceived of. . . . Nobody would have heard of the killer-ape concept if it hadn't been for Robert Ardrey, but in the same breath I will say it is certain that a majority of the people now interested in supporting early-man research would never have heard of early man. . . . [His book generated tremendous interest] and if the price was the killer-ape hypothesis, then I think it's cheap at the price. . . . [But it] should have been abandoned by everybody long ago.

Nevertheless, recent studies of chimpanzees in the wild have documented some horrific and brutal killings of male chimps by other males (some have been captured on videotape). Victims were individuals who either had invaded another group's territory or attempted to usurp another's place in their own group's dominance hierarchy. As documented in Harvard anthropologist Richard Wrangham and science writer Dale Peterson's book *Demonic Males* (1996), males may take turns beating a chimp to death, sometimes using sticks as clubs. Other documented incidents include groups of males who systematically kill all the infants in a group, or even exterminate all members of a subgroup over several years' time. Laying to rest a common belief that chimpanzees in the wild are gentle creatures, since 1971 observers have witnessed male chimpanzees carry out rape, border raids, brutal beatings, and warfare with rival territorial gangs.

See also AUSTRALOPITHECINES; BABOONS; BROOM, ROBERT; DART, RAYMOND ARTHUR; GORILLAS; HUNTING HYPOTHESIS; *QUEST FOR FIRE*

KIN SELECTION
Altruism and the "Selfish Gene"

Social insects, wrote Charles Darwin in *Origin of Species* (1859), present "one special difficulty, which at first appeared to me insuperable, and actually fatal to my whole theory." How, he wondered, could the worker castes of bees or ants have evolved if they are sterile and leave no offspring?

Darwin solved the problem by devising an idea known today as "kin selection," which was developed in the 1960s by W. D. Hamilton. Although the workers themselves do not reproduce, their seemingly altruistic actions perpetuate their close fertile relatives. The collective that is the hive society thus reproduces itself. As Darwin put it, "A well-flavoured vegetable is cooked, and the individual is destroyed; but the horticulturist sows seeds of the same stock, and confidently expects to get nearly the same variety."

Despite this important insight, which predated Mendelian genetics, Darwin's views on natural selection were usually interpreted in terms of an individual's struggle to pass on its own genes. Social Darwinists argued that "survival of the fittest" (a phrase coined by Herbert Spencer) eliminates altruism as unnatural and destructive of species.

Nevertheless, naturalists have recorded many instances of "altruism" in nature, and not just in social insects. Some birds risk their lives by crying out to warn of predators approaching the flock. Meerkats—small South African desert mongooses—live in social colonies in which sentries scan the skies for hawks and sound a loud alarm when they see one. Meerkat colonies also contain individuals that do not directly reproduce but act as "helpers," caring for several related youngsters while the parents are out hunting.

Students of Hamilton and of Edward O. Wilson's sociobiological approach argued that the meerkat's altruism, like that of the worker insects, can be viewed as a means of genetic survival for the group. The bird or meerkat who risks itself by warning its fellows protects close relatives that share many of the same genes, which will survive.

J. B. S. Haldane, the great British geneticist, was explaining natural selection to some friends in a pub and declared tongue-in-cheek that he would gladly sacrifice himself for his family if he knew that his genes would live on. "How many relatives would be enough?" asked his companion. Haldane seized the back of an envelope and did some hurried calculations. "I am willing to die," he announced, "for two brothers, four uncles, or eight cousins."

See also "SELFISH GENE"

> **"I am willing to die for two brothers, four uncles, or eight cousins."**
>
> —J. B. S. Haldane

KING KONG (1933)
Classic Ape Icon

Well, Denham," says the policeman, "the airplanes got him." Impresario Carl Denham, who has just watched his "prehistoric Giant Gorilla" fall off the top of the Empire State Building, knows better: "Oh no, it wasn't the planes. It was beauty killed the beast." So ends *King Kong*, the 1933 movie masterpiece that has been called Hollywood's most perfect realization of a terrifying dream.

Horrific and fierce though he is, Kong's kinship with ourselves is always painfully clear and evokes tremendous sympathy. He is not at fault for the nightmare that results from showman Denham's obsessive lust for gold and glory. In his arrogance, Denham has attempted to subdue an elemental force of nature for the amusement of paying crowds. His own downfall results, too, from refusal to acknowledge Kong's capacity for love—the ape's "humanity." Denham never imagined Kong would care more for Ann Darrow (Fay Wray) than for his own life or freedom.

Merian C. Cooper, the film's producer-director, had been a combat pilot in World War I, had filmed animals in remote jungles (with his partner Ernest B. Schoedsack), and aspired to become a great explorer. Among other accomplishments, he helped pioneer technicolor film, organized major airlines, and coproduced westerns with famed director John Ford.

Like his screen alter ego, Carl Denham, Cooper was tough, adventurous, stubborn, and a born showman with a sense of high drama. But his fascination with wild animals (especially baboons and gorillas) also made him something of an aspiring naturalist. While in Africa filming a "jungle adventure" (a very popular genre in the 1920s), he observed baboon troops near his locations, kept systematic notes daily for months, and wrote the first monograph about their behavior. (Unfortunately, the only manuscript was thrown out by a careless housemaid.)

Cooper lived a short distance from the American Museum of Natural History and became friendly with the wealthy explorer-naturalist W. Douglas Burden, one of the museum's trustees. Around 1925, Burden had financed and led a museum expedition to the remote and mysterious island of Komodo in the East Indies. Huge carnivorous lizards had been discovered there in 1912, soon after Sir Arthur Conan Doyle published *The Lost World*, a fantasy adventure set in an isolated land where ape-men and dinosaurs survived into the present.

After traveling 15,000 miles by ship, Burden, his crew, and young wife reached Komodo's "gnarled mountains . . . that bared themselves like fangs to the sky . . . as fantastic as the mountains of the moon. . . . We seemed to be entering a lost world." His first impression of a live Komodo dragon, he wrote in his 1927 book *Dragon Lizards of Komodo*, was of "a primeval monster in a primeval setting." (In his imagination, Cooper had dreamed of capturing giant gorillas in Africa; at the time, they were almost as little known as the dragons.)

Burden shot the first motion pictures of the dragons, killed a few (which are still on exhibit at the American Museum of Natural History), and even captured two live nine-foot specimens for the Bronx Zoo. In a 1964 letter to Burden, Cooper recalled:

> When you told me that the two Komodo Dragons you brought back to the Bronx Zoo, where they drew great crowds, were eventually killed by civilization, I immediately thought of doing the same thing with my Giant Gorilla. I had already established him in my mind on a prehistoric island with prehistoric monsters, and I now thought of having him destroyed by the most sophisticated thing I could think of in civilization, and in the most fantastic way . . . to place him on the top of the Empire State Building and have him killed by airplanes.

DREAMING OF PREHISTORIC MONSTERS, writer-producer Merian C. Cooper conjures a cinematic vision of his giant ape, Kong, in this 1933 publicity still. The Hollywood classic drew on Cooper's background as a combat pilot, wildlife photographer, filmmaker, and student of natural history.

KING KONG (1933)

During 1929–1930, Cooper wrote the first drafts of his movie masterpiece. He frequently consulted with Burden and visited him at the museum, where he was also inspired by the murals of prehistoric life by Charles R. Knight, the "Father of Paleoart." The idea of taking Fay Wray to Skull Island was inspired by Burden's wife, who had gone along on the dragon expedition. In Burden's book, he called his great lizard "King of Komodo," which Cooper transformed into "King Kong."

When David O. Selznick became head of RKO in 1931, he invited Cooper to assist him with a film called *Creation*, about prehistoric animals. It was an ambitious experiment in stop-motion animation by the talented Willis O'Brien, who had created similar scenes for *The Lost World* (1925).

The first test shot was Kong's death struggle with the pterodactyl. O'Brien's innovative crew, including the model-maker Marcel Delgado, became the core of a tradition. (In the scene where Kong wrecks the elevated trains, miniature storefronts bore the names of the artists—"O'Brien's Florist," "Delgado's Diner," etc.) Subsequently, many of the same artists and their students, including (on the later films) Ray Harryhausen, worked on *Son of Kong* (1934), *Mighty Joe Young* (1947), *The Beast from 20,000 Fathoms* (1953), and the *Sinbad* series that began in 1958.

For years, Cooper kept Kong's actual size a closely guarded secret. As late as 1964, at a Hollywood exhibit of classic special effects, visitors were still jarred by seeing the original Kong model on display—only about 15 inches high. Press photographers asked actress Fay Wray to hold up the miniature Kong, just as the giant ape had grasped her in the film. Miss Wray refused the pose, explaining quietly, "I could never do that to him."

Fay Wray (1907–2004), the original "girl in the hairy paw," was only 23 when she was chosen by Merian Cooper as the apple of Kong's eye. She was already a star of the silent screen, and during her lifetime she appeared in about 100 movies, but according to her *New York Times* obituary, "her fame is inextricably linked with the hours she spent struggling, helplessly screaming, in the eight-foot hand of King Kong." Ironically, her one criticism of the finished film was that the editor had inserted "too much screaming."

King Kong's roots are deep, drawing on rich traditions of both literature and natural history. Story elements came from the fairy tale "Beauty and the Beast," from the "fatal character flaw" of Greek tragedies (for both Denham and Kong), and from the nightmarish tales of gorillas popularized by hunter Paul du Chaillu. The film drew also on the filmmaker's explorer friends like Burden and his own glory-seeking stubbornness, on Charles Darwin's *Descent of Man* (1871) and Alfred Russel Wallace's *Malay Archipelago* (1869), on museum expeditions of the 1920s to exotic islands, the romance of jungle movies such as *Tarzan*, Conan Doyle's *Lost World*, and animator Willis O'Brien's long-standing obsession with prehistoric monsters.

Remote from these sources, producer Dino de Laurentis attempted a remake of *King Kong* in 1975, using a man in an ape suit in place of stop-motion animation; the World Trade Center towers replaced the Empire State Building as New York's tallest buildings. One scene featured a disastrous full-sized Kong robot, which cost over a million dollars to construct but ended up being used for only 10 seconds in the finished film. The dismal de Laurentis effort only reaffirmed the brilliance of the 1933 classic.

A 2005 remake of Kong was created by Peter Jackson, fresh from directing the *Lord of the Rings* trilogy. In the Jackson version, the giant ape's computer-enhanced movements were mimed by actor Andy Serkis, digitally wrapped in a virtual gorilla body. Just prior to shooting in 2004, Serkis spent weeks in the highlands of Rwanda observing and videotaping free-living gorillas.

A few months before her death, Jackson asked Fay Wray if she would appear in his film as an homage to the original, and speak the closing line, "It was beauty killed the beast," but she declined. She told him that any remake could not replace the original, because "the film I made was so extraordinary, so full of imagination and special effects, that it will never be equaled." In fact, Jackson achieved effects and digital animation that far surpassed the original's handmade technology. Unfortunately, however, he also shifted the story's focus from horror nightmare to four-handkerchief romance—a decision that was well intentioned but disastrous.

Perhaps he took his cue from Fay Wray herself, the "scream queen" who had developed a genuine soft spot in her heart for the big guy. Wray told a *New York Times* reporter in 1993 that, over the years, she felt that Kong had "become a spiritual thing to many people, including me." Whenever she was in New York, she said, "I look at the Empire State Building and feel as though it belongs to me, or is it vice versa?" At her death at the age of 96, the powerful floodlights that illuminate the building's tower at night were blacked out in tribute to Wray and her "tall, dark, leading man."

See also APES; GORILLAS; *JURASSIC PARK*; *LOST WORLD*; O'BRIEN, WILLIS

KNIGHT, CHARLES R. (1874–1953)
The Father of Paleo Art

Charles R. Knight spent his life creating paintings and sculptures of a world no man had ever seen: a younger Earth of monstrous dinosaurs, birds with teeth, and flying reptiles, all of which vanished millions of years ago. Combining skilled artistry with a mastery of anatomy and paleontology, Knight established a scientifically accurate vision of extinct creatures.

Born in Brooklyn, New York, on October 12, 1874, Charles was the only child of an American mother and English-born father who worked for the Morgan banking house, a connection that was to prove important. In later years, J. P. Morgan became a major patron and sponsor of Knight's murals for the American Museum of Natural History.

Knight's artistic talent and passion for animals showed itself early, but a childhood accident almost put an end to his career before it had begun. A playmate threw a pebble that struck his right eye, which remained damaged throughout his life. During his teens, he began full-time training in art at both the Metropolitan Art School and the Art Students League in New York City. In 1890, he got his first job, with a manufacturer of church windows, where he designed birds and animals for their stained-glass panels. Even while working full-time,

THE ARTIST WHO SAW THROUGH TIME, Charles R. Knight was the first to paint accurate, lifelike reconstructions of prehistoric animals, establishing modern images of the dinosaur. Knight created this *Stegosaurus* sculpture in clay as an aid to painting.

KNIGHT'S RENDITIONS of tyrannosaurs, *triceratops*, and brontosaurs (now called apatosaurs) became widely copied classics.

Knight's love of drawing animals led him often to the Central Park Zoo. After his father's death in 1892, he decided to turn to book and magazine illustration, and eventually began to get assignments from *McClure's* and *Harper's* magazines. To increase his knowledge of animal anatomy, he studied carcasses in the taxidermy department of the American Museum of Natural History and became friendly with staff scientists.

One day in 1894, one of his museum mentors mentioned that the fossil department needed a lifelike drawing of an extinct piglike mammal. Knight examined the fossil skeleton and made a skillful sketch of what the animal might have looked like. Delighted with the results, the paleontologists gave him further assignments.

His art soon came to the attention of the museum's autocratic president, Henry Fairfield Osborn (1857–1935), who was building its great dinosaur halls. Recognizing Knight's talent, he set him to work on a series of huge murals depicting prehistoric life. Osborn obtained sponsorship for the paintings from his uncle, the banker and financier J. P. Morgan, who had fortuitously employed Knight's father, George, as his longtime private secretary.

Osborn had also hired paleontologist William D. Matthew, who was preparing fossil skeletons in lifelike action poses, a novel idea at the time. After questioning Matthew and Osborn for their interpretations of an animal and studying the mounted skeleton, Knight would then sculpt a small clay model and cast it in plaster. Painting from his model, he could pose it at any angle and see exactly where highlights and shadows would fall. In addition, his accumulated knowledge of muscles, behavior, and expressions of living creatures enabled him to depict extinct animals with an unprecedented scientific accuracy.

Some of his most dramatic interpretations resulted from a brief, fruitful collaboration with Edward Drinker Cope, one of the founders of American paleontology. Cope had been Osborn's mentor, and Knight had always wanted to meet him. They spent several months together at Cope's fossil-crammed home in Philadelphia, planning a series of paintings based on Cope's interpretations of his dinosaurs. Cope died soon after, but Knight continued the paintings based on their discussions.

Although he always worked on a free-lance basis, Knight is closely identified with the American Museum of Natural History, which has an extensive collection of his paintings and murals. Excellent examples of his work can also be seen at the George C. Page Museum in Los Angeles and at the Field Museum in Chicago, where Knight did some of his finest work. Among the private commissions he accepted was a special request from President Calvin Coolidge to paint his beloved collies.

Even his mistakes were widely copied. In the 1890s, following a paleontologist's conjectures, he had painted *Agathaumus*, an unlikely mixture of dragon and rhinoceros. Imagined from fragments, it was later dropped by science. But that news failed to reach Hollywood; "Agathaumus" was animated by Willis O'Brien in his classic *The Lost World* (1925), along with Knight's other dinosaurs. Firmly established in popular culture, it continued to appear in children's books on dinosaurs until the 1960s. Knight's reconstructions of dinosaurs, mammoths, and saber-tooth cats were copied not only in scores of books, but have influenced comic book artists, toymakers, and filmmakers for more than a century.

See also COPE, EDWARD DRINKER; *KING KONG*; *LOST WORLD*

KOKO
First Signing Gorilla

During the 1970s a number of psychologists reported that they had successfully taught a modified form of American Sign Language to chimpanzees. Francine (Penny) Patterson, then a graduate student at Stanford University, decided to see if gorillas could do as well as their chimp cousins. She obtained permission to work with a young female gorilla, Koko, at the San Francisco Zoo in 1972.

Beginning when Koko was one year old, Patterson attempted to teach her sign language. Koko soon matched and then surpassed her chimpanzee rivals; by 1979, she used more than 400 signs. According to Patterson, Koko can use 1,000 signs based on American Sign Language and understand approximately 2,000 words of spoken English. Patterson also reports that Koko can communicate about the past or future and can argue, joke, and even tell lies. Patterson's work with Koko has continued for more than three decades.

A few years ago, when asked whether she was an animal or a person, Koko would sign, "Fine animal gorilla." Today, however, she responds, "Fine *person* gorilla"—which Patterson interprets as a significant shift in her self-image.

When a young male gorilla named Michael was moved into her quarters, Koko began signing to him. Michael lived with Koko for several years and learned 500 signs. Although he did not sign as frequently as Koko, Patterson says that he started telling stories—something Koko rarely does. Patterson had hoped that the pair would produce a signing infant, but they behaved as siblings and never mated. Michael died in 2000.

Despite Herbert Terrace's infamous "reconsideration" of his own language experiments with chimps [see APE LANGUAGE CONTROVERSY; NIM CHIMPSKY], Patterson has stuck to her guns about the reality of Koko's ability to speak through signs. Critics have fired away at her methods and controversial results, but she continues her daily pongid conversations with unflagging conviction and devotion.

Patterson asserts she has repeatedly demonstrated that Koko can use her thousand-word vocabulary in complex statements and questions. Most of the signs are standard American Sign Language (ASL), but some have been either invented or modified by Koko into what Patterson calls Gorilla Sign Language (GSL) or "Gorilla Speak."

Twenty years ago, journalist Emily Hahn visited Patterson and her gorillas for her book on women who investigate apes (*Eve and the Apes*, 1988). Hahn wrote that when she asked Patterson if she thought Michael could remember being taken from his mother in the African forest, Patterson replied, "Oh, yes, . . . he does. We know because we asked him. He said he was with her in the woods, and some bad men came and hit her *here* on the back of the neck. He saw the blood. After all, he was about two." Her matter-of-fact assertion that gorillas could report on past events left an astonished Hahn to "wrestle with the concept."

Patterson has objected that both she and Michael were misquoted. "He never said he remembers blood," she explained. "He said he saw red." In 2004, Michael was videotaped retelling the story of his capture in the PBS *Nature* television program *Conversations with Koko,* allowing viewers to decide for themselves whether Patterson's "translation" is scientific fact or wishful interpretation. Her work with gorillas, Patterson reiterates, is "forcing us to reconsider everything we thought we knew about animals."

See also ALEX; CLEVER HANS PHENOMENON

KOKO THE GORILLA and Penny Patterson have conversed daily in sign language for more than thirty years.
© Pete Von Sholly.

*Some of my colleagues don't believe we can really converse.

* That's alright. **None** of the gorillas believe it!

* translated from sign language

Penny and Koko discuss their critics.

KROPOTKIN, PRINCE PETER (1842–1921)
"Law of Mutual Aid"

During an era of industrial robber barons, militarism, laissez-faire economics, and colonial exploitation, Social Darwinism became a handy excuse for ruthlessness. In the words of Prince Peter Kropotkin, ideologues twisted Charles Darwin's conception of nature into "a world of perpetual struggle among half-starved individuals, thirsting for one another's blood. They made . . . *woe to the vanquished* . . . the last word of modern biology . . . [and] raised the 'pitiless' struggle for personal advantages to the height of a biological principle which man must submit to as well."

Born a prince in prerevolutionary Russia, Kropotkin railed against the social orders that gave him hereditary privileges and became an anarchist. His father, Prince Alexei Petrovich Kropotkin, had owned nearly 1,200 "souls" (male serfs) in three different provinces and large tracts of land.

The "anarchist prince," after holding a variety of diplomatic and military posts as a young man, became a writer-philosopher and sometime naturalist. In later years, he turned his energies to overthrowing his country's rigidly stratified social system, believing that "sociability is as much a law of nature as mutual struggle . . . mutual aid is as much a law of animal life as mutual struggle."

Kropotkin considered the two great movements of his time were "towards Liberty of the individual and social co-operation of the whole community"—goals that were summed up in his idealistic anarchist-communism.

His early travels for the military took him to Siberia, and he later conducted a geological survey of Manchuria. A careful observer of both wildlife and local villagers, he became convinced that—even in such a cold, harsh environment, where one might expect competition to be keenest—survival depended more on cooperation than on competition. He observed horses forming defensive circles to resist wolf attacks, and the cooperative hunting strategies of the wolves themselves.

Kropotkin published a remarkable book, *Mutual Aid* (1902), which was his corrective to the popular view of a "struggle for existence." Taking his cue from an 1880 lecture, "On the Law of Mutual Aid," by the Russian zoologist Karl F. Kessler, Kropotkin spent years building his case for the survival value of compassion, nurture, and altruism. It was not until almost 70 years later, with the rise of sociobiology, that the role of altruism in evolution was seriously examined. He clearly saw the implications for human politics and eugenics programs as well.

Kropotkin criticized Darwin's remarks in *The Descent of Man* (1871) about the "alleged inconveniences" of maintaining the "weak in mind and body" in society. Darwin, Kropotkin scolded, seemed to think advanced societies were burdened with too many "unfit" individuals, "as if thousands of weak-bodied and infirm poets, scientists, inventors, and reformers, together with other thousands of so-called 'fools' . . . were not the most precious weapons used by humanity in its struggle for existence by intellectual and moral arms." It was Darwin himself, said Kropotkin, who had shown that "sociability" conferred an important evolutionary advantage.

As he grew older, Kropotkin did everything he could to undermine the Russian social system, which he saw as unjust, inhumane, and "unnatural." If the corrupt political and economic institutions could be dismantled, he thought, mankind would return to its more "natural" state of harmony and cooperation.

See also KIN SELECTION; SOCIAL DARWINISM

TRAITOR TO HIS CLASS, Russian Prince Kropotkin advocated anarchy to bring down the corrupt, despotic artistocracy. After observing wildlife, he concluded that evolution favors cooperation in social animals.

LA BREA TAR PITS
Asphalt Fossil Trove

In Hancock Park, near a busy shopping district in Los Angeles, California, seven miles west of the civic center, natural asphalt pits contain one of the richest collections of fossil Pleistocene animals in the world. This bituminous outcrop of petroleum shale—clay, sand, and asphalt—contains remains of mastodons, bison, horses, camels, wolves, birds, and saber-toothed cats. About 40,000 years ago, they were the only residents of Los Angeles.

Although fossils from the pits weren't described until 1875, the site had been mined for years, and thousands of tons of fossiliferous asphalt had already been shipped to tar roofs and roads in San Francisco. During the 19th century, the site was part of Rancho La Brea, then on the western outskirts of the city.

University of California scientists began collecting fossils there in 1906, when the first saber-toothed cats (*Smilodon*) found anywhere in the world were discovered in the pits. Canine teeth of the saber-tooth resemble curved 10-inch knives and were used for stabbing and tearing prey. More than 700 skulls of the saber-toothed cat were found, and initial reports described densities of 20 saber-tooth and dire wolf skulls per cubic yard.

Most authorities believe that prey animals, such as the camels, horses, and bison, may have become mired in the tar, and their struggles attracted crowds of saber-toothed cats and wolves, which also were trapped in the sticky mass. A skeleton of an early American Indian was also found preserved under the bones of an extinct vulture.

TRAPPED IN ASPHALT POOLS, tens of thousands of animals left their bones in what is now Los Angeles, California. Saber-toothed cats, American lions, giant sloths, and mastodons succumbed. Mural by Charles R. Knight.

The 23-acre park is open to the public free of charge. A large cluster of fossil bones has been left unexcavated and undisturbed, so the visitor can see the natural state in which the profusion of fossils occurs in the asphalt. It is not unusual to see a sparrow or squirrel wander from the park into the pit today and become entrapped in the gooey tar, the sad spectacle of a fossil in the making.

The extensive collections from Rancho La Brea (more than 565 species) are stored and exhibited in the George C. Page Museum of La Brea Discoveries in the park. Opened in 1977, the Page Museum features life-sized outdoor sculptures of mastodons, seemingly trapped in the actual tar deposits. Visitors can also observe the museum's scientists and technicians as they meticulously extract fossil treasures from the ancient tar deposits.

LAETOLI FOOTPRINTS
Earliest Fossil Man-Tracks

FOOTPRINTS IN VOLCANIC ASH were made four million years ago by three upright hominids—possibly a male, female, and child. Other tracks nearby include those of saber-toothed cats.

Mary Leakey described it as "perhaps the most remarkable find I have made in my entire career." The veteran paleoanthropologist was referring neither to a fossil hominid skull nor a stone tool, but to a trackway of petrified footprints she had excavated in 1978 near an ancient volcano in Tanzania. When she first came across the hominid prints Leakey was sceptical, but later she became convinced that she had found the earliest prints of man's ancestors, evidence that hominids three-and-three-quarter million years ago walked upright with a free-striding gait, just as we do today.

These earliest human footprints were found at a site called Laetoli, in a wooded area about 25 miles south of Olduvai Gorge, where Mary Leakey, her husband, Louis, and son Richard had made so many important fossil discoveries. They were actually found by Paul I. Abell (1924–2004), a chemistry professor from the University of Rhode Island, who had a special interest in paleoclimates and for 17 years spent his sabbaticals helping the Leakeys search for hominid fossils. Working with Mary Leakey's team, he was the first to chance upon a hardened footprint in volcanic ash that turned out to be part of an 80-foot trail left by a pair of adult hominids and a child several million years ago.

Preserved in the hardened volcanic mud are tracks of various animals, including spring hares, guinea fowl, elephants, pigs, rhinos, buffaloes, hyenas, antelopes, baboons, and a saber-toothed cat. Among these are the tracks of three hominids—a large individual walking slowly north, a smaller one following behind, and a youngster. The young one seemed to have been following alongside them, at one point turning to look around to the left.

Like nearby active volcanoes in East Africa today, the ancient volcano Sadiman—very near the prints—occasionally belched out clouds of gray ash over the surrounding countryside. This ash sets hard as cement when it is first dampened slightly, then dried in the sun. A brief shower moistened the ash layer; tiny raindrop craters can be seen in its surface. Then the sun came out and hardened it, leaving this extraordinary record of an upright-walking hominid group, from almost four million years ago.

See also LEAKEY, MARY

LAMARCK, JEAN-BAPTISTE ANTOINE DE MONET, CHEVALIER DE (1744–1829)
Naturalist, Evolutionist

Pioneer evolutionist Jean-Baptiste Antoine de Monet (later known as the Chevalier de Lamarck) came from a long line of horse soldiers, imbued with honor, bravery, tenacity, and a desire for glory. When Lamarck traded a military career for one in science, he had simply found a new field of combat, and to this day "Lamarckians" remain embattled.

His war-weary father, determined to shield his 11th child from becoming cannon fodder, sequestered him with the Jesuits as a priest-in-training. But at 19, young Jean-Baptiste fled his school to join a regiment defending a German town at the start of the Seven Year War. Within a few days, Lamarck distinguished himself in the thick of battle. Seizing a field command, he refused to let his men withdraw—though they were isolated and outnumbered—until official orders to retreat came through the lines. Immediately thereafter, he received a regular officer's commission.

After years of distinguished service, Lamarck traded his sword to seek glory in the battle for men's minds. He decided to do no less than revolutionize knowledge in all of science by creating his own system of chemistry, meteorology, and biology (a term he coined).

Much of Lamarck's ambitious system was based on assumptions that were disproved during his lifetime. He took on the great experimental chemist Antoine Lavoisier (1743–1794), for instance, with "laws" of fluids and compounds based on logic and speculation rather than experiment. "In this respect," wrote his enemy Georges Cuvier, "he resembled so many others who spend their lives in solitude, who never entertain a doubt of the accuracy of their opinions, because they never happen to be contradicted."

After an unhappy stint as a bank clerk, Lamarck found he could sustain himself in cheerful poverty by writing encyclopedias of natural history. His *Dictionary of Botany* (1778) became quite popular, because (as Cuvier noted) he stole only from the best authors of the day.

For 11 years he published almanacs based on his meteorological system, predicting the year's weather in advance. The weather refused to cooperate, but people continued to buy a year's worth of inaccurate forecasts, just as they do today.

Lamarck also became a proficient botanist and published the first "keyed" field guide to French flowers, making identification easy for casual hikers. He also created a new system of animal classification based on the fundamental distinction between animals with backbones (vertebrates) and all those without (coining the term "invertebrates"), which still stands.

He struggled for years without a regular position, through four successive marriages that produced four children. Georges-Louis Leclerc, Comte de Buffon, one of the eminent naturalists of the day, recognized his abilities and hired him as tutor to his own son. For several years, Lamarck was able to pursue his natural history interests in that capacity, hoping one of the scarce positions in the then tiny scientific establishment might open up.

In 1794, when the dust from the French Revolution began to settle, the king's garden (Jardin des Plantes) was reorganized as the Museum of Natural History (Musée d'Histoire Naturelle), but most of the new posts were already promised to those who were professors and keepers under the old regime. Only the lowest, least desirable position was offered to Lamarck: keeper of insects, shells, and worms. He accepted.

Oddly enough, Lamarck's name has come to stand chiefly for an idea he did not originate: the development or atrophy of organs through "use or disuse" and their transmission to offspring who inherit these "acquired characteristics." Lamarck did not consider the idea of "inheritance of acquired characteristics" an original or important part of his evolutionary theory. Everyone in his day accepted it, and it continued to hold force until Mendel's ideas took hold around 1900. Darwin believed in the inheritance of acquired characteristics, although he realized that there were contradictory data, and he was never able to discover the mechanisms of inheritance.

Lamarck's classification work led him to several evolutionary speculations. First, he saw living things as tending to progress from simpler organisms to those with more complex nervous systems (man, of course, being the pinnacle of perfection). Second, he tried to address the problem of extinction at a time when few could imagine "gaps" in the natural order. Comparing fossil oysters in his collection with modern ones, he concluded that the ancient species had not really died out at all; they had simply changed into those of today.

Lamarck described no mechanisms for producing change, but he is remembered for asserting that bodies are shaped by habitual behavior caused by an animal's "needs" or "wants" (*besoins*). Thus, generations of giraffes keep stretching to reach ever-higher branches, and wading birds keep raising themselves from the mud until their legs become stilts.

Such examples brought Cuvier's ridicule and later Darwin's exasperation. Some of the criticism grew out of a semantic problem; the French word *besoin* can signify "lack," "need," or "want," depending on context. Darwin expressed

Lamarck's name has come to stand for an idea that he did not originate: the inheritance of acquired characteristics.

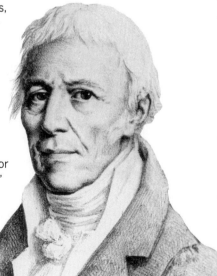

A PIONEERING FRENCH EVOLUTIONIST, Lamarck tried unsuccessfully to convince his colleagues of the truth of evolution and was not appreciated until after his death.

contempt for Lamarck's purported vision of trees or worms "willing" themselves to adapt or progress—an absurdity of which Lamarck was not guilty. It was only the more advanced creatures, he thought, that had "sensibility" and could strive to meet their needs. Lamarck believed plants and invertebrates changed because of unconscious physiological responses to environments.

Unlike most dissectors and collectors of his time, Lamarck always conceptualized organisms in relation to their behavior in nature and the challenges of changing environments. Rejecting accepted concepts of a young steady-state Earth, he believed geology pointed to gradual shifts in climate, land, and sea that took place over millions of years.

If species had been fixed at creation and remained static ever since, he realized, they could not survive environmental changes. Therefore, they must be constantly adapting, even if their appearance changed little. Further, he saw the development (evolution) of life as slow, smooth, and gradual—an approach similar to Darwinian gradualism.

Lamarck's ideas did not take hold in France, although after the Darwinian revolution he was reclaimed by French biologists. "It is not enough to discover and prove a useful truth," Lamarck wrote in *Zoological Philosophy* (1809). "It is necessary also to be able to propagate it and get it recognized." During his last years he continued to fight for his ideas even after he went blind, attending scientific meetings on the arm of a devoted daughter.

Lamarck had argued that species of blind fish found in dark caves had lost their eyes through disuse. In one of the meanest remarks ever recorded in science, Cuvier publicly taunted the aged Lamarck after he once again defended his evolutionary views. "Perhaps," said the baron, "your own refusal to use your eyes to look at nature properly has caused them to stop working." Even as his daughter led the frail genius from the room, she begged him not to despair: "Have no doubts, Father, posterity will honor you."

Although it is a misnomer and historically inaccurate, the outmoded doctrine of "use and disuse" inheritance continues to be known as "Lamarckian." The Russian Trofim Lysenko and other 20th-century "Lamarckian evolutionists" lifted this unimportant fragment of his thinking and extolled it as offering hope for the progressive improvement of societies.

See also NEO-LAMARCKISM

> " It is not enough to discover and prove a useful truth. It is necessary also to . . . get it recognized."
>
> —Jean-Baptiste Lamarck

LANKESTER, E. RAY (1846–1929)
Evolutionist, "Ghost-Buster"

Charles Darwin knew Ray Lankester's father, Edwin, the Kent county medical examiner who renovated London's sewers and rid the city of cholera. Edwin hoped one of his sons would become a great biologist and named them accordingly: Forbes (after Edward Forbes), Owen (after Richard Owen), and Ray (after John Ray).

By the time Ray was in his teens, he had met Darwin and Thomas Henry Huxley and was a dedicated, budding evolutionary biologist, eager to gain the approval of those two great naturalists. He graduated with a degree in zoology from University College London, where he had been Huxley's demonstrator in anatomy classes. Darwin called him "a rising star in Natural History."

Eager to please Darwin, who detested professional "spirit-mediums," the young graduate student hauled Henry Slade, London's latest celebrity conjuror of ghostly spirits, into police court. In 1876, Lankester charged Slade with criminal fraud and prosecuted him as a "common rogue." Darwin was so delighted that he sent Lankester a generous contribution, praising his exposure of the spiritualist as a "public benefit." [See SLADE TRIAL.]

Curious and energetic, Lankester had interests spanning a wide range of evolutionary biology. One now-forgotten area that caught his attention was the Degeneration Theory. Evolution, he thought, could lead in the direction opposite from progress, to decadence and degeneration. As he wrote in *Degeneration: A Chapter in Darwinism* (1880),

> Any new set of conditions occurring to an animal which render its food and safety very easily attained, seem to lead as a rule to degeneration; just as an active healthy man sometimes degenerates when he becomes suddenly possessed of a fortune; or as Rome degenerated when possessed of the riches of the ancient world. . . . Let the parasitic life once be secured, and away

go legs, jaws, eyes, and ears; the active, highly-gifted crab, insect, or annelid may become a mere sac, absorbing nourishment and laying eggs.

Accordingly, he warned his fellow members of the "English race" against allowing themselves to be "overtaken" in the struggle for existence through sheer laziness and consequent evolutionary degeneration.

Lankester wrote a weekly newspaper column, "Science from an Easy Chair," much of which was reprinted in two books by that title (1910 and 1912) and in *Diversions of a Naturalist* (1915). For eight years he served as director of the British Museum of Natural History and battled bureaucrats to secure its independence from its parent institution, the British Museum Library. More important, he managed to establish a dynamic Darwinian tradition at the museum, which had formerly been devoted mainly to classification of the thousands of specimens in its collections.

His classic children's book *Extinct Animals* (1905) became the first standard introduction to dinosaurs and ancient animals, illustrated with pioneering reconstructions by artists working under his direction. Sir Arthur Conan Doyle used many of Lankester's prehistoric creatures (often simply lifted without further change) as illustrations in *The Lost World* (1912), which became the first enduring dinosaur fantasy-adventure. In that story, Doyle has his fictitious Professor Challenger refer to the real-life Lankester as "my gifted friend."

According to evolutionist-historian Ernst Mayr, Lankester was responsible for rescuing the idea of natural selection from limbo in British science. Despite the efforts of Alfred Russel Wallace, Sir Joseph Hooker, Henry Walter Bates, and a few others to promote it, natural selection had became unpalatable to British scientists in the decades following Darwin's death. Confused about a plausible mechanism for heredity, they were unable to see how "descent with modification" could work.

Lankester was impressed by the new theories of August Weismann, the German geneticist who suggested that an independent "germplasm" that existed only within the sex organs was the basis of heredity. This germplasm could not be directly modified by behavior or environment, and therefore replaced the old "Lamarckian" idea of "use and disuse" as the mechanism of the transmission of traits from parents to offspring. Lankester invited Weismann to visit England and enthusiastically welcomed him into the scientific community.

At Oxford, Lankester founded a school of thought that combined Weismann's ideas about heredity with Darwinian natural selection—a forerunner of the Modern Synthetic Theory. In *The Growth of Biological Thought* (1982) Ernst Mayr concluded that Lankester's influence was "of decisive importance," since until that time not a single experimental biologist was pursuing research based on natural selection. Darwin's favorite "ghost-buster" had reinvigorated natural selection, the central proposition of his theory, pioneering a direction that was later followed by R. A. Fisher, J. B. S. Haldane, and Julian Huxley in shaping 20th-century biology.

When prehistoric archeology attracted his interest, Lankester coined many technical terms that are still used to describe stone tools. But his enthusiastic quest for human origins also led him to accept the "Piltdown Man" fossil forgery as genuine. Although at first skeptical of the concocted "ape-man" skull and jaw found in a Sussex gravel pit in 1912, within a few years he had joined his colleagues in hailing the unsuspected fraud as "the earliest Englishman."

Stone and bone evidence seemed reassuringly "solid" to Victorian scientists but in this instance they proved to be as thin as the "spirits" Lankester had discredited in Slade's parlor. The great debunker of hoaxers had been taken in. Perhaps, as anthropologist John Winslow suggested in the magazine *Science '83* (1983), Piltdown was a spiritualist's revenge on the British scientific community for Lankester's public prosecution (Conan Doyle called it "persecution") of Henry Slade.

See also DEGENERATION THEORY; *LOST WORLD*; PILTDOWN MAN (HOAX); WEISMANN, AUGUST

FEISTY, FEARLESS, AND STUBBORN, Dr. E. Ray Lankester continued the Darwin-Huxley tradition in England as director of the British Museum of Natural History and author of popular essays on evolution and natural history.

LASCAUX CAVE
Prehistoric Art Treasures

The decorated limestone caverns of Lascaux, France, contain masterpieces of prehistoric art, believed to have been painted 17,000 years ago. During the Pleistocene glaciations, our ancestors left a legacy of mysterious, magnificent paintings of mammoths, aurochs, wild horses, and woolly rhinos. These artworks show careful observation and intimate familiarity with the great mammals of the last Ice Age.

In September 1940, during the German occupation of France, four teenage boys, Marcel Ravidat, Jacques Marsal, Simon Coencas, and Georges Agniel, were hiking in the Lascaux Forest when Ravidat's dog Robot discovered a small opening in the ground. This "burrow" turned out to be an entrance to the Lascaux cave. Despite the war, the discovery of the prehistoric art gallery created a worldwide stir.

For months the boys camped out at the cave's entrance to guard their discovery. In 1943 Marsal, who was Jewish, was sent into forced labor in Austria, while Ravidat joined the local French Resistance. Coencas was held at the Drancy deportation camp before being saved by the French Red Cross; both his parents, however, had been killed. After the war, Marsal and Ravidat worked at the cave until its closing in 1963, and Marsal subsequently became the official caretaker.

Millions of visitors have viewed the extraordinary cave art, and unwittingly brought with them moisture, algae, and bacteria, which began to attack the walls. By 1960, alarmed custodians began to notice patches of green algae spreading and covering up the animal figures. Three years later, the cave was closed to the public and remains so, while conservators worked with temperature and humidity controls, antibiotics, and a new ventilation system that may have caused additional harm.

About 200 yards from the cave is the faux Lascaux II, a painstaking replica of the cave and its art, which was opened to the public in 1984. Within a huge blockhouse buried in a quarry, engineers recreated the contours of its walls. Artist Monique Peytral used the same pigments (red ochre, charcoal, sulfur) as Ice Age artists and took five years to reproduce the paintings in exacting detail. More than 300,000 visitors now visit Lascaux II annually, while experts work to preserve the nearby original.

LAURASIA
Northern Lands

Of the ancient former supercontinents, Laurasia was the northern land mass during the Mesozoic. When it broke apart, Laurasia formed North America, Greenland, northern and central Europe, and Asia (except India).
See also CONTINENTAL DRIFT; GONDWANALAND; *LYSTROSAURUS* FAUNA

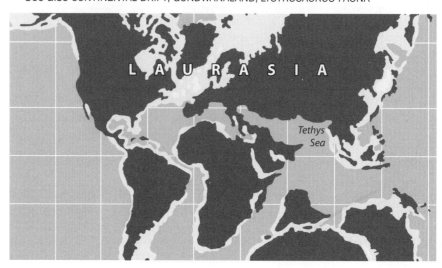

SUPERCONTINENT of Laurasia in the mid-Cretaceous, about 100 million years ago. Later, it broke up to form North America, Greenland, and parts of Asia and Europe.

LEAKEY, LOUIS (1903–1972)
Paleoanthropologist

Louis Seymour Bazett Leakey was a maverick, an independent fossil hunter, archeologist, and anthropologist who, with his wife, Mary, changed the entire picture of human prehistory. He is best known for his discovery of scores of fossils and stone tools relating to human evolution at Olduvai Gorge, Tanzania, where the Leakeys confirmed Charles Darwin's prediction that Africa was the probable "cradle of mankind."

Until then, most anthropologists thought that man had evolved in Asia about a million years ago. Since the Leakeys' pioneering investigations—continued by their son Richard—the focus has been on Africa, and the time depth has been pushed back beyond four million years.

Leakey was born in Kenya to Church of England missionaries, but the Africans they were "converting" in turn made a Kikuyu of young Louis. He was initiated into the tribe at 13 and often thought and dreamed in Kikuyu, which was his first language. He became known as the "White African," the title of his first book of reminiscences.

Early education came from his father's tutorials, as well as initiation into the culture and hunting methods of his African teachers. Plans to send him abroad for education were delayed by World War I, but at 16 he entered a school at Weymouth and eventually made his way to St. John's, Cambridge. There he persuaded dubious university officials to accept Kikuyu as his foreign language requirement, submitting as proof of his competence a dictated letter signed by a Kikuyu chief's thumbprint.

At 20, he took a year off to lead a fossil-hunting expedition to Tanganyika for the British Museum, after which he returned to Cambridge and took a degree in anthropology. His first independent "archeological expedition" to East Africa, consisting of himself and a friend, yielded many Late Stone Age bones and artifacts from the Nakuru-Naivasha area of Kenya, to which he would return. In 1931, he began his work at Olduvai Gorge, and in subsequent years searched Rusinga Island in Lake Victoria for Miocene ape fossils.

For more than 40 years, Leakey (later joined by Mary) worked in the African sun under harsh conditions to search for remains of early humans. He had to haul water hundreds of miles in his Land Rover to the site and often had to dodge charging rhinos and prowling lions to do his work. He found not only many early hominid remains, including *Homo habilis,* the still-debated fossil he considered the oldest known human ancestor, but also what he believed were early "living sites" strewn with Acheulean stone tools and a crude, early assemblage of worked pebbles from what he called the Oldowan culture.

During the 1950s, Kenya was in turmoil because of the Mau Mau uprising, a bloody revolution for freedom and independence from colonial rule. Two thousand Kikuyu who refused to take the Mau Mau oath were murdered, along with 32 white civilians. One of the victims was Leakey's cousin Gray Leakey, who was buried alive after having been forced to watch his wife strangled. Leakey campaigned for an end to the violence and terror, although he fully supported the Africans' goal of self-government.

When rebel leader Jomo Kenyatta, a Kikuyu and Western-trained anthropologist, was tried for conspiracy by the white government, Leakey was asked to serve as interpreter at the trial. Defense lawyers claimed his interpretation of certain Kikuyu words was inaccurate and prejudicial, and the episode created years of enmity between him and Kenyatta, who later emerged victorious as the first president of an independent Kenya. A price was put on Leakey's head by the Mau Mau. He tried to explain his position in two books, *Mau Mau and the Kikuyu* (1952) and *Defeating Mau Mau* (1954), in which he sought to "heal the mental wounds that have been inflicted on all races in Kenya."

In 1959, Mary discovered *Zinjanthropus* (later named *Paranthropus boisei*) at Olduvai while Louis was in camp with a fever. It was the first early hominid to which the new potassium-argon method of dating could be applied, and the resulting determination of 1.75 million years radically changed scientific views on the time scale of human evolution. Leakey's team also gathered and studied more than 2,000 stone tools and flakes at the site.

Leakey brought an unconventional, experimental approach to his theorizing about early

During Louis Leakey's fossil hunts at Olduvai Gorge, he had to haul in his own water and dodge prowling lions and charging rhinos.

humankind. When investigating the role of weapons in human survival, he attempted to subsist on small game caught with his bare hands alone, and succeeded. To determine the use and efficiency of stone tools, he skinned and butchered large game animals with them. He even became quite proficient in chipping his own replicas of stone tools.

Once he raised eyebrows at a graduate seminar in America by suggesting that students try gulping down whole mice, frogs, and lizards to see how their skeletons looked after they had passed through the human digestive tract. (He had found the remains of small creatures at prehistoric sites and wondered if they had got there by accident or as a food item of early man.)

In addition to his fossil hunting, he was also a fund-raiser, a founder of the Nairobi National Park, and the second curator of the Coryndon Museum, now the Nairobi National Museum, where his son Richard was later director.

His interest in the behavior of early humans also led him to select and encourage several women to conduct field studies of the great apes. They all made important contributions to understanding the social behavior of chimps, gorillas, and orangs in the wild.

During the last years of his life, slowed down by heart disease and arthritis, Louis followed with intense interest the career of his son Richard, who was making important discoveries of his own. After a few years of father-son feuding, the two reconciled, and Richard became a successor rather than a rival. Leakey died in 1972, but Richard and Mary continued his work. Richard's wife Meave and their daughter, Louise (named for her grandfather), further the family quest.

See also "APE WOMEN," LEAKEY'S; FOSSEY, DIAN; *HOMO HABILIS*; LEAKEY, MARY; LEAKEY, RICHARD

LEAKEY, MARY (1913–1996)
Archeologist, Paleoanthropologist

Mary Nicol Leakey was the great-great-great-granddaughter of John Frere, the first British prehistorian, who in 1800 published descriptions of flint tools and mammoth bones found together in a cave deposit in the Suffolk countryside. Frere attributed them to "a remote period, even beyond that of the present world," an interpretation far too radical for his contemporaries to consider.

Born in England, Mary trained as an artist-archeologist and went to France to work on early stone tool sites. In 1933, she met her future husband, Louis Leakey, and became his dedicated partner in the search for early humans in Africa. One of the spectacular finds usually attributed to her husband's efforts, the *Zinjanthropus* or "Nutcracker Man," was actually discovered by Mary and her dalmatian dogs in 1959. The "Zinj" skull has since been assigned to *Paranthropus boisei*, a hyper-robust hominid with heavy bones, powerful jaws, a ridge of bone atop the skull (the sagittal crest) that anchored heavy jaw muscles, and enormous molar teeth.

A talented and meticulous artist, Mary drew most of the illustrations of bone and tool specimens for Louis's books and scientific papers and published several volumes of her own. She spent much more time working at Olduvai Gorge than Louis did. While he was raising funds abroad, it was Mary who did most of the systematic development of the site.

Louis was excitable, impulsive, and intuitive, while Mary was careful, systematic, and logical. She recalled that he would get strong hunches about where fossils might be, and he was very often right—but very often wrong as well. Mary once opined that if they had both been the same kind of person, they wouldn't have accomplished as much.

In 1948, at Rusinga Island in Lake Victoria, Mary discovered a partial skull of an early apelike creature, dated at 18 million years. Louis named it *Proconsul* after Consul, a famous chimp at the Belle Vue Zoological Gardens, since he believed it to be ancestral to the chimpanzee. Louis never actually found a skull by himself. Mary, the African "hominid gang," and sometimes the Leakey children were the actual discoverers.

Between 1976 and 1978, at a site called Laetoli in Tanzania, 30 miles south

FINDING SPOTS to dig for early hominid fossils came easily to Mary Leakey and her dalmatians. Discoverer of the *Zinjanthropus* skull, Mrs. Leakey also unearthed three-million-year-old footprints of early hominids at Laetoli.
Photo © by Delta Willis.

of Olduvai Gorge, Leakey made what she considered the most exciting discovery of her career: preserved footprints of three hominid individuals who had left their tracks more than three million years ago in soft volcanic ash that solidified after a light rain. (Actually, they were found by an amateur member of her team, an American chemistry professor named Paul I. Abell.) It is a remarkable record of "fossilized" behavior, establishing that very ancient manlike creatures walked exactly as we do.

At a scientific meeting in Sweden in 1978, at which she was the honoree, Mary was scheduled to announce her discovery of a new fossil jaw at Laetoli, which she classified as *Homo habilis*. Shortly before her talk, the upstart Donald Johanson scooped her, not only describing her fossil but also declaring that it was not *Homo* but an australopithecine. He and Tim White thought it identical with his own *Australopithecus afarensis* ("Lucy") discovery, which came from Ethiopia—a thousand miles north of Leakey's site. Mary was outraged; thus began the bitter Johanson-Leakey feud that was to enliven paleoanthropology for years.

Mary was also, of course, mother and archeological mentor to Richard Frere Leakey, whom she had the foresight to tag with her prehistorian ancestor's family name.

See also FRERE, JOHN; LAETOLI FOOTPRINTS; LEAKEY, LOUIS; LEAKEY, RICHARD; "LUCY"

MARY LEAKEY PRESENTS her *Zinjanthropus* skull, dated at 1.75 million years, to President Nyerere of Tanzania in 1965.

LEAKEY, MEAVE E. (b. 1942)
Dynasty's Standard-Bearer

On March 22, 2001, the *New York Times* announced that paleoanthropologist Meave Leakey's discovery of *Kenyanthropus platyops* ("flat-faced man of Kenya") "threatens to overturn the prevailing view that a single line of descent stretched through the early stages of human ancestry." Almost immediately, Leakey's fossil of a 3.5-million-year-old skull and partial jaw was accepted as a previously unknown lineage of early hominids. The branching bush of human evolution had become a bit bushier.

Meave Epps, a native of London, attended convent and boarding schools, then pursued her early interest in science, taking degrees in zoology and marine zoology. While completing her doctorate in 1965, she went to work for Louis Leakey at his Tigoni Primate Research Centre, near Nairobi. There she met his son, Richard, who invited her to join his team of fossil hunters at Koobi Fora, near Kenya's Lake Turkana.

She married Richard in 1970, and the couple produced two daughters. One of them, Louise, born in 1972 and named for her grandfather, is now a respected paleoanthropologist who continues the family tradition.

In 1989, Meave Leakey became head of the National Museums of Kenya's paleontological field research in the Turkana Basin, where she focused on finding human fossils in 8-to-4-million-year-old deposits. In 1994, at a site called Kanapoi, her efforts paid off with the discovery of *Australopithecus anamensis*. Dated at 4 million years, it predates *A. afarensis*, or "Lucy," the oldest known hominid. *Anamensis* was not only a new species on the human family tree, but pushed the timeline for the evolution of hominids still further back.

LEAKEY, RICHARD ERSKINE FRERE (b. 1944)
Paleoanthropolgist, Wildlife Conservationist

When Richard Leakey was a boy in Kenya, he grew to hate the sight of old stones and bones. Since his parents, Louis and Mary, had dedicated their lives to the search for human origins, each new scrap of fossil bone became a rival for their attentions. When he found one, it was promptly snatched away. If there was one thing he was absolutely not going to be when he grew up, he decided, it was a paleoanthropologist.

Given the option of attending college or going to work when he was 17, he recalls in his book *One Life* (1981), "there was no doubt in my mind that I should avoid at all costs an academic life, and . . . determined to distance myself from my parents and their work."

RICHARD LEAKEY

Instead, he learned the ways of the bush, of organizing expeditions and tracking game, and decided to become a "white hunter," leading tourists on safaris. After a few years of success, he sickened at the senseless slaughter of wildlife and became an outspoken protector of African animals. Several times he organized equipment and vehicles for film crews covering his parents' work; he was impressed by the worldwide interest it attracted.

More like his father than he would admit (even to himself), the brilliant renegade from a family of brilliant renegades began to dream of following the track of early man. But it had to be on his own terms, not as his father's assistant or apprentice. "I wanted a show of my own," he later recalled.

Following his own hunches, he (and his friend, archeologist Glynn Isaac) prospected a few promising sites near Lake Turkana (then Rudolf). In 1968, he accompanied his father on a fund-raising visit to the National Geographic Society in Washington, D.C., and made his first bold step out of his father's shadow.

After Louis had received continued support for his and Mary's work at Olduvai Gorge, young Richard seized the floor to describe his own plans. Although untrained as an anthropologist or archeologist, he asked the startled board members to bankroll him at Lake Turkana. Startled, they agreed to sponsor one season. But they warned him that if he found nothing, he could never ask them for funding again.

His "Leakey luck" held, for he turned out to be a fossil-finder of the first rank. Early on, his assistant Kimoya Kimeu found the only known lower jaw of *Paranthropus boisei*, previously known only from Mary Leakey's find at Olduvai in 1959 (when it was called *Zinjanthropus*, or "Nutcracker Man").

But as Leakey's reputation grew, so did the tensions between him and his father, rival anthropologists, and government bureaucrats involved in administering the Kenya National Museums and archeological sites. His book *One Life* details the convoluted daily battles and infighting at which he proved to be proficient. In 1972, Leakey's team discovered the famous 1470 skull—a hominid with a large brain (750 cc) he identified as *Homo habilis*, a species of human rather than *Australopithecus*. His father, Louis, was delighted; it was the kind of evidence he had been seeking for an archaic type of human.

Unfortunately, shortly after that reconciliation, Louis died of a heart attack in London. With his mother's support, Richard became his successor in running the search for hominid fossils in East Africa. His team continued to make spectacular finds, including the remarkable skeleton of an adolescent *Homo erectus* (or *ergaster*) known as the "Turkana boy," and the "Black Skull" of a robust australopithecine.

Over the years, Leakey helped to overthrow simple linear schemes of hominid evolution and demonstrated that several distinct species of humans simultaneously inhabited ancient East Africa. He has championed *Homo habilis* as our earliest known ancestor, with all australopithecines as side branches. On this issue, Richard Leakey and his family were in remarkable agreement.

In 1979, Leakey almost died of kidney failure and was facing an end to his career. His brother, Philip, with whom he had never gotten along, offered to donate one of his own kidneys. Richard agonized over old resentments, the dangers to his brother, and the enormity of this gift of life, and finally accepted the organ transplant. Philip, a career politician, told a reporter, "Well, I can't really say I hate my brother's guts any more."

Leakey is fond of insisting he is not an anthropologist, a profession he swore he would never enter, and that he "has never been to a university—except to lecture." During his stint as director of the museum, he supervised projects in botany, zoology, geology, and many other fields, and has brought Africans into the scientific enterprise, which for so long was monopolized by Europeans.

In 1988, he accused high officials in the Kenyan government of being "soft" on elephant poachers, charging that they were profiting from wildlife slaughter. Political pressure forced him out of his posts as head of the museum and Department of Archeological Sites. Within a week, however, the president of Kenya personally intervened and reinstated Leakey. In

1989, he became Director of the Kenya Wildlife Service and launched a campaign to save the elephants from extermination by ivory hunters.

Declaring war on the poachers, Leakey assembled and trained a group of armed rangers so effective that some government officials began to fear his "private army." Highly placed ministers, some of whom were accepting cash from the poachers, wanted to clip Leakey's wings. In 1993, the bush plane he had flown for decades crashed during a short flight. Leakey survived, but lost both legs below the knees. After an investigation into possible sabotage, police declared it an accident.

A few years later, in 1995, Leakey ran for president of Kenya, heading a "Noah's Ark" coalition of multiracial ethnic groups who endorsed his ideals of a nontribal, pro-wildlife, education-fostering government. When he attempted to vote for himself at a polling place, he was attacked by hired thugs (linked by local journalists to the government), who pushed him off his prosthetic legs and whipped him until citizens came to his rescue. Five years later, President Moi appointed him director of the country's civil service, where he sought to clean up corruption and inefficiency. Leakey has since retired from public life to teach and tend his Kenyan farm.

See also *HOMO HABILIS*; JOHANSON, DONALD; LEAKEY, LOUIS; LEAKEY, MARY; TURKANA BOY

LEBENSBORN MOVEMENT
Attempt to Create a "Master Race"

Evolutionary studies inspired the eugenics movement, founded by Charles Darwin's cousin Sir Francis Galton in the 19th century. Galton proposed programs of selective breeding to "improve" the human species and discourage genetically "inferior" individuals from reproducing. But no government would attempt to evolve a new, "superior" breed of humans until the rise of Nazi Germany.

In 1936, Heinrich Himmler and his Storm Troopers (S.S.) founded an institution called *Lebensborn*, or "Fountain of Life." Its purpose was to create millions of blond, blue-eyed "Aryan" Germans as the genetic foundation of the new "master race." Lebensborn children would be raised to be obedient, aggressive, patriotic, and convinced that their destiny was to dominate or destroy all "inferior" races or nations. Galton's well-intentioned dream of human improvement had become a nightmare in reality.

The objective of Lebensborn was "to enlarge Germany's existing blood basis of 90 million to 120 million." This goal was to be achieved by forcibly taking "Aryan" or "Nordic" (blond, blue-eyed) children from their parents, wherever found, and raising them as Nazis. Another method was to mate Aryan-looking single women with genetically "superior" fathers. S.S. officers were assigned to perform this patriotic task. According to the Lebensborn Report for 1938, "832 valuable German women decided, despite their single state, to present the nation with a child." The report concluded that if each new child during its lifetime contributed 100,000 Reichsmarks to the economy, the Lebensborn program would have enriched Germany by 83 million Reichsmarks.

On his first inspection tour of occupied Poland, Himmler had been amazed by the Nordic appearance of many "Slavic" children and decided to kidnap them from their parents and send them to Lebensborn foster homes and orphanages in Germany. Two special S.S. agencies—the Volksdeutsche Liaison Office and the Race and Settlement Office—searched Europe for "human specimens considered suitable for Germanization." They also "cleared" vast areas of land of "inferior" Poles and Jews by killing or enslaving the inhabitants, and put to death thousands of Germans who were considered feebleminded, infirm, mentally ill, or crippled. (The gas chambers were originally invented to weed out "unfit" Germans and were later applied to the genocide of entire "undesirable races," particularly Gypsies and Jews.)

It was an article of faith that the blond, blue-eyed, "Nordic-looking" children would prove intellectually and morally superior and would "breed true" when mated. Neither assumption was correct. Instead of changing evolutionary history, the Lebensborn experiment produced a group of perfectly ordinary, confused orphans, who had to cope with the chaotic aftermath of a society gone mad.

See also "ARYAN RACE," MYTH OF; EUGENICS; GALTON, SIR FRANCIS

Nazis set up a human breeding program as the "scientific" foundation of a new "Master Race."

LESQUEREUX, LEO (1806–1889)
Heroic Paleobotanist

One of the most tragic and remarkable of earth scientists, Leo Lesquereux helped found the study of fossil plants in North America. His beautifully illustrated volumes of *American Coal Flora* (1879–1884) rivaled the best that were being published in Europe, yet he was unable to earn a living from teaching or research. A Swiss immigrant to America, he learned English after he became deaf and so never heard the language in which he wrote his books.

Lesquereux's family pursued a typically Swiss occupation: they manufactured parts for watches. As a boy, he loved to climb difficult mountains to gather rare flowers from their peaks. At the age of 10, a misstep took him over the edge of a cliff and he bounced down a mountainside, unconscious most of the way. No bones were broken, but he lay in a coma for two weeks. His hearing was permanently damaged; it would grow progressively worse over the years.

He tutored French, married a noblewoman who was one of his students, and became principal of the local high school. His spare time was spent studying mosses and collecting rocks in the mountains. He was the first to understand the nature of peat deposits, and he analyzed their stratification, the specific gravity of their different layers, the deformation of sphagnum mosses by pressure, and rates of growth and decay. His authoritative work was the first to determine the true causes and conditions of peat formation, which advanced understanding of the formation of coal.

Meanwhile, his hearing grew worse, and an operation in Paris intended to restore it instead extinguished it completely. He had to give up his career as an educator and followed his friend August Agassiz (brother of Louis, the Harvard paleontologist) to America. Lesquereux was 41 when he arrived in Boston in 1847. With a wife and five children to support, he spoke no English and was now totally deaf.

Agassiz hired him to classify a plant collection. Soon his reputation grew, prompting an invitation to Ohio to assist in studies of mosses. Other invitations and commissions followed, which allowed him to make important contributions, including definitive books on the Pennsylvanian coal deposits.

Scientific work did not provide a living for him. No university would hire a deaf professor, so Lesquereux worked as an unpaid assistant on various geological expeditions in order to pursue his passion for paleobotany. In between field trips, he set up a small watchmaking and jewelry business, and trained his three sons in that business.

When he could manage it, Lesquereux plugged away tirelessly at paleobotany, overcoming every obstacle. On expeditions, he sat day after day on waste slate heaps, turning over every rock in his search for the plant fossils that fascinated him. Cold winds, noonday heat, or pouring rain made no difference. Despite the formidable hardships he faced, he continued to be productive into his eighties.

Colleagues marveled at his abilities to overcome adversity. Writing in 1895, J. P. Lesley recalled: "I have been present when Lesquereux talked with three persons alternately in French, German, and English by watching their lips. The interview would begin by each one saying what language he intended to use." Then Lesquereux would answer each in their own tongue. However, since he had never heard the sounds of English, "his pronunciation of it was curiously artificial and original." As he told *Popular Science Monthly* in 1887, "My deafness cut me off from everything that lay outside of science. I have lived with Nature, the rocks, the trees, the flowers. They know me, I know them. All outside are dead to me."

Lesquereux's faith in revealed scripture was strengthened rather than weakened by the *Origin of Species* (1859). Charles Darwin, in a letter to his friend Joseph Hooker in 1865, expressed amazement at "an enormous letter from Leo Lesquereux on Coal Flora." Although Lesquereux had originally been a trenchant critic of the *Origin of Species*, Darwin wrote, "he says now after repeated reading of the book he is a convert! But how funny men's minds are; he says he is chiefly converted because my books make the Birth of Christ, Redemption by Grace, etc., plain to him!!"

> Although stone deaf, Lesquereux could converse with three people at once in French, German, and English by watching their lips.

LINNAEAN "SEXUAL SYSTEM"
Erotic Botanica

Carl Linnaeus has come down in history as a dry-as-dust classifier who gave thousands of Latin names to plants labeled in museums. What is often forgotten is that some of those labels created a sensation in the mid-18th century and brought their author both fame and infamy.

As a student, Linnaeus became fascinated with plant sexuality and peppered his botanical writing with erotic imagery. In his thesis, *Praeludia Sponaliarum Plantarum*, presented in 1729 at the University of Uppsala, he wrote of the pistils and stamens of flowers:

> The actual petals of a flower contribute nothing to generation, serving only as the bridal bed which the great Creator has so gloriously prepared, adorned with such precious bed-curtains and perfumed with so many sweet scents in order that the bridegroom and bride may therein celebrate their nuptials. . . . When the bed has thus been made ready, then is the time for the bridegroom to embrace his beloved bride and surrender himself to her.

Taking his cue from the earlier work of Rudolf Jakob Camerarius (*Epistola de sexu plantarum,* 1694) and Sébastien Vaillant ("Sermo de Structura Florum," 1718), Linnaeus had no hesitation in describing pollination as a sexual act. Pollen was sperm, seeds were ova, and there is no fertilization if the anthers are removed (castration).

By the time he brought out his *Systema Naturae* in 1735, Linnaeus had devised a complete "sexual system" for classifying plants. Flowering plants were grouped according to the number of stamens, while flowerless groups were defined by the form of the female organs.

Monandria (ginger, cardamom, arrowroot) is like "one husband in a marriage," *Diandria* (lilac, privet, nightshade) like "two husbands in the same marriage," and the *Polyandria*—a group that includes the linden and the poppy—"twenty males or more in the same bed with the female." The counting of stamens and pistils was a practical way of determining the group to which it belonged, and the erotic imagery a strong aid to remembering the family name.

The system proved to be very practical in identifying species and also greatly helped to popularize the study of plants and flowers. It even inspired Charles Darwin's grandfather Erasmus to write a controversial book of erotic poetry, *The Botanic Garden; or, Loves of the Plants* (1790). Illustrated with drawings of sensual human lovers, Erasmus's book-length poem celebrated the sexuality of vegetation with decorous lustiness.

Many of Linnaeus's fellow naturalists admired the system's usefulness and were amused by its explicitness, but others were outraged at his "botanical pornography." Long after the great taxonomist's death, the bishop of Carlisle wrote that "nothing could equal the gross prurience of Linnaeus' mind." Churchmen looked to nature for morality: the industrious ant, the persistent tortoise, the faithful dove. To equate plant fertilization with the variegated sex life of animals was to call attention to "polygamy, polyandry, and incest" in nature. Such botanical names as *Clitoria* (a pea plant) were, as the bishop of Carlisle wrote in 1808, "enough to shock female modesty." In 1820, even the great German poet-naturalist Goethe advocated censorship of the sexy botanical textbooks used by university students.

See also BINOMIAL NOMENCLATURE; LINNAEUS, CARL; LINNEAN SOCIETY; SPRENGEL, CHRISTIAN KONRAD

Linnaeus classifed plants by their manner of reproduction, which churchmen found shocking.

LINNAEUS, CARL (1707–1778)
Botanist, Founder of Taxonomy

In Sweden, Karolus (or Carl) Linnaeus is still a great name, as important as Shakespeare is to England or Dante to Italy. His system of classification of living things still forms the basic language of biology. During his lifetime, he was an international celebrity who dominated European biology, although churchmen were shocked by his "sexual system" for classifying plants.

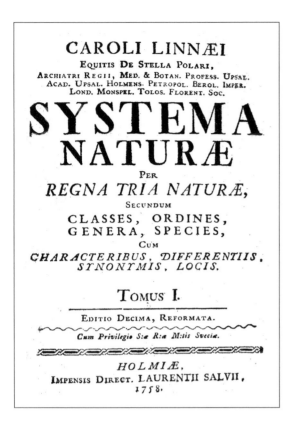

CAROLI LINNÆI
Equitis De Stella Polari,
Archiatri Regii, Med. & Botan. Profess. Upsal.
Acad. Upsal. Holmens. Petropol. Berol. Imper.
Lond. Monspel. Tolos. Florent. Soc.

SYSTEMA NATURÆ

Per
REGNA TRIA NATURÆ,
Secundum
CLASSES, ORDINES,
GENERA, SPECIES,
Cum
CHARACTERIBUS, DIFFERENTIIS,
SYNONYMIS, LOCIS.

TOMUS I.

Editio Decima, Reformata.

Cum Privilegio S:æ R:æ M:tis Sveciæ.

HOLMIÆ,
Impensis Direct. LAURENTII SALVII,
1758.

A GROUNDBREAKING work on classification, or taxonomy, Linnaeus's *Systema Naturae* revolutionized botany and zoology. Superior to previous systems, it later became the basis for evolutionary systematics.

Born in the southern Swedish province of Småland near a beautiful lake, he grew up surrounded by gardens and wildflowers. The surname Linnaeus was coined by his father Nils, after a lime tree near his birthplace. His parents had hoped he would follow his father into the church, but, to their disappointment, he showed neither interest nor talent for metaphysics and other priestly subjects.

He loved to work with plants, and a kindly professor steered him toward medical study. All physicians had to know some botany, since most drugs came from the rare plants they had to grow themselves in special gardens of *materia medica*.

Linnaeus was much more interested in the plants than in their medical uses and set out to be a dedicated botanist. After years of poverty and perseverance, he achieved his goal of becoming professor of botany at the University of Uppsala.

It was the perfect time to be a master of botany. As European colonization of far-flung lands reached its zenith, immense numbers of animal and plant specimens new to science were being sent back to England, Holland, and France. It became urgent to bring order to this rapidly expanding body of knowledge.

Since natural science grew out of the religion of his time, Linnaeus regarded each species as a distinct idea in the mind of God. He saw natural groups as divine "plans" of organization and thought his mission was to discern that plan. As he modestly put it, "God created, but Linnaeus classified."

Linnaeus co-invented his classification system with zoologist Peter Artedi (1705–1735), who helped to develop the binomial system. They decided to divide up the natural world, with Artedi taking the fishes while Linnaeus took plants, birds, and insects. Before Linnaeus and Artedi developed their system, natural history classification was in chaos. Without regard for systematic anatomical comparison, some put flying fish and birds together (because both fly) or turtles and armadillos (because both have shells.)

Artedi and Linnaeus vowed that if one should die, the other would take over his notes and finish his work. In 1735, on the way home from an evening out in Amsterdam, Artedi stumbled into a canal and drowned. True to his word, Linnaeus finished up his friend's book on fish, allowing Artedi to be remembered as the "father of ichthyology"—although few but fish experts know his name today.

Little by little, his wide knowledge, unceasing labor, and personal charisma brought Linnaeus to the top of his field. He developed unsurpassed collections of animals and plant specimens, containing thousands of species he had personally discovered and named. Many students came to study with him, and he soon had a clique of "disciples."

Established as a powerful and popular professor, he sent out fourteen of his students to the world's most remote wilderness areas to discover and collect thousands of animals, plants, and insects to add to the roster of known species. Eventually, he and his students devised scientific names for the roughly 4,400 species of animals and 7,700 plants that were then known. Botanical names published before 1763 have no standing unless they were adopted by Linnaeus in his *Species Plantarum* (1753) or *Genera Plantarum* (1737; 6th ed. 1764). His *System of Animate Nature* (1735) has gradually been expanded, until today it includes 350,000 plants and more than a million animals.

Linnaeus said that he was following Adam's example by spending his life giving names to all living things, though he was also interested in their behavior, reproduction, and distribution. His system of classification became so successful that for many years later naturalists became preoccupied with finding and naming new species to the exclusion of studying how plants and animals live in nature.

See also BINOMIAL NOMENCLATURE; LINNAEAN "SEXUAL SYSTEM"; LINNEAN SOCIETY

LINNEAN SOCIETY
British Science Organization

Carl Linnaeus (1707–1778), the inventor of the scientific classification system of living things, spent a lifetime amassing a priceless museum of specimens. Many still exist: 2,000 are in the Natural History Museum in Stockholm. But his personal collection—including all his books, manuscripts, correspondence, dried fishes, shells, insects, and herbarium—became a legacy for his widow. She would have preferred cash.

In 1784, Mrs. Linnaeus happily sold the collections to a wealthy English naturalist named James E. Smith. They were his pride and joy until he died 30 years later, at which time Mrs. Smith put them up for sale. She could not find an individual who wanted the entire collection, so she decided to break it up and sell it piecemeal.

Some of Linnaeus's English admirers thought it would be a tragedy to further disperse the collections. In 1829, a public-spirited group of gentlemen decided to form an association to keep what remained of Linnaeus's collection and library intact and out of the hands of mercenary widows. After raising the necessary contributions, they bought the great taxonomist's collection and founded the Linnean Society of London. At its headquarters at Burlington House, Piccadilly, the society maintains an extensive natural history library and keeps Linnaeus's preserved specimens as well as his correspondence in a temperature-controlled vault.

During the 19th century, the society became a major force in British science, where naturalists and botanists presented their new discoveries to the country's scientific elite.

In April 1858, the distinguished geologist Sir Charles Lyell and the botanist Sir Joseph Hooker "communicated" to the society a joint theoretical statement by Charles Darwin and Alfred Russel Wallace. Although their versions were separate (Darwin's had been written 14 years earlier), their theory proposed that species gradually evolved through the process of "natural selection."

That occasion marked the beginning of modern biology, but neither Darwin nor Wallace appeared at this famous non-event. It attracted little attention and no discussion at the society, whose namesake had championed the permanence of species. In the society's annual report for 1858, its president noted that it had been a dull year in biology, with no important new theoretical developments of the sort that occasionally revolutionize a scientific field.

In recent years, the Linnean Society has had a spectacular resurgence, leading the application of computer technologies to the systematics of living things, while preserving its invaluable historical archives of natural history. In 2004, led by its executive secretary, the late John Marsden, the organization successfully resisted attempts by the government to remove the "Linn Soc" from its venerable home in Burlington House, where it continues to serve as a major clearinghouse for many branches of the life sciences.

LONESOME GEORGE
Symbol of Endangered Species

There is something deeply affecting about a creature who has spent the better part of a century searching for a mate and has no way of knowing he is the very last of his kind. "Lonesome George," a 200-pound Galápagos giant tortoise, who resides in a spacious pen at the Charles Darwin Research Station on Santa Cruz Island, has became a symbol of endangered species. The San Diego Zoo has erected a bronze statue in his honor, and he has inspired thousands of contributions to the cause of threatened wildlife.

George is famous as the last of the saddle-backed species (*Geochelone elephantophus abingdoni*) originally native to Pinta Island, one of the driest in the Galápagos. An upturned shell shape enabled this species to raise their necks and reach high for cactus pads, unlike those from other islands whose shells are dome-shaped.

When he was discovered in 1971, no giant tortoises had been seen on Pinta for years and were believed to have been hunted to extinction. As the last of a doomed species, he was

> "When the last individual of a race of living things breathes no more, another heaven and another earth must pass before such a one can be again."
>
> —William Beebe

LONESOME GEORGE.
© Heidi Snell/Visual Escapes.

removed for safekeeping, and a $10,000 reward offered to anyone who could find a mate for him in any of the world's zoos or private collections. None was ever found.

Darwin visited the Galápagos in 1835 and was fascinated by its animals, including the tortoises. He was intrigued to learn that each of the 14 islands had its own species and that many islands had their own related varieties of mockingbirds and other creatures, which led him to conclude that the populations had diverged from common ancestors and were not the result of multiple creations.

Giant tortoises take their time to move, to breed, and to live—some for more than 200 years. Sailors, pirates, and whalers carried off more than 200,000 of them to help feed crews on long voyages. Hundreds were dumped, live, into ships' holds for months, awaiting their turn to become soup. So badly were they abused that sailors, perhaps to assuage their own guilt, believed that wicked naval officers were doomed to return as giant tortoises.

Those remaining on the islands were decimated by the goats, cats, pigs, dogs, and rats that had been introduced, which ate thousands of the tortoise eggs as well as their plant foods. In addition, seven scientific "collecting" expeditions invaded the islands between 1897 and 1906. The last of them, from the California Academy of Sciences, took 75,000 specimens, including 264 tortoises from 10 different islands. It is remarkable that any Galápagos wildlife managed to survive the earlier attentions of the scientific community.

In recent years, conservationists at the Charles Darwin Research Station have bred thousands of tortoises, which were restored to the various islands appropriate to their species. For now, George is a pampered prisoner, sharing his pen with two females of closely related species, in which he has shown no interest. Why does Lonesome George receive so much attention when his species is apparently a "lost cause"? As William Beebe, an American naturalist, wrote in *The Bird: Its Form and Function* (1906), "When the last individual of a race of living things breathes no more, another heaven and another earth must pass before such a one can be again."

But hope springs eternal. Despite the failed quest to find a female Pinta tortoise anywhere in captivity, in 2007 scientists made a promising discovery. After sampling the genes of tortoises on Isabela Island, biologists announced that they had detected some Pinta genes among the large tortoise population there. Now there's a new place to seek a female for George.

How could some Pinta tortoise genes have mixed into the population of another island? Biologists speculate that perhaps some sailors more than a century ago may have lightened their cargo load by dumping a few Pinta tortoises overboard near Isabela Island, where they made it to shore. Using Pinta DNA preserved in long-dead museum specimens for comparison, the scientists were able to recognize it when they found it among the present Isabela population. Humans, which came within a hair's breadth of exterminating the Pinta tortoise, may still have an opportunity—though an uncomfortably small one—to rescue it. If a female should eventually be discovered, it will then be up to the diffident and reclusive George as to whether or not he will remain eternally lonely.

See also DARWIN RESEARCH STATION; EXTINCTION; GALÁPAGOS ARCHIPELAGO

LORENZ, KONRAD (1903–1989)
Ethologist, Behavioral Evolutionist

Konrad Lorenz's entertaining account of his animal acquaintances, *King Solomon's Ring* (1952), is a natural history classic. According to myth, Solomon's magical ring granted him the gift of speaking with birds and beasts; in reality, Lorenz pioneered new scientific understanding of animal communication. A naturalist in the Darwinian tradition, he also attempted to understand how animal "displays" or social signals have evolved.

Born in Altenberg, Austria, Lorenz took a medical degree at Vienna, but spent most of his life pursuing a childhood fascination with animals. Working with creatures common about his country home, such as jackdaws and geese, he was able to unravel behavior patterns that had eluded generations of comparative psychologists, with their mazes and "puzzle boxes."

With Niko Tinbergen and Karl von Frisch, he founded ethology, a science of behavior that considered the whole animal. Ethologists were not interested in some arbitrary measure of "intelligence" or "learning ability," but in the animal's natural behavior: its courtship displays, social signals, species isolating mechanisms, nesting, and territorial behavior. They revolutionized the study of animal behavior and placed it back in the context of an animal's life cycle, ecology, and evolution. (By the 1990s, the terms "sociobiology" and "evolutionary psychology" had largely replaced "ethology," but all are focused on behavioral evolution.)

Lorenz's approach was to watch what animals did naturally, rather than stick them into preconceived experiments. Within a day or so after hatching, for instance, ducklings and goslings begin to follow their mothers to food and away from danger. They become, in Lorenz's terms, "imprinted" on the mother duck at a critical period of development.

Lorenz showed he could imprint the young waterfowl on something very unlike their natural mothers: for example, on himself. He raised broods of geese and ducks, leading them to water and showing them where to forage. Some scientists found his role of surrogate mother as bizarre as the birds' behavior. Lorenz realized that his own parental instincts were triggered by the duckling's "cuteness"—the round, big-eyed face common to the young of many vertebrate species. [See CUTENESS, EVOLUTION OF.]

While insisting that ethology must objectively describe observed behavior patterns, Lorenz found it difficult not to consider his study subjects little people, often attributing to them humanlike emotions and motives.

Pair formation among greylag geese, he wrote, contains all the emotions of human marriage. It "follows exactly the same course as with ourselves." In *The Year of the Greylag Goose* (1979), Lorenz spoke of one bird's "scorned mistresses," another's "unfaithful mate," and a goose's "dumbstruck grief" at the loss of his beloved. In *Natural History* magazine, a zoologist wondered, "How did this soap opera get into a book about geese?"

In attempting to reconstruct the evolution of behavior, Lorenz tried to compare "units" of behavior the same way 19th-century anatomists had compared bones and muscles. During courtship, for instance, waterfowl show a range of different, stereotyped behaviors peculiar to various species: One gives an exaggerated preening display, while another combines a few of the same movements with "ritualized" feeding or aggressive behavior.

Lorenz tried to arrange such "displays" into a family tree of behaviors, deducing how the more elaborate had evolved from the simpler. Although zoologists were intrigued with his conclusions, ultimately they were abandoned as conjectural and unprovable—much like Oxford's Pitt-Rivers Museum's 19th-century arrangements of human technology in evolutionary series. [See COMPARATIVE METHOD; PITT–RIVERS MUSEUM.]

With his fellow ethologists Tinbergen and von Frisch, Lorenz was awarded the 1973 Nobel Prize for Physiology or Medicine. But his brilliant, influential work was flawed by a biological determinism that lent itself to too-facile "jumps" from animals to man. It was an attractive bit of circular reasoning. After projecting human motivations onto animals, he jumped back from animals to man to prove his case. In *On Aggression* (1966), he argued that wars and political turmoil stemmed from an innate need for humans to attack and fight each other over mates and territory unmediated by "genetic mechanisms for social cooperation." Such fierceness was built into humans by evolution, he concluded, but not into what he considered the gentler carnivores—wolves and lions!

Even more upsetting to his supporters, it transpired that Lorenz, who had joined the National Socialist Party as a youth, had once bolstered Nazi racial theories with early writings on biological "purity of type." In a 1940 paper, "Disorders Caused by the Domestication of Species-Specific Behavior," Lorenz argued that domesticated animals lose their inborn preference to mate with the "pure wild type." Humans, he thought, had become "domesticated" by race-mixing in Europe, resulting in the loss of pure "types" and "degeneration" of the "higher" races—a situation that could only be corrected by state-controlled breeding and elimination of the "unfit." Only five years before, German Jews and gentiles were forbidden to intermarry based on similar pseudoscience. In a 1938 paper on the psychological effects of domestication in animals, Lorenz wrote:

> This high valuation of our species-specific and innate social behavior patterns is of the greatest biological importance. In it as in nothing else lies directly the backbone of all racial health and power. Nothing is so important for the health of a whole *Volk* [people] as the elimination of "invirent types": those which, in the most dangerous, virulent increase, like the cells of a malignant tumor, threaten to penetrate the body of a *Volk*. . . . [By studying animals], which are easier and simpler to understand, . . . [we] discover facts which strengthen the basis for the care of our holiest racial, *Volk*ish, and human hereditary values.

In a paper published in 1940, Lorenz suggested that when civilization causes "imbalance" within a population, "race-care must consider an even more stringent elimination of the ethically less valuable than is done today."

Some of Lorenz's defenders have claimed that he was politically naive or misunderstood. Certainly, he was appalled when, in 1943, he personally witnessed a group of Gypsies carted off to the death camps. In his acceptance speech for the 1973 Nobel Prize in Physiology, Lorenz rationalized his earlier views but stopped short of an apology: "Many highly decent scientists hoped, like I did, for a short time for good from National Socialism, and many quickly turned away from it with the same horror as I."

During the 1970s, there was an exaggerated reaction by some biologists and anthropologists to E. O. Wilson's book *Sociobiology* (1975), a new attempt to explain the biology of behavior from insect and fish to man. To Wilson's astonishment, scientific meetings became stormy and hostile; epithets of "biological determinist" quickly turned to "fascist" and "Nazi." It was, unfortunately, part of the legacy of Konrad Lorenz, who left an equally strong image as the charming papa of a line of baby ducks.

See also BIOLOGICAL DETERMINISM; ETHOLOGY; LEBENSBORN MOVEMENT; TINBERGEN, NIKO

LOST WORLD, THE (1912)
Doyle's Prehistoric Adventure

Sir Arthur Conan Doyle, the creator of Sherlock Holmes, had a love-hate relationship with evolutionary scientists. On the one hand, he was intrigued by their visions of ape-men and prehistoric monsters, but he was repelled by their narrow materialism and rejection of his Spiritualist religion.

His "boy's adventure novel," *The Lost World*, expressed both attitudes. Originally begun as a serial novel in the *Strand* magazine in April 1912, it introduced his raucous maverick scientist, Professor Challenger, who was to star in several subsequent adventures.

Challenger, a robust, dark-bearded explorer, leads an expedition to the deepest unknown jungles of South America, where his party stumbles into a "lost world," which had remained unchanged for millions of years. They encounter stegosaurs, flying reptiles, and tribes of red-haired ape-men in adventures Doyle admitted were inspired by the novels of Jules Verne.

When Challenger returns to civilization, he has a hard time convincing his fellow scientists he is not a liar. What evidence will satisfy them? A bone? A photograph? "A bone can be as easily faked as a photograph," says one of the characters—a sly reference to the controversy that surrounded Doyle's belief in "spirit photography."

In an appreciation of the book, novelist John Dickson Carr noted that its satiric descriptions of scientific debates are more entertaining than the dinosaurs. Conan Doyle paints a comic picture of "grave-bearded men of science, where some abstruse theory is concerned, behav[ing] exactly like temperamental prima donnas and . . . fully as jealous of each other. . . . The zoologists are as interesting as the zoo."

Doyle identified closely with his hero and even

donned a fake beard to pose as Challenger for the frontispiece when the story was published in book form.

In 1983, anthropologist John H. Winslow proposed in *Science '83* magazine that *The Lost World* contained hidden clues to the Piltdown hoax and suggested that Conan Doyle was the hoaxer. He noted that a map of the "Weald" in the novel was similar to the terrain of the Piltdown site, that there were references to "faked bones" and "prehistoric practical jokers," and that the ape-men were red-haired. (The Piltdown "man" was a human skull with the jaw of an orangutan planted by hoaxers.)

"THE LOST WORLD."
The Leader of the Explorers, with some of their Adventures.

Conan Doyle's motive, Winslow claimed, was to embarrass "materialist" scientists who denied a "spirit" role in human origins. Some Darwinian evolutionists had embarrassed Doyle's Spiritualist friends, claiming they were incapable of separating real evidence from fraud. Winslow singled out Edwin Ray Lankester, director of the British Museum of Natural History.

Lankester had become famous while a young man for exposing the medium Henry Slade as a swindler and prosecuting him as a "common rogue" in police court. It was the first time a scientist had charged a psychic with conducting fraudulent experiments in a court of law and did much to publicly discredit the Spiritualist movement.

Doyle never forgave Lankester and later spent much of his fortune promoting Spiritualism, which he believed was an "all important revelation" for mankind. Still, Doyle was impressed with Lankester's book *Extinct Animals* (1905) and drew heavily on its descriptions of prehistoric dinosaurs for *The Lost World*.

For years, Lankester's reaction to *The Lost World* was lost to history. However, in 2004 the British Library bought a box of papers from Conan Doyle's estate that contained the following letter, written by Lankester in 1912:

Dear Conan Doyle—

You are perfectly splendid in your story of the lost world mountain top. I feel proud to have had a certain small share in its inception as you indicate by quoting my book on Extinct Animals at the start—It is a delightful thing to imagine all those beasts and the ape-men—lingering together on this shut off peak—

Can your men escape by training a vegetarian pterodactyl to fly with them one at a time? Will some ape-woman fall desperately in love with the Professor and murder the leaders of her tribe to save him? Natural gas collected in a balloon made of Pterodactyl wings—might be a means of escape from the mountain. . . . [The ape-men] might be poisoned by alcohol or by some infective germ. . . . Sweep them all out with a Trypanosome brought up in the Professor's saliva!

Sincerely yours,
E. Ray Lankester

Although the *New York Times* headlined "Arthur Conan Doyle Is Piltdown Suspect" (1983) in reporting Winslow's views, Winslow was never able to muster more than circumstantial evidence that Doyle was the culprit. Other plausible suspects have been advanced, as summarized in *The Piltdown Inquest* (1986) by Charles Blinderman and *Unraveling Piltdown* (1996) by John Evangelist Walsh. The actual perpetrator seems to have been Charles Dawson, a local amateur archaeologist, whose many other "discoveries" of strange and unusual artifacts have since turned out to be fakes as well.

One of the first and most influential dinosaur adventure movies was made of *The Lost World* in 1925, starring Wallace Beery as a memorable Professor Challenger. Although the special effects, including one of the first uses of miniature models in stop-frame photography, are crude by current standards, they caused a sensation in 1925.

The process was so new that Conan Doyle created a stir by previewing the dinosaur sequences without explanation. At the American Club of Magicians in New York on June 2, 1925, Doyle announced that he would show the magicians a glimpse of something not exactly "psychic," but certainly "not nature as we can now observe it."

After building an atmosphere of expectation and mystery, he showed prehistoric iguanodons and brontosaurs fighting and rearing their young on the big screen.

Next day the *New York Times* headlined: "Dinosaurs Cavort in Film for Doyle. Spiritist Mystifies World-Famed Magicians with Pictures of Prehistoric Beasts—Keeps Origin a Secret."

Conan Doyle's resurrected monsters had been animated by a young genius in the new medium named Willis O'Brien. Some years later, O'Brien created and animated the models for the original *King Kong*.

See also *KING KONG*; KNIGHT, CHARLES R.; LANKESTER, E. RAY; O'BRIEN, WILLIS; PILTDOWN MAN (HOAX); SLADE TRIAL

LUBBOCK, SIR JOHN (LORD AVEBURY) (1834–1913)
Pioneering Prehistorian

When John Lubbock was a boy, his father, a wealthy banker, astronomer, and mathematician, told him that he had some wonderful news. Young Lubbock guessed he was going to get a pony. "Oh, it is much better than that," Lubbock Senior replied. "Mr. Darwin is coming to live at Down." "I confess I was much disappointed," Lubbock recalled, "though I came afterwards to see how right he was."

The Lubbocks lived on a grand estate called High Elms in the Kentish countryside, about a mile from Darwin's Down House. (The mansion burned down in 1967, and the estate grounds are now a public park and nature reserve.) "Insofar as one could be born and bred to Darwinism before 1858," writes historian George Stocking in *Victorian Anthropology* (1987), "John Lubbock was." He found in Darwin a teacher, mentor, and father figure who greatly influenced his life and career. At the celebrated Oxford Debate at which Huxley confronted Bishop Wilberforce, Lubbock gave a long, effective defense of Darwinism, using evidence from embryology.

An overachieving Victorian gentleman, he was talented and active in many fields. As a young man, he told Darwin his three goals: to be Lord Mayor of London, Chancellor of the Exchequer, and president of the Royal Society. Darwin said he could be any one if he gave up the other two. Lubbock ignored this advice and did not attain any of them.

Nevertheless, he made his mark in finance, politics, and science. While he dutifully followed his father into the banking business, his heart followed his mentor's interests into natural history. As a teenager, he discovered the first fossil musk-oxen in England, thus helping to establish the existence of a cold glacial period, which delighted Darwin. He published many original papers, but his most lasting contributions were in the fields of comparative psychology, prehistory, and the behavior of social insects.

One of the leaders of the "prehistoric movement," he focused on exploring a time period thousands of years before what was generally considered knowable "history." He was fascinated by the stone "hand axes" and other evidences of early humans that were coming to light, largely through the excavations of the French amateur archeologist Jacques Boucher de Perthes, whose work had been met with skepticism and disbelief for 25 years. Lubbock toured the Somme River gravels in 1860, escorting English geologist Sir Joseph Prestwich and others to see the prehistoric sites for themselves. All were impressed by the abundant evidence of extinct mammoths, woolly rhinos, and other cold-weather animals in the same layers with stone tools made by early humans. Lubbock published his landmark book, *Prehistoric Times*, in 1865.

Lubbock decorated the walls of High Elms with hundreds of primitive tools and weapons from ancient digs and current tribal peoples. He coined the terms for the divisions between the Old Stone Age (Paleolithic) and New Stone Age (Neolithic), which are still used today. As historian George Stocking pointed out, the full title of Lubbock's book, *Pre-historic Times, as Illustrated by Ancient Remains, and the Manners and Customs of Modern Savages*, spells out the comparative approach of sociocultural evolutionism, which was to dominate anthropology for a century. In 1867, Lubbock became the first president of the Royal Anthropological Institute, and he also served as president of the Linnean Society of London.

SIR JOHN LUBBOCK, M.P., F.R.S.

How doth the Banking Busy Bee
Improve his shining Hours
By studying on Bank Holidays
Strange Insects and Wild Flowers !

NATURALIST, PREHISTORIAN, POLITICIAN, AND BANKER, Sir John Lubbock became Charles Darwin's "scientific son."

While he admired the cleverness of early peoples, as reflected in their artifacts, his view of contemporary tribes was distorted by upper-class Victorian condescension. In general, he thought non-Western tribes had beastly manners and no "real" religion. Far from painting the savage as noble and free, Lubbock thought tribal man was "a slave to his own wants [and] passions . . . [suffering] from the cold by night and the heat of sun by day; ignorant of agriculture, living by the chase, and improvident in success, hunger always stares him in the face, and often drives him to the dreadful alternative of cannibalism or death." In his later book *Origin of Civilisation* (1870), however (published within a year of Darwin's *Descent of Man*), Lubbock revised his view of savages to allow for evolutionary progress.

One special room in Lubbock's home contained more than 30 ants' nests of many species. His classic *Ants, Bees, and Wasps* (1882) detailed many experiments on the behavior, social organization, and "mental activity" of social insects. Ants are mindless "automata," he thought, but occasionally show altruistic feelings for their fellows. Lubbock describes them "tenderly" carrying off wounded comrades. Indeed, his observations on social insects were sometimes kinder and more sympathetic than his descriptions of tribal peoples.

He was also intrigued by so-called slave-owning ants, which were so dependent on being fed by their "slaves" that their mouth parts degenerated. This proved, said the Liberal Lubbock, that slavery caused "degeneration" among slave-holding ants much as it did "among humans who become dependent on slave labour." (Recent studies of these species by entomologist Howard Topoff and others have shown that the biology of ant behavior is so dissimilar to human institutional slavery as to make such comparisons meaningless.)

Among Lubbock's other important discoveries, he was the first to document color vision in bees, which confirmed Darwin's view that flower forms and colors were adaptations to attract pollinating insects. [See DIVINE BENEFICENCE, IDEA OF.]

In the 1870s, Lubbock commissioned a series of watercolor paintings by Ernest Griset depicting mammoth hunts and other activities of early humans—the first series of reconstructions in the new genre, now known as "paleoart" (the paintings are now in the Bromley Museum in Orpington, Kent).

After several unsuccessful tries, Lubbock was elected to Parliament in 1869; his public list of supporters included John Stuart Mill and Charles Darwin. He established the first secular bank holiday in England—the first time workers were given a long weekend without the hypocrisy of a religious excuse. (It is the first Monday in August, which used to be popularly called "St. Lubbock's Day.") He also fought for shorter working hours, the introduction of scientific education into the schools, and laws protecting native birds and forests.

In later life, he wrote popular books on travel, free trade, botany, and economics, as well as *The Senses, Instincts, and Intelligence of Animals* (1888) and *The Pleasures of Life* (1887–1889), which included his celebrated list of the "One Hundred Best Books." He outlived many of his great scientific friends who were fellow members of the exclusive "X Club"—an intimate circle of Darwinian scientists—and served as a pallbearer at both Darwin's and Huxley's funerals.

During the 1890s, Lubbock devoted himself to rescuing a group of ancient earthen mounds at the village of Avebury, in Wiltshire, that once contained a great circle of 600 immense stones spread over several miles. He considered it "the finest megalithic ruin in Europe . . . older and much grander than Stonehenge." When the area was to be leveled for commercial development, he promptly bought up as much land there as possible so he could rescue the remaining 100 stones at the ancient site. He was rewarded for his public service with a peerage in 1899, and chose to call himself Lord Avebury, after his beloved prehistoric ruins.

See also CAVEMAN; INSECT SOCIETIES, EVOLUTION OF; INSECTIVOROUS PLANTS; LINNEAN SOCIETY; "X" CLUB

DARWIN'S PROTÉGÉ studies one of his glass-enclosed ant colonies at High Elms, the Lubbock mansion.

Lubbock's mansion housed collections of primitive tools, prehistoric artifacts, and glass cases filled with live ants, bees, and wasps.

"LUCY"
Early East African Hominid

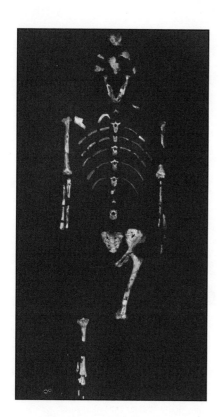

THE MOST COMPLETE SKELETON from 3.2 million years ago, "Lucy" was discovered by the Johanson-Taieb field team in the Afar region of northern Ethiopia in 1976.

According to anthropologist Donald Johanson's account *Lucy: The Beginnings of Humankind* (1981), this 3.2-million-year-old near-man (or near-woman) is "the oldest, most complete skeleton of any erect walking hominid found anywhere in the world." Discovered by Johanson at Hadar, in the Afar region of northern Ethiopia, on November 30, 1974, "Lucy" captured the attention of the world press.

While celebrating the find, Johanson's team sat around their evening campfire playing the Beatles tune *Lucy in the Sky with Diamond*s on a tape machine and nicknamed her Lucy. Her scientific name, *Australopithecus afarensis*, means "southern ape from the Afar" (the genus was first named from South African fossils), and Johanson has championed the species as our earliest known direct ancestors.

Although Lucy was proclaimed to be a 40 percent complete skeleton, in fact less than 25 percent of the complete skeleton was actually recovered (47 bones out of 206). Only by "doubling" many bones (with a mirror image of its known mate on the opposite side) did the specimen appear about 40 percent complete. Nevertheless, with long bones of arms and legs, ribs, sacrum, and half a pelvis, Lucy remains the most complete skeleton ever found from such an early deposit.

Despite excited press accounts, Lucy was not the first specimen to establish that australopithecines were bipedal and erect. Thirty years earlier, anatomist Robert Broom had found australopithecine leg bones and pelvises at Sterkfontein, showing that early hominids were upright walkers. But Lucy certainly provided dramatic, clinching confirmation. As one paleontologist put it, "The South African fossils are excellent, but they have not as good a salesman as a Leakey or Johanson."

Although the Lucy fossils were initially dated at three million years, Johanson announced them as 3.5 million because he said the species was "the same" as mandibles and teeth found by Mary Leakey at Laetoli, Tanzania. By proposing Mary Leakey's find as the "type specimen" for *Australopithecus afarensis*, he was identifying Lucy with another fossil 1,000 miles from the Afar and half a million years older! Mary thought the two not at all the same and refused to have any part of linking her specimen with *afarensis*. Convinced that her own specimen belonged to the genus *Homo*, she announced that she strongly resented Johanson's "appropriating" her find, her reputation, and the older date to lend authority to Lucy. Thus began the bitter, persistent feud between Johanson and the Leakeys.

Lucy, now believed to be 3.2 million years old, certainly is one of the most remarkable relics ever discovered. Upright walking had evolved in Lucy and other early hominids long before they had very large brain size. We did not "stand tall" as the result of some special human cleverness.

See also AFAR HOMINIDS; AUSTRALOPITHECINES; JOHANSON, DONALD; LEAKEY, MARY; TURKANA BOY

LYELL, SIR CHARLES (1797–1875)
Geologist, Darwin's Mentor

Charles Darwin wrote in an 1844 letter, "I always feel as if my books came half out of Lyell's brains. . . . [I see things] partially through his eyes." But Sir Charles Lyell, the great naturalist's friend and mentor, had a hard time returning the compliment. Although he privately encouraged Darwin's evolutionary work for years, Lyell could not bring himself to endorse his friend's theories in his own popular geology books. Much to Darwin's disgust, Lyell was a past master of the art of coming down squarely on both sides of an issue.

Lyell's father, Charles Lyell of Kinnordy, was a Scots laird who was torn between scientific and literary interests. He was a keen amateur botanist, but also managed to produce a well-known English translation of Dante's *Inferno*. The younger Charles was also divided between two professions; he started out as a lawyer, but his strong interest in geology finally won out. He had admired the works of James Hutton and explored volcanic rocks in Italy with the geologist William Buckland.

Just before Darwin was to leave on his five-year voyage aboard HMS *Beagle*, his Cambridge professor Reverend John Henslow recommended he take along the first volume of Lyell's *Principles of Geology* (1830), which had recently been published. Although he urged Darwin to study it, he warned the young naturalist: "On no account accept the views therein advocated." In 1832, during the voyage, Henslow sent Darwin the second volume.

In this founding document of modern geology, Lyell had argued (1) that the geologic past can best be understood in terms of natural processes we can observe today, such as rivers depositing layers of silt, wind and water eroding landscapes, glaciers advancing or retreating (actualism); (2) that change is slow and steady (gradualism), rather than quick and sudden; (3) that natural laws are constant and eternal, operating at about the same intensity in the past as they do today (uniformitarian—sometimes slower or faster, but averaging about the same in overall rate of change).

Darwin devoured the book, which was brilliantly written, thoroughly grounded in field-work, and seemed to place the study of geology on a new and sensible footing. Although Lyell believed living species were fixed and not related by common descent, he gave Darwin the means of seeing what wonders could be wrought in geology by slow, small forces operating over immense spans of time. "I am tempted," Darwin wrote, "to carry parts [of Lyell's views] to a greater extent, even than he does."

Although he inspired Darwin and became his lifelong friend and mentor, Lyell had great difficulty accepting "the descent of man from the brutes," because, he confessed to Darwin in 1863, "it takes away much of the charm from my speculations on the past relating to such matters." Darwin was frustrated with Lyell's refusal to support him in print, though he did so in private conversations. Lyell simply could not, as he put it, "go the whole Orang." Nevertheless, later in life he had to grudgingly acknowledge the growing evidence.

What was really peculiar to Lyell are two ideas rarely associated today with his *Principles of Geology*: that earth and water trade substances and shape each other, maintaining some kind of long-range balance (the steady-state Earth); and that time and life proceed in cycles. It was conceivable to Lyell that man and our familiar animals could all become extinct, only to be replaced by dinosaurs again in a subsequent creation, followed, in some distant age, by a "new creation" of man. Aside from historians of science, Lyell's belief in cyclic time has been all but forgotten.

Lyell wrote that only a few extinctions occurred at a time, which "gradually" added up, rather than wholesale extinctions. New species were "called into being" to replace them. Somewhat cynically, he refused to specify how or what he meant, leaving the interpretation open so as not to ruffle the feathers of the theologians.

Caution about antagonizing anyone and an extreme desire for social acceptance curbed Lyell's adventurousness in the realm of ideas. Darwin thought it ridiculous how Charles and Lady Lyell would spend hours poring over dinner invitations, making it a matter of great importance which to accept and which to decline. It was no accident that Lyell's determination to offend no one of importance resulted in his receiving a knighthood and, later, being named baronet. Darwin received no national honors during his lifetime.

Charles Darwin had Lyell in mind when he wryly remarked that scientific men should die at the age of 60, so their inflexible habits of mind could not interfere with the progress of the next generation. After finally including Darwin's theories at length in his *Geological Evidences as to the Antiquity of Man* (1863), Lyell asked Darwin if "now he might be allowed to live."

See also ACTUALISM; CATASTROPHISM; GRADUALISM; STEADY-STATE EARTH; UNIFORMITARIANISM

ARGUING HIS CASE for the "uniformitarian" view of geology came easily to Sir Charles Lyell, as he had been trained as a lawyer. Charles Darwin read Lyell's books during his voyage, and followed their principles.

LYSENKOISM
Ideological Genetics

During the 1930s, Trofim D. Lysenko rose to power in the U.S.S.R. by convincing government ideologues that he could create a genetic science consonant with their political philosophies. They wanted science to support the view that Soviet society could be literally transformed in a few generations, and that the Russian people were progressively evolving. It might take too long to create the socialist utopia if each generation had to be separately educated without cumulative inheritable improvements.

Lysenko proclaimed that Mendelian genetics was a "tool of bourgeois society" in teaching that mutation is random and that genes are usually passed on unchanged. His reading of Marxist principles was that man and nature are improvable and perfectible. Stalin gave him free rein, and within a few years his rivals (and any honest scientists who openly supported Mendelism) were out of work, imprisoned, or even executed.

Lysenko claimed he could alter species of wheat by changing their environment and that he intended to transform Soviet agriculture. He said he could change winter wheat into spring wheat merely by altering the temperatures at which they were grown. He even claimed that environmental manipulation could change wheat into rye in one generation!

According to Lysenko, Charles Darwin's concept of the "struggle for existence" was a bourgeois notion used to justify competition in a capitalist society. Nature was altruistic. Seeds should be planted in clusters, so that all except the most perfect and vigorous ones would sacrifice themselves for the good of the species.

Lysenko became so powerful he was never forced to produce evidence for his assertions. His policies retarded Soviet biology and genetics for 30 years and devastated the agricultural production of the country. Repeated crop failures and shortages finally caused his ouster in 1965.

After the death of Russian dictator Joseph Stalin in 1953, Lysenko revealed that the head of state had helped prepare his famous speech of 1948, in which he blasted "Mendelism." "Comrade Stalin found time even for detailed examination of the most important problems of biology," Lysenko declared in his eulogy for Stalin in *Pravda*. "He directly edited the plan of my paper, 'On the Situation in Biological Science,' in detail, explained to me his corrections, and provided me with directions as to how to write certain passages in the paper."

The term "Lysenkoism" is used in a restricted sense to describe his notions of environmental manipulation of genes, coupled with inheritance of acquired characters. In its broader meaning, Lysenkoism has come to symbolize the disastrous consequences of making science subservient to political ideology.

See also NEO-LAMARCKISM

> "Comrade Stalin found time even for detailed examination of the most important problems of biology."
>
> —Trofim D. Lysenko, 1953

LYSTROSAURUS FAUNA
Evidence for Ancient Continent

Lystrosaurus is a large mammal-like reptile (synapsid) of the Triassic period, originally known from South Africa, where it is associated with other characteristic fossils. Often it is found in the same rocks as another mammal-like reptile called *Thrinaxodon*, some small ancestral lizards (eosuchians), the little reptile *Procolophon,* and a particular type of amphibian—all collectively known as the *Lystrosaurus* fauna.

Squat, tusk-toothed *Lystrosaurus*—and the unique array of animals always found with it—helped prove that Antarctica, South Africa, and India were once joined together. Also, oceanographic mapmakers have found that details of their submerged coastal outlines today, at a depth of 1,000 fathoms, would form a remarkably close fit.

According to paleontologist Edwin Colbert, one of the discoverers of the *Lystrosaurus* fauna on all three continents, these animals once occupied a continuous range on ancient Gondwanaland. When the supercontinent broke up, one segment—once a land of lush forests and strange tropical animals—drifted southward to become frozen Antarctica.

See also CONTINENTAL DRIFT; GONDWANALAND; PLATE TECTONICS

THE UGLY TUSKED brute called *Lystrosaurus* helped to prove that the Earth's continents were once joined together; its bones are in Antarctica, India, and South America.

MALTHUS, THOMAS (1766–1834)
Inspiring Economist

Because of theorists like Thomas Malthus, who feared that overpopulation by the poorer classes would outstrip Europe's ability to produce food, economics came to be known as "the dismal science." Yet the dreary conclusions of Malthus's *Essay on the Principle of Population* (1798) had the unusual effect of giving hope to two great naturalists—Charles Darwin and Alfred Russel Wallace—by helping to inspire their ideas about natural selection.

Malthus, a British clergyman and political economist, was distressed by the slums of preindustrial England. Human misery, he perceived, arose from human fertility combined with irresponsibility. Taking a cue from plants and animals, he showed that reproductive potential far exceeds the ability of resources to support them if all offspring were to survive.

Biologists have since discovered that overproduction in nature goes far beyond what Malthus supposed. For instance, it is now known that every female sturgeon produces six million eggs per year, each mayworm 40 million, and each tapeworm 60 million. Without enormous mortality, the Earth would shortly be overrun by the descendants of a single breeding pair of almost any creature.

Even granting technological improvement, humans' food resources could only increase at an arithmetic rate, Malthus calculated, which could not keep up with the geometric increase in population. Therefore, Malthus concluded that humans will grow increasingly miserable as starvation becomes inevitable among the poorer social classes. His own solution was to advocate sexual abstinence for the poor and punitive laws against parents producing more children than they were able to support. Throughout most of the world where population growth is rapidly exceeding food supply, the Malthusian time bomb is still ticking.

In his *Autobiography* (1876), Darwin wrote:

In October 1838, that is, fifteen months after I had begun my systematic inquiry, I happened to read for amusement Malthus on Population, and being well prepared to appreciate the struggle for existence which everywhere goes on from long-continued observation of the habits of animals and plants, it at once struck me that under these circumstances favourable variations would tend to be preserved, and unfavourable ones to be destroyed. The results of this would be the formation of new species. Here, then, I had at last got a theory by which to work.

Wallace had also read Malthus and apparently made a similar connection as, in 1858, he was mulling over his observations on the competition for resources among tribal peoples of the Malay Archipelago. During a malarial fever, Wallace conceived the theory of natural selection, and realized it would apply to both human and animal populations.

Later, Social Darwinists would attempt to return the biological theory of natural selection to the human economic sphere, applying it to the conditions in urban slums that had been its point of departure. The whole episode is an impressive example of how social and biological theory can draw from each other, as well as a striking instance of independent invention by similarly prepared minds.

See also DARWIN, CHARLES; NATURAL SELECTION; SOCIAL DARWINISM; WALLACE, ALFRED RUSSEL

291

MAMMOTHS
Siberian Fossil Elephants

When great fossil elephants began to turn up in Europe and America in the late 1700s, there was a great deal of confusion about them. Those from Siberia were the first known to science; travelers since the early 1600s had been amazed at the quantities of fossil ivory found there. A Dutchman named Cornelius Witzen was the first to print the word "mammoth" in 1694, noting that "by the Inlanders [Russian settlers in Siberia] these teeth are called *mammouttekoos*, while the animal itself is called *mammount*."

Contradictory stories about the fabulous "mammount" trickled out of the north country for years, including debates over whether the animal was still alive. In the 1770s, when immense quantities of fossil bones and tusks were found, demand for them soon prompted a flourishing trade. During the 19th century, the tusks of perhaps 50,000 Siberian mammoths were unearthed and sold for ivory in Europe.

When elephant-like tusks and fossils were discovered in the Ohio Valley, President Thomas Jefferson took a keen interest in them and called them "mammoths." But he failed to notice, as the French expert Georges Cuvier pointed out, that the teeth of the American and European species were entirely different. While the Siberian mammoths had flat grooved grinders like those of modern elephants, the American fossils had many curvy, cone-shaped surfaces. Cuvier thus suggested calling the Ohio Valley animal a "mastodon," meaning "breast-shaped tooth."

Nevertheless, Jefferson (and almost everyone else) kept referring to the American fossil tuskers as "mammoths," and "mammoth fever" swept the country. Giant loaves of bread were advertised as "mammoth bread" in Philadelphia bake shops. Some of Jefferson's admirers in Massachusetts made a "mammoth cheese," which weighed 1,235 pounds, and shipped it to the White House. He put it on exhibit in the East Room, which he nicknamed "the Mammoth Room."

See also JEFFERSON, THOMAS; PEALE'S MUSEUM

MANTELL, GIDEON ALGERNON (1804–1892)
Discovered First Dinosaur

With the thousands of types of dinosaurs known today, it seems incredible that less than 200 years ago the only tales of ancient giant reptiles were about the dragons of myth. Dr. Gideon Mantell and his wife, Mary Ann, made the first discoveries of dinosaur bones—in Kent, England, in 1822. Of course, dinosaur fossils had been found long before by European quarrymen, as well as by Native Americans and Chinese, but they were usually considered "monsters" or "dragon bones." Mantell was the first Western scientist to recognize and interpret such fossils as extinct species of animals.

It was actually Mrs. Mantell who spotted the first huge fossil tooth as she accompanied her husband on one of his "geologising" walks through Tilgate Forest. Dr. Mantell was an avid amateur geologist, constantly searching for unusual relics of "antediluvian" creatures. When local authorities seemed baffled by his find, he sent it to the world's greatest fossil expert, Georges Cuvier, in Paris. The teeth belonged to a rhinoceros, Cuvier replied, and the bones were those of a hippopotamus.

But an English naturalist who saw the teeth disagreed. They were lizardlike, he thought, and convinced Mantell by comparing it with an iguana skull from the American tropics. Mantell concluded that he had found a giant lizard, named it *Iguanodon* ("iguana-tooth"), and read a paper describing it before the Royal Society of London in 1825. Cuvier, on second thought, admitted he was wrong and agreed that it was a previously unknown type of gigantic reptile. Mantell assembled the bone fragments in his private museum, where they became known as the "Mantell-piece."

PEALE'S MASTODON became world famous in 1799, symbolizing America's great size and vigor. Benjamin Franklin and Thomas Jefferson had encouraged Charles Willson Peale to unearth the fossil skeleton.

A EUROPEAN MAMMOTH'S GRINDER (above left) shows a series of long grooves and ridges. The American fossil elephant's molar's curving mounds inspired the name "breast-shaped tooth" or *Mastodon*.

Not long after, another dinosaur was found near Oxford by Reverend William Buckland (1784–1856), although the term "dinosaur" had yet to be invented by the British Museum's anatomist Richard Owen. Mantell redoubled his efforts at Tilgate Forest and was rewarded with a new saurian—this one bearing the cutting teeth of a meat-eater.

In 1832, Mantell dug up an armored *Hylaeosaurus* covered with bony plates and spikes. By now, Mantell, completely obsessed with his fossil hunting, was ignoring his medical practice. His house became so crammed with rocks and fossil bones that his wife moved out, probably regretting that she had sparked his obsession by finding that first tooth. Eventually, he had to sell off his dinosaurs to the British Museum in order to survive.

The official coat of arms of Maidstone, the county seat of Kent, features an iguanodon. Although it looks authentically Old English, the design was actually adopted by the town in 1949. It is the only municipality in the world with a dinosaur on its official seal.

See also DINOSAUR; IGUANODON DINNER

MARXIAN "ORIGIN OF MAN"
Dialectical Darwinism

In his 1876 essay "The Part Played by Labor in the Transition from Ape to Man," Friedrich Engels collaborated with Karl Marx in developing what they called "scientific socialism," which exalted labor as the definitive characteristic of *Homo sapiens*. "Labor . . . is the prime basic condition for all human existence," he wrote, "and this to such an extent that, in a sense, we have to say that labor created man himself. . . . First labor, after it and then with it, speech—these were the two most essential stimuli under the influence of which the brain of the ape gradually changed into that of man."

Marx had defined labor in *Capital* (1867) as a process in which man "mediates, regulates, and controls his material interchange with nature by means of his own activity. . . . Acting upon nature outside of him, and changing it, he changes his own nature also." Marx distinguished human labor from the work of spiders or bees, which he thought instinctual or automatic. To man alone he attributed a conscious purpose, a mental picture of the results of his labor. Since labor was the crucial factor in human evolution, an industrial system that "alienates" workers from their tools and labor is literally inhuman.

Engels discussed how chimpanzees build tree nests and shelters and even wield sticks and stones. But though ape anatomy is similar to human, "no simian hand has ever fashioned even the crudest stone knife." [See APE TOOL USE; KANZI.]

During the vast period of prehistory, Engels speculated, the hand, freed from locomotion, "could henceforth attain ever greater dexterity and skill. Thus . . . the hand is not only the organ of labor, it is also the product of labor. Only by labor, by adaptation to ever new operations" do muscles, ligaments, and bones adapt and improve until the hand can "conjure into being the paintings of a Raphael."

But when social organization diversified to where some individuals merely planned the labor that would be carried out "by other hands than [their] own," at that point bourgeois philosophy originated. "All merit for the swift advance of civilization was ascribed to the mind, to the . . . activity of the brain."

Essences, ideals, and religious thought dominated human cultures. The exalted status of thinking over actually making things "still rules them to such a degree that even the most materialistic natural scientists of the Darwinian school are still unable to form any clear idea of the origin of man, because under this ideological [bourgeois religious] influence they do not recognize the part that has been played therein by labor."

This Marxian idea of an early feedback loop (or dialectic) between the evolution of the hands and labor doubtless seems bizarre to non-Marxians, unaccustomed to seeing the loaded word "labor" used in discussions of human evolution. However, if one rereads Engels's passages and substitutes the phrase "tool-use" for "labor" every time it occurs, the theory becomes identical to the anthropological orthodoxy of the past few decades.

> "Darwin provides a basis in natural science for the class struggle in history."
>
> —Karl Marx to Friedrich Engels, 1861

KARL MARX STATUE in Chemnitz, Germany.

MATERIALISM
Everything's the Matter

In everyday speech, "materialism" means a special fondness for wealth, cars, houses, possessions—material "stuff" you can see and touch. In 19th-century science and philosophy, materialism was the belief that matter is the only "stuff" there is. Evolutionary theory—like classic chemistry and physics—is based on that materialist assumption. Science itself is often defined as "the study of the properties of matter."

The underlying postulate of materialism (and science) is the unprovable idea that everything—the Earth, the stars, animals, even our minds, dreams, and personalities—is a product of physical matter, which is everywhere the same. Although the idea goes back to the ancient Greeks, its acceptance in science really began in the 18th century with Newtonian physics.

Only a century ago, materialism was still a very hot topic, provoking passionate debate. Because its view of nature excludes "immaterial" demons, gods, ghosts, and fairies, materialism was considered by many to be an enemy of religion, morality, and ethical values. Scientists sought personality in brain rather than soul and focused on the creative properties of matter, leaving a creator God out of their theories. [See BELFAST ADDRESS.] Today a small but insistent group of creationists has been trying their best to bring deities back into science [see INTELLIGENT DESIGN], while some scientists and philosophers in this atomic age are not at all satisfied that reducing everything to "matter" is an adequate description of the natural world.

Most scientists who adopted the materialist postulate did not abandon Judeo-Christian values, and many professed a belief in God. Newton himself spent more time studying the book of Revelation than he did physics. Yet materialist scientists shocked Victorian lecture audiences by asserting that "thought is as much a secretion of brain as urine is of kidneys." Thomas Henry Huxley liked to compare the mind to the whistle on a steam engine—a noisy adjunct to the body, driven by the same force.

Opponents of materialism asked, "Is that all there is?" and complained that science was reducing the human being to a thing devoid of spirit. Vitalists insisted there must be an undiscovered "life force" pushing evolution forward; spiritualists argued for an invisible "guiding hand" from another dimension; while fundamentalists looked no further than "the Creator" for explanations of natural phenomena.

Although perhaps more comforting than "cold" materialism, these nonphysical approaches proved very poor tools for advancing biological understanding. Rejecting all these unseen, unmeasurable "forces," paleontologist George Gaylord Simpson wrote in his *Tempo and Mode in Evolution* (1944): "The progress of knowledge rigidly requires that no non-physical postulate ever be admitted in connection with the study of physical phenomena. . . . The researcher who is seeking explanations must seek physical explanations only." Simpson's proscription excludes not only consciousness, spirit, and God, but also Platonic ideals—patterns, types, archetypes—that such naturalists as Richard Owen and Lorenz Oken had believed to be as real as bones. [See ESSENTIALISM; *NATURPHILOSOPHIE*.]

Physical explanation as the only scientific one is a legacy of Newtonian physics, which by the 19th century was the accepted scientific model of the universe. However, modern physics has changed that picture drastically. Critic of anthropology William R. Fix wrote in *The Bone Peddlers: Selling Evolution* (1984): "Quantum physicists have been led more and more to consider models of consciousness and theories of perception as part of the 'stuff' that the new physics is about. Subatomic particle behavior, uncertainty principles, and other recent developments have led quantum physicists to describe 'reality' in terms that are often restatements of Buddhist metaphysics."

Fix's critique is that modern evolutionary biologists accept as their "reality" a "naive realism" based on outdated Victorian science: an old-fashioned materialism that was daring and appropriate in Darwin's day but is no longer the model used in physics. (In one of his notebooks on transmutation of species [1838], Darwin had written to himself: "Why is thought being a secretion of brain, more wonderful than gravity a property of matter? . . . Oh, you materialist!")

Thomas Henry Huxley, who fought to establish Darwinian theory, lamented in his 1871 essay "Bishop Berkeley and the Metaphysics of Sensation" that "there are numbers of highly cultivated and indeed superior persons to whom the material world is altogether contemptible; who can see nothing in a handful of garden soil, or a rusty nail, but types of the passive and the corruptible." Before they could appreciate Darwin's work, Huxley insisted, they would have to raise their estimation of matter:

> To modern science . . . the handful of soil is a factory thronged with swarms of busy workers [microorganisms]; the rusty nail is an aggregation of millions of particles, moving with inconceivable velocity in a dance of infinite complexity yet perfect measure; harmonic with like performances throughout the solar system. . . . these particles [and the energy that stirs them have always] existed and will exist. . . . Form incessantly changes, while the substance and the energy are imperishable.

In Huxley's view, matter was no less remarkable and mysterious than what some would call "consciousness." He wrote that what we call mind and matter "in our little speck of the universe are only two out of infinite varieties of existence . . . which we are not competent so much as to conceive—in the midst of which, indeed, we might be set down, with no more notion of what was about us, than the worm in a flower-pot, on a London balcony, has of the life of the great city."

Although Huxley has often been criticized for "reducing" everything to matter, in truth he never forgot that materialism was only a useful working assumption. He was always careful not to consider it an ultimate explanation or a provable fact, and it never diminished his sense of wonder. He resented that some scientists "talk as if [accepting matter] as the 'substance' of all things cleared up all the mysteries of existence. In point of fact, it leaves them exactly where they were."

One Victorian wit, despairing of ever finding a solution to the Mind-Matter question, summed up his perplexity in this epigram: "What is mind? No matter. What is matter? Never mind."

See also MECHANISM; SPIRITUALISM; THEORY, SCIENTIFIC

"[We have] no more notion of what [is] about us [in the universe] than the worm in a flower-pot on a London balcony has of the life of the great city."

—Thomas Henry Huxley, 1878

MATTHEW, PATRICK (1790–1874)
A Darwinian Predecessor

Eccentric Scots naturalist Patrick Matthew believed that good ideas must eventually surface, no matter where they appear. A tree expert and fruit grower by trade, he published a concise summary of natural selection 26 years before Charles Darwin and Alfred Russel Wallace, but his achievement was unnoticed and ignored. That was because, in 1831, Matthew chose to reveal his theory of evolution in an appendix to an obscure work titled *On Naval Timber and Arboriculture*, where it made no impact whatsoever.

After Darwin shook the scientific world with his epochal *Origin of Species* (1859), Matthew wrote a letter to the *Gardener's Chronicle* claiming priority not only for "the organic evolution law," but also for the idea of the "steam ram . . . and a navy of steam gun-boats as requisite in future maritime war"—other ideas for which he received no credit.

Upon reading Matthew's protest, Darwin replied that he had not heard of the earlier publication, but would now "freely acknowledge that Mr. Matthew has anticipated by many years the explanation which I have offered of the origin of species [by] natural selection." He also noted that, considering the strange place in which it was published, "I think that no one will feel surprised that neither I, nor apparently any other naturalist, had heard of Mr. Matthew's views." In subsequent editions of the *Origin*, however, Darwin duly credited Matthew and other forerunners in a historical preface.

Not satisfied with Darwin's acknowledgment, Matthew tried to belittle his achievement by noting that he had thought natural selection was perfectly obvious and could not see what all the fuss was about:

> To me the conception of this law of Nature came intuitively as self-evident fact, almost without an effort of concentrated thought. Mr. Darwin here seems to have more merit in the discovery than I have had—to me it did not appear a discovery . . . it was by a general glance at the scheme of Nature that I estimated this select production of species as . . . fact—an axiom, requiring only to be pointed out.

The difference, as Stephen Jay Gould has written, was that Matthew had indeed stated the theory first, but had failed to appreciate and work out its vast implications. In contrast, Darwin amassed huge amounts of data and tackled such diverse problems as plant and insect coadaptations, human evolution, variation in domesticated animals and plants, and hundreds of other topics. "He established a workable research program for an entire profession," notes Gould.

Darwin himself was somewhat amused when an even earlier discoverer of natural selection came to light. A scientist named William Charles Wells—famous for his "Essay on Dew"—had published in 1818 a statement of natural selection applied to the races of man. On learning of Wells in 1865, Darwin wrote to a friend, "So poor old Patrick Matthew is not the first, and he cannot, or ought not, any longer put on his title-pages 'Discoverer of the Principle of Natural Selection.'" Darwin acknowledged Wells in the fourth edition of the *Origin of Species* (1866), saying that he "distinctly recognizes the principle of natural selection, and this is the first recognition which has been indicated."

See also NATURAL SELECTION; WELLS, WILLIAM CHARLES

MAYR, ERNST (1904–2005)
Evolutionary Biologist, Ornithologist, Historian

On March 23, 1923, a young German medical student named Ernst Mayr chanced to spot a pair of very rare ducks—the first fateful "accident" of a brilliant, unplanned, and unexpected career. Recently graduated from the Dresden secondary school, he had thought his life was set: to follow the four-generation family tradition of successful physicians. He certainly did not expect to become an explorer, naturalist, ornithologist, philosopher-historian of science, Harvard professor, and one of the 20th century's greatest evolutionary biologists. The ducks changed everything.

Mayr had bicycled into the countryside for some bird-watching. When the two odd-looking ducks with brilliant red bills swam into view, he realized that they were red-crested pochards, a species not seen by anyone in nearly 80 years. No one believed him, and he poured out his heart to a new acquaintance, a pediatrician, who set up an introduction with the greatest ornithologist in Germany, Erwin Stresemann. Mayr journeyed to Berlin to meet the professor, but he received a rough reception. Stresemann quizzed him mercilessly, probed his knowledge of natural history, and asked to see his prior notes and journals of field observations. Satisfied at last, he published the sighting as genuine.

Stresemann invited the young man to work in the Berlin Museum as a volunteer, classifying bird specimens received from the tropics. Mayr thought he was "given the keys to heaven" and continued to work at the museum during breaks from medical school. Just before he was to receive his degree, Stresemann offered to send him to the tropics if he delayed his medical career and earned a doctorate in ornithology.

By age 21, Mayr had earned that doctorate, and Stresemann sent him to Lord Walter Rothschild, titular head of the wealthy European banking family. At his own private museum at Tring, in Hartfordshire, England, Rothschild was assembling the world's largest and most comprehensive collection of preserved birds and animals.

Again Mayr benefited from a well-timed accident of circumstance. Rothschild's staff naturalist in New Guinea had suddenly died after many years of service, and Rothschild was desperately seeking a new bird collector. Mayr was hired on the spot.

Within the year, he had traveled through six unexplored New Guinea mountain ranges, eventually collecting 3,400 bird skins and discovering 38 new species of orchids. In 1930, while suffering from malaria and dysentery in his mountain camp, he received an urgent invitation to join an expedition to the western South Seas sponsored by the American philanthropist Harry Payne Whitney.

Again Mayr was in the right place at the right time: a week before departure, the expedition had suddenly found itself without a leader. He accepted.

The Whitney South Seas Expedition was an epic scientific adventure, which made important contributions to biology, discovered scores of new species, and provided the American Museum of Natural History with the materials for a new hall. In 1931, after collecting in the Solomon Islands, Mayr was hired to come to New York and work with the bird specimens at the museum.

In his first year at the museum, Mayr published a dozen papers, describing scores of new species and subspecies. The following year, Rothschild's curator retired; Mayr was invited to take charge of the collection at Tring. There he worked up a series of related species from different islands, a convincing demonstration of geographic speciation.

In 1936, he invited the evolutionary geneticist Theodosius Dobzhansky to study the series. Mayr's study influenced Dobzhansky's important book *Genetics and the Origin of Species* (1937), the founding work of the Synthetic Theory of evolution.

During the 1930s and 1940s, Mayr collaborated with Dobzhansky, Julian Huxley, and George Gaylord Simpson to formulate the modern evolutionary synthesis, incorporating new discoveries by naturalists and population geneticists.

Mayr might have stayed on as curator at Tring but for another "great accident;" Rothschild was facing financial ruin and could no longer keep his hard-won collection. He had become involved with a married, titled woman who was now blackmailing him with threats of a family scandal. Her merciless and increasing demands for large sums of money forced Rothschild to cut back on staff and finally to sell off his beloved bird collection. New York's American Museum of Natural History purchased 280,000 bird skins from Rothschild, courtesy of the Whitney family, in the hardest year of the Depression. Mayr helped pack and ship 185 cases, each containing 7,000 skins of rare birds.

He then returned to the New York museum, where he continued his study of how species are formed that he had begun in the field, and was asked to give two lectures on speciation in animals at Columbia University, which grew into *Systematics and the Origin of Species* (1942), an influential classic that redefined species in terms of breeding populations.

Building on his decades of familiarity with island populations of birds, Mayr advanced a general theory (in 1954) of how species evolve. Either through the appearance of geographic barriers or by a few "founders" settling in a new area beyond the species' customary range, a very small population can become established—the first step toward reproductive isolation. Over time the little colony inbreeds, local conditions exert their selective pressures, and descendants become increasingly different from their ancestral population. If they are ever reunited, the two populations may no longer be capable of interbreeding or producing viable offspring.

TREKKING THROUGH NEW GUINEA in 1929 (above), Ernst Mayr, who became a Harvard evolutionary biologist, collected rare birds of paradise (top) for Lord Walter Rothschild. Bird painting courtesy of and © by Errol Fuller.

ERNST MAYR

This kind of rapid evolution at the edge of a species' range—what is technically called "peripatric evolution"—was emphasized by Mayr in the 1950s, and it was to become one of the foundations for the punctuated equilibrium theory of Niles Eldredge and Stephen Jay Gould in the 1970s. The apparent quickness of such evolutionary change, however, is only from the geologist's long perspective. In fact, the change is very gradual occurring over many thousands of generations.

Amused at the chancy, undirected path of his own career, Mayr describes it in terms of contingent history—a process that parallels evolution itself. Had Rothschild's former lover not blackmailed him, Mayr might have spent his life at Tring in England and never had contact with Columbia and Harvard, which eventually led him from his work as taxonomist (classifier) to historian of science, to philosopher of evolution. Had Rothschild's field collector not died, had Mayr not seen those rare ducks or met the pediatrician who knew the ornithologist, he claimed, he would not have had his remarkable career at all.

See also ALLOPATRIC SPECIATION; CONTINGENT HISTORY; DOBZHANSKY, THEODOSIUS; ROTHSCHILD, LORD WALTER; SPECIES, CONCEPT OF; SYNTHETIC THEORY; WAGNER, MORITZ

MECHANISM
Materialist Philosophy

Like all sciences, evolutionary biology is based on the idea that nature's workings, like those of a machine, can best be understood without reference to consciousness, gods, or spirits. Mechanism (and its companion concept, "materialism") has always had profound limitations as a mode of explanation, yet its application to scientific problems has yielded extraordinarily fruitful results. Materialism assumes that everything, living and nonliving, is formed of the same kind of measurable, physical matter.

Among the most important thinkers in establishing mechanistic explanations in science are the British physicist Sir Isaac Newton (1642–1727) and the French philosopher René Descartes (1596–1650). Newton's laws of motion and gravitation proved indispensable for building machinery in the industrial era and also encouraged the concept that the universe itself is like a machine.

Descartes believed there was a sharp division between mind or spirit *(esprit)*, possessed only by man, and body, shared by man and animals. Nonhuman creatures, in Descartes's view, were simply "automata," mindless machines like robots or what were then called "clockwork mannikins." (The title of Anthony Burgess's 1962 novel *A Clockwork Orange* is a reference to the human brain as a programmable machine encased in a soft, fruitlike rind.)

Religious philosophers were outraged at science's blind, "clockwork" universe. Where was the place in this design for a caring God who could intervene in human affairs? A mechanistic cosmos devoid of plan, purpose, or consciousness seemed devastatingly cold and unappealing. Although many scientists, including Newton himself, did not replace their personal religious beliefs with a mechanistic worldview, they nonetheless found it a necessity when doing science. There it was not a matter of comfort, hope, or morality; mechanistic models simply worked best at mimicking and predicting the behavior of natural phenomena.

Without Newtonian physics and the later advances of a "mechanistic, materialistic" chemistry and physiology, Darwin's evolutionary model would have been neither possible nor plausible. It is no accident that he was buried in Westminster Abbey just a few feet from the tomb of Sir Isaac Newton.

See also BELFAST ADDRESS; CARTESIAN DUALITY; MATERIALISM

MENDEL, ABBOT GREGOR (1822–1884)
Father of Genetics

The son of peasant farmers, Gregor Mendel was born in 1822 in the Silesian village of Heinzendorf, in what is now the Czech Republic. In an autobiographical sketch written in 1850, Mendel described his youth as "sorrowful," largely, as he says, because of his repeated failure to secure for himself the means to continue his education: "The distress occasioned by the disappointed hopes, and the anxious, dreary outlook which the future offered me, affected me so powerfully at that time that I fell sick, and was compelled to spend a year with my parents to recover."

This was apparently the first of a series of nervous breakdowns that Mendel suffered. To assure his future, he "felt compelled to step into a station of life" that would free him "from the bitter struggle for existence." This station was the Catholic Church, and in 1843 Mendel entered the Augustinian monastery in Brünn as a novitiate.

Freed from immediate necessity, Mendel began a career as a teacher, but was required to pass a state examination in order to receive a permanent appointment. He took the exam in 1850, but failed. The following year, at the monastery's expense, Mendel enrolled as a student at Vienna University to study natural sciences and continued there until the summer of 1853. In 1856, Mendel once more attempted the state examination, but this time could not even finish it. He suffered another nervous breakdown and in depression returned to Brünn, where he remained ill for some months. He never attempted the examination again, but worked for years as a temporary uncertified teacher.

Mendel's scientific fame rests on a single paper published in 1865 in the journal of the Brünn natural science society. The paper summarizes the results of his lengthy hybridization experiments on peas. It did not attract attention during his lifetime, and Mendel's few scientific correspondents saw nothing remarkable in his work, though it was, in fact, the long-sought key to understanding heredity.

"Mendel's Laws," though they were not explicitly stated in that 1865 paper, implied that parental characters do not blend in their progeny but are transmitted as discrete factors. Each trait is controlled by a pair of factors in which only one of the pair enters each mature sex cell or gamete. When joined with a gamete from the other parent, a new pair is formed: the genetic inheritance of the offspring.

Until recently, historians traditionally viewed Mendel as being scientifically isolated, discovering these "laws of heredity" by a set of brilliant experiments conducted in a small garden plot at the monastery, then spelling them out clearly. According to this tradition, his achievement was never grasped by his contemporaries and was ignored until 1900, when three researchers—Hugo de Vries, Carl Correns, and Erick von Tschermak—independently rediscovered the laws of heredity and Mendel's paper as well. They then brought the paper to the attention of the scientific community, where it was immediately recognized as the foundation document of the new science of genetics.

During the 1980s, this view of the history of genetics was challenged: The rules referred to as "Mendel's Laws of Heredity," it turned out, were never clearly formulated by Mendel but were read into his work by his rediscoverers. As plant physiologist Alan Bennett puts it, "There is no statement [in Mendel's paper] of the simple hypothesis that a given character is controlled by a pair of factors, and that a gamete carries only one member of the pair."

Mendel's experiments in plant hybridization lasted only about 15 years. After 1868, when he was elected abbot at the age of 46, he ceased teaching; his position as head of the monastery gave him administrative responsibilities that left little time for scientific work. Mendel died in 1884 of cardiac degeneration and kidney disease made worse by chronic nicotine poisoning—he was a heavy cigar smoker in the latter part of his life.

After his death, the new abbot had his experimental garden destroyed and all his notebooks, papers, and scientific records burned. Thus, Mendel's posthumous renown rested solely on his one paper in an obscure journal of natural history.

It was not until 16 years after his death that Mendel's ideas were resurrected, and

> Mendel's claim to fame rests on an obscure 1865 paper that went unnoticed until fifteen years after his death.

his key to understanding the age-old question of heredity was at last appreciated. Ironically, among the few relics of Mendel's life to survive is a poem from his youth, which concludes:

> May the might of destiny grant me
> The supreme ecstasy of earthly joy. . . .
> To see, when I arise from the tomb,
> My art thriving peacefully
> Among those who come after me.

See also MENDEL'S LAWS

MENDEL'S LAWS

In genetics, the principles of inheritance are known as Mendel's Laws, after the Austrian abbot who conducted thousands of breeding experiments with pea plants in the garden of the Brünn monastery. The two basic formulations are:

The Law of Segregation: Sometimes called "particulate inheritance," it states that a hybrid or heterozygote (Aa) transmits to each mature sex cell (gamete) only one factor (A or a)—not both—of the pair received from its parents.

The Law of Independent Assortment: Different characters (e.g., shape, Aa, and color, Bb, of peas) are recombined at random in the gametes (e.g., AB, aB, Ab, ab).

Although tradition attributes these laws to Gregor Mendel, he never actually formulated them. Instead they were read into his writings by his "rediscoverers" around 1900 and have been attributed to Mendel ever since. In Mendel's famous 1865 paper, independent assortment is only briefly alluded to, while segregation is never even mentioned—it is present as an assumption only.

See also MENDEL, ABBOT GREGOR

"Mendel's Laws" were never actually formulated by Mendel, but by his rediscoverers 35 years later.

MESOZOIC (GEOLOGICAL ERA)
Age of the Dinosaurs

Some paleontologists spend a lifetime fascinated by ancient fish or fossil snails but the museum-going public favors the dominant reptiles of the Mesozoic era, better known as dinosaurs.

The Mesozoic, or "time of the middle animals," began about 248 million years ago and lasted about 183 million years. For most of that time, dinosaurs of all shapes and sizes populated the planet—from creatures the size of a chicken to the largest land animals that ever walked the Earth. During the mid-Mesozoic, the first birdlike animals appeared, while small mammals lived somewhat inconspicuously throughout the era. Mammals did not "out-compete" the dinosaurs; but when the great saurians were wiped out in a mass extinction, they replaced them as the dominant land species.

There are three major divisions of the Mesozoic: the Triassic (248–213 million years ago), Jurassic (213–140 million years ago), and Cretaceous (140–65 million years ago). During the Triassic, small dinosaurs began their spectacular radiation into many families and species, adapting to many varied niches.

Sauropod dinosaurs and flying reptiles lived during the Jurassic, named for deposits from the Jura Mountains. The remarkable fossils of *Archaeopteryx* are found in the late division of the Mesozoic. During the late Cretaceous, small mammals evolved, but were inconsequential in the dinosaurian landscape.

A mass dying, or extinction, at the end of the Cretaceous wiped out not only the dinosaurs, but also seagoing plesiosaurs, mollusks, and ammonites in the oceans—indeed, the majority of all animal life. Despite a great deal of research, and even more speculation, what happened at the close of the Mesozoic is still a tantalizing mystery.

No traces of humans or other hominids are found in Mesozoic deposits; none appear until 63 million years later. Contrary to pop culture fantasies like *One Million B.C.*, *The Flintstones,* or Creationist comic books, humans and dinosaurs never walked the Earth at the same time.

See also DINOSAURS, EXTINCTION OF; GREAT DYINGS

METAPHORS IN EVOLUTIONARY WRITING

Jitney buses in modern Greece are called "metaphors," meaning they will carry you from one place to another. Poet Robert Frost thought every scientist could learn a practical lesson from poets—skill in the use of metaphors. A good writer, said Frost, knows just how far to extend a metaphor before it becomes strained. Imagery, unexpected comparisons, and seeing similarities in the dissimilar are the poet's stock-in-trade; they are also the hallmark of exciting scientific writing. Charles Darwin would certainly have agreed; the metaphors he included in his exposition of evolutionary theory retain their vigor long after his careful collections of "facts" became outdated.

Explaining natural selection, Darwin pictured Nature as a stock breeder, selecting variations that improve or adapt species and discarding others. But whereas a human breeder could only select characteristics he could see, like size or color, Nature could act on the composition of the blood, the size and shape of internal organs, and every part of the organism. "What might she not achieve?" Darwin wondered.

Another of his metaphors is the image of a "thousand wedges." Darwin pictured pressing in on every part of a plant or animal to fit it to its environment and way of life. The metaphor evokes not only the concept of many consistent and directional pressures, but also the idea that great effects can result from cumulative small causes.

The final paragraph of the *Origin of Species* (1859) contains Darwin's metaphor of the tangled bank of a river—a jumble of plants, animals, worms, and birds. It is a commonplace scene, observable in any wood or garden, yet confusing to the casual observer. Beneath the apparent disarray, Darwin sees orderly forces that evolve species in relation to others, each with its role to play in the ecology.

These life forms "most beautiful and most wonderful" are evolving "whilst this planet has gone cycling on according to the fixed law of gravity," which is Newton's metaphor of the universe as a machine. Thus, in his final sentence of the *Origin of Species,* Darwin ties the "tangled bank"—with all the beautiful messiness in life—to the logical, predictable order of a whirring engine.

Perhaps Darwin's most widely known metaphor is his depiction of evolutionary history as a tree of life branching from a broad common trunk to finer limbs, slender branches, and delicate twigs. Biologists have since slightly modified the metaphor into a branching bush to emphasize the fullness and bushiness of the delicate lineages and to minimize the inflexibility and verticality of the trunk.

Darwin's friend "Doubting" Thomas Huxley, premier agnostic, seemed better at creating metaphors for ignorance than for knowledge. In an 1868 essay on liberal education, his own dark image of man's relationship to Nature was that of the hidden chess player.

> The chessboard is the world, the pieces are the phenomena of the universe, the rules of the game are . . . the laws of Nature. The player on the other side is hidden from us. We know that his play is always fair, just, and patient. But also we know, to our cost, that he never overlooks a mistake, or makes the smallest allowance for ignorance. To the man who plays well, the highest stakes are paid, with that sort of overflowing generosity with which the strong shows delight in strength. And one who plays ill is checkmated—without haste, but without remorse.

Huxley's metaphor for himself was a tugboat, like those who bring the great ocean liners into harbor and help guide them out to sea. He thought his role as a scientist was to help bring in the new great ideas, and send the old ones out.

As for any real apprehension of the universe around us, Huxley wrote in 1871, we have as much understanding as a "worm in a flower-pot, on a London balcony, has of the life of the great city." In a letter of 1860, Darwin expressed the same thought about human ignorance

In Greece, any public vehicle that carries you from one place to another is called a *metaphor*.

of a cosmic plan or purpose with a different metaphor: "A dog might as well speculate on the mind of Newton."

See also BRANCHING BUSH; MATERIALISM; TANGLED BANK

METAPHYSICAL SOCIETY

One of the most remarkable collections of English scientists, theologians, priests, philosophers, and iconoclasts ever assembled, the Metaphysical Society was founded on April 21, 1869, 10 years after publication of Charles Darwin's *Origin of Species*. Its purpose was to explore and discuss all points of view on faith, belief, science, God, morality, miracles, truth, and the basis of "knowing" anything.

Founded by the poet Alfred, Lord Tennyson, the Reverend Charles Pritchard (an astronomer), and theologian Richard Holt Hutton, the group eventually grew to about 60 members. At various times, discussion participants included Prime Minister William Gladstone, Archbishop Manning, the Duke of Argyll, Sir John Lubbock, Dr. William B. Carpenter, Professor St. George Mivart, Henry Sidgwick, the Bishop of Gloucester, the Archbishop of York, Professor Thomas Henry Huxley, Walter Bagehot, John Ruskin, and Professor John Tyndall—a "who's who" of British intellectual and religious life.

Members sometimes read formal papers for discussion, on a wide variety of topics: Is God unknowable? What is Death? Has a Frog a Soul? On the words Nature, Natural, and Supernatural. What is Matter? How do we come by our Knowledge? The Verification of Beliefs. What is a Lie?

At the first meeting, a member remarked that "if we hung together for twelve months, it would be one of the most remarkable facts in history." In fact they continued to meet once a month for 12 years.

The Metaphysical Society finally disbanded not because of violent disagreements, but because the dialogue was no longer producing new truths. It had become a gathering of gentlemen who agreed to permanently disagree. "After twelve years of debating," founding member R. H. Hutton recalled in 1885, "there seemed little to be said which had not already been repeated more than once."

MILLER, HUGH (1802–1856)
Champion of Scriptural Geology

Which dangerous science did some Victorians believe frequently led sane men to madness and death?

What uplifting scientific pursuit, conducted out in the fresh air of the countryside, can lead a sane man to madness and death? In 1856, most literate people would quickly have answered, "Geology." For that year a remarkable Scots author, who had almost singlehandedly made studies of the rocks not only respectable but widely popular, came to a tragic end. His name was Hugh Miller, a writer of such remarkable power that he turned a book about fossil fish into an international bestseller.

Despite the 19th-century craze for natural history, geology was considered dangerous. Fossils of ancient life discovered in rock strata were telling a fascinating but disturbing story. Skeletons of monstrous reptiles and giant ground sloths not mentioned in scripture were piling up. Evidence of prehistoric events was steadily pushing back the Earth's age.

Although geologists reassured everyone there could be no disharmony between God's word and His record in the rocks, they fought each other tooth and nail over the meaning of each. Meanwhile, most people were bewildered and put off by the whole subject, and many clergymen recommended they avoid it.

Hugh Miller was a struggling writer before he was a geologist, but as the son of working-class people—blacksmiths and harness makers—he had to get a trade. Short, stocky, and muscular, he decided to try his hand at being a stonecutter at a quarry and hoped to also find time to write. As he went to work the first day, he dreaded becoming a drudge, too exhausted to pursue dreams of literature, "working to eat, and eating to work."

To young Miller's amazement, he enjoyed physical work in the wooded countryside, soon became fascinated with the rocks, and eventually was able to create a spectacular career writing geology books. His works, *The Old Red Sandstone* (1841), *Footprints of the Creator* (1847), and *Testimony of the Rocks* (1857), were admired by scientists and literary critics alike and gained him a tremendous readership on both sides of the Atlantic.

Miller exemplified an ideal of the Victorian age: the self-taught working man who could uplift himself and join the "aristocracy of merit." Gentlemen scientists of more privileged classes, like his idol, Louis Agassiz, welcomed Miller as a valued colleague.

His discoveries were plentiful, and his descriptions were stylish and entertaining. Above all, he conveyed a deeply felt piety, which his love for science seemed never to disturb.

He had no tolerance at all for pre-Darwinian theories of evolution or "development" of one species into another. Almost gleefully, he attacked the evolutionary writings of "Telliamed" (Benoît de Maillet), Jean-Baptiste Lamarck, and especially Robert Chambers's anonymously published *Vestiges of the Natural History of Creation* (1844). (Miller knew and liked Chambers, never suspecting he was secretly "Mr. Vestiges.")

Miller feared that the growing popularity of evolutionary ideas would lead to widespread atheism and immorality. Belief in evolution rather than special creation, he wrote in *Footprints of the Creator,* was the province of "sciolists and smatterers," by which he meant ignorant dabblers in pseudoscience.

Yet there was a dark, brooding side to this "meditative stonemason," whom historian Lynn Barber imagines "holding a Bible in one hand and a fossil fish in the other." Despite his valiant attempts to bolster biblical accounts of creation with geological evidence, it became an ever more difficult battle, which increasingly troubled him.

A MEDITATIVE STONE-MASON, Hugh Miller exemplified the unpretentious workingman as natural philosopher. His internationally known books popularized geology, while defending biblical creationism against the new theory of evolution.

In his early essays, he had insisted that there was no contradiction in the Earth having been created in six 24-hour days, but in his last, posthumously published work, *Testimony of the Rocks,* he had to admit the biblical "days" might each represent millions of years. Also, he had become increasingly morbid, convinced the record of the rocks pointed to inevitable extinction for mankind—part of a pattern no different from what had gone before.

As reported in the memorials to Miller that preface *Testimony,* he suffered from horrible dreams and visions while writing the book, awakening convinced that he had wandered the streets all night. (At such times, he insisted on checking his clothing for mud stains, but none were found.) He often wrote all night and day, with a knife and gun at his side to repel imagined burglars or intruders. There were searing headaches; he told his doctor that his mind was "giving way" and that he could not "put two thoughts together." (Yet the writing is perfectly lucid, showing no deterioration.)

Finally, on December 23, 1856, Miller used his pistol to take his own life, leaving this bizarre note worthy of a Victorian horror story:

> DEAREST LYDIA—My brain burns. I must have walked; and a fearful dream rises upon me. I cannot bear the horrible thought. God and Father of the Lord Jesus Christ have mercy upon me. Dearest Lydia, dear children, farewell. My brain burns as the recollection grows. My dear, dear wife, farewell.
>
> *Hugh Miller*

In our century, one could plausibly diagnose depression, nervous breakdown, chemical mood swings, migraines, or brain tumors. (Miller did suffer from the quarryman's disease, silicosis, caused by breathing in rock dust.) But public opinion had its own answer: Hugh Miller died of trying to reconcile fossils with scripture. His doctor ascribed his descent into madness and suicide as the result of "overworking the brain." Geology became "dangerous" once more, and ministers again warned their flocks against studying it.

A few years after Miller's death, in 1859, Charles Darwin published the *Origin of Species*, which also became a scientific bestseller. Miller's arguments against evolution, raised against the *Vestiges*, looked feeble in light of Darwin's massed evidence. Although he had once been acclaimed among the 19th century's most interesting and persuasive writers, no one—except historians of science—would ever read Hugh Miller's books again.

See also AGASSIZ, LOUIS; SCIOLISM; "TWO BOOKS," DOCTRINE OF THE; *VESTIGES OF CREATION*

MIMICRY
Imitating Poison Prey

Henry Walter Bates (1825–1892) is chiefly remembered for proving that nature can come up with some very deceptive packaging. One of the most gifted tropical naturalists of the 19th century, Bates was a good friend and sometime traveling companion of the remarkable Alfred Russel Wallace, Darwin's "junior partner" in the discovery of natural selection.

Bates was intrigued by Wallace's discovery that harmless insects were usually drably camouflaged, while poisonous, bad-tasting species were dressed in gaudy colors as a warning to predators. Upon investigating further, he noticed something else—that the colors of some harmless species copied or mimicked the poisonous ones. Mistaking them for noxious prey, predators might leave the mimics alone. Bates pointed out that this was not conscious imitation, but a gradually evolving resemblance between species brought about by natural selection. His writings on the phenomenon convinced other naturalists, who began to call it "Batesian mimicry."

Some critics scoffed at the idea that some species evolve elaborate markings and colors to imitate others, but Charles Darwin did not. "I am rejoiced that I passed over the whole subject in the *Origin*," Darwin wrote Bates in 1862, "for I should have made a precious mess of it. . . . Your paper is too good to be largely appreciated by the mob of naturalists without souls."

In recent times, experimenters have put Bates's idea to the test. One researcher put scrub jays in an enclosure with poisonous black-and-orange monarch butterflies and its harmless mimic, viceroys. Sure enough, inexperienced young birds gobbled up the viceroys. But birds that first ate what Alfred Russel Wallace called "the disgusting morsels" of monarchs quickly learned their bitter lesson and would not touch the harmless viceroy mimics. When various orange butterflies were introduced, even very approximate look-alikes went untouched by the birds—demonstrating that slight resemblances can confer survival advantage. Over generations of selection, the resemblance could become increasingly fine-tuned, as Bates had originally suggested.

The case of the poisonous monarch butterfly and its harmless viceroy mimic has been a textbook example of Batesian mimicry for more than a century. A 1991 study by David B. Ritland and Lincoln P. Brower of the University of Florida in Gainesville, however, revealed a surprise: when butterfly abdomens, without wings, were served up to birds, the predators found the viceroy just as unappetizing as the monarch.

Scientists, not birds, had been deceived. They assumed that the viceroy's orange warning colors were just a bluff, and that all foul-tasting butterflies had to acquire their poisons from food plants. It had been well known that monarchs must eat milkweed to gain their toxins, but viceroys, it turns out, manufacture noxious chemicals on their own.

Fritz Müller (1821–1897), another naturalist, first described how two or more equally distasteful butterfly species can gain greater protection from predators by evolving the same general appearance. He assumed that young, inexperienced predators learn to avoid certain prey through trial and error—by killing and eating some. If the foul-tasting species varied widely in appearance, predators would have to kill many of each before they learn which to avoid. But if the noxious prey came to share coloration, the predator would learn to avoid one basic pattern and the protection would be spread out over many species. The survival value of resemblance is that it cuts the loss to each population. Müllerian mimicry is common among poisonous tropical butterflies, both in Africa and the Amazon. So the viceroy may actually be a "Müllerian" mimic of the monarch, or it may have first evolved as a Batesian mimic, then "changed the rules of game" and developed its own chemical defenses as well.

> "Your paper is too good to be largely appreciated by the mob of naturalists without souls."
>
> —Charles Darwin to Henry Walter Bates, 1862

Another recent study has shown that the monarch butterfly gathers a second, totally different plant poison when its protective dose of milkweed cardenolides, or cardiac glycosides, has worn thin after its annual southern migration to Mexico. This second poison, a pyrrolizidine alkaloid, from the locally abundant *Senecio* flower, is effective against 35 out of 37 birds and most rodents in the area. However, one species of Mexican mouse, the scansorial black-eared mouse (*Peromyscus melanotis*), can dine without ill effects on monarchs that have imbibed it. With this selective advantage, it has become the most successful mouse on the mountainsides of central Mexico.

See also BATES, HENRY WALTER; CONSPICUOUS COLORATION

"MISSING LINK"
Mythical Ape-Man

The most widespread misconception about human evolution seems to be the myth of the "missing link." For 100 years, many have accepted the cliché that Charles Darwin's theory is not "proved" because "no one has yet found the missing link between ape and man."

The idea of "links" in a Great Chain of Being is derived from medieval theology. Church philosophers ranked all creatures as "higher" or "lower," with man at the top, the crowning glory of Creation. Above man were the angels, archangels, and other spiritual beings, leading up to Almighty God. The Chain of Being was also reflected in the earthly social order, with those of "low degree" at the bottom, an aristocracy above, and royalty at the apex of the social pyramid. That one species might develop or evolve into another was as unthinkable as a servant aspiring to move into the ruling class—a violation of the natural order.

A second source of the "missing link" idea lies in the comparative method developed by 19th-century naturalists. When Thomas Huxley wrote his famous essay *Evidence as to Man's Place in Nature* (1863), he made an exhaustive comparison of the anatomy of monkeys, apes, and humans. Humans and apes showed overwhelming, detailed similarities in structure, from which the kinship between them was inferred, but at the time there were no known fossil men, fossil apes, or near-men, with the exception of a few fragmentary Neanderthal crania.

Darwin was careful to say in *The Descent of Man* (1871) that humans are not descended from anything like a modern ape or monkey, but that these groups are related through common ancestry. He also guessed (correctly, it turned out) that ancient near-men fossils would be found in Africa, since it appears to be the ancestral home of the chimp and gorilla. Despite his caution, however, many of his readers imagined a half-man, half-ape as the ancestor of both humans and apes. Ernst Haeckel, the influential German evolutionist, went so far as to include this hypothetical "missing link" in his books, and he even gave it a Latin species name: he called it *Pithecanthropus alalus*, "ape-man without speech."

Since Haeckel's day, taxonomists have made a strict rule against giving scientific names to species before they've been discovered. Otherwise, unicorns and mermaids might return to catalogs of zoology. But Haeckel's insistence that *Pithecanthropus* existed and would be found had a remarkable effect. It inspired a Dutch army surgeon named Eugène Dubois to set out in search of *Pithecanthropus*—and he found it! In 1893, after several dedicated, difficult years, Dubois dug up the fossil hominid remains he eventually called *Pithecanthropus erectus*, popularly known as Java Man. It later turned out that they were not very "apish" after all, and were really an ancient species of human; whether or not they were "without speech" no one knows. (They have since been renamed *Homo erectus*, to reflect the conclusion that they were not apes at all.)

In 1912, a "missing link" hoax fooled British anthropologists, and it was not exposed for 40 years. An unknown prankster combined an ape jaw with a human skull and planted the forgery in Sussex, England, at a site called Piltdown. One reason it was eagerly accepted as authentic was that it certainly combined ape and human characteristics—half and half. One famous anthropologist, Sir John Lubbock, said he was impressed that it was "the most simian" fossil human yet found.

PITHECANTHROPUS ALALUS, meaning "ape-man without speech," was named by German zoologist Ernst Haeckel long before a fossil "ape-man" was ever found. Haeckel inspired this 1887 etching by Henri du Cleuziou.

The Piltdown forgery was not only a cruel hoax on the scientists who analyzed it, but a devastating comment on the simplistic concept of a "missing link."

There have been other famous "missing link" jokes that were not so malicious. When Charles Darwin went to Cambridge in 1877 to receive public acclaim for his life's work and an honorary doctorate, the students had decorated the hall with an effigy of a monkey hanging from several huge, interlocking hoops representing the "link." And 50 years later, at the conclusion of the Scopes "Monkey Trial" in Tennessee, lawyer Clarence Darrow—who had just lost his client's battle to teach evolution in the schools there—received a telegram from a friend in California who was trying to cheer him up. It read, "HAVE FOUND MISSING LINK. PLEASE WIRE INSTRUCTIONS."

Over the past few decades, paleoanthropologists have unearthed a bewildering variety of human and near-human fossils going back more than four million years. There are erect, bipedal near-men (australopithecines) with small brains and somewhat apelike teeth. There are larger, more robust hominids with huge molars and powerful jaw muscles. One spectacular recent discovery is the Toumai skull, *Sahelanthropus tchadensis* (2002), which is believed to represent the closest common ancestor of both apes and humans. The evolutionary relationship of various fossils to each other and to ourselves are still a mystery and may never be completely known, but fossil hominids and hominins are no longer "missing."

See also APES; AUSTRALOPITHECINES; DART, RAYMOND ARTHUR; DUBOIS, EUGÈNE; *HOMO ERECTUS*; PILTDOWN MAN (HOAX); TAUNG CHILD

MITOCHONDRIAL DNA
Genetic "Family Archives"

Genetic blueprints for an individual, coded in DNA, reside in the cell nuclei of most plants and animals: half from the individual's mother and half from the father. But some DNA also exists outside the nucleus, in tiny organelles called mitochondria, which are crucial to a cell's metabolism. Mitochondrial DNA (mtDNA) is passed on only through the female line; the father's genes do not affect it at all.

Recent discoveries about mitochondrial DNA have opened up new possibilities for tracing evolutionary lineages of living populations. Beginning in the 1980s with such diverse creatures as desert tortoises, red-winged blackbirds, eels, and humans, thousands of species have been studied on the assumption that complex similarities in mitochondrial DNA are attributable to shared matrilineal ancestors.

See also GENOGRAPHIC PROJECT; RACE

MIVART, ST. GEORGE J. (1827–1900)
"Heretical" Biologist

English biologist St. George J. Mivart, a devout convert to Catholicism, embraced the Darwinian theory of evolution with one qualification: After the human body evolved from apelike ancestors, God intervened to infuse it with a soul. In many ways a forerunner of Father Pierre Teilhard de Chardin (1881–1955), Mivart tried to bring scientific truth to the church and a religious perspective into biology. Also like Teilhard, he succeeded mainly in drawing heavy fire from both quarters.

Originally graduated as a barrister, Mivart returned to his boyhood interest in natural history and studied comparative zoology on his own. Eventually, he produced first-rate contributions, such as his monumental 557-page anatomy *The Cat* (1881), which guided generations of comparative anatomy students. For many years he lectured on biology at St. Mary's, a Catholic college, where he taught that evolution was entirely compatible with church dogma.

For almost a decade (1861–1869), Mivart was Thomas Huxley's devoted student, attending almost every anatomy lab and evolutionary lecture. (Huxley called him "my constant reader.") His mastery of the anatomy of newts and monkeys prompted Darwin to request information from him for his own books. But Mivart's increasing "theological fervor," fanned by a priestly colleague, eventually led him to adopt a strong anti-Darwinian stance. In 1869 he told Huxley, whose friendship he treasured, that he intended to write a strong critique on the "insufficiency of Darwinism." It was a painful, emotional confrontation, though Huxley neither became angry nor argued. As Mivart wrote in "Some Reminiscences of Thomas Henry Huxley" (1897), "As soon as I had made my meaning clear, his countenance became transformed as I had never seen it. Yet he looked more sad and surprised than anything else. He was kind and gentle as he said regretfully, but most firmly, that nothing so united or severed men as questions such as those I had spoken of."

Now Mivart determinedly churned out articles for Catholic journals on difficulties of the theory of natural selection and wrote a popular book, *On the Genesis of Species* (1871), as a rebuttal to Darwin's *Descent of Man* (1871). His stated object, he wrote in 1872, had been "to show that the Darwinian theory is untenable, and that natural selection is not *the* origin of species . . . upon scientific grounds only. My second object was to demonstrate that nothing . . . in evolution generally, was necessarily antagonistic to Christianity." He followed up with *Apes and Men* (1873), an exposition of antievolutionary analysis of comparative anatomy.

Ironically, it was his friend and scientific mentor Huxley who challenged Mivart's absurd arguments that evolution had been anticipated in church doctrine. Meeting him on his own grounds, Huxley quoted Catholic history and theology with the ease of a bishop to demonstrate that Mivart's claim was founded entirely on wishful thinking. Huxley would have liked to believe the church was congenial to evolutionary ideas, he wrote, but could no more condone "unfaithfulness to truth" by fudging ecclesiastical rulings than by stretching science beyond its proper boundaries.

One of Mivart's major arguments against natural selection—one that has never gone away—is the question of how complex adaptive structures could originate. What good is one-quarter of a wing or half an eye? Why would such incipient structures be selected before they were fully useful? Mivart thought this stumper proved the "logical insufficiency" of the theory of natural selection. [For one answer to Mivart's objection, see EXAPTATION.]

Darwin was perplexed by Mivart's private expressions of friendship even as he attacked him in the public press. In 1871, Darwin wrote to a friend: "You never read such strong letters Mivart wrote to me about respect, [and] begging I would call on him . . . yet [in his published articles] he . . . makes me the most arrogant, odious beast that ever lived. I cannot understand him; I suppose that accursed religious bigotry is at the root of it. . . . It has mortified me a good deal."

Although he was praised by cardinal-to-be John Henry Newman, who admired his competence in science, Mivart's increasingly liberal theological views led to his excommunication in 1900, shortly after which he died. Years later, his friends argued to church authorities that Mivart was not willfully heretical, but that the diabetes that caused his death had also unbalanced his mind. Accepting this "mechanistic" medical interpretation of Mivart's heresy, the church allowed him a Christian burial.

See also TEILHARD DE CHARDIN, FATHER PIERRE

> **Professor Huxley looked sad and "said regretfully . . . that nothing so united or severed men as questions such as [these]."**
>
> — St. George Mivart

MONAD
Particles of Life

Gottfried Leibniz (1646–1716), the German philosopher, proposed the monad as an elementary particle of life. Impressed by the new world of previously invisible creatures that was being revealed by the microscope, he defined them as the smallest living units, just as atoms were theoretically the smallest particles of matter.

Monads were thought to be "units of force" that could develop into plants or animals. From the first, the concept of monads as entities was confused with the simple microscopic organisms that were as yet unidentified.

Sir Charles Lyell thought monads might exist in great numbers at the present time, perhaps in warm ponds, where they could develop into new species to replace those few that become extinct from time to time. (He thought there was a slow, steady rate of species replacement and regarded the idea of rapid mass extinctions as a "catastrophist's" daydream.) Thus, monads were thought to be not only the original source of species, but an enduring resource that was continually renewing the Earth with species as needed in the "economy of nature."

By the end of the 19th century, the "monad" joined the "pangene" and the "homunculus" on the scrap heap of nonexistent theoretical entities.

See also MONISM

MONISM
Social Darwinist Philosophy

In late 19th- and early 20th-century Germany, monism (from Greek, meaning "one") emerged as a popular cultural and political expression of Social Darwinism. Its founder, Ernst Haeckel (1834–1919), the eminent German evolutionary biologist, characterized it as the union of matter and spirit, based entirely on the creative properties of matter.

Haeckel developed monism into a nationalistic, romantic, and anticlerical movement, which he contrasted with dualism, a doctrine that he thought made false distinctions between matter and spirit and between humans and nature. In *The Scientific Origins of National Socialism in Germany* (1971), historian Daniel Gasman characterizes Haeckel's agenda as a combination of Social Darwinism and deeply felt mystical nationalism.

Haeckel preached that it was necessary to "regenerate the German race," eliminate the church and all traditional religions, and conquer non-German peoples in order to fulfill a "higher evolutionary destiny." Darwinism for Haeckel was a harsh "struggle for existence," and the laws of nature had to become the laws of society as well.

In 1904, Haeckel proposed the formation of a Monist League at the International Free-thought Congress, which was meeting in Rome—an attempt to translate his version of Social Darwinism into political action. After the meeting, the pope was quoted as remarking that the Holy City was in need of cleansing after this "insult done to Almighty God"; one journalist quipped that the pope desired "a divine fumigation."

> Haeckel's international meeting of free-thinkers and atheists in Rome intentionally scandalized the pope.

Two years later, although he was then past 70, Haeckel established the Monist League in his university town of Jena. Within five years, the league had 6,000 members meeting in 42 cities and towns throughout Austria and Germany, published a weekly journal (*Monist Century*), and had developed enormous influence, both in the international Freethought Movement and in German politics.

Darwin himself was not a Social Darwinist and thought it a bitter joke that his scientific theories were being used, as he put it, to "prove Napoleon was right and every cheating tradesman is also right."

See also "ARYAN RACE," MYTH OF; HAECKEL, ERNST; SOCIAL DARWINISM

MOSAIC EVOLUTION
A Patchwork Process

Organisms do not change "all over" by gradual degrees. Sometimes one part of the system may remain stable over long periods, while another evolves rapidly. Many birds, for example, seem to be a "mosaic" of adaptations acquired at different times. Most show little variation in the ancient body plan, whereas wings, feet, or beaks have become specialized for different methods of obtaining food.

One of the most striking examples of mosaic evolution is our own species, *Homo sapiens*. Different parts of the human body did not evolve toward our present form at the same time or at the same rate; it is a mosaic, with different features added at a different times. First came the fully rotating shoulders and upper torso, which developed early (perhaps 15 million years ago) and which we have in common with our fellow hominoids. Anthropologists believe this arm-shoulder-chest complex evolved in a forest habitat, as an adaptation to moving through the trees before the hominid line diverged from the apes.

Next, about four to five million years ago, our double-arched feet, long legs, and basin-shaped pelvis became adapted to upright walking and running. This shift to a bipedal posture was accomplished by the time of *Australopithecus*, which was still small-brained but very like humans from the waist down.

The most recent phase of hominid evolution—the expansion of the brain and further reduction of the jaws—first appeared during the *Homo erectus* stage (half a million to 1.5 million years ago), perhaps associated with an increasing reliance on language, tools, and cultural traditions.

MULLIS, KARY B. (b. 1944)
Irrepressible Prince of PCR

Kary Mullis's invention in 1983 of the polymerase chain reaction (PCR)—a technique for rapidly copying large quantities of DNA from a tiny sample—revolutionized the study of the chemical basis of life. DNA is "the King of molecules," he wrote in his 1998 autobiography, *Dancing Naked in the Mind Field.* "The DNA molecules in our cells are our history, and they are the stuff of which our future will be crafted." Mullis's discovery provided new tools and research programs for evolutionary studies, medicine, forensics, and genetic engineering.

Raised by a conservative family in rural North Carolina, by the 1960s Mullis had become a disciple of "alternative" California culture, reveling in its psychedelic spiritual hedonism. In 1975, he had also managed to complete a doctorate in biochemistry at the University of California, Berkeley.

Mullis invented PCR during a stint as a hardworking, innovative laboratory scientist at the Cetus Corporation, at the same time completing his transition from bookish nerd to freewheeling surfer.

Mullis experienced his "aha moment" while driving along the twisty mountain roads of Mendocino County, California. "Natural DNA is a tractless coil, like an unwound and tangled audio tape on the floor of the car in the dark," he later wrote. "I had to arrange a series of chemical reactions, the result of which would represent and display the sequence of a stretch of DNA. The odds were long. Like reading a particular license plate out on Interstate 5 at night from the moon." However, before he had arrived at his mountain cabin, the solution flashed into his mind. Mullis knew "I would be famous. I would get the Nobel Prize."

Mullis claims he had merely envisioned a "simple" combination of existing techniques, just as Robert Fulton's steamboat involved the simple combination of a steam engine and a boat, both of which already existed. When Mullis first explained his proposed technique of replicating large quantities of DNA by means of a rapid chain reaction, his fellow biochemists at Cetus all agreed that his reasoning was correct—but they didn't think it would work.

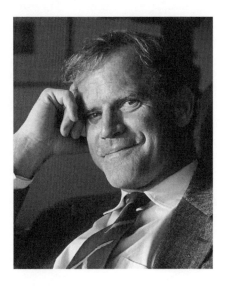

OPINIONATED AND UN-CONVENTIONAL biochemist Kary Mullis invented a technique that revolutionized genetic studies and forensic criminology.
Photo © by Paul Nestor.

Within a decade, however, all DNA labs were performing PCR with an apparatus that can be easily contained in a toaster-sized machine. Evolutionary biologists could now more easily compare the relationships between all living things based on tiny samples of DNA. Thanks to PCR, forensic investigators can use a fleck of blood or semen from a crime scene to yield an evidentiary quantity of DNA. Thousands of rapists and murderers have been convicted because crime labs can amplify tiny but unmistakable bits of their one-of-a-kind DNA. Minuscule skin or mouth scrapings are now routinely used to resolve paternity cases (often on daytime television). And when families of terrorist victims of New York's 9/11 Twin Towers disaster needed to identify a relative's fragmentary remains, they were asked to bring the deceased's hairbrush or toothbrush, so that a small DNA sample could be gathered for amplification before identification was attempted.

Even science fiction and crime stories have been influenced by PCR. Michael

Crichton's book *Jurassic Park* begins with the premise that living dinosaurs can be cloned from amplified dinosaur DNA taken from blood in fossilized dino-feeding mosquitoes. Popular television crime series now routinely hinge on cases built on DNA and PCR evidence.

On October 13, 1993, after being notified by telephone that he had won the Nobel Prize for chemistry, Mullis ducked reporters by surfing at a remote beach off the California coast, as if he were avoiding a subpoena. Next day, newspapers headlined: "SURFER WINS NOBEL PRIZE." After overcoming his initial reticence, Mullis was to become an irritating, controversial Nobel laureate.

"I was pretty happy working at Cetus and assumed, innocently, that if the [PCR] reaction worked big time, I would be amply rewarded by my employer," he wrote. "I was plenty wrong on that one." Instead, he received a $10,000 bonus from the company, and three years later Cetus sold the license to Hoffmann–La Roche for $300 million. Mullis later pondered, "Why couldn't they have charged $301 million, and given one million to me?"

Disillusioned by that episode, when Mullis was invited to give a lecture on PCR at a scientific conference, he spoke instead of dishonesty in the scientific establishment, claiming it was common practice to fudge data; he punctuated the lecture with color slides of nude women. A scandalized chairman cut him off in mid-speech and banned him from further invitations. Another time, he expressed his belief that "there is no scientific evidence that HIV is the probable cause of AIDS and that . . . people taking the drug AZT were being poisoned." When a large drug company abruptly reneged on one of his appearances in 1993, paying him a $6,000 cancelation fee, he joked that "this was the most money I had ever made specifically for not doing something." He wrote a satirical ad, "Have Slides, Will Stay at Home," in which he solicited other venues to pay him *not* to speak.

Mullis seemed convinced that the Nobel Prize gave him license (or even the duty) to speak "outside of the envelope" in public. In his autobiography he maintains his advocacy of LSD and marijuana, along with his beliefs in alien abduction and astral travel. He even tells of having had a conversation one evening with a "glowing raccoon" on his cabin porch. Like any good California surfer dude, he relied on astrology as his personal star guide. "I was born at 17:58 Greenwich Mean Time on December 28, 1944 in Lenoir, North Carolina," he wrote in his autobiography. "You can find out more about me from that than you can from reading this book."

Alfred Russel Wallace, the great Victorian naturalist and evolutionist, wrote in his 1874 "Defence of Spiritualism" that indeed "the pure dry air of California" was known to produce "powerful and . . . startling manifestations."

See also *JURASSIC PARK*; POLYMERASE CHAIN REACTION

MURCHISON, SIR RODERICK IMPEY (1792–1871)
Geologist

Perhaps no other geologist is more typical of his era than Roderick Murchison. Ambitious, tyrannical, possessed of seemingly limitless energy, and with a towering ego, Murchison could also be urbane and charming. He was, in fact, the perfect example of the "gentlemanly specialist," that curious breed of Victorian scientist who, having independent means, an inquisitive nature, and little formal training, did much to advance our knowledge of the world.

Murchison was born in 1792 into a wealthy Scots family and was educated at Great Marlow, an important military school—which perhaps accounts for his later pugnacity. His military ambitions were stifled, however, by a posting in Ireland, far from the field of valor during the Peninsular Campaign. He spent his early years traveling in Europe, then settled in a large country house and indulged his passion of fox hunting six days a week. In 1823, however, under the influence of Sir Humphrey Davy, president of the Royal Society, and his own wife, who was fed up with his lavish expenditures, Murchison sold his hounds and horses to his hunting cronies and decided to take up geology. His friends were much amused when one of his earliest discoveries, during a "geologising" trip to Germany, turned out to be a fossil fox.

Exploring the geology of the south of England with his wife, he devoted special attention to the rocks of northwest Sussex and presented his first scientific paper to the Geological Society in 1825. He then turned his attention to Continental geology and, with Charles Lyell, explored the volcanic region of Auvergne, parts of southern France, northern Italy, Tyrol, and Switzerland.

After studying for a time under Reverend William Buckland, the great geologist from Oxford, and then William Lonsdale, Murchison came under the spell of one of the preeminent scientists of the 19th century: the Reverend Adam Sedgwick. Together, they attacked the difficult problem of the geological structure of the Alps. Their joint paper giving the results of their study is one of the classics in the literature of Alpine geology. Sedgwick also led him into the study of Greywacke, a very ancient, distorted series of rocks about which little was known. Murchison took it up with enthusiasm and made the Upper Greywacke his personal domain. He named these strata the Silurian after an ancient kingdom of Britons. In 1839, he published his mammoth two-volume work *The Silurian System,* which he dedicated to Sedgwick. In it and his succeeding book, *Geology of Russia in Europe and the Ural Mountains* (1845), Murchison delineated the succession of Paleozoic rocks, sweeping aside the claims of other geologists. Thus was born the long-contested Cambrian-Silurian Question.

Although now all but forgotten, this knotty problem divided the scientific world for nearly two decades as Sedgwick, Murchison, Sir Henry Thomas De la Beche, and other key figures sought to determine the true succession of Europe's ancient rocks. Although the problem was eventually resolved, with Sedgwick's Cambrian, De la Beche's Devonian, and Murchison's Silurian all retaining their sovereignty, it was Murchison, as head of both the Geological Society and the British Geological Survey, who captured the public's imagination.

This episode taken from the *Midland Naturalist* of 1891 typifies his incredible popularity among the general public:

> During the year 1849 . . . an excursion was made to Dudley, when Sir Roderick Murchison, in the great cavern of the Castle Hill, briefly explained the system of strata to which he had given the name of "Silurian." . . . In proposing a vote of thanks to him, the Bishop of Oxford (Dr. Samuel Wilberforce) said that although Caractacus was an old king of [Siluria] . . . Sir Roderick should be acknowledged the modern king of Siluria. The Bishop, then taking a gigantic speaking-trumpet, called upon all present to repeat after him . . . one word at a time . . . Hail—King—of—Siluria! The vast assembly thrice [repeated the words] with stentorian voices and most hearty hurrahs, and ever afterwards, Sir Roderick was proud to be acknowledged "King of Siluria."

FAMOUS AS THE "KING OF SILURIA," Sir Roderick Impey Murchison gave up fox hunting for geology. He advanced understanding of early rock strata, but suppressed the work of other talented geologists.

Although he shied away from scientific controversies later in his life, Murchison retained his vigor and dictated his last presidential address to the Geological Society from his deathbed at the age of 79. The Murchison crater on the Moon and fifteen geographical locations on Earth are named after him.

In 1871, by his will, Murchison bequeathed £1,000 to the Geological Society for the founding of the Murchison Geological Fund, which, in addition to funding worthy scientists, was to present a bronze Murchison Medal every two years. The artwork, commissioned by its founder, features a bust of Murchison on the obverse; on the reverse are two crossed rock hammers, surrounded by fossil trilobites and brachiopod shells. Across the top is the single word "SILURIA."

See also CAMBRIAN-SILURIAN CONTROVERSY; SEDGWICK, REVEREND ADAM

NAPI
Native American Creation Story

According to one version of a Plains Indian creation story, in olden times the Sun was a great fiery chief who lived in his lodge in the sky. His principal servant was Napi, an immense being who did the Sun's bidding so he wouldn't be distracted from keeping the Earth warm.

Napi was usually occupied with the Sun's many tasks, but one day he sat down to smoke his pipe near a spring and noticed some damp clay. He made a great many little sculptures out of the clay, then let them all dry in the sun.

Finally, he picked one up, blew his breath on it, and said, "Go you now, my son. Be a Bighorn Sheep and live out on the plains." And the sheep galloped off.

Then Napi blew on the others, giving life to the Bear, the Antelope, the Beaver, the Badger, and many more. To each animal he gave a name and then sent them to where they were each supposed to live.

One strange little clay shape was left, one with two legs instead of four, and Napi smoked and looked at it for a long time. After a while, he blew the breath of life into it, and said, "Go you now, my son. Be a Man. Live with the wolves, and hunt meat on the plains."

PLAINS INDIAN LEGEND tells of Napi, the Sun's helper, forming people and animals out of clay. Many origin stories from all over the world begin with a deity creating humankind from common clay.

Napi thought he had done well, and that all the creatures would be happy. But a few days later, when he went to the spring again, all the animals came to complain. First, the Buffalo said, "Grandfather, I cannot live in the mountains where you sent me. The hills are too steep, the rocks break my hooves, and there is no grass."

The Bighorn Sheep complained that he could not live on the plains. "Grandfather," he said, "my hooves grow too fast there and curl up. There is no moss, no hills to climb, and my legs get weak." And the Antelope had similar complaints.

"All right, my sons," said Napi, "I will give you each a home suited to you. You, Bighorn Sheep, go up to the mountains, and take the Goat to live there, too. Bear, my son, you go and live among the forested hills; Cougar, you go there also. Buffalo, my son, go and take Antelope and live in the plains and eat the grass there. Badger and Prairie Dog, go also to the plains and dig burrows in the earth, where you will find food. And Wolf, you will share the meat of the plains with Man."

So all the animals listened to Napi and went where he told them to live, and they have lived there and been content ever since. All except Man, "who is never satisfied anywhere and always wants everything."

See also ORIGIN MYTHS (IN SCIENCE)

> **Every creature enjoys its place in nature, except for human beings, who are never satisfied anywhere and always want everything.**
>
> —Plains Indian creation story

NATURAL LAW
Formulation of Regularities

Natural law was one of the basic ideas that led to the ascendancy of science in Western thought. English physicist-astronomer Sir Isaac Newton was one of its founders, demonstrating that complex physical phenomena could be reduced to discoverable "rules," although he also believed that miracles were possible.

Most people assume that "natural" phenomena are those that are fixed, regular, and expected—arising from the properties of matter. Theologians, however, saw no logical difficulty in accepting the reality of supernatural events. If God could start the world spinning, he could temporarily stop it. If he could make one kind of beetle, why not 100,000? He was, by definition, infinite and beyond human comprehension.

If God had set natural law in motion, they argued, why should it be "unnatural" for him to intervene directly every now and then?

Emulating physicists and chemists, biologists realized that research was only possible with a commitment to finding regular "secondary causes." Such was the "revolution" of Charles Lyell in geology and Charles Darwin in biology.

See also *NATURPHILOSOPHIE;* "PHILOSOPHICAL" NATURALISTS; THEORY, SCIENTIFIC

NATURAL PRODUCTS CHEMISTRY
Evolutionary Pharmaceuticals

Living things have evolved thousands of chemical substances during 3.8 billion years in the world's oceans. Marine invertebrates like corals, sponges, and worms have had eons to develop chemical defenses against rivals and predators. "Some of these remarkable substances will also destroy viruses and cancers in humans," says Professor Robert Pettit of Arizona State University.

Dr. Pettit is part of an effort to develop "natural products chemistry," a field that began around 1812 in Europe. Biologists have searched molds, microorganisms, and plants for new medicines, leading to such discoveries as quinine, tetracycline, and penicillin. Many were derived from plants of the tropical rain forest, a repository of still-unknown medicinal chemicals that may be destroyed before anyone finds them. [See RAIN FOREST CRISIS.]

Since the 1960s, a small but dedicated group of scientists have turned to the ocean. There, in the nearly three-quarters of the planet covered by water, hundreds of thousands of animal and plant species still exist—a significant portion of the Earth's organisms. If only 10 percent of them produce potentially useful chemicals, 40,000 unique substances are waiting to be discovered.

Pettit's team discovered an effective cell growth inhibitor for mammals, which a marine worm uses for defense. Its biochemical structure is utterly unlike anything the researchers had expected. In 2004, a synthetic form of venom from the South Pacific cone snail was approved by the Food and Drug Administration to treat severe, chronic pain. The drug, called Prialt, controls pain in a new way—by blocking the calcium channels in nerve cells that transmit pain signals. It is 1,000 times more powerful than morphine and has become an alternative to opioid drugs like OxyContin and morphine.

Despite the great evolutionary distance between humans and these eyeless, limbless sea animals, their bodies produce substances that can have remarkable effects in our own systems—biochemical testimony to Charles Darwin's conclusion that all life is one great, related family.

NATURAL SELECTION
A Mechanism of Evolution

Charles Darwin complained that his critics said what was good in his theory was old and what was new in it was wrong. The "old" part was simply the *fact* of evolution: that species had developed over time and that all life is linked through common ancestry. The new part was how it worked: the mechanism of natural selection.

That living things evolve is as certain as a scientific fact can be. The evidence is overwhelming and continues to accumulate. Just *how* evolution occurs—entirely through natural selection, or in other ways as well—is still an open question. However, despite a century and a half of criticism and attack, natural selection continues to be one of the most fruitful organizing principles of biological research.

Natural selection starts with two observations:

1. There is vast overproduction of new individuals in nature. Every organism produces many more offspring (or eggs or seeds) than will survive to reproduce themselves, as anyone who walks through the woods can see.

2. There is a great amount of variation between individuals, which a casual observer may not see. All zebra foals or bullfrog tadpoles may look alike at first glance, but a naturalist who spends years studying them is struck by the wide range of variability within the same species.

Darwin used the term "natural" as opposed to what he called "artificial" selection, the deliberate breeding of varieties of domestic animals or plants by man. If pigeon breeders could select for fancy plumage and horse breeders for speed or disposition, he thought, surely organisms would be similarly shaped by relentless selection in nature.

In each generation, many individuals will not reproduce. Selection pressures may include predators; climate; other members of their own social group; competition for space, food, or mates; parasites; and disease. The popular notion that "survival of the fittest" means simply that strong "winners" kill off weak "losers" is ideology, not biology.

Creation of new life forms, according to current theory, requires only a source of genetic variation and a "sorting sieve," which lets some alleles through to the next generation and blocks others. (An allele is a variant of a gene.)

Paleontologist George Gaylord Simpson's famous image is a hat containing several sets of the 26 letters of the alphabet on slips of paper. If you were to draw out letters at random, chances are poor that they would spell C, A, and T, in that order. But if each time you picked a slip, you threw it away if it was not C, A, or T, and returned it to the hat if it was one of those three letters, soon you would have mostly Cs, As, and Ts in the hat. Your chances of drawing out C-A-T would keep improving, until eventually you would draw them in the proper order.

Darwin first wrote down his theory of natural selection in 1842; he sent a brief statement of it in an 1844 letter to his botanist friend Joseph Hooker. Bit by bit he assembled an enormous amount of data from geology, zoology, botany, animal husbandry, paleontology, biogeography, and scores of other disciplines. Patiently and brilliantly, he marshaled his

That living things evolve is as certain as a scientific fact can be. What drives evolution is still an open and contentious question.

evidence in *Origin of Species* (1859), giving his readers the feeling they had discovered the theory themselves, almost against Darwin's objections.

Another great naturalist, Alfred Russel Wallace, who was 14 years Darwin's junior, had come to the same conclusions. Working in the jungles of Malaysia, Wallace lay on a hammock with malarial fever thinking about Malthus's *Essay on the Principle of Population* (1798)—the same book that Darwin said led him to the theory—when he independently conceived the idea of natural selection. It is one of the most striking cases of parallel discovery in the history of science.

In 1858, Wallace mailed his own paper on natural selection to Darwin, sending the older naturalist into a panic to finish his "big book" on the subject. Wallace and Darwin received joint credit for the theory, which was communicated to the Linnean Society in 1858. Although Wallace's name is little known today, Darwin wrote him that he considered natural selection "just as much yours as mine."

Darwin was handicapped by the ignorance about inheritance in his day. Sometimes he spoke as if random variations were selected, eliminated, or preserved in populations; at other times he adopted the so-called "Lamarckian" theory that an individual's habitual use or disuse of organs would be passed on to offspring. (It wasn't just Lamarck's; practically everyone in his time believed in use-inheritance.)

So uncertain was Darwin on the question of inheritance that, under fire from critics, he began to retreat from natural selection as his main evolutionary mechanism. In later editions of the *Origin of Species*, he suggested possible alternatives and special cases. By the last (sixth) edition, Darwin had shrunk natural selection in importance to one of several possible mechanisms of evolution.

Meanwhile, his junior partner, Alfred Russel Wallace, continued to maintain the validity of their original vision, insisting that natural selection was the major driving force of evolution. Influenced by his spiritualist beliefs, however, Wallace could not see why early humans should have evolved a brain "so much better" than seemed strictly necessary for survival unless some "unknown agency" had intervened. Our physical bodies evolved by natural selection, he concluded, but at some point "the Unseen World of Spirit" had injected a higher consciousness and intellect into the human species. When he advanced that view in 1868, Darwin was saddened and disappointed. How could Wallace apply natural selection to every living creature except man? He was abandoning science to bring special creation in the back door. "I hope," Darwin wrote Wallace, "you have not murdered too completely your own and my child."

Natural selection has been misused as a political analogy, overextended into "just-so" origin stories, and undergone radical shifts in meaning as evolutionary theory broadened to take in population genetics and DNA. Yet despite more than 100 years of attempts to dislodge it, natural selection remains a central idea in biology, still generating new theories, observations, and fruitful research designs.

See also *ORIGIN OF SPECIES*; "SURVIVAL OF THE FITTEST"; "TERNATE PAPER"; WALLACE, ALFRED RUSSEL; "WALLACE'S PROBLEM"

> **Despite more than a century of attempts to dislodge it, natural selection remains a central idea in biology.**

NATURALISTIC FALLACY
Mother Nature Knows Best

When humans seek answers to questions of ethics, behavior, or custom, they sometimes appeal to observations of plants and animals to see what is "natural." That is the "naturalistic fallacy," the idea that what appears to be the case in nature is right or correct for humankind.

Aesop's fables, dating from ancient Greece, taught that the "industrious" ant was more to be admired than the "lazy" grasshopper, who fiddled all day and would have no food for the winter. Or that the tortoise proved that "slow and steady wins the race." (One might argue that a tortoise also shows that you can't get anywhere unless you stick your neck out.)

Medieval bestiaries detailed the supposed behavior of animals (largely inaccurately) and offered good or bad examples for mankind. The moralistic tradition was carried over into

19th-century science and still survives, despite great strides in accurate observation and the realization that nature can provide examples for any kind of behavior.

Some early anthropologists, after having watched chimps, argued that primitive promiscuity was natural to humans, while others pointed to the "monogamous gibbons." Among baboons, one male keeps a "harem" of females, while others have temporary "consorts."

Among animals there is every conceivable kind of system—from monogamy to socially cooperative mating and rearing of young. There is also frequent cannibalism and infanticide in nature, as well as "altruism" and parasitism. Ants have social "slavery" (though in no way resembling our meaning of the term); seagulls have "lesbian" female pairs, which court and nest together (though again, the human label is misapplied).

We don't even now what is "natural" for our own species. Every few years a new theory emerges on what is our "natural" diet, our "natural" life span, our "natural sexual practices, our "natural" social system, or our "natural "relationship with nature.

Nature is endlessly fascinating, but offers no "natural" way of life for humans to copy. Even in evolution, there is no "natural" tendency toward "progress," "perfection," or "ascent." Most of the time, we don't even know what is going on in nature.

NATURPHILOSOPHIE
German Idealist Movement

For almost a century, *Naturphilosophie*—a combination of Platonic idealism with a search for aesthetic purity—was the prevailing philosophy of natural history in Germany. Ushered in by the poet-philosopher Johann Wolfgang von Goethe (1749–1832), it was the search for archetypes; ideals of pure form and design.

Promulgated by a group of professors at Munich, especially Friedrich W. J. von Schelling, Lorenz Oken, and Ignatius Dollinger, it became extremely influential as students carried it all over the world.

Naturphilosophie aimed to encompass all nature in an absolute, unified system of ideas about plan, pattern, and type. The Creator and His designs were the ultimate reality, of which the actual animals and plants were derived manifestations. Many of its adherents made brilliant discoveries; among them were the German embryologist Karl von Baer, the English anatomist-paleontologist Richard Owen, and the Swiss-American Louis Agassiz, who elucidated the ice ages and fossil fish. All of them were staunch and influential opponents of Charles Darwin's view of form resulting from evolutionary history, rather than imperfect manifestations of an ideal type or plan.

See also AGASSIZ, LOUIS; ESSENTIALISM; OWEN, SIR RICHARD

NAZIS, EVOLUTIONARY PROGRAM OF

See "ARYAN RACE," MYTH OF; EUGENICS; HAECKEL, ERNST; INTELLIGENT DESIGN; LEBENSBORN MOVEMENT; MONISM

NEANDERTHAL MAN
First Known Fossil Man

A NEANDERTHAL FACE, reconstructed from a skull found in the French Dordogne region. Reconstruction courtesy of and © by Viktor Deak.

Fossil man does not exist!" declared French paleontologist Georges Cuvier in 1812. But in 1856—24 years after his death and a few years before Charles Darwin published *Origin of Species*—quarrymen working at a limestone cave deposit near Düsseldorf accidentally discovered Europe's first known human fossil. Science was still not prepared to accept it as such, but the public enthusiastically embraced the German "caveman" from the Neander Valley as an iconic prehistoric precursor.

A heavy skullcap and 15 parts of a skeleton, showing obvious differences from modern *Homo sapiens*, were found in the cave. Fortunately, the quarry owner called the local teacher, Johann Karl Fuhlrott, who sent the bones to Professor Hermann Schaaffhausen of

Bonn for analysis. Another skeleton was recovered, but it was lost before the First World War.

After careful study, the anatomist concluded that the strangely shaped skull did not belong to any German, but perhaps to a "wild northern tribe," far older than the Celts, that had been conquered in ancient times.

His opinion that the Neanderthaler was a normal adult of some kind of ancient man was essentially correct, but his colleagues insisted it was a pathological freak. Rudolf Virchow, Germany's most influential anatomist, opined that the skeleton was not an ancient man at all, but a recent cripple with a deformed skull who suffered from arthritis. Another anatomist pronounced it to be an abnormal idiot who had been a recluse living in the cave. Yet another learned professor concluded it was a Mongolian Cossack who had served in Napoleon's army 30 years before and had died during the retreat from Moscow.

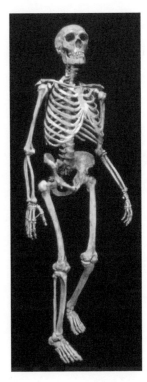

IN "STORMBOUND," artist Charles R. Knight envisioned Neanderthal folk (left) enduring harsh conditions in the deep freeze of the last Ice Age. Courtesy of and © by Rhoda Knight Kalt.

In 2004, Gary Sawyer and Blaine Maley of the American Museum of Natural history reconstructed a Neanderthal skeleton (right) by combining several specimens. Though similar to *Homo sapiens*, the Neanderthal has heavier legbones and pelvis, a barrel chest, and stockier build.

Over the next century, more and more remains of these strange "deformed" men and women with the heavy jaws and eyebrow ridges kept turning up. One had already come from Gibraltar, and several more were found in France; later, specimens popped up in the U.S.S.R., France, Israel, China, Italy, and North Africa. Soon scientists recognized that Neanderthals were a distinctive and wide-ranging population, which had inhabited much of the Old World, from Europe to the Near East, Africa, and Asia, during the last ice age. Neanderthal fossils are now known to span the period between about 200,000 and 30,000 years ago.

A few Neanderthal skeletons were discovered deliberately buried, some with animal horns and red ochre. One burial seemed to have been covered with flowers, as it contained a heavy concentration of pollen. Some anthropologists in the 1960s called them the first "flower children." However, later studies suggested that the flowers were more likely imported into the grave sites by hoarding rodents.

Neanderthals are associated at many sites with stone tools of a particular style, first discovered at Le Moustier in France and known as Mousterian. It was a distinctive style of making hand axes and stone scrapers that persisted almost unchanged for an incredible 150,000 years.

The name Neanderthal has an interesting but little-known history. A theologian, poet, and writer of hymns named Joachim Neander (1650–1680) lived near the picturesque Düssel River valley, where he went to compose his poems and also held popular religious services and theology classes in the wooded ravines. Ironically, his original family name was Neumann ("new man"), but his grandfather changed it to the Greek equivalent, Neander—a popular fad of the period. In the 19th century, a large limestone cave in the region was named Neanderhöhle in his honor, and eventually the valley itself became known as the Neander Valley (in German, *Neanderthal*). Later, a change in German orthography dropped the "h" to become Neandertal. Joachim Neander is thus the only man of the church to have had a fossil hominid named after him—albeit indirectly.

Neanderthals are now well known, at least in their physical characteristics. They had heavy skulls, with pronounced brow ridges and large faces. Although they were only five feet tall, they were strong-boned and muscular, with large hands. For years they were portrayed as violent and brutish creatures, who ambled along in a forward-slumping posture. Around 1960, however, scientists realized that the first Neanderthal skeleton found did indeed have arthritis, but that healthy individuals walked upright. Anthropologists still cannot agree on how Neanderthals were related to *Homo sapiens* populations, which apparently coexisted with them in the Levant for at least 60,000 years. At one time, Neanderthals were considered our ancestors; but it is now believed that they were contemporaneous with modern-looking people and are probably a separate species.

Ancient Neanderthal DNA (some of which, remarkably, has survived in a few fossils) supports the view that their genetic makeup differed significantly from our own. It seems possible, too, that they were specially adapted to cold climates, since they flourished during two periods when glaciers moved across Europe. Distinctive features of their nasal bones may have been adapted to humidify and warm very cold air.

A rich fossil trove of so-called "Pre-Neanderthals," dating from 400,000 years ago, has recently been unearthed in Atapuerca, northern Spain—the largest concentration of early human fossils in the world. The Pit of Bones at that cave site has thus far yielded the remains of about 30 individuals, and is by no means exhausted. While the skulls show some Neanderthal features, they are not identical with "classic" Neanderthals, but may be ancestral or related to them.

A final unsolved mystery is what became of the Neanderthals. Did they (as some scholars still believe) merge with the gene pool of *Homo sapiens*? Or did they disappear because we outcompeted or simply exterminated them? Whatever the case, the Neanderthal decline began about 40,000 years ago when the first "Cro-Magnons" entered Europe. By 28,000 years ago, *Homo sapiens* occupied almost the entirety of the European continent and the Neanderthals were gone forever.

See also ATAPUERCA; FOSSIL HUMANS

NEBRASKA MAN
A Swinish Missing Link

One of the most embarrassing incidents in the history of evolutionary science began in 1922, when a solitary molar tooth was found in Nebraska. First-rank paleontologists, anthropologists, and anatomists examined the cusp pattern, and all agreed with its discoverer that the tooth belonged to an ancient ape-man: a "missing link" of tremendous importance, to which they gave the scientific name *Hesperopithecus* or "western ape."

The tooth was certainly ancient and was embedded in Pliocene deposits, but what else could be said about it? For starters, Australian anatomist Sir Grafton Elliot Smith and a museum artist, Amédée Forestier, collaborated to produce a drawing of both male and female *Hesperopithecus* for the *Illustrated London News*. Their "reconstruction" featured full figures of a well-muscled, slope-browed pair in a prehistoric landscape complete with early horses and camels.

Professor Henry Fairfield Osborn, president of the American Museum of Natural History, welcomed the news. Antievolutionist William Jennings Bryan, the famous politician and lawyer who would prosecute John T. Scopes for teaching evolution in the 1925 "Monkey Trial" in Tennessee, was a Nebraskan, and Osborn rubbed it in. "The Earth spoke to Bryan from his own State," Osborn crowed. "This little tooth speaks volumes . . . in that it affords evidence of man's descent from the ape."

In 1927, an enthusiastic team of paleontologists returned to the Nebraska site where *Hesperopithecus* had been discovered five years earlier, determined to find more of this mysterious creature. To their joy, weathering had exposed parts of a jaw and skeleton on the precise spot. Eagerly, they brushed away dust and sand until the ancient fossil emerged to tell its truth, but they were not happy with what they found. The famous molar was about to become infamous: it had once belonged to an extinct pig!

See also "MISSING LINK"; PILTDOWN MAN (HOAX)

A RECONSTRUCTION DRAWING BY A. FORESTIER.

A PREHISTORIC COLUMBUS WHO REACHED AMERICA BY LAND?—AN ARTIST'S VISION OF HESPEROPITHECUS (THE APE-MAN OF THE WESTERN WORLD) AND CONTEMPORARY ANIMALS.

NEO-DARWINISM

"Hard" Inheritance

George Romanes, Charles Darwin's disciple in animal psychology, coined the term "Neo-Darwinism" in the 1890s to describe the theory of natural selection without the inheritance of acquired characters ("Neo-Lamarckism" or "soft" inheritance). Romanes was convinced that Gregor Mendel and August Weismann were correct about inheritance operating only through sex cells and not soma (body) cells. (The term "gene" was coined later, in 1909.)

Unlike most scientists of the time, who saw the new genetics as the death knell of Darwinism, Romanes expected the two theories to merge into an improved understanding of evolution, which in fact they eventually did. Alfred Russel Wallace fought Mendelism tooth and nail, believing it to be utterly incompatible with natural selection.

Some writers (incorrectly) use Neo-Darwinism as a more general term, to cover all of 20th-century biology, including the combination with population genetics in the 1930s that produced the Synthetic Theory of evolution.

See also DARWINISM; NATURAL SELECTION; ROMANES, GEORGE JOHN; SYNTHETIC THEORY

NEO–LAMARCKISM

Evolution by Conscious Effort?

From the late 19th century until well into the 20th, some biologists adopted the idea of evolution, but not the mechanism of "random variation" and "natural selection." As the novelist Samuel Butler had written, a mechanistic evolution without plan or purpose "banished mind from the Universe," which was difficult for many people to accept. Any intel-

AN EMBARRASSMENT to evolutionists, "Nebraska Man," or *Hesperopithecus*, was identified on too little evidence: a single molar tooth, which turned out to come from a fossil pig. The drawing above appeared in the *Illustrated London News*.

lectual or moral progress that was made by individuals would die with them, rather than be inherited by their offspring for the improvement of the species. Each generation had to learn anew—a depressing setback for those who hoped for incremental human progress. That was exactly the point that so rankled Soviet Communists that they fired "Mendelist" geneticists or shipped them off to gulag prisons.

Neo-Lamarckians saw themselves as bringers of hope and inspiration, the antidote to what physicist Sir John Herschel called Darwin's "law of higgeldy-piggeldy." Among them, over the years, were the novelist Samuel Butler, playwright George Bernard Shaw, and philosopher Henri Bergson. Perhaps the most famous Neo-Lamarckians of the 1930s were the Soviet ideologue Trofim Lysenko, who set Russian genetics and agriculture back 30 years, and the tragic Austrian experimentalist Paul Kammerer, who was accused of conducting fraudulent experiments and finally took his own life.

All believed Lamarckian inheritance to be the true basis of evolution and stood fast against the new Mendelian genetics. But Lamarck had proposed no theory of *how* acquired characteristics could be inherited; he simply shared the common belief of his time that changes in parents can be passed to offspring. Even Charles Darwin shared that belief and suggested a mechanism for how it might work: his futile theory of pangenesis.

Developments in 20th-century genetics have put the crude form of Neo-Lamarckism to rest, but the problem of inheritance of acquired characters remains important. The "Central Dogma of Genetics" says genetic information goes only one way—from DNA to expression in the organism, and never the reverse. But scientists continue to discover exceptions and complexities to this general rule.

Possible mechanisms for the inheritance of acquired characters include epigenetic regulators, which may alter embryonic development in response to certain environmental situations. Another unexpected phenomenon, discovered by the fiercely independent Cornell cytogeneticist Barbara McClintock (1902–1992), is that of "jumping genes," which may change positions on a chromosome and so help an organism adapt to its environment. Her early innovative work on maize in the 1920s was dismissed by her uncomprehending colleagues, but by the 1950s it was finally understood and accepted.

Among other "Neo-Lamarckian" developments, experiments conducted during the 1990s established that certain retroviruses can be acquired by an individual, enter the DNA, and become heritable by offspring.

These research directions have produced a "New" Lamarckism (as opposed to the dead-end Neo-Lamarckism). Scientists continue to explore plausible mechanisms by which acquired characters or responses to environment can be inherited.

See also BUTLER, SAMUEL; "CENTRAL DOGMA" OF GENETICS; LAMARCKIAN INHERITANCE; "NEW" LAMARCKISM; SHAW, GEORGE BERNARD

NEOTENY
Are Adults Big Babies?

INFANT APE AND ADULT HUMAN have strikingly similar facial proportions, such as high forehead, button nose, and flattened face (top), while mature apes develop low foreheads and protruding snouts. Neotenous species retain juvenile features into adulthood.

Neoteny is the retention of juvenile characteristics in the adult form of an animal. A classic example is the axolotl, a Central American salamander that spends its entire life underwater, never shedding the gill structures similar species outgrow when they develop lungs. The axolotl (its name is from a Nahuatl Indian word meaning "servant of water") remains in a permanent larval stage.

This curious phenomenon seems rare in nature and would be of little interest except that it may have been an important factor in human evolution: *Homo sapiens* has many characteristics of a fetal or infant ape. Adult chimps and gorillas, for instance, have elongated faces, heavy brow ridges, powerful jaws, and small braincase in relation to overall skull and other characteristic proportions. Baby apes have flat faces, rounded braincase, light brow ridges, proportionately smaller jaws, and many other bodily features strikingly like adult humans.

This idea—that man is in some way a fetalized ape—has tantalized students of evolution. Some have suggested it might help account for the apparently small genetic difference between humans and apes. One of the crucial differences may simply be genetic instructions to juvenilize an anatomical blueprint that is held largely in common.

See also CUTENESS, EVOLUTION OF

NEURAL DARWINISM
Theory of "Mind"

A new debate on the nature of "mind" began in 1987 with the publication of *Neural Darwinism: The Theory of Neuronal Group Selection* by Nobel laureate Gerald Edelman. After years of dissatisfaction with computer models of the brain, "black boxes," and "wiring diagrams," Edelman attempted to explain the workings of mind in terms of Darwinian selection.

In a long, difficult, and complex argument, which very few neurologists seem to grasp and have yet to confirm, Edelman sets forth his theory that perception, action, and learning are based on a selection process. Groups of cells that respond successfully to stimuli from the environment are preserved and their connections strengthened. Those that do not are eliminated; thus, a self-correcting process of adaptation continues throughout life.

Memories, rather than being stored in neat, local compartments, are continually reworked and recategorized. Every experience in our lives, from before birth up until death, alters and shapes our brains.

Edelman's "neural Darwinism" is an attempt to apply selection theory to one of the most intractable of all phenomena: the workings of the mind. Time and testing will tell whether it is scientifically productive or will itself be "selected out" in the struggle for existence of useful explanations.

As Ambrose Bierce wrote in *The Devil's Dictionary* (1906), the mind is "a mysterious form of matter secreted by the brain. Its chief activity consists in the endeavor to ascertain its own nature, the futility of the attempt being due to the fact that it has nothing but itself to know itself with."

NEUTRAL TRAITS
Nonadaptive Characters

The Darwin-Wallace theory of natural selection assumed "useful" variations would become established in a population, while all others would be eliminated. But some naturalists insisted that many traits in plants and animals had no demonstrable positive or negative advantage—they were nonadaptive or "neutral."

Although he easily demolished arguments for plan or "design" in nature, Charles Darwin seemed stumped by the critics who suggested evolution might be even more random than he had supposed. After the third edition (1861) of *Origin of Species*, he began to allow more room for a pluralistic view of evolution, until natural selection became only one mechanism among several.

Distressed by this turn of events, Alfred Russel Wallace published his own version of evolutionary theory in 1889, a few years after Darwin's death. Wallace urged a return to the original vision of their joint theory—a thoroughgoing selectionism. Ironically, he titled the book *Darwinism* and rejoiced when reviewers called him "more Darwinian than Darwin." (They really meant "more selectionist," more insistent on the importance of adaptation in shaping organisms. Wallace always assumed that persistent traits had some "selective advantage," even if their value was not immediately obvious.)

But Wallace did not have the last word. After his death, the 20th-century synthesis of Darwinism with Mendelian genetics sparked renewed interest in the possibility of neutral traits, especially among population geneticists.

By 1932, geneticist J. B. S. Haldane had concluded in *The Causes of Evolution* "that innumerable characters [of animals and plants] show no sign of possessing selective value, and moreover, these are exactly the characters that enable a taxonomist to distinguish one species from another." A few decades later, however, Oxford biologists proved the opposite—

that many characters then considered "neutral" do, in fact, have selective value, which could be revealed by experiment, not by speculation.

In the 1960s, with new discoveries about the enormous genetic variability within natural populations, a "neutralist school" arose once again. Researchers Jack L. King, Motoo Kimura, and others argued that if evolution was not like a preplanned journey with an intended destination, it might be more like a "random walk"—taking a turn in this direction or that, for no particular reason except the contingencies of its history. Both "chance" and directional selection shape organisms, but in what proportions? The debate and research continue.

See also ADAPTATION; GENETIC DRIFT

"NEW" LAMARCKISM
The Responsive Gene

The old idea that behaviors or experiences acquired in an individual's lifetime can be passed on to offspring became associated with the name of the French biologist Jean-Baptiste Lamarck, though he did not originate it. The rise of Mendelian genetics disproved this "old" Lamarckism because changes (mutations) in genetic material appeared unaffected by environment or individual experience.

According to the "central dogma" of genetics, protein synthesis can occur in one direction only: from DNA (the chemical code) to RNA (the messenger that interprets it as blueprint) to protein (body cells). The individual's genetic makeup (genome) remains fairly static, affected only by chance variations caused by replication mistakes or (at least in laboratory experiments) by exposure to radioactivity or certain chemicals.

However, as researchers probed more deeply into heredity during the 1970s, some became dissatisfied with the prevailing model of a static genome, totally unresponsive to environment except through the selection of chance mutations. Geneticist Barbara McClintock, working independently with maize for 30 years, discovered the unexpected phenomenon of "jumping genes"—genes that may abruptly change their positions in relation to others on the chromosome. Such changes in position, she found, can profoundly affect how the blueprint is expressed. In experiments made during the late 1980s, researchers were able to move genes to different positions in a fruit fly's chromosomes. The artificial repositioning resulted in a different eye color for the fly's offspring, which was subsequently inherited by descendants although the genes remained the same.

Genomes, it now appears, are not at all static. Within the ocean of genetic material there is constant movement and reorganization. Masses of material formerly labeled "junk DNA" (a confession of ignorance) do indeed have functions, which are only just being investigated. Whether some of this activity can be a direct response to environmental factors has recently become an open (and very controversial) question.

One research team during the 1980s reported an apparent responsiveness to environment in the genomes of hungry bacteria. Some seemed to evolve more rapidly when placed on lactose, and many produced lactose-digesting mutants. Moreover, some genes can be "switched" on and off by certain organisms in response to environmental changes.

In recent years, researchers have been astounded by bacteria that pass bits of genetic material (plasmids) between individuals, which can become incorporated into the recipient's own genome. Such "lateral gene-swapping" has been found to occur in hundreds of kinds of bacteria and algae.

The "central dogma" notwithstanding, in the early 1970s David Baltimore and Howard Temin found that certain viruses use RNA rather than DNA as genes and are able to reverse-transcribe the DNA from RNA. By 1987, reverse transcription had been found to be widespread, not only in these "retroviruses" (which infect RNA and travel to DNA to cause heritable diseases) but in yeasts, plants, and even mammals.

Some researchers, following the initial work of J. V. McConnell of the University of Michigan, are attempting to explore whether RNA may actually be changed by learning. If there were a mechanism for the changed RNA to transcribe back into DNA, a molecule with such easy access to the genome would play a major role in evolution. An idea so antithetical to the orthodox Synthetic Theory of evolution is not finding easy acceptance. Still, RNA as a mediator of evolution, responding to environment and experience, remains a working hypothesis that fascinates proponents of the "new" Lamarckism.

See also "CENTRAL DOGMA" OF GENETICS; EVO-DEVO; NEO-LAMARCKISM

NIM CHIMPSKY
Ape Communication Experiment

A few years after linguist Noam Chomsky of MIT had written that language is biologically unique to humans, an ape named Nim Chimpsky talked back.

Playfully named by psychologist Herbert S. Terrace of Columbia University, Nim was the subject of an experiment begun in December 1973. Its purpose was to teach American Sign Language to the two-week-old chimp, in hopes that he could learn a language if he didn't have to speak the words.

It was an idea that went back to Robert M. Yerkes, who had tried to teach chimps to talk in the 1930s, and concluded that they were better at mimicking visual signals than vocal ones. By 1975, Nim was putting two signs together, and a year later, combinations of three, such as "you-tickle-me" or "me-more-eat."

Nim was given lots of affectionate attention by human trainers, who treated him like a small child. Almost five years later he was returned to the Institute for Primate Studies in Norman, Oklahoma. At that point, it was thought that Nim understood 300 signs, could produce 125 of them, and had put thousands of "sentences" together.

Professor Terrace and Nim became well known for their achievements, but soon a controversy began to swirl around them. Was Nim really "speaking," or was he signing without comprehension to gain the approval of his handlers? Were his responses triggered by subtle cues from Terrace and his staff? Critics brought up the lesson of a famous German "talking" horse, Clever Hans.

In 1979, Terrace wrote a book, *Nim*, in which he disavowed his previous results. After reviewing the photos and videotapes, he concluded that he had vastly overrated Nim's language abilities. Word combinations had not increased in length, and much behavior could indeed be attributed to cues unconsciously given by humans.

During the 1990s, in a new series of experiments, a bonobo named Kanzi far surpassed Nim's language achievements—and once again the book on chimp language had to be rewritten. The extent to which "signing" apes can really master symbolic language is still a tantalizing question for research.

See also APE LANGUAGE CONTROVERSY; CLEVER HANS PHENOMENON; KANZI; KOKO

NOAH'S FLOOD
The Great Deluge

F or centuries, Noah's flood was offered as a serious explanation for the existence of all fossil-bearing rocks. Scriptural literalists believe fossils were deposited when millions of animals drowned in the 40-day downpour. Some carcasses sank to the bottom (lower strata), while others floated up and landed on higher ground.

Young Charles Darwin—no slouch as a geologist—was delighted to find a giant ground sloth skeleton in South American fossil beds and to see thousands of seashells high and dry on mountainsides. Such observations stimulated his interest in theories of geologic succession, which he tried to discuss with his friend Robert FitzRoy, the commander of the *Beagle*.

But Captain FitzRoy had a ready explanation. The shells had remained on the mountain when the waters of the worldwide Great Deluge had receded, and gigantic extinct animals had perished because they were too large to fit into the doorway of Noah's Ark and so drowned in the flood.

For decades, scriptural geologists tried to reconcile the record of the rocks with the Great Flood, but the growing mass of evidence just didn't fit. Some parts of the world showed no

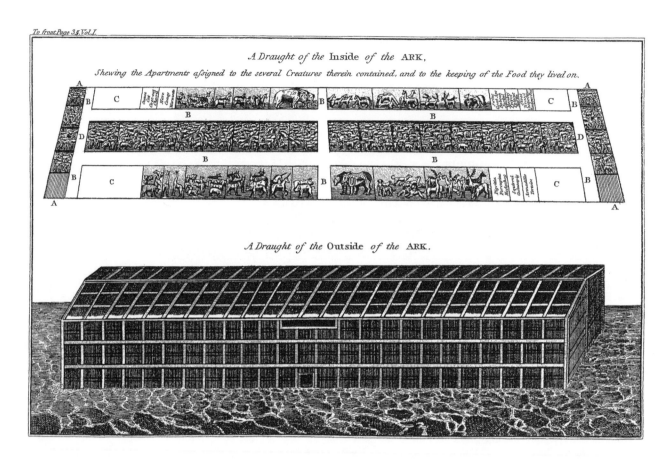

To front Page 35.Vol.I.

A Draught of the Inside of the ARK,

Shewing the Apartments afsigned to the several Creatures therein contained, and to the keeping of the Food they lived on.

A Draught of the Outside of the ARK.

THE CAPACITY OF NOAH'S ARK was calculated by Dr. Edward Wells in an 1801 work, *An Historical Geography of the Old Testament.* He concluded there was room for seven individuals of all "clean" (kosher) beasts, and two of each "unclean" beast, with enough hay, food, and water for all.

signs of having been inundated, while others had clearly been submerged not once, but many times. Even the great reconciler of the Bible with geology, Hugh Miller, was forced to conclude in his last book, *The Testimony of the Rocks* (1857), that perhaps Noah's Flood was locally confined to the Middle East.

If all animals had existed at the same time and their remains had been submerged in a single flood, a hodgepodge of life forms would be mixed together in the rocks, with no orderly pattern of succession. Smaller trilobites would have floated up while heavier ones remained on bottom, along with the large land animals, which would sink. But that's nothing like the actual record of the rocks.

Fossil layers (strata) everywhere show a pattern of succession that is completely incompatible with sorting by water action. Throughout the world, sea creatures and trilobites are in the lower rock strata, fish in the next higher, then amphibians, reptiles, birds, and mammals. Nowhere does one find reptiles on the bottom and trilobites on top, birds before reptiles, or primates mixed with dinosaurs.

It is rare for an individual animal's remains to become naturally preserved, yet the quantity of fossil skeletons is enormous. Such accumulations must have occurred over immense periods of time. The Karroo Formation in South Africa alone contains fossil remains of about 800 billion animals. If all the trillions of fossilized creatures had been alive at the same time, they would have blanketed the entire Earth.

Many older geology books (1800–1860) continued to accept the Great Deluge, even after the consistent patterns in fossil evidence had become well known. When the British Museum's famous anatomist Richard Owen coined the term "dinosaur" and commissioned the great life-size *Iguanodon* models for the Crystal Palace in 1853, neither he nor his sculptor, Benjamin W. Hawkins, had evolution in mind. Owen considered them "antediluvian monsters"—animals from before the Flood, perhaps from a "pre-Adamic Creation"—as did the Victorian public, which flocked to see them.

How many animals were there on Noah's Ark? In the 18th century, only hundreds of species were known, and there was no difficulty in imagining that all could fit on the Ark. But by early

Victorian times, zoologists recognized 1,000 species of mammals, 6,000 birds, and 1,500 reptiles and amphibians, and the numbers were rapidly increasing year by year. That would require that almost 20,000 animals, with enough food for more than a month, had to be crammed into Noah's Ark. Meat-eaters, like lions and tigers, would have to get by on vegetable food, or perhaps on an extended supply of herbivores. By the mid-19th century, any visitor to the London Zoo could see the insuperable difficulties of the story.

See also CREATIONIST MUSEUM; IGUANODON DINNER; MILLER, HUGH; "NOAH'S RAVENS"; PROGRESSIONISM

"NOAH'S RAVENS"
Birdlike Tracks in Stone

Reverend Edward Hitchcock, president of Amherst College during the mid-19th century, has a unique place in history. There is no doubt that, beginning in 1835, he conducted the first sustained research on dinosaur evidence in North America. But another man gets the credit as the first to study an American dinosaur, because Hitchcock had no idea what kind of animal he was studying.

When he first became aware of the abundant petrified three-toed trackways preserved in Connecticut Valley sandstone, dinosaurs were still virtually unknown to science. The footprints had been local curiosities, which farmers called "tracks of Noah's ravens."

Fascinated, Hitchcock spent most of his time and energy collecting, studying, and describing them for the next 30 years, and in 1858 published his splendid book *Ichnology of New England*. To the day of his death, he believed they were the footprints of giant birds. Years later, paleontologists linked them with Upper Triassic bipedal dinosaurs.

Joseph Leidy, who described *Hadrosaurus* teeth in 1856 and a skeleton two years later, is credited as first to discover a dinosaur in North America. Hitchcock was first to study petrified dinosaur footprints but thought they were made by birds, so Leidy was the first to describe a dinosaur *knowing* it to be a dinosaur. Current theories about the links between theropod dinosaurs and birds, however, may mean that Hitchcock was more right than wrong to consider them birds.

See also BIRD, ROLAND THAXTER

AMISH FARMERS in Pennsylvania often found giant "bird tracks" in stone, which they assumed were evidence of Noah's Flood. Reverend Edward Hitchcock collected them, unaware that they were dinosaur footprints.

NOBLE SAVAGE
Ideal of "Natural" Man

People have often been dissatisfied with themselves and their society, wishing to believe in another time and place where folks were pure, unspoiled, and "naturally" good. Scientists as well as philosophers have searched eagerly for man in the natural state: Adam and Eve before the Fall at a simpler "stage" of social evolution.

This ideal primitive was an obsession of philosopher Jean-Jacques Rousseau (1712–1778), the most influential and famous admirer of the "noble savage." An acute observer of artfully mannered, parasitic, and devious French aristocrats, he concluded that civilization corrupts human nature. When people lived in the wilderness, in simple tribes next to nature, he thought, they were straightforward, self-sufficient, democratic, and honest.

But where were these noble savages? With the exploration of the New World, some thought they had found them in the American Indian. Rousseau idealized the Indians as "children of the wilderness . . . nature's noblemen of the Plains," a Romantic notion that inspired European painters and sculptors to depict graceful, muscular superheroes, completely in tune with nature.

In his *Discourse on the Origin of Inequality* (1754), Rousseau describes a "Man of Nature" as he existed before society and history. He is solitary, "wandering in the forests, without industry, without words, without a home, without war, and without ties, without any need of his fellows and without any desire to harm them."

The 20th-century adventure hero "Tarzan of the Apes" is also very much Rousseau's noble savage—explicitly stripped of his European title, "Lord Greystoke," as well as his clothes. Reared by apes in an isolated jungle, uncontaminated by "corrupting institutions," he reverts to man's basic goodness and heroic virtue. Tarzan has remained popular for almost a century, incarnated in scores of books and movies.

This Romantic ideal of the uncorrupted natural man (and woman) was revived in full force by "hippies" during the 1960s. Like Rousseau, they openly opposed and disobeyed laws they considered unjust, thought the current war a perversion of human nature, and gave lip service to a purer, simpler way of life. Like them, two centuries earlier, Rousseau had thrown away his clock and had advocated natural foods, long hair for men, and public breast-feeding of babies.

But Rousseau also spoke of his state of nature as "a state which exists no longer, perhaps never existed, probably never will exist, of which none the less it is necessary to have just ideas, in order to judge well our present state." Utopian novelists and, later, science fiction writers have continued to explore his premise.

When Rousseau sent his *Discourse on Inequality* to Voltaire, the elder philosopher sarcastically replied: "One longs, in reading your book, to walk on all fours. But . . . I have lost that habit for more than sixty years. . . . Nor can I embark in search of the savages of Canada . . . because the example of our actions has made [them] nearly as bad as ourselves."

Nevertheless, Rousseau's ideas became so popular in the French court that sophisticated ladies began to breast-feed their infants at formal social functions.

Yearning for the noble savage spilled over into science, both in anthropologists' descriptions of tribal peoples and in their imaginative reconstructions of our far-distant ancestors. Many anthropologists have written of tribal peoples' alleged superiority in their respect for nature, purposely preserving the land and its creatures from permanent destruction even as they use it for sustenance. On close examination, this assumption seems to be Romantic thinking. Certainly, some tribal peoples use natural resources with understanding and wisdom, but most simply lacked the technology or population size to create destructive impact comparable with our own.

Recent studies, for example, show that Polynesian peoples exterminated scores of bird species before the arrival of Europeans, eating their way through the islands' faunas. Many agricultural tribes practiced wanton burning of forests (and still do), while some hunting peoples were needlessly destructive of wildlife habitats.

Oddly enough, when it seemed the idea of the noble savage had just about been abandoned by sadder but wiser anthropologists, animal behaviorists picked it up. Dian Fossey thought her gorillas were "superior" to humans in their gentleness, loyalty, and group solidarity. Jane Goodall once said chimpanzee mothers could teach humanity a thing or two about nurturing youngsters with devotion and tolerance. (She was shocked and disillusioned when, in 1974, she began to observe chimpanzee "warfare," infanticide, and cannibalism, concluding the apes were "more like humans than I at first supposed.")

Early humankind is often described now in similar terms: sharing food with members of their kin group, caring for cripples, cooperating in hunts, dividing labor between the sexes, and living in harmony with their environment. When a hoaxer in the 1960s offered a new vision of such an idyllic people—the so-called Tasaday tribe of the Philippines—they were celebrated worldwide as exemplars of the

THE ROMANTIC TRADITION lingers in this 1870 reconstruction of the Paleolithic hunter as "noble savage," inspired by stone tool discoveries in the French Dordogne region.

noble savage. It turned out there was no Tasaday tribe. Local villagers were recruited by corrupt government officials to shed their clothes and put on a charade for visiting journalists. Few recalled Rousseau's disclaimer that his vividly portrayed natural human "exists no longer, perhaps never existed, probably never will exist."

NOPCSA, BARON FRANZ (FRANZ BARON NOPCSA VON FELSŐ-SZILVÁS) (c. 1875–1933)

One of the most bizarre characters in the history of evolutionary science was the Transylvanian nobleman Franz Baron Nopcsa von Felső-Szilvás. While still a young student, Nopcsa (NOPE-tschah) contributed a brilliant paper on *Limnosaurus*, a new species of dinosaur he had discovered on his sister's estate. Dinosaurs would become his obsession, second only to his special passion for the land and culture of Albania, which he roamed at every opportunity.

Fascinated by the abundant remains of Cretaceous reptiles in his native Transylvania, Nopcsa discovered and described dozens of new species. His major book on fossil reptiles, *Die Familien der Reptilien* (1923), extended the pioneering work of H. G. Seeley and Friedrich von Huene in classifying dinosaurs. Seeley had been the first to establish the two major divisions of dinosaurs: the Saurischia (lizard-hipped) and Ornithischia (bird-hipped). A flamboyant polyglot, Nopcsa was able to read the paleontological literature in five languages.

Before World War I, when pockets of feudalism still persisted, Nopcsa was able to live like a baronial lord. As he drove through Hungary between his various estates, country folk bowed low and men snatched off their caps. (Later, in a changed Europe, the imperious Nopcsa was beaten and bloodied with pitchforks by these same humble peasants.)

Restless, with only his noble rank and scientific studies to occupy him, Nopcsa began having grander ambitions—like becoming king of his beloved Albania. He had long identified with the Albanian people, isolated in the Balkan mountains, and their backward, storybook culture. On frequent travels throughout the country, he made excellent ethnographic studies of its language and customs. When Albania was freed from Turkish rule in 1913, Austro-Hungarian rulers decided that it should have a new king who would be friendly to Vienna. Nopcsa confidently applied for the job.

Since he knew the dialects and culture of the country, he argued, he could take it over in short order. All he required from the central powers was a small army of about 500 soldiers with artillery. (He himself guaranteed to provide a couple of steamships and a white horse, on which he intended to ride triumphantly through the streets, arrayed in a splendid uniform.) He also proposed to bolster the country's economy by marrying an American millionairess—after he was king, of course. He seemed genuinely astonished when the Austro-Hungarian leaders turned him down and instead installed a minor Austrian prince, who had to flee the country six months later.

Nopcsa became an officer during World War I, serving as a spy along the Hungarian border. When peace came, most of his lands and estates were confiscated, since they were now inside the new borders of Rumania. For a while, Nopcsa headed the Hungarian Geological Survey, but his arrogant and flamboyant ways antagonized his colleagues. Disgusted, he gathered his last monies and set off with Bayazid Doda, his devoted male lover and secretary, on a motorcycle trip from Eastern Europe to Italy.

Unable to find a way to recapture the wealth or glories of his youth, Nopcsa sank into depression in his later years, crippled by "a complete breakdown of my nervous system." In April 1933, he put a pistol to his head and killed himself, but only after first murdering his longtime companion Doda while he slept. Wrote Nopcsa in his suicide note, "I did not want to leave him ill, miserable, and poor, for further suffering in this world."

BARON FRANZ NOPSCA

Dinosaurs were Baron Nopsca's obsession, but his greatest ambition was to become king of Albania.

O'BRIEN, WILLIS (1886–1962)
Special Effects Film Pioneer

STOP-MOTION ARTIST
Willis O'Brien never appeared before the cameras, but he was the real "star" of *King Kong* and *Mighty Joe Young,* imparting movements and personality to the "giant apes."

Restorations of prehistoric animals had begun with huge sculptures of dinosaurs at the Crystal Palace exhibition of 1853 and were carried forward 50 years later by the magnificent paintings of Charles R. Knight. It was inevitable that movies, the dominant art form of the 20th century, would bring the great creatures of the past to life on the screen. But how could extinct animals be made to move and appear real?

Willis O'Brien developed the answer: a special-effects technique called stop-motion animation. O'Brien's technique seems simple today, but when he started working on it in 1910, the process was unknown and revolutionary. First he constructed a flexible steel armature, with movable joints made of ball sockets, which became the core of the miniature model. The animal's body was sculpted in clay, and then cast in foam rubber, with the movable metal skeleton embedded within it.

Stop-motion animation is a slow, painstaking process. Each model is moved a fraction of an inch by hand and the scene photographed. When 24 frames per second are projected onto a screen, the illusion of movement is achieved.

O'Brien first developed his art for a picture called *Creation,* which was to depict scenes from the grand pageant of the evolution of life on Earth. He completed several scenes, but the film was discontinued. However, his work had impressed the studios, and they hired him to animate Sir Arthur Conan Doyle's *The Lost World* (1912), the story of a scientific expedition's adventures in an isolated jungle still populated by dinosaurs and ape-men.

After years of perfecting his technique, O'Brien was hired by Merian C. Cooper to create what was to become the masterpiece of stop-motion: the original *King Kong* (1933). Among the first scenes filmed were O'Brien's animations of Kong's fight with a pteranodon and the attack of a giant dinosaur on the expedition's raft.

Among the movie's innovations was the combination of live actors with the miniature models to create exciting multiple images. Today, such effects are achieved with great precision through the use of computers. Willis O'Brien did it all by hand.

See also IGUANODON DINNER; *KING KONG; LOST WORLD*

OCCAM'S RAZOR
Principle of Parsimony

William of Occam (c. 1285–1350) was a renowned scholastic philosopher, second in influence only to St. Thomas Aquinas. Since there is no reliable account of his life, it is uncertain whether he came from Ockham in Surrey or Ockham in Yorkshire. He was often embroiled in political controversies and once had to be rescued from the pope's wrath by the German emperor.

He is best remembered in science for the principle that became known as "Occam's Razor": "Entities are not to be multiplied without necessity." In other words, a well-constructed theory about nature is the simplest possible explanation consistent with the facts. The "razor" shaves off any unnecessary flourishes or complications.

Its usual formulation, as given above, does not appear in Occam's writings. Philosopher-historian Bertrand Russell cited the actual version in his *History of Western Philosophy* (1946): "It is vain to do with more what can be done with fewer." Or as modern minimalist designers would say, "Less is more."

Applied to geology and paleontology over the years, "Occam's razor" has raised (and resolved) key questions: Why postulate that God continually intervened to change climate, species, floods, etc., when it is simpler to assume uniform laws (proximate causes) that produce all subsequent events? Why assume, as some geologists did, that God planted fossils of creatures that never existed in order to test man's faith, when the simplified explanation is that they are the remains of animals that once lived? Why imbue life with a "vital force" or "innate drive toward progress," when such "multiplied entities" are neither demonstrable nor necessary to account for evolution?

In 1966, statistician Maurice G. Kendall proposed an update of Occam's principle specifically as applied to theories about how the mind works. "We should not invoke any entities or forces to explain mental phenomena," he wrote, "if we can achieve an explanation in terms of a possible electronic computer." Kendall called this special application "a kind of Occam's Electric Razor."

"OMEGA MAN"
Future Evolved Being

Father Pierre Teilhard de Chardin was a Jesuit biologist who looked back to the dawn of prehistory and forward to the next step in the evolution of mankind. Although he made many contributions to the study of Pleistocene mammals, early man, and paleolithic archeology, Teilhard's lifelong quest was to reconcile his beliefs in Christianity and evolution and to project their final synthesis. Christianity was the hope of making men better, and evolution was the means.

Teilhard believed there is a divine plan for Earth: Human destiny was to break out of the biosphere and into the "noosphere," the realm of higher thought, consciousness, and self-knowledge. "When for the first time in a living creature instinct perceived itself in its own mirror," he wrote, "the whole world took a pace forward." The next breakthrough, he thought, would be the conquering of this new dimension by "Omega Man," a future being who would surpass *Homo sapiens* both intellectually and spiritually.

Omega Man is quite unlike Nietzsche's self-contained and aloof "Superman," who evolves through ruthless competition and triumph of will. Instead, Teilhard projected Christianity's ideal of brotherly love, a mystical union of individuals in species-wide cooperation, for wise and loving stewardship of the Earth. Omega Man is an expression of cosmic optimism by a hopeful evolutionary philosopher.

See also TEILHARD DE CHARDIN, FATHER PIERRE; TELEOLOGY

OMPHALOS (1857)
Reconciling Fossils and Scripture

In the opinion of even his most sympathetic friends, Philip Gosse's *Omphalos* (1857), a unique defense of creationism, was a bizarre failure. A Victorian author of popular books on birds, insects, and ocean life, Gosse (1810–1888) was a respected naturalist who had corresponded with Charles Darwin on natural history subjects. But he was increasingly distressed by what he considered the conflict between evolutionary ideas in natural science and the biblical account of creation.

Gosse sought a theory that would reconcile the evidence of Earth's past with a literal interpretation of Genesis. The result was *Omphalos* (Greek for "navel"), which his son

Edmund, in his 1907 memoir *Father and Son,* called "this curious, this obstinate, this fanatical volume." His theory was that fossils—like Adam's navel—were evidence of a natural birth that never occurred.

This paradox was explained by Gosse's strange idea of "prochronism." Since nature is an ongoing, cyclical process, he argued, the act of creation had to start somewhere in the cycle. A chicken implies an egg in its past, just as an egg implies it was laid by a chicken. If God had created a chicken in an instant, we could imagine the egg from which it hatched, but that egg would exist outside time (a "prochronic" egg).

Fossils and other geologic evidence of Earth's past, Gosse argued, are prochronic. For example, just as Adam was created with a navel, there were growth rings on the trees in the Garden of Eden. They were brought into being "complete" by the creator, so they could be part of the ongoing natural process—evidence left by God to suggest a past that never existed. Thus, Gosse could accept that strata and fossils exist, yet still insist that the Earth was actually created in six days, only 4,004 years before Christ.

In effect, Gosse was actually maintaining that God had put fossils in the ground to fool geologists! "Who will dare to say," he asked, "that such a suggestion is a self-evident absurdity?"

> **Gosse attempted to reconcile scientists and churchmen, and ended up reviled by both.**

The answer to that was "everyone," friends as well as foes. The Reverend Charles Kingsley, himself a naturalist who wanted to believe in Genesis, was horrified. He simply could not give up what he knew of geology, he wrote Gosse, to "believe that God has written on the rocks one enormous and superfluous lie." Later, in a footnote to his 1859 edition of *Glaucus,* Kingsley publicly and sadly called *Omphalos* a "desperate" attempt ". . . more likely to make infidels than to cure them. For what rational man, who knows even a little of geology, will not be tempted to say—If Scripture can only be vindicated by such an outrage to common sense and fact, then I will give up Scripture, and stand by common sense?"

After the dismal reception given his magnum opus, Gosse retreated from the world, bitterly disappointed. After years of being the public's darling for his nature books, his son related, "He could not recover from amazement at having offended everybody by an enterprise which had been undertaken in the cause of universal reconciliation."

See also CREATIONISM; "SCIENTIFIC CREATIONISM"; "TWO BOOKS, DOCTRINE OF THE"

"ONTOGENY RECAPITULATES PHYLOGENY"

See BIOGENETIC LAW; FREUD, SIGMUND

ORANGUTAN
The Red Ape

Two hundred years ago, it was the red-haired Asian ape, the orangutan, and not the African chimpanzee that was considered man's closest kin.

During the 18th century, both Scots philosopher Lord Monboddo and the French naturalist Georges Buffon had marveled at the orang's approximation to human form, "lacking only in the art of speech." Samuel Taylor Coleridge, the poet, had accused evolutionist Erasmus Darwin (Charles's grandfather) of espousing "an Orang-utan theology." A generation later, geologist Sir Charles Lyell told Charles Darwin he simply couldn't "go the whole Orang" in accepting his theory of evolution.

One of the reasons Alfred Russel Wallace, Darwin's partner in founding evolutionary biology, went to Malaysia was in hopes of studying the orang and also finding fossil clues to early man. Many then believed Asia was the cradle of humanity, and Wallace was the first European to study a species of ape—the orang—in the wild.

Orangs figured prominently in the imagination of early-19th-century writers, as in this extraordinary passage from Henry Hallam's *Introduction to the Literature of Europe* (1838–39):

> Every link in the long chain of creation does not pass by easy transition into the next. There are necessary chasms, and, as it were, leaps from one creature to another. . . . If Man was made in the image of God, he was also made in the image of an ape. The framework of the body of

Him who has weighed the stars and made the lightning his slave, approaches to that of a speechless brute, who wanders in the forests of Sumatra. Thus standing on the frontier land between animal and angelic natures, what wonder that he should partake of both!

Orangutans in Malaysia today are dangerously close to extinction, as most of their original forest habitat has been cut down. For years they were never seen using tools spontaneously in the wild, although individuals reared in captivity take great interest in such human gadgets as latches or water pumps, which they easily master. An orangutan in the Basel zoo learned to make stone knives by chipping with another stone. [See ABANG.]

In 1994, Dutch primatologist Carel van Schaik, now at the Anthropological Institute and Museum of the University of Zurich, became the first anthropologist to document the use of tools among wild orangutans. He watched orangutans in the swampy Gunung Leuser National Park, in the northwest corner of Sumatra, and established that they prepare special twigs to poke termites out of their nests, to break into beehives, and to clean the prickly spikes off wild fruits. They also use leaves to make rain hats and leakproof roofs over their sleeping nests

Anatomical comparisons, along with DNA and serum albumin tests, indicate the orangutan is more distantly related to man than are the African apes and probably split off and went its own way millions of years before chimps and gorillas diverged.

"Orangutan" is a Dyak word meaning "old man of the forest"; tribal peoples of Malaysia believe orangs can speak but never do while people are around, for fear they will be put to work. When Alfred Russel Wallace shot, skinned, and prepared the skeletons of several orangutans for shipment to the British Museum, the local tribesmen were terrified that Wallace would go after their own heads and skins next.

See also APES, TOOL USE OF; BUFFON, COMTE DE; WALLACE, ALFRED RUSSEL

A DYAK HUNTER is attacked by an enraged orangutan in this illustration from Alfred Russel Wallace's book *The Malay Archipelago* (1872).

ORCHIDS, DARWIN'S STUDY OF
Analyzing Adaptations

During the mid-1800s, a craze for orchids swept England. Collectors braved tropical fever, snakebite, and hostile Indians to supply the greenhouses of gentlemen and ladies back home. (No one yet knew how to grow orchids from seeds.) In the midst of this popular mania, Charles Darwin chose orchids as the subject of his first book after the *Origin of Species* (1859). Published in 1862 and revised in 1877, *The Various Contrivances by which Orchids are Fertilised by Insects* was revolutionary, though it is almost unread today.

Orchids is devoted to dissecting flowers of many species, exploring their structure, and understanding the evolution of their special adaptations. Although most of the botanical details are technical, its major arguments are still crucial to understanding Darwin's thought and continuing influence.

Darwin's approach completely changed existing ideas about plant sexuality, while breaking new ground for understanding coevolution and pollination ecology. Previously (with the obscure exception of Christian Sprengel), naturalists thought of flowers as beautiful, pleasing, bizarre, or useful to man—examples of divine artistry. Darwin set out to show that the strange and intricate structures beyond "what the most fertile imagination of man could invent" were the result of sex and history. Structures were analyzed strictly as adaptations to the pollinating insects that evolved along with them.

In Darwin's words, "Nature tells us, in the most emphatic manner, that she abhors perpetual self-fertilisation." From field reports, experiments in his greenhouse, and data from botanist friends, he found "an almost universal law of nature that the higher organic beings require an occasional cross with another individual . . . no hermaphrodite fertilises itself for a perpetuity of generations." Since pollen from flowers on the same orchid plant was ineffectual, he demonstrated that "varied . . . contrivances have for their main object the fertilisation of the flowers with pollen brought by insects from a distinct plant."

Darwin documented case after case where flowers evolved in tandem with particular kinds of insects, in some instances to lure them inside a pollen chamber, in others to provide landing platforms. A century later, following his lead, even more amazing adaptations have been discovered. Some orchids give off the sexual smells of female wasps to attract the males, while others even mimic the body of the female wasp. Fooled into copulating with the flower, the wasp rolls in its pollen, which he will unwittingly carry to another flower.

So convinced was Darwin that the structures of flowers had coevolved with insects that he even predicted the discovery of a bizarre moth that no one believed could exist. The beautiful white Christmas Star orchids of Madagascar sport foot-long structures containing nectar only at the bottom. "What can be the use," Darwin wondered, "of a nectary of such disproportionate length?" Then, with Sherlock Holmesian logic, he concluded: "In Madagascar there must be moths with proboscides capable of extension to a length of between ten and eleven inches! This belief of mine has been ridiculed by some entomologists."

Forty years later, a night-flying moth with a 12-inch coiled tongue was discovered on the island. It was named *Xanthopan morganii praedicta*—Morgan's yellowish moth that was predicted! For years, no one observed it fertilizing the orchid, but in 2004 an entomologist using night vision video equipment filmed the event.

Darwin showed how orchids coevolved with their pollinating insects. One species even mimics a female wasp.

Darwin would have been amazed to see the great strides made in orchid raising during the past 150 years. About 18,000 wild species and 35,000 recorded hybrid crosses are now known. (No hybrids were known until 1856, when the first "orchaceous mule"—to use the wonderful Victorian term—was produced.) Growing them from seeds is easy now, since it was discovered some years ago that they will sprout in an agar-sugar mixture. In the wild, sugar is provided by a symbiotic fungus, a fact that was unknown to the Victorians.

Orchid commerce was revolutionized by meristem culture, a type of cloning developed by Frenchman Georges Morel in 1956. By growing bits of apical cells (from the tip of new shoots) in rich nutrients, an indefinite number of "copies" of an orchid can be produced. Although most of us imagine them as sensuous tropical exotics, there are plenty of tough, drab little orchids in the cooler climes. One of Darwin's hopes was that his book would stimulate interest in the homely little orchids of England; he derived great pleasure from an "Orchis Bank" of local species that grew near his home in rural Kent.

Currently, his great-great-grandson Randal Keynes is leading an effort to protect the orchis bank from development and to designate the entire village of Downe and environs as a UNESCO World Heritage Site.

See also COEVOLUTION; DIVINE BENEFICENCE, IDEA OF; *ORIGIN OF SPECIES*; SPRENGEL, CHRISTIAN KONRAD

ORIGIN MYTHS (IN SCIENCE)
Evolutionists as Storytellers

Origin myths are stories people tell about how they came to be who they are and where they are. The Zuni, a farming people of the American Southwest, claim to have emerged from a mystical hole in Mother Earth, establishing their special kinship with the land. A tale from the nomadic Plains Indians explains that Father Sun's helper created men from clay along with animals. [See NAPI.] When scientists put forth their best guesses about the origin of our species, they, too, become unintentional mythmakers.

While a student at Yale University in the early 1980s, anthropologist Misia Landau studied some of the older descriptions of human evolution written by such 19th-century evolutionists as Henry Fairfield Osborn, Sir Arthur Keith, and Sir Grafton Elliot Smith. Analyzing the passages as literature rather than for scientific content, she found they almost always took the form of a classic European "hero" myth, with its typically recurring elements.

The epic usually begins with a hero (a tree-dwelling ape, man-ape, early hominid, etc.) who lives contentedly in a stable environment (the trees, the forest) but is expelled from his happy home (climatic change, retreat of forest) and forced to set off on a dangerous journey. (What women were doing during this saga of "man" now seems conspicuously absent.) Then he must overcome a series of difficult challenges (new conditions, competing species) and emerge triumphant (develop language, tool use, intelligence). Often, he undergoes still more hardships (Ice Age), but finally emerges erect and victorious. Yet, as in classic myths, the scientist-narrator typically warns that if this hard-won success results in arrogance, the inevitable outcome will be self-destruction.

Other scientific origin myths sprang from archeologists' characterizations of various "stages" of human society, inferred from tools and artifacts. [See PALEOLITHIC.] For more than a century after Darwin, textbook writers spoke confidently of a Stone Age, Bronze Age, Iron Age, and so on. Different tribes and "races" of humans arrived at these stages at different times. Technologically "backward" peoples alive today were commonly described as "still living in the Stone Age." That was one rationale used by the British to slaughter Australian Aborigines and Tasmanians: They belonged to an earlier stage of evolution and should have become extinct anyway.

Landau's professor, paleontologist David Pilbeam, was quick to agree that "our theories about human origins have often said far more about the theorists than they have about what actually happened." For the theories, as he reported in "Major Trends in Human Evolution" (1980), "were relatively unconstrained by fossil data." Pilbeam should know. It was he who, with Elwyn Simons, built a reputation and a career by promoting a fragmentary Miocene jawbone of an early primate called *Ramapithecus* into a celebrated human "ancestor," then spun elaborate stories about the creature's presumed social organization and tool-using ability. When they subsequently found additional fossil fragments that did not fit their earlier reconstructions, Simons and Pilbeam had to "rethink" their conclusions. Eventually, the species was renamed and "demoted" to a possible apelike ancestor of orangutans.

Roger Lewin, in his book *Bones of Contention* (1987), cites another example: "The emphatic shift in theoretical stance between the 1950s and '60s, when the specter of Man the Hunter, Man the Killer Ape dominated paleoanthropology, and the 1970s and '80s, when peace and cooperation were stressed instead, with the emergence of Man the Social Animal." In the intervening years, Lewin points out, there was no new fossil evidence to support such a dramatic shift in reconstructions of human origins. The change, says Lewin, reflected a shift in social attitudes away from "a time when war was an acceptable instrument of international policy."

Assumptions about the abilities and attributes of men and women and the traditionally subordinate status of women have also influenced science's "origin myths." For one thing, woman was all but invisible in the male-dominated accounts of the origin of "man." Nineteenth-century male scientists depicted aggressive males who were mighty hunters; females were attracted to the strongest dominant males, who in turn selected the most pleasing mates. In primate studies, the troop was believed to be centered around domineering males, who competed for status with other males, while females vied for the males' protection. This view, already present in Charles Darwin's *Descent of Man* (1871), persisted for a century and was subsequently adopted by many leading writers and anthropologists.

In the 1970s, with the influx of women into anthropology (and, perhaps not coincidentally, with the rise of the feminist movement), a new theme sounded in reconstructions of early humankind. Adrienne Zihlman of the University of California, Santa Cruz, wrote of "Woman the Gatherer" as the economic mainstay of nomadic hunting tribes, providing the everyday roots, grubs, and vegetable foods, which were supplemented by the men's hunting.

After his embarrassment over the *Ramapithecus* debacle, David Pilbeam adopted an epigram, which he claims sums up what he has learned about scientists' pronouncements on the behavior and intelligence of fossil man (and woman): "We do not see things the way they are; we see them the way we are." When asked for the source of this wisdom, he used to tell his Yale students that this trenchant quotation comes from the Talmud. With his renewed devotion to truth, however, Pilbeam now admits he found it in a Chinese restaurant fortune cookie.

See also BABOONS; KILLER APE THEORY; PILTDOWN MAN (HOAX)

> **"We do not see things the way they are; we see them the way we are."**
>
> —David Pilbeam

DARWIN'S STUDY at Down House, where *Origin of Species* was written, is visited by thousands every year. Portraits of his mentors, botanist Joseph Hooker, geologist Sir Charles Lyell, and Josiah Wedgwood, helped inspire him.

ORIGIN OF SPECIES (1859)
Epochal Essay on Evolution

One of the great classics of science, Charles Darwin's *On the Origin of Species by Means of Natural Selection, or the Preservation of Favoured Races in the Struggle for Life* was published on November 24, 1859; it is a dense, difficult book, which even Darwin's best friends found rough going. Uncertain of its appeal to nonspecialists, he offered to cover his publisher's losses. To his amazement, however, the entire first printing of 1,250 copies was snapped up by booksellers the first day. The *Origin* continued selling through six revised editions over the next 17 years and has remained in print ever since.

Darwin described in his "one long argument" that the best way to account for thousands of facts about the world of life is by "descent with modification"—or what later came to be known as "evolution." (The word appears nowhere in the first edition; Darwin used it first in *The Descent of Man* [1871], and in the sixth edition of the *Origin* [1872]. However, "evolved" is the last word in all editions of the *Origin*.)

Similarities and differences between animal species, Darwin showed, pointed to common origins rather than separate, independent creations. He did not originate that idea (it had been stated before by, among others, the French biologist Jean-Baptiste Lamarck, Alfred Russel Wallace, and his own grandfather Erasmus Darwin), but Darwin marshaled the evidence that convinced the scientific world that evolution is a fact. Having done that, he went on to propose natural selection as its mechanism. The *Origin* sparked a revolution in thought and kindled intense controversies. In Thomas Huxley's words, it was considered a "decidedly dangerous book by old ladies of both sexes."

Previous evolutionary theories depended, in part, on destinies, goals toward which organisms must irresistibly develop, or untestable vital forces. Darwin approached living things as natural phenomena, shaped by natural causes that could be probed by experimental research and systematic field observations. For instance, Darwin doesn't merely speculate that certain plants might have reached far-flung oceanic islands as floating seeds; he soaks many kinds of seeds in barrels of salt water for months, then plants them to see which species will sprout. He does not merely speculate about whether crossing different varieties of domestic pigeons will improve the stock; he performs the breeding experiments. He invents a whole new science, then shows how to apply it.

Building his case in a disarming, indirect manner, Darwin brings the reader along almost in spite of himself. He constantly raises possible objections, readily admits there are many

Darwin's masterpiece sold out its first edition in 1859 and has remained continuously in print ever since.

difficulties, and appears to yearn for an alternate explanation. Yet he seems to be inexorably led by the evidence to his stated views, despite his best efforts to resist.

So convincing were Darwin's diffident arguments for evolution and natural selection that one contemporary reader, who was skeptical at first, wrote a friend: "I don't know what's so brilliant about your Mr. Darwin. If I had access to all the facts he does, I'm sure I would have come to exactly the same conclusions."

Having established evolution as the most plausible explanation for the facts of geology, paleontology, comparative anatomy, embryology, and other specialties, Darwin then proposes a mechanism for the way it works. Natural selection is a two-step process, involving (1) overproduction and variation within a species and (2) greater survival and reproduction of those individuals with any slight advantage over their fellows; "fitter" traits are preserved and accumulated in successive generations. "Multiply, vary, let the strongest live [and reproduce] and the weakest die [leaving few progeny]."

In addition, the *Origin* is also a founding work on the interpretation of fossils, distribution of plants and animals (biogeography), taxonomy or classification, comparative morphology, and many other fields of modern biology. However, it skirts the question of human evolution, which Darwin tackled a dozen years later, in *The Descent of Man*. In *Origin,* he allowed himself only the single remark that "light will be thrown on the origin of man and his history."

Darwin had been incubating the major ideas in the *Origin of Species* for about 21 years. He had opened his first "transmutation" notebook in 1837, a year after returning from the *Beagle* voyage, and continued recording ideas on the subject for the next several years while working on other projects. He wrote a 13-page "pencil sketch" of natural selection in 1842 and a much expanded summary (231 pages) in 1844; both were later submitted to the Linnean Society of London in 1858, to be presented along with Alfred Russel Wallace's independent invention of the theory.

Until Wallace forced Darwin's hand by mailing him the "Ternate" essay ("On the Tendency of Species to Depart Indefinitely from the Original Type," 1858) from the Malaysian jungle, Darwin had intended to set forth the theory in a massive "Species Book," bulging with overwhelming quantities of relevant facts. Although he had made some headway and was a strongly convinced evolutionist and natural selectionist by the 1840s, Darwin kept on collecting data for another 15 years. Then, springing into action to avoid being scooped by Wallace, he dashed off the 155,000-word *Origin* in only 13 months. Since it was intended only as a summary of the proposed longer work, he originally titled it *An Abstract of an Essay on the Origin of Species etc. etc.,* but his wise publisher cut the first three words. His "big book," of course, was never completed.

Scientific content alone cannot explain the special excitement generated by the *Origin of Species.* According to critic-historian Stanley Hyman, it is also "a work of literature, with the structure of tragic drama and the texture of poetry." It conveys the urgency of a personal testimony and an evangelical sense of mission.

Through Darwin's eyes we no longer see just a sparrow or a cactus, but a roiling drama of conflict and competition, a dynamic landscape of organic beings caught in a relentless struggle for existence. Darwin writes of nature as a frenetic, omniscient goddess, "daily and hourly scrutinising, throughout the world, every variation, even the slightest; rejecting [some], preserving [others]; silently and insensibly working" to choose and reject, to bestow favor or toss into the pit of extinction.

"With a book, as with a fine day," Darwin wrote to Huxley in 1863, "one likes it to end with a glorious sunset," as he had done with the *Origin*'s concluding paragraph:

> Thus, from the war of nature, from famine and death, the most exalted object which we are capable of conceiving, namely, the production of the higher animals, directly follows. There is grandeur in this view of life, with its several powers, having been originally breathed into a few forms or into one; and that, whilst this planet has gone cycling on according to the fixed law of gravity, from so simple a beginning endless forms most beautiful and most wonderful have been, and are being, evolved.

See also DARWIN, CHARLES; NATURAL SELECTION; VOYAGE OF HMS *BEAGLE;* WALLACE, ALFRED RUSSEL

"*Origin of Species* is [not only a scientific treatise] but a work of literature, with the structure of tragic drama and the texture of poetry."

—Stanley Hyman,
The Tangled Bank (1962)

ORTHOGENESIS
"Goal-Directed" Evolution

O rthogenesis (meaning an origin that is spelled out from the beginning) is the idea that evolution follows a preordained path, especially in the case of humans, guiding us on to a higher destiny. It is wonderfully compatible with many religious beliefs about the soul's inevitable progress, but causes much confusion to those who seek a biologist's understanding of evolution.

Some philosophers have imagined a guiding force or "vital principle," which acts like a spiritual magnet, drawing mankind toward "higher" grades or levels. These orthogenetic ideas are derived from the medieval Chain of Being, an ascent from lowly rungs upward toward the infinite.

Modern evolutionary theory holds that evolution is "opportunistic," in the word of paleontologist George Gaylord Simpson. At any point, it goes in the direction that is advantageous, often reshaping old structures for new uses. It does not know its destination, nor is it impelled to follow one particular direction. Sometimes the move is "sideways" into a different adaptive zone, as when horses shifted from forests to grasslands. Evolution can also move from complex to more simple, as in the development of certain parasites.

Humans cannot claim to be perched on an inevitable pinnacle, the ultimate "goal" of apes and near-men. We were shaped by a particular opportunistic and unique history. As Simpson put it, orthogenic theories "provide evidence that some scientists' minds tend to move in straight lines, not that evolution does."

See also CONTINGENT HISTORY; GREAT CHAIN OF BEING; HORSES, EVOLUTION OF; "OMEGA MAN"; TELEOLOGY

> Orthogenetic theories "provide evidence that some scientists' minds tend to move in straight lines, not that evolution does."
>
> —George Gaylord Simpson

OWEN, SIR RICHARD (1804–1892)
Zoologist, Paleontologist

V ictorian England's most prestigious zoologist and paleontologist, Sir Richard Owen held every scientific honor of his time. A master of classification, comparative anatomy, and paleontology, he disliked his nickname "the British Cuvier" because he considered himself far superior to the great French naturalist. Unfortunately, his fellow scientists were unanimous in their opinion of his character: Owen was the coldest, most arrogant, spiteful backstabber any of them had ever encountered. And he directed his most venomous, underhanded attacks at the young evolutionists Thomas Huxley and Charles Darwin, who had once admired him.

As superintendent of natural history at the British Museum, and later the first director of the British Museum of Natural History, Owen was the premier expert consulted by Her Majesty's government on everything imaginable: from evidence in murder cases to the authenticity of sea serpent sightings.

Young Darwin had sought Owen's friendship and advice and had even asked the senior naturalist to write the official description and classification of the fossil mammals he had collected on his voyage aboard HMS *Beagle*. Owen undertook the task, and his monographs, published between 1838 and 1840, occupied four volumes of *Zoology of the Voyage of HMS Beagle*. But, as Darwin later recalled, "After the publication of the *Origin of Species*, he became my bitter enemy, not owing to any quarrel between us, but as far as I could judge out of jealousy at its success. . . . His power of hatred was certainly unsurpassed."

Owen had a chilly politeness that masked his underhanded attacks. As historian Lynn Barber put it in *The Heyday of Natural History* (1980), "If Owen could steal the credit for someone else's achievements, he would always do so; if he could not, he would strive to discredit the achievement. Hugh Falconer, the elephant expert, warned Darwin that Owen was 'not only ambitious, very envious and arrogant, but untruthful and dishonest.'"

In addition to his 400 technical papers on anatomy, Owen is remembered for

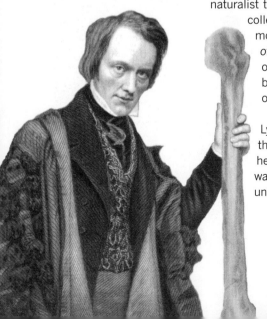

HOLDING THE BONE OF A MOA, anatomist Richard Owen initially identified a much smaller fragment as belonging to a monstrous bird. More complete fossils proved him correct.

History of The Art of War from the Earliest ages.

Stage 1.
The Primitive Age.

Stage 2.
Invention of weapons.

Stage 3.
Of a deadly nature.

The Lady She is too tired to dance

But who recovers in a rapid & unexpected manner

Anatomist Richard Owen, who coined the term "dinosaur," detested Darwin and his theory of evolution.

several noteworthy contributions to the reconstruction of extinct creatures, which, ironically, later helped establish the Darwinian theory he detested.

In 1839, a sailor brought him a puzzling six-inch fragment of bone from New Zealand, which Owen identified as belonging to an unknown, huge flightless bird. His colleagues were skeptical—could any bone so massive possibly have belonged to a bird? But Owen stuck by his original identification. A decade later, several crates of gigantic fossils reached the museum, filling out the skeleton of the same creature he had named *Dinornis*. The existence of New Zealand's giant extinct moa bird was confirmed.

Similarly, Owen was the first to correctly reconstruct the giant ground sloth *Mylodon* from South American fossils and was even able to deduce something about its way of life. But his most famous reconstructions were the life-sized prehistoric animals, including *Iguanodon*, which were sculpted under his direction by the artist Benjamin Waterhouse Hawkins. Originally created for the Crystal Palace exhibition in 1853, they remain in the gardens of Sydenham, near London, to this day. [See IGUANODON DINNER.]

Dinosaurs will always be linked with Richard Owen. He was the first to understand and publish the significance of the remarkable fossil reptile bones that had recently been discovered by Dr. Gideon Mantell and others. It was Owen who coined the term "dinosauria," meaning "terrible lizards," for this previously unknown group of extinct "monsters." He also described and named *Archaeopteryx*, the early reptile-like bird, when he purchased the world-famous specimen for the British Museum.

In his reviews of Darwin's books, Owen was consistently inaccurate and unfair in representing what had been said. Privately, he sniped that Charles "was just as great a goose as his grandfather." When shown advance proofs of the *Origin of Species* (1859), he complained it contained too many statements of Darwin's opinions, beliefs, or conjectures. When Darwin immediately offered to cut them out, Owen sternly advised him not to change a thing, or he would "spoil the charm" of his book. Then, in his review, Owen promptly singled out Darwin's use of such phrases as "I am convinced" or "I believe" and attacked them as unscientific. Darwin felt not only abused, but tricked and cheated as well.

Owen adhered to the idea of "archetypes" that he had learned from the German zoologist Lorenz Oken, a follower of the *Naturphilosophie* school. He clearly saw and studied in detail the widespread pattern of homologies in structure—the principle that a bat's wing, human foot, and whale's flipper are made of the same structural elements, bone for bone. He wrote a treatise *On the Archetype and Homologies of the Vertebrate Skeleton* in 1848, in which he illustrated homologies in the anatomy of diverse animal species.

But to Owen, these were variations of a divine plan, not transformations based on common descent with modification. Nevertheless, Owen was the first to make the crucial distinction between analogies—differently structured parts with similar functions, such as insects' wings and birds' wings—and homologies—the same underlying structures performing a wide variety of functions, such as horses' hooves corresponding to apes' third finger. Although the horse walks on an enlarged nail of a single digit, the other four digits are present as "splints" in the leg.

It was also Owen who prepared Bishop Samuel Wilberforce with the ammuni-

tion to attack Darwin at the Oxford debate in 1860. Nor was this the first of his "undercover" attacks on the evolutionists. Owen had anonymously published a scathing, spiteful review of the *Origin of Species* in the *Edinburgh Review* (1860), then lied to Darwin's face that he had not written it.

At other times, Owen seemed to want to take credit for developing an evolutionary perspective in biology. Darwin noted, with annoyance, that Owen later taught that perhaps "all birds are descended from one, and advances as his own idea that the oceanic wingless birds have lost their wings by gradual disuse. [But] he never alludes to me, or only with bitter sneers."

Owen, in fact, sometimes hinted that he had his own theory of evolution, which was better than Darwin's and which he had originated earlier—but he never revealed just what that idea might be. In effect, he was claiming evolution was wrong, but that, if it was right, he had invented it.

Despite all his efforts, however, Owen was unable to stem the Darwinian wave that swept the life sciences. He went on to establish and direct Britain's Natural History Museum in South Kensington, which opened in 1881, until his retirement at the age of 80. His arrogant prediction that Darwin's work "would be forgotten in ten years" followed him into old age, for it was his own theories that were discredited and forgotten. Although he outlived Darwin by several years, his archetypal ("essentialist") ideas about species had become as extinct as the dinosaurs.

See also ARCHAEOPTERYX; ESSENTIALISM; HOMOLOGY; *NATURPHILOSOPHIE*; OXFORD DEBATE

OXFORD DEBATE (1860)
"Darwin's Bulldog" vs. The Bishop

Soon after Charles Darwin's *Origin of Species* appeared in 1859, the theory of evolution was under attack by churchmen. The first confrontation of opposing forces took place at Oxford University on June 30, 1860, during the weeklong meetings of the British Association. "Darwin's Bulldog," Professor Thomas Henry Huxley, debated Samuel Wilberforce, the Bishop of Oxford, on the bishop's home ground.

Wilberforce was a wily and confident orator. Although he didn't know much science, he had been coached for several days by Richard Owen, the British Museum's paleontologist who was a rival of Darwin's and an enemy of Huxley's. Only a few days before, Owen and Huxley had clashed about the comparison of human and gorilla brains, with Huxley accusing the renowned Owen of not knowing what he was talking about. Now Owen was seeking revenge on the Darwinians by prepping the bishop for the Saturday session.

Huxley, after a week of meetings, intended to skip the last day. He knew Samuel Wilberforce was an accomplished speaker, capable of sliding around any argument or confrontation—which had earned him the nickname "Soapy Sam." (According to the bishop, he got the name because he was always in hot water and always came out of it with clean hands.) Besides, Huxley was not a scheduled speaker.

On Friday, Huxley ran into Robert Chambers, the author of *Vestiges of Creation* (1844), and told him he planned to miss the next day's session since he "did not see the good of giving up peace and quietness to be episcopally pounded." Chambers accused him of deserting, and Huxley replied, "Oh! If you are going to take it that way, I'll come and have my share of what is going on."

The new University Museum was crowded with 700 men and women who sensed an impending battle of giants. On the platform were the botanists J. S. Henslow and Joseph Hooker, Bishop Wilberforce, Dr.

John William Draper of New York, Sir John Lubbock, Sir Benjamin Brodie (the Queen's physician and president of the Royal Society), and Huxley.

Draper droned on for an hour with a mediocre paper discussing Darwin's views, and several clerics and other members of the audience rose to respond. Among them was Admiral Robert FitzRoy, captain of the *Beagle* 30 years before; some say he held a Bible over his head and denounced Darwin for having been a viper in his midst. Finally, there were calls for the bishop, who began with theatrical ruminations abut how disquieted he would be if anyone should prove he had a "venerable ape" in his family. According to Hooker's account, Wilberforce "spouted for half an hour with inimitable spirit, ugliness and emptiness and unfairness" without any grasp of the scientific issues involved.

Overconfident that the audience was his, Wilberforce pulled out all the stops and, with a turn to Huxley, demanded to know whether he was descended from an ape on the side of his grandfather or his grandmother? At this, Huxley turned to Brodie, slapped his knee, and whispered, "The Lord hath delivered him into mine hands!" The bishop had violated Victorian propriety by getting personal, attacking his opponent's family ancestry, and—worst of all—insulting womanhood!

Huxley, according to his own recollections in *Life and Letters* (1900), waited to be called to speak by the audience, then rose slowly and deliberately. When we talk of descent, he calmly explained, we mean through thousands of generations. He had listened carefully to the bishop, he said, but could not discover any new facts or new arguments—except the questioning of his own family tree. "It would not have occurred to me to bring forward such a topic as that for discussion myself," Huxley continued. But "if the question is put to me, would I rather have a miserable ape for a grandfather, or a man highly endowed by nature and possessed of great means and influence, and yet who employs these faculties and that influence for the mere purpose of introducing ridicule into a grave scientific discussion—I unhesitatingly affirm my preference for the ape."

Women fluttered their handkerchiefs and Lady Brewster fainted, while the room rebounded in laughter. Huxley continued with serious, sober arguments, and Lubbock and Hooker spoke after him. Hooker pointed out that, from the bishop's remarks, it was clear he could never have read Darwin's book and was (despite Owen's briefing) completely ignorant of the subject.

Writing to Francis Darwin in 1861, Huxley noted: "I happened to be in very good condition and said my say with perfect good temper and politeness—I assure you of this because all sorts of reports [have] been spread about, e.g. that I had said that I would rather be an ape than a bishop, etc." In fact several reports written later by participants vary in their account of exactly what was said that day.

The importance of the Oxford debate was that it was the first time open resistance was made to the church's authority over the question of human origins and science's right to pursue investigations of human origins. Instead of being crushed under ridicule, Huxley and his colleagues won a fair hearing for the new theories.

Huxley later reprised his Oxford performance many times, and eventually tired of it. According to the *Year-book of Facts in Science and Art* (1870), several clergymen spoke against evolution at the 1869 meeting of the British Association for the Advancement of Science. An Archdeacon Freeman insisted that "certain ideals pre-existed" in the mind of God, and that "after these the creatures were formed." Huxley replied that it was annoying to have to respond over and over to the same objections that had been dealt with satisfactorily many times. He said he sometimes thought it likely that "Abel was the first man of science and Cain the first theologian." He further opined that "theologians hang on to certain . . . doctrines till their fingers are burnt, and then, letting go, say it is of no importance."

Although after the Oxford meeting he treated the bishop with perfect courtesy, Huxley could not resist a cruel parting shot at the time of the bishop's death. In 1873, Wilberforce was riding through a field, when he was thrown from a horse and hit his head on a stone. "His end has been all too tragic for his life," Huxley wrote to John Tyndall. "For once, reality and his brains came into contact and the result was fatal."

See also CHAMBERS, ROBERT; FITZROY, CAPTAIN ROBERT; HOOKER, SIR JOSEPH DALTON; HUXLEY, THOMAS HENRY; OWEN, SIR RICHARD

> **"Would I rather have a miserable ape for a grandfather or a man [of great intellect who mocks scientific inquiry?]—I unhesitatingly affirm my preference for the ape."**
>
> —Thoms Huxley to Bishop Samuel Wilberforce, Oxford, 1860

PALEOLITHIC
Old Stone Age

Sir John Lubbock, Charles Darwin's neighbor and protégé, was a pioneer of what was known as the "Prehistory Movement," when discoveries of fossil man and his tools were not yet established. Lubbock brought England's top geologists to France to see for themselves the stone tool sites excavated by Jacques Boucher de Perthes.

Thousands of finely chipped flint tools were soon accepted as the work of early humans. In his book *Pre-historic Times* (1865), Lubbock named the earliest layers the Paleolithic, meaning "Old Stone Age" (100,000 to 1 million years ago), and later named a "Neolithic" age (9,000 years ago), from which came the polished stone axes associated with early agriculture. When early worked metals were found, a "Bronze Age" and an "Iron Age" were added, but these were later dropped as characterizing limited, local developments.

One of the fallacies associated with the Victorian labeling of "ages" was that they thought the tool types represented definite stages in human history. When tribal peoples were discovered with stone axes, it was said that they were "still in the Stone Age," which of course they were not. No people's history has to pass through some predetermined number of stages, and humans who use stone implements today are not necessarily anything like those of a million years ago in their language, thought, or social organization.

Also, what we find today is not necessarily the only technology used by ancient peoples. Where only stone tools survive, the people may also have used ropes, nets, wooden implements, and other perishable materials. The basic joke of the "Fred Flintstone" cartoons is that everything is made of stone—including books, vehicles, and furniture—because they're all supposed to be products of the Stone Age.

See also LUBBOCK, SIR JOHN; PITT-RIVERS MUSEUM

PALEY'S WATCHMAKER
Design Implies Deity

Everyone with a passing interest in natural history in the 19th century—which included the majority of educated Victorians—had heard of William Paley's famous "Watch-maker." If they did not read his widely studied book *Natural Theology,* first published in 1802 and reprinted almost annually thereafter, they "would have had its ideas dinned into them from the cradle up, and heard them expounded from every pulpit in the land," as Lynn Barber has phrased it.

Paley's book begins with "The Watch on the Heath," an analogy that illustrates the "argument from design," which he thought proved the existence of a creator:

> In crossing a heath, suppose I pitched my foot against a stone, and were asked how the stone came to be there. . . . [I could answer that] it had lain there forever. . . . But suppose I had found a watch upon the ground. . . . I should hardly think of [the same answer because] when we come to inspect the watch, we perceive . . . that its several parts are framed . . . to point out the hour

of the day. . . . The inference, we think, is inevitable, that the watch must have had a maker; that there must have existed, at some time, and at some place or other, an artificer . . . who formed it for the purpose . . . and designed its use.

According to natural theology, everything in nature reflects the beauties and perfection of the God who designed it. Paley marshaled the same kinds of facts Darwin later did: the construction of the human eye, the physiology of plants, the relationships between organs. But to Darwin, the intricate mechanism of nature came about through the gradual fitting and honing of natural selection. To Paley, all had been designed in one stroke by a "Master Designer"—and his watchmaker argument is still often advanced by today's creationists and proponents of "intelligent design."

See also BEECHER, HENRY WARD; DIVINE BENEFICENCE, IDEA OF; INTELLIGENT DESIGN; NATURAL SELECTION

PANDA'S THUMB
An Evolutionary "Contraption"

That zoo favorite, the giant panda, does not have a true opposable thumb, since its five toes long ago evolved into a bearlike paw. But when it adapted to feeding on tasty bamboo leaves, one of its wrist bones became an extra digit for holding and stripping the stalks. Thus, pandas have six digits—but the extra thumb is a made-over part. Evolution uses what is available.

Harvard paleontologist Stephen Jay Gould used the panda's "thumb" as the central image in a popular book of evolutionary essays (*The Panda's Thumb: Reflections in Natural History,* 1980). During a period when creationists were repopularizing the old idea that adaptation implies perfect design, Gould emphasized the sometimes quirky history of life.

Since the evolutionary pathway leading to opposable thumbs was not taken by the panda's ancestors, its thumb was "committed" to the paw structure. So the radial sesamoid bone developed into an efficient, but inelegant, thumb.

Gould reminds us that Charles Darwin consistently turned to those organic parts and geographic distributions that seemed to make the least sense, for in the "leftovers" and "contraptions" could be read a particular history.

The panda's thumb is Gould's favorite example that the strongest proof of evolution is not optimal design, but "odd arrangements and funny solutions . . . paths that a sensible God would never tread but that a natural process, constrained by history, follows perforce."

See also ADAPTATION; NATURAL SELECTION; PALEY'S WATCHMAKER

GIANT PANDA EVOLVED an "extra" thumb from one of its wrist bones, a secondary specialization for feeding on bamboo stalks. Its original first digit is immobilized in the bony structure of its paw.

PANGEA
Original Supercontinent

Pangea was a single, huge equatorial continent, which coalesced about 225 million years ago. Enormous plates moved together in conjunction with climatic shifts and other drastic geophysical changes. The result, apart from the formation of the single land mass, was the Permian extinction: the most extensive "great dying" in the Earth's history, during which thousands of species disappeared forever.

Eventually, the great moving plates began to tear apart, pulling Pangea into a northern and a southern half. The northern supercontinent, Laurasia, was formed by 65 million years ago and subsequently broke up to form North America, Europe, and Asia. Gondwanaland, the southern portion, later drifted apart to form the continents of South America, Australia, India, and Antarctica.

See also GONDWANALAND; LAURASIA; PLATE TECTONICS

PANGENESIS
Darwin's Wrong Guess

Because Charles Darwin realized evolutionary theory was incomplete without an understanding of heredity, he devised a theory of how parents pass on their characteristics to offspring. He called it pangenesis; it turned out to be wrong.

According to pangenesis, a trait acquired by a parent during his or her lifetime could be passed on to children (Lamarckian or "soft" inheritance). If a man worked to develop large muscles, for instance, the repeated habit of weightlifting would somehow leave a lasting record in the cells of his body. Darwin called hypothetical particles that carried this information "gemmules." They would migrate from all parts of the body to the sex cells, whereby they could be inherited by the offspring.

Within 20 years of Darwin's death, Mendelian genetics was belatedly recognized by science, and August Weismann demonstrated that what he called the "germplasm" is distinct from the rest of the body.

Sex cells, we now know, produce genetic material independently, without input from arms, intestines, or any other area of the body. Acquired characteristics, such as lopped-off fingers, trained muscles, or a mastery of French, are not passed on through "gemmules." However, Darwin was on the right track; he correctly surmised the existence of tiny biochemical structures that contain coded blueprints for the offspring's characteristics. We call them genes, a term coined by Wilhelm Ludwig Johannsen in 1909.

See also MENDEL, ABBOT GREGOR; WEISMANN, AUGUST

PANSPERMIA
Life from Outer Space

British astronomer Fred Hoyle and his colleagues first suggested in 1978 that the seeds of life may be constantly raining down on earth from outer space—a theory he called panspermia. Francis Crick, codiscoverer of DNA, supported Hoyle's theory.

According to Hoyle, the conditions for the origin of life may have been better among the vast amount of organic matter he believes floats through interstellar space. Unimaginably immense quantities of chemical molecules colliding in space might make the rare and improbable combinations more likely, almost inevitable. Simple life forms or amino acids may have journeyed to Earth on comets or meteors. Of course, this is no explanation for the origin of life; it simply moves the problem farther off to another time and place.

Although biologists ridiculed panspermia for years, in 1989 the idea acquired some supporting evidence when amino acids were found within meteor craters. And contrary to expectations, the molecules were not damaged by their rough landings. On learning of these discoveries, astronomer Carl Sagan commented: "The impact giveth and the impact taketh away!"

J. Craig Venter, the pioneer of human genomic research, believes that life is ubiquitous throughout the universe and that "life on our planet earth most likely is the result of a panspermic event." He has written:

> DNA, RNA and carbon based life will be found wherever we find water and look with the right tools. . . . In sequencing the genetic code of organisms that survive in the extremes of zero degrees C to well over boiling water temperatures we begin to understand the breadth of life, including life that can thrive in extremes of caustic conditions of strong acids to basic pH's that would rapidly dissolve human skin.

A species of bacteria called *Deinococcus radiodurans*, according to Venter, can survive millions of rads of ionizing radiation and complete desiccation for years or perhaps millennia, then repair any DNA damage within hours of being placed in water. "From the millions of genes that have been discovered in common bacteria," he concludes, "a finite number of themes are used over and over again and could have easily evolved from a few microbes arriving on a meteor or on intergalactic dust. Panspermia is how life spread throughout the universe and we are contributing to it from earth by launching billions of microbes into space."

See also BACTERIA

PASTRANA, JULIA (1834–1860)
Victorian "Gorilla Woman"

Early evolutionists thought a hirsute Mexican Indian dancer named Julia Pastrana was a "throwback" to an apelike stage of humanity. Charles Darwin, Ernst Haeckel, and Alfred Russel Wallace all wrote of her unfortunate genetic condition (hypertrichosis terminalis) as a scientific curiosity, and the poor woman was endlessly studied and measured.

Writing of her case in *The Variation of Animals and Plants under Domestication* (1868), Darwin referred to Pastrana, with his customary tact, as "a remarkably fine woman, but she had a thick masculine beard and a hairy forehead." He continued: "She had in both the upper and lower jaw an irregular double set of teeth, one row being placed within the other. . . . From the redundancy of teeth her mouth projected, and her face had a gorilla-like appearance." Darwin wondered if large teeth were correlated with hairiness: many hairless animals had small teeth or none. (He was thinking of armadillos and porpoises, but seemed to have forgotten that bare-skinned elephants and walruses grow the biggest teeth of all.)

Julia Pastrana became famous as the "gorilla woman"; she appeared in freak shows and sold photographs of herself. When she died, a heartless showman had her skin stuffed and mounted and placed in a glass case, along with that of her young child who had died in infancy. He did a brisk business charging the public to see them.

PEALE'S MUSEUM (1786–1845)
Early Natural History Museum

In July 1786, America's most famous portrait painter, Charles Willson Peale (1741–1827), put an ad in the *Pennsylvania Packet* asking for "Natural Curiosities" and "Wonderful Works of Nature." Artist, amateur scientist, entrepreneur, and showman, Peale was putting together the world's biggest and best private museum of natural history. (No such public institutions yet existed.)

To his delight, the public's interest matched his own, and he could barely keep up with the avalanche of specimens: minerals, shells, fossils, feathers, creatures dead and alive. Benjamin Franklin sent a French angora cat, and President George Washington a brace of pheasants.

It all started when a local naturalist asked the artist to illustrate some puzzling, gigantic bones he had found. After drawing these "curiosities," Peale decided to exhibit them among his portraiture. Thus, the first real museum of natural history grew out of an art gallery, rather than a science laboratory.

Peale's shifting interests were reflected in the naming of his sons. While the older boys were all named for artists—Raphael, Rembrandt, Titian, Rubens, and Vandyke—those born after he became interested in science were called Charles Linnaeus and Benjamin Franklin Peale. (Now best remembered as a philosopher-politician, Franklin was also famous as a scientist during his lifetime.)

By trial and error, Peale developed methods of preserving specimens (often dousing them in arsenic to keep away insects) and found an eager public willing to pay the admission fee to his establishment. It soon became one of the most celebrated in the world, the envy of European naturalists and a Philadelphia landmark.

Intrigued with the gigantic bones that had begun his natural history collection, Peale decided to find a complete skeleton of this unknown monster ("the great American incog-

FAMOUS BEARDED LADY, Julia Pastrana traveled with European circuses and freak shows. Her hairy body and other anomalies piqued the interest of both Charles Darwin and Alfred Russel Wallace.

PEALE'S MASTODON was discovered in a water-filled marl pit near Newburgh, New York, in 1799 (top). Charles Willson Peale (holding drawings) designed and built a bucket-wheel powered by treadmill. Spectators took turns running inside the wheel to bail out the site while workmen quarried the massive fossil skeleton, which was later exhibited at Peale's Natural History Museum (above).

nitum," as it was nicknamed) for his museum. After following several leads, he found what he was looking for at the farm of John Marsten, near Newburgh in New York State. Marsten had excavated some large fossil bones from a water-filled marl pit, and he sold them to Peale for $200.

Convinced the rest of the skeleton was buried there, Peale launched a full-scale excavation, rigging elaborate block-and-tackle system and a specially built pump, operated by a man-powered tread-wheel. Costing Peale $1,000, it was the first full-scale scientific expedition in American history and a complete success. Although a few bones (including the jawbone) were missing, Peale obtained an almost-complete skeleton of a huge mammal that had never before seen and had it shipped back to Philadelphia.

A few months later, his crew dug up a similar skeleton at another farm, and Peale pieced the two individuals together to form the first mounted mastodon (or "mammoth," as he called it). When Americans got their first look at the great extinct elephant that once roamed the present Hudson River valley, it became an immediate sensation. It was also a new symbol of national pride, since chauvinistic European naturalists had insisted that "everything is smaller in America."

For half a century, the word "mammoth" meant the biggest and best. (It was later supplanted in popular parlance by the name of P. T. Barnum's huge circus elephant, "Jumbo.")

Peale's discovery of the mastodon was an important and original contribution to science, though the French anatomist Georges Cuvier was first to accurately describe and classify it.

After the massive fossil bones had been mounted in the museum's Mammoth Room in December 1801, Peale and a dozen jubilant patriots celebrated by holding a formal dinner inside the skeleton, where they sang "Yankee Doodle." (Some years later, not to be outdone, British scientists and museum artists held an elegant dinner inside the world's first life-sized dinosaur model—an iguanodon at the Crystal Palace Exposition.)

Peale always claimed his main interest was scientific and took pride in being the first museum-keeper to arrange his large and diverse collection according to the Linnaean system of classification. But detractors were fond of pointing out that his exhibits also included such crowd pleasers as waxworks, stuffed monkeys wearing suits and gowns, a cow with five legs and two tails, a machine for drawing silhouettes, and clever mechanical dioramas that he called "moving pictures."

Raised among such wonders and oddities, one of Peale's sons, Rubens, became a shameless showman who built the museum into an even more popular (and unscientific) entertainment; another (Titian) became a serious naturalist; and still another (Rembrandt) became a fine painter who incorporated superbly observed flowers and animals into his compositions.

Part of Peale's legacy to the great natural history museums is their perennial problem of determining who should create the exhibits: scientists, showmen, or artists. In Charles Willson Peale—as in the best museum people since—these three distinct talents were highly developed in a single individual.

See also BARNUM, PHINEAS T.; IGUANODON DINNER; JEFFERSON, THOMAS; MAMMOTH

PEER REVIEW
Science's Referee System

Scientific innovations, when first presented, are often unrecognized by experts in the field. When Alfred Russel Wallace published his brilliant "Sarawak Law" on the regulation of species in 1855, it was almost completely ignored by zoologists. In 1858, the epochal Darwin-Wallace papers on evolution by natural selection were not greeted with acclaim by the majority of members when they were read at the Linnean Society of London. Some months later, the president of the society, in his annual report, lamented that no significant new theories had been presented that year in biology.

One objective of science is to generate novel breakthroughs: something never seen, done, or understood before. Yet scientists have a strong tendency to reject new facts or theories on those very grounds. Gregor Mendel's experiments on hybrid pea plants is a notorious example from evolutionary biology. Although published in a scientific journal and disseminated to the leading botanists of Europe, his work was not recognized as the foundation of a new science of genetics until 50 years later.

Charles Darwin was well aware that scientists are not convinced of new ideas so much as that younger men eventually replace their senior colleagues. All scientific men should have the grace to die at the age of 60, he wrote, so they could not "oppose all new doctrines" with their "inflexible" brains and impede the progress of the next generation.

Initially plagued by doubts as he began writing the *Origin of Species* (1859), Darwin first "fixed in my mind three judges, on whose decision I determined mentally to abide": the botanist Sir Joseph Hooker, the zoologist Thomas Huxley, and the geologist Charles Lyell. He would put aside his "awful misgivings" if they could agree with his approach and conclusions. Lyell alone of the three was afraid to "go the whole Orang," and his years of fence-sitting greatly upset Darwin.

After the initial acceptance by his self-chosen referees, Darwin mounted a personal campaign to convince about a dozen other top men in natural history of the truth of evolution. He even picked and targeted them and kept running lists of who was still "unconverted." If these colleagues could be won, he thought, "my theory will be safe."

Thomas Huxley early on had no illusions about the "objectivity" of peer review. He had written an original anatomical monograph but knew the leading journal was then controlled by Sir Richard Owen, the antievolutionary zoologist. Owen was the top expert in the field, but "he thinks natural history is his private preserve, and no poaching allowed," Huxley wrote a confidant.

Alfred Russel Wallace, the coauthor of the theory of evolution by natural selection, was exasperated by the difficulties geologists had convincing their colleagues that man-made stone tools had been found in sealed layers with the bones of extinct elephants and other Ice Age mammals. In his book *The Wonderful Century* (1898), Wallace wrote:

> In 1840 a good geologist confirmed these discoveries, and sent an account of them to the Geological Society of London, but the paper was rejected as being too improbable for publication! All these discoveries were laughed at or explained away. . . . These, combined with numerous other cases of the denial of facts on a priori grounds, have led me to the conclusion that, whenever the scientific men of any age disbelieve other men's careful observations without inquiry, the scientific men are always wrong.

"What a good thing . . . if every scientific man was to die when sixty years old . . . [so he could not] oppose all new doctrines."

—Charles Darwin, *Autobiography* (1876)

For a scientist, to acknowledge a novel discovery may be to admit that a colleague or rival is smarter, more methodical, a better observer—or even just luckier than oneself. If the new knowledge should come from an amateur, the resistance can be especially intense. [See CHAMBERS, ROBERT; SCIOLISM.]

At least Wallace and Mendel got their ideas published, even if they were ignored. One of the founders of the modern revolution in geology wasn't so lucky. In 1963, Canadian geophysicist Lawrence W. Morley, working independently, made a synthesis of current data and worked out one of the foundations of modern plate tectonics: The sea floor is continually spreading, causing continents to move, split apart, or crunch together.

Morley had been studying magnetic bands in rocks retrieved from the ocean floor, a record of alternating polarity reversals in the Earth's history. When he correlated them with matching evidence from inland rocks, he came up with a theory that seafloor spreading is the engine that moves the continental plates.

Morley was no weekend amateur, but chief of the Geophysics Division of the Geological Survey of Canada. In February 1963, he wrote a brief letter to the journal *Nature* outlining his theory, but was told they "did not have room" to print it. According to an account in William Glen's *Road to Jaramillo* (1982), he next tried the *Journal of Geophysical Research*, which sent Morley's paper to an anonymous referee, who advised that "such speculation makes interesting talk at cocktail parties, but it is not the sort of thing that ought be published under serious scientific aegis."

By the time the manuscript was returned to him, five months later, two other researchers had independently published the theory before him. (Fortunately, there was a happy ending; it did eventually become known as the Vine-Matthews-Morley hypothesis, one of the foundations of modern geophysics.)

In 1977, psychologist Michael J. Mahoney of Pennsylvania State University decided to test the objectivity of peer reviewers in judging scientific merit. He sent two specious papers on a disputed topic to 75 experts in the field whose opinions about this controversy were known. The vast majority praised the dummy paper whose conclusions agreed with their own position, while rejecting the one that appeared to offer evidence for the opposing view. When he published these results, Mahoney was threatened with loss of his job and called "unprincipled and unethical" for daring to experiment on the scientific system itself without the participants' knowledge. (Getting their permission, of course, would have made the study impossible.)

A few years later, two other psychologists, Douglas P. Peters of the University of North Dakota and Stephen J. Ceci of Cornell, decided to follow up on Mahoney's innovative research. Their strategy was a bit different. They took many papers that had already been published by top scientists from prestigious institutions, had them retyped, and substituted the names of unknown authors from low-prestige institutions. Then they sent them to the journals in which they had originally been published for peer review. The results: All of the papers by supposedly low-status authors were rejected, and none of the experts recognized that the same articles had previously appeared in the journal.

See also MENDEL, ABBOT GREGOR; "SARAWAK LAW"; TAUNG CHILD

PEKING MAN

See BEIJING MAN

PEPPERED MOTH
Controverted "Icon" of Evolution

A homely little insect living in a soot-blackened woods near Manchester, England, became one of the most hotly debated "icons" of evolution. During the 1930s, a group of British biologists seeking to demonstrate natural selection at work thought they had found their ideal subject in the peppered moth, *Biston betularia*.

This small, drab moth has three color phases: a whitish, a brownish mottled, and a melanistic (almost black) type. All three colors exist in the population, but in Victorian insect collections the lighter and mottled types predominate. Then, in the late 19th century, the formerly rare melanic phase began to appear more frequently, until by the 1920s the species was almost all black.

In the 1950s, British butterfly expert H. B. Kettlewell and his colleagues correlated the moth's color change with the industrialization of the Manchester area. Soot and sulfur dioxide from mills and factories coated tree trunks, turning them black, and killing mottled tree lichens. Lighter and mottled moths now stood out clearly against the trunks, making them easy prey for predatory birds, while the dark moths became hard to see.

As tree trunks darkened over the years, the formerly rare dark moths survived in greater numbers, reproduced, and dramatically increased their frequency in the population. Within a few decades, the majority of the moths were dark.

The change from light to dark moths became a textbook example of natural selection in action. Countless biology texts have recounted rapid evolution of color change in peppered moths as a classic demonstration of Darwinian selection observable within a human lifetime. For half a century, the peppered moth served as the standard bearer—partly because of the conspicuous absence of any other field studies of natural selection during that period. (Thanks to Rosemary and Peter Grant's field studies, Darwin's finches, with their rapidly evolving beaks, have now replaced the moths as textbook icons of observed natural selection in the field.)

But the resilient insect—and the truth about it—continues to evolve before our eyes. Because of antipollution laws, English skies have been clearing over the past few decades. Trees have been getting lighter again, and the lighter-colored moth has made a big comeback in the last few decades.

Laurence Cook, a biologist at the University of Manchester, published a study in 1986 of 1,825 moths collected throughout Great Britain, showing that the area dominated by the black moths was steadily shrinking toward England's northeast corner. Meanwhile, Sir Cyril Clark trapped moths for years near his home in Merseyside and reported in 1984 that dark peppered moths had made up 90 percent of the total before 1975, but had fallen to 60 percent 10 years later.

Did the white moth's coloration become protective again, as its environment became cleaner and lighter colored? Are birds the agents of selection? Can and do they now pick off the black ones much more easily than the white? If confirmed by field experiment, it would be a neat explanation of the color reversal, adding to the tidiness of the peppered moth's story.

During the past decade, however, Kettlewell's original studies increasingly came under fire, by both evolutionary biologists and creationist critics. Scientists uncovered serious flaws in the methodology of the original studies. Creationists protested that the moth study was a deliberately fraudulent attempt to prop up Darwinian theory. Kettlewell himself was attacked posthumously in a 2002 book, *Of Moths and Men: The Untold Story of Science and the Peppered Moth*, by Judith Hooper, who portrayed him as an unhappy man who was alternately abused by mean-spirited senior colleagues and plagued by self-doubt. His tormentors, who secretly blocked his admission into the Royal Society, built their own careers on the success of the peppered moth story. Kettlewell finally committed suicide in 1979.

One of the major scientific critics of Kettlewell's original study was Michael Majerus, an evolutionary geneticist at the University of Cambridge, who picked Kettlewell's methods apart in his 1998 book *Melanism: Evolution in Action*. Yet Majerus remains convinced that Kettlewell's basic conclusion will ultimately be proven correct: that differential predation by birds was indeed the key to the moths' color shift.

Majerus has been working for decades, with both lab-reared and wild-caught moths, to get "a definite answer." He believes he could now win the case for bird predation in a courtroom "miles beyond reasonable doubt, but it's not scientific proof." Time is running out for the studies, however. Continuing antipollution efforts have resulted in so many new "clean" tree trunks that within a few years the proportion of melanic moths is expected to decline to less than 1 percent.

The resilient insect, and the truth about it, continues to evolve before our eyes.

When asked in *Science* magazine (June 2004) what methodology he would recommend for asking questions of nature, Majerus suggested: "Find your organism and look at it incredibly closely, for a long time, in great detail, and see what questions it asks *you*."

See also DARWIN'S FINCHES; GENETIC VARIABILITY; NATURAL SELECTION

"PHILOSOPHICAL" NATURALISTS
Searching for "Nature's Laws"

Beginning in the late 18th century, when most zoologists were concerned with collecting or discovering new species, the more curious sought regularities or "laws." They referred to themselves as "philosophical naturalists."

Natural philosophy—a system of general principles describing the nonliving world—had already been established by physicists and astronomers. But "natural history," the study of rocks, plants, and animals, was still largely an exercise in description and classification. Philosophical naturalists may have enjoyed collecting, but they also shared a faith that the "laws" of life itself would soon be discovered.

They tried many approaches to find such laws. German poet Johann Wolfgang von Goethe thought he glimpsed a "unity of plan" behind living structures. In 1790, he analyzed flowering plants as variations on an "ideal" leaf and animal skulls as transformed vertebrae. Zoologist William Sharpe MacLeay invented a "quinary system" of classification (1819), based on his idea that all nature was built around the number five. In their day, these were worthy attempts by brilliant men, and there were many others.

All shared the conviction that discoverable, underlying principles would "make sense" of the facts of anatomy, geology, botany, animal distribution, and fossils. Without philosophical naturalism, there could have been no evolutionary biology.

See also BIOLOGY

PHRENOLOGY
Reading Character from Crania

READING BUMPS on the skull was a popular pseudoscientific fad of the 19th century. Despite excesses that now seem ridiculous, the phrenologists' system encouraged early research on localization of brain functions.

During the 19th century, thousands came to believe in phrenology, which purported to "read" human character with scientific accuracy from bumps on the skull. Some went so far as to base hiring decisions—or even marriage plans—on the results of analyzing shapes of heads or facial features.

Phrenology was the "sister" to physiognomy—the reading of facial features. Among the latter's enthusiasts was young Captain Robert FitzRoy, who almost rejected Charles Darwin as ship's naturalist because his nose lacked angular definition. "The Captain doubted," Darwin reported, "whether anyone with my nose could possess sufficient energy and determination for the voyage." (Later in life, Darwin was amused when a phrenologist told him he had a "bump of reverence developed enough for ten priests.")

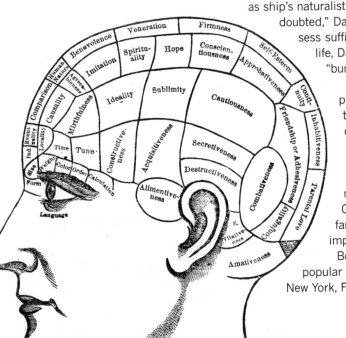

Franz Joseph Gall (1758–1828), an influential German physician and neuroanatomist, first developed the theory that scores of personality traits, talents, and attributes are detectable by relative sizes of their corresponding "brain centers." Phrenologists believed they had mapped the mental "organs" that controlled "Ideality," "Sublimity," "Hope," "Veneration," "Firmness," "Secretiveness," "Destructiveness," "Amativeness," and "Love of Home"—to name a few. Gall and his collaborator Johann Spurzheim had become famous by the 1830s, when their ideas were considered an important scientific breakthrough.

Books and magazines based on phrenology continued to be popular for almost 100 years. During the 1870s, a family firm in New York, Fowler and Welles, made a fortune offering individual head-readings, training programs for phrenologists, traveling lecturers, and publications. (The woman in that partnership, Lydia Folger Fowler, was also America's second woman doctor.) Their *Phrenological Journal* was popular and progressive, often carrying articles on anthropology, animal behavior, evolution, psychology, and social reform.

As a rationale for science, phrenology led to a great deal of social mischief. "Experts" gravely pronounced on "racial capabilities" on the basis of head shape; "craniologists" collected and measured skulls endlessly, with no useful result. Criminal "types" were identified and classified by facial features, which we now know to be irrelevant. Worst of all, people were judging one another by cranial "bumps" or ear shapes rather than by their actual deeds and real capacities.

By the beginning of the 20th century, biologists had rejected this pseudoscience so emphatically that they denied science owed anything to it, even as they made advances in localizing brain functions through electrode stimulation. But Darwin's codiscoverer of the principle of natural selection, Alfred Russel Wallace, staunchly defended phrenology.

"This very fact of the connection of certain definite brain-areas with muscular motion is no new discovery, as modern writers seem to suppose," Wallace wrote in *The Wonderful Century* (1898), "but was known to Dr. Gall himself, although he [lacked] modern appliances for the full experimental demonstration of it." Historians credit Wallace with being right: Phrenology turned out to be a necessary prelude to more sophisticated studies of localized brain function.

Unfortunately, Wallace had a misplaced faith that phrenology would make a dramatic comeback in the 20th century: "Phrenology will assuredly attain general acceptance [as] . . . the true science of mind. Its practical uses in education, in self-discipline, in the reformatory treatment of criminals . . . and the insane, will give it one of the highest places in the hierarchy of the sciences." He believed it was rejected because his scientific colleagues had "narrowminded prejudices" against his favorite ideas: phrenology, spiritualism, and mesmerism. But Wallace fought tooth and nail against widespread adoption of a medical technique he believed "absolutely useless" and a "dangerous delusion"—vaccination against smallpox.

See also BIOLOGICAL DETERMINISM; FITZROY, CAPTAIN ROBERT; SCIOLISM; WALLACE, ALFRED RUSSEL

PHYLA
The Systematist's Nightmare

When the great systematist Carl Linnaeus set up his system of classification in the 18th century, he included all living things in the "Empire Biotes," encompassing the animal and plant "kingdoms." Minerals, although they can grow and self-organize into shapes, occupied a separate, nonliving kingdom of their own. Within the living kingdoms are the phyla—the great groups of organisms with differing basic structures or organization.

Biologists today recognize a five-kingdom system, composed of unicellular and colonial organisms (Monera); protists, which include a quarter million species of protozoans and algae; fungi; and the million and a half plants and animals with which we are more familiar. Most biology students know about the phylum Chordata—the animals with backbones—because it includes ourselves. When we turn to "invertebrate" phyla, however, the ground soon becomes mushy. Tens of thousands of these species simply don't look or act like familiar plants or animals at all. The most contentious organisms to classify have been the Archaea, very ancient microbial life forms often placed in a sixth kingdom all their own.

Faced with the bewildering thousands of microscopic life forms, even the experts admit to their state of perpetual uncertainty and confusion. Some place the unicellular parasites, such as the trypanosomes that carry malaria, into a phylum all their own. Another phylum is represented only by one-celled creatures that digest wood and live in the guts of termites, where they have evolved in a closed universe for eons. Many seemingly unclassifiable microscopic creatures are considered plants by botanists and animals by zoologists. Some of them even have two names—a zoological and a botanical one.

SELECTION MACHINE groups balls by size, showing how some traits can ride "piggy-back" on those being sorted. Smallest balls are black, next smallest gray, while others are white. Although the sieve selects for size only, the balls end up sorted by color as well.

After E. Sober.

Scientists cannot agree on how many phyla exist, how they are related to each other, and which evolved from which. Adding to the confusion are a number of extinct phyla known only from fossils—such as some of the Ediacaran organisms, and the strange Burgess Shale creatures—whose relationship to other phyla and to living groups of organisms is still unknown. Even more puzzling is the question of how the various body plans originally evolved.

See also BURGESS SHALE; CLADISTICS; EDIACARAN FAUNA; EUKARYOTES

PIGGYBACK SORTING
Gene "Hitchhiking"

Sometimes natural selection favors a particular trait, and another, unrelated one gets carried along with it. The trait that comes along for the ride may seem puzzling from the standpoint of evolutionary "fitness"—it may appear to be nonadaptive or even harmful.

Often, genes are passed on as a "package deal," linked together on chromosomes. Natural selection acts on the whole "package," even if it should contain a deleterious gene mutation. The success of such linked genes has been compared to a football game in which the whole team wins the trophy, including the jerk who kept fumbling the ball.

An actual example of gene hitchhiking has been recently studied among a human population on the island of Sardinia. The islanders have a very high proportion of a particular type of color blindness; researchers wondered why it is so common in this somewhat isolated population. Certainly, defective vision confers no advantages according to the theory of natural selection.

Genetic studies showed this kind of color blindness is "linked" with genes that confer a hereditary immunity from malaria, which is the crucial selective factor in leaving more offspring.

PILTDOWN MAN (HOAX)
Famous Fossil Forgery

On December 18, 1912, newspapers throughout the world blared sensational headlines: MISSING LINK FOUND—DARWIN'S THEORY PROVED. The source of all the excitement was a gravel pit at Piltdown, Sussex, in the southern English countryside, where a local amateur archeologist had found "the earliest Englishman."

William Dawson, the discoverer, had been searching for fossils and stone tools for many years. Assisted by workmen at first, and later by the young French Jesuit priest Teilhard de Chardin, he found pieces of a skull (cranium) and jaw, along with remains of ancient mammals and some stone and bone tools. A new species, *Eoanthropus dawsoni*, "Dawson's Dawn Man," was named and authenticated by experts at the British Museum.

Remarkably, the skull appeared entirely human, but the jawbone fragment was apelike. A good deal of interpretation was necessary: The face was missing and so were the parts of the hinges that join the jaw to the skull. While a few scientists questioned whether the skull and jaw really belonged together, most rushed to embrace Piltdown man as a genuine intermediate between humans and apes—the "missing link."

Artists made imaginative reconstructions of his face, and statues of his presumed physique graced museums. In the U.S. there was even a popular comic strip in the Sunday papers called *Peter Piltdown*, precursor of *Alley Oop* and *The Flintstones*.

Most prestigious British anthropologists put their names and reputations on the line in authenticating Piltdown. When Sir Arthur Keith was challenged on his reconstruction, he gave a dramatic demonstration of his prowess. A known skull was smashed into pieces, and he correctly reconstructed its shape and cranial capacity from fragments. Others who championed the authenticity of the "great discovery" included anatomist Arthur Smith Woodward and Sir Ray Lankester, director of the British Museum of Natural History.

Forty years later, the famous bones again made world headlines: PILTDOWN APE-MAN A

FAKE—FOSSIL HOAX MAKES MONKEYS OUT OF SCIENTISTS. Back in 1911, someone had taken a human cranium and planted it at the gravel excavation together with a doctored orangutan jaw. The orang teeth had been filed to make them look more human, and the jaw was deliberately broken at the hinge, to obscure correct identification. All the fragments had been stained brown with potassium bichromate, which made them appear equally old.

Piltdown's authenticity had been repeatedly questioned over the years, but conclusive proof of fraud came in the 1950s, when the British Museum's Kenneth Oakley devised a new method for determining whether ancient bones were of the same age. His radioactive fluorine test proved the skull fragments were many thousands of years older than the jaw. They could not be from the same individual unless, as one scientist put it, "the man died but his jaw lingered on for a few thousand years."

This discrepancy led Oakley to closely reexamine the original material and to make other physical and chemical tests. While the skull was indeed an authentic *Homo sapiens* of ancient vintage, he realized that the jaw belonged to a modern orangutan and had been stained to match the skull. Moreover, he discovered that teeth had been filed to resemble human teeth, and the connections (condyles) between skull and jaw cleverly obliterated.

Critics pounced on Piltdown as an example of the weakness of evolutionary anthropology. Looking back, it appears that British scientists, fed up with news of sensational fossil men found in Germany and France, strongly craved an ancestor of comparable age. Besides, Sir Arthur Keith had theorized that the "big brain" came first in human evolution and was the hallmark of humanity. Piltdown filled the bill. Because of that bias, he had pooh-poohed Raymond Dart's australopithecine discovery in Africa—a creature with an ape-sized brain and "human" teeth and jaw, exactly the opposite of Piltdown.

Successful hoaxes generally share two characteristics: they prop up questionable but cherished beliefs while supporting local pride or patriotism.

In their enthusiasm, few thought it odd that the "earliest man" should have been found about 30 miles from the home of Charles Darwin or that a strange, paddlelike bone implement found in the pit resembled nothing so much as a prehistoric cricket bat—a bit too appropriate an artifact for the "first Englishman."

When the hoax was finally discovered, there were cheers and jeers. Creationists proclaimed all evolutionary science as phony. Anthropologists said the exposure proved their discipline is self-correcting and eventually roots out frauds. Advocates of the African australopithecine fossils felt vindicated in their view that hominids had small canines and walked upright very early on and only developed expanded brains much later.

A member of the British Parliament proposed a vote of "no confidence" in the scientific leadership of the British Museum. The motion failed to carry when another M.P. reminded his colleagues that politicians had "enough skeletons in their own closets."

The Piltdown hoax remains one of the most intriguing mysteries in the history of science. Who was the culprit, and what was his motive? For years, the finger of accusation pointed at William Dawson. As the "discoverer" of more than fifty antiquities, stone tools, and fossils that have since been exposed as fraudulent, Dawson remains the chief and most likely candidate. During the past decade, however, historians of the hoax have implicated others.

A minority view, advanced by anthropologist John Winslow, pins the real-life mystery on Sherlock Holmes's creator, Sir Arthur Conan Doyle, who lived a short distance from the

SKULLDUGGERY behind "Piltdown man," a human skull and ape jaw (below), fooled scientists for decades. Among them was Sir Arthur Keith (bottom, center), who examines a skull while "discoverer" William Dawson stands to his left.

RUSTING SIGN at a closed restaurant near Piltdown features a winking, leering, malevolent skull.

Piltdown site. A fanatic spiritualist, Doyle's motive might have been revenge against the scientists who ridiculed both his beliefs and his credulity in accepting the evidence of professional spirit-mediums. E. Ray Lankester, for instance, who later became director of the British Museum of Natural History, incurred Conan Doyle's lasting resentment as the first scientist to prosecute a "spirit-medium" for fraud in police court. [See SLADE TRIAL.] Winslow has also suggested that perhaps Conan Doyle wanted to demonstrate how easily scientific experts would uncritically accept evidence for their own beliefs, even as they scoffed at the authenticity of his photographs of "spirit-beings."

Other possible candidates have been offered, as have stories of elusive tape-recorded accusations by aged survivors and apocryphal anecdotes that muddy the waters.

Charles Blinderman, in *The Piltdown Inquest* (1987), furnished a detailed review of the full roster of about 10 of the "usual suspects." His tongue-in-cheek conclusion: The fictional "missing link" was planted by a fictional culprit, the Victorian supervillain Professor Moriarty. His motive? To discredit Conan Doyle, the creator of his nemesis, Sherlock Holmes!

See also LANKESTER, E. RAY; SPIRITUALISM; TEILHARD DE CHARDIN, FATHER PIERRE

PITHECANTHROPUS ERECTUS

See BEIJING MAN; *HOMO ERECTUS*

PITT-RIVERS MUSEUM
Evolution of Technology

Lieutenant-General Augustus Henry Lane Fox, a gentleman of good family but modest means, was a leader of the "prehistoric movement" within the Darwinian revolution. He believed that not only the races of man had evolved in "stages," but so had cultures and even tools and artifacts. An avid collector of bows, clubs, and stone tools, he dreamed of creating a museum where they could be exhibited in evolutionary series. It would show, step by step, the development of human technology.

Lane Fox got his opportunity in 1880, when he inherited the 29,000-acre estate of his great uncle, George Pitt, the second Baron Rivers. (Twelve intervening heirs had died through illness or accident.) Convinced he was carrying out the will of God to reveal the laws of evolution, he took the name Pitt-Rivers and began excavating numerous prehistoric sites on the estate.

Within a short time, the immense collection of prehistoric and tribal artifacts of now-Major General Lane Fox Pitt-Rivers had outgrown two small museums. He offered to donate the entire collection to the British Museum—with one catch. The museum must guarantee to keep his artifacts permanently exhibited in the "evolutionary series" he had arranged, according to "laws of technological evolution" he had discovered. Trustees refused to meet that condition, so the major general took his legacy to the world elsewhere.

In 1882, Oxford University agreed to accept the collection on Pitt-Rivers's terms, provided he underwrote construction of a separate building to house it, as an annex to the university museum. He also provided funds for the preparation and preservation of the collection and stipulated that the university establish regular lectures in anthropology.

Pioneer anthropologists E. B. Tylor and Baldwin Spencer helped arrange exhibits in the new Pitt-Rivers Museum, which was later directed by Henry Balfour. Balfour published *The Natural History of the Musical Bow* (1899), in which he chose specimens from the collection to demonstrate the evolution of a weapon into a musical instrument. One tragic result of Pitt-Rivers's obsession with evolutionary classification of tools was what historian George Stocking calls "the paleolithic equation." Stone tools from the ancient past of Europe were exhibited side by side with those of contemporary Tasmanians or Australians. The implication was that because flint axes, knives, and spearheads appeared similar, the modern tribes were "stone age people," biologically equivalent to long-extinct ancestral Europeans. As Stocking wrote in his book *Victorian Anthropology* (1987), "Left behind long since by the ancestors of the Europeans, [Tasmanians] had outlived their time by many thousands of years. . . . Not only did the paleolithic equation help to distance the horror of the Tasmanian's extinction; it seemed even to set the seal of anthropological science upon their fate."

In recent years, uncertainty and debate about the future have surrounded the Pitt-Rivers Museum. Some believe Oxford should have preserved the donor's Victorian "evolutionary" sequences of spears, bows, fire drills, "primitive" musical instruments, and stone axes intact, which was, after all, the pledge it made to Pitt-Rivers. However, a few years after his death, curators rearranged the exhibits and objects by type, destroying most of Pitt-Rivers's evolutionary sequences. The entire museum itself, revisionism and all, is now a Victorian cultural artifact of historical significance.

Others feel the exhibits are misleading, outdated, and racist, and should be rearranged according to geographical or cultural areas, along functionalist principles. The upper floors, now under renovation, will embark on yet another departure from the terms of Pitt-Rivers's bequest; they are being transformed into modern educational galleries. One tongue-in-cheek notice in the Royal Anthropological Institute's newsletter claimed that a group of "radical Museo-Marxists" want the artifacts to be classified "according to their means of production."

See also "VERCORS" (JEAN BRULLER); WELLS, H. G.

PLANET OF THE APES
Endless Ape Epic

May I kiss you?" a grateful Charlton Heston asks the female chimpanzee who saved his life. She offers her cheek, trying to suppress her obvious disgust at such intimate physical contact with a human. "I'm sorry," she says, "it's just that you're so ugly."

A wildly popular series of "Ape Planet" action movies began with a witty satirical novel by French author Pierre Boulle (who had already scored a major movie hit with *Bridge on the River Kwai*). His novella *Planet of the Apes* (1963) was a humorous parable, tied loosely to an action-adventure plot. The screen version shifted its major focus to the violent action, which continued through the numerous lucrative sequels.

Orangs, chimps, and gorillas in Boulle's original satire act not so much like different species as takeoffs on our own society. Chimps see matters clearly, try to act decently, and are flexible and innovative, which often lands them in trouble with the ape establishment. Orthodoxy in thought and policy is maintained by overly dignified orangutans, who serve on all scientific boards and committees. Gorillas are brutal, stupid, and follow orders well; they form a state police or military caste. Among the film's most striking images are the legions of armed, leather-clad gorillas mounted on horseback.

In *Planet*, naked humans are rounded up by the gorillas and caged for medical experimentation. When the astronaut Taylor (Charlton Heston) tries to protest, he finds that no ape will believe humans can speak, since most of the individuals they've seen have degenerated into mute animals.

During a jury trial to determine whether Taylor is capable of intelligence, the three ape judges form a "no-see, no-hear, no-speak" tableau, which may be homage to the Scopes "Monkey Trial" and its continuing aftermath.

In 1968, the first film version of *Planet of the Apes* was released, with a screenplay by Rod Serling (creator of the classic television series *The Twilight Zone*) and Michael Wilson, a screenwriter who had been blacklisted during the McCarthy era. Produced by Arthur Jacobs, the film starred Heston, Roddy McDowell, Kim Hunter, Maurice Evans, James Whitmore, and Linda Harrison.

Special prosthetic makeup for the apes' faces was created by John Chambers, who had originally developed the process to help disfigured war veterans. His masks were so successful (and so expensive) that the molds were used many times over in the sequels, thus amortizing their cost. The designs also became the basis of hundreds of popular toys, models, and other merchandise.

Beneath the Planet of the Apes, the first of four sequels, was released in

HUMAN LAB ANIMALS are captured for study by chimpanzee scientists in the fantasy-adventure *Planet of the Apes*, based on Pierre Boulle's anthropological satire. Charlton Heston and fellow specimens are brought to a lab by gorilla security forces.

1969. Although based on a flimsy premise (a spaceship sent to rescue the astronauts falls into the same time warp), it was one of the most successful of any genre sequels.

Despite puerile plots that are quickly overwhelmed by battle scenes and "action adventure," the popularity of these fantasies marks a change in mass perception of apes. No longer is the movie ape a monster (*King Kong*), a clown (Cheetah, Bonzo), or a "primitive man" (*2001: A Space Odyssey, Greystoke*). The apes hold social roles, but are also individuals—mostly conforming and brutal, but sometimes rebellious and heroic.

See also APES; *HOMO SAPIENS*, CLASSIFICATION OF; *KING KONG*

PLATE TECTONICS
Traveling Continents

As late as 1987, some scientists still doubted whether the "new geology" was here to stay. That Africa or North America moved around, migrating from one part of the world to another, seemed incredible. Yet evidence for the idea of continental drift kept piling up. And then a space-age technique convinced even the skeptics—by actually measuring the speed of plate movement with a "ruler" from beyond the stars.

Earth's crust is a thin outer shell, made up of about 20 tectonic plates. Each plate picks up molten volcanic rock from the Earth's deeper layers, which spews up through rifts on the ocean floor. As a layer of magma cools, its metallic particles align themselves with the Earth's magnetic poles, which change direction a couple of times every million years. Crustal layers, therefore, are made of alternating bands of rock with different magnetic directions, "freezing" a history of their formation. Thickness of the bands may indicate length of intervals or varying speed of Earth's movements.

Another line of evidence comes from the similarity of rocks and fossils that appear in continents that are widely separated today. For example, Edwin Colbert's discoveries of *Lystrosaurus* in South Africa, India, and Antarctica support the idea that these far-flung lands were once joined in a single supercontinent, now known as Gondwanaland.

Dozens of other studies added support to the theory. Then, the U.S. National Aeronautics and Space Administration sponsored a Crustal Dynamics Project that, in 1987, reported it had actually measured the rates at which continents move. Since continental movement is very slow, it was necessary to compare the continents' positions to a point that seemed almost perfectly fixed. James Ryan of Goddard Space Flight Center and his colleagues chose immensely distant quasars that emit radio noise from a far side of the universe. Huge radio antennas were built in Alaska, Hawaii, and Japan; the millisecond differences in their reception of the same signals gave a very precise way of measuring the antennas' distances from one another. After three years of tracking changing reception times, a clear pattern of continental movement emerged: Hawaii and Alaska are moving about 52.3 millimeters closer each year, while Hawaii is moving toward Japan at a rate of about 83 millimeters per year. The Pacific plate is moving much faster than its Atlantic counterpart; North America and Europe are moving apart by 17 millimeters each year.

Collisions between plates are thought to thrust up mountains. Relatively recent ramming of the Indo-Australian plate into Asia, for example, is believed to have created the Himalayan range. When the edge of one plate dives beneath the edge of another in a subduction zone, some crust returns to the mantle.

Earthquakes occur along "fault lines," which are now seen as meetings of plate boundaries. The famous San Andreas fault in California, responsible for the infamous San Francisco quake of 1906, is really a boundary between the North American and Pacific plates. With measurements from satellites and outer-space signals, it is now possible to calculate the precise rate of plate movement—an important step toward predicting the timing and severity of a quake.

Studies of plate tectonics have begun to explain many diverse phenomena, but some basic questions are still wide open: What drives the plates to move? Why does the Earth's magnetic field reverse itself? Do "pulses" of Earth activity cause changes in climate? As always in science, new answers bring with them further questions.

See also CONTINENTAL DRIFT; GONDWANALAND; LAURASIA; WALLACE'S LINE

The speed of Earth's plate movements was measured with a "ruler" from beyond the stars.

PLENITUDE
Biotic Saturation

One of the old, traditional ideas that Darwin inherited from natural theology was the notion of "plenitude," which went along with the Scala Naturae, the Great Chain of Being. In a world created by a beneficent God, every possible niche in nature was filled by a species that "belonged" there. Only 100 years before, belief in plenitude even ruled out the concept of extinction: no gaps or imperfections could exist in the natural order, since God's "ideas" (species) could neither be lost nor destroyed.

When fossil remains of giant sloths and mastodons turned up in 18th-century America, most naturalists assumed that some of the unknown creatures must be still alive in the wilderness. It was Georges Cuvier, the great French paleontologist, who first dared to establish that some species had actually disappeared from the face of the earth forever. However, even after extinction was accepted as a reality, it was believed that the roles occupied by species that had disappeared were replaced by new species, so that the "economy of nature" retained its balance.

Steven M. Stanley, in *The New Evolutionary Timetable* (1981), points out that the continuing influence of "plenitude" on Charles Darwin caused thorny problems with his theory of natural selection. First, how could species vary significantly in a world into which they were packed niche against niche, occupying every possible place in nature? There would be no room for natural selection to operate. And how could any species move into a new niche?

Given his belief in plenitude, Darwin thought perhaps small waves of natural selection might travel through the whole system, causing slow, "insensible" changes over long periods of time in very large populations. Stanley argues it was Darwin's commitment to plenitude that forced him to adopt that model of gradual change, in which nature has no gaps. In fact, recent studies seem to support Stanley's view that "the world is packed loosely enough with species that much variation is tolerated." Out of large worldwide populations, sometimes small local populations diverge rapidly into distinctive new species.

See also DIVINE BENEFICENCE, IDEA OF; GRADUALISM; GREAT CHAIN OF BEING

> "The world is packed loosely enough with species that much variation is tolerated."
>
> —Steven Stanley, 1981

POLYMERASE CHAIN REACTION
Amplifying Small Bits of DNA

Kary Mullis, a young California biochemist working for the Cetus Corporation in 1983, invented a new method for the rapid duplication of genetic material. Within a decade, his revolutionary technique for creating the polymerase chain reaction (PCR) has impacted everything from microbiology research to criminology and even Hollywood science fiction.

PCR capitalizes on DNA's natural method of replication, which copies quantities of itself from tiny samples. In the process, the two long strands of its double helix unwind, and shorter segments (primers) attach themselves to the ends of each strand. Then a specialized protein (polymerase) hitches itself to the ends of each strand and strings together a copy strand for each original segment.

Both halves of the original double-stranded molecule are copied, resulting in two identical molecules. To push the molecules to replicate rapidly, a DNA double helix is first heated until it unwinds. Synthetic primers and then polymerase enzymes are added to the mix, kicking off the rapid replication process. Each time the DNA molecule copies itself, both the original and the dupe are copied again by the polymerase. The quantity of DNA increases exponentially—from two strands to four to eight, up to as many as a million copies within a few hours time.

Many studies involving DNA require fairly large amounts of the molecule. In the early days of forensic workups, for instance, a blood-soaked sheet was needed for DNA identification, while a single hair follicle was useless. PCR changed that. The process not only produces substantial amounts of DNA from a tiny sample, but does it in less than a day. Thanks to Mullis's technique, microscopic bits of genetic material in a fleck of blood at a crime scene can be quickly amplified for comparison or identification. Minuscule skin or mouth scrapings are now routinely used to resolve disputed paternity cases.

In 1993, Mullis received the Nobel Prize in chemistry for his work, but the PCR technique garnered vastly wider fame from the film *Jurassic Park*. Author Michael Crichton based his imaginative scenario on what might happen if scientists were to amplify bits of dinosaur DNA that had been preserved in a fossil mosquito's gut.

Reconstituting live dinosaurs (or other extinct animals) is not possible at our present knowledge and technology. However, PCR has been used to capture other kind of monsters: rapists and murderers who, a few years ago, would have eluded punishment.

See also MULLIS, KARY

POPULATION THINKING
Shift in Biological Concepts

According to Harvard evolutionist and historian of biology Ernst Mayr, Darwinian evolution is based on a major shift of thought from "essentialism" to "population thinking." It is the difference between seeing a species as the "ideal type specimen" once favored by collectors or as the "reproductive population" observed by field naturalists.

The older view of "types" was compatible with divine creation, "archetypes," and a zoology based on study of museum-drawer specimens. But Charles Darwin and Alfred Russel Wallace had deliberately gone beyond being "mere collectors" (though they never forgot its allure) and shifted to trying to understand species in the wild–natural breeding populations. Variability among individuals was the natural condition of a species—the raw material of natural selection—not a "departure" from some ideal individual.

Darwin and Wallace were familiar with the studies of Belgian anthropologist Lambert Quetelet, who in 1842 had shown how normal frequency distribution curves applied to human population measurements. They were also inspired by Thomas Malthus's discussions of population growth and resources. Although the coauthors of natural selection theory were both untalented at mathematics, their shift of focus from studying "types" to understanding variability opened biological problems to mathematical analysis. Darwin's cousin Sir Francis Galton founded the biometrics movement, which eventually resulted in sophisticated studies of population dynamics and population genetics—the foundations of current research into biological evolution.

See also ESSENTIALISM; SPECIES, CONCEPT OF

POSITIVISM
The "Scientific" Religion

Positivist philosophy was a peculiar beast—on the one hand helping define science, and on the other, creating a screwball religion all its own. Its founder, the Frenchman Auguste Comte (1798–1857), and his mentor, Claude-Henri de Saint-Simon (1760–1825), had drunk deeply from the cup of the French Enlightenment philosophers. They believed in the inevitability of human progress and the irresistible advance of scientific knowledge—all tending to an enlightened, benevolent, egalitarian society in the near future. Saint-Simon had written in 1825 that "the golden age, which a blind tradition has hitherto placed in the past, is before us."

Human knowledge, Comte taught, must evolve through three stages. First, it is theological: Workings of nature are considered to be mysterious, the unpredictable will of God, the result of intervention by a First Cause. In the next stage of human understanding, philosophers look for nature's regularities or "second causes," although they may still cling to the idea of an underlying supernatural creator. Finally, Comte proclaimed people will refuse to speculate about any ultimate causes, which are incapable of proof. They should be concerned only with "positive facts" and "proven laws," which will be discerned in all phenomena and can form the only "positive" basis for human action and belief. In time, Comte believed, it would be possible to discern the "laws" of human social behavior, and sociology would acquire the precision of physics or chemistry. Social sciences would then become the crowning glory of all sciences and the best guide for human destiny. (Despite a modest shortfall on these aims, Comte did help to found and establish sociology.)

English logician and philosopher John Stuart Mill became interested, and wrote an influential book, *Auguste Comte and Positivism* (1865), which raised Comte's stock considerably among English intellectuals. As positivism gained ground, however, adherents of this "rational," "scientific" philosophy attempted to turn traditional religion inside out. Led and inspired by a kind of scientific priesthood, positivists were going to remake society from top to bottom. Humanity itself, as the source of truth, beauty, and knowledge, was to become the object of worship and veneration. Members would wear little statuettes of a "Goddess of Humanity" around their necks in place of the Virgin—familiar trappings "to ease the transition," as they put it.

Positivist leaders such as Richard Congreve and Frederic Harrison (both former Church of England theologians) became obsessed with detailing new rational rituals and secular sacraments. Master timetables were drawn up for the scientific improvement of society. In France, under Comte himself, the plan was to replace Napoleon with a positivist triumvirate, which would eventually empower a kind of positivist pope—all in the name of devotion to Humanity.

Thomas Henry Huxley, who had at first applauded positivist scientific values, quickly distanced himself from any attempts at a secular religion. With characteristic wit, he noted the positivists were simply "re-inventing Catholicism, minus Christianity."

Despite its initial splash, positivism lasted only about 25 years and never became a widely popular movement. Nevertheless, some of Comte's ideas on science and sociology continue to be taught in universities today—usually with no reference whatsoever to his role as high priest of the New Religion of Humanity.

POZZUOLI, PILLARS OF
Lyell's Icons of Geology

At Pozzuoli, Naples, in a little parklike setting near the Bay of Baiae, three 40-foot-high marble pillars erected 2,000 years ago remain standing in a pool of water. Famous as the ruins of the so-called Temple of Serapis, they were actually part of an entranceway to an ancient Roman marketplace.

The pillars would seem to be a minor scrap of architecture in this city of magnificent monumental ruins, yet they are a must-see mecca for every geologist and evolutionary biologist who visits Naples, for they hold a special place in the history of science. Not only did these pillars provide Charles Lyell with dramatic evidence for his new approach to geology, but their publication in his *Principles of Geology* (1830) turned them into icons. (*Principles* went through eleven editions during his lifetime.)

Lyell lamented that there was "an extreme reluctance [among other geologists] to admit that the land, rather than the sea, is also subject alternately to rise and fall." Using the pillars as touchstones, he argued that geological elevations and subsidences—without earthquakes—are common in some localities.

When Lyell first saw the pillars of Pozzuoli in 1828, during a tour of Europe, he realized that he could read them as "tide gauges" that recorded changes of shoreline level from the early Christian era to the present. Evidence of their multiple submersions in the sea is clearly marked by ringed zones on the pillars, made by small marine creatures.

There are thousands of perforations made by burrowing clams clustered in one zone, which is now near the top of the pillars, while another ring bears traces of sea worm activities. The pillars show that the level of the coast changed at least twice during the Christian

Led by a kind of scientific priesthood, Positivists intended to remake society from top to bottom.

THE ICONIC PILLARS of Pozzuoli appeared as the frontispiece of Charles Lyell's *Principles of Geology.* Bands of encrustations mark the rise and fall of the Neapolitan shoreline over 2,000 years. The pillars also appear on the prestigious Lyell Medal in geology (top).

era, each time by more than 20 feet. Following Lyell, subsequent visitors found evidence of several additional, smaller fluctuations on the pillars. Other geological evidence in the area showed that it was the land and not the sea that had risen and sunk.

Since Lyell thought there was abundant evidence throughout the world that land is as likely to rise and fall as ocean levels, he considered it unfortunate that poets had established rocks as symbols of the eternal and the sea as constantly changing. More important, his beloved pillars upheld the doctrine of gradualism—that modern causes, operating within presently observable rates, can explain the history of the Earth without the need for imagining a history of sudden, catastrophic earth changes. As Stephen Jay Gould wrote of them, "Lyell used these three pillars as a . . . uniformitarian antidote to the image of fiery [nearby] Vesuvius as a symbol for catastrophic global endings."

See also GRADUALISM; LYELL, SIR CHARLES; STEADY-STATE EARTH; UNIFORMITARIANISM

PRE-ADAMITE CREATIONS
Before the "Present World"

In the 18th and early 19th centuries, an immense body of detailed information accumulated on stratigraphy, glaciation, and fossil organisms. Such dramatic large vertebrates as plesiosaurs, pterodactyls, giant sloths, and seagoing ichthyosaurs became well known. When they could not reconcile new fossil discoveries to scripture, geologists assigned them to "previous creations" outside the scope of the Bible. All animals of the "present world," they believed, were created as fixed species at the same time as Adam and Eve. The others were "pre-Adamite." Scriptural geologists saw no conflict in accepting such creations as having come before the one described in Genesis. Professor D. Lardner of University College London, writing in 1856, claimed there is "no discordance between Scriptural history and geological discovery, [only] the most remarkable and satisfactory accordance with natural phenomena."

At least one artist of the day slyly wondered if the present creation would be succeeded by still another, in which post-Adamite giant reptiles would reign again. A popular cartoon shows a Professor Ichthyosaurus lecturing his reptilian students on a human skull. "With such small teeth and weak jaws," says the crocodilian-snouted professor, "it is altogether wonderful that this [human] creature was able to procure food."

PRESTWICH, SIR JOSEPH (1812–1896)
Confirmed Prehistoric Man

Clergyman-geologist Sir Joseph Prestwich did not discover any stone tools or fossil men and never publicly suggested mankind was older than church teachings acknowledged. Yet he is credited with establishing the fact that man coexisted with long-extinct mammals and that humans lived on Earth much earlier than anyone had previously suspected.

When Jacques Boucher de Perthes of Abbeville, France, announced in 1846 that he had excavated ancient man-made flint implements sealed with undisturbed bones of rhinos and mammoths in local gravel beds, no one paid him the slightest attention. Three years later he found many more tools with more remains of extinct mammals, and again he was ignored, even ridiculed. Only one French scientist came to see for himself; he became convinced, and then received the same treatment from his colleagues as Boucher de Perthes.

In 1858, Darwin's friend Dr. Hugh Falconer saw Boucher de Perthes's flint collections and wrote to Sir Joseph Prestwich, asking him to visit Abbeville. A respected churchman, Prestwich was renowned as a highly competent geologist who had done careful studies of English gravels. He also was known as a scrupulously honest observer who avoided conjecture, speculation, and theological controversy.

Prodded by Falconer, as well as by other gentlemen archeologists, Prestwich went to Abbeville and Amiens in 1859 to see the gravel beds. After examining the evidence, he said he was convinced (1) that the flints showed unmistakable human workmanship, (2) that they

Many doubted the antiquity of man until Joseph Prestwich—who was also a priest—confirmed seeing stone tools embedded with mammoth bones.

FOSSIL ANIMALS THAT DON'T FIT in the Bible were assigned to a previous creation, before Adam and Eve. This magnificent illustration is from *Pre-Adamite Man; or, The Story of Our Old Planet and Its Inhabitants, Told by Scripture and Science* by Isabella Duncan, 1860.

occurred in undisturbed sediments, (3) that they were associated with remains of extinct mammals, and (4) that the time period was late in Earth history, but before the formation of the present land surface.

Prestwich reported his observations, but made no attempt to account for them or explain their significance. As a devout Anglican, he abstained from drawing the obvious conclusion about the church's untenably short timetable for the appearance of humans on earth. He gave a truthful account of what he saw and left the inferences to others.

Nevertheless, only after Prestwich verified Boucher de Perthes's discoveries did anyone in England, France, or the rest of the world take them seriously. The impact of Prestwich's immense credibility as a clergyman-geologist illustrates the complex relationship that existed between science and religion in the 19th century. Science owes much to churchmen who developed geology, natural history, and genetics. Frequently, commitment to both religion and science coexisted in the same individuals.

See also BOUCHER DE PERTHES, JACQUES; LUBBOCK, SIR JOHN

PRIMATES
Lemurs, Monkeys, Apes, and Humans

Within the class of hairy, warm-blooded, milk-nurturing creatures known as mammals is the order Primates (Latin, pronounced pri-MATE-eez). So named by Linnaeus and meaning "the highest" or "first rank," it contains the creatures most closely resembling humans, which were considered "the pinnacle of creation." (The great classifier ranked all other mammals as "secondates," and the rest of the animal kingdom as "tertiates.")

There are 180 living species of primates, many of them in danger of extinction. South American monkeys, for instance, have evolved in their rain forests for 40 million years, but have lost vast ranges during the last 40 years—one-millionth of their history.

New World (American) monkeys evolved independently of Old World primates from very distant common ancestors. Old World monkeys, apes, and humans form a superfamily within the primates. With the exception of humans, most are confined to Asia and Africa. (The last "European" macaque monkeys still live on the Rock of Gibraltar.)

In size, primates range from two-ounce pygmy marmosets to 500-pound gorillas. Some are solitary species, while others are social; some are part-time meat-eaters, others full-time leaf-eaters; some are monogamous, others promiscuous; and their social organization

PRIMATE PERFORMERS regaled crowds by appearing in human clothes, skating, and bicycling. This late-19th-century poster claims that "Count," a macaque, is "living proof of the Darwin theory."

ranges from male-dominated "harems" to matriarchal alliances. Several primate species never leave the trees; some spend their whole lives on open plains, rocky cliffs, snowy mountains, or among ruins of ancient stone temples. The order Primates contains some of the shyest, most inoffensive creatures on Earth; it has also produced the one species that is alone capable of destroying all the rest.

See also BABOONS; BONOBOS; CHIMPANZEES; GORILLAS; ORANGUTANS

PROGRESSIONISM
Pre-Evolutionary Geological Idea

As geologists of the early 19th century dug more fossil animals out of the rocks, they began to notice a pattern in the strata or layers. At the bottom were mostly invertebrates, above them a gap, then mostly fish, a gap, then amphibians, reptiles, and finally mammals—always in that order. In their system, the idea of evolution did not yet exist, so they interpreted the series of animal groups (faunal succession) as a creation, followed by destruction, followed by a "higher" creation, again destruction, and so on.

Although the great French paleontologist Georges Cuvier had sidestepped the issue, some of his followers insisted that a brand-new creation had occurred after each "catastrophe." As world conditions changed, there was a series of separate creations, each one an improvement on the last—an idea known as "progressionism." It was different from later ideas of evolutionary "progress," because it did not maintain that all life through the ages is related and connected.

Most textbooks credit the geologist Charles Lyell, Darwin's mentor, with inventing a simple concept of "uniformitarianism," which Darwin then applied to living things. Actually, the history of that idea is much more complex. Lyell believed in recurring cycles of creation and destruction. In his view, after a series of creations the whole process was liable to repeat itself over again. For most of his life, he strongly resisted the idea of a single, unique history for all living things.

Louis Agassiz, the Harvard paleontologist who worked out the sequence of glaciers, staunchly opposed Darwinian evolution to the last. Like Cuvier, Agassiz taught a brand of progressionism that assumed each new creation reflected God's current conception, a step-by-step revision of all life.

"It did not occur to him what a blasphemy this interpretation really was," commented Harvard evolutionist and historian Ernst Mayr in *The Growth of Biological Thought* (1982). "It insinuated that God, time after time, had created an imperfect world, and that he completely destroyed it in order to do a better job the next time"—hardly a flattering picture of an omniscient Creator.

See also AGASSIZ, LOUIS; CATASTROPHISM; LYELL, SIR CHARLES; UNIFORMITARIANISM

Agassiz saw a progression of separate creations and destructions; God made and destroyed worlds each time to do a better job.

PUNCTUATED EQUILIBRIUM
Episodic Evolution

Ever since Charles Darwin's day, the fossil record posed a difficulty for evolutionists—and an arguing point for creationists—because it did not appear to confirm his notion of a slow, uniform development of species. Instead, some fossil organisms seem to persist through millions of years relatively unchanged, and then undergo a relatively rapid spurt of change. Sometimes (as with the dinosaurs) an entire group dies out, and is followed by the comparatively rapid appearance of many new forms (as with radiation of the mammals).

Darwin thought the rarity of complete evolutionary sequences could be explained because the fossil record was sketchy and incomplete. Conditions for preservation are so rare, he argued, that the rocks present only a fragmentary sampling of gradual transitions. Most of the "in-between" fossils are lost, giving the appearance of a jumpy picture. Antievolutionists interpreted the apparent "jumps" to mean that some species had perished in the Flood, or that there had been successive creations and destructions.

A century later, a great deal more is known about the fossil record, and its "incomplete-

The rocks reflect a real pattern: little change for long periods, followed by comparatively short spurts of rapid evolution.

ness" no longer seems convincing. In 1972, paleontologists Niles Eldredge and Stephen Jay Gould argued that the pattern is no "artifact" of chancy preservation, but that the rocks reflect a real pattern in the history of life: little change in species for long periods, followed by extinctions, followed by rapid change or branching radiations. They called that pattern of starts and stasis "punctuated equilibrium" (sometimes known as "punk eek").

Darwin himself was not always insistent on continuous gradualism. In the fifth edition of *Origin of Species,* he wrote that "the periods during which species have undergone modification, though long as measured in years, have probably been short in comparison with the periods during which they retain the same form." Thus punctuationalism in general is consistent with Darwin's conception of evolution, and with the natural selection of William Charles Wells, Patrick Matthew, and Alfred Russel Wallace.

But what known processes could produce such rapid divergence of species? Ernst Mayr of Harvard, one of the founders of the Synthetic Theory, had proposed a plausible mechanism in *Systematics and the Origin of Species* (1942). Mayr suggested that fairly rapid speciation could occur in small, isolated populations. Cut off from the larger gene pool by geographic barriers, a small sample of variation would be amplified by selection—the "founder effect." For years paleontologists ignored this effect as a special case of evolutionary process, rather than a major explanatory principle, but punctuationalism gave it new currency.

Over time, the real test of scientific theory is whether it explains previously puzzling data and leads to new discoveries. By those criteria, punctuated equilibrium is holding its own.

See also EXTINCTION; GENETIC DRIFT; GOULD, STEPHEN JAY; MAYR, ERNST

PURE LINE RESEARCH
Limits of Variability

Alfred Russel Wallace and Charles Darwin had insisted that through gradual, continuous change, species could (in Wallace's phrase) "depart indefinitely from the original type." Around 1900 came the first direct test of that proposition: the "pure line research" of Wilhelm Ludwig Johannsen (1857–1927).

What would happen, Johannsen wondered, if the largest members of a population were always bred with the largest, and the smallest with the smallest? How big or how small would they continue to get after a few generations? Would they "depart indefinitely" from the original type, or were there built-in limits and constraints?

Experimenting on self-fertilizing beans, Johannsen selected and bred the extremes in sizes over several generations. But instead of a steady, continuous growth or shrinkage, as Darwin's theory seemed to predict, he produced two stabilized populations (or "pure lines") of large and small beans. After a few generations, they had reached a specific size and remained there, unable to vary further in either direction despite continued selection.

Johannsen's work stimulated others to conduct similar experiments. One of the earliest was Herbert Spencer Jennings (1868–1947) of the Harvard Museum. He selected for size in *Paramecium* and found that after a few generations selection had no effect. Even after hundreds of generations, his pure lines remained constrained within fixed limits, "as unyielding as iron." One simply cannot breed a paramecium the size of a baseball.

Another pioneer in pure line research was Raymond Pearl (1879–1940), who experimented with chickens at the Maine Agricultural Experiment Station. Pearl took up a problem dear to farmers: Could you evolve a hen that lays eggs all day long?

He bred some super-layers, but an absolute limit was soon reached. Hens could lay only so many eggs, above which no amount of selective breeding could increase the quantity. In fact, Pearl found that egg production might actually be increased by relaxing selection—by breeding from "lower than maximum" producers.

Johannsen and other pure line researchers, who once seemed the nemesis of evolutionary theory, in fact ultimately rescued it from speculative fantasies. Johannsen's beans, Jenning's paramecia, and Pearl's hens established a new research tradition. Basic questions about selection and variation had been brought from the field into the laboratory.

See also GRADUALISM

One simply cannot breed a paramecium the size of a baseball.

QUEST FOR FIRE (1981)
Ultimate Hollywood Hominids

For 60 years, most Hollywood "caveman" films belonged to a sleazy genre—curvy starlets wearing scanty skins, and monosyllabic musclemen beaning each other with papier-mâché boulders on backlot sets. Sometimes rubber dinosaurs terrorized the tribe, despite science's conclusion that giant reptiles had vanished millions of years before humans appeared on Earth. Nevertheless, there has always been an audience hungry for images of how our ancestors might have lived.

In 1981, director Jean-Jacques Annaud completed *Quest for Fire*, a conscientious high-budget film about early man, against which all future efforts will be measured. An ambitious re-creation of life in the Pleistocene era, it was shot in harsh wilderness locations, featured real animals, and attempted to depict Ice Age language and culture as plausibly as possible.

Four types of hominids are shown coexisting in the same time and place. Three protagonists (played by Gary Schwartz, Nameer El-Kadi, and Franck-Olivier Bonnet) are men of the Ulam tribe—a wandering band of *Homo sapiens* who wear animal skins, take shelter in caves, and eat almost anything but human flesh. They are depicted nonchalantly munching on insects and leaves, but really get excited (and literally drool) when they see meat on the hoof.

While the Ulams know how to use fire for warmth and cooking, they do not know how to create it. If their carefully guarded fire goes out, they must find a smoldering tree that was struck by lightning or steal fire from another tribe. (There was no possibility in the film that they might buy or barter it.)

A second tribe, the Wagabou, use clubs made from bones. Covered with thick body hair, they are a different species—aggressive hunters who prey on weaker tribes. Near the start of the film, they stage a brutal dawn raid on the Ulams' cave shelter—bashing, raping, killing.

The Wagabou seem clearly to be based on the early "killer ape" reconstructions of Raymond Dart. They represent a more "primitive" type of hominid, something like what *Australopithecus* was imagined to be by its discoverers, Raymond Dart and

SHAGGY ANCESTORS roam a vast wilderness landscape in Jean-Jacques Annaud's *Quest for Fire*, a $20 million attempt to visualize the life of early humans.

Robert Broom, and by the dramatist Robert Ardrey in his bestseller *African Genesis* (1961). Ardrey described our common ancestor as a murderous man-ape born with a weapon in his hand and the mark of Cain on his brow. The Wagabou in Annaud's film are "semi-cannibalistic," resorting to human fare only when game is scarce.

A third group, the Kzamm, are fierce, cannibalistic Neanderthals. Human flesh is their preferred food, and it is from their camp that our intrepid heroes must steal their fire. At first, the hungry Ulams nibble at the remains of a Kzamm meal, but when they realize they are gnawing on human bones they recoil in disgust and horror.

While stealthily raiding the cannibals' camp, they rescue a young woman (Rae Dawn Chong) who was slated to be barbecued. She joins the Ulams and ultimately becomes their savior when they reach the land of her people, the Ivakas. Her tribe lives in settled villages of mud and straw huts and has mastered the art of making fire with a friction device. Ivakas wear little clothing, but paint their bodies with blue clay and wear reed masks. Throwing-sticks give their arms greater leverage and their spears longer flights; they also use animal skin containers. In general technology and appearence, the Ivakas seem to be modeled on Australian Aborigines, the "mud people" of New Guinea, and the Noubas of Africa. The Ulam leader takes the Ivaka girl as his mate, and they return to the Ulam, where she teaches the pathetic, shivering band how to make fire—causing them to erupt in primal screams of joy.

Scorning artificial studio sets, producer Michael Gruskoff shot the film in wilderness locations: the Badlands of Alberta, Bruce Peninsula of Oregon, Tsavo National Park in Kenya, and the Cairngorms of the Scottish Highlands. Masks and body costumes for the different characters were elaborate and expensive. Even more difficult was the task of "making up" the live animals to look like their ancient counterparts. Lions were fitted with plastic tusks to resemble saber-tooth cats, and a herd of Indian elephants was transported to Scotland to play mammoths. Each elephant was fitted with shoulder humps, curved tusks, and hundreds of pounds of wooly hair—truly a mammoth makeup job! (That was probably the last time lions and elephants will be subjected to movie makeup. With the advent of computer-generated imagery [CGI], caveman movies now rely on digital animals [e.g., *10,000 B.C.*, released in 2008].)

To create a credible communications system for the early men, the filmmakers hired as consultants zoologist-behavioral theorist Desmond Morris (author of *The Naked Ape*, 1967) and novelist-linguist Anthony Burgess (author of *A Clockwork Orange*, 1962). Morris and Burgess, in a unique collaboration, worked out a "primitive" language, which combined words, gestures, and primate communication signals.

Burgess developed his language for early man from "Indo-European," a supposedly ancestral language that linguists reconstructed by comparing words in various related language groups. For "fire," one of the most crucial words in the film, Burgess uses *atra,* which is tied to several related concepts in English, including "art," from the old meaning of scientific skill, as in making fire; and "hearth," the core of the home or camp, related in turn to "heart," the central, warming principle. For "animal," Indo-European *tir* is used, which became *Tier* in German and "deer" in English.

Morris fit gestures to Burgess's words, borrowing from tribal people's mimicry of animals and current field observations of living primates. His early men use monkey-ape signals for dominance (staring) and submission (looking away), lip-smacking and teeth-chattering, and practice social grooming by picking bugs from each other's hair.

If *Quest for Fire* ultimately does not present a true vision of our ancestors, it certainly offers a good picture to future historians of the state of late-20th-century theories about them.

Interestingly, anthropologists dismissed Annaud's vision of three different species of humans coexisting in time and space as an extravagant fantasy. Now, however—almost three decades since the film was made—paleoanthropologists have solid evidence that *Homo sapiens,* Neanderthals, *Homo erectus,* and possibly other hominins as well may indeed have frequently encountered one another.

See also DART, RAYMOND ARTHUR; ORIGIN MYTHS (IN SCIENCE)

RACE
Geographic Variability

In zoology, a "race" commonly means a variety or subspecies—a partially isolated breeding population that shows some genetic differences from closely related populations yet is still capable of interbreeding with the others.

When dealing with local races of differently pigmented frogs, no zoologist is inclined to praise the intellectual superiority of a "two-spotted" over the "three-spotted" population. In describing their fellow humans, however, scientists have historically propounded the most bizarre theories, based mainly on differences in culture, social traditions, and their own ideals of beauty or pigmentation, which they have confused with biology. Some anthropologists once considered various tribal peoples on a par with separate species; others confidently categorized particular "races" with such pseudoevolutionary terms as "primitive," "advanced," "degenerate," or "superior." While it may be painful for anthropologists to admit, they caused a lot of damage to many people, abetting racism by embracing crackpot theories purporting to be scientific.

Outmoded and discredited 19th-century evolutionary ideas about "race" continue to be tossed about. The fact is that scientists were never able to find the "pure races" they sought anywhere among human subpopulations. Anthropologists long ago gave up trying to make definitive classifications of "racial types," because it simply cannot be done. (Contentious learned professors "identified" between four and 400 races before they eventually abandoned the exercise as futile.) Human variability is fluid and complex, and its evolutionary significance remains largely unknown.

It was only around 1900 that anthropologists established the crucial distinction between a population's biological inheritance and its cultural tradition. For instance, "black music" is a rich African cultural tradition that must be learned by each generation, not a genetic ability that comes along with skin pigment. English is not a "white" language; anyone of any color can learn to speak it. A European child raised among East African Kikuyus, like the late paleoanthropologist Louis Leakey, will speak, think, and even dream in Kikuyu.

Charles Darwin thought human variability showed no discernible adaptive advantages, and so explained diverse skin colors and hair textures by his theory of "sexual selection." Local ideals of beauty, expressed in selection of mates, he believed, led to different appearances among diverse populations.

Over the past half-century, the traditional idea of "races of man" sputtered its last gasp. Recent genetic studies have shown that there is more variation within geographic populations than there is between them. Also, there are genetic characters (blood proteins, for instance) that vary independently of so-called racial groups. When plotted on maps, their distribution patterns (clines) cut right across populations with different skin colors or hair textures. Yet the frequencies of these hidden, biochemical traits are just as valid markers of gene flow as skin pigment or hair form.

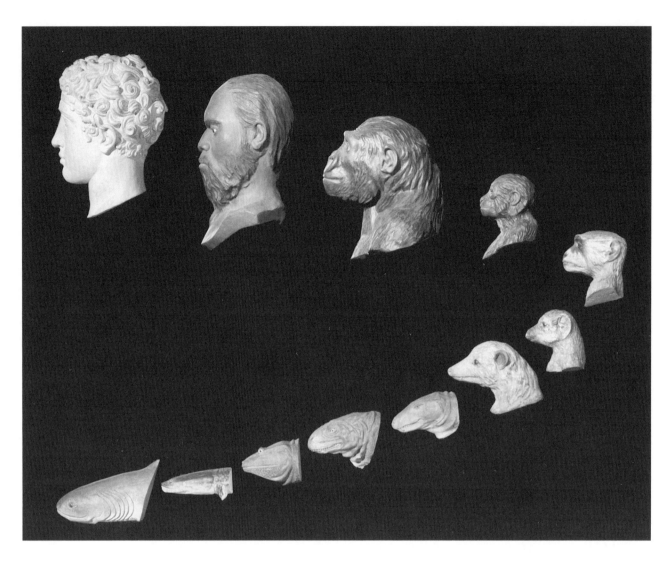

THIS FALLACIOUS LADDER at the American Museum of Natural History in the 1930s arranged living species in an "ascending" series. Widespread racism resulted in placing a black man "one rung below" a European—who was represented by a Greek god! (By the 1960s, the outmoded embarrassment was trashed.)

"There is as much variability within one small tribal group in New Guinea as there is between most of humankind."

—Richard Lewontin

Recent studies of mitochondrial DNA, traced through generations of a worldwide sample of females, indicate that the divergence and spread of human populations from Africa was comparatively recent, occurring just a few hundred thousand years ago—an eyeblink in evolutionary time. Human subpopulations have simply not been genetically isolated long enough to evolve very significant differences. Variations such as skin color or hair texture loom large because of our own social conditioning; biologically, they are insignificant.

Darwin and Huxley were "racists" by today's standards, although both were fervent abolitionists who were outspoken about the need to end slavery. They were raised to believe that light-skinned European males were the lords of the Earth, which easily translated into the "highest products of evolution." Women, whatever their color, were considered somewhat inferior beings, yet capable in their own limited spheres. Darwin, however, was impressed and puzzled by the educated, "Christianized" Fuegian Indians aboard his ship and by how different they had become from their rough tribal relatives.

Among evolutionists, the freest from racism was Alfred Russel Wallace, who neither carried a gun nor locked his doors during years among Amazon Indians, Malays, and Asian tribes. Wallace had developed "a high opinion of their character and morals."

Ideologists Comte Joseph de Gobineau in France and Houston Stewart Chamberlain in England, evolutionists Ludwig Woltmann and Ernst Haeckel in Germany, and paleontologist Henry Fairfield Osborn in America—all were entirely agreed on their own evolutionary superiority to all other peoples on Earth. They influenced immigration policies in America and lent respectability to genocidal activities in Germany.

Tribal peoples, to the Victorian English, were like the ancestors of Europeans of thousands of years ago. Since some used stone rather than metal for their tools, they were described as "still living in the Stone Age"—the so-called Paleolithic Equation, which equated living tribal peoples with prehistoric humans. Such thinking allowed the English to exterminate Australian Aborigines and Tasmanians as if they were subhumans.

One early, persistent controversy was whether all races were minor varieties within a single species (monogeny, or "single origin" theory) or distinct kinds of humans amounting to separate species (polygeny, or "multiple origin" theory). The nonracist view, holding for the unity of mankind, came from religious fundamentalists, who traced all humans back to the single creation of Adam and Eve. Early evolutionists, on the other hand, saw nonwhites as "primitive" peoples who were late to evolve or geographically specialized.

As late as the 1960s, the respected American physical anthropologist Carleton Coon vastly overemphasized group differences, which he confused with cultural behaviors, and imagined some kind of inevitability for "pre-human" populations to evolve into *Homo sapiens* wherever they might be. Can one similarly imagine "pre-moose" hoofed animals dispersing throughout the world, then each local population evolving to "cross the threshold of mooseness" at different times?

Harvard geneticist Richard Lewontin has demonstrated that there is more variability between individuals in a given local population than between the averages of widely separated populations. For instance, there is much more variability between black Africans from one end of the continent to the other, than between some kind of "average" African and "average" European. According to Lewontin's calculations, if most of the world's humans were wiped out, and the only surviving group was a thousand tribesmen in a remote mountain forest of New Guinea, that small, isolated population would contain 99 percent of the human genes and variations that exist.

See also COMPARATIVE METHOD; *DESCENT OF MAN*; EUGENICS; PITT-RIVERS MUSEUM

EXTREMES OF HUMAN VARIABILITY are exemplified by Bao Xi Shun and He Pingping, the world's tallest and shortest men, who met in 2007. Both were born in Inner Mongolia. Associated Press.

RAIN FOREST CRISIS
Contemporary Mass Extinctions

Tropical rain forests, the evolutionary wonderlands that so fascinated Charles Darwin and Alfred Russel Wallace, are being destroyed by man at the rate of 60 acres a minute. By the end of the 21st century, unless drastic measures are taken to reduce the damage, they will be almost gone—and with them many irreplaceable pages of the world's evolutionary history will disappear forever.

"Among the scenes which are deeply impressed on my mind," Darwin wrote as a young naturalist, "none exceed in sublimity the primeval forests undefaced by the hand of man. . . . No one can stand in these solitudes unmoved." In his *Journal of the Voyage of HMS Beagle* (1845), he recorded his delight in exploring a rain forest in 1832:

> Delight itself, however, is a weak term to express the feelings of a naturalist who, for the first time, has been wandering by himself in a Brazilian forest. . . . The elegance of the grasses, the novelty of the parasitical plants, the beauty of the flowers, the glossy green of the foliage . . . brings with it a deeper pleasure. . . . If the eye was turned from the world of foliage above, to the ground beneath, it was attracted to the [thick beds of ferns and mimosae] . . . wonder, astonishment, and devotion . . . fill and elevate the mind. . . . In these fertile climates, teeming with life, the attractions are so numerous, that [I was] scarcely able to walk at all.

Teeming with life, indeed. Within 25 acres of Bornean rain forest, botanists recently identified 700 different tree species—as many as exist in all of North America. One single tree studied in Peru was home to 43 different species of ants—more than are distributed throughout the British Isles. It took 60 million years for the present diversity of species to accumulate in these forests, and the various insects, frogs, snakes, and plants have evolved complex webs of interdependent relationships.

For example, certain fig trees have coevolved with a species of wasp that matures in its fruits and also pollinates the plant. Some species of ants live only in acacias; in return for food and shelter, they clear the trees of other insects and debris.

Great amounts of oxygen are generated by rain forests, and their vegetation holds and recycles vast quantities of water and nutrients. These forests also play a major role in creating and maintaining weather conditions for thousands of miles.

Oddly enough, the soil in which rain forests grow is neither thick nor rich. Most of the forest's valuable nutrients are trapped in the cycle of life and decay that goes on from just inches above the surface of the forest floor to the treetops or canopy area. When forests are leveled and cleared for farming, the soil itself is very poor, capable of supporting crops for only a few years. Rain then washes away the shallow topsoil, and erosion and floods follow.

Within the past 40 years, rain forests in Brazil, Indonesia, Central America, and Hawaii have shrunk from 5 million to 3.5 million square miles. More than 25,000 square miles are lost each year. If these forests disappear, as they will if destruction continues unchecked, thousands of species of plants and animals will disappear forever before they have even been discovered or studied. This could have dire consequences, for a variety of reasons:

1. Loss of species causes unpredictable disruptions in recycling of nutrients and production of clean, fresh air and water and triggers an increase in crop pests and disease-carrying insects.

2. Loss of such unglamorous species as soil bacteria and beneficial insects that are ecologically vital in maintaining the gas and nutrient cycles may affect climate and quality of life worldwide.

3. Rain forests are storage banks of genetic diversity or variability. One hundred acres of tropical forest contain more species than 1,000 square miles of Maine woods.

4. Each species, as the product of a long evolution, carries a special genetic code unique to its population. When a species becomes extinct, its genes are lost. And in this age of genetic engineering, such genes may be of crucial importance to humankind for agriculture and medicine.

5. When forests are burned, the released carbon builds up gases that tend to heat up the atmosphere. This "greenhouse effect" could cause a warming of temperate regions and accelerate the melting of polar ice, which would raise sea levels and submerge coastal areas.

6. Many tropical plants and animals could be an immediate source of foods or medicine if they were studied rather than destroyed. Coffee, quinine, ipecac, and reserpine have come from tropical rain forest plants. The Madagascar periwinkle, a rain forest flower, produces two drugs (vincristine and vinblastine) that have recently proved highly successful in treating Hodgkin's disease and leukemia.

Perhaps the greatest loss of all is knowledge. Tropical forests, which occupy only 7 percent of the planet's surface, may contain as many as half of all its 30 million life forms. Delicate systems of plants and animals that have evolved over millions of years can be wiped out overnight. It is a disaster comparable to the burning of a vast, rare library or the smashing of a complex mechanism before we even know what its parts are or how it works.

Local people usually believe they will reap large profits from the sale of timber and clearing for agriculture, but the short-term profits often pale beside long-term destruction by flooding and erosion. Leaders in destroying their forests are Brazil, India, Indonesia, Myanmar (Burma), Thailand, Vietnam, and the Philippines.

Edward O. Wilson, the Harvard sociobiologist and expert on social insects, says the "current reduction of diversity seems destined to approach that of the great natural catastrophes at the end of the Paleozoic and Mesozoic eras [the extinction of the dinosaurs]. In other words, the most extreme for 65 million years."

See also COEVOLUTION; ECOLOGY; EXTINCTION; GAIA HYPOTHESIS

(CHIEF) RED CLOUD (c. 1830–1880)
Ally of Dinosaur Hunters

Nineteenth-century fossil hunter Othniel C. Marsh and his arch-rival, Edward Drinker Cope, viewed the "Wild West" as the world's greatest graveyard of evolution. To those dedicated dinosaur diggers, the bloody war between whites and Native Americans was merely an annoyance to fieldwork. Marsh decided to negotiate his own separate peace with the Sioux Nation. The outcome was a museum full of dinosaurs and a major public scandal, the "Red Cloud Affair," which rocked the presidency of Ulysses S. Grant.

Red Cloud, an Oglala Sioux chief, led his warriors in scores of successful battles against the U.S. Army, until the government finally offered to trade land concessions for peace. According to a treaty of 1868, all of western South Dakota and the sacred Black Hills were reserved to the Sioux. But the chief couldn't know how badly white men would soon want what was under the lands: gold and dinosaur bones.

Red Cloud kept his word, but the U.S. government did not. Just three years after the treaty was signed, gold was discovered in the Black Hills. In 1874, General George Armstrong Custer led 1,000 men into the area on an "exploring expedition." Army forces protected the invading horde of gold-hungry prospectors instead of the Sioux, who had laid down their weapons.

Native Americans in the Badlands, near the Black Hills, were in a justifiably ugly mood. The Interior Department's "Red Cloud Agency," which administered the territory, was the first target of their discontent; they hacked the agency's flagpole to splinters.

Ignoring official warnings, Yale professor Othniel C. March entered the area with his field crew. In the midst of 12,000 Native Americans ready for battle, the paleontologist set up camp and asked to meet with Red Cloud and his council.

Marsh's request to dig up bones of extinct monsters on Sioux land was a novel one to Red Cloud, who was trying desperately to keep the peace. Still, the scientist's earnest quest for knowledge of ancient animals intrigued him. Other tribal leaders thought it was yet another ruse to dig for gold and that a government that cheated them with rotted rations and tattered blankets could not produce a truthful white man.

Marsh's earnest obsession won the day. He told the Sioux council that if they helped him get fossil bones, he would take all their grievances to the highest levels in Washington. Red Cloud took the gamble. With an escort party of Sioux warriors, Marsh and his workers set out for the Badlands, despite threats by hostile dissidents and dangers from other bands not bound to Red Cloud's word. After weeks of frantic labor, two tons of crated fossils were loaded onto wagons, and the field crew departed a day before a massive Indian war party swooped across their site.

Marsh now investigated Red Cloud's complaints; he was shocked at the rotten pork and beans, filthy flour, and scrawny beef the agency was providing. People were sick and starving because of profiteering within the Grant administration.

When Marsh tried to bring this to the attention of Washington officials, he was met with evasiveness and deceit. Patiently, he began to unravel the "Indian Ring" led by Secretary of the Interior Columbus Delano. Despite Delano's efforts to discredit him and Red Cloud, Marsh confronted President Grant and the cabinet, and finally took his case to the newspapers.

During a long public campaign, Marsh and Red Cloud swayed public opinion, and Marsh gained national attention. Editorial writers hailed him as a truthful scientist-reformer

CHIEF RED CLOUD'S warriors trounced the U.S. Army but protected Yale Professor Othniel C. Marsh's dinosaur diggers. Marsh in turn exposed corruption in the Bureau of Indian Affairs. In 1882, the two battlers posed in a New Haven, Connecticut, photo studio with a peace pipe symbolizing their friendship.

who had taken on the lying "big boys" and won—support and publicity that helped him gain an edge in his rivalry with fellow paleontologist Edward Cope. Chief Red Cloud and Marsh remained friends for life. "I thought he would do like all white men and forget me when he went away," said the old warrior. "But he did not. He told the Great Father everything, just as he promised he would, and I think he is the best white man I ever saw."

See also COPE, EDWARD DRINKER; COPE-MARSH FEUD

RESISTANT STRAINS
Evolving Super Bugs

Some evolutionary change is a direct threat to human well-being. Over the past century, science has developed insecticides to control insect pests, toxins to kill disease-bearing rats, and antibiotics to knock out infectious microbes. Yet in many cases, these organisms develop "resistant strains" or populations that have evolved immunity to our poisons. ALS-inhibitor herbicides, for example, widely adopted as a powerful weed-killer by crop farmers, seemed universally deadly when first used, but in the past few years more than 70 weed species have evolved a resistance to it.

So commonplace is this phenomenon that public health officials must routinely take it into account. For example, penicillin was widely used for a wide variety of infections, including sexually transmitted gonorrhea. After 20 years, many of these microorganisms had evolved resistant strains, and doctors found they had best rotate a variety of cures or penicillin might soon be useless.

Public health officials attempt "holding actions," as there's no way to win permanently against the evolution of noxious germs and insects.

Just as creatures in nature respond to predators with new defenses, so do targets of human medicine and pest control. Those who claim we never "see" evolution at work need only look at resistant strains, whose generations are usually very short compared to humans.

Two examples are the Indian meal moth, which consumes tons of stored grain, and the tropical blood parasite that causes malaria.

A microbial insecticide known as Bt (*bacillus thuringiensis*) had been used successfully for more than 20 years to control the Indian meal moth, which infests grain bins. It was thought to be the knockout punch for these destructive insects, but they rapidly developed resistance to pesticides. Experiments pointed to a single major (recessive) genetic factor, which had increased in frequency in the population.

Malaria was beaten down in many parts of the world by the 1960s, only to rebound sharply since then. There are now about 100 million cases a year, of which a million and a half are fatal. Most occur in tropical Africa, Asia, and South America, home of the *Anopheles* mosquito, which transmits the blood parasite by biting humans. Chloroquine had been the cheapest, most effective drug for combating malaria; it kills an early, asexual stage of the parasite. Resistant strains, however, began independently in Southeast Asia, Latin America, and Africa; they now occur globally.

New strategies are being developed to outfox the newfound immunity of malarial parasites, but these, too, will be effective only for a limited time. Public health officials now conduct "holding" actions rather than all-out warfare, for they have accepted the impossibility of absolute victory. The idea of a "cure" or medicine for every disease or a single substance that will exterminate pests has yielded to an evolutionary perspective.

See also DARWINIAN MEDICINE

RITUALIZATION
Evolution of Animal "Displays"

In *The Expression of the Emotions in Man and Animals* (1872), Darwin argued that actions that are originally voluntary may somehow become involuntary and "fixed" in a species' behavior. Half a century later, Konrad Lorenz and Niko Tinbergen based ethology—their attempt to understand animal behavior—on Darwin's premise.

Focusing on "displays," they observed that many kinds of animals repeatedly perform intense and (to humans) bizarre behaviors. During courtship, ducks, geese, and other waterfowl have a repertoire of seemingly aggressive moves, such as mock biting, and partial nesting behavior, like snatching clumps of vegetation. Some bob their heads, extend wings, ruffle feathers, entwine necks, dive down quickly and come up again, or vibrate their tails. There is a broad spectrum of possible courtship behaviors, but each species elaborates only a few.

Biologist Julian Huxley's field observations in 1906 of courtship in crested grebes first called attention to the phenomenon. The formalized, repetitive actions, which produce excitement and sometimes "ecstasy" in the birds, reminded Huxley of human rituals. He coined the term "ritualization" for the evolution of ordinary behaviors (feeding, preening, attacking, parenting) into social signals. Ritualization could involve abbreviating movements, speeding them up, slowing down, or incorporating them into a special patterned sequence that enhances their communicative function.

See also ETHOLOGY; HUXLEY, SIR JULIAN SORRELL; LORENZ, KONRAD; TINBERGEN, NIKO

ROMANES, GEORGE JOHN (1848–1894)
Comparative Psychologist

Canadian-born George J. Romanes, a graduate of Cambridge University (1870), was pursuing advanced studies in neurology and psychology in Europe when he read Charles Darwin's *Descent of Man* (1871), in which Darwin asserted that "there is no fundamental difference between man and the higher animals in their mental functions." Romanes determined to become "Darwin's disciple in animal behavior," applying evolutionary theory to comparative psychology.

Just as Darwin attempted to reconstruct the development of emotional expression (*The Expression of the Emotions in Man and Animals*, 1872), Romanes would trace "mental evolution" by comparing the intelligence and abilities of various organisms, including humans.

With the brashness of youth, the 25-year-old Romanes wrote Darwin and was thrilled when the naturalist invited him to visit. Darwin greeted him with "How glad I am that you are so young, Mr. Romanes!" for he foresaw a full lifetime's work ahead. When Romanes began compiling material on animal abilities, Darwin gave him his 40 years' collection of notes, files, and clippings on the subject. Darwin even offered his own essay on instinct, which was published posthumously as the final chapter in Romanes's *Mental Evolution in Animals* (1883).

Although not the first author to compile accounts of animal behavior, Romanes certainly founded the subject within the framework of Darwinian evolution, and was also the first to publish all the available information in English. His ambitious books *Animal Intelligence* (1881), *Mental Evolution in Animals* (1883), and *Mental Evolution in Man* (1888) established him as the "father of comparative psychology."

Oddly for an evolutionist, Romanes focused on the common domestic cat, whose mechanical abilities, he thought, indicated "a higher level of intelligence than any other animal except for monkeys or elephants." His own cat had mastered a series of complicated latches to let herself in and out of the house.

In 1881, he tested the common belief that cats have an uncanny sense of direction. He wrote Darwin, "I have got a lot of cats [from various homes, and will] drive them several miles into the country, and then let them out of their respective bags." Romanes stopped in the middle of a country road, released the cats, and climbed onto the coach roof to get a good view as they scurried away in all directions. Crestfallen, he reported the next day: "All the cats I have hitherto let out of their respective bags have shown themselves exceedingly stupid, not one having found her way back."

Romanes was tormented by continual flip-flops in his beliefs about science, religion, and even spiritualism. He had tried to broach the topic of spiritualism to Darwin in 1877. Romanes thought he had experienced "thought communication" with a supernatural intelligence and was impressed by the floating hands and luminous "disconnected faces" he had witnessed at seances. Darwin assured Romanes that his interest in Spiritualism would be their secret ("never mentioned to a human being"), but that he was unsympathetic, for "I fear I am a wretched bigot on the subject."

Comparative psychologist George Romanes thought his cat had "a higher level of intelligence than any other animal except for monkeys and elephants."

Frustrated by his mentor's attitude, Romanes called on evolutionist Alfred Russel Wallace, Darwin's friendly rival and a publicly avowed spiritualist, to discuss his secret belief in paranormal phenomena—a belief that could cost him his scientific reputation. Wallace recalled that he received Romanes graciously, promised to keep his confidence, and "gave him the best advice I could" about "the usual perplexities which beset the beginner."

A few years later, Romanes published a scathing review of Wallace's book *Darwinism* (1889). Its demonstration of natural selection in the animal world stops at human evolution, which Wallace thought was guided "by the unseen world of Spirit." Here, wrote Romanes, "We encounter the Wallace of spiritualism and astrology, the Wallace of vaccination and the land question, the Wallace of incapacity and absurdity."

Wounded, the usually gentle-mannered Wallace wrote Romanes: "There is [also] a Romanes 'of incapacity and absurdity!!' But he keeps it secret. He thinks no one knows it. He is ashamed to confess it to his fellow-naturalists; but he is not ashamed to make use of the ignorant prejudice against belief in such phenomena, in a scientific discussion with one who has the courage of his opinions, which he himself has not."

See also COMPARATIVE METHOD; *EXPRESSION OF THE EMOTIONS*; SPIRITUALISM; WALLACE, ALFRED R.

ROOSEVELT, THEODORE (1858–1919)
The Naturalist President

I sat at the feet of Darwin and Huxley," wrote President Theodore Roosevelt in 1918, recounting his lifelong interest in evolution and natural history. As a young boy, he recalled, he had loved to read about animals, birds and paleontology and—though handicapped by poor eyesight—tried to make field observations in the New York countryside near his family home.

When he was 14, Roosevelt's father, then a trustee of the American Museum of Natural History, encouraged his son's interests by giving him a shotgun and professional lessons in mounting birds. His taxidermist-tutor was John G. Bell, who had accompanied the famed John James Audubon on his collecting trips to the far West. When his wealthy father sponsored his field trips to Egypt, young Teddy learned he could make first-hand discoveries that were not to be found in his nature guidebooks.

As an adult, he continued traveling to wilderness areas throughout the world, including Africa and South America. His observations on behavior and ecology went into his books *African Game Trails* (1910) and (with Edmund Heller) *Life Histories of African Game Animals* (1914). He prided himself on his original observations, "which were really obvious, but to which observers hitherto had been blind . . . because sportsmen seemed incapable of seeing anything except as a trophy [and] partly because collectors had collected birds and beasts in precisely the spirit in which other collectors assemble postage stamps."

When he became the 26th president of the United States, Roosevelt was an active conservationist, the first to reserve millions of acres of lands for wilderness parks, national forests, and wildlife sanctuaries. Shortly after his unsuccessful race in the 1912 presidential election, he resumed making field trips all over the world to collect specimens for museums and to observe nature. His exploration of an unknown tributary of the Amazon, then called the Rio Misterioso, resulted in its being renamed the Rio Teodoro in his honor. On one trip to Africa, he shot a huge bull elephant for the American Museum of Natural History, which dominates the mounted herd in the Akeley African Hall.

Roosevelt could name dozens of genera and varieties of American mammals from their skulls alone and astounded professional naturalists with his skill. His passion for natural history, he said, "has added immeasurably to my sum of enjoyment in life."

In 1902, on a hunt in Mississippi, his servants and hounds treed a young bear. When called upon to kill it, Roosevelt indignantly declared the situation "unsporting" for a gentleman and refused to shoot. (Out of sight of reporters, however, he instructed one of the servants to kill it.) The president's apparent

EVOLUTIONIST-NATURALIST President Theodore Roosevelt changed from an avid hunter to visionary conservationist. When he refused to shoot a cornered bear, a craze for toy "Teddy Bears" swept the country.

mercy for his quarry made headlines all over the country, and Teddy's appealing bruin became a national mascot: the original Teddy Bear. Toy bears became a fad as a "teddy bear" craze swept the country.

Somewhat tongue in cheek, Roosevelt always credited his passion for natural history for saving his life from an assassin's bullet. When he was a boy stalking birds in the New York countryside, a twig caught his spectacles and flung them into the snow, leaving the near-sighted youth helpless and miserable. Ever after that painful incident, he recalled in 1919, "I never again in my life went out shooting, whether after sparrows or elephants, without a spare pair of spectacles in my pocket. After some ranch experiences I had my spectacle cases made of steel; and it was one of these steel spectacle cases which saved my life in after years when a man shot into me in Milwaukee."

ROTHSCHILD, LORD WALTER (1868–1937)
World's Greatest Bird Collector

Lord Walter Rothschild, scion of one of Europe's wealthiest banking families, spent a fortune amassing the world's largest and most complete collection of preserved birds. In 1889, for his 21st birthday, his father gave him a private museum on the family's estate in Tring, Hertfordshire, about thirty miles northwest of London. Eventually, it housed a quarter-million specimens of all kinds of animals. (Now a division of the London Natural History Museum, it continues to be a popular public attraction.)

Evolutionary biologists flocked to Tring to work on problems of species distribution, varieties, transitional forms, population variability, and classification. Rothschild's birds became legendary, along with their eccentric owner's attachment to his precious collection.

On March 10, 1932, Rothschild suddenly announced that most of the birds (including his especially beloved birds of paradise) would be sold. He was disconsolate; circumstances forced him, as he put it, "to tear one's being out by the roots." But he had no choice. His generous brother had died, the family cut off further funds, and he was pressed by creditors and tax debts.

He accepted the American Museum of Natural History's offer of $225,000 (about a dollar a bird) for a collection most scientists agreed was well worth $2 million. His only consolation was that his feathered treasure trove, which had taken 30 years to gather from all over the world, would be kept intact and remain accessible for the advancement of scientific studies.

Among his other eccentricities, the bird man of Tring had a strange way of filing his personal mail. For years, he had separated the letters into two piles. One pile was deposited, unopened, into large wicker laundry baskets. As the hampers filled up, he sealed them closed and placed them in a storeroom.

After his death in 1937, when they opened the hampers, Lord Rothschild's relatives made an astounding discovery. His aunt, Miriam Rothschild wrote in her memoir:

> It fell to the lot of his horrified sister-in-law to discover the existence of a charming, witty, aristocratic, ruthless blackmailer who at one time had been Walter's mistress, and, aided and abetted by her husband, had ruined him financially, destroyed his mind for forty years and eventually forced him to sell his bird collection . . . [He] seemed to shrink visibly in the period following the sale. . . . It was winter—his birds had flown.

See also BEETLES; MAYR, ERNST; PEALE'S MUSEUM

BANKING FAMILY SCION with a passion for natural history, Lord Walter Rothschild, shown here riding one of his Galápagos tortoises, tried unsuccessfully to maintain a carriage drawn by four zebras (top).

SALTATION
Evolutionary Leaps

The word "saltation," derived from the Latin, means jumping or leaping from place to place, and is used to describe the peculiar locomotion of grasshoppers and kangaroo rats.

In evolutionary studies, "saltation" means rapid change, where species seem to evolve by macromutations, rather than through a slow series of intermediate forms.

When Charles Darwin first expressed his theory of evolution, he adopted this timeworn cliché as part of the evolutionary process: *Natura non facit saltum* (Nature makes no leaps). His friend Thomas Huxley thought that was an unnecessary burden for the theory to carry. Although he was a staunch defender of the general truth of evolution, Huxley's reading of the fossil record presented some puzzles about evolutionary rates. Many species appeared to be stable, showing little change over long periods, while certain groups seemed to change and diverge fairly rapidly. Recent "punctuational" theorists incline more to Huxley's view.

Of course, from the vantage point of a human life span, evolution is excruciatingly slow—whether change takes place over millions of years or in mere thousands.

See also "HOPEFUL MONSTERS"; PUNCTUATED EQUILIBRIUM

"MY THINKING PATH" was how Charles Darwin described his Sandwalk, which he strolled several times a day, pondering his scientific problems.

SANDWALK
Darwin's "Thinking Path"

One of the first things Charles Darwin did when he and his wife settled in Downe village, in the Kentish countryside, was to construct a circular path through the fields and woods on his property. He called it the "Sandwalk," his "thinking path," and had the gardener sprinkle its length with sand.

This was to be no idle bit of landscaping, but an essential tool for his work. Each morning and each afternoon for over 40 years, he took his turns on the Sandwalk, sometimes accompanied by his little terrier. Scientific friends such as Thomas Huxley or Sir Joseph Hooker, when they visited, would join him for theoretical discussions on his walks, or to talk about, in Hooker's words, "old friends, old books, and things far off to both mind and eye."

Darwin was in the habit of placing a small pile of flints at the crossroad of the Sandwalk, the number of flints depending on the difficulty of the problem he was pondering. If it was a "three flint problem," he would knock a flint off with his walking stick each time he made a circuit; when the flints were gone, it was time to return home. (His method was strikingly similar to the "three pipe problems" of Sherlock Holmes.)

The Sandwalk has a dark and curvy side through the wood, and a straight and sun-lit stretch adjacent to the field. The alternation between straight and curved, shaded and bright, gives a stimulating rhythm to the walk.

One can go to Down House today (it is a Darwin museum, or, perhaps, shrine) and take the walk as Darwin did so many thousands of times. Some visitors report feeling a strong sense of the great naturalist's presence on the Sandwalk, as if he were just about to round the bend, swinging his walking stick, deep in thought.

See also DARWIN, CHARLES; DOWN HOUSE

"SARAWAK LAW" (1855)
Wallace's Evolutionary "Bombshell"

In 1854, naturalist Alfred Russel Wallace went to Sarawak, in northern Borneo, a tropical island of the Malay Archipelago. Stricken with malaria during the wet season, he was compelled to halt active field work and "look over my books and ponder over the problem which was rarely absent from my thoughts," the question of the origin of species. The result was his seminal paper known as "the Sarawak Law."

One of the most important short papers in the history of biology, Wallace's essay was published in the September 1855 issue of the *Annals and Magazine of Natural History* while he was still in Borneo. Its formal title was "On the Law Which has Regulated the Introduction of New Species"; its conclusion was that "every species has come into existence coincident both in space and time with a pre-existing closely allied species."

It was a closely reasoned argument pointing to the reality of evolution based on converging lines of evidence from the geographical distribution of living plants and animals, as well as the fossil record. Where Darwin had begun with an interest in varieties and the results of domestic breeding, Wallace was led to identical conclusions by the facts of zoogeography (animal distribution), a science he pioneered.

Though he was only 32 at the time and self-educated, Wallace displayed a sure grasp of botany, geology, geography, zoology, and entomology. He listed nine "well-known facts" to support his conclusion, but several were his own ingenious generalizations, then scarcely known at all. He also sketched a detailed metaphor of life as a branching tree, progressing "only in a general way" and diverging into side branches that were neither higher nor lower than their collaterals.

Wallace's Sarawak paper was getting awfully close to elaborating the theory on which Darwin had been laboring for 30 years. Lyell visited Darwin at Down House in 1856 and advised him to hurry up and publish something on evolution or Wallace would surely beat him to it. Darwin wrote Lyell on May 3, 1856, that "I rather hate the idea of writing for priority, yet I certainly should be vexed if any one were to publish my doctrines before me." Two years later, when Wallace published his Ternate Paper, Darwin wrote Lyell, "Your words have come true with a vengeance . . . that I should be forestalled."

Wallace, still in the Malaysian jungles, was entirely unaware of the anxiety and panic he was causing Darwin and his friends back in England by his apparent perversity in formulating and publishing "Darwin's theory" before Darwin.

In a letter to Darwin written September 27, 1857, Wallace thanks him for praising the Sarawak Law and is delighted that "my views on the order of succession of species [are] in accordance with your own; for I had begun to be a little disappointed that my paper had neither excited discussion nor even elicited opposition." He needn't have worried.

See also "DELICATE ARRANGEMENT"; LINNEAN SOCIETY; "TERNATE PAPER"; WALLACE, ALFRED RUSSEL

SCALA NATURAE

See GREAT CHAIN OF BEING

> "I rather hate the idea of writing for priority, yet I certainly should be vexed if any one were to publish my doctrines before me."
>
> —Charles Darwin to Charles Lyell, May 1856

SCIENCE
An Evolving System

Generations of students have been taught that scientific theories or facts may change, but science itself is rock-solid. It is a particular method for arriving at an objective understanding of nature, a self-correcting system that, given more facts, produces an ever-clearer picture of reality. Unlike religion, it is not blinkered by obedience to traditional dogma or faith, nor is it "subjective" in explaining the universe. Until recently, such was science's view of itself, repeated in countless textbooks.

Beginning in the 1960s, however, philosopher-historians of science began to discern a radically different picture. History shows that science is not at all a frozen method or system of thought, but is itself evolving. Oddly enough, though "scientific" discoveries have been made over thousands of years, the word "scientist" was not coined until 1809, by English philosopher William Whewell.

Change within science (previously called by various names, such as "natural philosophy") has not simply been a matter of correcting or rejecting previous theories and conclusions. "What is remarkable about what has occurred over the past few decades," wrote physicist-philosopher William B. Jones in 1988, "is the general recognition that the methodology and rules of science themselves change." It now appears there was never any such thing as a single, objective "scientific method," independent of its time and social context. And philosophers of science have struggled mightily (so far without success) to understand whether a discernible progress or pattern exists in the history of the scientific process itself.

In this new view of science, there can be no such thing as a "fact" uncontaminated by interpretation, expectation, or some kind of human bias. And "operationalism"—an optimistic attempt to define "facts" in terms of standardized operations by the scientist—has also failed to hold up as a method of distinguishing "meaning" from "pure facts." It has proven fruitless, for instance, to apply operational measurements to very diverse phenomena, which is why physicists usually fail when they address the "messier" problems of biology. But although they can explore Mars with their instruments, physicists cannot agree on such basic concepts as electrons, quarks, or the nature of gravity or light.

Science is also limited to what is "doable" within its narrow research strategies. Positivists once argued that whatever can be "scientifically proven" is necessarily meaningful or important. But it has become clear that since science can take on only certain kinds of problems—those that offer some hope of solution with known techniques—it often ignores possibly significant questions if they do not lend themselves to testing. Thus, published scientific work can (and does) "prove" an immense number of trivial, obvious conclusions. "Difficult problems rarely succumb to a frontal assault," notes Jones; "they are much more likely to yield to someone who appears on the scene with just the right tools, which were acquired for other reasons altogether."

Nor is science immune to personal animosities, priority disputes, and petty jealousies. Thomas Henry Huxley wrote to his sister in 1852: "You have no idea of the intrigues that go on in this blessed world of science. Science is, I fear, no purer than any other region of human acitivity; athough it should be. Merit alone is very little good; it must be backed by tact and knowledge of the world to do very much."

> **Neither science nor religion appears to be what we had supposed it to be.**

While there can be no doubt that since ancient times scientists have managed to reach some solid analytical conclusions, we are only beginning to realize how much of today's "advanced" knowledge is all mixed up with myths, fads, social biases, and religious values. The traditional conflict between science and religion appears increasingly irrelevant, for neither appears to be what we had supposed it to be. As physicist Werner Heisenberg wrote in *Physics and Philosophy* (1958), articulating his famous uncertainty principle, "Natural science does not simply describe and explain nature; it is part of the interplay between nature and ourselves; it describes nature as exposed to our method of questioning." And just as the history of an individual is that individual, so "the history of science is science itself."

See also PEER REVIEW; POSITIVISM; THEORY, SCIENTIFIC

"SCIENTIFIC CREATIONISM"
Elusive Research Program

During heated court battles over the teaching of evolution in public schools during the early 1980s, proponents of "balanced" presentations argued that "creation science" should be taught as an equally valid interpretation of the evidence.

They were not seeking to introduce sectarian religion into the public schools, argued Henry Morris, director of the Institute for Creation Research. "Scientific creationism," he wrote, has "no reliance upon biblical revelation, utilizing only scientific data to support and expound the creation model." Morris and his colleagues referred often to a growing body of scientific research supporting creationist interpretations in biology and geology.

At the height of the debate, Eugenie Scott and Henry Cole of the National Center for Science Education conducted a three-year search of 1,000 scientific and technical journals to survey the creationist research that might be taught in classrooms. "Nothing resembling empirical or experimental evidence for scientific creationism was discovered," they reported in 1985. A few prominent creationists who were professional scientists had published, but their topics were food processing, aircraft stress, and other unrelated areas.

Scott and Cole next sought to find how many manuscripts submitted by creationists had been rejected for publication by "establishment" science publications. After checking 68 journals, which had received more than 135,000 submissions during the three-year period, they found that only 18 articles were written by scientific creationists. Twelve of those were polemics on science education; the remaining six, mainly "refutations" of evolution, were rejected as incompetent and unprofessional.

There is, of course, the possibility that evidence for creationism would be "censored" by exclusion from orthodox, evolutionary-biased publications no matter how professional or well presented. But six articles submitted to 68 journals over three years hardly indicates a thriving research enterprise productive of new insights and discoveries.

Virtually the entire "creation science" literature consists of the books and tracts published by the Institute for Creation Research. Most are arguments against evolution, based on the logic that if evolutionary theory has weaknesses or cannot account for data, then creationism is proved correct. Their arguments assume that there are only two alternatives: creationism or Darwinian evolutionism.

Still, science teachers were pressured by public campaigns, some led by professors with science degrees, claiming there were abundant scientific studies based on creationist concepts. Many parents, politicians, and educators have assumed that a published body of scientific "creationist" research exists, when in fact it does not.

See also FUNDAMENTALISM; *INHERIT THE WIND*; INTELLIGENT DESIGN; SCOPES TRIAL

Creationists refer to a published body of scientific research that does not, in fact, exist.

SCIOLISM
An Extinct Insult

During the 19th century, the term "sciolism" referred to theories about natural or supernatural phenomena that the scientific establishment rejected as "false" or "pretended" knowledge. Spiritualism, phrenology, and astrology were among the pseudosciences attributed to "sciolists," which literally means "those who know only a little." (The word "science" is from Latin *scientia*, to know; *sciolus*, a smatterer, is the diminutive form.)

Hugh Miller, the most popular Victorian author on geology and the Earth's past, insisted the idea of evolution (or "development") belonged to sciolism. Reacting to early evolutionist Robert Chambers's anonymously published *Vestiges of Creation* (1844), Miller expressed outrage in his *Footprints of the Creator* (1847). Chambers had dared to cite fossils and other geological evidence to support evolution, claiming that professional geologists had misread the record of the rocks. Miller fumed:

> Can it mean, that he appeals from the only class of persons qualified to judge of his facts [professional geologists], to a class ignorant of these . . . that he appeals from astronomers

and geologists to low-minded materialists and shallow phrenologers—from phytologists and zoologists to mesmerists and phreno-mesmerists? . . . No true geologist holds by the development hypothesis;—it has been resigned to sciolists and smatterers;—and there is but one other alternative. [Species] began to be, *through the miracle of creation.*

See also CHAMBERS, ROBERT; MILLER, HUGH; PHRENOLOGY; SPIRITUALISM; *VESTIGES OF CREATION*

SCOPES TRIAL (1925)
Evolutionists vs. Creationists

Beginning on July 10, 1925, a small, sweltering courtroom in the American South became the focus of worldwide attention. High school biology teacher John T. Scopes of Dayton, Tennessee, was charged with violating the Butler Act, the state's new law against teaching human evolution or "any theory that denies the . . . Divine Creation of man" in the public schools. The famous "Monkey Trial" pitted fundamentalist politician William Jennings Bryan against the liberal lawyer Clarence Darrow in a classic courtroom drama, which filled the newspapers of the day and inspired books, plays, and movies for years afterward.

To many, the Scopes trial has come down as a triumph of science over religion, in which creationism was refuted by the evidence for evolution in open court. Scopes is depicted as an unwilling participant, arrested while going about his normal duties as a teacher. Bryan is said to have died soon after his loss to Darrow of a broken heart. And the town was said to have suffered the indignities of swarms of outsiders coming in to profiteer, turning the proceedings into a garish circus.

Influenced by the play *Inherit the Wind* (1955) by Jerome Lawrence and Robert E. Lee, these are all distortions and dramatizations of what really occurred. In fact, the 24-year-old Scopes was not a regular biology teacher, but an athletic coach who was subbing for a few weeks. His strongest belief was in freedom of expression, not evolution. The chain of events really began not in Dayton, but at the offices of the American Civil Liberties Union in New York. When the Butler Act passed, the ACLU sent out a press release to the newspapers

"MONKEY TRIAL" of teacher John T. Scopes in Dayton, Tennessee, captured worldwide attention in 1925. Famed attorney Clarence Darrow (in suspenders) defended Scopes, while fundamentalist politician William Jennings Bryan was prosecutor.

offering to pay for the defense of any teacher who violated the law by teaching evolution to a high school biology class.

Local merchants, rather than resenting the "circus atmosphere," gathered in F. E. Robinson's drugstore to instigate the case. They thought that promoting some excitement in the sleepy little town would be great for business, and asked Scopes whether he would be willing to provoke the test case of the new law.

Expert testimony by scientists about evolution vs. creationism was not permitted in the courtroom, nor was Bryan's planned closing speech, so the much-anticipated "great debate" never completely unfolded. The judge limited the case to whether or not Scopes had broken the law. Darrow, not Bryan, lost. Following a celebratory binge of overeating, Bryan died of diabetic complications, which doctors diagnosed as "apoplexy." Darrow maintained that "he didn't die of a broken heart but a busted belly."

Nevertheless, the public perception was that Darrow had won—and he had, in the public arena if not the courtroom. He had outfoxed his opponent, backed him into an intellectual corner, and made mincemeat of his most basic argument: that the Bible had one and only one clear meaning, which is not subject to human interpretation. Bryan had declared that "the one beauty about the Word of God is, it does not take an expert to understand it."

> **Darrow destroyed Bryan's argument that Scripture has only one clear meaning, not subject to human interpretation.**

Darrow then put Bryan on the stand as an expert witness on the Bible, and asked whether he believed that the sun stood still for Joshua so the day of battle would be lengthened. "I accept the Bible absolutely" was the reply. "Do you believe at that time the entire sun went around the Earth?" Darrow pressed. "No," Bryan admitted, "I believe that the Earth goes around the sun." Darrow pointed out that the biblical passage was written before it was known that the Earth rotates and orbits the sun. Even the descriptions of miracles depend on prevailing human assumptions about how nature works.

Bryan had a right to his own interpretation, Darrow stipulated, but any individual's understanding of the Bible had to be an interpretation. At any given time, hundreds of Christian and Jewish sects each claim to offer the one and only correct reading of divine revelation in the Bible.

A seasoned orator who had thrice run for president of the United States, Bryan pulled out all the stops. He told Darrow he was not interested in the age of rocks but in the Rock of Ages. "There is no place for the miracle in this train of evolution," he intoned,

> and the Old Testament and the New are filled with miracles. . . . [Evolutionists] eliminate the virgin birth . . . the resurrection of the body . . . the doctrine of atonement. [Scientists] believe man has been rising all the time, that man never fell; that when the Savior came there was not any reason for His coming. . . . [Outsiders] force upon . . . the children of the taxpayers of this state a doctrine that refutes . . . their belief in a Savior and . . . heaven, and takes from them every moral standard that the Bible gives us.

Darrow summed up his case with equal passion. Freedom cannot be preserved in written constitutions, he argued,

> when the spirit of freedom has fled from the hearts of the people. . . . Today it is the public school teachers, tomorrow the private. The next day the preachers . . . the magazines, the books, the newspapers. After a while, your Honor, it is the setting of man against man and creed against creed, until with flying banners and beating drums we are marching backward to the glorious ages of the sixteenth century, when bigots lighted fagots [torches] to burn the men who dared to bring any intelligence and enlightenment and culture to the human mind.

About 100 reporters had crowded into the tiny Dayton courtroom from all over the world to catch every word uttered by the two old war-horses, both in their late sixties. H. L. Mencken, an acerbic young reporter for the *Baltimore Evening Sun,* published a series of stinging satirical jibes at Bryan, whom he ridiculed as a "tinpot" Bible-thumper.

Editorial writers and cartoonists all over the country weighed in on the controversy. One of the more thoughtful editorials came from the *St. Louis Post-Dispatch,* who wrote that Bryan was putting religious faith on a very shaky

If Bryan's religious faith is based "on the assertion that Darwin . . . was wrong . . . What a gamble he makes of Christianity."

—*St. Louis Post-Dispatch*, 1925

basis. He was insisting, it said, "that the Christian religion cannot survive if one of the principal theories of modern science should prove to be true." The editorial continued:

> The gospel he brings to the people of Tennessee does not rest, as all sound faith must, on a continuing revelation in human experience. It rests instead on the assertion that Darwin or somebody else was wrong. . . . What a tiny base on which to erect a religion! What a gamble he makes of Christianity. . . . Is it fair, then, to tell the plain people in Tennessee that they must give up their whole religion, their consolation on earth and their hope of salvation hereafter, if the kinship of man with a lower order of animals is indisputably proved? . . . Who more than Mr. Bryan is threatening to rob devout people of their faith?

Of course, the drama of the battle for ideas was what held the world's attention. Everyone knew in advance that Scopes was guilty of breaking the law; he had admitted that from the beginning. He was convicted and fined $100, but the verdict was later overturned on a technicality. Tennessee did not attempt to retry the case.

After their hollow victory in Tennessee, fundamentalist activists shifted their focus to other states. In 1926 and 1927, they successfully lobbied for new legislation prohibiting the teaching of evolution in the public schools of Mississippi and Arkansas. These laws remained on the books—as did the Butler Act—for many years, though largely unenforced. Tennessee's antievolution law was repealed in 1967, and the following year its Arkansas counterpart was declared unconstitutional by the United States Supreme Court.

See also BRYAN, WILLIAM JENNINGS; BUTLER ACT; *CHRYSALIS*; FUNDAMENTALISM; *INHERIT THE WIND*; INTELLIGENT DESIGN; "SCIENTIFIC CREATIONISM"; SCOPES II

SCOPES II (1981)
"Creation Science" Trial

In 1981, 56 years after Tennessee's famous Scopes "Monkey Trial" had faded into history, another Southern courtroom was about to rehash the perennial American controversy about teaching evolution in the public schools.

After failing in several attempts to ban evolution from science classes, fundamentalist Christians had lobbied Arkansas legislators to pass Act 590, requiring "equal time" or "balanced treatment" for the theories of "creation science." If they had to put up with Darwin's theory, they said, they wanted equal time for "Moses' theory." Creationists described the trial, to be held in Little Rock, Arkansas, December 7–16, 1981, as "a test of fairness for the two scientific models." Scientists called it "Scopes II."

The challenge to the "equal time" law came from the Reverend Bill McLean in conjunction with the American Civil Liberties Union (ACLU), who regarded it as incompatible with America's constitutional separation of church and state. They were not suing any creationist group, but the Arkansas Board of Education, to enjoin it from enforcing an unconstitutional law.

A team of volunteer lawyers searched the country for expert witnesses on the subjects of science and religion. Among those who appeared for the ACLU were paleontologists Niles Eldredge and Stephen Jay Gould, philosopher of science Michael Ruse, biochemist Harold Morowitz, and theologians Langdon Gilkey and Father Bruce Vawter. Though science teaching was at issue, most of the plaintiffs, and half of the witnesses in the case, were priests, ministers, theologians, and historians of religion. Langdon Gilkey of the University of Chicago Divinity School later summed up his motives as a theologian for challenging the law: "There can be, I believe, no healthy, creative, or significant religious faith in a modern society unless [all] the forms of that faith are free. A politically enforced or supported religious faith becomes corrupt, dead, and oppressive."

Creationists came to argue that there was indeed a "scientific basis" for the scriptural version of human origins and that to exclude this "alternate model" from the classroom was a violation of academic freedom. They labeled evolutionary biology "a humanist religion" or "a fantasy based on guesses." The scientists explained that evolutionary biology was part of the complex tapestry of modern science, incorporating astronomy, geophysics, paleontology, biochemistry, genetics, anthropology, and more. It was a picture of interlocking disciplines, built up and tested, each feeding into the other. Theologian Gilkey, a committed Christian, was struck by the realization that "without this thesis of a universe in process over eons of time . . . there simply is no modern science."

On January 5, 1982, Judge William Overton released his 38-page ruling in favor of the plaintiffs, the ACLU. Overton ruled that "creation science" could not qualify as an alternative scientific explanation or theory. In the court's judgment, it is a religious doctrine concerning the biblical Christian God and the account of Creation given in Genesis. Act 590, Overton concluded, was therefore an attempt to establish religion in a state-supported school in violation of the First Amendment of the federal Constitution.

A similar test case of a "balanced treatment" law (Act 685) occurred in Louisiana in 1981–1982, which dragged on through several years of twists and turns, countersuits, and appeals. Finally, the U.S. Fifth Circuit Court of Appeals ruled on July 8, 1985: "The act's intended effect is to discredit evolution by counterbalancing its teaching at every turn with the teaching of creationism, a religious belief."

Still another replay of the trial and its Dayton predecessor took place in 2004 in Dover, Pennsylvania, when "creation science" was resurrected as "intelligent design." The Dover school board had decreed that students "will be made aware of gaps/problems in Darwin's theory and of other theories of evolution including, but not limited to, intelligent design." A group of parents mounted a legal challenge to the board's action, charging that members had assigned creationist textbooks, paid for by their church.

A conservative federal judge awarded victory to the plaintiff parents, chided the school board members for their thinly disguised attempt to force a particular religion into the public schools, and accused them of "breathtaking inanity" in wasting the court's time on an issue that had been repeatedly tried and settled. To borrow a phrase from the legendary ballplayer and philosopher Yogi Berra, "It was *déjà vu* all over again."

See also BRYAN, WILLIAM JENNINGS; FUNDAMENTALISM; INTELLIGENT DESIGN; "SCIENTIFIC CREATIONISM"; SCOPES TRIAL; SECULAR HUMANISM

> "A frog turning into a prince is called a fairy tale, but if you add a few million years, it's called evolutionary science."
>
> —Creationist pamphlet

SECULAR HUMANISM
Goodness without God

According to one authority on American English, William Safire, "Secular humanism may be defined as: (1) a philosophy of ethical behavior unrelated to a concept of God; (2) a characterization of an emphasis on individual moral choices as having the common denominator of atheism."

Many secular humanists define themselves as ethical atheists who try to behave well toward their fellow men because it is right, not because they fear punishment in an afterlife. Some who accept the label are churchgoers, and some do not consider themselves atheists, but agnostics. Many accept parts of the philosophy but refuse the label.

Although the ideas have been around for centuries, the term appears to go back only to about the 1960s. Corliss Lamont, in *The Philosophy of Humanism* (1957), defined it as "joyous service for the greater good of all humanity in this natural world and advocating the methods of reason, science, and democracy."

By that definition, a great many pre-Darwinian thinkers would qualify, including Voltaire, John Stuart Mill, Jeremy Bentham, and Thomas Jefferson, who had never used the phrase. Jefferson, for instance, looked on any form of supernaturalism as an offense against reason and the laws of nature. He believed that churchmen had so distorted the story of Jesus that "were he to return on earth, [he] would not recognize one feature." Accordingly, America's third president took scissors and paste and edited the New Testament, retitling it *The Life and Morals of Jesus*. Everything "contrary to reason or nature" was edited out of the "Jefferson Bible," which he always kept at his bedside.

Clerics rightly feared that he would dash their hopes of establishing official national churches, Jefferson wrote, "for I have sworn on the altar of God, eternal hostility against every form of tyranny over the mind of man." The irony of swearing to a God he doubted did not escape him; he wrote to a nephew in 1787: "Fix reason firmly in her seat. . . . Question with boldness even the existence of a God, because, if there be one, he must more approve of the homage of reason, than that of blind-folded fear."

Almost a century later, Karl Marx tried to wed reason, atheism, and humanity to a communist political ideology. American secular humanist organizations disavow any kinship with Marx's influence, but their adversaries try to link them. Evangelist James Kennedy defined secular humanism as a "Godless, atheistic, evolutionary, amoral, collectivist, socialistic, communistic religion." In the words of M.J. Rosenberg in the *Near East Report* (1985), that kind of rhetoric uses secular humanism as "the new label employed to indict anyone who opposes school prayer, believes in evolution, or disagrees with the religious right's views on abortion."

Free Inquiry magazine, published by the Council for Secular Humanism, affirms its

> commitment to the use of critical reason, factual evidence, and scientific methods of inquiry, rather than faith and mysticism, in seeking solutions to human problems . . . a search for viable individual, social, and political principles of ethical conduct, and . . . a philosophy called naturalism, in which the physical laws of the universe are not superseded by . . . demons, gods, or other "spiritual" beings outside the realm of the natural universe.

Secular humanists consider theirs a viewpoint "born of the modern scientific age and centered upon a faith in the supreme value and self-improvability of the human personality," as Warren Allen Smith defined it in 1951. And as the author Kurt Vonnegut expressed their credo, "We humanists try to behave well without any expectation of rewards or punishments in an afterlife. We serve as best we can the only abstraction for which we have any real familiarity, which is our community."

See also AGNOSTICISM; CREATIONISM; HUXLEY, THOMAS HENRY; JEFFERSON, THOMAS

> "In the struggle of life with the facts of existence, Science is a bringer of aid; in the struggle of the soul with the mystery of existence, Science is a bringer of light."
>
> —George H. Lewes, 1878

THE LIGHT OF REASON and knowledge, in this cartoon by Joseph Keppler (opposite), is represented by scientists Charles Darwin, Thomas Huxley, and John Tyndall, and secular "freethinking" forebears Voltaire, Thomas Jefferson, Ben Franklin, and others. Priests, popes, and (in center of group) the Reverend Henry Ward Beecher cringe in the dark shadows of mysticism, ignorance, and superstition. Keppler was a cofounder of *Puck*, the comic magazine in which his drawings appeared, in 1876.

REASON AGAINST UNREASON.

SEDGWICK, REVEREND ADAM (1785–1873)
Geologist

ONCE DARWIN'S MENTOR, the Reverend Adam Sedgwick turned on him after reading *Origin of Species*. "I have read your book with more pain than pleasure," he wrote. "You have deserted the true method of induction."

Darwin's teacher thought his theories were "as wild" and improbable as the idea of putting a man on the moon.

Like many of the foremost geologists of his day, Adam Sedgwick, professor of geology at Cambridge, had been ordained for the ministry. He had a reputation for being forthright and sincere and was called "the First of Men" by contemporaries—a play on his biblical first name. Despite his essentially generous and placid nature, however, Sedgwick became embroiled in one of the most acrimonious scientific controversies of Victorian England.

He had become friends with one of the young rising stars of British geology, Roderick Impey Murchison. Together they explored the complex rock formations of Devonshire and Wales. Sedgwick was the first to describe the fossils of the lower Greywacke strata, which he named the Cambrian system, after an ancient name for Wales. Eventually their studies led them to different levels of the Greywacke, where the mercurial and territorial Murchison claimed much of Sedgwick's domain for his newly founded Silurian system.

Inevitably, almost all of the members of the Geological Society were drawn into the fray, and when another geologist of the time, Sir Henry Thomas De la Beche, claimed part of the Greywacke for his Devonian period, the battle lines were drawn. For nearly a decade the Great Devonian Controversy, as it was called, raged on in the scientific journals. Although the dispute was eventually resolved with the Cambrian, Silurian, and Devonian each carefully delineated, Sedgwick and Murchison were to remain implacable foes for the rest of their careers.

As one of the foremost scientific figures of his time, Sedgwick, of course, was drawn into the controversy that revolved around the publication of the *Origin of Species* in 1859. Young Charles Darwin had accompanied him on a geologizing trip through Wales just before Darwin's historic voyage on HMS *Beagle*, and Sedgwick had written of him, "It is the best thing in the world for him that he went out on the Voyage of Discovery. There was some risk of his turning out an idle man, but his character will now be fixed, and if God spare his life, he will have a great name among the Naturalists of Europe."

Darwin sent Sedgwick a copy of his book when it first appeared, expecting the worst kind of criticism. He wasn't disappointed. The Reverend Sedgwick wrote:

> I have read your book with more pain than pleasure. Parts I laughed at till my sides were sore; others I read with absolute sorrow, because I think them utterly false and grievously mischievous. . . . You have deserted the true method of induction and started off in machinery as wild as Bishop Wilkin's locomotive that was to sail with us to the Moon.

Adam Sedgwick died in Cambridge in 1873, much beloved by many generations of geology students. No more fitting tribute could be made than when, in 1903, the collections that he had done so much to assemble were put on display in the newly opened Sedgwick Museum of Geology.

See also CAMBRIAN-SILURIAN CONTROVERSY; DARWIN, CHARLES; MURCHISON, SIR RODERICK IMPEY

"SELFISH GENE"
Unit of Selection?

English evolutionist Richard Dawkins's popular book *The Selfish Gene* (1976) emphasizes the genetic aspects of evolution. If reproductive success is the measure of "fitness," he argues, then evolution boils down to which genes continue to gain frequency in a population's gene pool. Genes "swarm in huge colonies," writes Dawkins,

> safe inside gigantic lumbering robots, sealed off from the outside world, manipulating it by remote control. They are in you and me; they created us body and mind; and their preservation is the ultimate rationale for our existence . . . we [the phenotypes of individuals] are their survival machines.

Selection favors those phenotypes that pass on the genes that produced them. Any behavior, any color or structure of a creature that increases the chance it will leave more of its genes will tend to be preserved and exaggerated. Within certain physical constraints, the chicken is really the egg's way of making another egg.

Understandably, field naturalists who had spent years observing animals were appalled at Dawkins' attempt to shift the focus of natural selection away from the organism's struggle to survive and reproduce. As Ernst Mayr put it, "the whole (potentially reproducing) individual," not the gene, is the target of selection.

Some years later Dawkins wrote a response in which he admitted his emphasis had been lopsided, but that he had sought to encourage a genetic, biometric, and biochemical perspective on fitness that was long overdue.

He was also promoting the sociobiological view that there is no real altruism in nature, where animals sacrifice themselves "for the good of the species." Adaptive behaviors, including apparent altruism and cooperation, are the result of an individual's "selfish" genes, whose only concern is to perpetuate themselves. Sometimes saving the lives of several close relatives can keep most of an individual's genes in circulation, even if he or she dies trying to rescue family members.

Like many current controversies and arguments over emphasis, the "selfish gene" is not new at all. Before the word "gene" had even been invented, one of the pioneers of genetics, August Weismann (1834–1914), took a similar view. An accomplished scientist who helped distinguish sex cells from body cells, Weismann is best remembered for his classic demonstration (by cutting the tails off generations of mice) that acquired modifications are not inherited. Around 1900, he wrote:

> From the point of view of reproduction, the germ cells [later called sex cells] appear the most important part of the individual, for they alone maintain the species, and the body sinks down almost to the level of a mere cradle for the germ cells, a place in which they are formed, nourished, and multiply.

Metaphors in science writing are always revealing. To Weismann's readers it was shocking enough to depict the human body as a "mere cradle," a passive, neutral incubator for genes. But Dawkins's "lumbering robot" not only harbors the genes; it lurches mindlessly through life, its every behavior mechanically controlled by them.

Critics thought Dawkins's imagery was reductionism carried to absurdity, a caricature of the biological study of behavior. In the 1980s, these critiques helped trigger a backlash against "biological determinism" and the newly established discipline of sociobiology and its spinoff, "evolutionary psychology."

See also KIN SELECTION; LYSENKOISM; METAPHORS IN EVOLUTIONARY WRITING; WEISMANN, AUGUST

SEXUAL SELECTION
Survival of the Flamboyant

Even Charles Darwin thought natural selection could not account for peacocks' tails or similar fantastic structures so prominent in courtship displays. On the contrary, elaborate appendages or tailfeathers could easily get in the way when animals had to escape enemies. Such secondary male sex characteristics, he thought, were actually selected by females.

Darwin first put forward this idea in *The Descent of Man and Selection in Relation to Sex* (1871), really two books in one. Almost a century earlier, his grandfather Erasmus Darwin had written that by consistently choosing stronger, better-formed males as their mates, women had gradually improved the human race. Although Charles read his grandfather's books, he seems to have reinvented the idea and doesn't mention Erasmus as a source. Rather, he came to it through the observation that among animal populations there is competition for mates, and often one male gets a larger share of females. If only a few males impregnate many females, the next generation contains a disproportionately large number of their offspring, carrying their traits.

With more emphasis on population thinking after Darwin, evolutionists realized the issue wasn't really "survival of the fittest"—it was differential reproductive success. The proportion of genes passed on to succeeding generations was the real battleground, not how much food or territory an individual could grab in a lifetime. Whatever characteristics make an individual more successful in leaving descendants would be intensified.

Strength or size might be decisive if males competed directly, but in many species they use courtship displays, which (in Darwin's words) "stimulate, attract, and charm" the opposite sex. Can the reality of sexual selection be demonstrated experimentally? During the 1980s, Swedish ethologist Malte Andersson showed that female birds of some species do indeed prefer males with exaggerated plumage.

Males of the East African widowbird have 18-inch-long black tailfeathers, which they whirl in spectacular displays visible for half a mile. Andersson cut some of the males' ornate feathers short and stuck the cut sections onto other males' tails. Some were glued right back on their rightful owners, to control for variables of capture and handling. (This experiment was made possible by the modern miracle of quick-setting superglue.) Results: Males with artificially extended tails attracted four times as many females as those with normal tails or those whose tails were cut short. Within their territories, the number of nests with eggs actually quadrupled.

Still, if elaborate plumage makes the birds more vulnerable to predators, why should evolution favor them, even if females do? Back in 1889, Alfred Russel Wallace suggested that birds with brighter plumage were generally stronger individuals that were really being selected for their vigor, not only their beauty. Sexual selection, he thought, was simply a special kind of natural selection, not a different process. Recent studies have followed his lead, exploring the relationship between exaggerated plumage, displays, and robust health.

Some years ago, English researchers proposed that there might be a correlation between species that were prone to blood parasites and males within that species displaying bright colors to advertise their health and vigor. In 1987, Andrew Read of Oxford University, after an exhaustive study of thousands of species of North American and European songbirds, found such a correlation. Species plagued by much higher levels of parasites also tended to be just those species with bright, displaying males—and the males with the brightest plumage tended to have the most resistance to parasites. Female birds may exercise a primitive sense of beauty, as Darwin thought, but selection seems to be working to produce vigor, as Wallace thought. If that is correct, then females pick mates who are literally "glowing with health."

See also DESCENT OF MAN; WALLACE, ALFRED RUSSEL

SHAW, GEORGE BERNARD (1856–1950)
Playwright of "Creative Evolution"

George Bernard Shaw was an evolutionist, but not a Darwinian. Appalled at the idea of a mechanistic universe devoid of purpose or design, the renowned Irish playwright argued that humankind could—indeed, should—consciously and deliberately take charge of its own evolution into a superior species of "Supermen."

A thoroughgoing Lamarckian in his view of inheritance, Shaw was convinced that conscious choice and aspirations to moral progress could be inherited to improve future generations. His play *Back to Methuselah* (1920) is based on the premise that the natural human life span should be hundreds of years, though it is usually cut short by unhealthy thoughts and habits.

A self-proclaimed foe of Darwinism, Shaw was fond of quoting Samuel Butler's remark that Darwin "had banished mind from the Universe." Both literary men agreed, too, that Darwin had stolen all the best parts of his theory from his grandfather Erasmus. What was new in it, the idea of natural selection, was not only wrong but morally offensive, leading to despair and abdication of responsibility.

GEORGE BERNARD SHAW FOCUSED some of his plays, particularly *Back to Methuselah* and *Don Juan in Hell*, on the implications of evolution. He thought humans could deliberately evolve into an ethically higher and more perfect species.

"When its whole significance dawns on you," Shaw wrote in his preface to *Back to Methuselah,* "your heart sinks into a heap of sand within you. . . . There is a hideous fatalism about it. . . . [The forces of nature] modify all things by blindly starving and murdering everything that is not lucky enough to survive in the universal struggle for hogwash."

Although disavowing natural selection, Shaw embraced the notion that evolution makes blood relatives of man, beast, and bird. Consequently, he proclaimed himself an antivivisectionist and vegetarian, and he remained so to the end of his life.

Shaw saw his mixture of socialism and evolutionism as a new religion; he thought artists and writers should develop new symbols and myths for it, to deliberately reshape world culture. This idea (or ideal) of "creative evolution," also developed by French philosopher Henri Bergson, is similar to ancient Asian philosophies, though Shaw thought it Western and modern—"the genuinely scientific religion for which all wise men are now anxiously looking."

According to creative evolution, change is driven by an *élan vital*—a life force—that pushes matter toward complexity and progress, upward toward the Infinite. It is the "ghost in the machine," with which humans had better cooperate or they will be swept away with mammoths, dinosaurs, and other "mistakes."

Shaw's parables of creative evolution include *Man and Superman* (1903), through which he hoped to give playgoers a sugar-coated indoctrination while they were being entertained. But, he complained, he had made the confection too good and "nobody noticed the new religion in the centre."

See also BUTLER, SAMUEL; NEO-LAMARCKISM; WEISMANN, AUGUST

> "Your heart sinks into a heap of sand within you. . . . There is a hideous fatalism about . . . the universal struggle for hogwash."
>
> —George B. Shaw,
> *Back to Methuselah*

SIMPSON, GEORGE GAYLORD (1902–1984)
Paleontologist, Evolutionist

During the 1930s, George Gaylord Simpson of the American Museum of Natural History helped to bring paleontology into the modern Synthetic Theory of evolution. While most paleontologists had become accustomed to thinking of a series of types "leading to" the modern forms, Simpson's attempt to approach the fossil record as a sampling of ancient breeding populations led to a revitalization of the field.

Simpson joined the museum as assistant curator of vertebrate paleontology in 1927, at the invitation of director Henry Fairfield Osborn, who had a knack for picking future "stars" in natural history. Simpson's productive field trips to the American West and to Patagonia added much new information on the evolution and distribution of extinct mammals of the New World.

Traditionally, fossil hunters had sought magnificent specimens for their museums and exhibited them as a series of individuals, like Othniel C. Marsh's famous linear "progression" of individual horse skeletons. In his classic book *Horses* (1951), Simpson exploded Marsh's "single-line" evolution of the horse from a fox-sized hoofless ancestor.

Instead, Simpson showed the complex and diverse branching of the horse's ancient relatives, not only through time, but over geographical areas, as early populations pushed into various habitats, adapting first to forests, then to open grasslands. Horses represented a complex, branching bush of diverging species—nothing like a line leading straight from *Eohippus* to old Dobbin.

A pioneer in tackling the problem of rates of evolution, Simpson was impressed with the pattern of long periods of stability in species, interspersed with relatively rapid change. Creationists had seen these "discontinuities" as evidence that no evolution had occurred, while Darwin considered them gaps in an imperfect fossil record. Employing Sewell Wright's idea of genetic drift, Simpson argued that important changes might occur fairly rapidly in very small populations, leaving little fossil evidence before they spread and stabilized in large numbers. In his book *Tempo and Mode in Evolution* (1944), he introduced the term "quantum evolution" for the phenomenon, a precursor of the theory of punctuated equilibrium. Thus, Simpson, as well as Ernst Mayr, laid the groundwork for the punctuated equilibrium theory that became widely popular during the late 1970s.

GEORGE GAYLORD
SIMPSON

He was also an engaging and popular writer: His journals of his travels and explorations attracted a wide readership (*Attending Marvels: A Patagonian Journal*, 1934). Simpson's lively and often definitive discussions of evolutionary theory and its history can be found in *The Major Features of Evolution* (1965), *Evolution and Geography* (1953) and his delightful *Book of Darwin* (1982), a personal guide to the life and works of the patriarch of evolutionary biology.

See also GENETIC DRIFT; HORSE, EVOLUTION OF; PUNCTUATED EQUILIBRIUM; SYNTHETIC THEORY

SINCLAIR DINOSAUR
Saurian Corporate Image

For 40 years, Harry Sinclair, chief executive of the Sinclair Oil and Refining Corporation, shared equal billing with a diplodocus. His thousands of gas stations sported the dinosaur sign, while ubiquitous billboards made it an inescapable part of the American landscape. But while Sinclair's "dino" sold gasoline, paleontologist Barnum Brown was selling Harry Sinclair real dinosaurs.

SINCLAIR GAS STATIONS promoted their petroleum products as the remains of dinosaurs (top). AERIAL FOSSIL HUNTERS Barnum Brown and his wife beside their plane, *Diplodocus*, provided by the Sinclair Oil Company to search for both oil fields and dinosaurs (bottom).

Brown courted and cajoled the oil magnate for months until he agreed to bankroll a decade of expensive bone-hunting expeditions in the American West. Throughout the 1920s and 1930s, Sinclair picked up the tab for Brown's field crew from the American Museum of Natural History, and each season's discoveries became part of a gasoline promotion campaign. Brown created popular dinosaur cards, stamp books, and maps that were given out at Sinclair gas stations.

After a while, Sinclair decided that the museum expeditions could really earn their keep by scouting for new oil fields. An acute field geologist as well as a dinosaur hunter, Brown pioneered the aerial survey in Sinclair's light plane, *Diplodocus.* Now he could search for dinosaurs and oil simultaneously. Rarely had there been a happier marriage of science and commerce.

On one of the expeditions to the Big Bend region of Texas, Brown and his chief assistant, R. T. Bird, unearthed an enormous fossil crocodile of previously unknown species. In honor of their benefactor, it has since borne the name *Phobosuchus sinclairi.*

In 1964, Sinclair Oil sponsored the creation of life-sized fiberglass dinosaur models at the World's Fair held in Flushing Meadow, New York. After the fair's run, they were donated to various parks and exhibits around the country. Several of the sculptures now have a permanent home in a small park near Glen Rose, Texas, where R. T. Bird and Barnum Brown found many dinosaur bones and footprint trackways of the real thing. Many dinosaur footprints can still be seen in the rocks of the nearby Paluxy River.

See also BIRD, ROLAND THAXTER; BROWN, BARNUM; *FANTASIA*

DARWIN'S GHOST-BUSTER, Dr. E. Ray Lankester, holds up a slate on which "medium" Henry Slade (far left) caused "spirit-writing" to appear. Backed by Charles Darwin, the 1876 prosecution of Slade was the first time a scientist had charged a psychic with criminal fraud.

SLADE TRIAL (1876)
Darwin vs. Wallace on Spiritualism

Charles Darwin and Alfred Russel Wallace—the two greatest naturalists of the 19th century—took opposing sides when the supernatural went on trial. The fascinating confrontation between the cofounders of evolutionary theory took place in November 1876, but it was downplayed by the Darwin family and almost lost to history. At issue was whether the famous American psychic Henry Slade was sincere about communicating with "departed spirits" or was merely a clever con man. Known as the "slate-writing medium," Slade's specialty was posing questions to the spirit of his dead wife and receiving mysteriously written answers on slates. Edwin Ray Lankester, a young zoology student of Thomas Henry Huxley's, had paid to attend a seance at Slade's with hopes of catching him in trickery. Boldly, he snatched the slate from the medium's hand in the darkened room and found an answer written before the question had been asked. Lankester hauled Slade into police court as a "common rogue."

Alfred Russel Wallace, a staunch believer in spiritualism, gladly appeared as star witness for the defense. Known for his honesty and as Darwin's codiscoverer of evolution by natural selection, Wallace testified that Slade was "as sincere as any investigator in a university department of natural science."

Darwin, on the contrary, was convinced that all "spirit-mediums" were "clever rogues," preying upon the credulous and bereaved. He wrote Lankester that he considered it a "public benefit" to put Slade out of business, and quietly contributed funds to the cost of prosecution.

Slade claimed at his trial that he didn't know how the writing was produced on the slate. John Nevil Maskelyne, the well-known stage magician, got up and performed a few tricks with slates to show how such effects might be produced, but the judge disallowed the performances in a vain attempt to keep the trial from becoming a circus. Finally, the judge announced that he had to rule "according to the ordinary course of nature" and convicted Slade under an old law against palmists and fortune-tellers.

Slade never served a day in jail. A few months later, his conviction was overturned on a technicality, and he fled England for Europe. He continued

Messrs. SLADE and LANCASTER,

In a Scene from the New Farce, "The Happy Medium; or, No Spirits should be above Proof."

his spirit-writing seances in Germany, where he impressed the renowned physicist Johann Zöllner with a demonstration of "odic force," and even convinced the local police chief of the authenticity of his powers.

However, after repeated exposures as a fraud, his reputation declined, and the faded celebrity ended up living in a run-down New York boarding house. On slow news days, city editors would initiate cub reporters by giving them the price of a seance and sending them to expose old Slade one more time.

One of Lankester's motives in pursuing Slade was to punish Wallace, who he thought had "degraded" scientific meetings by permitting a paper on "thought transference" to be read. Only a few days before Lankester's exposure of Slade, Wallace had chaired the anthropology section at the British Association for the Advancement of Science, and his vote had broken a bitter deadlock to allow a paper on psychic powers by physicist William Barrett. Despite his brilliance in science, Barrett had been taken in by a young woman who claimed to "see" events in London without ever leaving her village in Ireland.

The incident led to deep acrimony in the scientific community, for which some never forgave Wallace. Years later, when Darwin wrote his friend Sir Joseph Hooker asking help in securing a government pension for Wallace, Hooker refused. "The candidate is a leading and public Spiritualist," Hooker replied. But although they were at opposite poles on such questions as spirits and miracles, Darwin prevailed on his colleagues to approve Wallace's pension, in belated recognition of his extraordinary contributions to natural science.

See also LANKESTER, E. RAY; SPIRITUALISM; WALLACE, ALFRED RUSSEL

SMITH, SYDNEY (1911–1988)
Zoologist, Historian of Science

Cambridge zoologist Sydney Smith made important studies of trout embryos, led a chamber music group, and collected fine old wines. His imaginary ideal event, he said, would be to serve von Schubert wines "during pauses in the rehearsal of Franz Schubert's *Trout* Quintet." Historians of science may forget these passions of Smith's, but not his great labor of love: joining historian Frederick Burkhardt in 1974 to organize Charles Darwin's books, letters, and manuscripts, which had become a scattered and inaccessible jumble.

Darwin often cut up his notes, transferred them, sometimes destroyed or gave away pages of manuscripts, or made undated additions. Smith spent years helping to sort out the confusion, during which he acquired a deep understanding of Darwin's mind as he transcribed thousands of nearly illegible pages for their Darwin Correspondence Project.

> Frederick Burkhardt and Sydney Smith joined forces to gather and publish 15,000 Darwin letters.

Among the results are the definitive edition of Darwin's notebooks (1836–1844), published in 1987. (Smith's coeditors were Paul Barrett, Peter Gautrey, Sandra Herbert, and David Kohn.) With founding editor Burkhardt, he edited *A Calendar of the Correspondence of Charles Darwin, 1821–1882,* and in 1980 helped launch publication of the *Complete Correspondence of Charles Darwin,* which will ultimately include 15,000 Darwin letters.

Smith had to "badger, negotiate, charm, and coerce" owners of precious documents to part with them. The fruits of these labors now constitute the Darwin Archive in the Cambridge University Library—one of the world's outstanding resources for historians of science.

See also DARWIN, CHARLES, CORRESPONDENCE

SMITH, WILLIAM (1769–1839)
Discoverer of Strata

One of the most basic geological truths is that the Earth contains layers, or strata, that tell a consistent story of the past. William Smith, an English surveyor and self-taught engineer, learned how to read them. Orphaned at an early age and raised on a farm by an uncle, the rugged Smith was drawn to the outdoor life and made himself an expert on English terrain. After apprenticing to a surveyor, he earned a decent living as a practical expert in geology at a time when many canals were being dug for coal barges.

Canal companies hired him to survey, drain swamps, report on coal deposits, and recommend the best courses for canal routes. His livelihood depended on accurate knowledge and predictions of the composition of the land, and he was very good at his job.

While making his own surveys for maps, he found that each stratum contains its own characteristic minerals and fossils—a record of creatures and events. By understanding how to identify and use key fossils, "Strata" Smith, as he became known, taught himself to trace a given layer over long distances, even where it was eroded, discontinuous, or inverted. He could also work out its relationship to older and younger strata, discerning the sequence in which the deposits were laid down.

Before Smith, several geologists had noticed that there seemed to be a sequence in the strata. But they had looked only at the minerals or rocks, ignoring fossils as markers. Smith realized that each layer had a distinctive "signature," made up of the organisms it contained. Also, he noted that the oldest must be on the bottom, younger on top. Simple as that seems now, it was not at all obvious until Smith popularized his "doctrine of superposition." It also enabled him to recognize similar sequences that had become folded, overturned, or partially eroded.

Although he hated to write, he managed to publish *Delineation of the Strata of England and Wales* (1815), *Strata Identified by Organized Fossils* (1816), and *Stratigraphical System of Organized Fossils* (1817). He also succeeded in making the first geological maps of England. A dogged field worker who covered enormous distances over every kind of ground, he regretted in 1816 that "the theory of geology was in the possession of one class of men, the practice in another."

He declined to theorize, but bowed to the scientific climate of his times, which accounted for strange fossil creatures by conjecturing unknown or supernatural events. Nevertheless, he had opened the book of the Earth, and its story would soon be told.

Smith's efforts did not go unnoticed by men of science. In 1831, the polished gentlemen of the Geological Society of London awarded their first Wollaston Medal to the tough, weatherbeaten "Strata" Smith.

See also PROGRESSIONISM; UNIFORMITARIANISM

SOCIAL BEHAVIOR, EVOLUTION OF
A Widespread Adaptation

Social life is neither a human invention nor the special province of our primate relatives, but has evolved countless times in the natural world. Disproportionate attention paid to monkeys and apes because of their evolutionary kinship to ourselves has obscured an important fact: many kinds of creatures have evolved remarkable societies.

In the past, anthropologists, with the exception of Darwin's protégé Sir John Lubbock, never bothered to study such social insects as bees, ants, or wasps, summarily dismissing them as brainless automatons whose social behavior is instinctively programmed. Mere "herds" and "aggregations" of elephants or social birds did not hold a candle to primates. But over the last few decades, even as the world's wildlife is disappearing, dedicated field researchers have made astounding discoveries that are helping to place human social behavior in evolutionary perspective.

In 1968, anthropologists Michael and Barbara MacRoberts, who had previously studied free-ranging Gibraltar macaque monkeys, focused their attention on a small bird, California's acorn woodpecker. This remarkable woodpecker lives in complex colonies that communally store food in tree trunks, which they pack with thousands of acorns. Each acorn is hammered into a custom-carved compartment, carefully shaped by the birds for a snug fit. These "granaries" are worked on by scores of individuals and are passed on for the use of succeeding generations. After the MacRobertses' pioneering monograph appeared, scores of other researchers (zoologists rather than anthropologists) have continued to unravel the complexity of the woodpecker's world.

STORAGE GRANARIES pecked out of trees hold thousands of acorns for use by woodpeckers, who store enough for future generations. Their social behaviors seem as complex as those of monkeys. Photo © by M. H. MacRoberts.

HEADING INTO THE SALT MINES, elephant herds make difficult expeditions deep into Kitum Cave at Mt. Elgon in Kenya. Worn trails and tusk-gouged walls show that they have mined minerals there for thousands of years. Photo courtesy of and © by Ian Redmond.

With a brain no bigger than a thimble, acorn woodpeckers lead a social life as complex as any monkey's. They stake out and defend territories, which include nesting sites as well as food resources and storage granaries; exhibit elaborate communicative behaviors; and have a kin-based network of social relations. Also, they are "cooperative" breeders, with unmated "helpers" assisting parents in hatching eggs and rearing young.

Michael and Barbara MacRoberts could not find any distictive behaviors that would set societies of monkeys apart from those of other animals. It was heresy for anthropologists, though not an unusual point of view for some zoologists. George Schaller, for instance, studied gorillas, giant pandas, and tigers with an evenhanded approach, but anthropologists often treated primates as especially bright simply because they're our closest relatives.

Over the past thirty years, other scientists have begun to study social behavior in animals other than primates, discovering a wide range of complexity in their social lives.

Stephen Emlen (whose father, John Emlen, had trained George Schaller in zoology) made interesting discoveries about the societies of an African bird known as the bee-eater. Social bee-eaters live in sandy cliff "apartment houses," with each burrow scooped out by three to five "team" members who sleep together, cooperate in food-gathering, rearing and provisioning of young, and defense against predators. Teams sometimes allow other birds related to them (clan members) to sleep in their burrows, but will drive out individuals who belong to other clans.

Among mammals, recent studies of African hunting dogs, meerkats, and mongooses show hitherto unsuspected social complexity, learning, division of labor, teamwork, and manipulation of the environment. Meerkats and mongooses, for instance, take turns standing upright, performing "sentry duty" while the others feed, play, or mate. Good sentinels learn their roles from older individuals, who take on "apprentices." Usually they divide the task, with some searching the skies for hawks and eagles while the others scan the ground and horizon in all directions.

Like social birds, meerkats and mongooses help feed and rear their kinsmen's youngsters. Adults routinely risk danger to retrieve straying juveniles (not their own) to safety. One adult "babysitter" will patiently guard, nurture, and play with several of the colonies' youngsters for hours while their parents are out foraging.

Field research on African elephants continues to reveal unsuspected dimensions of social complexity, intelligence, and communication. Recent studies have proved their familiar trunk-raising and ear-spreading behavior enables herds to keep in touch with one another on the open plains, even though miles apart. Electronic equipment has picked up the low-frequency calls, inaudible to human ears, with which elephants can stay aware of the movements of distant groups in all directions.

Leadership by qualified, experienced matriarchs has become a well-known feature of elephant societies. Within the past decades, elephant "traditions" have been documented, including the remarkable trek of Kenyan herds to mountain caverns where they "mine" salts and minerals with their tusks to supplement their diets. Trails worn in the cave floors provide evidence of the hundreds of generations that have made the difficult, dangerous expedition. (A misstep in the cave can mean death, which occasionally happens.)

Scrutiny of primate behavior, from lemurs to monkeys and apes, has been going on now for more than 60 years. It was (and is) a worthwhile venture. But it has not yielded the sort of profound insights into the evolution of human behavior that latter-day Darwinians had expected. It appears that we share sociality with many kinds of animals, but evolution into the human niche was a unique historical event, not an inevitable development from primate societies.

See also APE LANGUAGE CONTROVERSY; BABOONS; CHIMPANZEES; INSECT SOCIETIES, EVOLUTION OF

SOCIAL DARWINISM
Political-Economic Ideology

What is usually called "Social Darwinism" was the wedding of evolutionary ideas to a conservative political program in the 1870s. It had a particular vogue among American businessmen, elevating traditional virtues of self-reliance, thrift, and industry to the level of "natural law." Proponents of the theory, which was based more on the writings of Herbert Spencer than of Charles Darwin, urged laissez-faire economic policies to weed out the unfit, inefficient, and incompetent.

One of Social Darwinism's leading spokesmen, William Graham Sumner of Princeton, thought millionaires were the "fittest" individuals in society and deserved their privileges. They were "naturally selected" in the crucible of competition. Andrew Carnegie and John D. Rockefeller agreed, espousing similar philosophies that they thought gave a "scientific" justification for the excesses of industrial capitalism. "This is a world," Sumner wrote in 1914 ("Reply to a Socialist"), "in which the rule is 'Root, hog, or die,' and . . . in which 'the longest pole knocks down the most persimmons.'"

From the 1870s through the 1890s, he waged a "holy war" against reformism, protectionism, socialism, and government intervention, writing several books, including *What Social Classes Owe to Each Other* (1883) and *The Absurd Effort to Make the World Over* (1894). One of his chapters surveying human customs had gotten too long, so he brought it out as a separate book, the classic *Folkways* (1906).

Sumner argued that it was not cruel to pit the strong against the weak in the economy, because the strong are "industrious and frugal," while the weak are "idle and extravagant."

Like other great truths, Darwinism seemed to lend itself to the most wildly conflicting political programs, depending on who was doing the interpreting. Edward Bellamy, the utopian social critic, thought the complete elimination of competition would hasten evolutionary perfection. Cooperation and socialism would develop through slow transitions; after all, as Darwin had taught, "nature makes no leaps." Sumner's reply was that socialism was "a plan for nourishing the unfittest and yet advancing in civilization"—a combination "no man will ever find."

Karl Marx wrote his friend Friedrich Engels that Darwin's theory was "the basis in natural history that we need" for the philosophy he called "scientific socialism." In Darwin's "materialism" he found ammunition against the "divine right" of kings and a social hierarchy supported by religion. And the idea that evolution is a history of competitive strife fit well with his ideology of "class struggle."

Marx sent Darwin a copy of his major work *Das Kapital* (1867), but the naturalist never read it (the pages remained uncut). Communists as well as capitalists claimed to be "Social Darwinists," though their reasons were very different. Engels eulogized Marx by claiming he had found the laws of human society, just as Darwin had found those of nature.

When Mendelian genetics came into vogue about 1900, the idea of discontinuous evolutionary jumps in nature suggested a basis for revolution in the social sphere. Some ideolo-

Both capitalists and communists have claimed to be the true "Social Darwinists."

gists seized upon it as the antidote to Darwin's notion of slow, steady changes. Yet after the Russian Revolution, "Mendelists" were reviled by doctrinaire Soviet scientists. Now a new society was to be achieved through "improvement" of the peasantry, producing cumulative genetic "progress." Under the tyranny of Lysenko's falsified "proof" of "use inheritance," they refused to believe each generation must be educated anew.

Another Darwinian social philosophy was claimed by the anarchists, whose spokesman was Peter Kropotkin, a Russian prince who despised the excesses of the nobility. Kropotkin took his cue from the cooperative social behavior of animals and certain passages in Darwin's *Descent of Man* (1871). He thought natural social cooperation was the true form of Social Darwinism.

In his book *Mutual Aid* (1902), Kropotkin argued that evolution had produced a great deal of social behavior in the natural world; survival often depended on individuals combining for their mutual benefit. His anarchist philosophy was not simply the absence of all rules and order, with everybody running wild. He deeply believed that if mankind was freed from oppressive and corrupt institutions, a natural, harmonious order would reassert itself. Cooperation for the common good in a classless society, he thought, was basic human nature in its natural state.

Theological liberals tied Darwinism to social progress as part of God's plan. Many Christians, such as the Jesuit paleontologist Pierre Teilhard de Chardin, found in evolution an inevitable "ascent" of humanity. Man was not a fallen angel but a risen ape, still progressing upward.

Reverend Henry Ward Beecher, once the most popular Protestant preacher in America, taught that God's plan is to constantly improve man. Moral progress toward a higher type of being lay ahead; sins were simply slips back to a more animalistic behavior. While some liberal Christian theologians were prepared to shed guilt and Original Sin, Social Darwinists like William Graham Sumner seemed just as driven by their grim duty to evolutionary "competition" as any Calvinist ever was by his duty to God.

Thomas Henry Huxley viewed evolution in nature as bloody and ruthless, but thought man is obliged to leave a gladiatorial view of nature behind and seek a better way. He taught that humans have the choice not to accept the "law of the jungle." Instead, we must struggle toward a compassionate and humane society.

Germany's leading evolutionist, Ernst Haeckel, on the other hand, thought man must "conform" to nature's processes, no matter how ruthless they may seem. The "fittest" must never become so foolishly "sentimental" that they stand in the way of evolutionary progress. In its extreme form, that social view was used in Nazi Germany to justify sterilization and mass murder of the "unfit" and "incompetent," and ultimately to the annihilation of "inferior races."

Darwin's politics were liberal (sometimes radical) for his day; he had too much compassion for the underdog to be a Social Darwinist in the Anglo-American sense. Writing to Charles Lyell in 1860, he laughed at a newspaper squib claiming "that I have proved 'might is right' and therefore that Napoleon is right, and every cheating tradesman is also right." He was passionately opposed to slavery, was known as a very lenient magistrate, campaigned against abusive child labor practices, and was locally admired for his philanthropies.

Yet he also was resigned to witnessing the subjugation of tribal peoples, whom most Englishmen considered "inferior." He had seen first-hand the extermination of South American Indians by the Argentine army and thought the slaughter of indigenous Australians and Tasmanians an inevitable outcome of the clash between "advanced" and "savage" races.

Sometimes empire-building Englishmen spoke of their "white man's burden": the duty of "civilized" nations to bring material and moral progress to the "backward" races. In an 1860 letter to Lyell, Darwin cynically remarked that thanks to the European's penchant for "improvement," many native populations were being "improv[ed] off the face of the earth."

See also BEECHER, HENRY WARD; CARNEGIE, ANDREW; HAECKEL, ERNST; HUXLEY, THOMAS HENRY; KROPOTKIN, PRINCE PETER; TEILHARD DE CHARDIN, FATHER PIERRE

SOCIOBIOLOGY

See ETHOLOGY; KIN SELECTION; "SELFISH GENE"

SPECIES, CONCEPT OF
From "Type" to Population

Few questions in evolutionary biology seem so simple, yet are really so complex as: What is a species? Surely any fool can tell a wolf from a tiger, or a bluebird from an ostrich. Creationists cite the biblical statement that God created each animal "after its kind," once and forever, as a fixed, unchanging "unit of creation."

Early naturalists, educated by the church, based their science on the congenial Platonic concept of "ideal types." Animals and plants were thought to be imperfect expressions of abstract forms or ideas in the mind of God. Any particular zebra partakes of the essence of Zebra.

In Carl Linnaeus's (1707–1778) influential system of classification, the naturalist chooses a "type specimen" when naming a new species. Imagine the perplexity of a naturalist from Mars, for instance, trying to identify a bulldog, chihuahua, dachshund, or great dane as members of the same species by comparing them to an arbitrarily chosen "type specimen" of *Canis familiaris*. Yet genomic studies show that all these varieties have descended from the wolf during the past 10,000 years.

Typological notions of species crumbled as naturalists confronted two pervasive facts. First is the tremendous variability within species. Even those that seem to be composed of almost identical creatures reveal, on closer examination, a wide range of individual differences. Second, mutations and hybrids that differ from their parents blur the boundaries of species—prompting even the great classifier Linnaeus, late in life, to question the typological species system he made famous.

In the late 18th century, some naturalists who saw the difficulties of defining species as an "essence" or "fixed type" declared that all we really see in nature are individuals. Grouping them in a class and naming that class "zebra" is an exercise of the human mind: It may be convenient, but it has no reality as a unit in nature. (This is known as the "nominalist" definition, after a medieval school of thought that held that reality was created by giving names to things.)

Charles Darwin began wrestling with the species concept during his voyage on HMS *Beagle* and by 1830 had jotted some important insights in his journal. During his sojourn in the Galápagos Islands, he noticed the great variability among different kinds of birds and reptiles, and he puzzled over whether they were true species or merely varieties of the same species. He finally decided to let the animals themselves decide the species question. Those that were attracted to one another and mated belonged to the same species. Darwin had observed courting displays and other behavioral modes of species recognition, which are known today as "isolating mechanisms."

Despite their apparent "repugnance" for individuals of other species, Darwin realized that at some future time such populations might mingle and produce hybrids. But he wrote that "until their instinctive impulse to keep separate" is overcome, "these animals are distinct species." He also noted that "species may be good ones and differ scarcely in any external character."

But 20 years later, when Darwin discussed species in the *Origin* (1859), it would almost seem, as Ernst Mayr has put it, "that one is dealing with an altogether different author." He now seems to regard species as a purely arbitrary label for the convenience of taxonomists and opines that only a naturalist of "sound judgment and wide experience" should attempt to identify one.

In a letter to Joseph Hooker in 1856, he commented on the "laughable" confusion over the definition of species. What goes on, he wondered, in naturalists' minds when they think of species? "In some, resemblance is everything and descent of little weight—in some, resemblance seems to go for nothing, and creation the reigning idea—in some, descent is the key—in some, sterility an unfailing test, with others it is not worth a farthing. It all comes, I believe, from trying to define the undefinable." His "solution" in the *Origin* is that species are fluid and change through time. Since they have no fixed essence, one need not worry too much about defining them.

The "biological species" concept championed by Harvard's Ernst Mayr strongly influenced two generations of biologists. Mayr characterized species as breeding populations occupying a specific niche in nature and reproductively isolated from one another by geography, ecology, or behavior. This concept, which Mayr believes young Darwin had glimpsed in the Galápagos, was lost for a century after he retreated from it and adopted the view that species were "undefinable." In recent years, however, Mayr's "biological species" concept has become controversial, and the reality of species remains an unsettled question.

See also ESSENTIALISM; HYBRIDIZATION; MAYR, ERNST; TRANSITIONAL FORMS

SPENCER, HERBERT (1820–1903)
Philosopher of Evolution

For his evolutionary theories of technology, ethics, nature, and society, English author Herbert Spencer was acclaimed as the greatest of Western philosophers. His works made him world famous by 1870, and in America his star rose higher than that of his countryman Charles Darwin. The majority of the public took little interest in Galápagos finches or fossil sloths, but were keen to know what policies science could recommend about the problems of the poor, industrialization, foreign wars, and the best way to conduct their lives. While Darwin confined his theories to organic evolution, Spencer took on the evolution of everything.

Applying his skills as an engineer and mathematician, Spencer set out to find the natural laws that govern all phenomena, from the evolution of societies to the "evolution" of stars. Seven years before the appearance of Darwin's *Origin of Species* (1859), Spencer published "The Development Hypothesis," which denied the biblically inspired notion of "special creation" and instead championed organic evolution. In his *Principles of Psychology* (1855), he maintained that "life under all its forms has arisen by a progressive, unbroken evolution; and through the immediate instrumentality of what we call natural causes." By evolution, he meant a steady change from homogeneous beginnings to a differentiation—whose branches might, at some point, fuse again in a new synthesis.

After writing *Principles of Biology* (1864), he published the three-volume *Principles of Sociology* (1876–1896), which established him as "the father of sociology." Spencer was much more interested than Darwin in understanding how society works and in seeking evolutionary "laws" that govern history, economics, and cultures. He viewed societies, like organisms, as entities whose structure and functions could be compared and whose evolution could be worked out.

Contrary to widespread distortions of his views, he did not believe that all societies evolved through the same stages toward some universal goal. Indeed, he deliberately used the word "evolution" rather than "progress" because he thought societies could also retrogress. Nevertheless, he worked out a general scheme of social evolution that some anthropologists notably, Robert Carneiro, at the American Museum of Natural History, still consider valid.

Despite his broad range of concerns, Spencer became best known for providing an ethical and scientific basis for laissez-faire capitalism. He coined the term "survival of the fittest," which Darwin later adopted for his descriptions of organic nature. Andrew Carnegie and other capitalists found in Spencer a rationale for ruthless elimination of competitors as part of the natural process that would eventually produce a harmonious and more perfect society. Indeed, some have said that "Social Darwinism" should have been known as "Social Spencerism."

In Spencer's vision of social evolution, the poor classes might eliminate themselves by their "unfitness," and the naturally strong and powerful would rise to the top and leave as many offspring as possible. He wrote in *Principles of Ethics* (part VI, 1892): "If left to operate in all its sternness, the principle of the survival of the fittest, which, as ethically considered, we have seen to imply that each individual shall be left to experience the effects of his own nature and consequent conduct, would quickly clear away the degraded."

Yet he was well aware of the contradictory fact that the uneducated underclass was reproducing alarmingly well, while the more cultivated and successful were producing comparatively few offspring. While he commiserated with the suffering of the poor, socialism was anathema to him. He wrote that he did not advocate "state-almsgiving," as it promoted

"SURVIVAL OF THE FITTEST" was a phrase coined by English philosopher of evolution Herbert Spencer, a polymath who dabbled in biology and economics and is sometimes known as "the father of Sociology." He also invented the paper clip.

idleness and dependency in recipients while victimizing those who were forced to fund it. In his often contradictory way, he also criticized the very capitalists who lionized him, particularly the American variety, for a pathological obsession with work that could lead them to avoidable breakdowns in mental and physical health. In his thousand-page *Autobiography*, he admitted that he was really preaching to himself, for he was a driven overachiever prone to neuroses and breakdowns who longed to be able to relax and enjoy life.

Spencer believed that war played an important role in shaping societies. When it came to warfare, his phrase "survival of the fittest" (which others used in a general and imprecise way) took on a specific meaning. Societies or tribes that had "outcompeted" others by force were objective examples of the principle in action. However, Spencer was deeply saddened by the persistence of war and wanted to believe that humankind had outgrown the need for it. When Europe turned toward militarism after 1870, Spencer despaired that true social progress seemed as elusive as ever. He was also troubled by the trend toward socialism, which he thought must lead to the total collapse of society.

Spencer's first theory of organic natural selection was published in *Social Statics* (1851), which contained numerous examples from the natural world. However, while he was brilliant at constructing theories, some critics among scientists complained that he seemed to have no interest in testing his ideas with experiments.

Darwin, for instance, admired Spencer's nimble theorizing but wrote in his *Autobiography* (1876) that any one of the ideas he was constantly throwing out "would be a fine subject for half-a-dozen years' work." To Darwin, Spencer's generalizations did not seem "to be of any strictly scientific use. . . . They do not aid one in predicting what will happen in any particular case . . . [and] have not been of any use to me."

In his personal life, Spencer's youthful rebellion against authority became crankiness as he grew older. He devoted himself completely to thinking and writing, and never married or romanced women. In later life, he was subject to nervous disorders, disliked social company, regularly used opium, and would stop his carriage in the midst of city traffic to take his pulse—a daily ceremonial event.

Even in middle age, his nervous condition caused him to stay awake all night if he became the least bit excited or upset. Therefore, he carried with him at all times a pair of ear pads connected by a spring that passed around the back of his head.

Sir Ray Lankester, the Darwinian zoologist who became director of the British Museum of Natural History, told Spencer's biographer, Hugh Elliot, of being asked by Spencer to provide him with some information on a biological question. When young Lankester answered the summons and appeared at Spencer's club, the philosopher immediately began to expound his own theories on the matter. Based on the facts he had gathered, Lankester tried to point out some difficulties and objections. "Spencer hastily closed the conversation by fitting on his ear-pads," Sir Ray later recalled to Elliot, "saying that his medical advisers would not allow him to enter into discussions."

Once when Thomas Huxley and other scientists were dining with him, Spencer announced he had just written a tragedy, as related in Francis Galton's *Memories of My Life* (1909). "Yes," Huxley replied, "I know the catastrophe." Spencer protested it had never been shown to anyone, but Huxley insisted he could reveal the plot. Spencer's idea of a tragedy, said Huxley, is "a beautiful theory, killed by a nasty, ugly little fact."

Spencer did have a very practical side, however. To organize the mass of manuscript pages on his writing desk, he invented the paper clip.

See also CARNEGIE, ANDREW; SOCIAL DARWINISM; "SURVIVAL OF THE FITTEST"

SPERM COMPETITION
Natural Selection of Gametes

Gametes are sex cells: ova (eggs) in the female and sperm in the male. When they combine to form an embryo, natural selection may operate on the developing fetus, or on the infant, the child, or the adult. But in 1970, a study of insect reproduction by Geoffrey Parker introduced an idea Darwin never considered: competition at the level of gametes.

Female insects sometimes mate with several males and store their sperm for days or, in some cases, years. "What determines which male's sperm will succeed in fertilizing her eggs?" Parker wondered.

He found that in some cases a "rival's" sperm was diluted or simply washed away; in others males would insert plugs in the female after mating—the insect version of a chastity belt. In recent years, the phenomenon has been studied in reptiles, amphibians, birds, and various mammals.

According to Allan Pacey of the University of Sheffield, sperm competition is like an evolutionary arms race. Pacey and his colleagues examined the semen of 12 different species of primates, and reported their results in *Nature Genetics* in 2004. In species where the females were most promiscuous, the males had developed several strategies to ensure they would be most likely to father offspring and pass on their genes. One of them has to do with the thickness and stickiness of the semen.

Chimpanzees, for example, which are a promiscuous species, have evolved stickier semen than that of gorillas, which tend to stay faithful to a partner for long periods of time. Bruce Lahn and his colleagues have shown that stickiness is determined by the gene SEMG2, which regulates the production of semen coagulum, a natural glue. In the most extreme cases, its effects were so pronounced that the semen became a solid plug, blocking entry of other sperm while fertilization was in process. Human semen is only moderately sticky, indicating perhaps that our species is less promiscuous than chimps but more so than gorillas.

See also SEXUAL SELECTION

SPIRITUALISM
Cosmic Evolution Controversy

SPIRITUALISM was a major antiscience movement for almost two centuries, fueled by con men and naive believers. Mimicking "spirit photography" (bottom), British illusionist John Nevil Maskelyne conjured departed souls as double exposures.

A SPIRIT-RAPPING SÉANCE!
WHAT FOXES WILL SAY, GEESE AND ASSES WILL BELIEVE

The spectacular rise of the spiritualist movement in 19th-century England and America posed a dramatic challenge to evolutionary science. Psychics claimed to offer "objective proof" of spiritual "phenomena," including communication with the dead—directly opposing scientific "materialism." And while they agreed humans had evolved, they proclaimed that evolution extended from the material body to cosmic spirit. Presented as a new philosophy, spiritualism was a reinterpretation of ancient Eastern beliefs, mixed with the Western desire to "secularize" the supernatural and establish it "scientifically."

A handful of respected savants, notably the chemist William Crookes, discoverer of the element thallium, physicist Sir Oliver Lodge, and Alfred Russel Wallace, codiscoverer of the theory of natural selection, believed they had contacted the spirit world. Wallace published a series of controversial essays supporting the existence of psychic phenomena. His book *Miracles and Modern Spiritualism* (1874) was endlessly quoted by mediums, who sought credibility through his scientific reputation. At the sensational trial of the "spirit-medium" Henry Slade in 1876, Wallace appeared as star witness for the defense. But most leading scientists, including Charles Darwin, physiologist William Carpenter, and botanist Sir Joseph Hooker, were implacable foes of spiritualism and resented Wallace's attempt to allow papers about "psychic phenomena" to be read at a scientific meeting.

Darwin expressed the majority opinion that "it was all trickery and deceit." Eventually, the sceptics were vindicated, for within the decade most of the top dozen mediums were caught in fakery and fraud. These exposures were usually made not by scientists, but by professional stage magicians and conjurers, who were experts in detecting how the uncanny "phenomena" were produced. Darwin detested mediums for preying on grief-stricken parents who had lost children (like himself), and carried on a secret campaign to expose them, which had not been known to historians until recently.

LIFE AFTER DEATH DECLARED PROVED BY EVOLUTION

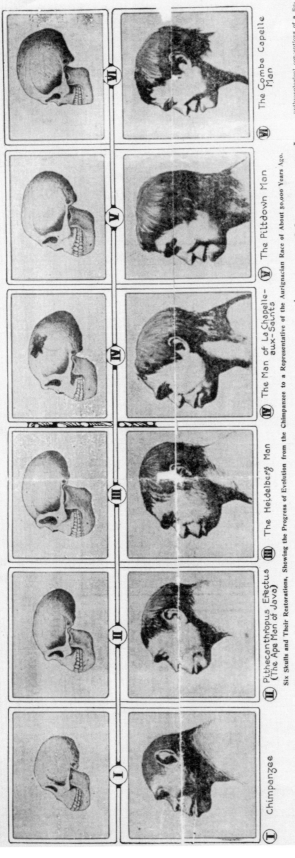

(I) Chimpanzee (II) Pithecanthropus Erectus (The Ape Man of Java) (III) The Heidelberg Man (IV) The Man of La Chapelle-aux-Saints (V) The Piltdown Man (VI) The Combe Capelle Man

Six Skulls and Their Restorations, Showing the Progress of Evolution from the Chimpanzee to a Representative of the Aurignacian Race of About 30,000 Years Ago.

Prominent Scientist Holds That Man's Ascent from Apes Means an Incessant Evolutionary Process for Humanity, Not Stopped by the Grave, and Proves the Existence of a Supreme Being.

By George MacAdam.

HOWEVER any of us might be willing to consider ourselves descended from an ape upon his father's side, no one would so demean his mother's memory as to imagine that she could possibly have shared in the descent."

That was the statement made by Bishop Wilberforce at the famous meeting of the British Association in 1860 at Oxford. The Bishop was the 'retiring President,' and, in accordance with the custom of the society, he gave a summary of the advance of science, especially during the preceding year. Darwin published his great work, "The Origin of Species," in 1859.

In shaping the sentence just quoted he got the better of him. But the bitterness of the Bishop undoubtedly got the better of him. But the spirit of his speech—a denunciation of a theory that would relegate the Divine creation of man in the Garden of Eden to the limbo of exploded beliefs and substitute in its place a simian ancestry—the spirit, this speech met with general approval. The theory attacked by Darwin and Wallace and soon championed by

many other old dogma by the evidence which science has recently brought to light has led to a profound reaction in thought during the past thirty or forty years.

"With many thinking people it has led to an almost complete renouncement not only of religious dogmas but of the very fundamentals of religion itself.

"So, too, in conversing with many scientific men on the relation of modern scientific thought to those deeper problems involving the idea of a Supreme Being and the continuity of life after death, one often encounters what is apparently the same attitude of mind.

"In reality these scientific critics are criticising images of dogma which

It is just that way. At the present moment, however, I wish to take up another point, and will come to my answer to your question later.

"There is a perfectly legitimate and logical point of view from which it may be said that man has always existed. If by that we mean that the appearance of the first unicellular organism, the first form of organic life, made the final appearance of organic life inevitable. And it is perfectly legitimate to think that the life impulse working in those first organic forms did make the final appearance of man inevitable.

"So viewed, the whole long line of ascent from amoeba to man may be regarded as the human line of evolution. Somewhere, always keeping to

danger to him, was originally part of a more extended digestive area.

"It is perfectly well known that the human embryo in its development passes through the entire evolutionary process of the vertebrates.

"A further and convincing proof I have recently worked out, establishing man's relationship to the anthropoid apes. This evidence is the fact that there are three distinct and strongly marked types of teeth which are found in all races of mankind, ancient and modern, savage and civilized, and also in the anthropoid apes. This is regarded by Prof. Haeckel, author of 'The Riddle of the Universe' and 'The Evolution of Man,' as a new and convincing proof of the origin of man and of the anthropoid apes from a common

his place in the evolutionary scale as shown by the size of his own skull and other parts of his organism.

"Thus we see that his rise in the evolutionary scale is entirely due to the slow triumph of mind over matter. As man rises in the scale we see a gradual increase in his brain capacity.

"Let us glance at a few facts about these skulls:

"In 1891 Dr. Eugene Dubois, Professor of Geology in the University of Amsterdam, found on the banks of the Bengawan, at Trinil, Java, the top of the skull, two molar teeth, and the thigh bone of a creature presenting characteristics which placed him in a position about midway between the highest apes and the lowest human beings heretofore known. The esti-

tained that this could not be the skull of a normal human being. But the subsequent discovery of three other twenty skulls of the same general type proved that Virchow was quite wrong. The discovery of so many skulls of one ancient type, extending as they do over a great stretch of time is one of the most important features of our knowledge of primitive man. Let us compare three or four of these specimens.

A Remarkable Skull

"We may take that wonderful skull known as 'The Man of La Chapelle-aux-Saints' as typical of the race. We see at a glance that we have here to deal with something altogether different from the modern type of man. This outline of

anthropological conceptions of a Supreme Being."

Asked for his own conception of the Supreme Being or of his purpose in carrying on the evolutionary process, Dr. Williams replied:

"I can sum up and express my whole idea when I say that looking over the evolutionary power, involving logically inevitably from one stage of life to another, chance is simply unthinkable. I am forced to accept the conclusion that back of it is an Infinite Intelligence.

"Farther than that I do not attempt to go. Finite conceptions must necessarily have finite aspects and limitations. But whatever conceptions we are able to form of the workings of an Infinite Intelligence we must keep as far away as is finitely possible from all that limits such Infinite Intelligence.

"I think that there is much evidence that the thinking world is weary of two things—dogmatic theology and dogmatic materialism. What the thinking man of to-day needs most to do is to keep a completely open mind toward all the great problems of life.

"It is perfectly well known that the range of human consciousness, les-

IN 1913, the *New York Times* lent credibility to a claim by a Columbia University professor that human evolution is "not stopped by the grave" and "proves the existence of a Supreme Being."

After years of inconclusive attempts by the Association for Psychic Research to establish the truth of spiritualism, the organization dissolved in disarray. Eventually, the wonders of a "materialist" science—telephone, radio, television—eclipsed the marvels of mysterious rapping noises and self-playing accordions in darkened rooms. The movement declined after 1930, and mediums were reduced to carnival tricksters.

However, in the late 1970s spiritualism reemerged in a different form. Mediums were now called "channelers" but they still conveyed messages from the departed, and sometimes took on their personalities. Instead of prosecuting them in court as frauds, we now, inexplicably, give them national television shows.

SPRENGEL, CHRISTIAN KONRAD (1750–1816)
Botanist

> "It may be doubted whether [botanist] Robert Brown ever planted a more fruitful seed than in putting such a book into such hands."
>
> —Francis Darwin on his father being given Sprengel's book.

In 1841, Charles Darwin read a book that amazed him—a German book that had gone almost completely unknown and unread since its publication a half-century earlier. Its title, roughly translated, was *The Secret of Nature Revealed in the Structure and Fertilization of Flowers* (1793). His friend the botanist Robert Brown, of Kew Gardens, had sent it to him. "It may be doubted," wrote Francis Darwin, "whether Robert Brown ever planted a more fruitful seed than in putting such a book into such hands."

Christian Sprengel had been rector of Spandau, but was fired for neglecting his official duties for his beloved plants. The "secret" revealed in his book, of course, was pollination by insects—not then generally known or appreciated. For years a "pollen controversy" had raged among botanists: Was pollen really the "sperm" of plants? And if so, were lovely flowers at the mercy of vulgar visiting insects for their survival and reproduction? Only 200 years ago, this was still an open question.

Sprengel's book helped launch Darwin on a research program that would last the rest of his life, drawing him into thousands of experiments and observations in his greenhouse and garden. Perhaps its greatest culmination was his botanical classic *On the Various Contrivances by Which Orchids Are Fertilised by Insects* (1862).

See also INSECTIVOROUS PLANTS; ORCHIDS, DARWIN'S STUDY OF

STEADY-STATE EARTH
Change without Direction

For thousands of years, common observation showed the Earth was constantly in a state of change. Floods eroded hillsides, weather cracked and crumbled rocks, avalanches changed the shape of mountainsides. But such changes were not thought to modify the world in any particular direction, from one point in time to another. As the geological theorist James Hutton (1726–1797) had written, it was a steady, ongoing process with "no vestige of a beginning, no prospect of an end."

The idea, common in Elizabethan times, was expressed in Shakespeare's sonnet 64:

When I have seen the hungry ocean gain
Advantage on the kingdom of the shore,
And the firm soil win of the watery main,
Increasing store with loss and loss with store

This notion, which the Bard called an "interchange of state," has been likened to the Hindu "cosmic dance." Everything changes, yet the total remains the same.

Since Sir Charles Lyell's (1797–1895) "uniformitarian" geology strongly influenced Charles Darwin, it is commonly—and mistakenly—assumed that Lyell believed the Earth itself "evolved," undergoing progressive or directional change. Oddly enough, Lyell never questioned the very old idea of a "steady-state" Earth and wove it into his "uniformitarian" doctrine.

His famous *Principles of Geology* (1830–1833) appears in hindsight as a mixed bag of seemingly irreconcilable ideas. Lyell actually believed that time moved in great cycles,

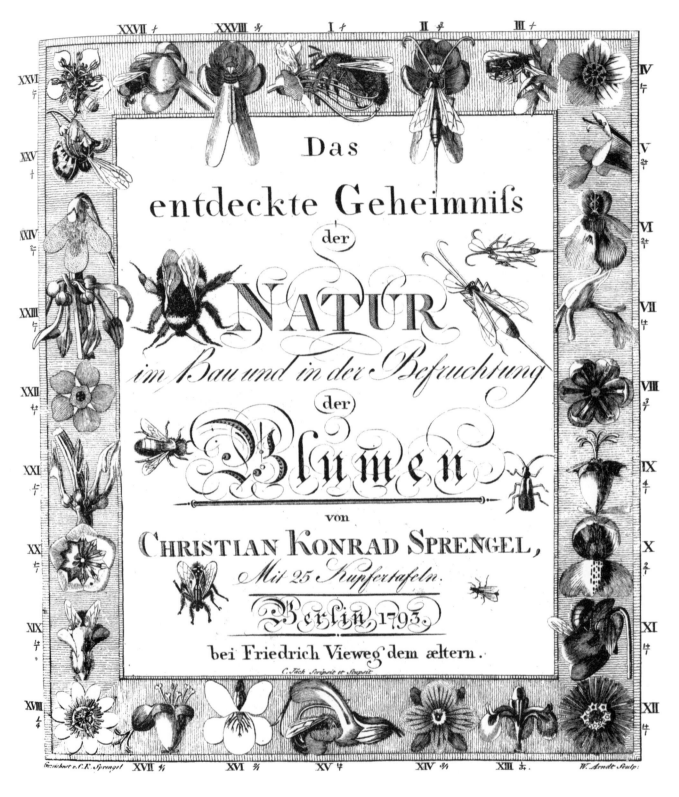

THE SECRET OF NATURE REVEALED (1793), the first book to describe pollination of plants by insects, helped to inspire Charles Darwin's botanical experiments. Its ornate title page admirably exemplfies the often close relationship between art and natural history.

resisted Darwin's demonstrations that evolution had taken place, and thought that perhaps one day the flying reptiles and dinosaurs would return.

See also ACTUALISM; CATASTROPHISM; GRADUALISM; LYELL, SIR CHARLES; UNIFORMITARIANISM

STOPES, MARIE C. (1880–1958)
Paleobotanist and Sexual Reformer

Marie Stopes was a fearless and relentless pioneer in trying to change public attitudes about sex. Her manual *Married Love* (1918) was reprinted in many languages, and in 1921 she and her husband established Britain's first birth control clinic, which was also the first in the world. After pressure and protests by religious authorities, however, her book was banned in England and the United States, although that did not prevent people from obtaining and reading it.

Stopes's thousands of letters, questionnaires, and notes from patients in her clinic compose a legacy that has still not been fully utilized. They are a treasure trove of attitudes and clinical problems about sex, the bulk of which have been sealed until 2026. One book that has already mined some of Stopes's letters, *Hidden Anxieties* by Lesley Hall, has demonstrated how valuable these primary sources remain.

Stopes's interest in birth control was heavily influenced by her background as a paleontologist with a strong interest in evolution. Her studies of fossil plants had made her aware of the fragility of species. Like other early eugenicists, she thought that the poorer classes must be made to stop breeding so prolifically, lest they bring ruinous dysfunction to society. Despite considerable opposition, she persevered in both areas. She also invented a novel birth control device, the cervical cap, which she believed was much healthier than a diaphragm, and publicly campaigned for its adoption.

Around 1900, when Marie Stopes explored Japan searching for 70-million-year-old flowers, women in academic work were still a novelty. In geology, long dominated by men (with the exception of the remarkable Mary Anning), they were considered almost an impropriety. Nevertheless, the Scotch-English paleobotanist didn't mind being the first foreigner, first woman, and first Western scientist most Japanese had ever seen.

In addition to her work on petrified plant remains from the Cretaceous rocks of Hokkaido, she wrote the charming *Journal from Japan* (1910) describing her adventures. Before 1920, her treatise *Cretaceous Flora* was published by the British Museum, and her nontechnical book *Ancient Plants* successfully popularized fossil botany. Stopes's *Monograph on the Constitution of Coal* (1918), a classic in its field, solved the puzzle of the formation of so-called coal balls. Stopes demonstrated that the structures are fossils of densely matted swamp vegetation, which often contain small marine animals that live on the plants.

Stopes was interested in the past life on Earth and in its future. A champion of women's rights along with Margaret Sanger and Victoria Woodhull, she became notorious for her views at a time when such matters were simply not discussed in public, especially by women.

To withstand the pressures of being a feminist, social reformer, scientist, and sexual guru, Stopes developed a unique personality. As one of her acquaintances put it:

> Her vanity was so colossal, so uninhibited, and so unashamed, as to be positively endearing. . . . Having done so much for other people's personal relationships, she mismanaged all her own. She was compassionate, headstrong, tactless, public-spirited, humourless, intellectually distinguished and wholly lacking in aesthetic taste. I count it a privilege to have known her.

Her legacy includes Britain's Marie Stopes International Organization, which is the country's leading reproductive health care charity, helping over 130,000 women and men each year. Its partner organizations throughout the world provide sexual and reproductive health services to nearly five million people in 38 countries.

Fossil plant specialist W. T. Gordon of King's College, London, once enlisted Stopes's aid in raising funds for a university charity. As part of his series of "science week" lectures, he

COAL BEDS AND MARRIAGE BEDS were among Marie Stopes's favorite topics. While her geological studies won scientific applause, her advocacy of legal and social equality for women made her a social pariah. Stopes founded the first birth control clinics in England.

THE UNIVERSITY OF MANCHESTER

commemorates
MARIE STOPES
1880-1958
Palaeobotanist and
Pioneer of Family Planning
Lecturer in Palaeobotany
1904-1907

announced a rare speaking appearance by the notorious Dr. Marie Stopes. It was during the height of her infamy as an author who challenged accepted sexual practices, and the lecture hall was packed with paying curiosity seekers (mostly men). This expectant audience was treated to an hour's talk by the irrepressible Stopes—on the formation of coal.

"SUE" THE TYRANNOSAURUS
Bones of Contention

The most complete fossil of a *Tyrannosaurus rex* ever found has an appropriate nickname, "Sue," considering the lawsuits it has inspired, but the spectacular fossil was really named for its discoverer, paleontologist Susan Hendrickson. (No one has yet established whether the 67-million-year-old *T. rex* was male or female.)

While prospecting for the Black Hills Institute of Geological Research, a commercial fossil company, on a private ranch in southern South Dakota in 1990, Hendrickson spotted the "king of the tyrant lizards" skeleton eroding from a cliff. After many months of excavation and preparation, the skeleton was fully revealed—and it was a stunner.

Most of the 22 known *T. rex* fossils were fragmentary, but Sue is over 90 percent complete. At more than 40 feet long, standing 13 feet tall at the hip, and with an estimated live weight of six to eight tons, it is the largest known tyrannosaur. The 5-foot-long skull sported rows of daggerlike teeth, with serrated edges, like steak knives. Huge olfactory cavities probably enabled the tyrant lizard to smell decaying carcasses from miles away; it would have been an adept scavenger as well as possibly a hunter.

The custody battle over Sue's remains soon became intense and acrimonious. Paleontologists, attorneys, Native Americans, ranchers, and the federal government engaged in a protracted tug of war over ownership. Fossil dealer Peter Larson, who employed Susan Hendrickson, had paid $5,000 to the rancher for the right to take fossils off his land. After Sue was found, however, the rancher demanded more money. While his suit was pending, the Cheyenne River Sioux claimed that he was merely a leaseholder who had no right to profit; the skeleton had been found on tribal land that was held in trust by the government.

As the court case dragged on, Larson and his crew spent two years preparing and preserving the fossils. Then the federal government intervened, originally to protect the Native Americans' interests, but later contending that Sue was a national treasure and could not be privately owned. In 1992, in a sudden raid on Larson's museum in Hill City, South Dakota, 35 FBI agents and 20 armed National Guardsmen seized the premises, crated up the many tons of Sue's remains, and carted them away to locked storage.

More suits, countersuits, and indictments escalated the battle, as Sue began to disintegrate in storage. At one point, Larson was briefly imprisoned as an example to "illegal" collectors. Months later, Sue's remains were finally awarded to Maurice Williams, the rancher, who promptly auctioned the fossils at Sotheby's. The top bidder at more than $8 million (with funds from donor corporations) was Chicago's Field Museum, where Sue was put on display in May 2000 and instantly became its star attraction. An obscure Darwinian lyricist summed up the case:

A tyrannosaurus named Sue
Was seized by an FBI crew
When her owners resisted
The lawyers insisted
We'll sue to give Sue to the Sioux.

SULFUR-BASED ECOSYSTEMS
Life Evolving in Sunless Worlds

Until the late 20th century, biologists believed that all life on Earth relied on energy from the sun. All animals, it was thought, depended on plants that manufacture nutrients in the presence of sunlight. During the past few decades, however, scien-

Self-contained
worlds of life have
evolved in sulfu-
rous caves and
deep-sea vents,
beyond the reach
of the sun's rays.

tists have discovered self-contained worlds that have evolved beyond the reach of the sun's rays, deep in sulfur-rich caves and around hydrothermal vents—openings in the oceanfloor that spew boiling liquid and molten magma.

These sulfur-based ecosystems were first encountered in 1977, when biologists aboard the deep-diving submersible *Alvin* located the first known hydrothermal vents. A decade later in the South Pacific, "black smokers" were found some 7,200 feet beneath the sea. Occurring where superheated (6000° F) magma erupts from underground volcanic fissures and hits 35° F seawater, the chimneys form when streams of dissolved minerals precipitate and solidify. Near the black smokers, scientists collected bizarre organisms that included sulfur-based microbes, sulfur worms, tube worms, and white crabs. Ghostly creatures new to science, they live in total darkness and somehow thrive under crushing pressure, extreme toxicity, and alternately frigid and scalding temperatures. Other thermal vents in the ocean floor support similar white worms and microbes, as well as Archaea, or blue-green algae. Among the most primitive organisms on Earth, Archaea are genetically as different from bacteria as bacteria are from humans.

When scientists sought similar ecosystems on land in the 1980s, they found them *under* the land—in the Movile Cave in Dabrogea, Romania (near the Black Sea), and the Cueva de Villa Luz in Mexico. Movile Cave contains 48 species of cave-adapted creatures—making it the most biologically diverse closed-cave system known. Despite almost total isolation from solar energy, the cave's level of food production is startlingly high, comparable to that of above-ground lakes.

Cueva de Villa Luz, a vast sulfur-spring cave system, is located near the town of Tapijulapa, at the edge of highland Chiapas. When veteran cave explorer James Pisarowicz first explored it in 1987, he was struck by its unpleasant, otherworldly quality. Yellow sulfur was everywhere, and the air was thick with the "rotten egg" odor of hydrogen sulfide. Clear gypsum crystals abounded, and the walls were stained with patches of multicolored slime. Hanging from the ceilings were strange rubbery deposits that dripped sulfuric acid—massive, oozy colonies of sulfur-oxidizing organisms. They looked like stalactites made of mucus, so Pisarowicz dubbed them "snottites."

Sulfur-eating bacteria in Cueva de Villa Luz are the base of the food chain. Other microbes ingest them, and are in turn eaten by tiny midges that break down the sulfur compounds. Spiders and microscopic worms prey upon the midges, and are in turn eaten by blind cave fishes, which are annually harvested by local villagers in a spectacular candle-lit ceremony.

Some researchers believe that if life exists on Mars, a sulfur-rich planet full of caves, it may be limited to an underground world. According to cave scientist Louise Hose of Chapman University in California, "Many researchers now believe that life on Earth may have originated with chemoautotrophic bacteria in caves, underground fissures, or seafloor smokers."

SUNDAY LEAGUE
Victorian "Sabbatarian" Controversy

Nineteenth-century London was a dreary place on its sacred Sundays. Shops were padlocked, recreations and concerts forbidden, and the great public cultural institutions—zoos, art galleries, museums, and libraries—were closed. In his novel *Little Dorrit* (1856), Charles Dickens described Sunday in sooty London as "gloomy, close, and stale," with "maddening church bells of all degrees of dissonance":

Sundays in London
had "no pictures,
no unfamiliar
animals . . . nothing
to see but streets,
streets, streets."

—Charles Dickens, 1865

> Everything was bolted and barred that could . . . furnish relief to an overworked people. No pictures, no unfamiliar animals, no rare plants or flowers, no natural or artificial wonders of the ancient world. . . . Nothing to see but streets, streets, streets. . . . Nothing for the spent toiler to do, but to compare the monotony of his seventh day with the monotony of his six days, [and] think what a weary life he led.

In 1831, conservative "sabbatarians" had formed the Lord's Day Observance Society to preserve the "sanctification of the Christian Sabbath" as "one of the few remaining threads

"A DREAM OF THE FUTURE" was the title of this 1885 cartoon lampooning the bitter fight to open London's libraries and museums on weekends. A drunken, loutish working-class mob is shown entering the British Museum on Sunday, and emerging refined and gentrified, studying books and carrying the banner of "Sweetness and Light."

on which [England's] destiny is awfully suspended." Then, in 1853, a jeweler named Robert Morrell countered with a new organization, the National Sunday League, designed to promote "elevating recreation" on that day. He founded a journal, the *Sunday Review,* which was sponsored by many liberal thinkers, literary men, and scientists, including such notables as Dickens, Thomas Henry Huxley, and Charles Darwin.

After a long and bitter fight between the two factions, the Sunday League won its first victory in 1856, when Lord Palmerston allowed the Guards' bands to play in the public parks. Outraged, the archbishop of Canterbury proclaimed that "unless the Sunday band concerts cease, I can be no longer responsible for the religion of the country." Palmerston maintained that "innocent intellectual recreations, combined with fresh air and healthy exercise [are not] at variance with the soundest and purest sentiments of religion." But he had to yield to conservative pressure, and the bands were withdrawn.

Some time later, Thomas Henry Huxley, who was unabashedly evangelical about science and Darwinism, presented the first scientific lecture ever given on a Sunday. To conform with traditional sabbath practices, the program included some singing before and after the talk. The lecture was attended by an enthusiastic audience of more than 2,000 on the evening of January 7, 1866, at St. Martin's Hall in Long Acre.

The event, though a great success, was followed by demands from sabbatarians for Huxley's arrest. Puzzled police could not find a law banning a "secular service" and in desperation charged him under an old law for "keeping a disorderly house." To no one's great surprise, the case was thrown out of court. Later, Huxley entitled a book of his collected lectures *Lay Sermons* (1879).

See also FUNDAMENTALISM; SECULAR HUMANISM

"SURVIVAL OF THE FITTEST"
Evolutionary Slogan

Nowhere in the first edition of Charles Darwin's masterpiece *On the Origin of Species* (1859) does he use the phrase "survival of the fittest." It was actually coined by English philosopher Herbert Spencer, in *Principles of Biology* (1864).

Although Spencer wrote a volume on biological evolution, he was no naturalist; evolution interested him as a "universal principle." His vague notion of "the fittest" meant those best able to foster general progress and improvement for their society or species.

Alfred Russel Wallace, the cofounder of evolutionary theory, was struck by "the utter inability of many intelligent persons" to understand what he and Darwin meant by natural selection and suggested they substitute Spencer's phrase. When Darwin obliged him by using "survival of the fittest" in later editions, readers were still confused; everyone seemed to have his own interpretation of what was meant by "the fittest."

However, the phrase caught the public's imagination and became completely associated with Darwin. Critics said it was a meaningless tautology—a proposition that simply repeats itself. Since the fit are the individuals who survive, they argued, wasn't it another way of saying "survival of the survivors"?

Evolutionary biologists have long been aware of that pitfall. Fitness, it turns out, is a relative term. Organisms that are the "fittest" in one environment, may be completely unsuccessful in another. Or they may be extremely successful for millions of years—as the dinosaurs were—only to be suddenly wiped out when conditions change.

In populational terms, fitness simply means reproductive success. The race does not always go to the strong or swift, but to those who manage, by whatever means, to produce the largest number of surviving offspring. Sometimes the "fittest" may be those who attain high social status (and more matings) with bold bluffs or subterfuges, rather than with prowess or strength. Other methods of outcompeting rivals in production of offspring include persistent sexiness, extraordinary tail plumage, or throwing your neighbor's eggs over a cliff.

Social Darwinists seized the phrase as a slogan to advocate a totally unregulated economy. Robber barons of the Gilded Age—James J. Hall, John D. Rockefeller, Andrew Carnegie—often told journalists their cutthroat business practices were, in the long run, helping society evolve. Elimination of weak and inefficient competitors was the road to progress, with future benefits for all.

But Thomas Henry Huxley knew full well that scoundrels were appropriating biology to exalt themselves. As he pointed out in his essay "On Providence" (1892), "We commonly use 'fittest' in a good sense, with an understood connotation of 'best' . . . [which we are] apt to take in its ethical sense. But the 'fittest' which survives in the struggle for existence may be, and often is, the ethically worst."

See also SOCIAL DARWINISM; SPENCER, HERBERT; SPERM COMPETITION

Robber barons claimed their cutthroat business practices were helping society evolve to greater efficiency and perfection.

SYNTHETIC THEORY
Reformulation of Evolution

Around 1900, Darwinism was in a scientific limbo. Hugo de Vries and Thomas Hunt Morgan had brought macromutation into prominence, relegating natural selection to a minor sorting role. Other geneticists had shown that most mutations were harmful and that the random shuffling of genes makes little change in the population. Paleontologists were talking about "straight line" evolution, and philosophers were looking for "vital forces" that guided evolution to a predetermined goal.

During the 1920s and 1930s, however, the rise of population genetics brought evolutionary studies full circle back to Darwin. Population concepts, shaped by new mathematical tools, dealt with changing gene frequencies in populations—and reaffirmed natural selection as a major force in evolution. Mathematical studies also lent new weight to "genetic drift," or sampling error associated with small breeding populations.

One of the leaders in reintegrating evolutionary science was Theodosius Dobzhansky, a Russian émigré geneticist who continued his work with the fruit fly *Drosophila* in America. His controlled laboratory experiments with the flies, which reproduce every 10 days, enabled him to observe evolution directly and to study adaptation as an experimental science. In *Genetics and the Origin of Species* (1937)—one of the founding documents of what would come to be known as the Modern Synthesis, or the Synthetic Theory—Dobzhansky showed how minor changes among a few flies in his small populations could greatly influence a large number of descendants.

Another seminal book in shaping the modern synthesis gave the movement its name. Biologist Julian Huxley, the grandson of "Darwin's Bulldog" Thomas Henry Huxley, published *Evolution: The Modern Synthesis* in 1942. At about the same time, paleontologist George Gaylord Simpson applied populational thinking and genetics to the study of fossils in his *Tempo and Mode in Evolution* (1944). Simpson also dealt with varying rates of evolution, and discredited attempts to view fossils in terms of "straight-line" predetermined evolutionary paths or goals.

Other major contributors to the Synthetic Theory included ornithologist Ernst Mayr, biologist J. B. S. Haldane, botanist G. Ledyard Stebbins, biometrician R. A. Fisher, and geneticist Sewall Wright. Fisher's book *The Genetical Theory of Natural Selection* (1930) provided a new empirical, logical, and mathematical basis for the theory. Difficult and technical, it is one of the most important and influential scientific books of the 20th century.

Through these scientists' efforts, a new understanding of Darwinism was reached that integrated the results of genetics, mathematics, paleontology, and especially populational thinking, with Darwin's original principle of natural selection. The Synthetic Theory became the basis of biology and prompted new experimental research on populations.

One change of emphasis in the theory was the importance of genetic recombination. Natural selection can only select among the existing variations in a population; it does not create new varieties. However, existing genes can recombine to produce new traits.

In the 1970s, Stephen Jay Gould, Niles Eldredge, Steven Stanley, and others criticized the Synthetic Theory for promoting an unsupported "gradualist" view of evolutionary change. Darwin's idea that most species undergo slow changes at a steady rate had become orthodoxy in the Synthetic Theory. Gould and Eldredge believed, instead, that fossil and genetic evidence points to widely varying rates of change. In 1972, they proposed the "punctuational" model of evolution: long periods of stability interrupted by species-forming episodes of rapid change. Defenders of Synthetic Theory accused them of having set up "gradualism" as a straw man. The Synthetic Theory, they argued, has always included the idea that evolution can proceed at different rates.

See also ESSENTIALISM; GENETIC DRIFT; PUNCTUATED EQUILIBRIUM

When the future of Darwinism looked bleak, the invention of population genetics gave it a new lease on life.

TANGLED BANK
Darwinian Metaphor

Darwin's *Origin of Species* (1859) is not only a scientific argument, but a work of literature as well—full of rich imagery, metaphors, and appeals to the reader's imagination. In the last paragraph of the book, Darwin offers his lyrical musings on the "tangled bank." (He once said that a good book, like a fine day, should end "with a glorious sunset"):

> It is interesting to contemplate a tangled bank, clothed with many plants of many kinds, with birds singing on the bushes, with various insects flitting about, and with worms crawling through the damp earth, and to reflect that these elaborately constructed forms, so different from each other, and dependent on each other in so complex a manner, have all been produced by laws acting around us.

In *The Tangled Bank: Darwin, Marx, Frazer, and Freud as Imaginative Writers* (1962), Stanley E. Hyman shows how several influential thinkers clothed their abstract theories in poetic language and imagery. The tangled bank expresses Darwin's special view of nature as a web of interrelationships binding various plants and animals into a community. Not only is the visible vegetation entangled; the lives of different species are also intertwined "in so complex a manner."

Ecological interdependence was one of the hallmarks of Darwin's contribution to modern thought, though the word "ecology" had yet to be invented (by German evolutionist Ernst Haeckel). "The metaphor of entanglement," Gillian Beer has written, "enacts what often remains latent in his argument: the extent to which evolution is a lateral rather than simply an onward movement, whose power lies in multiple relationships as much as in selecting out."

Why was this "tangled bank" so "interesting to contemplate?" Because beneath its illusory surface confusion, science could discover underlying regularities and relationships. Darwin's later work established the painstaking research methods that could eventually "untangle" his beloved tangled bank.

Is the metaphor based on a particular place? Historian Randal Keynes, a Darwin descendant, found family documents that mentioned Darwin's fondness of Down Bank, in Kent, and a hillock called the "Orchis Bank." The latter was well within the range of Darwin's daily walks, and Keynes believes Darwin was describing this place, which still exists. On the other hand, the tangled bank could be any English stream, familiar and commonplace—like the banks of the River Severn where Darwin played as a child.

Darwin was well aware that evolution's mysteries could be observed not only in the exotic rain forests of South America, but in one's own country garden as well—which, after returning from his adventures in the tropics, is exactly what he did.

See also EARTHWORMS; ECOLOGY; METAPHORS IN EVOLUTIONARY WRITING

TARZAN OF THE APES
Fictional Lord of the Jungle

Tarzan, the jungle hero raised by apes, became one of the most popular fictional creations of the 20th century. According to zoologist Jane Goodall, who has lived with wild chimpanzees for 30 years, her early exposure to Tarzan stories helped inspire her interest in African animals.

This fantasy adventure series by Edgar Rice Burroughs first appeared in the pulp magazine *All-Story* in 1912. Burroughs claimed he took the germ of his story from the classical myth of Romulus and Remus, the twin brothers who were abandoned, nursed by a she-wolf, and grew up to found the city of Rome. He was also strongly influenced by the romantic idea of a "noble savage," first celebrated by the 18th-century French philosopher Jean-Jacques Rousseau and later embedded in American popular art and literature. Part of Tarzan's appeal stems as well from the post-Darwinian appreciation of man's kinship with chimpanzees and gorillas. Even after the movie Tarzan establishes a treetop household with Jane and Boy, Cheetah the chimp remains an inseparable member of his family.

In the original episode of the story, Lord and Lady Greystoke are marooned on the African coast but survive for a year in a hut, where their son is born. One year later he is orphaned when his mother dies of fever and his father is killed by Kerchak, a fierce giant ape who rules the neighborhood. Adopted by Kala, a gentle female ape who is mourning her own dead infant, Tarzan grows up at home in the trees, thinking himself an ape until he meets other humans.

Burroughs posed the question of whether the child of English aristocrats could survive in a "state of nature." His answer was that his hero not only survived, but became benevolent ruler over all other creatures. Despite the crudity of his speech and manners, Tarzan has an innate nobility of character that leads him to perform heroic deeds.

When the American Jane Porter visits the jungle, she is abducted and brutalized by Kerchak. Tarzan enters into a life-or-death struggle to rescue her and avenges his father by killing the murderous ape. Jane falls in love with Tarzan, and they begin their life together in a treetop retreat.

JUNGLE MAN raised by apes, Tarzan was Edgar Rice Burrough's post-Darwinian version of the "noble savage." Johnny Weissmuller's film portrayals (top), from 1932 to 1948, were widely popular. The pongid Madonna scene (above) is from Hollywood's more recent *Greystoke* (1984).

More than 25 Tarzan books sold hundreds of millions of copies worldwide. In the 1930s, as one writer recalled it, "Millions of boys took to leaping from limb to limb and from tree to tree. The nation resounded with the Tarzan yell and the snapping of collarbones."

Tarzan's adventures have been popular on radio and television and in movies, cartoons, comic books, and a Broadway musical. The first of the many movie Tarzans, in 1918, was Elmo Lincoln, who appeared in three silent films. Olympic champion swimmer Johnny Weissmuller became the most enduring film Tarzan, although he was the ninth actor to play the role. Now pop classics, his films were shot in California and Florida—never in Africa—during the 1930s and 1940s. Asiatic elephants were used because they are easier to handle than their African cousins. Sometimes the films showed jungle monkeys hanging from their tails, although no African species can do that. (Monkeys with grasping or prehensile tails are found only in Central and South America.)

Tarzan's famous movie yell was a combination of several different sounds, including

Weissmuller yelling, a soprano singing, a note played on a violin, and hyena growls played backward. Eventually, Weissmuller himself was able to give a piercing rendition of the sound. During his terminal illness in 1979, newspapers reported that Weissmuller became sadly confused about his identity. He had to be moved from the Actor's Hospital because he frightened other patients with blood-curdling Tarzan calls in the middle of the night.

Burroughs insisted the sound of the name "Tarzan" was an important factor in his character's success, and had tried hundreds of sound combinations before settling on it. A lover of the "music" of words, he also invented an entire "ape language" for his fictional anthropoids, which were never specified as chimpanzees or gorillas in the original Tarzan stories.

As he grew wealthy from book and movie royalties, Burroughs bought a ranch in the San Fernando Valley, near Hollywood, where he lived happily for many years. There he founded a town and named it after his imaginary ape-man of Africa: Tarzana, California.

See also "APE WOMEN," LEAKEY'S; NOBLE SAVAGE

TAUNG CHILD
Dart's Fabulous Fossil

In November 1924, as he was dressing in tie and tails for a wedding, anatomy teacher Dr. Raymond Dart received three crates of rocks from a friend. With his collar still unfastened, and the groom and guests waiting, Dart opened the boxes and pulled out a small natural rock cast of a braincase, and the back of the skull into which it fitted. Immediately he knew he had something extraordinary, and the groom finally had to yank him away to be best man. The young anatomist at the University of Witwatersrand in South Africa had received the first known fossils of the ancient African man-apes.

Intrigued by the baboon skulls a student had shown him from a limeworks at Taung, 200 miles from his Johannesburg home, Dart had requested to see further fossils from the site. A geologist friend, Dr. R. B. Young, who was visiting the quarry, packed up some odd chunks from the day's blasting and sent them along. Dart's full-dress reception of them turned out to be strangely appropriate.

Over the next three months, Dart worked constantly to separate the face from the limestone in which it was embedded, using improvised tools, including his wife's knitting needles. What he uncovered was a six-year-old child's skull, of a creature no one had ever seen before, along with the mineral impression of its braincase. The Taung child lived about 2.5 million years ago.

"DART'S BABY," found in a South African limestone quarry in 1924, was the first known australopithecine skull. This small-brained primate with large, human-like teeth opened a new chapter in the search for human origins.
© Margo Crabtree.

Dart named the hominid *Australopithecus africanus*, meaning southern ape of Africa. Its large braincase, small teeth (compared to apes), rounded palate, and position of the hole at the base of the skull (foramen magnum) indicated an upright hominid with many similarities to humans. But when Dart announced his discovery in 1925, four of Europe's leading anthropologists scoffed. Sir Arthur Keith thought it was merely an aberrant chimpanzee; other anatomists proclaimed it an ancient gorilla. The jury of experts gave their verdict, and it proved to be totally wrong.

While anthropologists turned their attention to Beijing Man in the 1930s, Dart stuck to his guns, and reminded them of Darwin's half-forgotten prediction, made in 1871, that the most ancient human ancestors would probably be found in Africa, home of the chimp and gorilla. Some years later, after his colleague Robert Broom began to find adult skulls of *Australopithecus* in the southern Transvaal, the tide began to shift in Dart's favor. But it took another 40 years before he was fully vindicated; many subsequent discoveries have finally brought scientific opinion into agreement with his first assessment of the Taung child's importance. "It's no good being in front if you're going to be lonely," he once remarked—but in science that often comes with the territory.

See also AUSTRALOPITHECINES; DART, RAYMOND ARTHUR

TEILHARD DE CHARDIN, FATHER PIERRE (1881–1955)
Philosopher, Paleontologist

From his early boyhood among nine brothers and sisters in the rural French province of Auvergne, Pierre Teilhard de Chardin was remarkable for his religious devotion and scientific curiosity. If he had developed only one of these talents, the lanky, brilliant young man might have enjoyed a comfortable, uncomplicated life. Instead, his persistent attempts to straddle both science and religion brought years of anguished struggle. After spending his childhood at Jesuit schools, at 18 Teilhard entered the order, then was sent to teach physics and chemistry at a Jesuit college in Cairo for several years. On returning to France, his lifelong curiosity about nature led him to the Museum of Natural History in Paris, where he studied human paleontology with the great Marcellin Boule. There, at the Institute of Human Paleontology, he also met another mentor and lifelong friend, the Abbé Henri Breuil, a churchman-archeologist who pioneered the study of Ice Age cave paintings and stone tools.

Teilhard took a doctorate in paleontology at the Sorbonne, where he found acceptance and recognition as a scientist. His first experience with the excitement of a fossil man dig, oddly enough, was with Charles Dawson in Sussex, England, at the Piltdown site, where he found a celebrated tooth. (Decades later, it transpired that the gravels had been "salted" with doctored bones and artifacts by still-unknown hoaxers.) Between visits to England, he served as a paramedic in World War I, then returned to Paris, where he hoped to pursue a career of writing, teaching, and research.

But it was not to be. While Teilhard's science had led him to the fact of evolution, his devotion fastened on an impossible dream, upon which he refused to compromise: that the Catholic Church embrace the reality of evolution as part of the divine plan.

A scientific quest for the truth about human origins, he believed, would surely lead back to the same Creator worshiped by the church. Refusing him any kind of hearing, the hierarchy branded him a dangerous, heretical rebel who needed to be muzzled.

They sent him to China as a missionary and permitted him to attend excavations at remote sites—anything to keep him from speaking and writing in France. Paradoxically, his long stay in China resulted in his participation in one of the great discoveries in paleoanthropology: the excavation of the limestone caves at Zhoukoudian where Beijing Man (*Homo erectus*) had left a number of spectacular fossil skulls.

Holding him to the vow of strict obedience he had taken 30 years earlier, church officials forbade Teilhard to print any of his philosophical books or articles during his lifetime.

CATHOLIC EVOLUTIONIST Father Pierre Teilhard de Chardin pursued a dual career as paleontologist and theologian, but his church muzzled him for attempting to synthesize science and religion.

However, after his death in 1955, friends published the manuscripts, beginning with his grandest work, *The Phenomenon of Man (Le Phénomène humaine)*, which he had completed in 1941. An instant sensation, for several years it enjoyed a tremendous vogue, and was widely discussed by philosophers, psychologists, social scientists, and theologians (although some biologists dismissed it as an "an incoherent rhapsody").

Father Teilhard viewed humans as a phenomenon—a species that is not static, but in the process of being and becoming, for which he coined the term "hominisation." In the future, humankind would continue to evolve toward a distant "Omega point," drawn to a perfected spirituality that transcends its primate and hominid origins.

He also coined the term "noösphere" to denote the realm of developing mind or culture, which others have called "the superorganic" or "psychocultural evolution." Whereas organic evolution previously took place only in the biosphere, or world of life, the future of earthly life would be decided in this noösphere, the arena of human perceptions, thoughts, and values.

One of Teilhard's favorite notions was that special creativity results when an evolutionary "splitting-off" (divergence) is followed by a "coming together" of lines that have begun to splay out on different paths. He cited such "convergent integration" in his discussion of world "races" and cultures—pursuing separate histories for a time, but not for so long they could not then "melt" back together, enriching all. His great hope was that the divergent streams of scientific and religious thought, too, would one day merge, producing an enlarged understanding of the "phenomenon of man."

> **Teilhard de Chardin's attempts to reconcile evolution and religion were too mystical for scientists and too scientific for his church.**

Though Teilhard won the personal admiration and friendship of distinguished philosophers, scientists, and theologians, they never integrated his grand synthesis into the mainstream of Western thought. His work remained, as it had during his lifetime, too mystical for scientists and too scientific for religionists.

A small, dedicated band of admirers continues to cherish his memory and publish several journals in which to discuss his ideas. To these Friends of Teilhard Societies, he is a martyred visionary, an uncanonized saint.

Exiled from France and denied the satisfaction of seeing his philosophical works in print, Father Teilhard spent his final years in a small New York City apartment. He grew ill and alienated, living on the philanthropy of the Wenner-Gren Fund, whose president, Paul Fejos, was a longtime admirer. He died on Easter Sunday, 1955.

"How is it possible," he wrote during his last days, "after descending from the mountain and despite the glory that I carry in my eyes, I am so little changed for the better, so lacking in peace, so incapable of passing on to others through my conduct, the vision of the marvelous unity in which I feel myself immersed? . . . As I look about me, how is it I find myself entirely alone of my kind? . . . Why am I the only one who sees?"

See also BEIJING MAN; "OMEGA MAN"; PILTDOWN MAN (HOAX)

TELEOLOGY
Search for Design in Nature

Before the quest to understand evolution became their central focus, naturalists—many of whom were also churchmen—tried to find divine design and purpose in plants and animals. Since they believed the Creator had devised a plan for every creature, they named the focus of their study from the Greek *telos*, which means "purpose."

Early zoologists marveled at the supposedly "perfect design" of organisms to fit their environments, such as the structure of eagle's wings or the optics of the human eye. Ignoring abundant evidence of imperfect or quirky structures in nature, they focused on the seeming marvels of adaptation.

Teleologists argue backwards. Sir Thomas Browne praised God's wisdom for putting living things that need heat and sunshine in the tropics and those that thrive on cold in the Arctic. Since Darwin's day, we take it for granted that the vegetation has evolved to suit the climate, and not the other way round—a view that would have been considered shockingly "mechanistic" a few hundred years ago.

In Voltaire's *Candide* (1758), teleology is satirized through Dr. Pangloss, a caricature of the philosopher Leibniz, who continually asserts that God made everything for the best in "this best of all possible worlds." At one point, the optimistic doctor remarks how wonderful it is that the nose is midway between the eyes, for God knew it would be the perfect position to support eyeglasses. Voltaire was joking, but zoologist Richard Owen was not when he suggested that the gap in horse's teeth was "designed" to allow room for the bit, so men could ride them.

To some, Darwin at first seemed to "rescue" teleology by giving it a new basis. After all, his experiments with orchids revealed the unsuspected "purpose" of bizarre structures and colors. Oddly shaped nectaries of orchids, for instance, could be viewed as "designed" to accommodate the heads or tongues of pollinating insects. But Darwin also demonstrated how the structures were shaped by the coevolution of insects and orchids—parts of a system whose components have evolved together, creating a "fit" between them.

He also showed that most of the expected "perfection" in nature did not hold up under systematic examination. In fact, most life forms are more like odd contraptions, evolutionary makeovers that resemble the work of a make-do tinkerer, not a "Master Planner."

See also ADAPTATION; GRAY, ASA; ORCHIDS, DARWIN'S STUDY OF; PALEY'S WATCHMAKER; PANDA'S THUMB

TEMPLE OF NATURE, THE (1803)
Erasmus Darwin's Evolutionary Epic

After a lifetime of pondering the phenomena of nature and humankind's place in it, 18th-century philosopher Erasmus Darwin expressed his mature views in an epic poem rather than a scientific treatise. His masterpiece, *The Temple of Nature* (1803), was published a year after his death.

Heroic and grandiose in scope, musical in its language, *The Temple of Nature* is a remarkable presentation of the pioneer evolutionist's view of the Earth's creation, the rise of life from microscopic "filaments" in the seas, and the radiation of plants and vertebrates, including humans. Combining science and art, Erasmus Darwin footnoted references to the raw material on which his mind fed—horticulture, medicine, anatomy, botany, chemistry. Indeed, he parades before the reader surprisingly more "facts" than one might suppose from his reputation as a speculative visionary.

It is an extraordinary work, little read or studied today, but widely popular at the turn of the 19th century, when it was praised by philosophers and literary men alike. Its excellence and rarity justify reprinting the following excerpt:

Erasmus Darwin published an evolutionary treatise before Charles was born—and wrote it all in verse.

> Then, whilst the sea, at their coeval birth
> Surge over surge, involv'd the shoreless earth;
> Nurs'd by warm sun-beams in primeval caves,
> Organic Life began beneath the waves.
>
> First forms minute, unseen by spheric glass,
> Move on the mud, or pierce the watery mass;
> These, as successive generations bloom,
> New powers acquire, and larger limbs assume;
> Whence countless groups of vegetation spring,
> And breathing realms of fin, and feet, and wing.
>
> Next, when imprison'd fires in central caves
> Burst the firm earth, and drank the headlong waves;
> And, as new airs with dread explosion swell,
> Form'd lava-isles, and continents of shell;
> Pil'd rocks on rocks, on mountains mountains rais'd,
> And high in heaven the first volcanoes blaz'd.
>
> In countless swarms an insect-myriad moves
> From sea-fan gardens, and from coral groves;
> Leaves the cold caverns of the deep, and creeps

On shelving shores, or climbs on rocky steeps
Cold gills aquatic, for respiring lungs,
And sound aerial flow from slimy tongues.

Thus the tall Oak, the giant of the wood,
Which bears Britannia's thunders on the flood;
The Whale, unmeasured monster of the main,
The lordly Lion, monarch of the plain,
The Eagle, soaring in the realms of air,
Whose eye undazzled drinks the solar glare;—

Imperious man, who rules the bestial crowd,
Of language, reason, and reflection proud
With brow erect, who scorns this earthly sod,
And styles himself the image of his God;
Arose from rudiments of form and sense,
An embryon point, or microscopic ens!

The elder Darwin summarizes his major thesis in one of the poem's footnotes: "The Great Creator of all things has infinitely diversified the works of his hands, but has at the same time stamped a certain similitude on the features of nature, that demonstrates to us that the whole is one family of one parent." Half a century later, Charles Darwin's words in *Origin of Species* (1859) unmistakably echo his grandfather: "There is grandeur in this view of life . . . having been originally breathed into a few forms or into one, and that whilst this planet has gone cycling on . . . from so simple a beginning endless forms most beautiful and most wonderful have been, and are being evolved."

See also DARWIN, ERASMUS; *ORIGIN OF SPECIES*; ORTHOGENESIS

> "The whole [world of life] is one family of one parent."
>
> —Erasmus Darwin, 1803

TENNYSON'S *IN MEMORIAM* (1850)
Evolutionary Requiem

When her beloved husband Prince Albert died, Queen Victoria told the poet Alfred, Lord Tennyson (1809–1892), that "next to the Bible, *In Memoriam* is my comfort." So popular was the 131-stanza poem that within a few months of its publication in June 1850, Tennyson was named poet laureate of England.

This poem, to which so many Victorians turned for comfort during dark days, was partially inspired by Robert Chambers's *Vestiges of Creation* (1844), an important work on evolution that appeared well before Darwin's *Origin of Species* (1859).

In Memoriam was Tennyson's tribute to the memory of his beloved friend Arthur Hallam, who died at the age of 22 in 1833. Tennyson, who had been tutored by the great William Whewell, had an intense interest in science and wove his reactions to scientific developments into his work. Near the middle of the long poem, Tennyson despairs, faced with what seems to be the meaninglessness of the natural world, according to his reading of Charles Lyell and Chambers:

Are God and Nature then at strife,
    That Nature lends such evil dreams?
    So careful of the type she seems,
So careless of the single life;

That I, considering everywhere
    Her secret meaning in her deeds,
    And finding that of fifty seeds
She often brings but one to bear,

I falter where I firmly trod,
    And falling with my weight of cares
    Upon the great world's altar-stairs
that slope thro' darkness up to God . . .

> "So careful of the type?" but no,
>   From scarped cliff and quarried stone
>   She cries, "A thousand types are gone:
> I care for nothing, all shall go."

Toward the end of his poem, Tennyson regains his optimism through a faith in an evolutionism that seeks progress. His friend Hallam, he believes, was a premature example of a higher type of man.

> A soul shall draw from out the vast
> And strike his being into bounds,
>
> And, moved thro' life of lower phase
>   Result in man, be born and think,
>   And act and love, a closer link
> Betwixt us and the crowning race . . .

ALFRED, LORD TENNYSON

Among the phrases that became common in the language are the description of "Nature, red in tooth and claw," which "shriek'd against" the creed of divine love. And of course, the poem's core message, which is often wrongly attributed to Shakespeare: " 'Tis better to have loved and lost / Than never to have loved at all." One beautiful stanza could serve as an epitaph for all naturalists, past and future:

> My love has talk'd with rocks and trees;
>   He finds on misty mountain-ground
>   His own vast shadow glory-crown'd;
> He sees himself in all he sees.

The Victorians loved the poem and quoted it endlessly. But paradoxically, they failed to notice that its philosophy ran counter to their most cherished religious teachings. Since "its hope comes from evolutionism and its prospect of progress to a race of supermen like Hallam," writes historian-philosopher Michael Ruse in *The Darwinian Revolution* (1979), "one suspects that deep down many Victorians did not much care whether organic evolutionism . . . or the doctrinal niceties of conventional Christianity" were really true.

"What they did care about," according to Ruse, "was the frightening rapidity of change in their lifetimes and the essential lack of security of their society—a society supported and surrounded by the vast underprivileged, frequently starving masses. When Tennyson held out the hand of hope and progress to a better state, they grasped it thankfully, not bothering about details."

"TERNATE PAPER" (1858)
Wallace's Evolution Theory

One of the most important—and most obscure—founding documents of evolutionary biology is "The Ternate Paper" (1858), written in a tropical jungle by Alfred Russel Wallace. Entitled "On the Tendency of Varieties to Depart Indefinitely from the Original Type," it is a clear, concise statement of Wallace's theory of evolution by natural selection, worked out independently of Charles Darwin.

Wallace was collecting specimens on the island of Ternate in the Moluccas (now Indonesia). Like Darwin, he conceived his theory after reading Thomas Malthus's *Essay on the Principle of Population* (1798), a striking example of historical parallelism. Although he and Darwin had corresponded, he did not know that the senior naturalist had been working secretly for 20 years on the same theory.

Innocently, Wallace sent the paper off to Darwin on March 9, 1858, from Ternate, asking him if he thought it was worthy of publication. It carried such topic headings as "The Struggle for Existence," "Adaptation to the Conditions of Existence," "Useful Variation Will Tend to Increase; Useless or Hurtful Variations to Diminish." Wallace concluded the paper saying that "there is a tendency in nature to the continued progression of certain classes of varieties further and further from the original type." This explanation of evolutionary process, he concluded, will "agree with all the phenomena presented by organized beings, their extinction and succession in past ages, and all the extraordinary modifications of form, instinct, and habits which they exhibit."

When Darwin received the Ternate Paper, he was thrown into a panic. He feared he had been "forestalled . . . so that all my originality . . . has gone for naught." He was astounded that Wallace's topics "could stand as my chapter headings" and appealed to his friends Charles Lyell and Joseph Hooker for help.

There followed a series of events still debated by historians, culminating with the joint publication of the "Darwin-Wallace" theory in the *Journal of the Linnean Society of London* in 1858. Darwin then frantically set to work on his long-delayed book and produced the *Origin of Species* (1859) in 13 months. When Wallace returned from the Moluccas three years later, he was gracious about deferring to Darwin's priority for the discovery. If he had not, evolutionary theory might be known as "Wallaceism."

See also DARWIN, CHARLES; "DELICATE ARRANGEMENT"; NATURAL SELECTION; WALLACE, ALFRED RUSSEL

THEORY, SCIENTIFIC
"Truth" and Uncertainty

Critics often charge that evolution by natural selection is "only a theory" and "cannot be proved." Charles Darwin would have agreed with them—not because he didn't believe in evolution, but because he was a subtle philosopher of science. He well understood that "induction" (going from facts to general principles) could not guarantee a final, absolute truth. But he was among the first to realize the enormous power and value of "provisional" truth in advancing human understanding.

Darwin caused a great uproar, not only by the substance of his theory, but also by his redefinition of science itself. Some were more upset at his insistence that there is no absolute truth in science than with his belief in man's kinship with the apes. A good scientific explanation, he thought, is simply one that accounts for the most facts at a given time. It does not need to be proven beyond all doubt as "true" forever. If a more productive and comprehensive explanation is devised, the theory is superseded or becomes a "special case."

His stated procedure was to try a probable explanation on a group of facts and—if it seemed to solve problems—to attempt a broader application. It was a modest, realistic, and revolutionary concept of the limitations of science as well as its strengths. Darwin rarely used the word "law" as in physics, preferring the more probabilistic terms "doctrine," "principle," "theory," or "explanation."

His frankness about the uncertainty of his most deeply held beliefs was not welcome news to a shaky society looking to science or religion for unambiguous answers. But Darwin, and his advocate Thomas Henry Huxley, knew that truth-seekers would have to learn to live with a new acceptance of uncertainty—in both religion and science.

Darwin explained in a private letter (1838) that his theory was based on four "general considerations": (1) the "struggle for existence" or competition in nature; (2) "the certain geological fact [from fossils] that species do somehow change"; and (3) "the analogy of change under domestication by man's selection." And most important of all, (4) the theory connects and makes intelligible "a host of facts."

He matter-of-factly admitted that he could not "prove" any living species had changed or that supposed changes are beneficial. Nor could he explain "why some species have changed and others have not," and he was baffled by how the variability he observed in species might be produced.

If he could not prove with certainty that evolution was true, how could Darwin expect any thinking person to adopt his theory? The answer is simply that it is productive; it works. Applying it produced a torrent of discoveries, insights, new information. Connections arose to bridge formerly diverse disciplines: comparative psychology, geology, botany, paleontology. Working scientists found it solved many puzzles; and formerly inexplicable phenomena made sense within a coherent larger picture.

Darwin wrote a friend in 1861, "The change of species cannot be directly proved, and . . . the doctrine must sink or swim according as it groups and explains [disparate] phenomena. It is really curious how few judge it in this way, which is clearly the right way." A few years later he wrote that he was "weary of trying to explain" the point; most people could not grasp it.

See also MATERIALISM; MECHANISM; SCIENCE

TINBERGEN, NIKO (1907–1988)
Pioneer Ethologist

Although Charles Darwin had been a great field naturalist, acceptance of evolution had the strange effect of sending several generations of scientists indoors to their laboratories. Ethologists brought the study of animal behavior out of the labs and back into the fields.

Dutch ethologist Niko Tinbergen—along with Konrad Lorenz and bee expert Karl von Frisch—took their studies back to the woodlands and rivers. After watching animals in natural settings, they devised simple experiments out in the open that would help elucidate and analyze what the creatures were doing.

As a student in the Netherlands, Tinbergen had distinguished himself as a hockey player and pole-vaulter, but he was an indifferent scholar and did poorly on his zoology exams. Whenever possible, he fled the classrooms for long, solitary hikes across fields and beaches, where he became increasingly curious about the behavior of birds and animals.

Rather than seek out rare or exotic creatures, Tinbergen focused on common species, ignored by many naturalists. He wondered about little things, like the function of bright red splotches on the beaks of herring gulls, or how digger wasps locate their narrow burrows after flights of several miles.

While still a student, he caught several wasps and painted them with colored dots, for identification. Next, he removed pebbles, grass clumps, or pine cones from around their burrows. As he had suspected, the airborne wasps recognized those landmarks near their holes. When he removed them, the wasps were not able to locate their homes. Moving the objects to another place—but keeping them in the same configuration—fooled returning wasps into heading for where the burrow entrance "ought" to be.

Using such simple methods, Tinbergen was able to discover how wasps can recognize their own homes among many. More important, he demonstrated their unsuspected ability to quickly learn and memorize new patterns of objects in the environment.

Tinbergen's classic study of herring gulls revealed how their bright red bill splotches function. Hungry chicks, he discovered, will peck at anything red. When they target the mother's beak, they receive regurgitated food. But Tinbergen showed they will also peck at a cardboard cutout, or a stick, that has been painted with a red splotch, and even prefer it to a real beak on which the red spot has been covered over. His investigations of bird behavior culminated in such classics as *The Study of Instinct* (1950) and *The Herring Gull's World* (1953).

Tinbergen was held for two years as a prisoner of war by the Nazis during World War II, which afterward led to some friction with Konrad Lorenz, his colleague in pioneering the study of animal behavior. A member of the Nazi Party during his youth, Lorenz had interpreted "survival of the fittest" in a way that meshed with Nazi "race hygiene" policies. When Tinbergen was 76, he shared the Nobel Prize in Physiology or Medicine with Lorenz and von Frisch for establishing ethology, their new science of animal behavior. At the awards ceremony, Lorenz apologized for having used his science to support the Nazi racial agenda.

See also ETHOLOGY; LORENZ, KONRAD

TOOL–USING, EVOLUTION OF
Former "Human" Hallmark

For centuries philosophers believed that using tools was one trait that separated mankind from animals. From the simple stone implements found in prehistoric sites, the human species has developed a tool kit that includes computers, scuba underwater breathing gear, and rocket ships. Because these tools allow humans to enter new environments, some anthropologists speculated that once humans became tool-using creatures, they may have liberated themselves from further biological evolution.

Although humans are the tool users par excellence, field studies have produced a growing list of other creatures that use simple tools to manipulate or modify their environments. When Egyptian vultures want to crack open ostrich eggs, for instance, they pick up stones and fling them at the eggs until the shells break. Another bird, the green heron, has been observed "bait-fishing." The heron drops a small object into the water, which attracts small fish to the surface, where it can grab them.

On Darwin's beloved Galápagos Islands, the woodpecker finch uses twigs and cactus spines held in its beak to wrest insects out of small holes in trees. After selecting a suitable twig, the bird uses it to poke and pry grubs from underneath bark. It may transfer the twig to a foot, hold it while getting a better stance, then seize it again in its beak. Sometimes finches carry the tools with them to use on another foray. Not all members of the species have mastered the trick, indicating that it may well be a learned tradition. (Darwin himself never witnessed this remarkable behavior during his brief stay on the Galápagos.)

Another simple tool is used by the sea otters that swim among the kelp beds off the coast of California. They often float on their backs and crack open abalone shells with a hammerstone used against another flat stone balanced on their chests as anvils. One was observed using the thick part of a Coca-Cola bottle as a hammer, in place of the usual stone used for that purpose.

Some anthropologists supposed that it was tool-making, not merely using, that was unique to humans. That objection was shattered forever when Jane Goodall discovered how chimpanzees in the African forest not only use but also make simple tools. Chimps take some care with selecting twigs and sticks, biting them to just the right size, stripping them of leaves, and chewing them into shape for simple jobs, such as gathering termites, prying up ant nests, or extracting honey from beehives.

When anthropologist Louis Leakey first heard of his protégé Jane Goodall's discoveries, he sent her a telegram: "Now we must redefine tool, redefine man, or accept chimpanzees as humans." A few years later, anthropologists decided to reclassify apes into the same family as humans: Hominidae.

But some naturalists think the accomplishments of the Galápagos woodpecker finch—with a brain a mere fraction of the size of a chimp's—show up the apes as underachievers. "Given their hands and huge brains, it's amazing apes and monkeys don't do a lot more tool-using. They're incredibly stupid," says anthropologist and naturalist Michael MacRoberts in a personal communication. MacRoberts, who has conducted field studies of both free-ranging macaque monkeys and acorn-storing woodpeckers, further argues:

> If Leakey had seen the Galápagos finch prying and stabbing hidden grubs with cactus spines, or watched California woodpeckers chisel trees into collective "granaries" for storing acorns, would he say we would have to change the definition of man—or birds? No, because primatologists are like doting parents. Anything "their" monkeys or apes do is remarkably clever, because they expect them to be bright. And anything other animals do is "just instinct," because they're supposed to be far removed from man.

After interviewing Jane Goodall on television, comedian Jay Leno remarked, "I don't believe chimps are similar to humans just because they use tools. What would really make them like humans would be if they *borrowed* the tools—*and never gave them back.*"

See also ABANG; APES, TOOL USE OF

TOUMAI SKULL

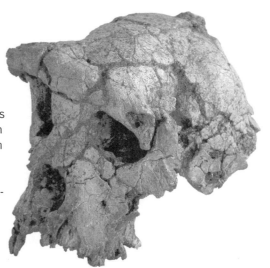

Paleoanthropologist Michel Brunet was so moved he could hardly get the words out. "It's a lot of emotion to have in my hand the beginning of the human lineage," Brunet, of the University of Poitiers, France, told reporters in 2001. "I have been looking for this for 25 years, and I knew I would one day find it." The skull he held so tenderly is known as Toumai ("hope of life" in Goran, a language spoken in southern Chad) or by its scientific name, *Sahelanthropus tchadensis,* meaning "man from the Sahel region of Chad."

Brunet's long-sought prize was found by members of his team in the western Djurab Desert of northern Chad, thus opening a "third geographic window" on early hominid evolution in central-east Africa, thousands of miles west of the established fossil hominid treasure troves in Kenya and Ethiopia's Afar region. Toumai also opened a new window into time. While the australopithecines, like "Lucy," are dated as far back as four million years, Toumai lived three million years before that. Hundreds of fossils of distinctive extinct relatives of hippos, pigs, and elephants that were found near the site confirm the skull's association with a six- to seven-million-year-old fauna.

Nature, the science magazine of record, called it "the most important fossil discovery in living memory," rivaling that of the Taung baby, the first known australopithecine, in 1924. According to Brunet, *Sahelanthropus tchadensis* shows a mosaic of apelike and more human features. It has heavy brow ridges, flattened facial structure (rather than a protruding, apelike snout), and hominid-like teeth, yet it has "an apelike brain size and skull shape."

No jawbone, arms, limbs, or torso bones were found with the partially distorted skull, but a computer image revealed that the angle at which it once perched on the spinal column is consistent with an upright biped. Walking upright, rather than large brain size, has become the defining characteristic of hominids.

Toumai's braincase, 350 cc in volume, is one-quarter the size of a modern human's, and about the size of a chimp's. But there's nothing apish about its teeth, which appear to be hominid. Also, it does not sport large projecting canines, as do modern gorillas and chimpanzees.

Brunet believes that *S. tchadensis* is the oldest known common ancestor of both chimps and humans. Of course, discoverers always want their "children" to be the long-sought holy grail, a "missing link" that joins humans to our primate family tree. Almost immediately, much to Brunet's annoyance, envious detractors among anthropologists were ready to proclaim it an ape, without having studied the fossil. Even if Toumai turns out not to be a direct ancestor of the human lineage, however, it would still be the oldest generalized relative of apes and humans known, falling in time between the earliest known ancestral apes (9 million years ago) and australopithecines like Lucy (4 million years ago). For many decades, that enigmatic 5-million-year-long period was a total blank in the fossil record, known as "the Miocene gap." Of course, *Sahelanthropus* was probably not the only Miocene ape or protohominid; there may well have been several species wandering Africa, whose remains have yet to be found.

A year and a half prior to the discovery, Brunet had accurately predicted to his talented Chadian student Ahounta Djimadoumalbaye, "It is you who will find it." According to Djimadoumalbaye's recollection of the discovery, shortly after he lifted the precious cranium from the surface of a sand dune, he shouted to his Chadian colleague, "We have what we seek. We are victorious."

AFTER SEARCHING for 25 years, Michel Brunet's team found in Chad a seven-million-year-old skull of a species that may have lived near the junction of the ape and human lineages.

TRACKWAYS

See BIRD, ROLAND T.; "NOAH'S RAVENS"

TRANSITIONAL FORMS
Links between Species

Where is the fish on its way to becoming a lizard or the lizard turning into a cat? Creationists claim that there are no transitional forms, or "missing links," to document the evolution of species and of major groups of living things.

While it is true that "gaps" in the fossil record have tantalized evolutionists since Darwin, numerous fossil transitions between families or orders are becoming known. In *Evolution: What the Fossils Say and Why It Matters* (2007), paleontologist Donald Prothero chronicles newly discovered "links" that join bears, seals, and walruses in a single group or clade, forming a remarkably complete series. Cretaceous fossil snakes with legs and hip bones have recently been found in the Middle East. Prothero's impressive catalog demonstrates that many presumed "missing links" are no longer missing.

The term "transitional form" is slippery, however, because it is impossible to state with certainty that fossil A is a direct ancestor of fossil B and also of living animal C. Therefore, paleontologists speak of animals or plants with transitional *features*. [See CLADISTICS.]

For years, a transitional link between dinosaurs and birds was thought to be *Archaeopteryx*, an ancient winged creature with a seemingly lizardlike skeleton. First discovered in 1861, it caused a sensation with its fully developed feathers, in conjunction with a full set of teeth, claws on its wings, long tail vertebrae, unfused back vertebrae, and other "reptilian" traits. In the past few years, however, scores of earlier theropods with feathers have been found in China. Feathered dinosaurs, it turns out, may have already evolved into a successful group millions of years before *Archaeopteryx* appeared. It is now viewed as a relative latecomer that still retained some features of its dinosaurian ancestors—just as today's platypuses, which are milk-giving mammals, retain their reptilian ability to lay eggs.

In comparing living organisms, zoologists have found that many "kinds" or "species" are not at all fixed or well defined. Intermediate variant populations are common and often defy categorization. Thus, there is no "gap" between thrushes and wrens, between lizards and snakes, or between sharks and skates. A complete gamut of intermediate species runs from the great white shark to the butterfly ray, and each step in the series is a small one, corresponding to the slight differences that separate very similar species. Living plants, too, have many transitional and hybrid forms. Attempting to classify species of willows or grasses, for instance, is a botanist's nightmare.

Sometimes, as Darwin well knew, classification is wholly arbitrary. The distinction between early mammal-like reptiles and reptile-like mammals, for instance, is blurred. But a long series of intermediate fossils exhibits a clear trend: two bones that formed the articulation for the upper and lower jaws in reptiles became modified as middle ear bones (the malleus and incus, or hammer and anvil) of mammals. So if the lower jaw contains these bones in sequence, no matter what the animal looks like, scientists call it a reptile. If the lower jaw consists of only the dentary bone, however, it's a mammal.

Evolution is not a ladder but a branching bush, sending off dozens of shoots in many directions. One usually finds sequences of groups with intermediate structures rather than straight-line sequences of ancestors and descendants. We may find fossils of only a twig here, a bough there—usually from periods of stability, since most major changes occur in short bursts in small populations. In William H. Calvin's metaphor, in his book *The River That Flows Uphill* (1986): "If you were an archeologist digging up a parking garage, nearly all the cars you'd uncover would be parked on one level or another; few would actually be on a ramp between levels."

Recent discoveries of spectacular transitions that were formerly "missing" include the "walking whales" of Pakistan and the long-sought tetrapod *Tiktaalik*, a "walking fish" that shows clear skeletal evidence of limbs and toes within its fins. In 2004, Ted Daeschler and his team at the Philadelphia Academy of Natural Sciences discovered *Tiktaalik roseae* on Ellesmere Island, more than 600 miles north of the Arctic Circle—the most compelling evidence yet of an animal that was on the verge of the transition from water to land 375 million years ago.

TIKTAALIK is a transitional form between fish and four-legged walkers, or tetrapods. Discovered in the Canadian Arctic in 2006, it is a fish with simple foot bones inside its fins that lived 385 million years ago.

When we turn to humanlike primate fossils going back more than four million years, we find abundant evidence of evolutionary transitions. Asked about the supposed lack of "missing links" for humans, paleontologist Stephen Jay Gould cited the "unmistakable temporal sequence displaying a threefold increase of brain size, and corresponding reduction of jaws and teeth" over the past five or six million years. "What more could you ask," he wrote, "from a record of rare creatures living in terrestrial environments that provide poor opportunity for fossilization?"

See also *ARCHAEOPTERYX*; BRANCHING BUSH; CREATIONISM; DINOSAURS, FEATHERED; ESSENTIALISM; HYBRIDIZATION; "MISSING LINK"; WHALE, EVOLUTION OF

TREE OF LIFE
Branching Evolutionary Lineages

In a famous passage in the *Origin of Species* (1859), Charles Darwin spoke of "the great Tree of Life," establishing the arboreal metaphor as one of the central themes of evolutionary biology. "The affinities of all the beings of the same class have sometimes been represented by a great tree," he wrote. "As buds give rise by growth to fresh buds, and these, if vigorous, branch out and overtop on all sides many a feebler branch, so by generation I believe it has been with the great Tree of Life, which fills with its dead and broken branches [fossils of extinct animals] the crust of the earth, and covers the surface with its ever branching and beautiful ramifications [all living organisms]."

Since the time of Linnaeus (1707–1778), living things have been classified by how closely they resemble one another. Shared anatomical attributes of foxes, dogs, and wolves convinced taxonomists to group them in the dog family (Canidae), just as lions, tigers, and leopards are classified as cats (Felidae), and both groups, as meat-eaters, belong to the

PHYLOGENY OR TREE OF LIFE compiled by Elisabeth Vrba of Yale University incorporates discoveries in paleontology and molecular comparisons: a cladistic hypothesis of evolutionary history. Arrows point to earliest known fossil for each lineage. Artist: Susan Hochgraf.

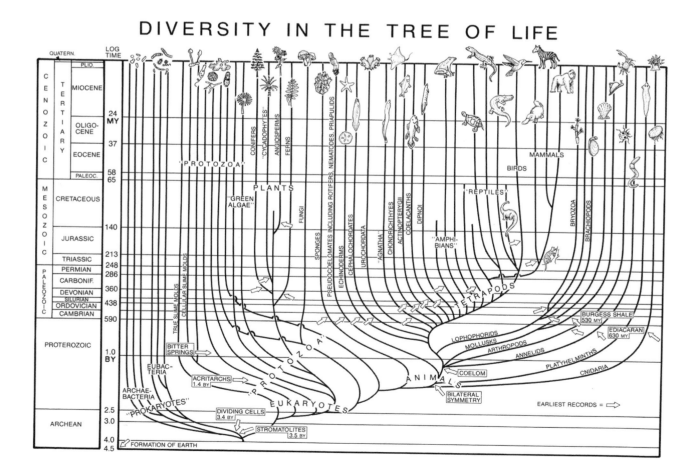

DIVERSITY IN THE TREE OF LIFE

order Carnivora. Since all carnivores are warm-blooded and furry, they are included (with ungulates, rodents, etc.) in the class of mammals (Mammalia).

Darwinian evolution gave this taxonomic system a new dimension: time. As evolutionary thought revolutionized biology, it became clear that species not only appeared to share certain resemblances; they were actually related because they descended from common ancestors. Evolutionary classification therefore should reflect how (and possibly when) various lineages split off from common ancestors

New methods in biochemistry are making inroads into the search for the tree's limbs and branches, which had long been the province of bone experts (paleontologists) and anatomists. If the test tube scientists are correct, an animal's microscopic molecules tell as much or more about its evolutionary history than the shape of its paws or dental patterns. Using comparisons of DNA, proteins, serum albumin reactions, and other techniques, researchers have refined classifications based on genetic closeness or distance.

To the dismay (and sometimes despite objections) of paleontologists and anatomists, biochemical researchers sometimes refuse to take into account the appearance of an animal or its supposed ancestors. Fossil bone fragments or certain kinds of anatomical similarities can be misleading, they claim; biochemistry of genomes is more accurate. Some cladists (from the Greek *klados*, a branch) even refuse to infer anything about timing of the branching; they believe science can only show closeness or distance of lineages with any high level of certainty. Others believe that nonadaptive proteins evolve at a known, steady rate, and rely on these "molecular clocks" to date the branching of lineages.

A flood of new information, from whole-genome sequences to detailed structural information to inventories of Earth's biota, is transforming 21st-century biology. Along with comparative data on morphology, fossils, development, behavior, and interactions of all forms of life on Earth, the growing body of genomic sequencing data has sparked a worldwide scientific effort to build up a phylogeny—the genealogical map for all lineages of life on Earth. The Tree of Life Web Project (ToL) is an international collaborative effort of biologists to chart that phylogeny for all 1.7 million described species.

So far, ToL has created more than 3,000 Web pages, compiling up-to-date information about the diversity of organisms on Earth and their evolutionary history (phylogeny). Each page contains information about a particular group of organisms (e.g., echinoderms, tyrannosaurs, phlox flowers, cephalopods). The pages are linked one to another hierarchically, starting with the root of all life on Earth and moving out along diverging branches to individual species.

Recently, however, the basic Darwinian concept of a tree of life has undergone some radical changes. Bacteria and retroviruses of different lineages, it turns out, can swap genetic material from one "branch" to another—horizontally, rather than vertically. In fact, such genetic exchanges between species from different lineages are going on all the time. As the geneticist Richard Lewontin wrote in *It Ain't Necessarily So* (2001):

> The branches of the tree of life come together again. . . . The basic cellular architecture that is shared by yeasts and humans is a consequence of ancient fusions of independent life forms . . . [An] evolutionary "tree of life" seems the wrong metaphor. Perhaps we should think of it as an elaborate bit of macramé.

See also BRANCHING BUSH; CLADISTICS; LINNAEUS

TURKANA BOY
Most Complete *Erectus*

In the summer of 1984, Richard Leakey's "hominid gang" unearthed the 1.6-million-year-old remains of a gangly adolescent boy—the most complete skeleton of a *Homo erectus* ever found. (Several years later, anthropologists referred it to a closely allied species, *Homo ergaster*.) Discovered by Kimoya Kimeu, the foreman of a team led by anatomist Alan Walker and Richard Leakey, the fossil boy came from a site called Nariokotome, near the western shore of Lake Turkana in Kenya. Perhaps the most complete fossil hominid skeleton ever found (aside from members of our own species, *Homo sapiens*), it is missing only a few arm bones, the feet, and some neck vertebrae, as well as tiny fragments from the skull and jaws. Every tooth is present, the spine and rib cage are complete, and there are hand bones, arm bones, a complete pelvis, and both legs.

In contrast with the modern human cranial capacity of approximately 1,300–1,400 cubic centimeters, the boy's braincase held about 840 cc. His height was five feet four inches (considered tall for an early hominid), and he had humanlike body proportions. Canine milk teeth had not yet been shed and the long bones weren't fully grown, indicating that he was less mature than a *sapiens* his age would be. As an adult, his brain would have reached about 882 cc—65 percent of what adults have today—and his body would have reached a stature of six feet. Also, his spinal cord is much narrower than that of modern humans.

Back in 1891, Eugene Dubois found the first remains of *Homo erectus,* which was popularly known as the "Java ape-man," as he thought it represented the "missing link" between apes and men. But spectacular as his finds were at the time, they consisted of only a fragmentary skullcap and some molar teeth. Thirty-five years later, more complete skulls and long bones were discovered at Zhoukoudian in China and nicknamed "Beijing Man." Since then, *Homo erectus* skulls and fragments have been found in various parts of Africa, Europe, and Asia—including several skulls excavated at Olduvai by the Leakeys.

But never before the Turkana boy was a specimen of such completeness found, with so few gaps left to the imagination. Along with "Lucy," a 40 percent complete skeleton of a female australopithecine from Ethiopia, the Turkana boy is one of the most important fossil hominid discoveries in the history of paleoanthropology.

See also HOMO ERECTUS; "LUCY"

TURKANA MOLLUSKS

During the mid-1970s, Harvard paleontologist Peter G. Williamson worked up the best-documented fossil sequence supporting the idea of punctuated equilibrium. But his series of mollusk layers taken from Lake Turkana, in northern Kenya, also demonstrates that "suddenness" in geology may really be a very long time.

Using the rate of reproduction of living relatives of the snails as a measure, Williamson calculated the rate of change in his fossil mollusk series. The "abrupt" discontinuities or "revolutionary changes" turn out to have taken 5,000 to 50,000 years—which would be about 20,000 generations of the creatures.

In a 1981 article in *Nature,* J. S. Jones of University College London translated this fossil record into the terms of an experimental geneticist. The time in which the major changes occurred in these species, he figured, would be the equivalent of 1,000 years of breeding fruit flies, 6,000 years of mouse generations, or 40,000 years of breeding dogs. (Most dog breeds we have today were developed over only a few thousand years from wolf ancestors.) Williamson's snails are therefore much slower-changing than they look to a geologist who reads thick strata punctuated by "rapid" evolution. Ordinary Darwinian selection can account for the seemingly fast changes; they seem fast only to a geologist, not to a geneticist.

See also GOULD, STEPHEN JAY; PUNCTUATED EQUILIBRIUM

> "Rapid" evolution seems fast only to a geologist, not to a geneticist.

TWAIN, MARK (1835–1910)
On Human Evolution

Samuel Clemens (Mark Twain) took a particular delight in deflating pomposity, whether of churchmen or scientists. His satirical essays on evolutionists, written around 1900, were later collected and published in *Letters from the Earth* (1938). Still fresh almost a century later, they offer insight into the public perception of major issues in 19th-century evolutionary theory.

In 1903, Alfred Russel Wallace, codiscoverer of natural selection, marshaled evidence for what he believed was the "guiding force" or "purpose" behind human evolution. His book *Man's Place in the Universe* concluded that Earth occupied a central position in the

EVOLUTIONARY SPOOFS were a staple in Mark Twain's satiric arsenal. Here a monkey wanders into his hotel room and destroys one of his books, whereupon Twain hails him as "An Honest Critic."

> **"[The oyster thought] that the nineteen million years was a preparation for *him*; [oysters are] the most conceited animal there is, except man."**
>
> —Mark Twain

universe and that certain special conditions were necessary to make the planet habitable for humans.

Despite Wallace's fresh approach and compilation of current data, scientists saw in it the old church doctrine of a preordained plan and purpose for everything in nature, with man at the center of things, like a spoiled child. Twain found it an irresistible target.

In "Was the World Made for Man?" (c. 1904) he gave his own version of Wallace's anthropocentric ("man-centered") theory. "It was foreseen," he began, "that man would have to have the oyster . . . [which] you cannot make . . . out of whole cloth, you must make the oyster's ancestor first. This is not done in a day."

Before the oyster, Twain wrote, there had to be an array of invertebrates so disgusting that you wouldn't feed them to the cat, and most would have to be failures and go extinct.

When the oyster was finally evolved, he continued, it started to think about how and why it got there:

> An oyster has hardly any more reasoning power than a scientist has; and so it is reasonably certain that this one jumped to the conclusion that the nineteen million years was a preparation for *him*; but that would be just like an oyster, the most conceited animal there is, except man.

Next, there had to be fish, Twain explained, so man could come along later and catch them. And great fern forests had to rise up, then die and turn to coal so man could fry the fish. Mastodons, giant sloths and Irish elk wander around ducking moving ice sheets, getting soaked when the continents submerge and burned by exploding volcanoes.

> At last came the monkey, and anybody could see that man wasn't far off now. . . . The monkey went on developing for close upon five million years, and then turned into a man—to all appearances.

In another satirical essay, "The Lowest Animal," also written around 1900, Twain argues that man is descended from the "Higher Animals," like snakes—a takeoff on the then-popular Degeneration Theory. Twain says he made the experiment of putting seven calves in with an anaconda and observed that it ate one and spared the rest. On the other hand, an English earl went out on the plains, shot 72 buffalo, ate part of one, and left the rest to rot. This proved, Twain wrote, "that the earl is cruel and the anaconda isn't."

> We have descended and degenerated, from some far ancestor—some microscopic atom wandering [in] a drop of water . . . —insect by insect, animal by animal, reptile by reptile, down the long highway of smirchless innocence, till we reached the bottom stage of development—namable as the Human Being. Below us—nothing. Nothing but the Frenchman.

Twain was an internationally popular speaker during the 1870s, and he played an engagement in London. It is not known whether he used some of his evolutionary spoofs in that program. We do know, however, that in 1879 the greatest American humorist of his day made a personal appearance in the little village of Downe to call on one of his devoted readers: Charles Darwin.

See also DEGENERATION THEORY; HAPPY FAMILY; SPIRITUALISM; "WALLACE'S PROBLEM"

"TWO BOOKS," DOCTRINE OF THE
God's Word vs. God's Works

Sir Francis Bacon (1561–1626), who took "all knowledge to be my province," helped create the modern scientific attitude, though he was neither an experimenter nor a systematic observer. As philosopher as well as counsel to King James I, Bacon used his nimble mind as a great arranger and compromiser. Approving chroniclers call him "a master of the balance of power among the different faculties" of the human mind, while critics have described his complexity as a mixture of "the soaring angel and the creeping snake."

One of the tasks he set himself was to make the pursuit of natural science acceptable in an age that took Scripture as revealed authority on all matters. In 1605, he published *Advancement of Learning,* with its doctrine of the "Two Books," which set the tone for the next two hundred years of natural philosophy:

> Let no man, upon a weak conceit of sobriety or an ill-applied moderation, think . . . that a man can search too far or be too well studied in the book of God's word or in the book of God's works, divinity or philosophy; but rather let men endeavor an endless progress or proficience in both; only let men beware . . . that they do not unwisely mingle or confound these learnings together. . . .
>
> . . . Our Saviour saith, "You err, not knowing the Scriptures, nor the power of God"; laying before us two books or volumes to study . . . first, the Scriptures, revealing the will of God, and then the creatures expressing his power. . . . The latter is a key unto the former . . . opening [both] our understanding . . . [and] our belief, in drawing us into a due meditation of the omnipotency of God, which is chiefly signed and engraven upon his works.

"Previously philosophers seldom drew the sharp distinction" between the lessons of God's work and God's words, writes philosophical historian James R. Moore in his essay "Geologists and Interpreters of Genesis in the Nineteenth Century" (1986). But "for Bacon the book of God's works is 'a key' to the book of God's word; students of nature may therefore instruct interpreters of the Bible." Cleric-naturalists, who might be tempted to explain nature in scriptural terms, tried to honor Bacon's dictum not to "confound these learnings together." This Baconian compromise, according to Moore, formed the basis of congenial relations between naturalists and churchmen, encouraging the growth of scientific research and observation.

The metaphor of the "Two Books" became an almost unconscious convention, embedded in the language of naturalists. Geologists today still speak of "reading the record in the rocks" as they did a century ago. Darwin himself spoke of the fossil record containing missing pages, and even whole chapters. On living plants and animals, however, he could see their evolutionary history "stamped in plain letters on almost every line of their structure." Bacon's definition of the "Two Books" was chosen by Darwin to face the title page of his own greatest book, the *Origin of Species* (1859).

2001: A SPACE ODYSSEY (1968)
Evolutionary Epic

Dawn on the ancient African plains a million years ago. A group of man-apes awakens and begins the daily search for food. Within a few minutes, they act out an image of our ancestral nature and evolving behavior as it was understood by anthropologists in the mid–20th century. These are killer apes brought to life on the Super Panavision screen, guided by the writings of paleoanthropologist Raymond Dart, animal behaviorist Konrad Lorenz, and best-selling popularizer Robert Ardrey, author of *African Genesis* (1961) and *The Territorial Imperative* (1966).

The film is the enormously influential *2001: A Space Odyssey,* directed and produced by Stanley Kubrick and costing $10.5 million. Based on a novella by science fiction author Arthur C. Clarke, it begins with the man-apes (played by costumed French mimes) and depicts an evolutionary journey that takes humans to the far planets aboard space vehicles, burrows inside the psychedelic vortex of his own brain, and locks him into a "struggle for existence" with the conscious machines he has created. ("Open the pod bay door, Hal!")

Eventually, man is reborn as a "starchild": an embryo floating in space, about to evolve into a higher, more advanced being. Each step of his evolutionary odyssey is heralded (and perhaps accelerated) by a mysterious "monolith," which first appears among the African man-apes, then reappears on the moon and again in outer space.

First announced by MGM with the title *From the Ocean to the Stars,* the original version was to start with unicellular life and trace in detail the evolutionary history of life on Earth

One must be "well studied [both] in the book of God's word [and] in the book of God's works."

—Sir Francis Bacon

Kubrick transformed a tossed bone tool into a space station within a few seconds of film, symbolizing the origins of a limitless technology.

before arriving at the man-apes. When Stanley Kubrick was brought in, he changed the concept and wrote the screenplay with Clarke, based on the latter's story *The Sentinel*, published in 1947. (The original tale did not include the apes, though it prominently featured the mysterious monoliths.)

The ape sequence begins with two groups fighting over which is to have access to a waterhole, the killing of a tapir with a rock, and the murder of one hominid by another, using a heavy bone as weapon. In a classic juxtaposition, which has become part of the language of cinema, the bone tool is tossed in the air, turning end over end in slow motion—then dissolves into a shot of an elongated space station, rotating slowly amidst the star-studded blackness. One tool implies the unfolding of a limitless technology.

Kubrick's film received five Academy Award nominations and won the Oscar for Special Effects. A sequel, *2010* (1984), written by Arthur C. Clarke and Peter Hyams, failed to generate excitement comparable to the original.

See also KILLER APE THEORY; *KING KONG*; O'BRIEN, WILLIS; *QUEST FOR FIRE*

TYNDALL, JOHN (1820–1893)
Physicist, Chemist, Natural Philosopher

With Charles Darwin and Thomas Huxley, the Irish physicist John Tyndall is inseparably connected with the mid-19th-century battle to bring science into the mainstream. He became for many the ideal professor of physics, whose powers of communication about his investigations into natural phenomena had enormous influence, in both England and America.

Tyndall mastered many sciences. He conducted groundbreaking studies of magnetism and polarity, the properties of heat, and the "magneto-optic properties of crystals" and their molecular arrangement. His friends and teachers included such scientific eminences as chemist Edward Frankland and physicists G. H. Magnus and Michael Faraday.

Not content with observing nature in the laboratory, in 1856 he climbed the Swiss Alps with Thomas Henry Huxley to investigate the movements and structure of glaciers. Three years later, Tyndall climbed to the Montanvert, one of the three great glaciers of Mont Blanc, to observe their winter movements. As he recounts in *The Glaciers of the Alps* (1860), he was impressed not only with their grandeur and immensity, but also by the beauty of the extraordinarily large, crystalline snowflakes that began to fall:

> The air became quite still, and the snow underwent a wonderful change. Frozen flowers . . . fell in myriads . . . the flakes were wholly composed of these exquisite blossoms entangled together. On the surface of my woollen dress they were soft as down, while my coat was completely spangled with six-rayed stars. And thus prodigal Nature rained down beauty, and had done so here for ages unseen by man. . . . Whence those frozen blossoms? Why for aeons wasted?

Tyndall was particularly awestruck by how both the crystals he had studied and snowflakes could organize themselves into such pleasing geometric patterns. Like Darwin and Huxley, he was one of those rare individuals who could welcome scientific understanding of nature without losing sight of her beauty or wonder.

In 1874, he delivered the Presidential Address at the British Association for the Advancement of Science. According to one 19th-century writer, that lecture "gave rise to a wide variety of opinion as to the conclusions expressed in it. "Tyndall's intent was clear. While trying to avoid giving offense, he had stated as diplomatically as possible that the church's days of restricting scientific inquiry into human nature and origins were over.

He was a member of the "X" Club, where he was influential in appointing scientists to various posts and committees. When Darwin was buried in Westminster Abbey, Tyndall served as one of the honored pallbearers.

See also BELFAST ADDRESS; "X" CLUB

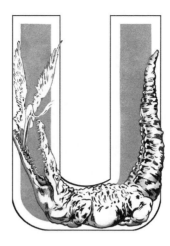

UNIFORMITARIANISM
Slow, Steady Change

In the early 19th century, the top geologists of England and France, among them the great Georges Cuvier, were convinced catastrophists. They believed the geology of the Earth could be explained by such biblical catastrophes as the Great Flood, or "Noachian Deluge" as they called it. Some even attempted to calculate the dimensions of Noah's Ark; Captain Robert FitzRoy of the *Beagle*, for instance, held a pet theory that mammoths became extinct because the door of the ark was too small to admit them!

Charles Lyell (1797–1895) published a revolutionary book, *Principles of Geology* (three volumes, 1830–1833), in which he theorized that the great features of the Earth had been produced by small causes working at a uniform rate over immense periods of time. These could still be observed at work today, such as water carrying sediments or wearing down rocks.

When Charles Darwin left on his voyage aboard HMS *Beagle,* he took the newly published first volume of Lyell's *Principles* with him. It had a profound effect on his geological observations. In 1832, when the ship stopped over at Montevideo, on the Río de la Plata, he received the second volume by mail. "I am become a zealous disciple of Mr. Lyell's views, as known in his admirable book." Darwin wrote a friend in 1835, "Geologising in South America, I am tempted to carry parts to a greater extent even than he does."

Actualism, the concept that ordinary present processes operated in the past, is the keystone of what we usually call uniformitarian thinking; it was not original with Lyell, though he made it widely popular. (In the mid-18th century Buffon, for instance, had written that "in order to judge what has happened, or even what will happen, one need only examine what is happening.")

Few noticed that Lyell spliced actualism with other ideas that seemed to be logical extensions but, in fact, were not. Gradualism and other theories were tied onto actualism like tin cans to a dog's tail. They had no necessary connection or unity, but Lyell's skillful presentation made them an accepted part of the package later called uniformitarianism. (A master of argument, Lyell had trained as a barrister before switching to geology.)

Modern geology is uniformitarian in accepting the actualist notion that the study of processes observable today can tell us what happened in the past, in postulating an immense age for the Earth, and in concluding that many great geologic features are the products of slow, steady forces causing gradual change over very long periods.

However, geology is also catastrophic in deducing radical changes in the atmospheric gases, in attributing global mass extinctions to fairly rapid shifts in climate, and in tracing some of these in turn to meteoric impacts. There has also been a shift toward the discontinuous, or jumpy, view of evolutionary events known as punctuationalism.

Today's earth scientists claim Lyell's *Principles of Geology* as their founding document but view it as a mixed bag of catastrophic and uniformitarian elements.

See also ACTUALISM; GRADUALISM; LYELL, SIR CHARLES; STEADY-STATE EARTH

USSHER-LIGHTFOOT CHRONOLOGY
Biblical Timekeepers

In 1650, Archbishop James Ussher of Armagh, Ireland, published the result of his calculations of the age of the Earth and humankind. Counting backward through all the "begats" in the Old Testament, he estimated the number of generations since Adam. Creation of Adam and Eve and all other creatures, he concluded, took place in 4004 B.C., giving the planet an age of about 6,000 years. His figures became the standard against which inquiries about the Earth's early history were measured.

Ussher's calculations were refined by another 17th-century churchman, Dr. John Lightfoot, vice-chancellor of Cambridge University, who computed that "man was created by the Trinity on 23rd October, 4004 B.C., at nine o'clock in the morning." An eminent Hebrew scholar, Dr. Lightfoot had also concluded that "heaven and earth, centre and circumference, and clouds full of water, were created all together, in the same instant." A convenient feature of Lightfoot's chronology was that the time of the month of all creation coincided exactly with the annual beginning of the academic year at Cambridge.

The Ussher-Lightfoot chronology was widely accepted for many years and was challenged only when evidence of a much greater age for the Earth began to accumulate from geological strata and fossils in the mid–18th century. But it continued to stand as dogma until the 19th century, when the rise of geology and evolutionary biology stretched the vistas of geologic time back hundreds of millions of years. Most people, including Charles Darwin, had thought the Ussher-Lightfoot chronology was part of the original scripture itself, since it was usually printed in the margins of most Bibles for 150 years.

See also CREATIONISM; FOUR THOUSAND AND FOUR B.C.; SMITH, WILLIAM

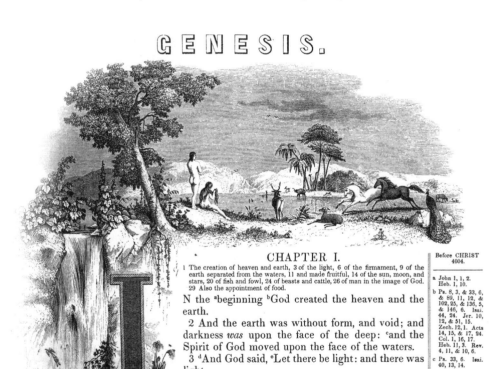

ARCHBISHOP USSHER'S chronology was inserted into Bibles around 1750. Most people, including Charles Darwin, thought it was an integral part of original Scripture. This *Illuminated Bible* is from 1843, published by Harper and Brothers, New York.

VEBLEN, THORSTEIN (1857–1929)
Sociologist, Economist

By the late 19th century, "survival of the fittest" had become the slogan of American free enterprise. Andrew Carnegie and John D. Rockefeller justified ruthless competition as the price of evolutionary progress. Sociologist Thorstein Veblen's classic *The Theory of the Leisure Class* (1899) outraged Social Darwinists by turning their theories about capitalism upside down.

Veblen concluded that successful entrepreneurs were far from "the fittest" members of the human species. Drawing on the bizarre evolutionary theories of Italian criminologist Cesare Lombroso, he lumped them with lower-class thieves and murderers as "atavistic" degenerates. In Veblen's words, the powerful industrialists shared with criminals "predatory aptitudes and propensities carried over . . . from the barbarian past of the race . . . with the substitution of fraud and . . . administrative ability" for naked violence.

Capitalist businessmen, in Veblen's view, were genetic "throwbacks" to early plundering, barbarian tribes—a lingering hindrance to social progress.

See also CARNEGIE, ANDREW; SOCIAL DARWINISM; "SURVIVAL OF THE FITTEST"

"VERCORS" (JEAN BRULLER) (1902–1991)
French Author, Illustrator

Talented writer-artist Jean Bruller ("Vercors") wondered what might happen if a tribe of primitive hominids had survived to the present day. Evolutionary kinship of species, he realized, raises ethical questions that have never been asked, let alone answered.

If "bigfoot" or "yeti" were actually found, would it be murder to shoot one—or merely cruelty to animals? How "close" to humans does an apelike creature have to be to deserve "human rights"? Vercors tackled these themes in *Les Animaux Dénaturés* (1952), published in English as *You Shall Know Them* (1953) and later retitled *The Murder of the Missing Link.*

Vercors set his story in Australia, where an elusive tribe of hominids is discovered. Scientists recognize them as australopithecines, a humanlike species from a million years ago. When word gets out, a factory owner sets out to capture them for exploitation as unpaid laborers. If they are not humans, he could legally treat them as domestic animals; but if they're defined as persons, it would be slavery.

When attempts to protect the tribe fall on deaf ears, Vercors's hero takes a desperate course. He gets a female pregnant, then announces he has killed her infant with a lethal injection. Now the legal system must determine his punishment. If he is to be convicted of murder, the jury must first define the man-apes as humans, which would rescue them from enslavement. Vercors's hero puts his own life on the line to establish the tribe's humanity and win their freedom.

With elegant Gallic logic, Vercors manages to rescue both his hero and the hom-

THE MINDLESS MOVIE
Skullduggery, adapted from Vercors's novel, mangled the book's central question: How closely must a hominid resemble ourselves before we grant it full human rights?

inids. After prolonged deliberations, the jury declares the creatures are indeed human and any future killing of them will be considered murder. Since they were not legally human at the time of the killing, however, the jury decides the hero has committed no crime. Like the great Swedish classifier Carl Linnaeus, Vercors concludes that man himself defines who is human and who is an animal. (Instead of formally describing the human species zoologically, Linnaeus had written: "Man, know thyself.")

Unfortunately, Vercors's unusual, thoughtful novel was made into an uncommonly cheesy Hollywood film, *Skullduggery* (1970), starring Burt Reynolds, which trivialized the story, and left viewers with a grotesque image of full-figured women in hairy body suits.

See also ANIMAL RIGHTS; *HOMO SAPIENS*, CLASSIFICATION OF

VESTIGES OF CREATION (1844)
A Victorian Sensation

In the early 19th century, "The treasures of the whole world of nature" were pouring into Europe from the colonized tropics, evolutionist Alfred Russel Wallace recalled in *The Wonderful Century* (1898), "and there was a general impression that . . . we must spend at least another century in collecting, describing, and classifying" before science could hope to tackle that "mystery of mysteries," the origin of species. "The need of any general theory of how species came into existence was hardly felt."

Robert Chambers (1802–1883) of Edinburgh was neither a scientist nor philosopher, but a writer-publisher of encyclopedias, biographies, and reference books. Most of Chambers's many publications, such as his *Dictionary of Eminent Scotsmen* and *Chambers's Cyclopaedia of English Literature,* were popular and thoroughly noncontroversial. In 1844, however, he created a furor with the first widely read book championing evolution, *Vestiges of Creation*, published 15 years before Charles Darwin's *Origin of Species* (1859).

Wallace credits Chambers with being the first to set down "the vague ideas of those who favored evolution . . . with much literary skill and scientific knowledge." By 1860, the *Vestiges* had gone through 11 editions.

Beginning with astronomer Pierre-Simon Laplace's Nebular Hypothesis, which had recently gained acceptance, Chambers summarized evidence for the evolution of solar systems. Given a primitive Earth and immense periods of time, he wrote, "an impulse . . . was imparted to the forms of life, advancing them in definite lines, by generation, through grades of organization terminating in the highest plants and animals."

> **Wallace claimed that Chambers's *Vestiges of Creation* convinced him that evolution took place by means of ordinary sexual reproduction.**

Though he argued for the "reasonableness" of modification through ordinary reproduction, rather than special creation, Chambers never tackled the how and why of evolution. "The book," Wallace said in 1898, "was what we should now call mild in the extreme." Its tone was reverential, its language respectful and even religious. Yet even as the public snapped up one printing after another, most scientific, religious, and literary critics greeted it "with just the same storm of opposition and indignant abuse which assailed Darwin's work fifteen years later."

Wallace claimed it was Chambers's *Vestiges of Creation* that convinced him that evolution of species took place by means of the ordinary process of reproduction, inspiring him to gather more evidence in support of the idea. (The initial result of that effort, known as Wallace's "Sarawak Law," was published in 1855.) Thomas Huxley, before he became an ardent evolutionist, had a quite different reaction—to his later regret, he published a nasty, dismissive review of the *Vestiges*.

Chambers's main objective was to extend the conception of the province of law in the universe and to establish the Theory of Development (evolution). All living organisms were connected, he concluded, obeying a system of law ordained by God. Change came about gradually and continuously, proceeding in the direction of increasing progress for all life—now spearheaded by humans, whose continuing progress was certain to continue.

Despite its conciliatory, reverent language, Chambers correctly anticipated the scorn and abuse it would provoke and took elaborate precautions to disguise his authorship, which was not revealed until long after his death.

See also CHAMBERS, ROBERT; "SARAWAK LAW"; WALLACE, ALFRED RUSSEL

VOYAGE OF HMS *BEAGLE*
Darwin's Journey of Discovery

Young Charles Darwin could hardly wait until December 27, 1831, when he would board HMS *Beagle* for a voyage around the world as ship's naturalist. "My second life will then commence," he wrote Captain FitzRoy two months before departure, "and it shall be as a birthday for the rest of my life."

During the surveying vessel's mission to make accurate coastal maps and charts for the Admiralty, Darwin collected animals, plants, fossils, and rocks from all over the world—and observed earthquakes, wars, and volcanic eruptions. He wandered the Brazilian rain forest, rode the backs of giant tortoises in the Galápagos, and confronted painted "savages" at Tierra del Fuego.

Autocratic Captain Robert FitzRoy (1805–1865), only four years older than young Darwin, was already a seasoned commander, who had once before (1826–1830) steered this ship through the treacherous waters around Cape Horn. He had sought a well-educated naturalist and companion who could share his meals and cabin. Darwin was not the first to apply, and his father hated the idea. But thanks to a recommendation from his Cambridge professor, Rev. John Henslow, he got the post.

FitzRoy conducted an intensive coastal survey of South America, though Darwin also made overland trips through Brazil, Chile, and Argentina. After two unsuccessful attempts to navigate around the Horn, during which the vessel almost capsized, FitzRoy steered through the Strait of Magellan. After visiting the Galápagos Islands, the *Beagle* struck out across the Pacific Ocean.

She visited Tahiti and the Society Islands, continued on to New Zealand, Australia, Tasmania, Keeling Island, and Mauritius in the Indian Ocean, and the Cape Colony in South Africa. Then, instead of heading north to England (to the chronically seasick Darwin's dismay), the *Beagle* traced the eastern coast of South America once more. Finally, on October 2, 1836, she returned to Falmouth, England.

For a good part of the voyage, Darwin was cramped in a chart room in the forward poop deck, where he had to be perfectly organized—there was no room for the slightest disarray. Although he grumbled about the tiny space, he later admitted it forced him to concentrate, complete one task at a time, and arrange his materials in impeccable order, a habit that was to stay with him for life. (Friends even suggested that his little study at his home in Downe, jammed with books, papers, and specimens, was an attempt to re-create the poop cabin.)

Young Darwin of the *Beagle* was different from the older, semi-invalid philosopher of Down House, who was easily tired and had daily bouts of headaches, abdominal pain, and vomiting. As a young man, he thought nothing of riding with "sinister" gauchos on the pampas, trekking 400 miles through wilderness, excavating fossils by hand, and climbing mountains.

He met his share of danger—more from people than from the wild animals he sought. In Argentina, he found himself in the midst of a bloody war of extermination against the Pampas Indians led by the ruthless General Juan Manuel Rosas, who at first mistook him for a spy. Among Brazilians, he observed the slavery of Africans first-hand: It sickened and infuriated him. (Years later, he still had nightmares about a screaming serving girl, whose fingertips were crushed in screw-vises by her mistress for small infractions.) Everywhere—among people as among birds and beasts—he saw competition, waste of life, struggle for survival.

He was impressed, also, by the fossil evidence of the "former inhabitants" of South America, giant sloths and armadillos, which bore obviously close resemblances to the modern small sloths and armadillos living there. It struck him that they might be ancestors and descendants. And some of the nearly naked hunters and gatherers he encountered made him think of what humans must have been like before the dawn of history.

Another striking fact was that the various volcanic islands in the Galápagos contained differently adapted species that appeared very closely related, as though they had all arisen from common mainland ancestors. (He nearly missed this important point at first, but bird

> ## "I could not have believed how wide was the difference between savage and civilized man."
> —Charles Darwin, 1839

expert John Gould, after studying his collections of finches back in London, set him on the right track.)

On the last leg of the voyage, Darwin focused on the formation of coral reefs and atolls, which became the subject of his first scientific book, *The Structure and Distribution of Coral Reefs* (1842).

Even before it was published, Darwin won fame with his travel narrative of discovery and adventure, generally known as *Voyage of the Beagle*. It was first issued as the last of a three-volume report, with the full title *Journal of Researches into the Geology and Natural History of the Various Countries Visited by HMS Beagle under the Command of Captain FitzRoy, R.N., from 1832 to 1836.*

The first volume, written by Captain Philip King, described the ship's first voyage. Volume two was written by FitzRoy, who was somewhat irritated when Darwin's volume on the natural history gained instant public attention and popularity. By Victorian standards, it was a bestseller, while the other volumes were ignored.

In *Through the Magic Door* (1907) Sir Arthur Conan Doyle, himself no stranger to adventure books, called Darwin's *Voyage of the Beagle* one of the two best books of "the romance of travel and the frequent heroism of modern life." (His other choice was Alfred Russel Wallace's *Malay Archipelago,* 1872.) Doyle admired Darwin's "gentle and noble firmness of mind" and devotion to his naturalist's quest:

Nothing was too small and nothing too great for [his] alert observation. One page is occupied in the analysis of some peculiarity in the web of a minute spider, while the next deals with the evidence for the subsidence of a continent and the extinction of a myriad animals. . . . Darwin rode the four hundred miles between Bahia and Buenos Ayres, when even the hardy Gauchos refused to accompany him. Personal danger and a hideous death were small things to him compared to a new beetle or an undescribed fly.

The excitement and stimulation of the *Beagle* voyage were enough to last Darwin for the rest of his life. Although he always maintained that he owed all later accomplishments to his strenuous voyage, he wrote his sister Carolyn in 1836 that his seafaring days were over: "I am convinced that it is a most ridiculous thing to go round the world, when by staying quietly [in one place], the world will go round with you."

See also (HMS) *BEAGLE;* CORAL REEFS; DARWIN, CHARLES; FITZROY, CAPTAIN ROBERT; *ORIGIN OF SPECIES*

WAGNER, MORITZ (1813–1887)
Founder of Isolation Theory

Famous German explorer, geographer, and naturalist Moritz Wagner was troubled by a common objection to Darwinian selection: Domesticated varieties, if set free from man's control, would be absorbed in a few generations back into the wild populations. Besides, while collecting beetles in Algeria in 1837 (more than 20 years before the Darwin-Wallace theory), it struck Wagner that different but closely related species appeared whenever he crossed a river.

Wagner created quite a stir with his "law of migration": Only when small populations are geographically isolated and prevented from crossing with the ancestral population can new species arise. Isolation, he wrote, is "the necessary condition for natural selection. . . . Organisms which never leave their ancient area of distribution will never change."

Most naturalists realized isolation was a common condition of speciation—particularly on remote islands—but were not willing to admit its general importance in the formation of new species. August Weismann, for instance, noted that in fossil beds at Steinheim Lake, new species were found in the same beds as older, closely related fossil species. Some believed Wagner's view was "deprived of all foundation" by Weismann's argument that sexual species must have evolved from original hermaphrodites "on the same territory." Darwin himself objected that Wagner's theory could not explain widespread adaptation among the majority of species spread out over vast areas.

In applying his "law" to the origin of man, Wagner speculated that one or several pairs of protohumans were driven away from their tropical forest homelands, where survival had been easy, while in less hospitable northern latitudes their return was cut off when the great Old World mountain chains were thrust up. Trapped in the harsher environment, they had to work and develop skills and tools to survive. (Wagner and his colleagues failed to see how closely this "just-so" story retells the familiar biblical account of the expulsion from the Garden of Eden, after which Adam had to live by the sweat of his brow.)

Despite such flights of fancy, Wagner's notion was rooted in sound observations of the distribution of natural populations relative to geographic barriers. Revived by Ernst Mayr 75 years later, the idea that small, isolated populations fostered speciation became part of the Synthetic Theory. This time it was anchored in a mechanism: "genetic drift" or "founder effect," which could be demonstrated in experimental populations.

Although Mayr had reestablished the theory on the basis of his studies of bird species in the 1930s, isolation was still not considered of great general importance by biologists until the 1970s, when paleontologists Stephen Jay Gould and Niles Eldredge tied it to punctuated equilibrium. Today, reproductive isolation of small populations is considered a major "cradle" for the origin of new species.

See also DIVERGENCE, PRINCIPLE OF; GENETIC DRIFT; HAWAIIAN RADIATION: MAYR, ERNST; PUNCTUATED EQUILIBRIUM; WEISMANN, AUGUST

> "Isolation of small populations is the necessary condition for natural selection."
>
> —Moritz Wagner

WALLACE, ALFRED RUSSEL (1823–1913)
Codiscoverer of Natural Selection

After publication of the *Origin of Species* in 1859, evolution by natural selection, biology's great unifying concept, became famous as "Darwin's theory." First announced and published jointly the previous year, it is actually the Darwin-Wallace theory. Nevertheless, Charles Darwin often called it "my theory," while Alfred Russel Wallace, his partner and coauthor, graciously insisted, "It [is] actually yours and yours only."

Wallace carried modesty to extremes, even calling his own book on evolution *Darwinism* (1889). Had he been more ambitious and less generous, evolutionary science might have become known as "Wallaceism."

An explorer, zoologist, botanist, geologist, and anthropologist, Wallace was a brilliant man in an age of brilliant men. Famous not only as cocreator of the natural selection theory, he was the discoverer of thousands of new tropical species, the first European to study apes in the wild, a pioneer in ethnography and zoogeography (distribution of animals), and author of some of the best books on travel and natural history ever written, including *A Narrative of Travels on the Amazon and Rio Negro* (1853) and *The Malay Archipelago* (1869). Among his remarkable discoveries is "Wallace's Line," a natural faunal boundary between islands (now known to coincide with a junction of tectonic plates) separating Asian-derived animals from those evolved in Australia.

Born in 1823 in Usk, England, a small town near the Welsh border, Wallace was raised in genteel poverty. His first employment was helping his brother John survey land parcels for a railroad. While still in his twenties, he served a stint as a schoolmaster in Leicester, where he met young Henry Walter Bates, who shared his passion for natural history. On weekend bug-collecting jaunts, the would-be adventurers discussed such favorite books as the *Voyage of HMS Beagle* (1845) and dreamed of exploring the lush Amazon rain forests of Charles Darwin's ecstatic descriptions.

BRILLIANT, ECCENTRIC, and utterly his own man, Alfred Russel Wallace independently developed the theory of evolution by natural selection.

Another book also inspired them: Robert Chambers's anonymously published *Vestiges of Creation* (1844), a controversial, literary treatise on evolution. Scorned by scientists, *Vestiges* championed the idea that new species originate though ordinary sexual reproduction rather than by spontaneous creation. Wallace and Bates decided they would comb the exotic jungles to collect evidence that might prove or disprove this exciting "development hypothesis" (only later known as evolution). When Darwin had embarked on his own voyage of discovery some 20 years earlier, he had had no such clear purpose in mind.

Science was not yet a well-established profession, and naturalists were often dedicated amateurs from wealthy families. When Darwin went on his circumglobal voyage, his father paid all expenses, even providing a servant to assist with his work. Wallace's achievements are all the more remarkable, for he had to finance his expeditions by selling thousands of natural history specimens, mainly insects, for a few cents apiece. When his exploring and collecting days were over, Wallace struggled to support his family on author's royalties and by grading examination papers. (He said in *My Life* (1905) that the "capability of a man in getting rich is in an *inverse* proportion to his reflective powers in in *direct* proportion to his impudence.")

Bates and Wallace reached Pará, at the mouth of the Amazon, in May 1848; they collected and explored the surrounding regions for several months, then decided to split up. Wallace went up the unknown Rio Negro, leaving Bates to explore the upper Amazon regions. From 1848 until 1852, Wallace collected, explored, and made numerous discoveries despite malaria, fatigue, and the most meager supplies.

When he finally returned to rejoin Bates downriver, he found that his beloved younger brother had traveled across the world to join the adventure and had just died of yellow fever in Bates's camp. Grief-stricken, exhausted, and suffering from malaria himself, Wallace boarded the next ship for England. With him went his precious notebooks and sketches, an immense collection of preserved insects, birds, and reptiles, and a menagerie of live parrots, monkeys, and other jungle creatures.

In the middle of the North Atlantic, as Wallace suffered a new attack of malaria, the ship suddenly burst into flames. He wrote in *My life,* "I began to think that almost all the reward of my four years of privation and danger was lost." He was able to rescue only a few notebooks as he dragged himself into a lifeboat; everything else burned or sank beneath the waves. In *Travels on the Amazon and Rio Negro,* he recalled:

> How many times, when almost overcome by the ague, had I crawled into the forest and been rewarded by some unknown and beautiful species! How many places, which no European foot but my own had trodden, would have been recalled to my memory by the rare birds and insects they had furnished to my collection! . . . And now I had not one specimen to illustrate . . . the wild scenes I had beheld!

The measure of Wallace's enormous courage and resilience showed itself shortly after returning to England. With the insurance money he received for part of his lost collections, he immediately set out on a new expedition—this time to the Malay Archipelago (1854–1862).

Wallace mastered Malay and several tribal languages, for he was intensely interested (as Darwin never was) in "becoming familiar with manners, customs and modes of thought of people so far removed from the European races and European civilization." A self-taught field anthropologist, he made pioneering contributions to ethnology and linguistics and developed "a high opinion of the morality of uncivilized races." He later recalled with satisfaction that while he lived among them he never carried a gun or locked his cabin door at night.

In the Moluccas he tracked orangutans through the deep forest, shot several for the British Museum's collection, and raised an orphaned infant orang in his field camp. Since local tribesmen regarded the red-haired apes as "men of the woods," they were horrified when he shot and skinned them, convinced he would next want to add their own skulls to his collection.

Wallace collected natural history specimens with an extraordinary passion. As he recounts in *The Malay Archipelago* (1869),

NATURALIST, EXPLORER, anthropologist, and founder of zoogeography, Alfred Russel Wallace was also the first European to observe orangutans in the forest.

> I found . . . a perfectly new and most magnificent species [of butterfly]. . . . The beauty and brilliancy of this insect are indescribable, and none but a naturalist can understand the intense excitement I experienced. . . . On taking it out of my net and opening the glorious wings, my heart began to beat violently, the blood rushed to my head, and I felt . . . like fainting . . . so great was the excitement produced by what will appear to most people a very inadequate cause.

Wallace came to the idea of evolution not through artificial selection of domestic animals, as Darwin did, but through his observations of the natural distribution of plants, animals, and human tribal groups and their competition for resources. Like Darwin, he was influenced by Thomas Malthus's *Essay on the Principle of Population* (1798), which he had read some years before.

In 1855, while in Sarawak, he composed "my first contribution to the great question of the origin of species." Combining his knowledge of plant and animal distribution with Sir Charles Lyell's account of "the succession of species in time," he came up with a conclusion about when and where species originate. ("The how," he wrote, "was still a secret only to be penetrated some years later.") His paper, titled "On the Law Which Has Regulated the Introduction of New Species," stated that "every species has come into existence coincident

YOUNG WALLACE set off for the Brazilian rain forest on a quest to find evidence for or against the idea of evolution.

both in space and time with a pre-existing, closely-allied species." This preliminary conclusion, he knew, "clearly pointed to some kind of evolution."

Published in an English natural history journal in September 1855, Wallace's "Sarawak Law" was generally ignored by the scientific world. When he expressed his disappointment in a letter to Darwin, "He replied that both Sir Charles Lyell and Mr. Edward Blyth, two very good men, specially called his attention to it." Writing years later, Thomas Huxley said, "On reading it afresh I have been astonished to recollect how small was the impression it made."

In February 1858, Wallace was living on Ternate, one of the Moluccan Islands, and was suffering from a sharp attack of intermittent malarial fever, which forced him to lie down for several hours every afternoon. From his combined accounts in a 1903 article and in *My Life*, his 1905 autobiography, here are Wallace's recollections about his independent discovery of natural selection:

> It was during one of these fits, while I was thinking over the possible mode of origin of new species that somehow my thoughts turned to the "positive checks" to increase among savages and others described . . . in the celebrated *Essay on Population* by Malthus . . . I had read a dozen years before. These checks—disease, famine, accidents, wars, etc.—are what keep down the population. . . . [Then] there suddenly flashed upon me the idea of the survival of the fittest . . . that in every generation the inferior would inevitably be killed off and the superior would remain.
>
> Considering the amount of individual variation that my experience as a collector had shown me to exist . . . I became convinced that I had at length found the long-sought-for law of nature that solved the problem of the origin of species. . . . On the two succeeding evenings [I] wrote it out carefully in order to send it to Darwin by the next post.

It was this article, "On the Tendency of Varieties to Depart Indefinitely from the Original Type" (1858), that sent Darwin into a panic, convinced his friend Charles Lyell's warning that he would be "forestalled" by Wallace "had come true with a vengeance."

Lyell and Sir Joseph Hooker, attempting to rescue their friend's threatened prior claim, arranged to have Wallace's paper published along with some of Darwin's early drafts. The announcement of the Darwin-Wallace theory of evolution by means of natural selection was read at the Linnean Society and published in its journal in 1858; the following year Darwin completed the *Origin of Species* and rushed it into print.

Wallace was informed of these developments while still in the Moluccas, and he wrote that he happily and graciously approved. When he returned to England in 1862, Darwin was still anxious about Wallace's reaction, and was relieved to discover his "noble and generous disposition." Later Wallace maintained that even if his only contribution was getting Darwin to write his book, he would be content. But the fact remains that Wallace was not given an opportunity to exercise his nobility or generosity, since the joint publication was decided without anyone consulting him.

In addition to the chronicles of his travels, Wallace turned out a remarkable series of books, all landmark studies in evolutionary biology: *Contributions to the Theory of Natural Selection* (1870), *Geographical Distribution of Animals* (1876), *Island Life* (1880), and *Darwinism* (1889). In *The World of Life* (1910), he describes the living Earth as a single, complex system, an idea that seems, in some sense, to have foreshadowed the Gaia hypothesis:

> There are now in the universe infinite grades of power, infinite grades of knowledge and wisdom, infinite grades of influence of higher beings upon lower. . . . This vast and wonderful universe, with its almost infinite variety of forms, motions, and reactions of part upon part, from suns and systems up to plant life, animal life, and the human living soul, has ever required and still requires the continuous co-ordinated agency of myriads of such intelligences.

Unlike the cloistered, tactful Darwin, in his later years Wallace was imprudently outspoken about his religious and political beliefs. Outraged colleagues wanted to dismiss him as a "senile crank" for his strong advocacy of utopian socialism, pacifism, wilderness conservation, women's rights, psychic research, phrenology, and spiritualism, as well as his campaign against vaccination. Wallace replied he was not "brain-softening" with age, but had held many of these beliefs for 30 years.

Spiritualism strongly influenced his ideas on human evolution, causing him to differ with Darwin in 1869 on whether natural selection could explain "higher intelligence" in man. Wallace thought the human mind was supernaturally injected into an evolved ape from "the unseen world of Spirit." He also rejected Darwin's concept of "sexual selection," which he dismissed as merely a special case of natural selection. Although the two men remained friendly and mutually respectful, they never really understood each other's perspective. [See SPIRITUALISM; "WALLACE'S PROBLEM."] Nevertheless, Wallace was called upon to be an honored pallbearer at Darwin's funeral at Westminster Abbey.

In 1876, Wallace helped introduce a Spiritualist paper at the British Association's scientific meetings, which apparently touched off the notorious Slade affair. [See SLADE TRIAL.] He testified for the defense at the trial of Henry Slade and often defended other professional "spirit-mediums" who were accused of conducting fraudulent "psychic experiments." In 1881, Wallace joined the Society for Psychic Research.

He headed the Land Nationalisation Society in 1882 and openly declared himself a Socialist in 1890. Some of his admirers had recommended he be appointed director of the proposed new park at Epping Forest, but Wallace immediately lost the position by stating that he would keep the woodland exactly as it was for future generations, allowing no restaurants, hotels, or other concessions.

When Darwin started a petition among scientists to get Wallace a civil pension, botanist Sir Joseph Hooker and others objected to appealing for government funds on behalf of "a public and leading spiritualist." However, Darwin and Huxley prevailed and Wallace got his pension. (Huxley, though differing with Wallace on many issues, assured him in 1866 that he would never seek "a Commission of Lunacy against you"!)

In his last book, *Social Environment and Moral Progress* (1913), Wallace cataloged the horrors of the urban poor, colonial exploitation, and unchecked greed: "It is not too much to say that our whole system of society is rotten from top to bottom, and the Social Environment as a whole, in relation to our possibilities and our claims, is the worst that the world has ever seen." He was deeply saddened and outraged, as he wrote in *The Wonderful Century* (1898), by "reckless destruction of the stored-up products of nature, which is even more deplorable because more irretrievable."

He was furious when apologists for the status quo told him society needed no safety net for its poor or infirm, since, according to the "law" of natural selection, they ought to be eliminated. "Having discovered the theory," he fumed in his 1913 book, "it is rather amusing to be told . . . that I do not know what natural selection is, nor what it implies." Eugenicists who sought to regulate human breeding for selective improvement he considered "dangerous and detestable," and he warned that lawmakers were "sure to bungle disastrously" any legislation on the subject.

Influenced by the socialist Henry George, Wallace urged a policy of land nationalization and an economy in which "all shall contribute their share either of physical or mental labor, and . . . every one shall obtain the full and equal reward for their work. [Then] the future progress of the race will be rendered certain by the fuller development of its higher nature acted on by a special form of selection which will then come into play."

What "special form of selection" might be the salvation of humanity? Wallace argued that human populations produce many more males than females, but in his day young men were dying by the millions. Alcoholism, dangerous occupations, and particularly the frequent wars left Europe with a huge proportion of unattached women. But under a just and nonmilitaristic social system, Wallace predicted, the number of males would rise dramatically, until they greatly outnumbered women: "This will lead to a greater rivalry for wives, and will give to women the power of rejecting all the lower types of character among their suitors." The well-educated, enfranchised, responsible "women of the future [will be] the regenerators of the entire human race . . . in accordance with natural laws."

Wallace's special hope for the salvation of mankind, then, was none other than "sexual selection," one of Darwin's favorite mechanisms for explaining the evolution of man—which Wallace had always insisted did not exist! However, Wallace added a twist to Darwinian sexual selection: an explicit acknowledgment of the large evolutionary effects of a slight change in sex ratio, a surprisingly modern way of thinking about populations.

During the 1970s and 1980s, Alfred Russel Wallace become a hero among disaffected academics and independent scholars. They saw in him a brilliant scientist, working outside the establishment, scrabbling for a living, snubbed by those with wealth and position, persecuted for unpopular social views—possibly even deprived of his rightful place in history. Yet Wallace was morally triumphant as a great human being and fearless truthseeker, cheerful, optimistic, and productive into his ninetieth year.

In 1985, the British Entomological Society, of which Wallace was once president, launched a series of major expeditions to study the insects of the world's tropical rain forests. They called it "Project Wallace."

See also BATES, HENRY WALTER; BEETLES; GAIA HYPOTHESIS; HAMPDEN, JOHN; PHRENOLOGY; "SARAWAK LAW"; SEXUAL SELECTION; *VESTIGES OF CREATION*; WALLACE'S LINE

WALLACE'S LINE
Landmark in Zoogeography

Alfred Russel Wallace (1823–1913), the talented English naturalist who codiscovered the theory of evolution by natural selection, has often been the forgotten man in the Darwin-Wallace theory. But there is another monument to his brilliance that stands alone, and can still be seen today on every geologist's and biologist's map of the world: Wallace's Line.

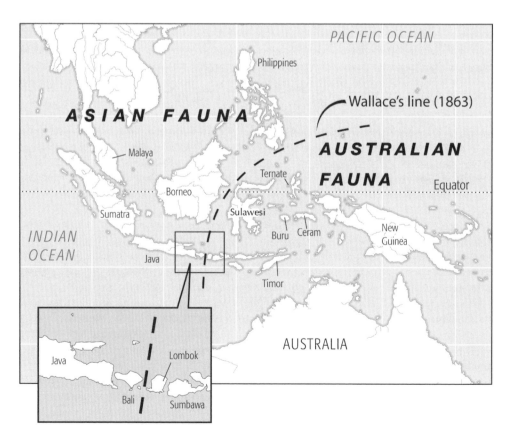

WALLACE'S LINE is an inferred natural boundary between animals evolved from Asian precursors (western side) and those of Australian (eastern side). A century after Wallace proposed it, geologists confirmed that his line is near the edge of the Indo-Australian plate.

After years in the Amazon and Malaysia observing and collecting wildlife, Wallace attempted to make sense of the facts of animal geography. While he was exploring the vast 2,500-mile chain of islands known as the Malay Archipelago, he noticed similarities and differences in the distribution of plants and animals that suggested a pattern. Also, he noted an ever-diminishing number of species as he traveled farther from the mainland peninsula. He wrote in *The Malay Archipelago* (1869), "I have arrived at the conclusion that we can draw a line among the islands, which shall so divide them that one-half shall truly belong to Asia, while the other shall no less certainly be allied to Australia."

Wallace's Line runs along a narrow strait between the islands of Bali and Lombok, and between Borneo and Celebes (now Sulawesi). To the east of the line are animals that evolved on the Australian continent, while west of it the fauna is derived from Asia.

In *The Geographical Distribution of Animals* (1876), a classic of zoogeography, Wallace presented evidence for his theory. Islands west of the line once must have been connected, but even then the eastern islands had been isolated from the western by a deep channel.

Wallace had no way to study the geology of the seafloor directly, but from his observations of wildlife species on either side of the uncommonly deep channel he deduced that the eastern islands must have been separated from the western groups for much longer than any individual islands were separated from each other. More than a century later, geologists and oceanographers, armed with new technologies, found proof that he was correct: the major biogeographical boundary he discovered does indeed reflect an ancient separation between tectonic plates. The geophysical evidence that supports his conclusion is deep beneath the sea. Wallace's Line follows the edge of a submerged beach near the Asian plate; the Australian plate lies somewhat farther east.

Today, so many of the area's native species have been exterminated, decimated, or artificially carried across the channels that Wallace's Line has been blurred. During the intervening century and a half since he made his observations, the distinctions and boundaries he recorded between natural populations have been nearly obliterated.

See also PLATE TECTONICS; WALLACE, ALFRED RUSSEL

"WALLACE'S PROBLEM"
Evolution of Human Brain

Alfred Russel Wallace, Charles Darwin's "junior partner" in discovering natural selection, had a disturbing problem: He did not believe their theory could account for the evolution of the human brain.

In the *Origin of Species* (1859), Darwin had concluded that natural selection makes an animal only as perfect as it needs to be for survival in its environment. But it struck Wallace that the human brain seemed to be a much better piece of equipment than our ancestors really needed.

After all, he reasoned, humans living as simple tribal hunter-gatherers would not need much more intelligence than gorillas. If all they had to do was gather plants and eggs, and kill a few small creatures for a living, why develop a brain capable not merely of speech, but also of composing symphonies and doing higher mathematics? Why, indeed, evolve a brain capable of formulating a theory of evolution?

Wallace did not share in the racist views of his time that held that tribal peoples had brains intermediate between apes and "civilized" peoples. During his travels through Malaysia and South America, he had gained great respect for the character and intelligence of nonliterate peoples. "Natural selection," he wrote in 1869, "could only have endowed the savage with a brain a little superior to that of an ape, whereas he actually possesses one but very little inferior to that of the average member of our learned societies."

But since he thought the human brain is everywhere "an instrument . . . developed in advance of the needs of its possessor," Wallace postulated a "spiritual" dimension in human evolution. While our bodies had been shaped by natural selection, he thought, at a crucial moment there had been divine intervention to expand the human brain.

There exists in man, Wallace wrote in *Darwinism* (1889), "something which he has not derived from his animal progenitors—a spiritual essence or nature, capable of progressive development under favorable conditions." Although he insisted we evolved from apelike creatures through natural selection, yet he contended that human intellectual and moral faculties "can only find adequate cause in the unseen universe of Spirit." (Not a conventionally religious man, Wallace was a staunch believer in spiritualism who attended séances.)

In 1869, when Wallace expressed his beliefs to Darwin, he responded, "I differ grievously from you; I can see no necessity for calling in an additional and proximate cause in regard to man." In another letter Darwin expressed his displeasure in stronger terms: "I hope you have not murdered too completely your own and my child." In Darwin's view, natural selection was a sufficient explanation for the emergence of the human mind, given slow, gradual evolution.

Since Wallace's time, we have additional evidence to ponder. Both the australopithecines—bipedal hominids with brains a third the size of our own—and the larger-bodied *Homo erectus,* with intermediate-sized brains, have been discovered. Nevertheless, Wallace's problem remains unsolved; the emergence of the human mind is still a mystery. Stephen Jay Gould offered the hypothesis that many "higher" functions of the brain came along with its adaptation to more complex social behavior, including language. If you design a computer to handle business accounting, he argued, its very structure may also be capable of word processing or working out musical harmonies—side effects of its basic design for other functions.

WASHOE
First Signing Chimp

Allen and Beatrice Gardner, a married team of behavioral psychologists at the University of Nevada, were convinced chimpanzees were capable of language. Past attempts to teach them spoken words had failed, but the Gardners thought they could succeed with visual symbols. So they taught a young female chimp to communicate in American Sign Language (Amslan) and compared her progress to that of deaf children.

Named for the Nevada county in which she was raised, Washoe had been captured as an infant in Africa and reached the Gardners in June 1966. Her trainers took care never to speak around her, except through signs and gestures; they taught her names of things by showing the object, then immediately arranging her hands in the appropriate sign.

By the end of three years, Washoe could make 132 signs and had used three or more in 245 combinations, such as "hug me good" and "you tickle me." She had also become something of an ape celebrity, inspiring more research into apes' language ability.

Dr. Roger Fouts, who had been the Gardners' chief assistant, took Washoe with him to the Institute for Primate Studies in Norman, Oklahoma, in 1971. There, Washoe saw other chimpanzees, whom she called (in signs) "black bugs," and a swan, or "water bird."

After losing two infants, Washoe adopted Loulis, a one-year-old male chimp, and began signing to him. Loulis learned the gestures quickly. Fouts called it "the first case of cultural transmission of a language between generations" of chimpanzees. But other researchers were not convinced. A few years later, psychologist Herbert Terrace questioned whether the chimps' responses constituted real language behavior, or an ambiguity produced by flawed research design. The "ape language controversy" had begun.

See also ALEX; APE LANGUAGE CONTROVERSY; KOKO; NIM CHIMPSKY

WATER BABIES, THE (1863)
Classic Evolutionary Parody

Reverend Charles Kingsley, a clergyman, naturalist, poet, social reformer, and novelist (*Westward Ho*, 1855) penned *The Water Babies: A Fairy Tale for a Land-Baby* in 1862 as a present for his youngest son, then four years old. The tale of Tom, a child chimney sweep who is magically transformed into a "water baby," became a children's classic, reprinted for more than a century. But few modern readers (or their children) realize that it is also a spoof of some of the major issues of Victorian science and Darwinian evolution.

When Tom is transformed into a "water baby" (four inches long with external gills), he warns the reader not to dismiss such evolutionary "degeneration" as impossible. Scientific "experts," he complained, are supposed to cultivate open minds toward natural phenomena, yet they are always imposing their own limitations of imagination on nature.

Issues of science, belief, and dogma were on Kingsley's mind as he wrote his children's story. No scientist, however formidable his reputation, was qualified to deny that "water babies"—or unknown animal species, or "immaterial" human souls—could exist:

> How do you know that? Have you been there to see? And if you had been there to see, and had seen none, that would not prove that there were none. . . . And no one has a right to say that no water babies exist till they have seen no water babies existing, which is quite a different thing, mind, from not seeing water babies.

If they really exist, the young reader objects, someone would have caught one, put a story in the newspaper, "or perhaps cut it into two halves, poor little thing, and sent one to Professor [Richard] Owen, and one to Professor [Thomas Henry] Huxley, to see what they would each say about it." The author replies,

> You must not say that this cannot be, or that is contrary to nature. You do not know what nature is, or what she can do; and nobody knows, not even Sir Roderick Murchison, or Professor Owen, or Professor Sedgwick, or Professor Huxley, or Mr. Darwin. . . . They are very wise men; and you must listen respectfully to all they say. But even if they should say, which I am sure they never would, "That cannot exist. That is contrary to nature," you must wait a little, and see; for perhaps even they may be wrong.

The joke was that each of them had indeed made contradictory pronouncements on what was "contrary to nature." Kingsley also advanced his favorite theory of degeneration—that evolution doesn't necessarily imply progress.

At the same time, Kingsley makes it clear that there are limits to his own credulity, which did not extend to the then-current fad for spiritualism. Here he aligned himself with Darwin and Huxley, stating that a natural wood table is in itself more wonderful "than if, as foxes say, and geese believe, spirits could make it dance, or talk to you by rapping on it."

An illustration in the book shows Professors Huxley and Owen examining the bottled water baby with huge magnifying glasses. When Huxley's grandson Julian was five years old, in 1892, he saw that picture and requested an expert opinion on this matter of natural history:

> DEAR GRANDPATER—Have you seen a Waterbaby? Did you put it in a bottle? Did it wonder if it could get out? Can I see it some day?—Your loving
>
> JULIAN

Huxley sent back a laboriously hand-lettered reply:

> MY DEAR JULIAN—I never could make sure about that Water Baby.
> I have seen Babies in water and Babies in bottles; but the Baby in the water was not in a bottle and the Baby in the bottle was not in water.
> My friend who wrote the story of the Water Baby was a very kind man and very clever. Perhaps he thought I could see as much in the water as he did—There are some people who see a great deal and some who see very little in the same things.
> When you grow up I dare say you will be one of the great-deal seers, and see things more wonderful than Water Babies where other folks can see nothing.

DISGUISED ALLEGORY about science, Rev. Charles Kingsley's children's book *The Water Babies* remained popular for almost a century. In one of the original illustrations by Edward Linley Sambourne, Professors Richard Owen and Thomas Henry Huxley (with magnifying glass) are depicted as sinister zoologists.

Julian did grow up to be a distinguished evolutionary biologist, carrying on his grandfather's work in philosophy as well as science. His landmark book *Evolution: The Modern Synthesis* (1942) was one of the founding documents of the modern Synthetic Theory of evolution, even coining the name by which the theory has become known.

Strangely enough (or perhaps inspired by his childhood love for the Water Babies), Julian Huxley's first biological experiments were on the axolotl, a salamander that is a real-life counterpart to Kingsley's imaginary creature. Like Tom, the axolotl never becomes an adult, but retains the external gills of a juvenile all its life. Huxley induced development in adults by feeding hormones to the creatures. They sprouted limbs and lost their gills, inspiring wild newspaper stories about the prospect of artificially speeding up human evolution.

See also HUXLEY, SIR JULIAN; HUXLEY, THOMAS HENRY; NEOTENY; OWEN, SIR RICHARD

WATSON, JAMES D.

See DNA; GENOMIC PROJECT

WEGENER, ALFRED

See CONTINENTAL DRIFT; PLATE TECTONICS

WEIDENREICH, FRANZ

See BEIJING MAN

WEISMANN, AUGUST (1834–1914)
Father of Somatic Mutation Theory

August Weismann pioneered scientific knowledge of sex, death, embryology, and aging: some of the most profound questions in biology. A German evolutionist of enormous accomplishments, he was philosopher, experimentalist, naturalist, cellular physiologist, and musician. Yet, like the farmer's wife in the nursery rhyme, he is chiefly remembered today for cutting off the tails of mice.

After 20 generations of amputations had no effect on the length of mouse tails, Weismann considered his point made. If Jean-Baptiste Lamarck (and most naturalists, including Charles Darwin) had been correct about acquired characters being inherited, eventually some mice should have been born with shorter tails or none at all. Weismann's classic demonstration established once and for all that changes inflicted by environment on the bodies of individuals could not be passed on to their descendants.

These famous mutilated mice also supported Weismann's idea that sex (or germ) cells are separate from body (or somatic) cells. If any change (mutation) takes place, it has to occur first on the genetic level in order to be passed on.

In the turn-of-the-century confusion over mechanisms of inheritance, some biologists ("neo-Lamarckians") insisted that Weismann's theory was incompatible with Darwinism. However, Weismann saw no contradiction; in fact, independence of body cells from sex cells, he realized, gave crucial support to the idea of natural selection.

Among the critics of Weismann's experiments was the Irish playwright George Bernard Shaw, who insisted "any fool could have told him beforehand" that the mice's tails would not become reduced in their offspring. Neo-Lamarckian Shaw, with his "scientific religion" of Creative Evolution, claimed Weismann need never have taken a blade to the innocent rodents. After all, he wrote in *Back to Methuselah* (1921),

> his experiment had been tried for many generations in China on the feet of Chinese women without producing the smallest tendency on their part to be born with abnormally small feet . . . [not to mention systematically performed] mutilations, the clipped ears and docked tails, practised by dog fanciers and horse breeders on many generations of the unfortunate animals they deal in.

"It is the quest after . . . truth, not its possession, . . . that gladdens us, fills up the measure of our life, nay! hallows it."

—August Weismann

Shaw argued that the proper experiment would have been to hypnotize the mice "into an urgent conviction that the fate of [their] world depended" on losing their tails. Soon, there would be a few mice born with little or no tail, and these would gain more food and mates, while "the tailed mice would be put to death as monsters by their fellows." Shaw had wit on his side, but history has vindicated Weismann's methods of doing science. Having established that sex cells kept repeating and recombining themselves—and were almost immortal—Weismann wondered why the body cells (soma) had to grow old and die. Where philosophers and most biologists had seen in death a necessary corollary of life, Weismann declared in 1882 that the mortality of individuals was "not a primary necessity, but . . . it has been secondarily acquired as an adaptation" to protect the nearly immortal genes. "The unlimited existence of individuals would be a luxury without a corresponding advantage" to the germplasm.

Weismann's advanced ideas about cooperative cell colonies evolving specialized functions, his correlation of individual longevity with the species reproductive cycle, and his sophisticated hypotheses on the origin of sex, aging, and death were based on a lifetime of careful experiment and observation. (He peered through microscopes until he went blind, but continued his investigations with the help of his wife and sighted assistants.)

Once, in pondering the question of how life arose out of nonliving matter, he conceded the problem appeared to be, "at least for the present, insoluble." But preoccupation with the great questions of biology did not produce despair in Weismann. "In fact," he wrote in *On Life and Death* (1884), "it is the quest after perfected truth, not its possession, that falls to our lot, that gladdens us, fills up the measure of our life, nay! hallows it."

Despite the stale textbook tradition of caricaturing Weismann as a mere clipper of mice tails, new generations of biologists exploring evolutionary mysteries will have to confront these deeper dimensions of his work.

See also LANKESTER, E. RAY; NEO-DARWINISM; "SELFISH GENE"; SHAW, GEORGE BERNARD

WELLS, H. G. (1866–1946)
Science Fiction Novelist

More than any other writer, H. G. Wells is the father of modern science fiction. His first "science romance," *The Time Machine* (1895), established a new popular genre, while his second, *War of the Worlds* (1898), is credited with having created literature's first true extraterrestrials. Humanoids from other planets were commonplace, according to science fiction historian Philip Klass, but Wells's Martians were "the first intelligent creatures who were clearly the product of an alien evolution and an alien planetary environment."

Wells was trained in science as a student of Professor Thomas Henry Huxley, the combative champion of Darwinian biology. After he had won fame as an author, Wells often acknowledged Huxley's influence; he was inspired not only by his teacher's enthusiasm for scientific ideas, but also by Huxley's skill at expressing them. In his *Experiment in Autobiography* (1934), Wells fondly recalled a lab session with "Darwin's bulldog":

> [Professor Huxley] spoke in clear completed sentences in words so well chosen that only afterwards did you realize how much that quiet leisurely voice had said and how swiftly it had covered the ground. . . . Before him was a dead rabbit. . . . We were going to see . . . how it was made . . . how it was related to other [creatures]. That little limp furry body was the key by which we were to make our way towards the understanding of the whole incessant network of life.

Huxley's evolutionism colored Wells's literary works for years to come; for instance, his time travelers encountered monsters accurately described according to paleontology. But much about the idea of evolution frightened him—and his audience.

Anxiety and reservations about Victorian science's quest for the laws of evolution led to *The Island of Doctor Moreau* (1896). A mad scientist invents a serum that speedily "evolves" other mammalian species into humans. Setting himself up as their God, Dr. Moreau becomes Giver of the Law to the beast-men he has created ("Not to go on all-fours; that is the Law.

BEAST-MEN were part of speeded-up evolution in the science fiction thriller *The Island of Doctor Moreau*, written by H.G. Wells in 1896. This scene is from the 1977 Hollywood version, which starred Burt Lancaster as Moreau.

Are we not Men?"). In the end, the doctor cannot control his world, and the island of evolutionary monsters is consumed in fiery self-destruction.

In *Man of the Year Million* (1893), Wells saw future evolution producing humans with huge heads and eyes, delicate hands, and much reduced bodies—an image appropriated countless times since by less imaginative writers. Wells also produced a popular account of the rise and fall of civilizations (*An Outline of History*, 1920) to make history accessible to "the ordinary man." A few years later, he teamed with his old professor's grandson, Julian Huxley, to write a popular compendium on evolutionary biology, *The Science of Life* (1925).

In 1929, Wells joined the editorial board of an independent literary magazine, *The Realist*, which lost money but made reputations. His fellow editors were Harold Laski, Arnold Bennett, Rebecca West, and the brothers Julian and Aldous Huxley.

Wells is perhaps best remembered for *War of the Worlds*, which endured as an Orson Welles radio play (1938) and Hollywood film (1953 and 2005). A terrifying fantasy of planetary conquest, it features strange Martian creatures arriving in advanced spaceships equipped with hi-tech weapons. The story's power has its roots in 19th-century history, when European colonial powers invaded far-flung lands, subduing native peoples with their superior technology and weapons. Some day soon, Wells was warning, "civilized" men might be on the receiving end of the terror.

His analogy touched a deep nerve in Western culture. Even 40 years after the novel first appeared, when the Orson Welles version was broadcast in America (Halloween, 1938), thousands called the station to ask if the alien invasion was real. One and a half million people fled their homes in panic, convinced that the day of reckoning had finally arrived.

An idealist as well as futurist, Wells could never resolve the conflict between his utopian dreams and Darwinian reality. In one of his last books, *Mind at the End of Its Tether* (1945), he warned that humankind is doomed unless it can adapt to its own advancing technology of destruction. "We must stop war," he said, "before it stops us."

In *All Aboard for Ararat* (1941), Wells allegorized the aloneness of modern humans in a Darwinian universe. There is to be a new cataclysm, and God asks Noah to build another

H.G. Wells wrote *War of the Worlds* as the ultimate "survival of the fittest" story.

ark. But Noah agrees only on one condition: This time God must ride as a noninterfering passenger, while man takes charge of his own (and the planet's) destiny.

See also HUXLEY, SIR JULIAN; HUXLEY, THOMAS HENRY; *LOST WORLD*

WELLS, WILLIAM CHARLES (1757–1817)
Early Natural Selectionist

Charles William Wells, the son of Scots immigrants to Charleston, South Carolina, came up with the idea of natural selection while Charles Darwin was still a young child. In 1813, four years before his death, Wells read a paper on the subject to the Royal Society of London, applying it to the origin of the human races.

After completing medical studies at Edinburgh, Wells had returned to Charleston, where he pursued a practice and developed his interests in botany and other scientific subjects. The American Revolution sent him packing to London in 1784; soon he became established as a physician-scientist and was elected to the Royal Society.

He won recognition for his classic explanations of binocular vision (*An Essay upon Single Vision with Two Eyes*, 1792) and why leaves were wet in the mornings (*An Essay on Dew*, 1814). His explanation of the formation of dew was published posthumously in 1818, together with the strange paper that contains his ingenious natural selection theory.

In this "Account of a Female of the White Race of Mankind, Part of Whose Skin Resembles That of a Negro," published posthumously in 1818, Wells described patches of darkly pigmented skin on one of his women patients. Black skin in Africans, he concluded, "is no proof of their forming a different species from the white race." Also, he thought this example showed that—contrary to common opinion—a hot climate was not necessary to produce black skin. (It was thought that "Negro" skin had been deeply tanned under tropical sunlight and the accumulated color passed on to succeeding generations.)

Wells's ideas seem amazingly ahead of his time. He knew, for instance, that Africans were resistant to certain tropical diseases that killed Europeans easily and quickly. Perhaps, he speculated in a paper to the Royal Society in 1813, such resistance was correlated with dark skin:

> Of the accidental varieties of man, which would occur among the first few and scattered inhabitants of [Central] Africa, some one [race] would be better fitted than the others to bear the diseases of the country [and might also be dark.] This race would consequently multiply, while the others would decrease, not only from their inability to sustain the attacks of disease, but from their incapacity of contending with their more vigorous neighbors. . . . [A] darker and a darker race would in the course of time occur, and as the darkest would be the best fitted for the climate, this would at length become the most prevalent, if not the only race, in the particular country.

Wells also compared artificial selection, as practiced by domestic animal breeders, with selection by natural forces. In addition, he suggested that "accidental peculiarities" or "varieties" could become established in local populations by inbreeding, isolation, and geographical barriers.

Ingenious as he was, Wells continued writing on many topics, but never sought to apply his very important insights to the data of zoology, botany, or paleontology. His limited development and application of natural selection made no impact on scientific thought, though it remains a striking precursor of the Darwin-Wallace theory.

Wells's paper was brought to Darwin's attention some years after he published *Origin of Species* (1859), and he acknowledged it in later editions. Yet another early theorist, Patrick Matthew, was an expert on tree farming. He had published a version of the theory and was constantly insisting he deserved priority as the founder of evolutionary theory. When the story of Wells's even earlier version came to light, Darwin was delighted: "So poor old Patrick Matthew is not the first!" he wrote to Hooker in 1865.

See also MATTHEW, PATRICK; NATURAL SELECTION; PIGGYBACK SORTING

William Wells published the idea of natural selection when Darwin was still a toddler.

WHALE, EVOLUTION OF
Origins of Earth's Largest Animal

Extremes of adaptation—such as the 90-foot-long blue whale—provoke wonder about how such a creature could have evolved. Sometimes larger than a whole herd of elephants, this intelligent mammal feeds on tons of tiny plants and animals (plankton) that it strains from seawater. Since it is air-breathing, warm-blooded, and milk-producing, the blue whale and 75 other species of cetaceans (whales, dolphins, and porpoises) appear to have evolved from land animals that returned to the sea. But 150 years ago, who could imagine how such a spectacular and seemingly improbable transformation could have come about?

Charles Darwin could. He had noticed in a traveler's account that an American black bear was seen "swimming for hours with widely open mouth, thus catching, like a whale, insects in the water." If this new food-getting habit became well established, Darwin wrote in the *Origin of Species* (1859): "I can see no difficulty in a race of bears being rendered, by natural selection, more and more aquatic in their structure and habits, with larger and larger mouths, till a creature was produced as monstrous as a whale."

"Preposterous!" snorted zoologists. Such an example, they thought, sounded so wild and far-fetched it must brand Darwin as a teller of tall tales. Professor Richard Owen of the British Museum prevailed on him to leave out the "whale-bear story," or at least tone it down. Darwin cut it from later editions, but privately regretted giving in to his critics, as he saw "no special difficulty in a bear's mouth being enlarged to any degree useful to its changing habits." Years later he still thought the example "quite reasonable."

Unfortunately, for the next century and a half no evidence surfaced to back up Darwin's conjecture. During the mid–19th century, fossil hunters found skeletons of *Basilosaurus* (misnamed "king lizard" and mistaken for a sea serpent) in the Upper Eocene rocks of Egypt and the United States. It had front flippers for steering and a completely flexible backbone, yet retained hind limbs that were too small to be functional. But *Basilosaurus* only served to demonstrate that as far back as 40 million years ago, whales were still whales.

During subsequent evolution, the whale's hind limbs became further reduced, finally losing toes and kneecaps until they became shrunken vestiges, located inside the bodies of

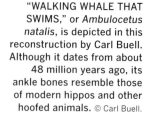

"WALKING WHALE THAT SWIMS," or *Ambulocetus natalis*, is depicted in this reconstruction by Carl Buell. Although it dates from about 48 million years ago, its ankle bones resemble those of modern hippos and other hoofed animals. © Carl Buell.

modern whales. Of course, tiny, nonfunctional legs and feet suggest that ancestral whales once walked on land, but remains of any intermediate terrestrial ancestors were nowhere to be found. Such an apparent gap in the fossil record prompted creationists to adopt whales as an exemplar of an animal created by divine fiat, without any intermediate forms. If whales really evolved from land animals, they argued, where were the "missing links" to back up that supposition, and to what group of living animals are whales most closely related?

Beginning in the 1960s, paleontologists found fossils of various four-footed ancestral whales in the Pakistan desert. During the 1990s, Philip Gingerich of the University of Michigan and Hans Thewissen of Northeastern Ohio University's College of Medicine described several spectacular fossil protowhales. The oldest known whale ancestor, *Pakicetus*, was found in riverine deposits in the Lower Eocene of Pakistan. While the skull is whalelike, its teeth resemble those of the grazing animals from which it evolved.

Next along the time line, they found *Ambulocetus*, whose remains, also from Pakistan, are found in shallow marine deposits rather than those formed by rivers. Its full scientific name, *Ambulocetus natalis*, means "walking whale that swims"; it seems to have behaved more like a seal or otter, alternating between water and land. Its feet were large and suitable for paddling, although it still had small hoofed toes. Its ankle joints are strikingly like those of its modern hoofed cousins. *Ambulocetus* had a highly flexible back, enabling it to move up and down rapidly, as whales do in diving and swimming.

In 2007, Thewissen announced that his team had discovered the cetacean "missing link," which he named *Indonyus*. "We've found the closest extinct relative to whales . . . closer than any living relative," he claimed. While it looked nothing like a whale, *Indonynus* resembles early whales in the structure of its ear and skull, as well as the thickened bones common to aquatic mammals. It belongs to the artiodactyls, an ancient order of hoofed mammals that had two or four toes on each foot, whose modern representatives include camels, pigs, sheep, cows, and hippopotamuses— a relationship confirmed by modern DNA comparisons. Several other fossil cetaceans, including *Indocetus* and *Rodhocetus*, were also found early in the 21st century in Pakistan, making that part of the world the richest source of protowhale remains.

Recently won knowledge about whale evolution has moved Moby Dick from the creationist poster animal to a well-documented example of macroevolution, or large-scale evolutionary change. Charles Darwin's "whale-bear" story now seems less like the product of a runaway imagination and more like a demonstrable example of the extreme plasticity of life forms adapting to new environments over vast periods of time.

See also ADAPTATION; EXAPTATION; OWEN, RICHARD; TRANSITIONAL FORMS

Formerly the poster animal of creationists, discoveries of "walking whales" have turned cetaceans into exemplars of evolution.

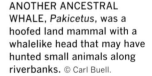

ANOTHER ANCESTRAL WHALE, *Pakicetus*, was a hoofed land mammal with a whalelike head that may have hunted small animals along riverbanks. © Carl Buell.

WHIG HISTORY
Selective Hindsight

Sir Charles Lyell, the founder of uniformitarian geology, predicted that dinosaurs would someday reappear on Earth. Charles Darwin, the architect of natural selection, believed in the "Lamarckian" inheritance of acquired characteristics. And Alfred Russel Wallace saw no difficulty in fusing natural selection and Spiritualism to explain human evolution.

Science is full of dead ends, strange combinations of ideas that later seem incompatible, and concepts that die and reemerge in totally different contexts. It is influenced by social, political, and artistic environments and quirks of personality no less than other human endeavors.

"Whig history" is the fallacy of relating past ideas to those of the present without regard for context, culture, or the state of knowledge at the earlier time. The phrase was coined by British historian Herbert Butterfield in *The Whig Interpretation of History* (1931) to describe colleagues who viewed all constitutional history as a progressive movement toward their own political values. Every prominent figure over the centuries was seen as either a "forward-looking" liberal (Whig) or "backward" conservative on the road inevitably leading to representative government.

Butterfield himself defined it as "the tendency in many historians to write on the side of Protestants and Whigs, to praise revolutions provided they have been successful, to emphasize certain principles of progress in the past, and to produce a story which is the ratification if not the glorification of the present."

"Whiggish" history of science rates each scientist according to his "contribution" to currently held concepts, ignoring all other aspects of his work. More often than not, similarity to modern thought is only superficial; the work had a totally different meaning in its time and place. Ernst Mayr, Stephen Jay Gould, and Niles Eldredge all complain of too much "Whig history" in science writing and insist that earlier biological theorists must be considered with reference to the context of their ideas, state of knowledge, and cultural environment.

WOLTMANN, LUDWIG (1871–1907)
German "Racial" Anthropologist

Ludwig Woltmann, physician and author, was a very influential writer on the subjects of eugenics, racial anthropology, and Social Darwinism around 1900. He was the most important German representative of Count de Gobineau's theory of the Nordic race and its agenda for Aryan supremacy. Woltmann was also a disciple of famed biologist Ernst Haeckel's evolutionary religion.

In 1900, Woltmann submitted an essay titled "Political Anthropology" to a contest of Social Darwinist writing sponsored by the industrialist Alfred Krupp. Although Professor Haeckel was one of the judges, the first prize went to another of his disciples. When Woltmann was given only third prize, because of many blatant mistakes in biology, he angrily withdrew from the contest and refused the money. That caused a permanent break with Haeckel, his old mentor.

As a result of the publicity surrounding this dispute, Woltmann became famous and published the essay as a book in 1903. A year earlier, in 1902, he had begun the *Political Anthropology Revue*, in which he campaigned to prevent the "deterioration of the Nordic race" and to maintain its supremacy—by force, if necessary.

Woltmann used evolution as a rationale for advocating a worldwide "racial" war for Aryan supremacy.

During the first few years of the 20th century, Woltmann published a series of books on race and politics that gained a large audience in Germany. Although he criticized Haeckel's bias against socialism, he accepted his teacher's basic view that a universal law of evolution operates in society as well as in nature.

In his attempt to wed the ideas of Haeckel with those of Karl Marx, he transformed the Marxist class struggle into one of worldwide racial conflict. Germans, he thought, were the highest species of mankind, whose "perfect physical proportions" expressed a heightened spirituality and inner superiority. It was their duty to avoid the "biological deterioration" that would come of "mixing" with other races.

According to historian Daniel Gasman, Woltmann followed Haeckel closely in teaching that "life was a constant struggle for existence and racial purity, and sought to forearm Germany against biological decay." The extent of Woltmann's influence is impressive, since he died prematurely at the age of 36 and never lived to see many of his racial ideas adopted and implemented by the Nazis a few decades later.

See also "ARYAN RACE," MYTH OF; HAECKEL, ERNST; MONISM; SOCIAL DARWINISM

WRIGHT, SEWALL (1889–1988)
Pioneering Population Theorist

After Charles Darwin, the day of the field naturalist was eclipsed for many years. Now it was the turn of lab men and mathematicians (Darwin hated mathematics) to put evolutionary theory on a new foundation. Three brilliant men, working separately, were the architects of the new understanding: British biometricians J. B. S. Haldane and R. A. Fisher and the American Sewall Wright.

Natural selection, as Darwin well knew, was incomplete without a theory of inheritance. But the rise of Mendelian genetics 20 years after his death was at first thought to disprove Darwinian ideas of continuously varying traits. Mendelists insisted there could only be large, discrete mutations, combining in established ratios.

Wright had studied with an experimental geneticist, William E. Castle, whose breeding experiments with hooded rats showed that inheritance in small, inbred populations was a lot more complicated than just sorting discrete one-trait genes. For instance, the presence of "modifier genes" altered the expression of other genes. Also, there was the phenomenon of "epistasis"—genes at one position (locus) on the chromosome showing varying effects, depending on what genes are present at other loci.

All this "extra variation" posed a challenge to Wright. Was there a way, he wondered, to mathematically describe complex systems of gene interaction? In 1920, he developed and published a mathematical model for such a system; it became one of the founding documents of population genetics. Darwinism was not, it turned out, incompatible with genetics; it simply needed to be rescued from the "beanbag" dilemma.

"Beanbag genetics" is a nickname for the original Mendelian theory of particulate inheritance. According to that theory, each gene was assumed to be a discrete unit. Even when jumbled together in chromosomes, each gene remains separate and independent, like beans in a beanbag. Each was assumed to carry instructions for a particular trait. But that model seemed incompatible with Darwin's idea of "continuous variations." An advantageous trait would be swamped before it could become established in a population.

Wright continued to attack the problem of how genes behaved in populations and published his influential paper "Evolution in Mendelian Populations" in 1931. Individuals in large populations, he thought, are not likely to mate randomly with any other individual in the entire range. Instead, there are smaller populations ("demes") within the larger where effects of variation can be intensified.

Working with breeders of short-horned cattle, Wright found rapid evolution is more likely to occur in these small subpopulations; they provide an effective source of continuous variation for the larger population. Wright showed mathematically how selection pressure could cause adaptive variations in a small deme to spread rapidly through the larger population. He also pictured populations as "adaptive landscapes," with their gene frequencies distributed in "peaks" and "valleys."

Although Wright's name is linked in the history of science with those of Haldane and Fisher, the three did not view themselves as members of the same team. Each came at the problems of population genetics with a different perspective, and their bitter quarrels and fights still echo through the literature. At one point, even their respective groups of students would not speak to one another.

Wright labored for years on a definitive textbook, *Evolution and the Genetics of Populations* (1968–1978); the first volume appeared when he was 79. He continued to write theoretical papers up to the time of his death in 1988, two years short of his 100th birthday.

See also GENETIC DRIFT; HALDANE, JOHN BURDON SANDERSON; SYNTHETIC THEORY

"X" CLUB
Victorian Scientific Elite

Between 1864 and 1893 in London, nine friends met regularly for dinner at six o'clock on the first Thursday of the month. Evolutionist Thomas Henry Huxley was the genial ringleader of this extraordinary dining club, composed entirely of brilliant scientists, philosophers, and mathematicians. Most had a strong interest in evolution, and were also friends of Charles Darwin, who rarely ventured into the city.

Members included physicist John Tyndall, director of Kew Gardens Sir Joseph Hooker, organic chemist Sir Edward Frankland, philosopher Herbert Spencer, and prehistorian-naturalist-banker Sir John Lubbock. Since they had no name for the group, they called themselves simply "The 'X' Club"—the unknown quantity.

Their dinners were usually held directly before meetings of the Royal Society. "But what do they do?" asked a curious journalist, according to William Irvine in *Apes, Angels, and Victorians* (1955). "They run British science," a professor replied, "and on the whole, they don't do it badly." Huxley insisted that the group was purely social, but during its existence the "X" Club provided three successive presidents of the Royal Society and six presidents and several officers of the British Association. Collectively, they profoundly influenced the priorities of British science.

Once a year, they went to the countryside together on jovial excursions to which their spouses were invited. It became known as "the weekend for x's and their yv's."

XANTHOPAN MORGANII PRAEDICTA
Moth That Darwin Predicted

Darwin's first book after the *Origin of Species* was about orchids and the insect pollinators on which they depend to reproduce. He realized that insects had, in a real sense, selected the scents and colors of flowers. Also, he noticed that the orchids' bags of nectar and the insects' mouthparts had coevolved a lock-and-key fit.

A NIGHT-FLYING MOTH with 14-inch proboscis feeds on nectar at the Christmas Star orchid in Madagascar.

When he observed that the white Christmas Star orchid of Madagascar sports foot-long containers that have nectar only at the bottom, he wondered, "What can be the use of a nectary of such disproportionate length?" He concluded that "in Madagascar there must be moths with proboscides capable of extension to a length of between ten and eleven inches! This belief of mine has been ridiculed by some entomologists." Alfred Russel Wallace, in *Contributions to the Theory of Natural Selection* (1870), agreed with Darwin and urged naturalists to "search for it with as much confidence as astronomers searched for the planet Neptune—and I venture to predict they will be equally successful!" Twenty years after Darwin's death, a night-flying moth with a foot-long tongue was indeed discovered in Madagascar—a subspecies of Morgan's African sphinx moth—named *Xanthopan morganii praedicta*: the moth that was predicted.

See also BENEFICENCE, DIVINE; COEVOLUTION; ORCHIDS

YANG ZHONGJIAN (1897–1979)
Father of Chinese Vertebrate Paleontology

YANG ZHONGJIAN

Yang Zhongjian (also known as C. C. Young) began his career in paleontology working with European and American scientists; there were no professional Chinese paleontologists. After graduating as a geologist from Beijing University in 1923, Yang gained his doctorate at the University of Munich and learned German, English, Latin, and Russian. During the 1930s he worked with Pierre Teilhard de Chardin, Davidson Black, and Franz Weidenreich on Triassic reptiles, late Cenozoic mammals, and early human fossils—particularly those of *Homo erectus*, or Beijing Man, at the hillside site of Zhoukoudian.

Beginning in 1929, at the Cenozoic Research Laboratory in Beijing, Yang effectively founded vertebrate paleontology in China. The facility grew to become the Institute of Vertebrate Paleontology and Paleoanthropology. Yang was appointed director in 1954 and continued to guide its development for 25 years. Under his tutelage, generations of students discovered hundreds of important fossil animals, including *Yunnanosaurus*, a bird-hipped herbivorous "grazing" dinosaur, the gigantic sauropod *Mamenchisaurus*, and a series of previously unknown stegosaurs. Unfortunately, he died before the remarkable troves of feathered dinosaurs came to light during the 1990s, as Chinese paleontology achieved its full florescence. Yang's cremated remains are buried at Zhoukoudian, near the cave where he had worked on the spectacular Beijing Man excavations that had sparked his early career.

YOUMANS, EDWARD LIVINGSTONE (1821–1887)
American Promoter of Spencer

While essayist John Fiske and botanist Asa Gray led the campaign to make evolution respectable among intellectuals in the United States, the American Edward Livingstone Youmans became its self-appointed salesman.

An author of chemistry textbooks, Youmans convinced D. Appleton and Company to publish cheap editions of Herbert Spencer and Charles Darwin and, in 1867, to launch a magazine that, a few years later, became *Popular Science Monthly*. Still published today, it focused on the latest technology; its original emphasis was on popularizing evolutionary ideas.

Youmans also created the red-covered International Scientific Series, which included titles by Darwin, Thomas Huxley, George Romanes, Edward B. Poulton, and other evolutionists. More than a century and a half later, they are sought-after collectibles.

Appleton prospered riding the crest of interest in evolution, and Youmans's fortunes rose too. He promoted Spencer's American visit, during which the philosopher became more famous there than he was in England. In 1871, Youmans, Spencer's literary agent, wrote him: "Things are going here furiously. I have never known anything like it. Ten thousand *Descent of Man* have been printed and I guess they are nearly all gone. . . . The progress of liberal thought is remarkable. Everybody is asking for explanations. The clergy are in a flutter."

See also CARNEGIE, ANDREW; SPENCER, HERBERT

ZOONOMIA (1794)
Pioneering Evolutionary Treatise

Between 1794 and 1796, Dr. Erasmus Darwin, grandfather of Charles, published a massive two-volume treatise, *Zoonomia, or the Laws of Organic Life*, that tapped a lifetime of medical practice and expounded his views on animal life. Its stated aim was to "unravel the theory of diseases," and it did exert wide influence on the treatment of fevers and insanity. But *Zoonomia* owes its secure place in the history of science to Erasmus Darwin's views on evolution, which in many ways anticipated those of his grandson.

As part of his attempt to establish "the laws of organic life," Darwin presented evidence to show that species do change, using examples from domestic breeds of horses, dogs, and pigeons. He discussed the function of many special organs among animals and how they might have arisen as adaptations for feeding or defense. Most remarkable, he even proposed ideas similar to what his grandson would call sexual selection and natural selection. For instance, Erasmus suggested that the antlers of male stags were formed "for the purpose of fighting [other stags] for the exclusive possession of the females. . . . The final cause of this contest amongst the males seems to be, that the strongest and most active animal should propagate the species, which should thence become improved."

Erasmus Darwin also proposed that in the "millions of ages before the commencement of the history of mankind . . . all warm-blooded animals have arisen from one living filament [with the] faculty of continuing to improve . . . and of delivering down those improvements by generation to its posterity, world without end!"

Desmond King-Hele, in his biography of Erasmus Darwin, claims *Zoonomia*'s profound influence on Charles has been seriously undervalued. Charles abetted the neglect, for he never presented himself as building on his grandfather's work. In later life, however, he published a small memoir honoring Erasmus. "My scrutiny of the evidence," says King-Hele, "drives me to conclude that the Darwinian theory of evolution is very much a family affair, in which the shares of Erasmus and his grandson Charles are more nearly connected, and more nearly equal, than is usually supposed."

See also DARWIN, ERASMUS; *TEMPLE OF NATURE*

> Erasmus Darwin's writings show that "the Darwinian theory of evolution is very much a family affair."
>
> —Desmond King-Hele

ZOOPHARMACOGNOSY
Animal Self-Medication

Seeking out natural remedies for illness is widespread throughout the animal kingdom, and may far predate the evolution of hominids and humans.

In the early 1980s, Richard Wrangham of Harvard University (a former student of Jane Goodall's) published his observations of chimpanzees eating leaves of the *Aspilia* plant, a relative of the sunflower. Although its leaves are rough and bitter, the chimps ate them whole, often grimacing as they swallowed. Since the leaves were often excreted intact, Wrangham deduced that the apes weren't getting much nutrition from them. He had a bio-

452

chemical analysis done, and found that the plant contains thiarubrine-A, a substance that kills parasitic worms, fungi, and viruses, and can also kill tumorous cancer cells.

The chimps, it seemed, were forcing down the distasteful plant because it was good for them. They roll the leaves around under their tongues before swallowing, apparently absorbing the chemicals before they can be destroyed by stomach acids. Wrangham concluded that the chimps were self-medicating and, with Eloy Rodriguez of Cornell, coined a tongue-twister to describe the phenomenon: zoopharmacognosy ("animal knowledge of medicines").

Primatologist Michael Huffman of the Primate Research Institute at Kyoto University, Japan, independently observed self-medication among East African chimps. While watching an obviously ill female, he saw her seek out and chew the pith of the bitter mjonso tree (*Vernonia amygdaline*), spitting out its fibrous material and swallowing the juice. On the following day, the ape's health was restored.

An African game officer told Huffman that his people, the WaTong we, use the same plant to treat their own gastrointestinal disorders. His grandfather was a traditional herbal doctor, who sometimes copied animals' use of native plants for his patients.

Huffman has advanced a "Velcro theory" of why the chimps may swallow rough and bristly leaves, which they also neither chew nor digest. As the leaves pass through their intestines, he suggests, their rough, textured surfaces pick up parasitic worms and carry them out of the body. Pursuing his guess, Huffman examined chimpanzee stools and found the undigested leaves with live worms trapped in the bristles.

In Venezuela, ecologist John Robinson of the Wildlife Conservation Society observed puzzling behavior in a group of capuchin monkeys (*Cebus olivaceus*). Upon finding a millipede, the monkeys rubbed it all over their bodies, their eyes appeared to glaze over, and they frequently placed the arthropod in their mouths while drooling. Three or four often shared a single millipede, passing it from one to the other. Afterward the monkeys rubbed up against one another, apparently to spread the millipede's secretions. Up to ten bouts of this frenetic, ecstatic anointing could take place within a single day.

In the 1990s, Robinson's student Ximena Valderrama, now an anthropologist, then at Columbia University, collected some of the millipedes (*Orthoporus dorsovittatus*) and sent them to chemical ecologists at Cornell University. After "milking" the millipede for their secretions, which are produced by glands along their sides, the chemists determined that they contained benzoquinones, which are powerful insecticides and disinfectants.

Because the monkeys rub themselves with millipedes only during the rainy season, when they are vulnerable to infection by mosquito-borne botfly larvae, they seem to be using the benzoquinones as protection against biting insects and parasites. Although millipede secretions contain other toxins and even carcinogens, they appear to cause the monkeys no harm. Of course, one cannot know whether the monkeys are acting with conscious intent to immunize themselves.

Similarly, several species of birds in various parts of the world swoop down on ant nests, pick up ants, and rub them all over their feathers—a behavior known as "anting." The distressed insects release formalin, a powerful chemical repellent that kills various small parasites in the bird's plumage.

Navajos in America claim that they learned of an antiparasitic plant by observing grizzly bears. The bears chew roots of certain lovage plants and rub the juice into their fur to keep away ticks, fleas, and lice.

Baboons, bonobos, and other primates apparently medicate themselves as well. Robert Sussman of Washington University in St. Louis observed pregnant long-tailed macaque females eating a lantana plant that is used by the Mauritian people as an antiparasitic, especially by pregnant women. (Not native to Mauritius, the monkeys were introduced there 450 years ago.)

How do animals recognize a medicine? Is the behavior instinctive or learned? Do the animals connect the plant with feeling better, or with another entirely different stimulus? As the world's rare animals and medicinal plants continue to disappear at an accelerating pace, the race is on to gain as much knowledge as possible about these remarkable relationships between animals and medicinal plants before the chance is lost forever.

Chimps apparently force themselves to eat a distasteful plant because it kills parasitic worms.

APPENDIX
Darwin Lyrics

DARWIN'S NIGHTMARE

Lyrics © 2007 by Richard Milner and John Woram
(In homage to W. S. Gilbert and Sir Arthur Sullivan)

I was lying awake with a dismal headache,
And my sleep was put off by anxiety,
For I feared that my plan of explaining how man
Had evolved would provoke notoriety.

Though I'd fled London town for the village of Down
And a home that is quiet and regal,
Yet I get no repose, and I can't even doze
Without dreaming I'm back on the *Beagle.*

We are rounding the Horn in a furious storm
And our progress is measured in inches,
Then we're rolling around 'til the crew's almost drowned
And they scatter like terrified finches.

Cap'n FitzRoy's in a mood, and he's coming unglued
And cannot say where our next port is.
I fear he's unwell, for he's sprouted a shell,
And turns into a monstrous tortoise.

He ignores the enormity of this deformity,
Carapace, scales, and the rest of it,
And vows his complexion is close to perfection,
And as for my fears, makes a jest of it.

After ten thousand miles—the Galápagos Isles,
Now I'm tense and exceedingly wary.
For I fear that this place could result in disgrace
When I use it to further my theory.

Then I scramble ashore as the sea lions roar,
And I hear a contemptuous snicker,
I glance up in a tree, and there what do I see?
Seven chimps and an Anglican vicar.

It's not hard to tell that they're all mad as hell,
So I ask, "What's the source of your fury?"
His Grace's retort is "Silence in Court!
For I am your judge and your jury."

With a barrister's smile, he declares I'm on trial
And tells me the cause for his action.
"You cannot escape, for you said 'man's an ape,'
And my clients demand satisfaction."

To challenge my fitness, he calls his first witness,
An albatross down from Guiana.
But the bird flicks its tail and turns into a snail,
Then uncoils and becomes an iguana.

The judge is perplexed, and I say, "Don't be vexed,
We've been changing since Adam's arrival.
The monk in his cloister descends from an oyster,
The fittest have mastered survival."

His Honor says, "Quiet! The court will not buy it,
You mock, Sir, the truth of Creation,
You've played fast and loose, now your neck's in a noose,
And you're off to eternal damnation."

I've given up hope as he tightens the rope
And the chimps start to make a commotion
Then a wave hits the shore with a volcanic roar
And the island sinks under the ocean.

With a shriek I awake, and it's all a mistake,
The iguana is really a kitten,
No chimps are in sight on this miserable night,
And no wonder—I'm back in Great Britain.

I'm a regular wreck with a crick in my neck,
My anxiety's hardly diminished.
And the night's been too long, ditto-ditto my song,
And thank goodness—they're both of them finished!

Illustrations by Pete Von Sholly

WHY DIDN'T I THINK OF THAT?

Lyrics © 1992 by Richard Milner, Music by Jacques Semmelman
(When Darwin's friend Thomas Henry Huxley first read *Origin of Species* in 1859,
he exclaimed, "How extremely stupid not to have thought of that!")

Of course! Of course! It must be so.
I should have seen it long ago.

'Twas adaptive radiation that produced the mighty whale
His hands have grown to flippers and he has a fishy tail.
Selection's made him streamlined for his liquid habitat.
Why didn't I think of that?

There was an ancient mammal that could hop and leap around,
But with webbing 'twixt his fingers, he could fly right off the ground.
And so this mousy creature evolved into a bat.
Why didn't I think of that?

There are fossils in the ground, protozoa in the sea
All these unrelated facts
Made a monkey out of me.
But now I see how species were selectively defined.
How could I have been so ruddy blind?

There was an ancient monkey with a long and curly tail
This ape evolved into a man
(He's teaching now, at Yale)
A chimp could pass for upper class
In gloves and a cravat
Why didn't I think of that?

The struggle for survival lies outside the jungle, too.
Just take a look at Parliament, it's better than a zoo!
We're at each other's throats just like the bulldog and the cat.

But why didn't I, why didn't I—
Your ideas on evolution will create a revolution!
Why didn't I think of that?

BIBLIOGRAPHY

A wealth of primary sources may be consulted at http://darwin-online.org.uk (*The Complete Work of Charles Darwin Online,* which includes multiple editions of published works and more than 20,000 private papers) and at www.darwinproject.ac.uk (the Darwin Correspondence Project).

ABANG

Desmond, Adrian John. *The Ape's Reflexion.* London: Blond & Briggs, 1979.

Parker, Sue Taylor, Robert W. Mitchell, and H. Lyn Miles, eds. *The Mentalities of Gorillas and Orangutans: Comparative Perspectives.* Cambridge: Cambridge University Press, 1999.

ACTUALISM

Bowler, Peter J. *Evolution: The History of an Idea.* 3d ed. Berkeley: University of California Press, 2003.

Gould, Stephen Jay. *Time's Arrow, Time's Cycle.* Cambridge, Mass.: Harvard University Press, 1987.

Greene, John C. *The Death of Adam: Evolution and Its Impact on Western Thought.* Ames: Iowa State University Press, 1959.

ADAM AND EVE

Baetzhold, Howard G., and Joseph B. McCullough, eds. *The Bible According to Mark Twain: Writings on Heaven, Eden, and the Flood.* Athens: University of Georgia Press, 1995.

Gosse, Phillip Henry. *Omphalos: An Attempt to Untie the Geological Knot.* 1857; repr. Woodbridge, Conn.: Ox Bow Press, 1998.

Oppenheimer, Stephen. *The Real Eve: Modern Man's Journey Out of Africa.* New York: Carroll & Graf, 2003.

Twain, Mark. *The Diaries of Adam and Eve.* Edited by Don Roberts. San Francisco: Fair Oaks Press, 2002.

ADAPTATION

Grant, Verne. *The Origin of Adaptations.* New York: Columbia University Press, 1963.

Provine, William B. "Adaptation and Mechanisms of Evolution after Darwin: A Study in Persistent Controversies." In *The Darwinian Heritage,* edited by David Kohn and Malcolm J. Kottler,

pp. 825–866. Princeton: Princeton University Press, 1985.

Williams, George C. *Adaptation and Natural Selection: A Critique of Some Current Evolutionary Thought.* Princeton: Princeton University Press, 1966.

AFAR HOMINIDS

Johanson, Donald, and Maitland Edey. *Lucy: The Beginnings of Humankind.* New York: Simon & Schuster, 1981.

Johanson, Donald, and James Shreeve. *Lucy's Child: The Discovery of a Human Ancestor.* New York: William Morrow, 1989.

Kalb, Jon. *Adventures in the Bone Trade: The Race to Discover Human Ancestors in the Ethiopian Afar Depression.* New York: Copernicus Books, 2001.

Leakey, Richard, and Roger Lewin. *Origins Reconsidered: In Search of What Makes Us Human.* New York: Doubleday, 1992.

Reader, John. *Missing Links: The Hunt for Earliest Man.* London: Collins, 1981.

White, Tim D. "Human Evolution: The Evidence." In *Intelligent Thought: Science versus the Intelligent Design Movement,* edited by John Brockman, pp. 65–81. New York: Vintage Books, 2006.

AGASSIZ, LOUIS

Agassiz, Elizabeth. *Louis Agassiz: His Life and Correspondence.* 2 vols. Boston: Houghton Mifflin, 1885.

Agassiz, Louis. *Geological Sketches.* 2 vols. Boston: Houghton Mifflin, 1886.

———. *Methods of Study in Natural History.* Boston: Ticknor & Fields, 1863.

Barber, Lynn. *The Heyday of Natural History.* Garden City, N.Y.: Doubleday, 1980.

Lurie, Edward. *Louis Agassiz: A Life in Science.* Chicago: University of Chicago Press, 1960.

AGNOSTICISM

Budd, Susan. *Varieties of Unbelief: Atheists and Agnostics in English Society, 1850–1960.* London: Heinemann Educational Books, 1977.

Coley, Noel George, and Vance M.D. Hall. *Darwin to Einstein: Primary Sources on Science and Belief.* Harlow, Essex: Longman in association with Open University Press, 1980.

Huxley, Leonard, ed. *Life and Letters of Thomas Henry Huxley.* London: Macmillan, 1900.

AKELEY, CARL

Akeley, Carl Ethan. *In Brightest Africa.* Garden City, N.Y.: Doubleday, Page & Co., 1924.

Akeley, Mary L. Jobe. *Carl Akeley's Africa: The Account of the Akeley-Eastman-Pomeroy African Hall Expedition of the American Museum of Natural History.* New York: Dodd, Mead & Co., 1930.

Bodry-Sanders, Penelope. *African Obsession: The Life and Legacy of Carl Akeley.* Jacksonville, Fla.: Batax Museum Publishing, 1998.

———. *Carl Akeley: Africa's Collector, Africa's Savior.* New York: Paragon House, 1990.

ALEX

Hillix, William A., and Duane M. Rumbaugh. *Animal Bodies, Human Minds: Ape, Dolphin, and Parrot Language Skills.* New York: Kluwer Academic/Plenum Publishers, 2004.

Pepperberg, Irene M. *The Alex Studies: Cognitive and Communicative Abilities of Grey Parrots.* Cambridge, Mass.: Harvard University Press. 2000.

———. *Alex and Me: How a Scientist and a Parrot Uncovered a Hidden World of Animal Intelligence.* New York: Collins, 2008.

ALLOPATRIC SPECIATION

Coyne, Jerry A., and H. Allen Orr. *Speciation.* Sunderland, Mass.: Sinauer Associates, 2004.

Dieckmann, Ulf. *Adaptive Speciation.* Cambridge: Cambridge University Press, 2004.

Weiner, Jonathan. *The Beak of the Finch: A Story of Evolution in Our Time.* New York: Alfred A. Knopf, 1994.

ANIMAL RIGHTS

Linzey, Andrew, and Paul A.B. Clarke. *Animal Rights: A Historical Anthology.* New York: Columbia University Press, 2004.

Singer, Peter. *Animal Liberation.* New York: Avon Books, 1975.

Sunstein, Cass R., and Martha Craven Nussbaum. *Animal Rights: Current Debates and New Directions.* Oxford: Oxford University Press, 2004.

Vyvyan, John. *The Dark Face of Science.* London: Michael Joseph, 1971.

ANNING, MARY

Clarke, Nigel J. *Mary Anning: A Brief History.* Lyme Regis, Dorset, Eng.: Clarke Publications, 1998.

Hawkins, Thomas. *The Book of the Great Sea-Dragons, Ichthyosauri and Plesiosauri.* London: W. Pickering, 1840.

Pierce, Patricia. *Jurassic Mary: Mary Anning on the Primeval Monsters.* Stroud, Gloucestershire, Eng.: Sutton, 2006.

APE LANGUAGE CONTROVERSY

Desmond, Adrian. *The Ape's Reflexion.* London: Blond & Briggs, 1979.

Linden, Eugene. *Silent Partners: The Legacy of the Ape Language Experiments.* New York: Times Books, 1986.

Rogers, Lesley J., and Gisela Kaplan. *Songs, Roars, and Rituals: Communication in Birds, Mammals, and Other Animals.* Cambridge, Mass.: Harvard University Press, 2000.

Terrace, Herbert. *Nim: A Chimpanzee Who Learned Sign Language.* New York: Alfred A. Knopf, 1979.

"APE WOMEN," LEAKEY'S

Jahme, Carole. *Beauty and the Beasts: Woman, Ape, and Evolution.* New York: Soho Press, 2001.

Fossey, Dian. *Gorillas in the Mist.* Boston: Houghton Mifflin, 1983.

Galdikas, Biruté M. F. *Reflections of Eden: My Years with the Orangutans of Borneo.* Boston: Little, Brown, 1995.

Goodall, Jane van Lawick. *In the Shadow of Man.* Boston: Houghton Mifflin, 1971.

———. *Through a Window: My Thirty Years with the Chimpanzees of Gombe.* New York: Houghton Mifflin, 1990.

Montgomery, Sy. *Walking with the Great Apes: Jane Goodall, Dian Fossey, and Biruté Galdikas.* New York: Houghton Mifflin, 1991.

APES, TOOL USE OF

Berthelet, A., and Jean Chavaillon. *The Use of Tools by Human and Non-Human Primates.* Oxford: Clarendon Press, 1993.

Köhler, Wolfgang. *The Mentality of Apes* (1917). Translated from the 2d rev. ed. by Ella Winter. London: Routledge & Kegan Paul, 1973.

Matsumura, Shuichi. "Orangutan Sociality, Tool Use, and Human Culture": Review of *Among Orangutans: Red Apes and the Rise of Human Culture,* by Carel van Schaik. *Current Anthropology* 47 (2006): 878.

McGrew, William Clement. *Chimpanzee Material Culture: Implications for Human Evolution.* Cambridge: Cambridge University Press, 1992.

———. *The Cultured Chimpanzee: Reflections on Cultural Primatology.* Cambridge: Cambridge University Press, 2004.

Moura, A. C. de A., and P. C. Lee. "Capuchin Stone Tool Use in Caatinga Dry Forest." *Science* 306 (2004): 1919.

Tomasello, Michael, and Josep Call. *Primate Cognition.* New York: Oxford University Press, 1997.

ARCHAEOPTERYX

Chambers, Paul. *Bones of Contention: The Archaeopteryx Scandals.* London: John Murray, 2002.

Chatterjee, Sankar. *The Rise of Birds: 225 Million Years of Evolution.* Baltimore: Johns Hopkins University Press, 1997.

Chiappe, Luis M. *Glorified Dinosaurs: The Origin and Early Evolution of Birds.* Hoboken, N.J.: John Wiley, 2007.

Fedducia, Alan. *The Origin and Evolution of Birds.* 2d ed. New Haven: Yale University Press, 1999.

Shipman, Pat. *Taking Wing: Archaeopteryx and the Evolution of*

Bird Flight. New York: Simon & Schuster, 1998.

ARISTOGENESIS

Hellman, Geoffrey. *Bankers, Bones, and Beetles: The First Century of the American Museum of Natural History.* Garden City, N.Y.: The Natural History Press, 1968.

Osborn, Henry Fairfield. *Man Rises to Parnassus.* Princeton: Princeton University Press, 1927.

ARTIFICIAL SELECTION

Clutton-Brock, Juliet. *A Natural History of Domesticated Mammals.* Cambridge: Cambridge University Press, 1987.

Darwin, Charles. *On the Origin of Species: The Illustrated Edition.* Edited by David Quammen. New York: Sterling, 2008.

———. *The Variation of Animals and Plants under Domestication.* 2 vols. London: John Murray, 1868.

Richards, Richard A. "Darwin, Domestic Breeding and Artificial Selection." *Endeavour* 22, no. 3 (1998): 106–109.

"ARYAN RACE," MYTH OF

Biddiss, Michael Denis. *Father of Racist Ideology: The Social and Political Thought of Count Gobineau.* New York: Weybright & Talley, 1970.

———. *The Age of the Masses: Ideas and Society in Europe since 1870.* Atlantic Highlands, N.J.: Humanities Press, 1977.

Chamberlain, Houston Stewart. *The Foundations of the Nineteenth Century.* Translated from the original German by John Lees. New York: John Lane, 1914.

Frawley, David. *The Myth of the Aryan Invasion of India.* New Delhi: Voice of India, 1994.

Grant, Madison. *The Passing of the Great Race; or, The Racial Basis of European History.* New York: Charles Scribner, 1916.

Poliakov, Léon. *The Aryan Myth: A History of Racist and Nationalist Ideas in Europe.* London: Chatto & Windus/Heinemann, 1974.

Spring, G. M. *The Philosophy of the Count de Gobineau.* Washington, D.C.: Scott-Townsend Publishers, 1996.

ATAPUERCA

Arsuaga, Juan Luis. *The Neanderthal's Necklace: In Search of the First Thinkers.* Translated by Andy Klatt. New York: Four Walls Eight Windows, 2002.

Arsuaga, Juan Luis, and Ignacio Martínez. *The Chosen Species: The Long March of Human Evolution.* Translated by Rachel Gomme. Malden, Mass.: Blackwell, 2005.

AUSTRALOPITHECINES

Constantino, Paul, and Bernard Wood. "The Evolution of *Zinjanthropus boisei.*" *Evolutionary Anthropology: Issues, News, and Reviews* 16, no. 2 (2007): 49–62.

Grine, Frederick E., ed. *Evolutionary History of the "Robust" Australopithecines.* New York: Aldine de Gruyter, 1988.

Gundling, Tom. *First in Line: Tracing Our Ape Ancestry.* New Haven, Conn.: Yale University Press, 2005.

Johanson, Donald, and Maitland Edey. *Lucy: The Beginnings of Humankind.* New York: Simon & Schuster, 1981.

Leakey, Mary D. *Olduvai Gorge: My Search for Early Man.* London: Collins, 1979.

Walker, Alan, and Pat Shipman. *The Wisdom of the Bones: In Search of Human Origins.* New York: Alfred A. Knopf, 1996.

BABOONS

Altmann, Jeanne. *Baboon Mothers and Infants.* Chicago: University of Chicago Press, 2001.

Cheney, Dorothy L., and Robert M. Seyfarth. *Baboon Metaphysics: The Evolution of a Social Mind.* Chicago: University of Chicago Press, 2007.

Sapolsky, Robert M. *A Primate's Memoir.* New York: Scribner, 2001.

Strum, Shirley. *Almost Human: A Journey into the World of Baboons.* New York: Random House, 1987.

BACTERIA

Dworkin, Martin, and Stanley Falkow. *The Prokaryotes: A Handbook on the Biology of Bacteria.* New York: Springer, 2006.

Dyer, Betsey Dexter. *A Field Guide to Bacteria.* Ithaca, N.Y.: Cornell University Press, 2003.

Sachs, Jessica Snyder. *Good Germs, Bad Germs: Health and Survival in a Bacterial World.* New York: Hill & Wang, 2007.

BANKS, SIR JOSEPH

Beaglehole, J. C., and Phillip Edwards, eds. *The Journals of Captain Cook.* London: Penguin Classics, 1999.

Chambers, Neil, ed. *The Letters of Sir Joseph Banks: A Selection, 1768–1820.* London: Imperial College Press, 2000.

Gascoigne, Tom. *Science in the Service of Empire: Joseph Banks, the British State, and the Uses of Science in the Age of Revolution.* Cambridge: Cambridge University Press, 1998.

Moyal, Ann Mozley. *"A Bright and Savage Land": Scientists in*

Colonial Australia. Sydney: Collins, 1986.

O'Brian, Patrick. *Joseph Banks: A Life.* Boston: D. R. Godine, 1992.

Rice, Tony. *Voyages of Discovery: Three Centuries of Natural History Exploration.* New York: Clarkson Potter, 1999.

Withey, Lynne. *Voyages of Discovery: Captain Cook and the Exploration of the Pacific.* Berkeley: University of California Press, 1989.

BARLOW, LADY NORA

Barlow, Lady Nora, ed. *Charles Darwin and the Voyage of the "Beagle."* New York: Philosophical Library, 1945.

———. ed. *Charles Darwin's Diary of the Voyage of HMS "Beagle."* Cambridge: Cambridge University Press, 1932.

———. *Darwin and Henslow: The Growth of an Idea.* London: John Murray, 1967.

BARNACLES

Darwin, Charles. *A Monograph of the Sub-Class Cirripedia.* 4 vols. London: Ray Society, 1850–1858.

Milner, Richard. "Mr. Darwin Meets Mr. Arthrobalanus." *Journal of the Linnean Society of London* 20, no. 3 (2004): 32–37.

Stott, Rebecca. *Darwin and His Barnacles.* New York: W. W. Norton, 2003.

BARNUM, PHINEAS T.

Barnum, Phineas Taylor. *The Life of P. T. Barnum, Written by Himself.* Urbana: University of Illinois Press, 2000.

———. *Struggles & Triumphs; or, Forty Years' Recollections.* Buffalo: Warren, Johnson & Co., 1872.

Harris, Neil. *Humbug: The Art of P. T. Barnum.* Boston: Little, Brown, 1973.

Kunhardt, Philip B. Jr., Philip B. Kunhardt III, and Peter W. Kunhardt. *P. T. Barnum: America's Greatest Showman.* New York: Alfred A. Knopf, 1995.

Saxon, A. H. *P. T. Barnum: The Legend and the Man.* New York: Columbia University Press, 1989.

BATES, HENRY WALTER

Bates, Henry Walter. *In the Heart of the Amazon Forest: Great Journeys.* London: Penguin Books, 2007.

———. *The Naturalist on the River Amazon.* London: John Murray, 1863.

Beddall, Barbara G., ed. *Wallace and Bates in the Tropics: An Introduction to the Theory of Natural Selection.* New York: Macmillan, 1969.

Wallace, Alfred Russel. *A Narrative of Travels on the Amazon and Rio*

Negro, with an Account of the Native Tribes, and Observations on the Climate, Geology, and Natural History of the Amazon Valley. New York: Greenwood Press, 1969.

(HMS) BEAGLE

Chancellor, Gordon. "The *Beagle* Paintings of John Chancellor (1925–1984)." In "Survival of the Fittest: Celebrating the 150th Anniversary of the Darwin-Wallace Theory of Evolution." *The Linnean,* special issue no. 9 (2008): 49–60.

Chancellor, Rita, and Austin Hawkins. *John Chancellor's Classic Maritime Paintings.* Newton Abbot, Devon, UK: David & Charles, 1989.

Darwin, Charles. *The Voyage of the "Beagle."* New York: Harper, 1959.

Keynes, Richard Darwin. *The "Beagle" Record: Selections from the Original Pictorial Records and Written Accounts of the Voyage of HMS "Beagle."* Cambridge: Cambridge University Press, 1980.

Marquardt, Karl Heinz. *HMS "Beagle": Survey Ship Extraordinary.* London: Conway Maritime Press, 1997.

Thomson, Keith. *HMS "Beagle": The Story of Darwin's Ship.* New York: W.W. Norton, 1995.

BEECHER, HENRY WARD

Abbott, Lyman, and S.B. Halliday. *Henry Ward Beecher: A Sketch of His Career.* Whitefish, Mont.: Kessinger, 2003.

Applegate, Debby. *The Most Famous Man in America: The Biography of Henry Ward Beecher.* New York: Doubleday, 2006.

Beecher, Henry Ward. *Evolution and Religion.* Parts I and II. New York: Fords, Howard & Hulbert, 1885.

Fox, Richard Wightman. *Trials of Intimacy: Love and Loss in the Beecher-Tilton Scandal.* Chicago: University of Chicago Press, 1999.

BEETLES

Evans, Arthur V., and Charles L. Bellamy. *An Inordinate Fondness for Beetles.* Berkeley: University of California Press, 2000.

Grimaldi, David A., and Michael S. Engel. *Evolution of the Insects.* Cambridge: Cambridge University Press, 2005.

Peck, Stewart B. *The Beetles of the Galápagos Islands, Ecuador: Evolution, Ecology, and Diversity (Insecta: Coleoptera).* Ottawa: NRC Research Press, 2006.

BEIJING MAN

Aczel, Amir D. *The Jesuit and the Skull: Teilhard de Chardin, Evolution, and the Search for Peking*

Man. New York: Riverhead Books, 2007.

Boaz, Noel T., and Russell L. Ciochon. *Dragon Bone Hill: An Ice-Age Saga of Homo Erectus.* New York: Oxford University Press, 2004.

Lanpo, Jia, and Huang Weiwen. *The Story of Peking Man: From Archaeology to Mystery.* New York: Oxford University Press, 1990.

Van Oosterzee, Penny. *Dragon Bones: The Story of Peking Man.* Cambridge, Mass.: Perseus, 2000.

BELFAST ADDRESS

Otis, Laura, ed. *Literature and Science in the Nineteenth Century: An Anthology.* New York: Oxford University Press, 2002.

Tyndall, John. *Fragments of Science for Unscientific People.* New York: Appleton, 1898.

Weber, Alan S., ed. *Nineteenth-Century Science: An Anthology.* Orchard Park, N.Y.: Broadview Press, 2000.

BELT, THOMAS

Belt, Thomas. *The Naturalist in Nicaragua.* London: John Murray, 1874.

Maslow, Jonathan. *Footsteps in the Jungle.* Chicago: Ivan R. Dee, 1997.

BENGA, OTA

Bradford, Phillips Verner, and Harvey Blume. *Ota: The Pygmy in the Zoo.* New York: St. Martin's Press, 1992.

Bridges, William. *A Gathering of Animals: An Unconventional History of the New York Zoological Society.* New York: Harper & Row, 1974.

Sifakis, Carl. *American Eccentrics.* New York: Facts on File, 1984.

BERINGER, JOHANNES

Jahn, Melvin E., and Daniel J. Woolf. *The Lying Stones of Dr. Johann Bartholomew Adam Beringer, Being His "Lithographiae Wirceburgensis."* Berkeley: University of California Press, 1963.

Gould, Stephen J. *The Lying Stones of Marrakech: Penultimate Reflections in Natural History.* New York: Harmony Books, 2000.

Silverberg, Robert. *Scientists and Scoundrels: A Book of Hoaxes.* Lincoln: University of Nebraska Press, 2007.

BINOMIAL NOMENCLATURE

Dunn, Rob R. *Every Living Thing: Man's Obsessive Quest to Catalog Life, from Nanobacteria to New Monkeys.* New York: Collins, 2009.

Farber, Paul Lawrence. *Finding Order in Nature: The Naturalist*

Tradition from Linnaeus to E.O. Wilson. Baltimore: Johns Hopkins University Press, 2000.

Heller, John Lewis. *The Early History of Binomial Nomenclature.* Pittsburgh: Hunt Botanical Library, Carnegie Institute of Technology, 1964.

Winston, Judith E. *Describing Species: Practical Taxonomic Procedure for Biologists.* New York: Columbia University Press, 1999.

BIOGENETIC LAW

Baer, Karl Ernst von, and George Sarton. *The Discovery of the Mammalian Egg and the Foundation of Modern Embryology.* London: St. Catherine Press, 1931.

Freeman, Brian. "The Myth of the Biogenetic Law." *American Biology Teacher* 63, no. 2 (2001): 84.

Gould, Stephen Jay. *Ontogeny and Phylogeny.* Cambridge, Mass.: Harvard University Press, Belknap Press, 1977.

Haeckel, Ernst. *Last Words on Evolution: A Popular Retrospect and Summary.* Translated by Joseph McCabe. London: A. Owen, 1906.

BIOLOGICAL DETERMINISM

Caplan, Arthur, ed. *The Sociobiology Debate.* New York: Harper & Row, 1978.

Gould, Stephen Jay. *The Measure of Man.* New York: W.W. Norton, 1981.

Lewontin, Richard C., Steven P.R. Rose, and Leon J. Kamin, eds. *Not in Our Genes: Biology, Ideology, and Human Nature.* New York: Random House, 1984.

Segerstråle, Ullica. *Defenders of the Truth: The Sociobiology Debate.* New York: Oxford University Press, 2001.

Singer, Peter. *The Expanding Circle: Ethics and Sociobiology.* New York: New American Library, 1981.

BIOLOGICAL EXUBERANCE

Bagemihl, Bruce. *Biological Exuberance: Animal Homosexuality and Natural Diversity.* New York: St. Martin's Press, 1998.

Roughgarden, Joan. *Evolution's Rainbow: Diversity, Gender, and Sexuality in Nature and People.* Berkeley: University of California Press, 2004.

Wolfe, Linda D. "Human Evolution and the Sexual Behavior of Female Primates." In *Understanding Behavior: What Primate Studies Tell Us about Human Behavior,* edited by James D. Loy and Calvin B. Peters. New York: Oxford University Press, 1991.

BIOLOGY

Burkhardt, Richard W. *The Spirit of System: Lamarck and Evolution-*

ary Biology. Cambridge, Mass.: Harvard University Press, 1995.

Mayr, Ernst. *The Growth of Biological Thought: Diversity, Evolution, and Inheritance.* Cambridge, Mass.: Harvard University Press, Belknap Press, 1982.

Serafini, Anthony. *The Epic History of Biology.* New York: Plenum, 1993.

Webster, Stephen. *Thinking about Biology.* Cambridge: Cambridge University Press, 2003.

BIOSPHERE, EVOLUTION OF

Haselton, Aaron, Lynn Margulis, and Clifford N. Matthews. *Environmental Evolution: Effects of the Origin and Evolution of Life on Planet Earth.* Cambridge, Mass.: MIT Press, 2000.

Lovelock, James. *The Revenge of Gaia: Earth's Climate in Crisis and the Fate of Humanity.* New York: Basic Books, 2006.

Margulis, Lynn, and Dorion Sagan. *Slanted Truths: Essays on Gaia, Symbiosis, and Evolution.* New York: Copernicus, 1997.

Myers, Norman, and Jennifer Kent. *The New Atlas of Planet Management.* Berkeley: University of California Press, 2005.

Schneider, Stephen Henry. *Scientists Debate Gaia: The Next Century.* Cambridge, Mass.: MIT Press, 2004.

Smil, Vaclav. *The Earth's Biosphere: Evolution, Dynamics, and Change.* Cambridge, Mass.: MIT Press, 2002.

Thomashow, Mitchell. *Bringing the Biosphere Home: Learning to Perceive Global Environmental Change.* Cambridge, Mass.: MIT Press, 2002.

Vernadsky, V.I., and Mark A. McMenamin. *The Biosphere.* New York: Copernicus, 1998.

BIRD, ROLAND THAXTER

Bird, Roland T. *Bones for Barnum Brown: Adventures of a Dinosaur Hunter.* Fort Worth: Texas Christian University Press, 1985.

———. *Dinosaur Valley State Park, Glen Rose, Texas: Where the Big Ones Ate and Ran.* Shirleysburg, Pa.: Laughingstock Ranch, 1978.

Colbert, Edwin H. *Men and Dinosaurs: The Search in Field and Laboratory.* New York: E.P. Dutton, 1968.

BONOBOS

Diamond, Jared M. *Why Is Sex Fun? The Evolution of Human Sexuality.* New York: HarperCollins, 1997.

Furuichi, Takeshi, and Jo Thompson. *Bonobos: Ecology, Behavior, and Conservation.* Springer Verlag, 2008.

Levine, Mathea, and Marian Brickner. *I'm Lucy: A Day in the Life of* a Young Bonobo. St. Louis: Blue Bark Press, 2008.

Sandin, Jo. Bonobos: Encounters in Empathy. Milwaukee: Zoological Society of Milwaukee, 2007.

Savage-Rumbaugh, E. Sue. Apes, Language, and the Human Mind. Oxford University Press, 1998.

de Waal, Frans, and Frans Lanting. Bonobo: The Forgotten Ape. Berkeley: University of California Press, 1998.

Wrangham, Richard W. Chimpanzee Cultures. Harvard University Press, 1996.

BOUCHER DE PERTHES

Brodrick, Alan. Father of Prehistory: The Abbé Henri Breuil. New York: William Morrow, 1963.

Newton, William M. Light on Palaeolithic Flint Figures, and Boucher De Perthes. N.p., 1910.

BRANCHING BUSH

Cracraft, Joel, and Michael J. Donoghue. Assembling the Tree of Life. Oxford: Oxford University Press, 2004.

BRIDGEWATER TREATISES

Babbage, Charles. The Ninth Bridgewater Treatise: A Fragment. London: John Murray, 1838.

Bridgewater, Francis Henry Egerton. The Bridgewater Treatises on the Power, Wisdom, and Goodness of God, as Manifested in the Creation. Treatises I–VIII. London: W. Pickering, 1834.

Topham, J. "Science and Popular Education in the 1830s: The Role of the Bridgewater Treatises." British Journal for the History of Science 25 (1992): 397–430.

BROOM, ROBERT

Broom, Robert. Finding the Missing Link. London: Watts, 1950.

Dart, Raymond A. Robert Broom: His Life and Work. Johannesburg: Hortors, 1951.

Findlay, George. Dr. Robert Broom, F.R.S., Palaeontologist and Physician, 1866–1951: A Biography, Appreciation, and Bibliography. Cape Town: A.A. Balkema, 1972.

Reader, John. Missing Links: The Hunt for Earliest Man. London: Collins, 1981.

BROWN, BARNUM

Brown, Frances R. Let's Call Him Barnum. New York: Vantage Press, 1988.

Brown, Lilian. Bring 'Em Back Petrified. New York: Dodd, Mead, 1956.

Dingus, Lowell, and Mark A. Norell. "The Bone Collector." Discover, March 2007.

Preston, Douglas. Dinosaurs in the Attic: An Excursion into the American Museum of Natural History. New York: St. Martin's Press, 1986.

BRYAN, WILLIAM JENNINGS

Bryan, William Jennings, and Mary Baird Bryan. The Memoirs of William Jennings Bryan. Philadelphia: John C. Winston, 1925.

Cherny, Robert W. A Righteous Cause: The Life of William Jennings Bryan. Norman: University of Oklahoma Press, 1994.

Kazin, Michael. A Godly Hero: The Life of William Jennings Bryan. New York: Alfred A. Knopf, 2006.

BUFFON, COMTE DE

Buffon, Georges-Louis Leclerc. Buffon's Natural History of the Globe, and of Man, Beasts, Birds, Fishes, Reptiles, and Insects. Corrected and enlarged by John Wright. London: T. Tegg, 1831.

Roger, Jacques. Buffon: A Life in Natural History. Ithaca, N.Y.: Cornell University Press, 1997.

BURGESS SHALE

Briggs, Derek E.G., Douglas H. Erwin, and Frederick J. Collier. The Fossils of the Burgess Shale. Washington, D.C.: Smithsonian Institution Press, 1994.

Conway Morris, Simon. The Crucible of Creation: The Burgess Shale and the Rise of Animals. New York: Oxford University Press, 1998.

Gould, Stephen Jay. Wonderful Life: The Burgess Shale and the Nature of History. New York: W.W. Norton, 1989.

Whittington, Harry. The Burgess Shale. New Haven, Conn.: Yale University Press, 1985.

BURIAN, ZDENĚK

Augusta, Josef, and Burian, Zdeněk. The Age of Monsters. London: Paul Hamlyn, 1966.

———. Prehistoric Animals. London: Paul Hamlyn, 1960.

———. Prehistoric Man. London: Paul Hamlyn, 1960.

BURNET, REVEREND THOMAS

Burnet, Thomas. The Sacred Theory of the Earth. Carbondale: Southern Illinois University Press, 1965.

Lynch, John M. Creationism and Scriptural Geology. Bristol, Eng.: Thoemmes Press, 2002.

BUTLER, SAMUEL

Butler, Samuel. Evolution, Old and New; or, The Theories of Buffon, Dr. Erasmus Darwin, and Lamarck, as Compared with Those of Mr. Charles Darwin. London: Hardwick & Bogue, 1879.

———. Luck, or Cunning, as the Main Means of Organic Modification? London: Trubner, 1887.

Raby, Peter. Samuel Butler: A Biography. Iowa City: University of Iowa Press, 1991.

Willey, B. Darwin and Butler: Two Versions of Evolution. London: Chatto & Windus, 1960.

BUTLER ACT

Murray, A. Victor. The School and the Church: The Theory and Practice of Christian Education under the Butler Act. London: S.C.M. Press, 1944.

BUTTON, JEMMY

Bridges, Lucas. The Uttermost Part of the Earth. London: Hodder & Stoughton, 1951.

Hazelwood, Nick. Savage: The Life and Times of Jemmy Button. New York: St. Martin's Press, 2001.

Marks, Richard Lee. Three Men of the "Beagle." New York: Alfred A. Knopf, 1991.

CAMBRIAN-SILURIAN CONTROVERSY

Rudwick, M.J.S. The Great Devonian Controversy: The Shaping of Scientific Knowledge among Gentlemanly Specialists. Chicago: University of Chicago Press, 1985.

Secord, James A. Controversy in Victorian Geology: The Cambrian-Silurian Dispute. Princeton: Princeton University Press, 1986.

CANNIBALISM CONTROVERSY

Arens, W. The Man-Eating Myth. New York: Oxford University Press, 1979.

Berglund, Jeff. Cannibal Fictions: American Explorations of Colonialism, Race, Gender and Sexuality. Madison: University of Wisconsin Press, 2006.

Donnelly, Mark, and Daniel Diehl. Eat Thy Neighbour: A History of Cannibalism. Stroud, Gloucestershire, Eng.: Sutton, 2006.

Goldman, Laurence. The Anthropology of Cannibalism. Westport, Conn.: Bergin & Garvey, 1999.

Krieger, Michael. Conversations with the Cannibals: The End of the Old South Pacific. Hopewell, N.J.: Ecco Press, 1994.

CARDIFF GIANT

Feder, Kenneth L. Frauds, Myths, and Mysteries: Science and Pseudoscience in Archaeology. Mountain View, Calif.: Mayfield, 1999.

Perrin, Pat, and Wim Coleman. The Mystery of the Cardiff Giant. Logan, Iowa: Perfection Learning, 2004.

CARNEGIE, ANDREW

Carnegie, Andrew, and John Charles Van Dyke. Autobiography of Andrew Carnegie. Boston: Houghton Mifflin, 1920.

Nasaw, David. Andrew Carnegie. New York: Penguin Books, 2006.

Rea, Tom. Bone Wars: The Excavation and Celebrity of Andrew Carnegie's Dinosaur. Pittsburgh: University of Pittsburgh Press, 2001.

Wall, Joseph F. Andrew Carnegie. New York: Oxford University Press, 1970.

CARTESIAN DUALITY

Garber, Daniel. Descartes Embodied: Reading Cartesian Philosophy through Cartesian Science. Cambridge: Cambridge University Press, 2001.

CATASTROPHISM

Eiseley, Loren. Darwin's Century: Evolution and the Men Who Discovered It. New York: Doubleday, 1958.

Gould, Stephen Jay. Time's Arrow, Time's Cycle. Cambridge, Mass.: Harvard University Press, 1987.

Greene, John C. The Death of Adam: Evolution and Its Impact on Western Thought. Ames: Iowa State University Press, 1959.

Palmer, Trevor. Controversy—Catastrophism and Evolution: The Ongoing Debate. New York: Kluwer Academic/Plenum Publishers, 1999.

CAVEMAN

Clark, Constance Areson. God—or Gorilla: Images of Evolution in the Jazz Age. Baltimore: Johns Hopkins University Press, 2008.

Lubbock, John. Pre-historic Times: As Illustrated by Ancient Remains, and the Manners and Customs of Modern Savages. London: Williams & Norgate, 1865.

McCaughey, Martha. The Caveman Mystique: Pop-Darwinism and the Debates over Sex, Violence, and Science. New York: Routledge, 2008.

Milner, Richard. "Portraits of Prehistory: Imaging Our Ancestors." In The Last Human: A Guide to Twenty-two Species of Extinct Humans, edited by G.J. Sawyer and Viktor Deak, pp. 237–250. New Haven, Conn.: Yale University Press, 2007.

"CENTRAL DOGMA" OF GENETICS

Crick, Francis. What Mad Pursuit: A Personal View of Scientific Discovery. New York: Basic Books, 1988.

CHAMBERS, ROBERT

Chambers, Robert. "Vestiges of the Natural History of Creation" and Other Evolutionary Writings. Edited by James A. Secord. Chicago: University of Chicago Press, 1994.

Milhauser, Milton. Just before Darwin: Robert Chambers and "Vestiges." Middletown, Conn.: Wesleyan University Press, 1959.

Secord, James. Victorian Sensation: The Extraordinary Publication, Reception, and Secret Authorship of "Vestiges of the Natural History of Creation." Chicago: University of Chicago Press, 2001.

CHAUVET CAVE

Chauvet, Jean-Marie, et al. *Dawn of Art: The Chauvet Cave, the Oldest Known Paintings in the World.* New York: Harry N. Abrams, 1996.

Clottes, Jean. *Chauvet Cave: The Art of Earliest Times.* Salt Lake City: University of Utah Press, 2003.

———. "Rhinos and Lions and Bears (Oh My!)." *Natural History* 104, no. 5 (1995): 30–35.

Packer, Craig, and Jean Clottes. "When Lions Ruled France." *Natural History* 109, no. 9 (2000): 52–57.

CHIMPANZEES

Goodall, Jane. *The Chimpanzees of Gombe: Patterns of Behavior.* Cambridge, Mass.: Harvard University Press, Belknap Press, 1986.

———. *In the Shadow of Man.* Boston: Houghton Mifflin, 1971.

Stanford, Craig B. *Chimpanzee and Red Colobus: The Ecology of Predator and Prey.* Cambridge, Mass.: Harvard University Press, 2001.

de Waal, Frans. *Chimpanzee Politics: Power and Sex among Apes.* Rev. ed. Baltimore: Johns Hopkins University Press, 2000.

———. *Our Inner Ape: A Leading Primatologist Explains Why We Are Who We Are.* New York: Riverhead Books, 2005.

Wrangham, Richard W., and Dale Peterson. *Demonic Males: Apes and the Origins of Human Violence.* Boston: Houghton Mifflin, 1996.

Wrangham, Richard W., W.C. McGrew, Frans B.M. de Waal, and Paul G. Heltne, eds. *Chimpanzee Cultures.* Cambridge, Mass.: Harvard University Press, 1996.

CHINA, EVOLUTION IN

Chiappe, Luis M., and Lawrence M. Witner. *Mesozoic Birds: Above the Heads of Dinosaurs.* Berkeley: University of California Press, 2002.

Norrell, Mark A., and Mick Ellison. *Unearthing the Dragon: The Great Feathered Dinosaur Discovery.* New York: PI Press, 2005.

Shih, Hu. *Living Philosophies.* New York: Simon & Schuster, 1931.

CLADISTICS

Hull, David. *Science as a Process: An Evolutionary Account of the Social and Conceptual Development of Science.* Chicago: University of Chicago Press, 1990.

Lecointre, Guillaume, and Hervé Le Guyader. *The Tree of Life: A Phylogenetic Classification.* Cambridge, Mass.: Harvard University Press, Belknap Press, 2006.

O'Brien, Michael J., and R. Lee Lyman. *Cladistics and Archaeology.* Salt Lake City: University of Utah Press, 2003.

CLEVER HANS PHENOMENON

Sebeok, Thomas A., and Robert Rosenthal. "The Clever Hans Phenomenon: Communication with Horses, Whales, Apes, and People." *Annals of the New York Academy of Sciences* 364 (1981): 1–311.

Yule, George. *The Study of Language.* Cambridge: Cambridge University Press, 2006.

COEVOLUTION

Futuyma, Douglas J., and Montgomery Slatkin, eds. *Coevolution.* Sunderland, Mass.: Sinauer Associates, 1983.

Levy, Charles Kingsley. *Evolutionary Wars: A Three-Billion-Year Arms Race.* New York: W.H. Freeman, 2000.

COMPARATIVE METHOD

Darwin, Charles. *The Expression of the Emotions in Man and Animals* (1872). Edited by Paul Ekman. Oxford: Oxford University Press, 1998.

Lorenz, Konrad. *Evolution and Modification of Behavior.* Chicago: University of Chicago Press, 1965.

COMPOSITE PHOTOGRAPHS

Burson, Nancy, Richard Carling, and David Kramlich. *Composites: Computer-Generated Portraits.* New York: Beech Tree Books, 1986.

Galton, Sir Francis. *Inquiries into Human Faculty and Its Development.* London: Macmillan, 1883.

CONGO

Morris, Desmond. *The Biology of Art.* London: Methuen, 1962.

CONSPICUOUS COLORATION

Cott, H.B. *Adaptive Coloration in Animals.* London: Methuen, 1940.

Poulton, E.B. *The Colors of Animals: Their Meaning and Use, Especially Considered in the Case of Insects.* London: Kegan Paul, Trench, Trübner, 1890.

CONTINENTAL DRIFT

Glen, William. *The Road to Jaramillo: Critical Years of the Revolution in Earth Science.* Stanford: Stanford University Press, 1982.

Oreskes, Naomi, and H.E. LeGrand. *Plate Tectonics: An Insider's History of the Modern Theory of the Earth.* Boulder, Colo.: Westview Press, 2001.

Wegener, Alfred. *The Origin of Continents and Oceans.* 1915; repr. New York: Dover, 1966.

CONTINGENT HISTORY

Gould, Stephen Jay. *The Flamingo's Smile: Reflections in Natural History.* New York: W.W. Norton, 1985.

———. *Wonderful Life: The Burgess Shale and the Nature of History.* New York: W.W. Norton, 1989.

CONVERGENT EVOLUTION

Conway Morris, Simon. *The Crucible of Creation: The Burgess Shale and the Rise of Animals.* Oxford: Oxford University Press, 1998.

Dawkins, Richard. *The Ancestor's Tale: A Pilgrimage to the Dawn of Evolution.* Boston: Houghton Mifflin, 2004.

COPE, EDWARD DRINKER

Cope, Edward Drinker. *The Origin of the Fittest: Essays on Evolution and the Primary Factors of Organic Evolution.* New York: Appleton, 1887.

———. *Tertiary Vertebrata.* Washington, D.C.: U.S. Geological Survey, 1884.

Osborn, Henry F. *Cope: Master Naturalist.* Princeton: Princeton University Press, 1931.

COPE-MARSH FEUD

Davidson, Jane Pierce. *The Bone Sharp: The Life of Edward Drinker Cope.* Philadelphia: Academy of Natural Sciences, 1997.

Howard, Robert W. *The Dawnseekers: The First History of American Paleontology.* New York: Harcourt, Brace, 1975.

Jaffe, Mark. *The Gilded Dinosaur: The Fossil War between E.D. Cope and O.C. Marsh and the Rise of American Science.* New York: Three Rivers Press, 2000.

Plate, Robert. *The Dinosaur Hunters: Marsh and Cope.* New York: McKay, 1964.

Wallace, David Rains. *The Bonehunters' Revenge: Dinosaurs, Greed, and the Greatest Scientific Feud of the Gilded Age.* New York: Houghton Mifflin, 1999.

Warren, Leonard. *Joseph Leidy: The Last Man Who Knew Everything.* New Haven: Yale University Press, 1998.

COPROLITE INDUSTRY

Grove, Richard. *The Cambridgeshire Coprolite Mining Rush.* Cambridge: Oleander Press, 1976.

O'Connor, Bernard. *The Dinosaurs of Coldham's Common: The Story of Cambridge's Coprolite Industry.* Cambridge: Bernard O'Connor, 1998.

CORAL REEFS

Darwin, Charles. *The Structure and Distribution of Coral Reefs.* London: Smith, Elder, 1842.

Dobbs, David. *Reef Madness: Charles Darwin, Alexander Agassiz, and the Meaning of Coral.* New York: Pantheon Books, 2005.

Milner, Richard. "Seeing Corals with the Eye of Reason." *Natural History* 118, no. 1 (2009): 18–23.

Stoddart, David R. "Darwin and the Seeing Eye: The Iconography and Meaning in the *Beagle* Years." *Earth Sciences History* 14, no. 1 (1995): 3–22.

"CRADLE OF HUMANKIND"

Leakey, Richard, and Roger Lewin. *Origins.* New York: E.P. Dutton, 1977.

Reader, John. *Missing Links: The Hunt for Earliest Man.* London: Collins, 1981.

CREATIONISM

Eldredge, Niles. *The Triumph of Evolution and the Failure of Creationism.* New York: W.H. Freeman, 2001.

Forrest, Barbara, and Paul R. Gross. *Creationism's Trojan Horse: The Wedge of Intelligent Design.* Oxford: Oxford University Press, 2004.

Godfrey, Laurie, ed. *Scientists Confront Creationism.* New York: W.W. Norton, 1983.

Leone, Bruno, *Creationism vs. Evolution.* San Diego, Calif.: Greenhaven Press, 2002.

Moore, James R. "The Creationist Cosmos of Protestant Fundamentalism." In *Fundamentalisms and Society: Reclaiming the Sciences, Education, and the Family*, edited by Martin E. Marty and R. Scott Appleby, pp. 42–72. Chicago: University of Chicago Press, 1993.

Numbers, Ronald L. *Creationism in Twentieth-Century America.* Hamden, Conn.: Garland, 1995.

Toumey, Christopher P. *God's Own Scientists: Creationists in a Secular World.* New Brunswick, N.J.: Rutgers University Press, 1994.

CREATIONISM, AMERICAN POLL ON

Kalthoff, Mark A., ed. *Creation and Evolution in the Early American Scientific Affiliation.* Vol. 10 of *Creationism in Twentieth-Century America.* New York: Garland, 1995.

Larson, Edward J. *Trial and Error: The American Controversy over Creation and Evolution.* New York: Oxford University Press, 2003.

Webb, George E. *The Evolution Controversy in America.* Lexington: University Press of Kentucky, 1994.

CREATIVE EVOLUTION

Bergson, Henri. *Creative Evolution* (1907). Edited by Keith Ansell-Pearson, Michael Kolkman, and Michael Vaughan. Basingstoke, Eng.: Palgrave Macmillan, 2007.

CUTENESS, EVOLUTION OF

Gould, Stephen Jay. "Biological Homage to Mickey Mouse." In *The Panda's Thumb: More Reflections on Natural History*, pp. 95–107. New York: W.W. Norton, 1992.

Hinde, R.A., and L.A. Barden. "The Evolution of the Teddy Bear." *Animal Behaviour* 33 (1985): 1371–1373.

Lorenz, Konrad. "Part and Parcel in Animal and Human Societies." In *Studies in Animal and Human Behavior*, 2: 115–95. Cambridge, Mass.: Harvard University Press, 1971 (originally published 1950).

CUVIER, BARON GEORGES

Cuvier, Baron Georges. *The Animal Kingdom*. London: W.H. Allen, 1886.

——. *Essay on the Theory of the Earth* (1817). Edinburgh: William Blackwood, 1827.

Rudwick, M.J.S., trans. and ed. *Georges Cuvier, Fossil Bones, and Geological Catastrophes: New Translations and Interpretations of the Primary Texts*. Chicago: University of Chicago Press, 1997.

DART, RAYMOND ARTHUR

Dart, Raymond. *Adventures with the Missing Link*. New York: Viking Press, 1959.

Dart, Raymond A., and Richard Stanton Rimanoczy. *Tools—Creator of Civilization: The Development of Man as Revealed by the Discoveries of Raymond A. Dart*. [Cleveland]: American Economic Foundation, 1965.

Wheelhouse, Frances, and Kathaleen S. Smithford. *Dart: Scientist and Man of Grit*. Sydney: Transpareon Press, 2001.

DARWIN, CHARLES ROBERT

Brent, Peter. *Charles Darwin: A Man of Enlarged Curiosity*. New York: Harper & Row, 1981.

Browne, Janet. *Charles Darwin: Voyaging*. New York: Alfred A. Knopf, 1995.

——. *Charles Darwin: The Power of Place*. New York: Alfred A. Knopf, 2002.

Clark, Ronald. *The Survival of Charles Darwin*. New York: Random House, 1984.

Darwin, Charles. *Autobiography*. Edited by Nora Barlow. New York: Harcourt, Brace, 1958.

Darwin, Francis. *Life and Letters of Charles Darwin*. London: John Murray, 1887.

Desmond, Adrian, and James Moore. *Darwin: The Life of a Tormented Evolutionist*. London: Michael Joseph, 1991.

Hodge, Jonathan, and Gregory Raik. *The Cambridge Companion to Darwin*. Cambridge: Cambridge University Press, 2003.

Huxley, Julian, and H.B.D. Kettlewell. *Charles Darwin and His World*. London: Thames & Hudson, 1965.

Irvine, William. *Apes, Angels, and Victorians: The Story of Darwin, Huxley, and Evolution*. New York: McGraw-Hill, 1955.

Keynes, Randal. *Annie's Box: Charles Darwin, His Daughter, and Human Evolution*. London: Fourth Estate, 2001.

Milner, Richard. *Charles Darwin: Evolution of a Naturalist*. New York: Facts on File, 1993.

Thomson, Keith. *The Young Charles Darwin*. New Haven: Yale University Press, 2009.

Ward, Henshaw. *Charles Darwin: The Man and His Warfare*. New York: Bobbs-Merrill, 1927.

DARWIN, CHARLES, DESCENDANTS

Chapman, Matthew. *Trials of the Monkey: An Accidental Memoir*. New York: Picador, 2001.

Darwin, Emma. *The Mathematics of Love*. New York: William Morrow, 2006.

Keynes, Randal. *Annie's Box: Charles Darwin, His Daughter, and Human Evolution*. London: Fourth Estate, 2001.

Milner, Richard. "Descent with Modification: Interview with Richard Darwin Keynes." *Natural History* 115, no. 4 (2005): 30–34.

Padel, Ruth. *Tigers in Red Weather: A Quest for the Last Wild Tigers*. New York: Walker & Co., 2006.

DARWIN, CHARLES, ILLNESS OF

Colp, Ralph, Jr. *Darwin's Illness*. Gainesville: University Press of Florida, 2008.

Milner, Richard. "Darwin's Shrink: A Noted Darwin Historian Probes the Naturalist's Inner Life." *Natural History* 114, no. 9 (2005): 42–44.

DARWIN, CHARLES, MAGISTRATE

Milner, Richard. *Charles Darwin: Evolution of a Naturalist*. New York: Facts on File, 1993.

——. "Charles Darwin: Ghostbuster, Muse, and Magistrate." In "Survival of the Fittest: Celebrating the 150th Anniversary of the Darwin-Wallace Theory of Evolution." *The Linnean*, special issue no. 9 (2008): 49–60.

DARWIN, EMMA

Healey, Edna. *Emma Darwin: The Inspirational Wife of a Genius*. London: Headline, 2001.

Litchfield, Henrietta Emma. *Emma Darwin: A Century of Family Letters, 1792–1896*. London: John Murray, 1915.

DARWIN, ERASMUS

Darwin, Erasmus. *The Botanic Garden*. Dublin: J. Moore, 1790.

——. *The Temple of Nature or, the Origin of Society*. Baltimore: Bonsall & Niles, 1804.

King-Hele, Desmond. *Erasmus Darwin: A Life of Unequalled Achievement*. London: Giles de la Mare, 1999.

——. ed. *Charles Darwin's Life of Erasmus Darwin*. Cambridge: Cambridge University Press, 2002.

Uglow, Jenny. *The Lunar Men: Five Friends Whose Curiosity Changed the World*. New York: Farrar, Straus & Giroux, 2002.

DARWIN COLLEGE

Keynes, Margaret. *A House by the River*. Published privately, 1984.

DARWIN CORRESPONDENCE PROJECT

Burkhardt, Frederick, ed. *Origins: Selected Letters of Charles Darwin, 1822–1859*. Cambridge: Cambridge University Press, 2008.

Burkhardt, Frederick, Alison Pearn, and Samantha Evans, eds. *Evolution: Selected Letters of Charles Darwin, 1860–1870*. Cambridge: Cambridge University Press, 2008.

DARWIN FISH

Caiazza, John. *The War of the Jesus and Darwin Fishes: Religion and Science in the Postmodern World*. New Brunswick, N.J.: Transaction, 2007.

Gibson, Sheila. "Trawling for the Truth behind the Bumper Battle Known as the 'Fish Wars.'" *Skeptic* 8, no. 3 (2000): 7–9.

DARWIN STATE MUSEUM OF NATURAL HISTORY (MOSCOW)

Ladygina-Kohts, N.N. *Infant Chimpanzee and Human Child: A Classic 1935 Comparative Study of Ape Emotions and Intelligence*. Edited by Frans B.M. de Waal, translated by Boris Vekker. New York: Oxford University Press, 2004.

DARWIN RESEARCH STATION

Watt, Nigel. *In the Steps of Darwin: Galápagos*. Hong Kong: Dan Arts, 1998.

DARWINIAN MEDICINE

Nesse, Randolph M., and George C. Williams. *Evolution and Healing*. London: Weidenfeld & Nicolson, 1994.

——. *Why We Get Sick: The New Science of Darwinian Medicine*. New York: Times Books, 1984.

DARWINISM

Kohn, David, ed. *The Darwinian Heritage*. Princeton: Princeton University Press, 1985.

Rachels, James. *Created from Animals: The Moral Implications of Darwinism*. Oxford: Oxford University Press, 1990.

Wallace, Alfred Russel. *Darwinism: An Exposition of the Theory of Natural Selection, with Some of Its Applications*. London: Macmillan, 1889.

DARWINISM, NATIONAL DIFFERENCES IN

Clark, Constance Areson. *God—or Gorilla: Images of Evolution in the Jazz Age*. Baltimore: Johns Hopkins University Press, 2008.

Glick, T.F., ed. *The Comparative Reception of Darwinism*. Austin: University of Texas Press, 1974.

Kohn, Marek. *A Reason for Everything: Natural Selection and the English Imagination*. London: Faber & Faber, 2004.

"DARWIN'S BULLDOG"

Hellman, Hal. *Great Feuds in Science: Ten of the Liveliest Disputes Ever*. New York: John Wiley, 1998.

Huxley, Thomas Henry. *Autobiography and Selected Essays*. Edited by Ada L.F. Snell. New York: Houghton Mifflin, 1909.

DARWIN'S FINCHES

Grant, Peter R., and B. Rosemary Grant. *Why Species Multiply: The Radiation of Darwin's Finches*. Princeton: Princeton University Press, 2007.

Lack, David. *Darwin's Finches: An Essay on the General Biological Theory of Evolution*. Cambridge: Cambridge University Press, 1947.

Podos, Jeffrey, and Stephen Nowicki. "Beaks: Adaptation and Vocal Evolution in Darwin's Finches." *BioScience* 54, no. 6 (2004): 501–510.

Sulloway, Frank J. "Darwin and His Finches: The Evolution of a Legend." *Journal of the History of Biology* 15 (1982): 1–53.

Weiner, Jonathan. *The Beak of the Finch: A Story of Evolution in Our Time*. New York: Alfred A. Knopf, 1994.

"DARWIN'S LIZARDS"

Losos, Jonathan B., and Kevin de Queiroz. "Darwin's Lizards—Leaping Lizards, Sprinting Lizards, Lounging Lizards: The Anoles of the Greater Antilles Adapted Quickly to Life in the Trees." *Natural History* 106, no. 12 (1997): 34.

DAVID GRAYBEARD (CHIMPANZEE)

Goodall, Jane, and Hugo van Lawick. *In the Shadow of Man*. Boston: Houghton Mifflin, 2000.

DAWKINS, RICHARD

Dawkins, Richard. *The Blind Watchmaker*. New York: Norton, 1986.

——. *The God Delusion*. Boston: Mariner Books, 2008.

——. *The Selfish Gene*. Oxford: Oxford University Press, 1976; 30th anniversary ed., 2006.

Grafen, Alan, and Mark Ridley, eds. *Richard Dawkins: How a Scientist Changed the Way We Think—Reflections by Scientists, Writers, and Philosophers*. Oxford: Oxford University Press, 2006.

Sterelny, Kim. *Dawkins vs. Gould: Survival of the Fittest*. New ed. [New York]: Totem Books, 2007.

DEAN, BASHFORD

Dean, Bashford. *Fishes, Living and Fossil: An Outline of Their Forms and Probable Relationships*. New York: Macmillan, 1895.

——. *Helmets and Body Armor in Modern Warfare*. New Haven, Conn.: Yale University Press, 1920.

Gregory, W. K. "Memorial of Bashford Dean." In *Bashford Dean Memorial Volume: Archaic Fishes*, ed. E. W. Gudger. New York: American Museum of Natural History, 1930.

DECEPTION, EVOLUTION OF

Byrne, Richard, and Andrew Whiten, eds. *Machiavellian Intelligence: Social Expertise and the Evolution of Intellect in Monkeys, Apes, and Humans*. Oxford: Oxford University Press, 1988.

Dugatkin, Lee. *Cheating Monkeys and Citizen Bees: The Nature of Cooperation in Animals and Humans*. New York: Free Press, 1999.

DEGENERATION THEORY

Lankester, E. Ray. *Degeneration: A Chapter in Darwinism*. London: Macmillan, 1880.

"DELICATE ARRANGEMENT"

Bedall, Barbara. "Wallace, Darwin, and the Theory of Natural Selection." *Journal of the History of Biology* 1 (1968): 261–323.

Brackman, Arnold C. *A Delicate Arrangement: The Strange Case of Charles Darwin and Alfred Russel Wallace*. New York: Times Books, 1980.

Shermer, Michael. *In Darwin's Shadow: The Life and Science of Alfred Russel Wallace*. New York: Oxford University Press, 2002.

Wallace, Alfred Russel. *The Wonderful Century: Its Successes and Its Failures*. New York: Dodd, Mead, 1898.

DE MAILLET, BENOÎT

Maillet, Benoît de. *Telliamed: Or, Conversations between an Indian Philosopher and a French Missionary on the Diminution of the Sea, the Formation of the Earth, the Origin of Men and Animals, etc.* (1750). Translated and edited by Albert V. Carozzi. Urbana: University of Illinois Press, 1968.

DESCENT OF MAN, THE

Campbell, Bernard, ed. *Sexual Selection and the Descent of Man, 1871–1971*. Chicago: Aldine, 1972.

Darwin, Charles. *The Descent of Man, and Selection in Relation to Sex* (1871). London: Penguin Books, 2004.

Desmond, Adrian, and James Moore. *Darwin's Sacred Cause: How a Hatred of Slavery Shaped Darwin's Views on Human Evolution*. Boston: Houghton Mifflin Harcourt, 2009.

DE VRIES, HUGO

Dunn, L. C. *A Short History of Genetics*. New York: McGraw-Hill, 1965.

Mayr, Ernst. *The Growth of Biological Thought: Diversity, Evolution, and Inheritance*. Cambridge, Mass.: Harvard University Press, Belknap Press, 1982.

Meijer, Onno G. "Hugo de Vries No Mendelian?" *Annals of Science* 42 (1985): 189–232.

Olby, Robert C. *Origins of Mendelism*. 2d ed. Chicago: University of Chicago Press, 1985.

Provine, William B. *The Origins of Theoretical Population Genetics*. Chicago: University of Chicago Press, 1971.

DEWEY, JOHN

Appleman, Philip, ed. *Darwin: A Norton Critical Edition*. 3d ed. New York: W. W. Norton, 2001.

Dewey, John. *The Influence of Darwin on Philosophy and Other Essays on Contemporary Thought*. New York: William Holt, 1910.

Westbrook, Robert B. *John Dewey and American Democracy*. Ithaca, N.Y.: Cornell University Press, 1991.

DIGIT

Fossey, Dian. *Gorillas in the Mist*. Boston: Houghton Mifflin, 1983.

DIMA

Guthrie, R. Dale. *Frozen Fauna of the Mammoth Steppe: The Story of Blue Babe*. Chicago: University of Chicago Press, 1990.

Stone, Richard. *Mammoth: The Resurrection of an Ice Age Giant*. Cambridge, Mass.: Perseus, 2001.

DINOSAURS

Colbert, Edwin. *Dinosaurs: An Illustrated History*. Maplewood, N.J.: Hammond, 1983.

Fastovsky, David E., and David B. Weishampel. *Dinosaurs: A Concise Natural History*. New York: Cambridge University Press, 2009.

——. *The Evolution and Extinction of the Dinosaurs*. 2d ed. New York: Cambridge University Press, 2005.

Glut, Donald. *The New Dinosaur Dictionary*. Secaucus, N.J.: Citadel Press, 1982.

Horner, Jack, and James Gorman. *Digging Dinosaurs*. New York: Workman, 1988.

Long, John A. *Dinosaurs in Australia and New Zealand and Other Animals of the Mesozoic Era*. Cambridge, Mass.: Harvard University Press, 1998.

Norell, Mark, Eugene Gaffney, and Lowell Dingus. *Discovering Dinosaurs in the American Museum of Natural History*. New York: Alfred A. Knopf, 1995.

Norman, David. *The Illustrated Encyclopedia of Dinosaurs*. New York: Crescent Books, 1985.

Paul, Gregory. *Predatory Dinosaurs of the World*. New York: Simon & Schuster, 1988.

Wallace, Joseph. *The Rise and Fall of the Dinosaurs*. New York: W. N. Smith, 1987.

Weishampel, David B., Peter Dodson, and Halszka Osmólska, eds. *The Dinosauria*. 2d ed. Berkeley: University of California Press, 2004.

DINOSAURS, EXTINCTION OF

Desmond, Adrian. *The Hot-Blooded Dinosaurs: A Revolution in Palaeontology*. London: Blond and Briggs, 1975.

Hsu, Kenneth J. *The Great Dying: Cosmic Catastrophe, Dinosaurs, and the Theory of Evolution*. New York: Ballantine/Random House, 1986.

Powell, James. *Night Comes to the Cretaceous: Dinosaur Extinction and the Transformation of Modern Geology*. New York: W. H. Freeman, 1998.

Wilford, John Noble. *The Riddle of the Dinosaur*. New York: Random House, 1985.

DINOSAURS, FEATHERED

Currie, Philip J., Eva B. Koppelhus, Martin A. Shugar, and Joanna L. Wright, eds. *Feathered Dragons: Studies on the Transition from Dinosaurs to Birds*. Bloomington: Indiana University Press, 2004.

Dingus, Lowell, and Timothy Rowe. *The Mistaken Extinction: Dinosaur Extinction and the Origin of Birds*. New York: W. H. Freeman, 1997.

Norrell, Mark A., and Mick Ellison. *Unearthing the Dragon: The Great Feathered Dinosaur Discovery*. New York: PI Press, 2005.

DIVERGENCE, PRINCIPLE OF

Bowler, Peter J. *Charles Darwin: The Man and His Influence*. Oxford: Blackwell, 1990.

Kohn, David. "Darwin's Keystone: The Principle of Divergence." In *The Cambridge Companion to the "Origin of Species,"* edited by Michael Ruse and Robert J. Richards, pp. 87–108. Cambridge: Cambridge University Press, 2009.

DIVINE BENEFICENCE, IDEA OF

Darwin, Charles. *The Various Contrivances by Which Orchids Are Fertilized by Insects*. London: W. Pickering, 1988.

Paley, William, Matthew Eddy, and David M. Knight. *Natural Theology; or, Evidence of the Existence and Attributes of the Deity, Collected from the Appearances of Nature*. 1802; repr. Oxford: Oxford University Press, 2008.

Ray, John. *The Wisdom of God Manifested in the Works of the Creation* (1691). Anglistica & Americana, 122. Hildesheim: Georg Olms, 1974.

DMANISI HOMINIDS

Agustí, Jordi, and David Lordkipanidze. *Del Turkana al Cáucaso: La evolución de los primeros pobladores de Europa*. Barcelona: RBA Libros, 2005.

Rosas, A., and J.M. Bermúdez de Castro. "On the Taxonomic Affinities of the Dmanisi Mandible (Georgia)." *American Journal of Physical Anthropology* 107 (1998): 145–62.

DNA, DISCOVERY OF

Carroll, Sean B. *The Making of the Fittest: DNA and the Ultimate Forensic Record of Evolution*. New York: W. W. Norton, 2005.

Judson, H. F. *The Eighth Day of Creation: The Makers of the Revolution in Biology*. London: Jonathan Cape, 1979.

Nelkin, Dorothy, and Susan Lindee. *The DNA Mystique: The Gene as a Cultural Icon*. New York: W. H. Freeman, 2000.

Watson, James. *The Double Helix: A Personal Account of the Discovery of the Structure of DNA*. London: Weidenfeld, 1981.

DNA IDENTIFICATION

Billings, Paul R. *DNA on Trial: Genetic Identification and Criminal Justice*. Plainview, N.Y.: Cold Spring Harbor Laboratory Press, 1992.

Genge, Ngaire. *The Forensic Casebook: The Science of Crime Scene Investigation*. New York: Ballantine Books, 2002.

Lazer, David. *DNA and the Criminal Justice System: The Technology of Justice*. Cambridge, Mass.: MIT Press, 2004.

DOBZHANSKY, THEODOSIUS

Dobzhansky, Theodosius. *Genetics and the Origin of Species*. New

York: Columbia University Press, 1937.

———. *Mankind Evolving.* New Haven, Conn.: Yale University Press, 1962.

Dobzhansky, T., F.J. Ayala, G.L. Stebbins, and J.W. Valentine. *Evolution.* San Francisco: W.H. Freeman, 1977.

DODO

Fuller, Errol. *The Dodo: From Extinction to Icon.* London: Collins, 2002.

Quammen, David. *The Song of the Dodo: Island Biogeography in an Age of Extinctions.* New York: Scribner, 1996.

DOLLY

Kolata, Gina Bari. *Clone: The Road to Dolly and the Path Ahead.* Ringwood, Vic.: Penguin Books, 1998.

Wilmut, Ian, and Roger Highfield. *After Dolly: The Uses and Misuses of Human Cloning.* New York: W.W. Norton, 2006.

Wilmut, Ian, Keith Campbell, and Colin Tudge. *The Second Creation: Dolly and the Age of Biological Control.* New York: Farrar, Straus & Giroux, 2000.

DOWN HOUSE

Atkins, Sir Hedley. *Down: Home of the Darwins.* London: Phillmore, 1976.

Milner, Richard. "Keeping Up Down House." *Natural History* 105, no. 8 (1996): 54–58.

Morris, Solene, and Louise Wilson. *Down House: The Home of Charles Darwin.* [London]: English Heritage, 2000.

Raverat, Gwen. *Period Piece: A Cambridge Childhood.* New York: W.W. Norton, 1953.

DOYLE, SIR ARTHUR CONAN

Booth, Martin. *The Doctor and the Detective: A Biography of Sir Arthur Conan Doyle.* New York: St. Martin's Minotaur, 2000.

Doyle, Arthur Conan. *The Lost World.* New York: Looking Glass Library, 1959.

Lycett, Andrew. *The Man Who Created Sherlock Holmes: The Life and Times of Sir Arthur Conan Doyle.* London: Free Press, 2007.

Stashower, Daniel. *Teller of Tales: The Life of Arthur Conan Doyle.* New York: Henry Holt, 1999.

DRAGONS

Mayor, Adrienne. *The First Fossil Hunters: Paleontology in Greek and Roman Times.* Princeton: Princeton University Press, 2001.

———. *Fossil Legends of the First Americans.* Princeton: Princeton University Press, 2005.

Norrell, Mark A., and Mick Ellison. *Unearthing the Dragon: The Great Feathered Dinosaur Discovery.* New York: PI Press, 2005.

Van Oosterzee, Penny. *Dragon Bones: The Story of Peking Man.* Cambridge, Mass.: Perseus, 2000.

Watson, Lyall. *The Dreams of Dragons: Riddles of Natural History.* New York: William Morrow, 1987.

DRYOPITHECINES

Walker, Alan, and Pat Shipman. *The Ape in the Tree: An Intellectual and Natural History of Proconsul.* Cambridge, Mass.: Harvard University Press, 2005.

DUBOIS, EUGÈNE

Shipman, Pat. *The Man Who Found the Missing Link: Eugène Dubois and His Lifelong Quest to Prove Darwin Right.* New York: Simon & Schuster, 2001.

EARTHWORMS AND THE FORMATION OF VEGETABLE MOLD

Darwin, Charles. *On the Formation of Vegetable Mould, through the Action of Worms.* Introduction by Stephen Jay Gould. 1881; repr. Chicago: University of Chicago Press, 1985.

ECOLOGY

Krebs, John R., and N.B. Davies. *An Introduction to Behavioral Ecology.* Oxford: Blackwell Scientific Publications, 1987.

May, Robert M. *Theoretical Ecology.* Philadelphia: W.B. Saunders, 1976.

Odum, E.P. *Ecology.* New York: Holt, Rinehart & Winston, 1963.

EDIACARAN FAUNA

Fortey, A.R., and R.H. Thomas, eds. *Arthropod Relationships.* London: Chapman and Hall, 1997.

Gould, Stephen Jay. "Of Embryos and Ancestors." In *The Lying Stones of Marrakech: Penultimate Reflections in Natural History,* pp. 317–332. New York: Harmony Books, 2000.

McMenamin, Mark A.S., and Dianna L. Schulte McMenamin. *The Emergence of Animals: The Cambrian Breakthrough.* New York: Columbia University Press, 1990.

Schopf, J. William. *Cradle of Life: The Discovery of Earth's Earliest Fossils.* Princeton: Princeton University Press, 1999.

ENDOSYMBIOSIS

Margulis, Lynn. *Symbiotic Planet: A New Look at Evolution.* New York: Basic Books, 1998.

Overmann, Jörg, ed. *Molecular Basis of Symbiosis.* Berlin: Springer, 2006.

ENGLAND, DARWINISM IN

Kohn, Marek. *A Reason for Everything: Natural Selection and the English Imagination.* New York: Faber and Faber, 2005.

Mitchell, Sally. *Daily Life in Victorian England.* Westport, Conn.: Greenwood Press, 1996.

Young, Robert M. *Darwin's Metaphor: Nature's Place in Victorian Culture.* Cambridge: Cambridge University Press, 1985.

ESSENTIALISM

Desmond, Adrian. *Archetypes and Ancestors: Paleontology in Victorian London, 1850–1875.* Chicago: University of Chicago Press, 1982.

Mayr, Ernst. *Evolution and the Diversity of Life: Selected Essays.* Cambridge, Mass.: Belknap Press, 1982.

———. *The Growth of Biological Thought: Diversity, Evolution, and Inheritance.* Cambridge, Mass.: Belknap Press, 1982.

ETHOLOGY

Alcock, John. *The Triumph of Sociobiology.* New York: Oxford University Press, 2001.

Barash, David P., and Judith Eve Lipton. *The Myth of Monogamy: Fidelity and Infidelity in Animals and People.* New York: W.H. Freeman, 2001.

Burkhardt, Richard W. *Patterns of Behavior: Konrad Lorenz, Niko Tinbergen, and the Founding of Ethology.* Chicago: University of Chicago Press, 2005.

Caplan, Arthur, ed. *The Sociobiology Debate: Readings on Ethical and Scientific Issues.* New York: Harper & Row, 1978.

Lorenz, Konrad. *Evolution and Modification of Behavior.* Chicago: University of Chicago Press, 1965.

Segerstråle, Ullica. *Defenders of the Truth: The Sociobiology Debate.* New York: Oxford University Press, 2001.

Thorpe, W. *The Origins and Rise of Ethology.* London: Heinemann, 1979.

Tinbergen, Niko. *The Study of Instinct.* Oxford: Clarendon Press, 1969.

Wilson, Edward O. *Sociobiology: The New Synthesis.* Cambridge, Mass.: Harvard University Press, 2000.

EUGENICS

Galton, Sir Francis. *Hereditary Genius.* London: Macmillan, 1869.

Jones, Greta. *Social Darwinism and English Thought.* Atlantic Highlands, N.J.: Humanities Press, 1980.

Kevles, Daniel. *In the Name of Eugenics: Genetics and the Uses of Human Heredity.* Berkeley: University of California Press, 1985.

Moore, James R. "Good Breeding." *Natural History* 114, no. 9 (2005): 45–46.

EUKARYOTES

Knoll, Andrew H. *Life on a Young Planet: The First Three Billion Years of Evolution on Earth.* Princeton: Princeton University Press, 2003.

Williams, R.J.P., and J.J.R. Fraústo da Silva. *The Chemistry of Evolution: The Development of Our Ecosystem.* Amsterdam: Elsevier, 2006.

EVO-DEVO

Carroll, Sean B. *Endless Forms Most Beautiful: The New Science of Evo Devo and the Making of the Animal Kingdom.* New York: W.W. Norton, 2005.

Laubichler, Manfred D., and Jane Maienschein. *From Embryology to Evo-Devo: A History of Developmental Evolution.* Cambridge, Mass.: MIT Press, 2007.

EVOLUTION

Bowler, Peter J. *Evolution: The History of an Idea.* 3d rev. ed. Berkeley: University of California Press, 2003.

Dawkins, Richard. *The Blind Watchmaker: Why the Evidence of Evolution Reveals a Universe without Design.* New York: W.W. Norton, 1996.

Futuyma, Douglas. *Science on Trial: The Case for Evolution.* New York: Pantheon Books, 1983.

———. *Evolutionary Biology.* Sunderland, Mass.: Sinauer Associates, 1998.

Gould, Stephen Jay. *The Panda's Thumb: More Reflections on Natural History.* New York: W.W. Norton, 1980.

———. *The Structure of Evolutionary Theory.* Cambridge, Mass.: Harvard University Press, Belknap Press, 2002.

Larson, Edward J. *Evolution: The Remarkable History of a Scientific Theory.* New York: Modern Library, 2004.

Mayr, Ernst. *What Evolution Is.* New York: Basic Books, 2001.

Milner, Richard, ed. "Darwin and Evolution." Special report. *Natural History* 114, no. 9 (2005): 35–68.

Morris, Richard. *The Evolutionists: The Struggle for Darwin's Soul.* New York: Holt, 2002.

Prothero, Donald. *Evolution: What the Fossils Say and Why It Matters.* New York: Columbia University Press, 2007.

Reader, John. *The Rise of Life.* New York: Alfred A. Knopf, 1988.

Ridley, Mark. *Evolution.* Oxford: Oxford University Press, 1997.

Zimmer, Carl, and Richard Hutton. *Evolution: The Triumph of an Idea.* New York: Harper Perennial, 2002.

EVOLUTIONARY PROGRAMMING
De Jong, Kenneth A. *Evolutionary Computation: A Unified Approach.* Cambridge, Mass.: MIT Press, 2006.

EXAPTATION
Gould, Stephen Jay, and Elisabeth S. Vrba. "Exaptation: A Missing Term in the Science of Form." *Paleobiology* 8, no. 1 (1982): 4–15.

EXPRESSION OF THE EMOTIONS
Darwin, Charles. *The Expression of the Emotions in Man and Animals* (1872). Edited by Paul Ekman. Oxford: Oxford University Press, 1998.
Ekman, Paul, ed. *Darwin and Facial Expression: A Century of Research in Review.* New York: Academic Press, 1973.
———. *Emotions Inside Out: 130 Years after Darwin's "The Expression of the Emotions in Man and Animals."* New York, N.Y.: New York Academy of Sciences, 2003.

EXTINCTION
Clark, Tim. *Averting Extinction: Reconstructing Endangered Species Recovery.* New Haven, Conn.: Yale University Press, 1997.
Ehrlich, Paul, and Anne Ehrlich. *Extinction: The Causes and Consequences of the Disappearance of Species.* New York: Random House, 1981.
Stanley, Steven M. *Extinction.* New York: Scientific American Books, 1987.
Stearns, Beverly Peterson, and Stephen C. Stearns. *Watching, from the Edge of Extinction.* New Haven, Conn.: Yale University Press, 1999.

FANTASIA
Rossbach, Thomas Joseph. "*Fantasia* and Our Changing Views of Dinosaurs." *Journal of Geological Education* 44, no. 1 (1996): 13.
Schickel, Richard. *The Disney Version: The Life, Times, Art, and Commerce of Walt Disney.* 3d ed. Chicago: Ivan R. Dee, 2007.

FITNESS
Huxley, Julian. *Evolution: The Modern Synthesis.* London: George Allen & Unwin, 1942.
Wallace, Alfred Russel. *Darwinism: An Exposition of the Theory of Natural Selection with Some of Its Applications.* London: Macmillan, 1889.

FITZROY, CAPTAIN ROBERT
Gribbin, John, and Mary Gribbin. *FitzRoy: The Remarkable Story of Darwin's Captain and the Invention of the Weather Forecast.* New Haven, Conn.: Yale University Press, 2004.

Mellersh, H. E. L. *FitzRoy of the "Beagle."* London: Mason & Lipscomb, 1968.
Moorehead, Alan. *Darwin and the "Beagle."* New York: Harper & Row, 1969.
Nichols, Peter. *Evolution's Captain: The Dark Fate of the Man Who Sailed Charles Darwin around the World.* New York: HarperCollins, 2003.

FLAT-EARTHERS
Michell, John F. *Eccentric Lives, Peculiar Notions: True Tales of Flat-Earthers, Head Drillers, Ufologists, Frantic Lovers, Welsh Druids, Finders of Lost Tribes, and Other Obsessed Individuals.* New York: Black Dog & Leventhal, 2003.
Rowbotham, Samuel Birley. *Zetetic Astronomy: Earth Not a Globe—An Experimental Inquiry into the True Figure of the Earth.* London: J. B. Day, 1873.

FLINT JACK
Richardson, Michael, and Tom Stamp. *Flint Jack.* Whitby, Eng.: Caedmon of Whitby, 1985.
Stephens, Joseph. *Flint Jack: A Short History of a Notorious Forger of Antiquities.* Reading: Printed at the Blagrave Street Steam Printing Works, 1894.

FLORES MAN
Morwood, Mike, and Penny Van Oosterzee. *A New Human: The Startling Discovery and Strange Story of the "Hobbits" of Flores, Indonesia.* New York: HarperCollins, 2007.

FLY ROOM
Allen, Garland E. *Thomas Hunt Morgan: The Man and His Science.* Princeton: Princeton University Press, 1978.
Schwartz, James. *In Pursuit of the Gene: From Darwin to DNA.* Cambridge, Mass.: Harvard University Press, 2008.

FOSSEY, DIAN
Fossey, Dian. *Gorillas in the Mist.* Boston: Houghton Mifflin, 1983.
Hayes, Harold. *The Dark Romance of Dian Fossey.* New York: Simon & Schuster, 1990.
Mowat, Farley. *Woman in the Mists: The Story of Dian Fossey and the Mountain Gorillas of Africa.* New York: Warner Books, 1987.

FOSSIL HUMANS
Delson, Eric, ed. *Ancestors: The Hard Evidence.* New York: Alan R. Liss, 1985.
Johanson, Donald, and Maitland Edey. *Lucy: The Beginnings of Humankind.* New York: Simon & Schuster, 1981.

Johanson, Donald, and Blake Edgar. *From Lucy to Language.* Rev. ed. New York: Simon & Schuster, 2006.
Lewin, Roger. *Bones of Contention: Controversies in the Search for Human Origins.* New York: Simon & Schuster, 1987.
Reader, John. *Missing Links: The Hunt for Early Man.* London: Collins, 1981.
Sawyer, G. J., and Viktor Deak, eds. *The Last Human: A Guide to Twenty-two Species of Extinct Humans.* Text by Esteban Sarmiento, G. J. Sawyer, and Richard Milner. New Haven, Conn.: Yale University Press, 2007.
Stringer, Chris, and Peter Andrews. *The Complete World of Human Evolution.* London: Thames & Hudson, 2005.
Tattersall, Ian. *The Fossil Trail: How We Know What We Think We Know about Human Evolution.* 2d ed. New York: Oxford University Press, 2008.
———. *Becoming Human: Evolution and Human Uniqueness.* New York: Harcourt Brace, 1998.
Tattersall, Ian, and Jeffrey Schwartz. *Extinct Humans.* New York: Westview/Nevraumont, 2000.
Walker, Alan, and Pat Shipman. *The Wisdom of the Bones: In Search of Human Origins.* New York: Alfred A. Knopf, 1996.

FOSSILS
Delisle, Richard G. *Debating Humankind's Place in Nature, 1860–2000: The Nature of Paleoanthropology.* Upper Saddle River, N.J.: Pearson Prentice Hall, 2007.
Thomson, Keith Stewart. *Fossils: A Very Short Introduction.* Very Short Introductions, 138. Oxford: Oxford University Press, 2005.

FOUR THOUSAND AND FOUR B.C.
Lewis, Cherry, and Simon J. Knell. *The Age of the Earth from 4004 B.C. to A.D. 2002.* Geological Society special publication, no. 190. London: Geological Society, 2001.

FOX-FARM EXPERIMENTS
Budiansky, Stephen. *The Covenant of the Wild: Why Animals Chose Domestication.* New Haven, Conn.: Yale University Press, 1999.
Truit, Lyudmjila. "Early Canid Domestication: The Farm-Fox Experiment." *American Scientist* 87, no. 2 (1999): 160–169.

FRANKENFOOD
Charles, Daniel. *Lords of the Harvest: Biotech, Big Money, and the Future of Food.* New York: Perseus, 2001.
Lambrecht, Bill. *Dinner at the New Gene Cafe: How Genetic Engineering Is Changing What We Eat, How

We Live, and the Global Politics of Food.* New York: St. Martin's Press, 2001.
Miller, Henry, and Gregory Conko. *The Frankenfood Myth: How Protest and Politics Threaten the Biotech Revolution.* New York: Praeger, 2004.
Pence, Gregory E. *Designer Food: Mutant Harvest or Breadbasket for the World?* Lanham, Md.: Rowman & Littlefield, 2002.

FRANKLIN, BENJAMIN
Chaplin, Joyce E. *The First Scientific American: Benjamin Franklin and the Pursuit of Genius.* New York: Basic Books, 2006.
Franklin, Benjamin. *Poor Richard's Almanac* (1758). New York: Ballantine Books, 1977.
Malthus, Thomas Robert. *An Essay on the Principle of Population.* Edited by Antony Flew. 1798; repr. Harmondsworth, Eng.: Penguin Books, 1982.

FRANKLIN, ROSALIND
Anderson, Lara. *Rosalind Franklin.* Chicago, Ill.: Raintree, 2008.
Maddox, Brenda. *Rosalind Franklin: The Dark Lady of DNA.* New York: HarperCollins, 2002.
Sayre, Anne. *Rosalind Franklin and DNA.* New York: W. W. Norton, 2000.

FREEMAN, RICHARD BROKE
Freeman, Richard. *Charles Darwin: A Companion.* Folkestone, Eng.: William Dawson/Hamden, Conn.: Archon Books, 1978.
———. *The Works of Charles Darwin: An Annotated Bibliographical Handlist.* 2d ed. Folkestone, Eng.: William Dawson/Hamden, Conn.: Archon Books, 1977.

FREUD, SIGMUND
Gould, Stephen Jay. "Freud's Evolutionary Fantasy." In *The Richness of Life: The Essential Stephen Jay Gould,* edited by Steven Rose, pp. 473–485. New York: W. W. Norton, 2007.
Sulloway, Frank. *Freud, Biologist of the Mind: Beyond the Psychoanalytic Legend.* New York: Basic Books, 1979.

FULLER, ERROL
Fuller, Errol. *Dodo: From Extinction to Icon.* London: HarperCollins, 2002.
———. *The Great Auk.* London: Errol Fuller, 1999.
———. *The Lost Birds of Paradise.* Shrewsbury, Eng.: Swan Hill Press, 1995.
Milner, Richard. "Remembrance of Auks Past: A Naturalist Pursues His Obsession with Vanished Species of Birds." *Natural History* 108, no. 7 (1999): 88.

FUNDAMENTALISM

Brasher, Brenda E. *Encyclopedia of Fundamentalism*. New York: Routledge, 2001.

Marsden, George M. *Fundamentalism and American Culture: The Shaping of Twentieth-Century Evangelicalism, 1870–1925*. New York: Oxford University Press, 2006.

Moore, James R. *The Post-Darwinian Controversies: A Study of the Protestant Struggle to Come to Terms with Darwin in Great Britain and America, 1870–1900*. New York: Cambridge University Press, 1979.

GAIA HYPOTHESIS

Lovelock, J.E. *The Ages of Gaia: A Biography of Our Living Earth*. New York: W.W. Norton, 1988.

——. *Gaia: A New Look at Life on Earth*. Oxford: Oxford University Press, 1979.

——. *The Revenge of Gaia: Earth's Climate in Crisis and the Fate of Humanity*. New York: Basic Books, 2006.

GALÁPAGOS ARCHIPELAGO

De Roy, Tui. *Galápagos: Islands Born of Fire*. Toronto: Warwick, 1998.

Larson, Edward J. *Evolution's Workshop: God and Science on the Galápagos Islands*. New York: Basic Books, 2002.

Steadman, David, and Steven Zousmer. *Galápagos: Discovery on Darwin's Islands*. Washington, D.C.: Smithsonian, 1988.

Woram, John. *Charles Darwin Slept Here: Tales of Human History at World's End*. Rockville Centre, N.Y.: Rockville Press, 2005.

GALTON, SIR FRANCIS

Forrest, Derek William. *Francis Galton: The Life and Work of a Victorian Genius*. New York: Taplinger, 1974.

Gillham, Nicholas W. *A Life of Sir Francis Galton: From African Exploration to the Birth of Eugenics*. New York: Oxford University Press, 2001.

GALTON'S POLYHEDRON

Gould, Stephen Jay. "A Dog's Life in Galton's Polyhedron." In *Eight Little Piggies: Reflections in Natural History*. New York: W.W. Norton, 1993.

GARNER, RICHARD LYNCH

Garner, R.L. *Apes and Monkeys: Their Life and Language*. Boston and London: Ginn, 1900.

——. *Gorillas & Chimpanzees*. London: Osgood, McIlvaine, 1896.

GENETIC DRIFT

Allendorf, Frederick William, and Gordon Luikart. *Conservation and the Genetics of Populations*. Malden, Mass.: Blackwell, 2007.

Depew, David J., and Bruce H. Weber. *Darwinism Evolving: Systems Dynamics and the Genealogy of Natural Selection*. Cambridge, Mass.: MIT Press, 1995.

GENETIC ENGINEERING

Avise, John C. *The Hope, Hype, and Reality of Genetic Engineering: Remarkable Stories from Agriculture, Industry, Medicine, and the Environment*. Oxford: Oxford University Press, 2004.

Yoxen, Edward. *The Gene Business: Who Should Control Biotechnology?* London: Pan Books, 1983.

——. *Unnatural Selection? Coming to Terms with the New Genetics*. London: Heinemann, 1986.

GENETIC VARIABILITY

Ausubel, Kenny. *Seeds of Change: The Living Treasure*. San Francisco: Harper, 1994.

Schwartz, Anne. "Banking on Seeds to Avert Extinction." *Audubon* 90, no. 1 (1988): 22–27.

Shell, Ellen Ruppel. "Seeds in the Bank Could Stave Off Disaster on the Farm." *Smithsonian* 20, no. 10 (1990): 95–105.

GENOGRAPHIC PROJECT

Wells, Spencer. *Deep Ancestry: Inside the Genographic Project*. Washington, D.C.: National Geographic, 2006.

——. *The Genographic Project*. Alexandria, Va.: National Geographic, 2005.

GENOME

Davies, Kevin. *Cracking the Genome: Inside the Race to Unlock Human DNA*. New York: Free Press, 2001.

Lewontin, Richard. *It Ain't Necessarily So: The Dream of the Human Genome and Other Illusions*. New York: New York Review of Books, 2000.

——. *The Triple Helix: Gene, Organism, and Environment*. Cambridge, Mass.: Harvard University Press, 2000.

Watson, James D., and Andrew Berry. *DNA: The Secret of Life*. New York: Alfred A. Knopf, 2003.

GERTIE THE DINOSAUR

Mitchell, W.J.T. *The Last Dinosaur Book: The Life and Times of a Cultural Icon*. Chicago: University of Chicago Press, 1998.

Sanz, José Luis. *Starring T. Rex! Dinosaur Mythology and Popular Culture*. Bloomington: Indiana University Press, 2002.

"GHOST" SPECIES

Barlow, Connie. *Ghosts of Evolution: Nonsensical Fruit, Missing Partners, and Other Ecological Anachronisms*. Foreword by Paul S. Martin. New York: Basic Books, 2000.

Janzen, Daniel H., and Paul S. Martin. "Neotropical Anachronisms: The Fruits the Gomphotheres Ate." *Science* 215 (1982): 19–27.

Martin, Paul S., and David Burney. "Bring Back the Elephants!" *Wild Earth* 9, no. 1 (1999): 57–64.

GOBI DESERT EXPEDITIONS

Andrews, Roy Chapman. *On the Track of Ancient Man*. New York: Putnam, 1926.

Man, John. *Gobi: Tracking the Desert*. New Haven: Yale University Press, 1999.

Preston, Douglas. *Dinosaurs in the Attic: An Excursion into the American Museum of Natural History*. New York: St. Martin's Press, 1986.

GONDWANALAND

Macdougall, J.D. *A Short History of Planet Earth: Mountains, Mammals, Fire, and Ice*. New York: John Wiley, 1996.

Rich, Pat Vickers, and Thomas H.V. Rich. *Wildlife of Gondwana: Dinosaurs and Other Vertebrates from the Ancient Supercontinent*. Bloomington: Indiana University Press, 1999.

GORILLAS

Du Chaillu, Paul. *Adventures in Equatorial Africa*. New York: Harper, 1871.

Fossey, Dian. *Gorillas in the Mist*. Boston: Houghton Mifflin, 1983.

Schaller, George B. *Year of the Gorilla*. Chicago: University of Chicago Press, 1964.

Taylor, Andrea B., and Michele L. Goldsmith, eds. *Gorilla Biology: A Multidisciplinary Perspective*. Cambridge Studies in Biological and Evolutionary Anthropology, 34. Cambridge: Cambridge University Press, 2003.

GOULD, STEPHEN JAY

Gould, Stephen Jay. *The Mismeasure of Man*. New York: W.W. Norton, 1981.

——. *Ontogeny and Phylogeny*. Cambridge, Mass.: Harvard University Press, 1977.

——. *The Panda's Thumb: More Reflections on Natural History*. New York: W.W. Norton, 1980.

——. *The Richness of Life: The Essential Stephen Jay Gould*. Edited by Paul McGarr and Steven Rose. New York: W.W. Norton, 2007.

——. *The Structure of Evolutionary Theory*. Cambridge, Mass.: Belknap Press, 2002.

——. *An Urchin in the Storm: Essays about Books and Ideas*. New York: W.W. Norton, 1987.

——. *Wonderful Life: The Burgess Shale and the Nature of History*. New York: W.W. Norton, 1989.

Milner, Richard. "Farewell, Fossil-face: A Memoir of Stephen J. Gould (1941–2002)." *Skeptic* 9, no. 4 (2002): 30–35.

GRADUALISM

Eldredge, Niles. *Life Pulse: Episodes from the Story of the Fossil Record*. New York: Facts on File, 1987.

Stanley, Steven M. *The New Evolutionary Timetable*. New York: Basic Books, 1981.

GRAY, ASA

Dupree, A. Hunter. *Asa Gray*. New York: Atheneum, 1968.

Gray, Asa. *Darwiniana: Essays and Reviews Pertaining to Darwinism*. Edited by A. Hunter Dupree. 1876; repr. Cambridge, Mass.: Belknap Press, 1963.

GREAT CHAIN OF BEING

Lovejoy, Arthur. *The Great Chain of Being: A Study of the History of an Idea*. Cambridge, Mass.: Harvard University Press, 1936.

Tillyard, E.M.W. *The Elizabethan World-Picture: The Idea of Order in the Age of Shakespeare, Donne and Milton*. New York: Vintage/Random House, 1942.

GREAT DYINGS

Benton, M.J. *When Life Nearly Died: The Greatest Mass Extinction of All Time*. New York: Thames & Hudson, 2003.

Glen, William. *The Mass-Extinction Debates: How Science Works in a Crisis*. Stanford: Stanford University Press, 1994.

Hallam, Tony. *Catastrophes and Lesser Calamities: The Causes of Mass Extinctions*. New York: Oxford University Press, 2004.

Ward, Peter D. *Under a Green Sky: Global Warming, the Mass Extinctions of the Past, and What They Can Tell Us About Our Future*. New York: HarperCollins, 2007.

GROUP SELECTION

Gilpin, Michael E. *Group Selection in Predator-Prey Communities*. Princeton: Princeton University Press, 1975.

Kummer, Hans. *Primate Societies: Group Techniques of Ecological Adaptation*. Chicago: Aldine Atherton, 1971.

Williams, George C. *Adaptation and Natural Selection: A Critique of Some Current Evolutionary Thought*. Princeton: Princeton University Press, 1966.

Wynne-Edwards, V.C. *Animal Dispersion in Relation to Social Behavior*. Edinburgh: Pliver & Boyd, 1962.

HAECKEL, ERNST

Bölsche, Wilhelm. *Haeckel: His Life and Work*. Translated by Joseph McCabe. London: T. Fisher Unwin, 1906.

Gasman, Daniel. *The Scientific Origins of National Socialism in Germany: Social Darwinism in Ernst Haeckel and the German Monist League*. London: Macdonald, 1971.

Gould, Stephen Jay. *Ontogeny and Phylogeny*. Cambridge, Mass.: Harvard University Press, 1977.

Haeckel, Ernst. *The History of Creation, or the Development of the Earth and Its Inhabitants by the Action of Natural Causes*. 2 vols. New York: D. Appleton, 1876.

HALDANE, J.B.S.

Clark, Ronald W. *JBS: The Life and Work of J.B.S. Haldane*. New York: Coward-McCann, 1969.

Dronamraju, K.R. *Haldane and Modern Biology*. Baltimore: Johns Hopkins University Press, 1968.

Haldane, J.B.S. *The Causes of Evolution*. London: Longman, 1932.

———. *Heredity and Politics*. London: George Allen & Unwin, 1938.

HAMPDEN, JOHN

Wallace, Alfred Russel. *My Life*. vol. 2, ch. 39. New York: Dodd, Mead, 1905.

HAPPY FAMILY, THE

Barnum, P.T., and James W. Cook. *The Colossal P.T. Barnum Reader: Nothing Else Like It in the Universe*. Urbana: University of Illinois Press, 2005.

Harris, Neil. *Humbug: The Art of P.T. Barnum*. Boston: Little, Brown, 1973.

HAWAIIAN RADIATION

Carlquist, Sherwin. *Island Biology*. New York: Columbia University Press, 1974.

———. *Island Life: A Natural History of the Islands of the World*. Garden City, N.Y.: Natural History Press, 1965.

Pratt, H. Douglas. *The Hawaiian Honeycreepers: Drepanidinae*. Oxford: Oxford University Press, 2005.

Wagner, Warren L., and V.A. Funk, eds. *Hawaiian Biogeography: Evolution on a Hot Spot Archipelago*. Washington, D.C.: Smithsonian Institution Press, 1995.

HAWKINS, BENJAMIN WATERHOUSE

Bramwell, Valerie, and Robert M. Peck. *All in the Bones: A Biography of Benjamin Waterhouse Hawkins*. Philadelphia: Academy of Natural Sciences, 2008.

Marshall, Nancy Rose. "'A Dim World, Where Monsters Dwell': The Spatial Time of the Sydenham Crystal Palace Dinosaur Park." *Victorian Studies* 49, no. 2 (2007): 286–301.

Preston, Douglas. *Dinosaurs in the Attic: An Excursion into the American Museum of Natural History*. New York: St. Martin's Press, 1986.

HENSLOW, JOHN STEVENS

Barlow, Lady Nora, ed. *Darwin and Henslow: The Growth of an Idea—Letters, 1831–1860*. Berkeley: University of California Press, 1967.

Russell-Gebbett, Jean P. *Henslow of Hitcham: Botanist, Educationalist, and Clergyman*. Lavenham, Suffolk: T. Dalton, 1977.

Walters, S.M., and E.A. Stow. *Darwin's Mentor: John Stevens Henslow, 1796–1861*. Cambridge: Cambridge University Press, 2001.

HOMININ

Johanson, Donald, and Blake Edgar. *From Lucy to Language*. Rev. ed. New York: Simon & Schuster, 2006.

Kingdon, Jonathan. *Lowly Origin: Where, When, and Why Our Ancestors First Stood Up*. Princeton: Princeton University Press, 2003.

Sawyer, G.J., and Viktor Deak, eds. *The Last Human: A Guide to Twenty-two Species of Extinct Humans*. Text by Esteban Sarmiento, G.J. Sawyer, and Richard Milner. New Haven, Conn.: Yale University Press, 2007.

Tattersall, Ian, and Jeffrey H. Schwartz. *Extinct Humans*. New York: Westview/Nevraumont, 2000.

Tobias, Phillip V., ed. *Hominid Evolution: Past, Present, and Future*. New York: Alan R. Liss, 1985.

HOMO ERECTUS

Boaz, Noel T., and Russell L. Ciochon. *Dragon Bone Hill: An Ice-Age Saga of Homo Erectus*. New York: Oxford University Press, 2004.

Delson, Eric, ed. *Ancestors: The Hard Evidence*. New York: Alan R. Liss, 1985.

Mayr, Ernst. "Taxonomic Categories in Fossil Hominids." *Cold Spring Harbor Symposium on Quantitative Biology* 15 (1950): 109–118.

Reader, John. *Missing Links: The Hunt for Early Man*. London: Collins, 1981.

Shapiro, Harry. *Peking Man*. New York: Simon & Schuster, 1974.

Swisher, Carl C. III, Garniss H. Curtis, and Roger Lewin. *Java Man: How Two Geologists Changed Our Understanding of Human Evolution*. Chicago: University of Chicago Press, 2000.

HOMO HABILIS

Leakey, Richard, and Roger Lewin. *Origins: What New Discoveries Reveal about the Emergence of Our Species and Its Possible Future*. New York: E.P. Dutton, 1977.

Tobias, Phillip V. *The Skulls, Endocasts, and Teeth of Homo Habilis*. Vol. 4 of *Olduvai Gorge*. Cambridge: Cambridge University Press, 1991.

HOMO SAPIENS, CLASSIFICATION

Blunt, Wilfred. *The Compleat Naturalist: A Life of Linnaeus*. New York: Viking Press, 1971.

HOMOLOGY

Desmond, Adrian. *Archetypes and Ancestors: Palaeontology in Victorian London, 1850–1875*. Chicago: University of Chicago Press, 1982.

Owen, Richard. *On the Archetype and Homologies of the Vertebrate Skeleton*. London: John van Voorst, 1848.

HOOKER, SIR JOSEPH DALTON

Allan, Mea. *Darwin and His Flowers: The Key to Natural Selection*. New York: Taplinger, 1977.

———. *The Hookers of Kew, 1785–1911*. London: Michael Joseph, 1967.

Hooker, Joseph D. *Himalayan Journals*. 2 vols. London: John Murray, 1854.

Huxley, Leonard, ed. *Life and Letters of Sir Joseph Dalton Hooker*. 2 vols. London: John Murray, 1918.

Scourse, Nicolette. *The Victorians and Their Flowers*. London: Croom Helm, 1983.

Turrill, W.B. *Joseph Dalton Hooker*. London: Scientific Book Club, 1963.

HOOTON, EARNEST ALBERT

Garn, Stanley M., and Eugene Giles. *Earnest Albert Hooton, 1887–1954: A Biographical Memoir*. Washington, D.C.: National Academy Press, 1995.

Hooton, Earnest Albert. *Man's Poor Relations*. New York: Doubleday, 1942.

———. *Up from the Ape*. 2d ed. New York: Macmillan, 1946.

"HOPEFUL MONSTERS"

Dawkins, Richard. *The Blind Watchmaker: Why the Evidence of Evolution Reveals a Universe without Design*. New York: W.W. Norton, 1996.

Goldschmidt, Richard. *The Material Basis of Evolution*. New Haven: Yale University Press, 1940. (Reissued with an essay by Stephen Jay Gould, 1982.)

HORSE, EVOLUTION OF

MacFadden, Bruce. *Fossil Horses: Systematics, Paleobiology, and Evolution of the Family Equidae*. New York: Cambridge University Press, 1992.

Simpson, George Gaylord. *Horses: The Story of the Horse Family in the Modern World through Sixty Million Years of History*. Oxford: Oxford University Press, 1951.

HUNDREDTH MONKEY PHENOMENON

Watson, Lyall. *Lifetide: The Biology of the Unconscious*. New York: Simon & Schuster, 1979.

HUNTING HYPOTHESIS

Ardrey, Robert. *African Genesis: A Personal Investigation into the Animal Origins and Nature of Man*. New York: Atheneum, 1961.

———. *The Hunting Hypothesis: A Personal Conclusion concerning the Evolutionary Nature of Man*. New York: Atheneum, 1976.

HUXLEY, ALDOUS

Clark, Ronald W. *The Huxleys*. New York: McGraw-Hill, 1968.

Huxley, Aldous. *Ape and Essence*. New York: Harper, 1948.

———. *Brave New World*. Garden City, N.Y.: Doubleday, Doran, 1932.

———. *Ends and Means: An Inquiry into the Nature of Ideals*. New York: Harper, 1937.

HUXLEY, SIR JULIAN SORELL

Clark, Ronald W. *The Huxleys*. New York: McGraw-Hill, 1968.

Harrison, G.A., and M. Keynes, eds. *Evolutionary Studies: A Centenary Celebration of the Life of Julian Huxley*. London: Macmillan, 1989.

Huxley, Julian. *Essays of a Biologist*. Cambridge: Cambridge University Press, 1912.

———. *Evolution: The Modern Synthesis*. London: George Allen & Unwin, 1942.

———. *Religion without Revelation*. New York: Harper, 1927.

Waters, C. Kenneth, and Albert Van Helden. *Julian Huxley: Biologist and Statesman of Science*. Houston: Rice University Press, 1992.

HUXLEY, LEONARD

Clark, Ronald W. *The Huxleys*. New York: McGraw-Hill, 1968.

Huxley, Leonard, ed. *Life and Letters of Sir Joseph Dalton Hooker*. 2 vols. London: John Murray, 1918.

———. ed. *The Life and Letters of Thomas Henry Huxley*. London: Macmillan, 1900.

HUXLEY, THOMAS HENRY

Ayres, Clarence. *Huxley*. New York: W.W. Norton, 1932.

Bibby, Cyril, ed. *The Essence of T.H. Huxley*. New York: Macmillan, 1967.

Clark, Ronald W. *The Huxleys*. New York: McGraw-Hill, 1968.

Desmond, Adrian. *Huxley: From Devil's Disciple to Evolution's High Priest*. Reading, Mass.: Addison-Wesley, 1997.

DiGregorio, Mario. *T.H. Huxley's Place in Natural Science*. New Haven, Conn.: Yale University Press, 1984.

Huxley, Leonard, ed. *The Life and Letters of Thomas Henry Huxley*. London: Macmillan, 1900.

Huxley, Thomas H. *Collected Essays*. 9 vols. London: Macmillan, 1893–1894.

———. *Evidence as to Man's Place in Nature*. London: Williams & Norgate, 1863.

———. *On a Piece of Chalk* (1868). Edited and with an introduction by Loren Eiseley. New York: Scribner, 1967.

Irvine, William. *Apes, Angels, and Victorians: The Story of Darwin, Huxley, and Evolution*. New York: McGraw-Hill, 1955.

HYBRIDIZATION

Levin, Donald A. *Hybridization: An Evolutionary Perspective*. Stroudsburg, Pa.: Dowden, Hutchinson & Ross, 1979.

Mallet, James. "Hybridization, Ecological Races, and the Nature of Species: Empirical Evidence for the Ease of Speciation." *Philosophical Transactions of the Royal Society* 363 (2008): 2971–2986.

ICE AGE

Hadingham, Evan. *Secrets of the Ice Age: A Reappraisal of Prehistoric Man*. New York: Walker, 1979.

Marshack, Alexander. *The Roots of Civilization: The Cognitive Beginnings of Man's First Art, Symbol, and Notation*. Rev. ed. Mount Kisco, N.Y.: Moyer Bell, 1991.

Stanley, Steven. *Children of the Ice Age: How a Global Catastrophe Allowed Humans to Evolve*. New York: Harmony Books, 1996.

Sutcliffe, Antony J. *On the Tracks of Ice Age Mammals*. Cambridge, Mass.: Harvard University Press, 1985.

White, Randall. *Dark Caves, Bright Visions: Life in Ice Age Europe*. New York: W. W. Norton, 1986.

ICEMAN

Dubowski, Cathy. *Ice Mummy: The Discovery of a 5,000-Year-Old Man*. New York: Random House, 1999.

Fowler, Brenda. *Iceman: Uncovering the Life and Times of a Prehistoric Man Found in an Alpine Glacier*. New York: Macmillan, 2001.

IGUANODON DINNER

Barber, Lynn. *The Heyday of Natural History*. Garden City, N.Y.: Doubleday, 1980.

INFANTICIDE

Hausfater, Glenn, and Sarah B. Hrdy. *Infanticide: Comparative and Evolutionary Perspectives*. New York: Aldine, 1984.

Hoogland, John L. *The Black-Tailed Prairie Dog: Social Life of a Burrowing Mammal*. Chicago: University of Chicago Press, 1995.

Koenig, W. D, et al. 1995. "Patterns and Consequences of Egg Destruction among Joint-Nesting Acorn Woodpeckers." *Animal Behavior* 50 (1995): 607–621.

Van Schaik, Carel, and Charles H. Janson, eds. *Infanticide by Males and Its Implications*. New York: Cambridge University Press, 2000.

INHERIT THE WIND

Lawrence, Jerome, and Robert Edwin Lee. *Inherit the Wind*. New York: Random House, 1955.

INSECT SOCIETIES, EVOLUTION OF

Dugatkin, Lee. *Cheating Monkeys and Citizen Bees: The Nature of Cooperation in Animals and Humans*. New York: Free Press, 1999.

Gordon, Deborah M. *Ants at Work: How an Insect Society Is Organized*. New York: W. W. Norton, 2000.

Hölldobler, Bert, and Edward O. Wilson. *The Ants*. Cambridge, Mass.: Harvard University Press, 1990.

———. *The Superorganism: The Beauty, Elegance, and Strangeness of Insect Societies*. New York: W. W. Norton, 2008.

Topoff, Howard. "Slave-making Queens." *Scientific American*, Nov. 1999, 84–91.

Tschinkel, Walter. "Colonies in Space." *Natural History* 110, no. 3 (2001): 64-65.

Wilson, Edward O. *The Insect Societies*. Cambridge, Mass.: Harvard University Press, Belknap Press, 1971.

INSECTIVOROUS PLANTS

Allan, Mea. *Darwin and His Flowers: The Key to Natural Selection*. New York: Taplinger, 1977.

Darwin, Charles. *Insectivorous Plants*. London: John Murray, 1875.

Juniper, B., Richard J. Robins, and D.M. Joel. *The Carnivorous Plants*. London: Academic Press, 1989.

INTELLIGENT DESIGN

Brockman, John, ed. *Intelligent Thought: Science versus the Intelligent Design Movement*. New York: Vintage Books, 2006.

Dawkins, Richard. *The Blind Watchmaker: Why the Evidence of Evolution Reveals a Universe without Design*. New York: W. W. Norton, 1996.

Dembski, William A. *Intelligent Design: The Bridge between Science and Theology*. Downers Grove, Ill.: InterVarsity Press, 1999.

Forrest, Barbara, and Paul R. Gross. *Creationism's Trojan Horse: The Wedge of Intelligent Design*. New York: Oxford University Press, 2004.

Johnson, Phillip E. *Darwin on Trial*. Downers Grove, Ill.: InterVarsity Press, 1991.

Larson, Edward J. *Trial and Error: The American Controversy over Creation and Evolution*. New York: Oxford University Press, 1985.

Milner, Richard, and Vittorio Maestro, eds. "Intelligent Design or Evolution?" Special report. *Natural History* 111, no. 3 (2002): 73–80.

Perakh, Mark. *Unintelligent Design*. Amherst, N.Y.: Prometheus Books, 2004.

Scott, Eugenie C. *Evolution vs. Creationism: An Introduction*. Westport, Conn.: Greenwood Press, 2004.

Shanks, Niall. *God, the Devil, and Darwin: A Critique of Intelligent Design Theory*. New York: Oxford University Press, 2004.

Shermer, Michael. *Why Darwin Matters: The Case against Intelligent Design*. New York: Times Books, 2006.

Young, Mark, and Taner Edis, eds. *Why Intelligent Design Fails: A Scientific Critique of the New Creationism*. New Brunswick, N.J.: Rutgers University Press, 2004.

ISOLATING MECHANISMS

Mayr, Ernst. *Animal Species and Evolution*. Cambridge: Belknap Press, 1963.

JARAMILLO EVENT

Glen, William. *The Road to Jaramillo: Critical Years of the Revolution in Earth Science*. Stanford: Stanford University Press, 1982.

Lawrence, David M. *Upheaval from the Abyss: Ocean Floor Mapping and the Earth Science Revolution*. New Brunswick, N.J.: Rutgers University Press, 2002.

Oreskes, Naomi, and H. E. LeGrand. *Plate Tectonics: An Insider's History of the Modern Theory of the Earth*. Boulder, Colo.: Westview Press, 2001.

JEFFERSON, THOMAS

Cohen, I. Bernard. *Science and the Founding Fathers: Science in the Political Thought of Jefferson, Franklin, Adams, and Madison*. New York: W. W. Norton, 1995.

Malone, Dumas. *Jefferson the Virginian*. Vol. 1 of *Jefferson and His Time*. Charlottesville: University of Virginia Press, 2006.

JOHANSON, DONALD

Johanson, Donald, and Maitland Edey. *Lucy: The Beginnings of Humankind*. New York: Simon & Schuster, 1981.

Johanson, Donald, and Blake Edgar. *From Lucy to Language*. Rev. ed. New York: Simon & Schuster, 2006.

Johanson, Donald, and James Shreeve. *Lucy's Child: The Discovery of a Human Ancestor*. New York: William Morrow, 1989.

Lewin, Roger. *Bones of Contention: Controversies in the Search for Human Origins*. New York: Simon & Schuster, 1987.

Reader, John. *Missing Links: The Hunt for Early Man*. London: Collins, 1981.

JURASSIC PARK

Crichton, Michael. *Jurassic Park*. New York: Alfred A. Knopf, 1990.

Horner, John R., and Edwin Dobb. *Dinosaur Lives: Unearthing an Evolutionary Saga*. New York: Harcourt, Brace, 1997.

Shay, Don, and Jody Duncan. *The Making of "Jurassic Park."* New York: Ballantine Books, 1993.

"JUST-SO" STORIES

Gould, Stephen Jay. "Sociobiology: The Art of Storytelling." *New Scientist* 80 (1997): 530–533.

Kipling, Rudyard. *A Collection of Rudyard Kipling's Just-So Stories*. Cambridge, Mass.: Candlewick Press, 2004.

Ricketts, Harry. *Rudyard Kipling: A Life*. New York: Carroll & Graf, 2000.

Schlinger, Henry D., Jr. "How the Human Got Its Spots: A Critical Analysis of the Just-So Stories of Evolutionary Psychology." *Skeptic* 4, no. 1 (1996).

Whitnall, Harold O. *A Parade of Ancient Animals*. New York: Thomas Y. Crowell, 1936.

KAMMERER, PAUL

Kammerer, Paul. *The Inheritance of Acquired Characteristics*. New York: Boni & Liveright, 1924.

Koestler, Arthur. *The Case of the Midwife Toad*. London: Hutchinson, 1971.

KANZI

Burling, Robbins. *The Talking Ape: How Language Evolved*. New York: Oxford University Press, 2005.

Savage-Rumbaugh, E. Sue, and Roger Lewin. *Kanzi: The Ape at the Brink of the Human Mind*. New York: John Wiley, 1994.

Segerdahl, Pär, William Fields, and Sue Savage-Rumbaugh. *Kanzi's Primal Language: The Cultural Ini-*

tiation of Primates into Language. Basingstoke, Eng.: Palgrave Macmillan, 2005.

KELVIN, LORD

Burchfield, Joe D. *Lord Kelvin and the Age of the Earth.* New York: Science History Publications, 1975.

Gould, Stephen Jay. *The Flamingo's Smile: Reflections in Natural History.* New York: Norton, 1985.

Gray, Andrew. *Lord Kelvin: An Account of His Scientific Life and Work.* New York: E.P. Dutton, 1908.

KENNEWICK MAN

Chatters, James C. *Ancient Encounters: Kennewick Man and the First Americans.* New York: Simon & Schuster, 2001.

Downey, Roger. *Riddle of the Bones: Politics, Science, Race, and the Story of Kennewick Man.* New York: Copernicus, 2000.

Thomas, David Hurst. *Skull Wars: Kennewick Man, Archeology, and the Battle for Native American Identity.* New York: Basic Books, 2000.

KIDD, BENJAMIN

Jones, Greta. *Social Darwinism and English Thought.* Atlantic Highlands, N.J.: Humanities Press, 1980.

Kidd, Benjamin. *Social Evolution.* New York: Macmillan, 1894.

KILLER APE THEORY

Ardrey, Robert. *African Genesis: A Personal Investigation into the Animal Origins and Nature of Man.* New York: Atheneum, 1961.

———. *The Hunting Hypothesis: A Personal Conclusion concerning the Evolutionary Nature of Man.* New York: Atheneum, 1976.

Lorenz, Korad. *On Aggression.* London: Methuen, 1966.

Rensberger, Boyce. "The Killer Ape Is Dead." *Alicia Patterson Foundation Newsletter,* Dec. 4, 1973.

Wrangham, Richard W., and Dale Peterson. *Demonic Males: Apes and the Origins of Human Violence.* Boston: Houghton Mifflin, 1996.

KIN SELECTION

Haldane, J.B.S., and Krishna R. Dronamraju. *Haldane and Modern Biology.* Baltimore: Johns Hopkins University Press, 1968.

Hamilton, W.D. "The Genetical Theory of Social Behavior." *Journal of Theoretical Biology* 7 (1964): 1–52. Repr. in *The Sociobiology Debate: Readings on Ethical and Scientific Issues,* edited by Arthur L. Caplan. New York: Harper and Row, 1978.

KING KONG

Annan, David. *Ape-Monster of the Movies.* New York: Bounty Books, 1975.

Goldman, Orville, and George E. Turner. *The Making of "King Kong."* New York: Ballantine Books, 1975.

Gottesman, Ronald, and Harry Geduld. *The Girl in the Hairy Paw: King Kong as Myth, Movie, and Monster.* New York: Avon, 1976.

Morton, Ray. *King Kong: The History of a Movie Icon, from Fay Wray to Peter Jackson.* New York: Applause Theatre and Cinema Books, 2005.

KNIGHT, CHARLES R.

Bogart, Michele. "Lowbrow/Highbrow: Charles R. Knight, Art Work, and the Spectacle of Prehistoric Life." In *American Victorians and Virgin Nature,* edited by T.J. Jackson Lears, pp. 64–79. Boston: Isabella Stewart Gardner Museum, 2002.

Czerkas, Sylvia, and Don Glut. *Dinosaurs, Mammoths, and Cavemen: The Art of Charles R. Knight.* New York: E.P. Dutton, 1982.

Knight, Charles R. *Animal Drawing: Anatomy and Action for Artists.* New York: McGraw-Hill, 1947.

———. *Before the Dawn of History.* New York: McGraw-Hill, 1935.

———. *Life through the Ages.* New York: Alfred A. Knopf, 1946.

———. *Prehistoric Man: The Great Adventurer.* New York: Appleton-Century-Crofts, 1949.

Milner, Richard. *Charles R. Knight: Father of Paleoart.* With a preface by Rhoda Knight Kalt. New York: Harry N. Abrams, in press.

KOKO

Hahn, Emily. *Eve and the Apes.* New York: Weidenfeld & Nicolson, 1988.

Patterson, Francine, and E. Linden. *The Education of Koko.* New York: Holt, Rinehart & Winston, 1981.

KROPOTKIN, PRINCE PETER

Kropotkin, Petr Alekseevich. *Mutual Aid: A Factor of Evolution.* New York: Garland, 1972.

Woodcock, George, and Ivan Avakumovic. *Peter Kropotkin: From Prince to Rebel.* Montreal: Black Rose Books, 1990.

LA BREA TAR PITS

Harris, John, and George Jefferson. *Rancho La Brea: Treasures of the Tar Pits.* Los Angeles: Los Angeles County Museum, 1985.

Jang, Allen W., and William S. Weston. *The La Brea Tar Pits: A Field Trip and Self-Study Guide—Understanding the Past and Critical Thinking.* Vancouver, B.C.: Fifth Province Media, 2006.

LAETOLI FOOTPRINTS

Agnew, Neville, and Martha Demas. "Preserving the Laetoli Foot-

prints." *Scientific American,* Sept. 1998, 44–55.

Meldrum, D. Jeffrey, and Charles E. Hilton. *From Biped to Strider: The Emergence of Modern Human Walking, Running, and Resource Transport.* New York: Kluwer Academic/Plenum, 2004.

LAMARCK, JEAN BAPTISTE DE

Lamarck, J.B. *Zoological Philosophy: An Exposition with Regard to the Natural History of Animals* (1809). Chicago: University of Chicago Press, 1984.

Packard, Alpheus S. *Lamarck, the Founder of Evolution.* New York: Longmans Green, 1901.

LANKESTER, E. RAY

Lankester, E. Ray. *Diversions of a Naturalist.* London: Methuen, 1915.

———. *Extinct Animals.* London: Constable, 1905.

———. *Science from an Easy Chair.* London: Methuen, 1910.

———. *Science from an Easy Chair.* Second series. London: Adlard and Son, 1912.

Lester, Joe, and Peter J. Bowler, eds. *E. Ray Lankester and the Making of Modern British Biology.* [Great Britain]: British Society for the History of Science, 1995.

LASCAUX CAVES

Bahn, Paul G., and Jean Vertut. *Images of the Ice Age.* New York: Facts on File, 1988.

Clottes, Jean. *Cave Art.* New York: Phaidon Press, 2008.

Félix, Thierry, and Philippe Bigotto. *The Secret of the Forest of Lascaux.* Foreword by Yves Coppens. Sarlat, France: Dolmen Editions, 1990.

Gurney, Kenneth P. *In the Caves of Lascaux.* Shorewood, Wis.: Hodge Podge Press, 1996.

LAURASIA

Deacon, Richard. *Atlas: Gondwanaland and Laurasia.* Oslo: Kunstnernes Hus, 1990.

LEAKEY, LOUIS

Leakey, L.S.B. *Adam's Ancestors: Evolution of Man and His Culture.* New York: Harper & Row, 1960.

———. *By the Evidence: Memoirs, 1932–1951.* New York: Harcourt Brace Jovanovich, 1974.

———. *White African: An Early Autobiography.* Cambridge, Mass.: Schenkman, 1966.

Morell, Virgina. *Ancestral Passions: The Leakey Family and the Quest for Humankind's Beginnings.* New York: Simon & Schuster, 1955.

LEAKEY, MARY

Leakey, Mary. *Disclosing the Past: An Autobiography.* Garden City, N.Y.: Doubleday, 1984.

———. *Olduvai Gorge: My Search for Early Man.* London: Collins, 1979.

Reader, John. *Missing Links: The Hunt for Early Man.* London: Collins, 1981.

Willis, Delta. *The Hominid Gang.* New York: Viking Press, 1989.

LEAKEY, MEAVE E.

Leakey, Meave, and John M. Harris, eds. *Lothagam: The Dawn of Humanity in Eastern Africa.* New York: Columbia University Press, 2001.

LEAKEY, RICHARD ERSKINE FRERE

Leakey, Richard. *The Making of Mankind.* New York: E.P. Dutton, 1981.

———. *One Life: An Autobiography.* London: Michael Joseph, 1981.

LEBENSBORN MOVEMENT

Carlson, Elof Axel. *The Unfit: A History of a Bad Idea.* Cold Spring Harbor, N.Y.: Cold Spring Harbor Laboratory Press, 2001.

LESQUEREUX, LEO

Andrews, Henry. *The Fossil Hunters: In Search of Fossil Plants.* Ithaca, N.Y.: Cornell University Press, 1980.

Lesley, J.P. "Memoir of Leo Lesquereux, 1806–1889." *National Academy of Sciences, Biographical Memoirs,* 3: 187–212. Washington, D.C.: National Academy of Sciences, 1895.

Tidwell, William D. *Common Fossil Plants of Western North America.* Washington, D.C.: Smithsonian Institution Press, 1998.

LINNAEAN "SEXUAL SYSTEM"

Blunt, Wilfrid. *Linnaeus: The Compleat Naturalist.* Citations from app. 1, by William T. Stearn. Princeton: Princeton University Press, 2002.

LINNAEUS, CARL

Blunt, Wilfred. *The Compleat Naturalist: A Life of Linnaeus.* New York: Viking Press, 1971.

Jackson, Benjamin D. *Linnaeus.* London: H.F. & G. Witherby, 1923.

Linnaeus, Carl. *Systema naturae. Regnum animale.* 1758; facsimile repr. London: British Museum (Natural History), 1956.

LONESOME GEORGE

Nicholls, Henry. *Lonesome George: The Life and Loves of a Conservation Icon.* London: Macmillan, 2006.

LORENZ, KONRAD

Lerner, Richard M. *Final Solutions: Biology, Prejudice, and Genocide.* University Park: Pennsylvania State University Press, 1992.

Lorenz, Konrad. *Evolution and Modification of Behavior.* Chicago: University of Chicago Press, 1965.

——. *King Solomon's Ring: New Light on Animal Ways.* New York: Thomas Crowell, 1952.

——. *On Aggression.* London: Methuen, 1966.

——. *The Year of the Greylag Goose.* New York: Harcourt, Brace, Jovanovich, 1979.

Singer, Peter. *The Expanding Circle: Ethics and Sociobiology.* New York: New American Library, 1981.

LOST WORLD, THE

Carr, John Dickson. *The Life of Sir Arthur Conan Doyle.* New York: Harper, 1949.

Doyle, Arthur Conan. *"The Lost World" of Arthur Conan Doyle.* Edited by John R. Lavas. Auckland, N.Z.: John R. Lavas, 2002.

——. *The Annotated "Lost World": The Classic Adventure Novel.* Edited by Roy Pilot and Alvin Rodin. Indianapolis: Wessex Press, 1996.

LUBBOCK, SIR JOHN

Hutchinson, Horace. *Life of Sir John Lubbock, Lord Avebury.* 2 vols. London: Macmillan, 1914.

Lubbock, Sir John. *Ants, Bees, and Wasps.* London: Kegan Paul, Trench, 1882.

——. *Origin of Civilisation and the Primitive Condition of Man.* London: Longmans, Green, 1870.

——. *Pre-historic Times: As Illustrated by Ancient Remains, and the Manners and Customs of Modern Savages.* London: Williams & Norgate, 1865.

Lubbock, Lyulph. "From the Origin of Species to the Origin of Civilization: A Perspective from a Corner of Kent." In "Survival of the Fittest: Celebrating the 150th Anniversary of the Darwin-Wallace Theory of Evolution." *The Linnean,* special issue no. 9 (2008): 61–87.

Stocking, George W., Jr. *Victorian Anthropology.* New York: Free Press, 1987.

"LUCY"

Johanson, Donald, and Maitland Edey. *Lucy: The Beginnings of Humankind.* New York: Simon & Schuster, 1981.

Johanson, Donald, and Blake Edgar. *From Lucy to Language.* Rev. ed. New York: Simon & Schuster, 2006.

Johanson, Donald, and James Shreeve. *Lucy's Child: The Discovery of a Human Ancestor.* New York: William Morrow, 1989.

Lewin, Roger. *Bones of Contention: Controversies in the Search for Human Origins.* New York: Simon & Schuster, 1987.

Reader, John. *Missing Links: The Hunt for Early Man.* London: Collins, 1981.

LYELL, SIR CHARLES

Bailey, Edward. *Charles Lyell.* London: Thomas Nelson & Sons, 1962.

Eiseley, Loren. *Darwin's Century: Evolution and the Men Who Discovered It.* New York: Doubleday, 1958.

Gould, Stephen Jay. *Time's Arrow, Time's Cycle.* Cambridge: Harvard University Press, 1987.

Greene, John C. *The Death of Adam: Evolution and Its Impact on Western Thought.* Ames: Iowa State University Press, 1959.

Lyell, Charles. *On the Geological Evidence of the Antiquity of Man.* London: John Murray, 1863.

——. *The Principles of Geology.* 3 vols. London: John Murray, 1830–1833; repr. Chicago: University of Chicago Press, 1990–1991.

LYSENKOISM

Joravsky, David. *The Lysenko Affair.* Cambridge, Mass.: Harvard University Press, 1970.

Zirkle, Conway. *Death of a Science in Russia: The Fate of Genetics as Described in "Pravda" and Elsewhere.* Philadelphia: University of Pennsylvania Press, 1949.

LYSTROSAURUS FAUNA

Lucas, Spencer G. *Chinese Fossil Vertebrates.* New York: Columbia University Press, 2001.

MALTHUS, THOMAS

Chase, Allan. *The Legacy of Malthus: The Social Costs of the New Scientific Racism.* New York: Alfred A. Knopf, 1977.

Malthus, Thomas. *An Essay on the Principle of Population* (1798). Edited by Geoffrey Gilbert. Oxford: Oxford University Press, 1999.

MAMMOTHS

Cohen, Claudine. *The Fate of the Mammoth: Fossils, Myth, and History.* Translated by William Rodarmor. Chicago: University of Chicago Press, 2002.

Haynes, Gary. *Mammoths, Mastodons, and Elephants.* New York: Cambridge University Press, 1993.

Silverberg, Robert. *Mammoths, Mastodons, and Man.* New York: McGraw-Hill, 1970.

Thomson, Keith. *The Legacy of the Mastodon: The Golden Age of Fossils in America.* New Haven, Conn.: Yale University Press, 2008.

MANTELL, GIDEON ALGERNON

Colbert, Edwin H. *Men and Dinosaurs: The Search in Field and Laboratory.* New York: E. P. Dutton, 1968.

Mantell, Gideon Algernon. *Medals of Creation; or, First Lessons in Geology, and in the Study of*

Organic Remains. London: Henry G. Bohn, 1844.

——. *The Wonders of Geology; or, A Familiar Exposition of Geological Phenomena.* 2 vols. New Haven, Conn.: A.H. Maltby, 1838.

MARXIAN "ORIGIN OF MAN"

Engels, Friedrich. *The Origin of the Family, Private Property, and the State.* New York: Pathfinder, 1972.

——. *The Part Played by Labor in the Transition from Ape to Man.* New York: International Publishers, 1950.

Marx, Karl. *Capital: An Abridged Edition.* Edited by David McLellan. Oxford: Oxford University Press, 2008.

MATTHEW, PATRICK

Dempster, W. J. *Natural Selection and Patrick Matthew: Evolutionary Concepts in the Nineteenth Century.* Edinburgh: Pentland Press, 1996.

Gould, Stephen Jay. *The Flamingo's Smile: Reflections in Natural History.* New York: Norton, 1985.

MAYR, ERNST

Dobzhansky, Theodosius. *Genetics and the Origin of Species.* New York: Columbia University Press, 1937.

Mayr, Ernst. *Animal Species and Evolution.* Cambridge, Mass.: Harvard University Press, 1963.

——. *The Growth of Biological Thought: Diversity, Evolution, and Inheritance.* Cambridge, Mass.: Harvard University Press, 1982.

——. *One Long Argument: Charles Darwin and the Genesis of Modern Evolutionary Thought.* Cambridge, Mass.: Harvard University Press, 1993.

——. *Systematics and the Origin of Species.* New York: Columbia University Press, 1942.

MENDEL, ABBOT GREGOR

Mawer, Simon. *Gregor Mendel: Planting the Seeds of Genetics.* New York: Harry N. Abrams, 2006.

MENDEL'S LAWS

Bateson, William, and Gregor Mendel. *Mendel's Principles of Heredity.* Cambridge: Cambridge University Press, 1913.

Mendel, Gregor, Alain F. Corcos, and Floyd V. Monaghan. *Gregor Mendel's Experiments on Plant Hybrids: A Guided Study.* New Brunswick, N.J.: Rutgers University Press, 1993.

Monaghan, F.V., and A. Corcos. "The Origins of the Mendelian Laws." *Journal of Heredity* 75 (1984): 67–69.

Stubbe, Hans. *History of Genetics: From Prehistoric Times to the*

Rediscovery of Mendel's Laws. Cambridge, Mass.: MIT Press, 1972.

MESOZOIC

Fastovsky, David E., and David B. Weishampel. *The Evolution and Extinction of the Dinosaurs.* Cambridge: Cambridge University Press, 1996.

METAPHORS IN EVOLUTIONARY WRITING

Beer, Gillian. *Darwin's Plots: Evolutionary Narrative in Darwin, George Eliot, and Nineteenth-Century Fiction.* London: Routledge & Kegan Paul, 1983.

Hyman, Stanley E. *The Tangled Bank: Darwin, Marx, Frazer, and Freud as Imaginative Writers.* New York: Atheneum, 1962.

Landau, Misia. *Narratives of Human Evolution.* New Haven, Conn.: Yale University Press, 1991.

Leatherdale, W.H. *The Role of Analogy, Model, and Metaphor in Science.* Amsterdam: North-Holland; New York: American Elsevier, 1974.

MILLER, HUGH

Barber, Lynn. *The Heyday of Natural History.* Garden City, N.Y.: Doubleday, 1980.

Miller, Hugh. *The Footprints of the Creator; or, The Asterolepis of Stromness.* With memoir of the author by Louis Agassiz. Boston: Gould & Lincoln, 1858.

——. *The Old Red Sandstone; or, New Walks in an Old Field.* Edinburgh: J. Johnstone, 1841.

——. *The Testimony of the Rocks; or, Geology in Its Bearings on the Two Theologies, Natural and Revealed.* Boston: Gould & Lincoln, 1857.

MIMICRY

Owen, Denis. *Camouflage and Mimicry.* Chicago: University of Chicago Press, 1982.

Wickler, Wolfgang. *Mimicry in Plants and Animals.* Translated by R.D. Martin. New York: McGraw-Hill, 1968.

MITOCHONDRIAL DNA

Ball, Edward. *The Genetic Strand: Exploring a Family History through DNA.* New York: Simon & Schuster, 2007.

Jones, Martin. *The Molecule Hunt: Archaeology and the Search for Ancient DNA.* New York: Arcade, 2002.

Lane, Nick. *Power, Sex, Suicide: Mitochondria and the Meaning of Life.* Oxford: Oxford University Press, 2005.

MIVART, ST. GEORGE J.

Gruber, J.W. *A Conscience in Conflict: The Life of St. George*

Jackson Mivart. New York: Columbia University Press, 1960.

Hull, David L., ed. Darwin and His Critics. Cambridge, Mass.: Harvard University Press, 1973.

Mivart, St. George Jackson. On the Genesis of Species. London: Macmillan, 1871.

MONISM

Bachli, Andreas, and Klaus Petrus. Monism. Frankfurt: Ontos, 2003.

Gasman, Daniel. The Scientific Origins of National Socialism: Social Darwinism in Ernst Haeckel and the German Monist League. London: Macdonald; New York: American Elsevier, 1971.

Romanes, George John. Mind and Motion and Monism. London: Longmans, Green, 1895.

MOSAIC EVOLUTION

Nevo, Eviatar. Mosaic Evolution of Subterranean Mammals: Regression, Progression, and Global Convergence. Oxford: Oxford University Press, 1999.

MULLIS, KARY B.

Grace, Eric S. Biotechnology Unzipped: Promises and Realities. Rev. 2d ed. Washington, D.C.: Joseph Henry Press, 2006.

Mullis, Kary. Dancing Naked in the Mind Field. New York: Random House, 1998.

Rabinow, Paul. Making PCR: A Story of Biotechnology. Chicago: University of Chicago Press, 1996.

MURCHISON, SIR RODERICK IMPEY

Geikie, Archibald. The Life and Letters of Sir Roderick Impy Murchison. 2 vols. London: John Murray, 1875.

Murchison, Roderick I. Siluria: The History of the Oldest Known Rocks Containing Organic Remains. London: John Murray, 1854.

Secord, James A. Controversy in Victorian Geology: The Cambrian–Silurian Dispute. Princeton: Princeton University Press, 1986.

NATURAL PRODUCTS CHEMISTRY

Krieg, Margaret B. Green Medicine: The Search for Plants That Heal. New York: Rand McNally, 1964.

Plotkin, Mark J. "The Outlook for New Agricultural and Industrial Products from the Tropics." In Biodiversity, edited by E.O. Wilson and Frances M. Peter, pp. 106–116. Washington, D.C.: National Academy Press, 1988.

NATURAL SELECTION

Gould, Stephen Jay. The Structure of Evolutionary Theory. Cambridge, Mass.: Belknap Press, 2002.

Haldane, J.B.S. "Natural Selection." In Darwin's Biological Work: Some

Aspects Reconsidered, edited by P.R. Bell, pp. 101–149. Cambridge: Cambridge University Press, 1959.

Kottler, Malcolm. "Charles Darwin and Alfred Russel Wallace: Two Decades of Debate over Natural Selection." In The Darwinian Heritage, edited by David Kohn, pp. 367–432. Princeton: Princeton University Press, 1985.

Sober, Elliot. The Nature of Selection. Cambridge: MIT Press, 1984.

Williams, George C. Adaptation and Natural Selection: A Critique of Some Current Evolutionary Thought. Princeton: Princeton University Press, 1966.

NATURPHILOSOPHIE

Mauch, Christof, ed. Nature in German History. New York: Berghahn Books, 2004.

Oken, Lorenz. Elements of Physiophilosophy. Translated by Alfred Tulk. London: printed for the Ray Society, 1847.

NEANDERTHAL MAN

Shreeve, James. The Neandertal Enigma: Solving the Mystery of Modern Human Origins. New York: William Morrow, 1995.

Stringer, Christopher, and Clive Gamble. In Search of the Neanderthals: Solving the Puzzle of Human Origins. New York: Thames & Hudson, 1993.

Tattersall, Ian. The Last Neanderthal: The Rise, Success, and Mysterious Extinction of Our Closest Human Relatives. New York: Macmillan, 1995.

Trinkaus, Erik, and Pat Shipman. The Neandertals: Changing the Image of Mankind. New York: Alfred A. Knopf, 1993.

NEBRASKA MAN

Gregory, William K. "Hesperopithecus: Apparently Not an Ape nor a Man." Science 66 (1927): 579–581.

Gregory, William K., Milo Hellman, and William Diller Matthew. "Notes on the Type of Hesperopithecus haroldcookii Osborn." American Museum Novitates, no. 53 (1923): 1–16.

NEO-DARWINISM

Bowler, Peter J. Evolution: The History of an Idea. Berkeley: University of California Press, 1984.

Mayr, Ernst. The Growth of Biological Thought: Diversity, Evolution, and Inheritance. Cambridge, Mass.: Harvard University Press, Belknap Press, 1982.

NEO-LAMARCKISM

Cook, George M. "Neo-Lamarckian Experimentalism in America; Origins and Consequences."

Quarterly Review of Biology 74 (1999): 417–437.

Hitching, Francis. The Neck of the Giraffe: Where Darwin Went Wrong. New Haven, Conn.: Ticknor & Fields, 1987.

NEOTENY

Gould, Stephen Jay. Ontogeny and Phylogeny. Cambridge, Mass.: Harvard University Press, Belknap Press, 1977.

NEURAL DARWINISM

Edelman, Gerald M. "Neural Darwinism: Population Thinking and Higher Brain Function." In How We Know, edited by Michael Shafto and Gerald M. Edelman, pp. 1–29. New York: Harper & Row, 1985.

———. "Group Selection as the Basis for Higher Brain Function." In The Organization of the Cerebral Cortex, edited by F.O. Schmitt et al., pp. 535–563. Cambridge, Mass.: MIT Press, 1981.

NEUTRAL TRAITS

Kimura, Motoo. The Neutral Theory of Molecular Evolution. Cambridge: Cambridge University Press, 1983.

King, J.L., and T.H. Jukes. "Non-Darwinian Evolution." Science 164 (1969): 788–798.

"NEW" LAMARCKISM

Sheldrake, Rupert. A New Science of Life: The Hypothesis of Formative Causation. Los Angeles: J.P. Tarcher, 1981.

Steele, E.J. Somatic Selection and Adaptive Evolution: On the Inheritance of Acquired Characters. Chicago: University of Chicago Press, 1981.

Wills, Christopher. The Wisdom of the Genes: New Pathways in Evolution. New York: Basic Books, 1989.

NIM CHIMPSKY

Chomsky, Noam. Language and Mind. New York: Harcourt, Brace, Jovanovich, 1968.

Terrace, Herbert. Nim: A Chimpanzee Who Learned Sign Language. New York: Alfred A. Knopf, 1979.

NOAH'S FLOOD

Custace, Arthur. The Flood: Local or Global? Grand Rapids, Mich.: Academic Books, 1979.

Godfrey, Laurie R., ed. Scientists Confront Creationism. New York: W.W. Norton, 1983.

"NOAH'S RAVENS"

Hitchcock, Edward. Ichnology of New England: A Report on the Sandstone of the Connecticut Valley, Especially Its Fossil Footmarks (1858). New York: Arno Press, 1974.

NOBLE SAVAGE

Ellingson, Terry Jay. The Myth of the Noble Savage. Berkeley: University of California Press, 2001.

Rousseau, Jean-Jacques. Discourse on the Origin of Inequality (1754). Translated by Franklin Philip. Oxford: Oxford University Press, 1994.

NOPCSA, BARON FRANZ

Colbert, Edwin. Men and Dinosaurs: The Search in Field and Laboratory. New York: E.P. Dutton, 1968.

Elsie, Robert. "The Viennese Scholar Who Almost Became King of Albania: Baron Franz Nopcsa and His Contribution to Albanian Studies." East European Quarterly 33 (1999): 327–337.

O'BRIEN, WILLIS

Archer, Steve. Willis O'Brien: Special Effects Genius. Jefferson, N.C.: McFarland, 1993.

Kinnard, Roy. The Lost World of Willis O'Brien: The Original Shooting Script of the 1925 Landmark Special Effects Dinosaur Film, with Photographs. Jefferson, N.C.: McFarland, 1993.

OCCAM'S RAZOR

Gordon, Malcolm S., and Soraya M. Bartol. Experimental Approaches to Conservation Biology. Berkeley: University of California Press, 2004.

Leinfellner, Werner, and Eckehart Köhler, eds. Developments in the Methodology of Social Science. Dordrecht: D. Reidel, 1974.

"OMEGA MAN"

Teilhard de Chardin, Pierre. The Phenomenon of Man. New York: Harper & Row, 1959.

OMPHALOS

Gosse, Edmund. Father and Son: A Study of Two Temperaments. 1907; repr. Stroud, Gloucestershire: Nonsuch, 2005.

Gosse, Phillip Henry. Omphalos: An Attempt to Untie the Geological Knot. 1857; repr. Woodbridge, Conn.: Ox Bow Press, 1998.

Thwaite, Ann. Glimpses of the Wonderful: The Life of Philip Henry Gosse. London: Faber & Faber, 2002.

ORANGUTAN

de Boer, Leobert E.M., ed. The Orang-utan: Its Biology and Conservation. The Hague: W. Junk, 1982.

Galdikas, Biruté M.F. Reflections of Eden: My Years with the Orangutans of Borneo. Boston: Little, Brown, 1995.

Peterson, Dale. The Deluge and the Ark: A Journey into Primate Worlds. Boston: Houghton Mifflin, 1989.

Schwartz, Jeffrey H. *The Red Ape: Orangutans and Human Origins.* Cambridge, Mass.: Westview Press, 2005.

ORCHIDS, DARWIN'S STUDY OF
Allan, Mea. *Darwin and His Flowers: The Key to Natural Selection.* New York: Taplinger, 1977.
Darwin, Charles. *On the Various Contrivances by Which Orchids are Fertilized by Insects.* London: John Murray, 1862; 2d ed., rev., 1877.
Scourse, Nicolette. *The Victorians and Their Flowers.* London: Croom Helm, 1983.

ORIGIN MYTHS (IN SCIENCE)
Darwin, Charles. *The Descent of Man, and Selection in Relation to Sex.* Edited by Carl Zimmer. New York: Plume, 2007.
Landau, Misia. *Narratives of Human Evolution.* New Haven, Conn.: Yale University Press, 1991.
Lewin, Roger. *Bones of Contention: Controversies in the Search for Human Origins.* Chicago: University of Chicago Press, 1997.
Long, Barry. *The Origins of Man and the Universe: The Myth That Came to Life.* London: Routledge & Kegan Paul, 1984.

ORIGIN OF SPECIES
Darwin, Charles. *Charles Darwin's Natural Selection, Being the Second Part of His Big Species Book Written from 1856 to 1858.* Edited by R.C. Stauffer. Cambridge: Cambridge University Press, 1975.
———. *On the Origin of Species by Means of Natural Selection.* London: John Murray, 1859.
———. *On the Origin of Species: The Illustrated Edition.* Edited by David Quammen. New York: Sterling, 2008.
Darwin, Francis, ed. *The Foundations of the Origin of Species by Charles Darwin.* (Includes Charles Darwin's earlier sketches of the theory.) Cambridge: Cambridge University Press, 1909.
Hyman, Stanley E. *The Tangled Bank: Darwin, Marx, Frazer, and Freud as Imaginative Writers.* New York: Atheneum, 1962.
Jones, Steve. *Darwin's Ghost: The Origin of Species Updated.* New York: Random House, 2000.
Simpson, George Gaylord. *The Book of Darwin.* New York: Washington Square Press, 1983.

ORTHOGENESIS
Simpson, George Gaylord. *The Meaning of Evolution: A Study of the History of Life and of Its Significance for Man.* 2d ed. New Haven: Yale University Press, 1967.

OWEN, SIR RICHARD
Barber, Lynn. *The Heyday of Natural History.* Garden City, N.Y.: Doubleday, 1980.
Desmond, Adrian. *Archetypes and Ancestors: Palaeontology in Victorian London, 1850–1875.* Chicago: University of Chicago Press, 1986.
Owen, Richard. *On the Archetype and Homologies of the Vertebrate Skeleton.* London: J. Van Voorst, 1848.
Rupke, Nicolaas A., *Richard Owen: Victorian Naturalist.* New Haven, Conn.: Yale University Press, 1994.

OXFORD DEBATE
Gould, Stephen Jay. "Knight Takes Bishop." In *Bully for Brontosaurus: Reflections in Natural History,* pp. 79–93. New York: W.W. Norton, 1991.
Irvine, William. *Apes, Angels, and Victorians: The Story of Darwin, Huxley, and Evolution.* New York: McGraw-Hill, 1955.
Tuckwell, Rev. William. *Reminiscences of Oxford.* London: Cassell, 1901.

PALEOLITHIC
Gamble, Clive. *The Palaeolithic Societies of Europe.* Cambridge: Cambridge University Press, 1999.
Johanson, Donald, and Blake Edgar. *From Lucy to Language.* Rev. ed. New York: Simon & Schuster, 2006.
Klein, Richard. *The Human Career: Human Biological and Cultural Origins.* Chicago: University of Chicago Press, 1999.
Lubbock, John. *Pre-historic Times: As Illustrated by Ancient Remains, and the Manners and Customs of Modern Savages.* London: Williams & Norgate, 1865.
Mellars, Paul. *The Emergence of Modern Humans: An Archaeological Perspective.* Ithaca, N.Y.: Cornell University Press, 1990.

PALEY'S WATCHMAKER
Barber, Lynn. *The Heyday of Natural History.* Garden City, N.Y.: Doubleday, 1980.
Dawkins, Richard. *The Blind Watchmaker: Why the Evidence of Evolution Reveals a Universe without Design.* New York: W.W. Norton, 1996.
Paley, William. *Natural Theology; or, Evidences of the Existence and Attributes of the Deity, Collected from the Appearances of Nature.* London: Baynes, 1802.

PANDA'S THUMB
Gould, Stephen Jay. *The Panda's Thumb: More Reflections in Natural History.* New York: W.W. Norton, 1980.

PANGEA
LeTourneau, Peter M., and Paul Eric Olsen. *The Great Rift Valleys of Pangea in Eastern North America.* New York: Columbia University Press, 2003.
Macdougall, J.D. *A Short History of Planet Earth: Mountains, Mammals, Fire, and Ice.* New York: Wiley, 1996.

PANSPERMIA
Crick, Francis. *Life Itself: Its Origin and Nature.* New York: Simon & Schuster, 1981.
Hoyle, Fred, and N.C. Wickramasinghe. *Evolution from Space: A Theory of Cosmic Creationism.* New York: Simon & Schuster, 1981.

PASTRANA, JULIA
Gylseth, Christopher Hals, and Lars O. Toverud. *Julia Pastrana: The Tragic Story of the Victorian Ape Woman.* Stroud, Gloucestershire: Sutton, 2003.
Tromp, Marlene. *Victorian Freaks: The Social Context of Freakery in Britain.* Columbus: Ohio State University Press, 2008.

PEALE'S MUSEUM
Barber, Lynn. *The Heyday of Natural History.* Garden City, N.Y.: Doubleday, 1980.
Sellers, Charles C. *Mr. Peale's Museum: Charles Willson Peale and the First Popular Museum of Natural Science and Art.* New York: W.W. Norton, 1980.

PEER REVIEW
Glen, William. *The Road to Jaramillo: Critical Years of the Revolution in Earth Science.* Stanford: Stanford University Press, 1982.
MacRoberts, Michael H. and Barbara R. "The Scientific Review System." *Speculations in Science and Technology* 3, no. 5 (1980): 573–578.

PEPPERED MOTH
Hooper, Judith. *Of Moths and Men: The Untold Story of Science and the Peppered Moth.* New York: W.W. Norton, 2002.
Kettlewell, Bernard. *The Evolution of Melanism: The Study of a Recurring Necessity; with Special Reference to Industrial Melanism in the Lepidoptera.* Oxford: Clarendon Press, 1973.
Majerus, M.E.N. *Melanism: Evolution in Action.* Oxford: Oxford University Press, 1998.

"PHILOSOPHICAL" NATURALISTS
Rehbock, Philip. *The Philosophical Naturalists: Themes in Early Nineteenth-Century British Biology.* Madison: University of Wisconsin Press, 1983.

PHRENOLOGY
De Giustino, David. *Conquest of Mind: Phrenology and Victorian Social Thought.* London: Croom Helm, 1975.
Martinez, Severiano. *Phrenology.* Lincoln, Neb.: Writers Club Press, 2001.

PHYLA
Margulis, Lynn, and Karlene V. Schwartz. *Five Kingdoms: An Illustrated Guide to the Phyla of Life on Earth.* New York: W.H. Freeman, 1998.
Valentine, James W. *On the Origin of Phyla.* Chicago: University of Chicago Press, 2004.

PILTDOWN MAN
Blinderman, Charles. *The Piltdown Inquest.* Buffalo, N.Y.: Prometheus Books, 1986.
Spencer, Frank. *Piltdown: A Scientific Forgery.* New York: Oxford University Press, 1990.
———. *The Piltdown Papers, 1908–1955: Correspondence and Other Documents Relating to the Piltdown Forgery.* New York: Oxford University Press, 1990.
Walsh, John Evangelist. *Unraveling Piltdown: The Scientific Fraud of the Century and Its Solution.* New York: Random House, 1996.
Weiner, J.S. *The Piltdown Forgery.* Oxford: Oxford University Press, 2003.

PITT-RIVERS MUSEUM
Bowden, Mark. *Pitt Rivers: The Life and Archaeological Work of Lieutenant-General Augustus Henry Lane Fox Pitt Rivers.* Cambridge: Cambridge University Press, 1991.
Stocking, George W. *Victorian Anthropology.* New York: Free Press, 1987.

PLANET OF THE APES
Boulle, Pierre. *Planet of the Apes: A Novel.* Translated by Xan Fielding. New York: Vanguard Press, 1963.
Greene, Eric. *Planet of the Apes as American Myth: Race and Politics in the Films and Television Series.* Jefferson, N.C.: McFarland, 1996.

PLATE TECTONICS
Kious, W. Jacquelyne, and Robert I. Tilling. *This Dynamic Earth: The Story of Plate Tectonics.* Washington, D.C.: U.S. Geological Survey, 1996.
Prothero, Donald R., and Robert H. Dott, Jr. *Evolution of the Earth.* 7th ed. New York: McGraw-Hill, 2004.

PLENITUDE
Sax, Dov F., John J. Stachowicz, and Steven D. Gaines. *Species Invasions: Insights into Ecology,*

Evolution, and Biogeography. Sunderland, Mass.: Sinauer Associates, 2005.

Stanley, Steven M. *The New Evolutionary Timetable: Fossils, Genes, and the Origin of the Species.* New York: Basic Books, 1981.

POLYMERASE CHAIN REACTION

Mullis, Kary B., François Ferré, and Richard Gibbs. *The Polymerase Chain Reaction.* Boston: Birkhäuser, 1994.

Rabinow, Paul. *Making PCR: A Story of Biotechnology.* Chicago: University of Chicago Press, 1996.

POPULATION THINKING

Crow, J.F., and M. Kimura. *An Introduction to Population Genetic Theory.* New York: Harper & Row, 1970.

Hutchinson, G. Evelyn. *An Introduction to Population Ecology.* New Haven, Conn.: Yale University Press, 1978.

Mayr, Ernst. "Typological versus Population Thinking." In *Conceptual Issues in Evolutionary Biology: An Anthology,* edited by Elliott Sober, pp. 157–160. Cambridge, Mass.: MIT Press, 1994.

POSITIVISM

Ayer, A.J. *Logical Positivism.* Glencoe, Ill.: Free Press, 1959.

Smith, Warren S. *The London Heretics, 1870–1914.* New York: Dodd, Mead, 1968.

POZZUOLI, PILLARS OF

Gould, Stephen Jay. "Pozzuoli's Pillars Revisited." *Natural History,* May 1999, pp. 24, 81–91.

Lyell, Charles. *Principles of Geology.* 3 vols. 1830–1833; repr. Chicago: University of Chicago Press, 1990–1991.

PRESTWICH, SIR JOSEPH

Prestwich, Joseph. *Geology: Chemical, Physical, and Stratigraphical.* Oxford: Clarendon Press, 1886.

Woodward, Horace B. *Joseph Prestwich.* Washington: G.P.O., 1898.

PRIMATES

Beard, Chris. *The Hunt for the Dawn Monkey: Unearthing the Origins of Monkeys, Apes, and Humans.* Berkeley: University of California Press, 2004.

Bourne, Geoffrey H. *Primate Odyssey.* New York: Putnam, 1974.

Cheney, Dorothy L., and Robert M. Seyfarth. *How Monkeys See the World: Inside the Mind of Another Species.* Chicago: University of Chicago Press, 1990.

Dunbar, Robin, and Louise Barrett. *Cousins: Our Primate Relatives.* London: Sean Moore, 2000.

Fleagle, John G. *Primate Adaptation and Evolution.* 2d ed. San Diego: Academic Press, 1998.

Groves, Colin. *Extended Family: Long Lost Cousins. A Personal Look at the History of Primatology.* Arlington, Va.: Conservation International, 2008.

Haraway, Donna. *Primate Visions: Gender, Race, and Nature in the World of Modern Science.* New York: Routledge, 1989.

Hartwig, Walter. *The Primate Fossil Record.* Cambridge: Cambridge University Press, 2002.

Kavanagh, Michael. *A Complete Guide to Monkeys, Apes, and Other Primates.* New York: Viking Press, 1983.

Peterson, Dale. *The Deluge and the Ark: A Journey into Primate Worlds.* Boston: Houghton Mifflin, 1989.

PROGRESSIONISM

Bowler, P.J. *Fossils and Progress: Paleontology and the Idea of Progressive Evolution in the Nineteenth Century.* New York: Science History Publications, 1976.

Coleman, William. *Georges Cuvier, Zoologist: A Study in the History of Evolution Theory.* Cambridge, Mass.: Harvard University Press, 1964.

Gohau, Gabriel, Albert V. Carozzi, and Marguerite Carozzi. *A History of Geology.* New Brunswick, N.J.: Rutgers University Press, 1990.

Huggett, Richard. *Catastrophism: Asteroids, Comets, and Other Dynamic Events in Earth History.* 2d ed. London: Verso, 1997.

PUNCTUATED EQUILIBRIUM

Eldredge, Niles. *Life Pulse: Episodes from the Story of the Fossil Record.* New York: Facts on File, 1987.

———. *Time Frames: The Rethinking of Darwinian Evolution and the Theory of Punctuated Equilibria.* New York: Simon & Schuster, 1985. (Includes reprint of the 1972 Eldredge and Gould paper, "Punctuated Equilibria: An Alternative to Phyletic Gradualism.")

Stanley, Steven. *The New Evolutionary Timetable.* New York: Basic Books, 1981.

QUEST FOR FIRE

Brach, Gérard. *Quest for Fire: Screenplay.* London: Scripts Ltd., 1980.

De Paolo, Charles. *Human Prehistory in Fiction.* Jefferson, N.C.: McFarland, 2003.

Sanello, Frank. *Reel v. Real: How Hollywood Turns Fact into Fiction.* Lanham, Md.: Taylor Trade Publications, 2003.

RACE

Blakely, Robert, and Judith Harrington. *Bones in the Basement: Post-Mortem Racism in Nineteenth-Century Medical Training.* Washington, D.C.: Smithsonian Institution Press, 1997.

Dobzhansky, Theodosius. *Mankind Evolving.* New Haven, Conn.: Yale University Press, 1962.

Gould, Stephen Jay. *The Mismeasure of Man.* New York: W.W. Norton, 1981.

Lewontin, Richard C. *Human Diversity.* New York: Scientific American Library, 1995.

Montague, Ashley. *Man's Most Dangerous Myth: The Fallacy of Race.* 5th ed. Oxford: Oxford University Press, 1974.

Shipman, Pat. *The Evolution of Racism: Human Differences and the Use and Abuse of Science.* New York: Simon & Schuster, 1994.

Stanton, William. *The Leopard's Spots: Scientific Attitudes toward Race in America, 1815–1859.* Chicago: University of Chicago Press, 1960.

RAIN FOREST CRISIS

Alameda, Frank, and Catherine M. Pringle, eds. *Tropical Rainforests: Diversity and Conservation.* San Francisco: California Academy of Sciences, 1988.

Caufield, Catherine. *In the Rainforest: Report from a Strange, Beautiful, Imperiled World.* Chicago: University of Chicago Press, 1991.

Lowman, Margaret D. *Life in the Treetops: Adventures of a Woman in Field Biology.* New Haven, Conn.: Yale University Press, 1999.

Wilson, E.O., and Frances M. Peter, eds. *Biodiversity.* Washington, D.C.: National Academy Press, 1988.

(CHIEF) RED CLOUD

Hyde, George E. *Red Cloud's Folk.* Norman: University of Oklahoma Press, 1937.

Plate, Robert. *The Dinosaur Hunters: Marsh and Cope.* New York: McKay, 1964.

RESISTANT STRAINS

Farmer, Paul. *Infections and Inequalities: The Modern Plagues.* Berkeley: University of California Press, 1999.

Häusler, Thomas. *Viruses vs. Superbugs: A Solution to the Antibiotics Crisis?* London: Macmillan, 2006.

RITUALIZATION

Burkhardt, Richard W. *Patterns of Behavior: Konrad Lorenz, Niko Tinbergen, and the Founding of Ethology.* Chicago: University of Chicago Press, 2005.

Darwin, Charles. *The Expression of the Emotions in Man and Animals* (1872). Edited by Paul Ekman. Oxford: Oxford University Press, 1998.

Lorenz, Konrad. *King Solomon's Ring: New Light on Animal Ways.* Translated by Marjorie Kerr Wilson. New York: Meridian, 1997.

Tinbergen, Niko. *Animal Behavior.* New York: Time-Life Books, 1965.

ROMANES, GEORGE JOHN

Boakes, Robert. *From Darwin to Behaviorism: Psychology and the Minds of Animals.* Cambridge: Columbia University Press, 1984.

Romanes, Ethel Duncan, ed. *The Life and Letters of George John Romanes.* London: Longmans, Green, 1896.

Romanes, George John. *Mental Evolution in Animals, with a Posthumous Essay on Instinct by Charles Darwin.* London: Kegan Paul, Trench, 1883.

ROOSEVELT, THEODORE

Cutright, Paul Russell. *Theodore Roosevelt the Naturalist.* New York: Harper & Bros., 1956.

Roosevelt, Theodore. *Life Histories of African Game Animals.* New York: Scribner, 1914.

ROTHSCHILD, LORD WALTER

Rothschild, Miriam. *Dear Lord Rothschild: Birds, Butterflies, and History.* Philadelphia: Balaban, 1983.

———. *Walter Rothschild: The Man, the Museum, and the Menagerie.* London: Natural History Museum, 2008.

SANDWALK

Hosler, Jay. *The Sandwalk Adventures: An Adventure in Evolution Told in Five Chapters.* Columbus, Ohio: Active Synapse, 2003.

"SARAWAK LAW"

McKinney, H. Lewis. *Wallace and Natural Selection.* New Haven, Conn.: Yale University Press, 1972.

Wallace, Alfred Russel. "On the Law Which Has Regulated the Introduction of New Species." In *Alfred Russel Wallace: An Anthology of His Shorter Writings,* edited by Charles E. Smith, pp. 220–231. New York: Oxford University Press, 1991.

SCIENCE

Coley, Noel, and Vance Hall, eds. *Darwin to Einstein: Primary Sources on Science and Belief.* London: Longman, 1980.

Jones, William B. "Scientific Explanations and Religious Accounts: A Historical and Critical Exposition for the Modern Pilgrim." *Dialogue and Alliance* 1, no. 4 (1987–88): 4–9.

Kuhn, Thomas S. *The Structure of Scientific Revolutions.* Chicago: University of Chicago Press, 1970.

Nagel, Ernest. *The Structure of Science: Problems in the Logic of*

Scientific Explanation. New York: Harcourt, Brace & World, 1961.

Wolpert, Lewis. *The Unnatural Nature of Science.* Cambridge, Mass.: Harvard University Press, 1993.

"SCIENTIFIC CREATIONISM"

Kitcher, Philip. *Abusing Science: The Case against Creationism.* Cambridge, Mass.: MIT Press, 1982.

Scott, Eugenie, and Henry Cole. "The Elusive Scientific Basis of Creation 'Science.'" *Quarterly Review of Biology* 60 (1985): 21–30.

SCIOLISM

Chambers, Robert. *"Vestiges of the Natural History of Creation" and Other Evolutionary Writings.* Edited by James A. Secord. Chicago: University of Chicago Press, 1994.

Miller, Hugh. *The Footprints of the Creator; or, The Asterolepis of Stromness.* With memoir of the author by Louis Agassiz. Boston: Gould & Lincoln, 1858.

SCOPES TRIAL

Caudhill, Edward, Edward Larson, and Jesse Fox Mayshark. *The Scopes Trial: A Photographic History.* Knoxville: University of Tennessee Press, 2000.

Darrow, Clarence, and William J. Bryan. *The World's Most Famous Court Trial: Tennessee Evolution Case.* Cincinnati: National Book Company, 1925; repr. Union, N.J.: Law Book Exchange, 1999.

De Camp, L. Sprague. *The Great Monkey Trial.* Garden City, N.Y.: Doubleday, 1968.

Larson, Edward J. *Summer for the Gods: The Scopes Trial and America's Continuing Debate over Science and Religion.* New York: Basic Books, 1997.

Lienesch, Michael. *In the Beginning: Fundamentalism, the Scopes Trial, and the Making of the Antievolution Movement.* Chapel Hill: University of North Carolina Press, 2007.

Scopes, John T. *Center of the Storm: Memoirs of John T. Scopes.* New York: Holt, Rinehart & Winston, 1967.

SCOPES II

Chapman, Matthew. *40 Days and 40 Nights: Darwin, Intelligent Design, God, Oxycontin, and Other Oddities on Trial in Pennsylvania.* New York: Collins, 2007.

Futuyma, Douglas. *Science on Trial: The Case for Evolution.* New York: Pantheon Books, 1983.

Gilkey, Langdon. *Creationism on Trial: Evolution and God at Little Rock.* Minneapolis: Winston Press, 1985.

Godfrey, Laurie, ed. *Scientists Confront Creationism.* New York: W. W. Norton, 1983.

Humes, Edward. *Monkey Girl: Evolution, Education, Religion, and the Battle for America's Soul.* New York: Ecco, 2007.

Milner, Richard. "Darwin in Court." *Natural History* 116, no. 5 (2007): 28–32.

Slack, Gordy. *The Battle over the Meaning of Everything: Evolution, Intelligent Design, and a School Board in Dover, PA.* San Francisco: Jossey-Bass, 2007.

SECULAR HUMANISM

Lamont, Corliss. *The Philosophy of Humanism.* New York: Continuum, 1990.

Kurtz, Paul. *What Is Secular Humanism?* New York: Prometheus Books, 2007.

SEDGWICK, REVEREND ADAM

Clark, John, and Thomas Hughes. *Life and Letters of Reverend Adam Sedgwick.* 2 vols. Cambridge: Cambridge University Press, 1890.

Rudwick, M.J.S. *The Great Devonian Controversy: The Shaping of Scientific Knowledge among Gentlemanly Specialists.* Chicago: University of Chicago Press, 1985.

Woodward, Horace B. *History of Geology.* New York: Putnam, 1911.

"SELFISH GENE"

Dawkins, Richard. *The Selfish Gene.* New York: Oxford University Press, 1976.

Weismann, August. *The Germ-Plasm; A Theory of Heredity.* Translated by W. Newton Parker and Harriet Rönnfeldt. New York: Scribner, 1893.

SEXUAL SELECTION

Andersson, Malte. "Female Choice Selects for Extreme Tail Length in a Widowbird." *Nature* 299 (1982): 818–820.

Bateson, Patrick, ed. *Mate Choice.* Cambridge: Cambridge University Press, 1983.

Cronin, Helena. *The Ant and the Peacock: Altruism and Sexual Selection from Darwin to Today.* New York: Cambridge University Press, 1992.

Darwin, Charles. *The Descent of Man, and Selection in Relation to Sex* (1871). London: Penguin Books, 2004.

SHAW, GEORGE BERNARD

Henderson, Archibald. *George Bernard Shaw: His Life and Works—A Critical Biography.* Whitefish, Mont.: Kessinger, 2004.

Shaw, George Bernard. *Back to Methuselah: A Metabiological Pentateuch.* London: Constable, 1922.

———. *Man and Superman: A Comedy and a Philosophy* (1903). Edited by Dan H. Laurence. London: Penguin Books, 1957.

SIMPSON, GEORGE GAYLORD

Laporte, Léo F. *George Gaylord Simpson: Paleontologist and Evolutionist.* New York: Columbia University Press, 2000.

Simpson, George G. *Attending Marvels: A Patagonian Journal.* New York: Macmillan, 1934.

———. *The Major Features of Evolution.* New York: Columbia University Press, 1953.

———. *Tempo and Mode in Evolution.* New York: Columbia University Press, 1944.

———. *This View of Life: The World of an Evolutionist.* New York: Harcourt, Brace & World, 1964.

SLADE TRIAL

Milner, Richard. "Charles Darwin and Associates: Ghostbusters." *Scientific American*, Oct. 1996, 96–100.

———. "Darwin for the Prosecution, Wallace for the Defense: How Two Great Naturalists Put the Supernatural on Trial." *North Country Naturalist* 2 (1990): 19–35.

———. "Darwin for the Prosecution, Part II: Spirit of a Dead Controversy." *North Country Naturalist* 2 (1990): 37–50.

SMITH, SYDNEY

Burkhardt, Frederick, and Sydney Smith, eds. *A Calendar of the Correspondence of Charles Darwin, 1821–1882.* 2d ed. Cambridge: Cambridge University Press, 1994.

Darwin, Charles. *Charles Darwin's Notebooks, 1836–1844: Geology, Transmutation of Species, Metaphysical Enquiries.* Edited by Paul H. Barrett, Peter J. Gautrey, Sandra Herbert, David Kohn, and Sydney Smith. Cambridge: Cambridge University Press, 1987.

Virgin, Peter. *Sydney Smith.* London: HarperCollins, 1994.

SMITH, WILLIAM

Eiseley, Loren. *Darwin's Century: Evolution and the Men Who Discovered It.* New York: Doubleday, 1958.

Morton, John L. *Strata: How William Smith Drew the First Map of the Earth in 1801 and Inspired the Science of Geology.* Franklin Park, Ill.: Arcadia, 2001.

Smith, William. *Stratigraphical System of Organized Fossils.* London: W. Arding, 1816.

Wigley, Peter, Peter Dolan, Tom Sharpe, and Hugh Torrens. *"Strata" Smith: His Two Hundred Year Legacy.* London: Geological Society, 2005.

Winchester, Simon. *The Map That Changed the World: William Smith and the Birth of Modern Geology.* New York: HarperCollins, 2001.

SOCIAL BEHAVIOR, EVOLUTION OF

Bonner, John Tyler. *The Evolution of Culture in Animals.* Princeton: Princeton University Press, 1980.

Chance, Michael, and Clifford Jolly. *Social Groups of Monkeys, Apes, and Men.* London: E. P. Dutton, 1970.

Douglas-Hamilton, Ian, and Oria Douglas-Hamilton. *Among the Elephants.* New York: Viking Press, 1975.

MacRoberts, M.H., and Barbara MacRoberts. *Social Organization and Behavior of the Acorn Woodpecker in Central Coastal California.* Ornithological Monographs, 21. Washington, D.C.: American Ornithologists Union, 1976.

Moynihan, Martin H. *The Social Regulation of Competition and Aggression in Animals.* Washington, D.C.: Smithsonian Institution Press, 1998.

Trivers, R.L. *Social Evolution.* Menlo Park, Calif.: Benjamin/Cummings, 1985.

de Waal, Frans, ed. *Tree of Origin: What Primate Behavior Can Tell Us about Human Social Evolution.* Cambridge, Mass.: Harvard University Press, 2001.

SOCIAL DARWINISM

Bannister, Robert C. *Social Darwinism: Science and Myth in Anglo-American Social Thought.* Philadelphia: Temple University Press, 1979.

Hofstadter, Richard. *Social Darwinism in American Thought, 1860–1915.* 2d rev. ed. New York: Braziller, 1959.

Jones, Greta. *Social Darwinism and English Thought: The Interaction between Biological and Social Theory.* Atlantic Highlands, N.J.: Humanities Press, 1980.

SPECIES, CONCEPT OF

Eldredge, Niles. *Unfinished Synthesis: Biological Hierarchies and Modern Evolutionary Thought.* New York: Oxford University Press, 1985.

Mayr, Ernst. *Populations, Species, and Evolution.* Cambridge, Mass.: Harvard University Press, 1970.

Otte, Daniel, and John A. Endler, eds. *Speciation and Its Consequences.* Sunderland, Mass.: Sinauer, 1989.

Salthe, Stanley N. *Evolving Hierarchical Systems: Their Structure and Representation.* New York: Columbia University Press, 1985.

SPENCER, HERBERT

Elliot, Hugh. *Herbert Spencer*. New York: Henry Holt & Co., 1917.

Francis, Mark. *Herbert Spencer and the Invention of Modern Life*. Ithaca, N.Y.: Cornell University Press, 2007.

Galton, Francis. *Memories of My Life*. London: Methuen, 1908.

Jones, Greta, and Robert A. Peel. *Herbert Spencer: The Intellectual Legacy*. London: Galton Institute, 2004.

SPERM COMPETITION

Birkhead, Tim. *Promiscuity: An Evolutionary History of Sperm Competition*. Cambridge, Mass.: Harvard University Press, 2000.

Birkhead, T.R., and A.P. Møller, eds. *Sperm Competition and Sexual Selection*. San Diego: Academic Press, 1998.

Smith, Robert L. *Sperm Competition and the Evolution of Animal Mating Systems*. Orlando, Fla.: Academic Press, 1984.

SPIRITUALISM

Kottler, Malcolm J. "Alfred Russel Wallace, the Origin of Man, and Spiritualism." *Isis* 65 (1974): 145–192.

Podmore, Frank. *Modern Spiritualism*. 2 vols. London: Methuen, 1902.

Washington, Peter. *Madame Blavatsky's Baboon: A History of the Mystics, Mediums, and Misfits Who Brought Spiritualism to America*. New York: Schocken, 1996.

Weisberg, Barbara. *Talking to the Dead: Kate and Maggie Fox and the Rise of Spiritualism*. New York: HarperOne, 2005.

SPRENGEL, CHRISTIAN KONRAD

Darwin, Charles, and Francis Darwin. *The Various Contrivances by Which Orchids Are Fertilised by Insects*. Chicago: University of Chicago Press, 1984.

Sprengel, Christian Konrad. *The Secret of Nature in the Form and Fertilization of Flowers Discovered*. 1793; repr. Washington, D.C.: Saad Publications, 1975.

STEADY-STATE EARTH

Lyell, Charles. *Principles of Geology*. 3 vols. 1830–1833; repr. Chicago: University of Chicago Press, 1990–1991.

Repcheck, Jack. *The Man Who Found Time: James Hutton and the Discovery of the Earth's Antiquity*. Cambridge, Mass.: Perseus, 2003.

STOPES, MARIE C.

Duffield, Arabella. "Marie Stopes, Eugenics, and the Birth Control Movement." *Journal of Biological Science* 30 (1998): 135–144.

Hall, Lesley A. *Hidden Anxieties: Male Sexuality, 1900–1950*. Cambridge: Polity Press, 1991.

Stopes, Marie Carmichael. *Ancient Plants: Being a Simple Account of the Past Vegetation of the Earth and of the Recent Important Discoveries Made in This Realm of Nature Study*. 1910; repr. Dehra Dun, India: International Book Distributors, 2004.

"SUE" THE TYRANNOSAURUS

Fiffer, Steve, and Robert T. Baker. *Tyrannosaurus Sue: The Extraordinary Saga of the Largest, Most Fought Over T Rex Ever Found*. New York: W.H. Freeman, 2001.

Horner, John R. and Don Lessem. *The Complete T. Rex: How Stunning New Discoveries Are Changing Our Understanding of the World's Most Famous Dinosaur*. New York: Touchstone Books, 1993.

SULFUR-BASED ECOSYSTEMS

Hose, Louise D. "Cave of the Sulfur-Eaters." *Natural History*, April 1999, pp. 54–61.

Schopf, J. William. *Cradle of Life: The Discovery of Earth's Earliest Fossils*. Princeton: Princeton University Press, 1999.

Taylor, Michael Ray. *Dark Life: Martian Nanobacteria, Rock-Eating Cave Bugs, and Other Extreme Organisms of Inner Earth and Outer Space*. New York: Scribner, 1999.

Tyson, Peter. "Neptune's Furnace." *Natural History*, June 1999, pp. 42–47.

"SURVIVAL OF THE FITTEST"

Darwin, Charles. *On the Origin of Species: The Illustrated Edition*. Edited by David Quammen. New York: Sterling, 2008.

Northcutt, Wendy. *The Darwin Awards 3: Survival of the Fittest*. New York: E.P. Dutton, 2003.

Spencer, Herbert. *The Principles of Biology*. New York: D. Appleton, 1898.

Sumner, William Graham. *The Challenge of Facts and Other Essays*. New Haven: Yale University Press, 1914.

SYNTHETIC THEORY

Ceccarelli, Leah. *Shaping Science with Rhetoric: The Cases of Dobzhansky, Schrödinger, and Wilson*. Chicago: University of Chicago Press, 2001.

Dobzhansky, Theodosius. *Genetics and the Origin of Species*. New York: Columbia University Press, 1937.

Eldredge, Niles, and Stephen J. Gould. "Punctuated Equilibria: An Alternative to Phyletic Gradual-

ism." In *Models in Paleobiology*, edited by Thomas J.M. Schopf, pp. 82–115. San Francisco: Freeman, Cooper and Co., 1972.

Gould, Stephen Jay. *The Structure of Evolutionary Theory*. Cambridge, Mass: Harvard University Press, Belknap Press, 2002.

Huxley, Julian. *Evolution: The Modern Synthesis*. London: George Allen & Unwin, 1942.

Mayr, Ernst. *Systematics and the Origin of Species*. New York: Columbia University Press, 1942.

Mayr, Ernst, and William B. Provine, eds. *The Evolutionary Synthesis: Perspectives on the Unification of Biology*. Cambridge, Mass.: Harvard University Press, 1998.

Simpson, George G. *Tempo and Mode in Evolution*. New York: Columbia University Press, 1944.

TANGLED BANK

Hyman, Stanley E. *The Tangled Bank: Darwin, Marx, Frazer, and Freud as Imaginative Writers*. New York: Atheneum, 1962.

TARZAN OF THE APES

Berglund, Jeff. *Cannibal Fictions: American Explorations of Colonialism, Race, Gender, and Sexuality*. Madison: University of Wisconsin Press, 2006.

Burroughs, Edgar Rice. *Tarzan of the Apes* (1914). New York: Penguin Books, 1990.

Utz, Richard J., and Elizabeth Sharpe. *Investigating the Unliterary: Six Essays on Burroughs' "Tarzan of the Apes."* Regensburg, Ger.: Verlag Ulrich Martzinek, 1995.

TAUNG CHILD

Thomas, Herbert. *Human Origins: The Search for Our Beginnings*. New York: Harry N. Abrams, 1995.

Tobias, Phillip V. *Dart, Taung, and the "Missing Link."* Johannesburg: Witwatersrand University Press, 1984.

TEILHARD DE CHARDIN, PIERRE

Aczel, Amir. *The Jesuit and the Skull: Teilhard de Chardin, Evolution, and the Search for Peking Man*. New York: Riverhead Books, 2007.

Lukas, Mary, and Ellen Lukas. *Teilhard: The Man, the Priest, the Scientist*. New York: Doubleday, 1977.

Teilhard de Chardin, Pierre. *The Phenomenon of Man*. Translated by Bernard Wall. New York: Harper & Row, 1975.

TELEOLOGY

Dawkins, Richard. *The Blind Watchmaker: Why the Evidence of Evolution Reveals a Universe without Design*. New York: W.W. Norton, 1996.

Gould, Stephen Jay. *The Panda's Thumb: More Reflections on Natural History*. New York: W.W. Norton, 1980.

TEMPLE OF NATURE

Darwin, Erasmus. *The Temple of Nature; or, The Origin of Society*. London: J. Johnson, 1803.

King-Hele, Desmond. *Doctor of Revolution: The Life and Genius of Erasmus Darwin*. London: Faber & Faber, 1977.

TENNYSON'S *IN MEMORIAM*

Roppen, Georg. *Evolution and Poetic Belief: A Study in Some Victorian and Modern Writers*. Oslo: Oslo University Press, 1956.

Ruse, Michael. *The Darwinian Revolution: Science Red in Tooth and Claw*. Chicago: University of Chicago Press, 1979.

"TERNATE PAPER"

Brooks, John L. *Just Before the Origin: Alfred Russel Wallace's Theory of Evolution*. New York: Columbia University Press, 1984.

Wallace, Alfred Russel. "On the Tendency of Varieties to Depart Indefinitely from the Original Type" (1858). In *Infinite Tropics: An Alfred Russel Wallace Anthology*, edited by Andrew Berry, pp. 52–62. New York: Verso, 2002.

THEORY, SCIENTIFIC

Ghiselin, Michael T. *The Triumph of the Darwinian Method*. Berkeley: University of California Press, 1969.

Giere, Ronald N. *Understanding Scientific Reasoning*. New York: Holt, Rinehart & Winston, 1979.

TINBERGEN, NIKO

Baerends, Gerard, Colin Beer, and Aubrey Manning, eds. *Function and Evolution in Behaviour: Essays in Honour of Professor Niko Tinbergen*. Oxford: Clarendon Press, 1975.

Fuller, Ray, ed. *Seven Pioneers of Psychology: Behaviour and Mind*. London: Routledge, 1995.

Tinbergen, Niko. *Animal Behavior*. New York: Time-Life Books, 1980.

TOOL-USING, EVOLUTION OF

Bradshaw, John L., and Lesley J. Rogers. *The Evolution of Lateral Asymmetries, Language, Tool Use, and Intellect*. San Diego: Academic Press, 1993.

Gibson, Kathleen Rita, and Tim Ingold. *Tools, Language, and Cognition in Human Evolution*. Cambridge: Cambridge University Press, 1993.

Goodall, Jane. "Learning from the Chimpanzees: A Message Humans Can Understand." *Science* 282 (1998): 2184–2185.

Ridley, Matt. *The Agile Gene: How*

Nature Turns on Nurture. New York: Harper Perennial, 2004.

TOUMAI SKULL
Boyd, Robert, and Joan B. Silk. *How Humans Evolved.* New York: W.W. Norton, 2002.
Gibbons, Ann. *The First Human: The Race to Discover Our Earliest Ancestors.* New York: Doubleday, 2006.

TRANSITIONAL FORMS
Futuyma, Douglas J. *Science on Trial: The Case for Evolution.* Sunderland, Mass.: Sinauer Associates, 1995.
Martin, Robert A. *Missing Links: Evolutionary Concepts and Transitions through Time.* Boston: Jones & Bartlett, 2004.
Prothero, Donald. *Evolution: What the Fossils Say and Why It Matters.* New York: Columbia University Press, 2007.
Shubin, Neil. *Your Inner Fish: A Journey into the 3.5-Billion-Year History of the Human Body.* New York: Pantheon Books, 2008.

TREE OF LIFE
Eldredge, Niles. *Darwin: Discovering the Tree of Life.* New York: W.W. Norton, 2005.
Lewontin, Richard. *It Ain't Necessarily So: The Dream of the Human Genome and Other Illusions.* 2d ed. New York: New York Review of Books, 2001.
Tree of Life Project: http://tolweb .org.

TURKANA BOY
Leakey, Richard, and Roger Lewin. *Origins Reconsidered: In Search of What Makes Us Human.* New York: Anchor Books, 1993.
Walker, Alan, and Richard Leakey, eds. *The Nariokotome Homo Erectus Skeleton.* Cambridge, Mass.: Harvard University Press, 1993.
Walker, Alan, and Pat Shipman. *The Wisdom of the Bones: In Search of Human Origins.* New York: Alfred A. Knopf, 1996.

TURKANA MOLLUSKS
Jones, J.S. "An Uncensored Page of Fossil History." *Nature* 293 (1981): 427–428.
Williamson, Peter G. "Evidence for the Major Features and Development of Rift Palaeolakes in the Neogene of East Africa from Certain Aspects of Lacustrine Mollusc Assemblages." In *Geological Background to Fossil Man: Recent Research in the Gregory Rift Valley, East Africa,* edited by Walter W. Bishop, 507–527. Edinburgh: Scottish Academic Press, 1978.
———. "Palaeontological Documentation of Speciation in Cenozoic Molluscs from Turkana Basin." *Nature* 293 (1981): 437–443.

TWAIN, MARK
Twain, Mark. *The Hidden Mark Twain: A Collection of Little-Known Mark Twain.* Edited by Anne Ficklen. New York: Crown Books, 1984.
———. *Letters from the Earth.* Edited by Bernard DeVoto. New York: Harper & Row, 1938.

"TWO BOOKS," DOCTRINE OF THE
Bacon, Francis. *The Advancement of Learning* (1605). Edited by G.W. Kitchin. Philadelphia: Paul Dry Books, 2001.
Eiseley, Loren. *The Man Who Saw through Time.* New York: Scribner, 1973. (Originally published in 1961 as *Francis Bacon and the Modern Dilemma.*)
Moore, James R. "Geologists and Interpreters of Genesis in the Nineteenth Century." In *God and Nature: Historical Essays on the Encounters Between Christianity and Science,* ed. David C. Lindberg and Ronald L. Numbers, pp. 322–350. Berkeley: University of California Press, 1986.

2001: A SPACE ODYSSEY
Clarke, Arthur Charles. *2001: A Space Odyssey.* New York: New American Library, 1968.
Stork, David G. *HAL's Legacy: "2001"'s Computer as Dream and Reality.* Cambridge, Mass.: MIT Press, 1997.

TYNDALL, JOHN
O'Flynn, Barry. *John Tyndall: A Biographical Note.* Kilcoole, Ire.: Cló Ceó Draoidheachta, 1999.
Otis, Laura, ed. *Literature and Science in the Nineteenth Century: An Anthology.* New York: Oxford University Press, 2002.
Tyndall, John. *Fragments of Science for Unscientific People: A Series of Detached Essays, Lectures, and Reviews.* London: Longmans, 1871.
Yamalidou, Maria. "John Tyndall, the Rhetorician of Molecularity. Part Two. Questions Put to Nature." *Notes and Records of the Royal Society of London* 53 (1999): 319–331.

UNIFORMITARIANISM
Gould, Stephen Jay. *Time's Arrow, Time's Cycle: Myth and Metaphor in the Discovery of Geological Time.* Cambridge, Mass.: Harvard University Press, 1987.
Lyell, Charles. *Principles of Geology.* 3 vols. 1830–1833; repr. Chicago: University of Chicago Press, 1990–1991.
Rudwick, M.J.S. *The Meaning of Fossils: Episodes in the History of Paleontology.* London: Macdonald, 1972.

USSHER-LIGHTFOOT CHRONOLOGY
Ford, Alan. *James Ussher: Theology, History, and Politics in Early-Modern Ireland and England.* Oxford: Oxford University Press, 2007.
Ussher, James. *The Annals of the World: Deduced from the Origin of Time.* London: Printed by E. Tyler, 1658.

VEBLEN, THORSTEIN
Jorgensen, Elizabeth Watkins, and Henry Irvin Jorgensen. *Thorstein Veblen: Victorian Firebrand.* Armonk, N.Y.: M.E. Sharpe, 1999.
Veblen, Thorstein. *The Theory of the Leisure Class.* Oxford: Oxford University Press, 2007.

"VERCORS" (JEAN BRULLER)
Dicks, Nathaniel. "The Essence of Man as Reflected in Jean Bruller-Vercor's Writings: 'Les Animaux Dénaturés' and 'Sylvia.'" Ph.D. diss., Atlanta University, 1973.
Vercors. *You Shall Know Them.* Translated by Rita Barisse. Boston: Little, Brown, 1953.

VESTIGES OF CREATION
Chambers, Robert. *"Vestiges of the Natural History of Creation" and Other Evolutionary Writings.* Edited by James A. Secord. Chicago: Chicago University Press, 1994.
Lynch, John M., ed. *"Vestiges" and the Debate Before Darwin.* Bristol, Eng.: Thoemmes Press, 2000.
Secord, James A. *Victorian Sensation: The Extraordinary Publication, Reception, and Secret Authorship of "Vestiges of the Natural History of Creation."* Chicago: University of Chicago Press, 2000.

VOYAGE OF HMS *BEAGLE*
Barlow, Lady Nora, ed. *Charles Darwin's Diary of the Voyage of HMS "Beagle."* Cambridge: Cambridge University Press, 1932.
Brosse, Jacques. *Great Voyages of Discovery.* New York: Facts on File, 1983.
Darwin, Charles. *Charles Darwin and the Voyage of the "Beagle."* Edited by Lady Nora Barlow. New York: Philosophical Library, 1945.
———. *Journal of Researches into the Geology and Natural History of the Various Countries Visited by HMS "Beagle."* London: H. Colburn, 1839.
Keynes, Richard. *The "Beagle" Record: Selections from the Original Pictorial Records and Written Accounts of the Voyage of HMS "Beagle."* Cambridge: Cambridge University Press, 1979.
———. *Fossils, Finches and Fuegians: Darwin's Adventures and Discoveries on the "Beagle."* New York: Oxford University Press, 2003.
Moorehead, Alan. *Darwin and the "Beagle."* New York: Harper & Row, 1969.

WAGNER, MORITZ
Mayr, Ernst. "Darwin and Isolation." In *Evolution and the Diversity of Life: Selected Essays.* Cambridge, Mass.: Harvard University Press, Belknap Press, 1982.
———. *The Growth of Biological Thought: Diversity, Evolution, and Inheritance.* Cambridge, Mass.: Harvard University Press, Belknap Press, 1982.
Wagner, Moritz. *The Darwinian Theory and the Law of the Migration of Organisms.* Translated by J.L. Laird. London: E. Stanford, 1873.

WALLACE, ALFRED RUSSEL
Berry, Andrew. *Infinite Tropics: An Alfred Russel Wallace Anthology.* London: Verso Books, 2002.
Brooks, John. *Just before the Origin: Alfred Russel Wallace's Theory of Evolution.* New York: Columbia University Press, 1984.
Marchant, James. *Alfred Russel Wallace: Letters and Reminiscences.* 2 vols. London: Cassell, 1916.
McKinney, H. Lewis. *Wallace and Natural Selection.* New Haven, Conn.: Yale University Press, 1972.
Quammen, David. *The Song of the Dodo: Island Biogeography in an Age of Extinction.* New York: Scribner, 1997.
Raby, Peter. *Alfred Russel Wallace: A Life.* Princeton: Princeton University Press, 2001.
Shermer, Michael. 2002. *In Darwin's Shadow: The Life and Science of Alfred Russel Wallace.* New York: Oxford University Press.
Smith, Charles H., and George Beccaloni, eds. *Natural Selection and Beyond: The Intellectual Legacy of Alfred Russel Wallace.* New York: Oxford University Press, 2008.
Wallace, Alfred Russel. *Alfred Russel Wallace: An Anthology of His Shorter Writings.* Edited by Charles E. Smith. New York: Oxford University Press, 1991.
———. *The Alfred Russel Wallace Reader: A Selection of Writings from the Field.* Edited by Jane Camerini. Baltimore: Johns Hopkins University Press, 2002.
———. *Darwinism: An Exposition of the Theory of Natural Selection with Some of Its Applications.* London: Macmillan, 1889.
———. *Island Life: Or the Phenomena and Causes of Insular Faunas.* New York: Harper, 1881.
———. *The Malay Archipelago: The Land of the Orang-utan and the*

Bird of Paradise. New York: Macmillan, 1872.

———. *Miracles and Modern Spiritualism.* London: Nichols, 1874.

———. *My Life.* 2 vols. New York: Dodd, Mead, 1905.

———. *A Narrative of Travels on the Amazon and Rio Negro.* London: Reeve, 1853.

———. *Social Environment and Moral Progress.* New York: Cassell, 1913.

———. *The Wonderful Century: Its Successes and Its Failures.* New York: Dodd, Mead, 1898.

WALLACE'S LINE

Daws, Gavan, and Marty Fujita. *Archipelago: Islands of Indonesia.* Berkeley: University of California Press, 1999.

Van Oosterzee, Penny. *Where Worlds Collide: The Wallace Line.* Ithaca, N.Y.: Cornell University Press, 1997.

Wallace, Alfred Russel. *The Geographical Distribution of Animals, with a Study of the Relations of Living and Extinct Faunas as Elucidating the Past Changes of the Earth's Surface.* New York: Harper & Bros., 1876.

"WALLACE'S PROBLEM"

Wallace, Alfred Russel. Review of Sir Charles Lyell's new edition of *Principles of Geology. Quarterly Review* (April 1869); repr. *Alfred Russel Wallace: An Anthology of His Shorter Writings,* edited by Charles E. Smith, pp. 31–34. New York: Oxford University Press, 1991.

———. *Social Environment and Moral Progress.* New York: Cassell, 1913.

WASHOE

Fouts, Roger, and Stephen Tukel Mills. *Next of Kin: My Conversations with Chimpanzees.* New York: Quill, 2003.

Proffer, Keiran. *Adam, Darwin, and Washoe: Genesis and the Talking Chimpanzee.* London: Underhill Management, 2002.

WATER BABIES, THE

Kingsley, Charles. *The Water Babies: A Fairy Tale for a Land-Baby* (1863). New York: Penguin Books, 2008.

WEISMANN, AUGUST

Weismann, August. *Essays upon Heredity and Kindred Biological Subjects.* 2 vols. Edited by E. B. Poulton et al. Oxford: Oxford University Press, 1891–1892.

———. *The Evolution Theory.* Translated by J. A. Thomson. 2 vols. London: Edward Arnold, 1904.

———. *The Germ Plasm: A Theory of Heredity.* Translated by W. Newton Parker and Harriet Rönnfeldt. New York: Scribner, 1893.

WELLS, H.G.

Coren, Michael. *The Invisible Man: The Life and Liberties of H. G. Wells.* Toronto: McArthur, 2005.

Klass, Philip. "Welles or Wells: The First Invasion of Mars." *New York Times Book Review,* October 31, 1988.

Wells, H.G. *Experiment in Autobiography: Discoveries and Conclusions of a Very Ordinary Brain (Since 1866).* New York: Macmillan, 1934.

———. *The Island of Dr. Moreau* (1896). New York: Modern Library, 1996.

———. *The Time Machine: An Invention* (1895). Cambridge, Mass.: R. Bentley, 1971.

———. *The War of the Worlds* (1898). New York: NYRB Classics, 2005.

Wells, H.G., Julian Huxley, and G. P. Wells. *The Science of Life: A Summary of Contemporary Knowledge about Life and Its Possibilities.* Garden, City, N.Y.: Doubleday, 1929.

WELLS, WILLIAM CHARLES

Wade, Nicholas J. *Destined for Distinguished Oblivion: The Scientific Vision of William Charles Wells.* New York: Kluwer Academic/Plenum Publishers, 2003.

Wells, William Charles. *An Essay on Dew and Several Appearances Connected with It.* 1814; repr. Whitefish, Mont.: Kessinger, 2007.

WHALE, EVOLUTION OF

Burnett, D. Graham. *Trying Leviathan: The Nineteenth-Century New York Court Case That Put the Whale on Trial and Challenged the Order of Nature.* Princeton: Princeton University Press, 2007.

Zimmer, Carl. 1998. *At the Water's Edge: Macroevolution and the Transformation of Life.* New York: Free Press, 1998.

WHIG HISTORY

Butterfield, Herbert. *The Origins of Modern Science.* London: G. Bell & Sons, 1957.

———. *The Whig Interpretation of History.* New York: W.W. Norton, 1931.

WOLTMANN, LUDWIG

Cornwell, John. *Hitler's Scientists: Science, War, and the Devil's Pact.* New York: Viking, 2003.

Gasman, Daniel. *The Scientific Origins of National Socialism.* London: Macdonald, 1971.

Mosse, George L. *The Crisis of German Ideology: Intellectual Origins of the Third Reich.* New York: Grosset & Dunlap, 1964.

Woltmann, Ludwig. *Politische Anthropologie.* Leipzig, 1903.

WRIGHT, SEWALL

Provine, William. *Sewall Wright and Evolutionary Biology.* Chicago: University of Chicago Press, 1986.

Wright, Sewall. *Evolution and the Genetics of Populations.* 4 vols. Chicago: University of Chicago Press, 1968–1978.

"X" CLUB

Barton, Ruth. "'Huxley, Lubbock, and Half a Dozen Others': Professionals and Gentlemen in the Formation of the X Club, 1851–1864." *Isis* 89 (1998): 410-444.

Irvine, William. *Apes, Angels, and Victorians: The Story of Darwin, Huxley, and Evolution.* New York: McGraw-Hill, 1955.

Strick, James Edgar. *Sparks of Life: Darwinism and the Victorian Debates over Spontaneous Generation.* Cambridge, Mass.: Harvard University Press, 2000.

YOUMANS, EDWARD LIVINGSTONE

Fiske, John. *Edward Livingston Youmans: Interpreter of Science for the People.* New York: D. Appleton, 1894.

ZOONOMIA

Darwin, Erasmus. *Zoonomia, or the Laws of Organic Life.* 2 vols. London: Johnson, 1794–1796.

King-Hele, Desmond. *Doctor of Revolution: The Life and Genius of Erasmus Darwin.* London: Faber & Faber, 1977.

———. *Erasmus Darwin: A Life of Unequalled Achievement.* London: Giles de la Mare, 1999.

ZOOPHARMACOGNOSY

Engel, Cindy. *Wild Health: How Animals Keep Themselves Well and What We Can Learn from Them.* New York: Houghton Mifflin, 2002.

Wrangham, Richard W., and Jane Goodall. 1989. "Chimpanzee Use of Medicinal Leaves." In *Understanding Chimpanzees,* edited by Paul G. Heltne and Linda A. Marquardt, pp. 22–37. Cambridge, Mass.: Harvard University Press, 1989.

ACKNOWLEDGMENTS

Three friends initially inspired this book: Carl Sifakis, innovative author of entertaining reference books; Harvard paleontologist and boyhood chum Stephen Jay Gould; and Facts on File CEO Edward Knappman, who originated the idea for my *Encyclopedia of Evolution*, the precursor to the present work, in 1985 and published it in 1990.

I am grateful to other friends (or enablers) who have provided tangible support and encouragement over the years: Susan and Bob Wilder, Gerry Ohrstrom, Miriam and Ira Wallach, Martin and Kate Cassidy, Norman Shaifer, Bob Adelman, Henry and Gloria Jarecki, and the late C. A. Tripp. Mary Kaye Linge produced the 1993 Holt paperback edition.

Science editors Blake Edgar and Chuck Crumly of the University of California Press are responsible for its present incarnation, maintaining enthusiasm, patience, and forbearance during the eons it took to evolve the *Encyclopedia* into *Darwin's Universe*. In addition to revisions and updates throughout, *Darwin's Universe* features hundreds of new essays and photos. Blake's expert knowledge of fossil man saved me from many gaffes about current paleoanthropology. Agent Russ Galen skillfully helped to navigate some of the turbulence during the book's transmutations.

Editor Rose Vekony at UC Press gave the text a thorough overhaul, update, and fact-check with admirable dedication, intelligence, and meticulous care. She was ably supported by editorial assistant Lynn Meinhardt and copyeditor Anne Canright. The text has benefited immeasurably from all their efforts.

The talented designer Rick DeMonico created the original layouts for this edition, which were expertly adapted by UC Press art director Lia Tjandra. Compositors at BookMatters managed to maintain the layouts while making numerous revisions to art and text. Computer genius and friend R. J. Cote has for years rescued me from the perils and pitfalls of digital technology.

Friends and colleagues at the American Museum of Natural History have been indispensable: Ian Tattersall, curator of anthropology, and physical anthropologists Will Harcourt-Smith and Gary Sawyer. Thanks also to *Natural History* magazine editor-in-chief Vittorio Maestro, former chief Peter Brown, and publisher Charles Harris.

I'm grateful to my friends at the Linnean Society of London, especially to Archivist Emerita Gina Douglas; Brian Gardiner, a past president of the Society; and my late, sorely missed friend, former Executive Secretary John Marsden. My association with the "Linnsock," as they call it, has been a joy and an education. My thanks to Executive Secretary Ruth Temple and her staff: Elaine Shaughnessy, Victoria Smith, and Kate Longhurst. Thanks especially to librarian Lynda Brooks and assistant librarian Ben Sherwood, conservator Janet Ashdown, and the *Linnean Journal*'s wonderful editor, Mary Morris. I'm also grateful to scholars at the Charles Darwin Correspondence Project in Cambridge, England, for their kindnesses, and particularly to its late founder and longtime editor-in-chief Frederick Burkhardt, who wrote me, in a treasured letter, that he kept the previous edition of this book on his night table and enjoyed it from cover to cover.

Special thanks to the Darwin family, especially to Randal Keynes, Richard Darwin Keynes, Sara Darwin Vogel, and Matthew Chapman, as well as to Lyulph Lubbock and Michael Huxley, descendants of Darwin's closest friends, who have generously shared family photos and archives. Brian Gardiner contributed the sketchbook of Richard Owen, from which the dour Victorian zoologist's unexpectedly lighthearted cartoons are published here for the first time.

Thanks also to the staffs of the Colindale Newspaper Library, New York Public Library, Charles Darwin Manuscript Library at Cambridge, British National Archive, and Maidstone Archive. Thanks to curators Adrian Green and Marie-Louise Kerr at the Bromley Museum in Orpington; Elisabeth Silverthorne, formerly of the Bromley Library; and Peter Gautrey and Philip Titheradge, former conservators of Down House. I am particularly grateful to the staff of the American Museum of Natural History Research Library, especially to its knowledgeable, indefatigable archivist Barbara Mathé.

Antiquarian booksellers are an undervalued, knowledgeable, and fiercely individualistic breed. Their passion for old books and letters has preserved much cultural history that might otherwise have been lost. David Bergman (Dave's Books), the late John Chancellor (Kew Books), Susan and Michael Lowell (Sue Lowell Natural History Books), and London bookseller extraordinaire Eric Korn have all opened many windows into the past.

Kudos to the wonderful artists who worked on this book: Pete Von Sholly, Viktor Deak, Bob Ziering, and, for the extraordinary new jacket art, Rosamond and Dennis Purcell. Thanks also to the other artists and photographers whose images appear: Margo Crabtree, Nancy Burson, Errol Fuller, Alexis Rockman, Kenneth Love, Johnny Hart, Carl Buell, Michael Rothman, Paul Nestor, Tui de Roy, Frans Lanting, Heidi Snell, and to Gordon Chancellor and the family of maritime painter John Chancellor for his magnificent re-creations of HMS *Beagle*.

I'd also like to thank the many who provided expertise, images, vetting, critiques, comments, emotional support, and encouragement: Ned Barnard, Judith Berman, Jean Thompson Black, Carmen Collazo, Ralph Colp Jr., Eric Delson, Jeannine Frank, Lindsay Fulcher, Daniel Gasman, Peter Gautrey, Fred and Nancy Golden, Eliot Goldfinger, Stephen Jay Gould, Sarah Granato, Richard Lancelyn Green, Colin Groves, Dean Hannotte, Leonard Horner, Judith Jennings, Donald Johanson, Rhoda Knight Kalt, Gene Keiffer, David Kohn, Malcolm Kottler, Gene Kritsky, Mike Levin, Michael and Barbara MacRoberts, James Mallet, John and Hazel Marsden, Ernst Mayr, Ivory and Abe Mertens, Jude Milner, Jennifer Minichello, William Montgomery, James R. Moore, Doug Murray, Gareth Nelson, Susan Parker, Roy Pinney, Steven Poser, Michael Price, Nina Root, Donald Stone Sade, Richard Seyfarth, Michael Shermer, Rob Susman, Anthony J. Sutcliffe, Howard Topoff, Melvin Van Peebles, Elisabeth Vrba, Alan Walker, Stephen Webster, Andrew Wilson, John Woram, and Richard Wrangham.

I am particularly indebted to the succession of talented, bright young assistants I think of as "Darwin's Angels": Rebecca Araya, Sara Arias, Heather Bloch, Elizabeth Donohue, Kathleen Letchford, Marisa Macari, Whitney Reiner, Kirsten Weir, Kimberly Wong, and Lily Xie. In imagination, I have awarded each of them a diamond-encrusted pin shaped like an evolved fish that has sprouted not only ruby feet, but angel wings of spun gold. Thanks to all of you for your valued presence in my book and in my life.

ILLUSTRATION CREDITS

AMERICAN BIBLE SOCIETY LIBRARY: 428

AMERICAN INTERNATIONAL PICTURES: 444

AMERICAN MUSEUM OF NATURAL HISTORY: 15 left (negative no. 34314), 15 right (311654), 16 top (211501), 18 left (325486), 24 (325097), 25 (271655), 28 top and bottom, 33, 40 top (315447), 40 bottom (315286), 42 (299134), 49 (2A3077), 55 top (321623), 55 bottom (34216), 56 top (19508), 57 (318651), 80 top, 80 bottom (310500), 87 top (© R. MICK-ENS), 91 (104670), 103 top (326799), 114 top (326665), 115 (326699), 120 (117120), 125 bottom, 158 top (110666), 158 bottom (338971), 200 top (410950), 200 bottom (410927), 216 (325487), 226 (35383), 263 (327667), 264 left (35827), 264 right (35799), 267 (39442), 289, 319 (2A 17487-A), 324, 344 top (310511), 351 top (362404A), 351 bottom (109353), 366 (318686), 387 (123978), 388 (bottom), 399

ASSOCIATED PRESS: 235 (photo by MARTIN MEISSNER), 367, 380

AUTHOR: 109 (top, second from right), 116 (left), 147 (bottom), 190 (bottom), 237 (top and bottom), 334, 352, 374. Animal Alphabet (at the head of each letter section) © Richard Milner, 2009.

AUTHOR'S COLLECTION: frontispiece, 2, 35 (top), 35 (bottom), 38 (bottom), 44, 58, 60, 67, 70, 83, 98, 104 (top right), 106, 117, 125 (top), 148 (top and bottom), 153, 155, 166, 172, 193 (left and right), 203 (top), 207, 213, 214, 231, 233 (top right), 243, 257 (left, middle), 261, 275, 276, 280, 284, 285 (top and bottom), 286 (bottom), 292 (top and bottom), 293, 303, 305, 326, 327, 331, 337 (bottom), 343, 348, 350, 378, 379, 389 (top and bottom), 396, 398 (top and bottom), 402 (bottom), 415, 429, 431 (all), 434, 435, 436 (top), 437 (top and bottom), 441 (top and bottom)

CHARLES F. BADLAND: 242

EDWARD S. BARNARD: 123 (second from top), 199 (top)

LA BELLE EPOQUE VINTAGE POSTERS, INC: 360

BELSTAR PRODUCTIONS: 363

MARIAN BRICKNER: 39 (top), 50, 51

BROMLEY MUSEUM SERVICE: 94, 95, 111

CARL BUELL: 446, 447

NANCY BURSON: 86

CARNEGIE LIBRARY OF PITTSBURGH: 69

GORDON CHANCELLOR: 37, 432 (bottom)

MATTHEW CHAPMAN: 109 (bottom right)

JAMES CHATTERS: 257 (top left). Copyright by James Chatters.

JAMES CHATTERS AND TOM MCCLELLAND: 257 (right). Copyright by James Chatters and Tom McClelland.

CHETHAM'S LIBRARY, MANCHESTER: 150

MATT CIOFFI: "Gangsta Chimp," 259

CIRCUS WORLD MUSEUM, BARABOO, WISCONSIN, with permission from RINGLING BROS. AND BARNUM AND BAILEY® THE GREATEST SHOW ON EARTH®: 203 (bottom)

JEAN CLOTTES/FRENCH MINISTRY OF CULTURE: 76 (right), 77 (left)

COLCHESTER AND IPSWICH MUSEUM SERVICE: 217

MARGO CRABTREE: 29 (top and bottom), 410

CREATION MUSEUM: 99

EMMA DARWIN; photo by RODERICK FIELD: 109 (right, middle)

DAVE'S BOOKS: 52, 64, 65, 73 (bottom), 146, 311 (top and bottom), 325, 357 (top and bottom), 359, 388 (top)

VIKTOR DEAK: 11, 143, 176, 179, 268, 316

TUI DE ROY/ROVING TORTOISE PHOTOS: 123 (top, bottom three), 189

THE DIGIT FOUNDATION: 135

DISNEY ENTERPRISES, INC.: 26, 100 (top), 168

E.C. PUBLICATIONS INC.: 105

MICK ELLISON: 139 (top and bottom)

FENIMORE ART MUSEUM, COOPERSTOWN, N.Y.: 68

ERROL FULLER NATURAL HISTORY IMAGES: 20, 56 (bottom), 118 (top and bottom), 149, 167 (top and bottom left and right), 186 (top and bottom), 215, 271, 297 (top), 373 (top and bottom)

GILCREASE MUSEUM: 257 (bottom). John Wesley Jarvis, *Black Hawk and Son Whirling Thunder*. From the Collection of Gilcrease Museum, Tulsa, Oklahoma.

STEPHEN J. GOULD: 60 (photo © by ALAN C. KEMP), 89 (right), 165

GRYPHON PRODUCTIONS LTD., © PETER VON PUTTKAMER: 18 (right)

ERNST-HAECKEL-HAUS, FRIEDRICH-SCHILLER-UNIVERSITÄT, JENA: 53, 209, 210

JOHN L. HART FLP AND CREATORS SYNDICATE, INC.: 21

THE HUNTINGTON LIBRARY, SAN MARINO, CALIFORNIA: 108 (left and middle). Reproduced by permission of the Huntington Library, San Marino, California.

MICHAEL HUXLEY: 206, 232 (top three), 233 (bottom two)

INSTITUTE OF HUMAN ORIGINS: 12, 249, 288

INTERNATIONAL CENTER OF PHOTOGRAPHY: 224 (right). © 1994, International Center of Photography, New York. Bequest of Wilma Wilcox.

ITV/GRANADA: 140

RHODA KNIGHT KALT: 317 (left)

RANDAL KEYNES: 108 (bottom right), 109 (top, left and second from left)

KING FEATURES SYNDICATE: 100 (bottom)

M.E. KORN: 239 (bottom)

STANLEY KRAMER PRODUCTIONS: 241 (top and bottom)

VLADIMIR KRB: 61 (left and right)

GENE KRITSKY: 233 (top left)

FRANS LANTING/MINDEN PICTURES: 76 (left)

JOE LEMONNIER: 88, 201 (bottom), 272, 438

LIBERTY FILMS: 89 (left)

LINNEAN SOCIETY OF LONDON: 286 (top), 337 (top)

PAT LINSE/SKEPTIC.COM: 205 (top)

LONDON OBSERVER: 136

KENNETH LOVE: 23, 78 (left and right)

LUBBOCK FAMILY ARCHIVE: 287. Copyright by Lubbock Family Archive.

MICHAEL H. MACROBERTS: 391

JAY H. MATTERNES: 220

ARCHIVES OF THE ERNST MAYR LIBRARY OF THE MUSEUM OF COMPARATIVE ZOOLOGY, HARVARD UNIVERSITY: 298

JOE MCDONALD: 87 (middle and bottom)

METRO PICTURES: 73 (top)

METRO-GOLDWYN-MAYER: 409 (top)

MISSION PALÉOANTHROPOLOGIQUE FRANCO-TCHADIENNE: 419 (top and bottom). © MPFT.

DOUG MURRAY ARCHIVES: 328

MUSÉE DE L'HOMME, PARIS: 101, 269, 411

MUSEUM OF COMPARATIVE ZOOLOGY, HARVARD UNIVERSITY: 13 (top)

NATIONAL MARITIME MUSEUM, GREENWICH, LONDON: 432 (top)

NATIONAL ZOOLOGICAL PARK, WASHINGTON, D.C.: 341

NATURAL HISTORY MAGAZINE, INC.: 297 (bottom)

GARETH NELSON and JORGE LLORENTE: 81. © 1989 Jorge Llorente.

PAUL NESTOR: 309

NEVRAUMONT PUBLISHING COMPANY: 11, 176

NEW YORK PUBLIC LIBRARY PICTURE COLLECTION: 38 (top), 185, 191, 224 (left), 228, 229, 248, 266, 336, 338, 339, 386, 404, 424, 436 (bottom)

RUTH PADEL; photo by ROBERT CARPENTER TURNER: 109 (top, second from right)

PARAMOUNT PICTURES: 116 (right)

PEABODY MUSEUM OF ARCHAEOLOGY AND ETHNOLOGY, HARVARD UNIVERSITY: 34

PEABODY MUSEUM OF NATURAL HISTORY, YALE UNIVERSITY: 92 (all), 369

JOYCE PENDOLA: 154

PENNSYLVANIA ACADEMY OF THE FINE ARTS, PHILADELPHIA: 344 (bottom). Gift of Mrs. Sarah Harrison (The Joseph Harrison Jr. Collection).

IRENE PEPPERBERG; photo by ARLENE LEVIN-ROWE: 16 (bottom)

STUART PIVAR: 9

MICHAEL E. PRICE: 232 (bottom)

BRUCE J. RAMER: 401

IAN REDMOND: 392

RKO RADIO PICTURES: 262

ALEXIS ROCKMAN: 180

ROSLIN INSTITUTE: 147 (top)

MICHAEL ROTHMAN: 199 (bottom)

ROYAL MAIL, U.K.: 107 (top)

ROYAL NAVAL COLLEGE, GREENWICH, LONDON: 170

GARY J. SAWYER, DIVISION OF ANTHROPOLOGY, AMNH: 317 (right)

WILLIAM SCHOPF: 43

NEIL SELKIRK/TIME & LIFE PICTURES/GETTY IMAGES: 175

ROSE A. SEVCIK, LANGUAGE RESEARCH CENTER, GEORGIA STATE UNIVERSITY: 254

HEIDI SNELL/VISUAL ESCAPES: 119, 282

STANFORD UNIVERSITY LIBRARY: 13 (bottom)

STEVE STOLLMAN COLLECTION: 383

MARIE STOPES INTERNATIONAL: 402 (top)

CHARLES G. SUMMERS, JR., WILD IMAGES: 77 (right)

JOHN TELFORD-TAYLOR: 72

20TH CENTURY FOX: 353

UNIVERSAL PICTURES: 251 (top and bottom)

UNIVERSITY OF THE WITWATERSRAND, JOHANNESBURG: 102

PETE VON SHOLLY: 31, 39 (bottom), 112, 167 (bottom middle), 265, 290, 420, 454, 455, 488. Copyright by Pete Von Sholly.

ELISABETH VRBA: 421. Phylogeny chart authored by Elisabeth Vrba, drawn by Susan Hochgraf.

WARNER BROS.: 409 (bottom)

DELTA WILLIS: 204, 205 (bottom), 274

JOHN WORAM: 190 (top)

INDEX

Page numbers in bold refer to encyclopedia entries.

selectionism, 321
"selfish gene," 260, **384–85**
Selfish Gene, The (Dawkins), 125
Selznick, David O., 262
Sereno, Paul C., 202
Serkis, Andy, 262–63
Serling, Rod, 353
sex, 33, 45, 46, 50, 141, 185, 259, 291, 316, 385, 402, 442; botanical, 84, 141–42, 279, 331–32, 400; and hybridization, 234; and infanticide, 240; and isolating mechanisms, 246; and isolation theory, 433; and Mendelian genetics, 299, 300; and natural selection, 430, 434; origin of, 163, 443; and pangenesis, 342; sperm competition, **397–98**
sexual selection, 10, 132, 365, **385–86,** 437–38
Shakespeare, William, 400, 415
Shapiro, Harry, 41
Shapiro, Robert, 74
Shaw, George Bernard, 63, 120, 191, 320, **386–87,** 442–43
Sherman (chimpanzee), 21–22
Shermer, Michael, 130
Silurian period, 66, 311, 384
Simon, Helmut, 238
Simons, Elwyn, 333
Simpson, Edward, 172–73
Simpson, George Gaylord, 99, 225, 227, 230, 294, 314, 336, **387–88,** 407
Sinanthropus pekinensis, 40, 176, 219
Sinclair, Harry, 49, 388
Sinclair dinosaur, **388**
Singer, Peter, 19
Sinodelphis szalazi, 79
Sinosauropteryx, 79
Sivapithecus, 150
Skullduggery (movie), 429, 430
Slade, Henry, 270, 271, 285, 389–90, 398
Slade trial, **389–90**
"slave-making" ants, 128, 287, 316
slavery, 128, 132–33, 170, 287, 394, 431
sloths, giant, 199, 248, 267, 337, 355, 358, 431
Smellie, William, 141
Smith, Sir Grafton Elliot, 319, 332
Smith, James E., 281
Smith, John Maynard, 156
Smith, Sydney, 116, **390**
Smith, William, **390–91**
Smithson, Robert, 168
social behavior, evolution of, **391–93**
Social Darwinism, 27, 58, 62, 69, 187, 233, **393–94,** 406, 429, 448; Kropotkin's critique of, 266; and monism, 308; and Spencer, 396
"social selection," 46
sociobiology, 45, 47, 85, 103, 158, 385
sociology, 357, 396
somatic mutation theory, 442–43
Soviet Union, 118, 121, 145, 254, 290, 320, 394
speciation, 161, 225
species, concept of, 103, 234, 235, 246, **395–96**

Spencer, Baldwin, 352
Spencer, Herbert, 69–70, 105, 121, 161, 260, 393, **396–97,** 451; "survival of the fittest" coined by, 406; and "X" Club, 450
Spencer, Timothy, 144
sperm competition, **397–98**
Spielberg, Steven, 250, 251
spiritualism, 234, 271, 294, 352, 377, **398–400,** 441; and Conan Doyle, 284, 285; and Romanes, 371–72; and Slade trial, 389–90; and Wallace, 75, 148, 310, 315, 349, 372, 398, 447–48, 449
Spoor, Fred, 220
Sprengel, Christian Konrad, 141, 331, **400, 401**
Spurzheim, Johann, 348
Stalin, Josef, 290
Stanley, Steven M., 355, 407
State Darwin Museum of Natural History (Moscow), **117–19**
steady-state Earth, **400, 402**
Stebbins, G. Ledyard, 407
Stegodons, 173, 177
Stegosaurus, 137, 263
Stewart, Patrick, 256, 257
Stocking, George, 286, 352–53
Stoddard, Lothrop, 27
Stokes, Capt. Pringle, 36, 112
Stokowski, Leopold, 168
Stopes, Marie C., **402–3**
"straight line" evolution, 407
strata, geological, 390–91
Stresemann, Erwin, 296
Strong, Augustus Hopkins, 187
Structure and Distribution of Coral Reefs, The (Darwin), **93–95,** 105, 107, 113, 153
"struggle for existence," 84, 85, 290, 308, 448–49
Strum, Shirley, 30–31, 193
subspecies, 365
Suess, Eduard, 48, 201
"Sue" the Tyrannosaurus, **403**
sulfur, ecosystems based on, **403–5**
Sulloway, Frank, 122, 185
Sultan (chimpanzee), 23
Sumner, William G., 121, 393, 394
Sunday League, **405–6**
supercontinents, 272, 290, 341–42
"survival of the fittest," 69, 155, 161, 162, **406,** 436; coined by Spencer, 396, 397; Huxley's view of, 233; meaning of fitness, 169; and sexual selection, 386; and Social Darwinism, 260, 429
Sussman, Robert, 453
Swinton, W.E., 137
symbiosis, 84
Synthetic Theory, 145, 156, 157, 271, 297, **407,** 433, 442; and neo-Darwinism, 319; and "new" Lamarckism, 323; and paleontology, 387; and punctuated equilibrium, 362; and variation, 160
Systema Naturae (Linnaeus), 279, 280

Taieb, Maurice, 11, 12, 249
"tangled bank" metaphor, **408**

Tarzan of the Apes stories, 262, **409–10**
Tasaday tribe (hoax), 259
Tattersall, Ian, 28, 219
"Taung child" skull, 29, 55, 102, **410–11**
taxidermy, 14
taxonomy, 47, 103, 157, 421–22
technology, evolution of, 352–53
Teddy bear, 100
Teilhard de Chardin, Pierre, 306, 329, 394, **411–12**
teleology, **412–13**
Telliamed (de Maillet), 131, 303
Temin, Howard, 322
Temple of Nature, The (E. Darwin), **413–14**
Tenniel, John, 9
Tennyson, Alfred Lord, 155, 302, 414–15
"terminator" seeds, 181
"Ternate Paper" (Wallace), 335, 375, **415–16**
Terrace, Herbert, 21, 265, 323, 440
Tertiary period, 95
theory, scientific, **416–17**
Thewissen, Hans, 447
Thistleton-Dyer, William, 223
Thomas Aquinas, Saint, 97, 328
Thornhill, Randy, 158
Three Ages (movie), 73, 74
Tinbergen, Niko, 85, 125, 157, 229, 283, 370, **417**
tool use, 293–94, 333; of apes, **23–24;** of bonobos, 50; of chimpanzees, 17, 23–24, 78; evolution of, **418;** of gorillas, 23; hominids and, 12; of orangutans, 8, 23; and prehistory movement, 340
Topoff, Howard, 287
tortoises, giant, 119, 123, 189–90, 281–82, 373, 431
Toth, Nick, 255
Toumai skull, **419**
Tracy, Spencer, 241
transitional forms, 161, 163, **420–21**
"tree of life," **421–22**
Triassic period, 137, 290, 300
Triceratops, 137
Tripp, C.A., 116
Tschermark, Erick von, 299
Tschinkel, Walter, 242
Tuilefano, George, 144
Turkana Boy, 143, 219, 276, **422–23**
Turkana mollusks, **423**
Twain, Mark, 9, 214, **423–24**
Tweed, "Boss" William, 18
"two books," doctrine of, **424–25**
2001: A Space Odyssey (movie), 38, 354, **425–26**
Tylor, Edward, 85, 352
Tyndall, John, 41, 302, 383, **426,** 450
Tyrannosaurus rex, 56, 79, 137, 168, 204, 251–52, 264; kinship with birds, 139; "Sue" fossil, **403**

uniformitarianism, 8, 59, 72, 94, 101, 120, 289, 361, 400, **427**
United States: Darwinism in, 121; dinosaur fossils in, 137; eugenics movement, 158–59, 191–92; fun-

damentalist religion in, 187; immigration policies, 366; sterilization laws, 129, 159, 191
Ussher, Archbishop James, 179, 428
Ussher-Lightfoot chronology, **428**

Vaillant, Sébastien, 279
Valderrama, Ximena, 453
variability in organisms, 157, 356
variation, 105, 118, 134, 145, 157, 160, 162, 164, 192, 194, 296, 314, 319, 436; anatomical, 224; and "founder effect," 362; and hybridization, 234; and plenitude, 355; and race, 207, 365
Variation of Animals and Plants under Domestication, The (Darwin), 26, 107, 114, 194, 206
Vawter, Father Bruce, 381
Veblen, Thorstein, **429**
vegetable mold, Darwin's study of, 107, 114, **152–53**
Venter, J. Craig, 342–43
"Vercors" (Jean Bruller), **429–30**
Vereshchagin, Nikolai, 136
Vernadsky, Vladimir, 48
Verne, Jules, 284
Verner, Samuel, 42
Vestiges of Creation (Chambers), 54, 74–75, 303, 304, 338, 377, 414, **430,** 434
Victoria, Queen, 155, 170, 414
Victorian Anthropology (Stocking), 286, 352–53
Vigl, Eduard Egarter, 238
Vine, Fred, 247
Vine-Matthews-Morely hypothesis, 346
Virchow, Rudolf, 151, 317
Vogel, Sarah Darwin, 109
Vogelsang, Ulrich, 149
Voliva, Wilbur Glenn, 172
Voltaire, François, 60, 383, 413
Volterra, Vito, 84
Vonnegut, Kurt, 382
Voyage of HMS Beagle, 33, 64, **431–32,** 434
Vrba, Elizabeth, 165
Vries, Hugo de, **133–34,** 299, 407

Waal, Frans de, 50
Wagner, Moritz, 234, **433**
Wagner, Richard, 27
Walcott, Charles D., 60, 89
Walker, Alan, 422
Wall, Joseph Frazer, 70
Wallace, Alfred Russel, 60, 121, 234, 256, 334, 343, 430, **434–38;** and Bates in Amazon, 35; as beetle collector, 39, 104; on conspicuous coloration, 86, 87; on divergence, 139, 140; and flat earth bet, 212–13; on human origins, 95–96; "joint" publication with Darwin, 129–30, 161, 281, 335, 416; in Malay Archipelago, 129, 130, 150, 291, 315, 330, 331, 375, 415–16, 435–36; and Malthus, 182, 291, 356, 435, 436; on Mendelian genetics, 319; on mimicry in insects, 304; and natural

**Art Direction: Richard Milner
Designer: Rick DeMonico**

**Contributing Artists: Pete Von Sholly, Viktor Deak
Compositor: BookMatters, Berkeley
Indexer: Alexander Trotter
Text and display: Benton Gothic
Printer/Binder: Thomson-Shore, Inc**